TERRARVM ORBIS

8

TERRARVM ORBIS

Histoire des représentations de l'espace : textes, images
History of the Representation of Space in Text and Image

Collection dirigée par
General Editor of the Series
Patrick Gautier Dalché

2011
BREPOLS

Angelo Cattaneo

Fra Mauro's Mappa Mundi and Fifteenth-Century Venice

2011
BREPOLS

Cover illustration: a combination of the image of the celestial and sublunary world from the upper left corner of Fra Mauro's *mappa mundi* with the text of a legend in which Fra Mauro declares that it is only by undertaking a critical approach to, and reading of the *Geography* that he can remain faithful to the teachings of Ptolemy: "Io no(n) credo derogar a Tolomeo | se io non seguito la sua cosmographia, | perchè se hauesse uoluto obs(er)var i sui | meridiani ouer paralleli ouer gradi | era necessario quanto a la demo(n)st(r)a|tion de le parte note de questa circu(m)|ferentia lassar molte proui(n)cie de le | qual Tolomeo no(n) ne fa me(n)tio(n), ma | per tuto maxime i(n) latitudine çoè tra | ostro e tramo(n)tana dice terra i(n)cognita, | e q(ue)sto p(er)chè al suo te(m)po no(n) li era nota." A translation into English is provided at the beginning of Chapter IV.

Graphics: Angelo Cattaneo and Alberto Martini

This publication is based on a revised version of an European University Institute PhD thesis defended at the Department of History and Civilization in 2005.
This book has been published with financial subsidy from the

European University Institute

and

Centro de História de Além-Mar
(Faculdade de Ciências Sociais e Humanas - Universidade Nova de Lisboa)

D/2011/0095/10
ISBN 978-2-503-52378-1

FOREWORD

He had all the doggedness of a Flat-earther.
These people had to find the hard way.

William Boyd, *The Blue Afternoon*

At first sight this book might seem redundant, but it is not. Indeed, the valuable edition published in 2006 (*Terrarum Orbis*, 5) is remarkable because it has made the text of the Fra Mauro's world map accessible to a wider public and has offered scholars a complete view of this exceptional work.

Even if historians see the map as a landmark of cartography, it has always been interpreted as a characteristic product of the late Middle Ages. By mixing facts and fantasy, Fra Mauro was supposedly trying – for some readers "heroically" – to associate on one map irreconcilable traditions: the "traditional" or "outmoded" or "conventional" authority of the Antiquity, the "symbolic" values of medieval Christianity, and the "field experience" acquired from travel, commerce and contact with different cultures. So, his enterprise was necessarily marked by conflict, producing the representation of an irremediably confused space, in other words, a failure.

Whether Fra Mauro has succeeded or failed is of little importance. These judgments – because these are judgments, not analyses – are too widespread to be footnoted; they reproduce the tired opposition between the "Middle Ages" and "Renaissance" invented by post-romantic history, whose vigour still seems inextinguishable in the history of cartography, despite the strong objections voiced over the last decades, in different fields, by scholars of cultural and intellectual history, texts and facts in hand. It is precisely in dealing with the fifteenth-century documents that mark the turning point between the two periods and whose value is purely pedagogical that this type of historiography shows its weaknesses. *Quattrocento* authors are often presented as belonging to two culturally opposed worlds and unable to elaborate coherent works. But who is judging their coherence ? The modern historian, with his ideology, his unconscious and uncritical *epistémè* – as some would needlessly say. This is the "overhanging knowledge" once denounced by the phenomenologists, which has no close understanding of the intellectual productions transmitted from the past.

The present work does not fall into this trap. The author takes the document at face value and, with a learned and unbiased eye, identifies Fra Mauro's intentions. He analyses the sources not from a closed perspective, but by situating them in the context of the theological, philosophical and cosmographic speculations known to Fra Mauro. These meticulous analyses take into account the cultural determination of the milieux that produced this knowledge, and thus offer a precise understanding of how the cartographer was meeting the expectations of his contemporaries. Finally, the reception of this monument is studied in extended time, at the Portuguese court, in Medici Florence (during the "Renaissance"…), in the *Serenissima* of the Enlightenment and in imperial England, during the triumph of orientalism.

Here is an example of intellectual and cultural history with broad perspectives, confirmed by scrupulous scholarship. It is a method one hopes will be applied to other objects belonging to what is called the "history of cartography".

The Editor

per Agnese e Raffaele

TABLE OF CONTENTS

II
PATRONAGE. MARKETPLACES. THE HISTORY OF CARTOGRAPHY

ACKNOWLEDGMENTS

This book is the result of over ten years of research, work, and writing. I first saw the *mappa mundi* of Fra Mauro in Christmas of 1995 in the valuable facsimile edition of 1956, in the study of my father Sandro, a lithographer who is among the few who still know print using arrays of stone. A collector of books, he bought it because its color plates were accounted at the time they were created a true example of virtuoso lithography. I was then working in a research center at the University of Milan which was focused on historical narratives and new media. The *mappa mundi*, covered by a teeming mass of streets, cities, mountains, and ships, enchanted me. The luminous monumentality of the whole and the painted details seemed to me perfect for a historical narrative both textual and visual, which one then called (with great emphasis and mystery) "multimedia." The 1992 Columbus quincentennial celebrations had just finished, and I prepared a project on the map which was never considered. A few years later it became the proposal for a doctoral thesis accepted by the Department of History at the European University Institute, which gave me a wonderful opportunity to study, write, and live. This time multimedia played no role at all. I could see, then in a confused way, the opportunity to begin a study on the history of culture in the context of "the great fact of civilization which is the fifteenth century," as Eugenio Garin wrote.

Over the years that led to the elaboration of the thesis defended in September 2005, titled *La mappa mundi di Fra Mauro Camaldolese. Venezia, 1450*, Diogo Ramada Curto, Patrick Gautier Dalché, Marica Milanesi, and Anthony Molho have followed my work, believed in me, and read, commented on, and corrected my pages. Many of the ideas and insights that emerged in my research I owe to the discussions and exchanges which I had with them. Marica Milanesi welcomed me into her house in Genoa one afternoon in June of 1997 and I came out with a mountain of books, yet when we did not really know each other. I will be forever grateful to her for knowing how to guide, and restrain my enthusiasm. Anthony Molho and Patrick Gautier Dalché made me realize that philology, analysis, and comparison all come from the same burning intellectual fire. Marica Milanesi and Patrick Gautier Dalché also gave me the courage to go beyond what is found easily in supermarkets of "fashionable ideas" and in the cultural climate of the moment. Diogo Ramada Curto has taught me that the *longue durée* and an interest in the big picture are not necessarily inconsistent with micro-historical analysis, and that we must also learn to dare. I am also grateful to Kirti Chaudhuri who wanted me among his students. At the Institute I learned to write in two languages, and read many others. If I do so clumsily, the fault is my own. In the academic year 1999-2000, David Woodward, who unfortunately has since passed away prematurely, advised me for a semester at the University of Wisconsin at Madison. From weekly meetings in his office I realized that the road then I was following was wrong. His knowledge, dedication and humility are a great lesson. Norman Fiering, director of the John Carter Brown Library in Providence (RI) for the first time made me really feel part of a humanistic circle.

I started to prepare the final draft of this book, which is a thorough revision of my Ph.D. dissertation, while I was a post-doctoral fellow at the Biblioteca Nacional de Portugal, in Lisbon, sponsored by the Fundação para a Ciência e a Tecnologia. Joaquina Esteves Feijão, in charge of the Map Room, Inês Cordeiro, vice-director of the Biblioteca Nacional, and my friends and colleagues João Carlos Garcia, André Ferrand Almeida, and Renata Malcher de Araujo have all contributed to make my stay in Lisbon very rich and profitable. Later, at the invitation of Patrick Gautier Dalché, I was able to significantly deepen my knowledge of manuscript culture at the Institut de recherche et d'histoire des textes (I.R.H.T.) of the Centre national de la recherche scientifique in Paris, where they have unparalleled research tools. To Anne-Marie Eddé, director of the I.R.H.T., Jean-Pierre Rothschild, director of the Section latine, Adriano Oliva, Monica Calma, and Marie-Jeanne Gaudoin, I am grateful for the openness and warmth of their welcome. The Centro de História de Além-Mar (CHAM) of the Faculdade de Ciências Sociais Humanas at the Universidade Nova of Lisbon, where I have been working since October 2008, has allowed me to continue my research and to follow new paths. I am grateful to its director, João Paulo Oliveira e Costa, and my colleagues.

During recent years, the following institutions and foundations have helped fund my research : The American Friends of the J. B. Harley Research Fellowships, Inc., the Centro Nazionale della Ricerca, the Centre national de la recherche scientifique (C.N.R.S.), the European University Institute, the Fundação para a Ciência and Technology (F.C.T.), the Gulbenkian Foundation, the J. B. Harley Research Trust, and the Korea Foundation. To the respective trustees I wish to express my deep gratitude, for it is thanks to their support that I have come to know many scholars with whom I am now exchanging ideas.

I alone am responsible for any errors in this book, but my colleagues and former supervisors have enabled me to avoid many embarrassments. Marica Milanesi and Patrick Gautier Dalché have continued to discuss with me and more importantly to read drafts of this work, offering learned commentary and helpful corrections, and generously sharing with me their own research. Giovanna Murano contributed accuracy and grace to the Italian version of this work and helped me continuously in the examination of manuscript sources. Her friendship and generosity are a gift. For their intelligent and engaging conversation about many aspects of this book I am extremely grateful to Peter Barber, Edoardo Barbieri, Zoltán Biedermann, Wulf Bodenstein, Nathalie Bouloux, Cécile Caby, Christine Gadrat, Alfred Hyatt, Lucia Merolla, Alessandro Scafi and Vladimiro Valerio.

Father Ugo Fossa of the archive of the Monastery of Camaldoli has greatly facilitated my research in the monastery archives, a place of peace and beauty. Paola Pirolo, director of the Manuscripts Room of the Biblioteca Nazionale in Florence has given me the privilege of exploring the Library's magnificent manuscript collections. Her nobility and kindness made my days of study at the Library pleasant, and helped me discover the cloister of Brunelleschi from another point of view. Alexandra Curvelo has read and commented on many parts of this book with critical attention, intellectual generosity and great care and offers me new perspectives to understand and explore the *imago mundi* with new eyes and sensitivity.

The translators who have accompanied me in the preparation of this book over the years deserve special mention : Megan Metters, Molly McIlwrath, Eliza la Rocca, Sheila Barker, and Chet Van Duzer. Sheila and Chet in particular have honored me with their dedication, care, and skill, and now with their friendship. Many of the ideas discussed here have been refined thanks to their painstaking work and our exchanges during the preparation of the translation. Francis Herbert, an untiring reader, read a recent draft of this book. His friendly dedication, calm competence, and generosity have removed many inconsistencies.

Two men deserve my particular thanks because without them this book would not exist. Luc Jocqué, publishing manager at Brepols, has consistently but patiently urged the project forward, patient and confident that the book would be ready. Patrick Gautier Dalché not only offered me the opportunity to publish this work, but also gave me the strength and the motivation to finish it, finally.

My thanks go to my mother Marisa and my father Sandro, for their closeness, their work ethic, and their understanding. Finally, my special gratitude to Barbara Cimolato for the continuous help and support she has given me during these years of life together, including in the writing of this book – and especially for her calm care (*cura*).

This book is dedicated to our children, Agnese and Raffaele, for their love.

Florence, September 2010

INTRODUCTION

> Nei luoghi letterari qualcosa pur sempre rimane. Sta avvolto definitivamente nel nostro sguardo di lettori.[1]
>
> Claudio Magris

This book has its origins in a doctoral thesis written over the course of seven years and defended in September of 2005 before the Department of History and Civilization at the European University Institute, located in San Domenico di Fiesole, near Florence. Its goal is to analyze the way in which the *imago mundi* was imagined, represented and systematized, within the context of Venetian civilization and culture in the mid fifteenth century, an ecumenical crossroads of men, knowledge, and cultures. The theme will be brought into focus by means of a study of the world map of Fra Mauro, prepared around 1450 at the Camaldolese monastery of San Michele di Murano by Fra Mauro, a Camaldolese *converso* (a term denominating a lay member of the religious community who has not taken monastic vows, and who is usually responsible for humble but necessary duties within the monastery). Active in the first half of the fifteenth century and living perhaps until October of 1459 (and certainly not beyond 1464), he was honored by his contemporaries with the sobriquet "*cosmographus incomparabilis*."

Fra Mauro's world map, drawn on parchment laid down onto a wooden support, is contained within a square, gilded wood frame measuring 223 × 223 cm. Set within this square frame is a circular frame with a diameter of about 196 cm. Dispersed throughout the surface are three thousand inscriptions (two hundred of these being lengthy, detailed legends, while the rest are toponyms) ; these inscriptions comprise of a total of more that 115,000 characters. In addition, Fra Mauro's world map also includes hundreds of images, representing cities, temples, funerary monuments, streets, and ships, as well as a scene in the lower left corner representing Earthly Paradise. By means of both the words and images, Fra Mauro's work synthesizes and recapitulates aspects of Aristotelian-Christian cosmology ; Aristotelian-Ptolemaic cosmography of the sublunar world ; the chorography and geography of peoples and commerce ; nautical and Ptolomaic cartography ; fourteenth- and fifteenth-century travel narratives (above all, but not limited to, those of Marco Polo and Niccolò de' Conti) ; hearsay based on the oral accounts of travelers (from the Coptic monks that drew for him the maps of Ethiopia, to the ambassador João Fernandes da Silveira from the Portuguese court of Afonso V, who relayed information gathered during the first Portuguese navigational explorations along the coasts of Africa) ; descriptions of the principal routes for the global commerce of spices, gold, and silver ; as well as the histories of the ancient conquests of Alexander the Great and the recent ones of Tamerlane. All these frames of reference, and others besides, will be analyzed in the course of this study. According to Fra Mauro's own writings, this *mappa mundi* was drawn *a contemplation di questa*

[1] "In places touched by literature something is always left behind. It is forever fixed up in the looks of us readers."

Illustrissima Signoria, that is to say, for the benefit of the Republic of Venice ; this statement suggests that he might have intended to present his map to the government of Venice.[2] Nevertheless, the map did not leave Fra Mauro's monastery until 1811. In the years immediately following Fra Mauro's death in Venice, his work seems to have fallen into obscurity, eclipsed by the maps of a Venetian cosmographer who at the time enjoyed far greater fame : Don Antonio Leonardi. Fra Mauro's maps did, however, achieve renown abroad. In 1457, the Portuguese court of Afonso V commissioned a copy, which was dispatched to Lisbon in April of 1459. Then, around 1480, the Medici family in Florence – which at that time was headed by Lorenzo "the Magnificent" – commissioned some painters to visit to San Michele in order to make a copy. All the legends in the copy made for the Medici were translated into Latin by Pietro Dolfin, who at that time was prior of San Michele di Murano, and who later would become general of the Camaldolese Order. In the same years, Dominican friar Felix Fabri, who had traveled to Venice in order to embark on a ship that ferried pilgrims between Europe and the Holy Land, saw Fra Mauro's world map during his stay in that city. Fabri made note of it in his travel diary entitled, *Evagatorium in Terrae Sanctae, Arabyae et Egypti peregrinationem* (*c*. 1480-1483), including it among his list of the marvels that awaited visitors to Venice, a list otherwise comprised mostly of relics, churches and monasteries. In 1559, Giovanni Battista Ramusio (1485-1557) described the world map as one of the "miracles" of Venice, dedicating an entire, densely written page to it in the introduction to his edition of Marco Polo's *Milione* included in the second volume of *Navigazioni e viaggi*. Notably, with these reflections, Ramusio provided some of the earliest literary evidence of the way in which *mappae mundi* were studied and interpreted in the sixteenth century.[3] Altogether, this illustrious history of reception demonstrates that Fra Mauro's map enjoyed – from the very start – a remarkable degree of 'visibility' (that is to say, it was known, sought out, and discussed) even though it remained at the monastery of San Michele. Moreover, it was also known of through the two above-mentioned copies, displayed in palaces in two different European courts.

Thus, Fra Mauro's world map was famous and authoritative ; moreover, in comparison with other *mappae mundi*, it stands out for giving rise to a robust branch of scholarship and for having inspired the most noteworthy early facsimile reproductions : to my knowledge, no other ancient cartographic enterprise has been the focus of as much interest and attention over the past two hundred years as Fra Mauro's world map. By the end of the eighteenth century, four Camaldolese monks at San Michele di Murano – namely, the abbot Giovanni Benedetto Mittarelli (1708-1777) and Anselmo Costadoni (1714-1785), both co-authors of the *Annales Camaldulenses*, followed later by Mauro Cappellari (1765-1846, later Pope Gregory XVI) and Placido Zurla (1769-1834) – had redacted the very first studies of Fra Mauro's *mappa mundi* ; their work culminated in the publication of one of the first monographs ever to be dedicated entirely to a single cartographic work : *Il mappamondo di Fra Mauro camaldolese*, pubished by Placido Zurla in Venice in 1806. Between 1804 and 1807, William Vincent, the deacon of

[2] Marica Milanesi notes : "La carta (…) non risulta essere stata commissionata dalla Repubblica ; ma potrebbe essere stata fatta per esserle offerta, come era d'uso con i Principi, in vista di una ricompensa." M. Milanesi, "Cartografia per un principe senza corte : Venezia nel Quattrocento," in *Micrologus. Natura, scienze e società Medievali / Nature, Sciences and Medieval Societies* 16 (2008), pp. 189-216 (p. 206).

[3] For a complete bibliography of the cited works, see Chapter I below and its Documentary Appendix.

Westminster Abbey, believed that he had discovered in Fra Mauro's creation the world map of Marco Polo, and so, with the support of a group of British aristocrats involved in the administartion of the colonies and the patronage of the East India Company, commissioned a manuscript copy of the same dimensions as the original, and they placed it on deposit in the British Museum (now it is part of the collection of the British Library, formerly a division of the British Museum). This copy, in turn, served as the model for a published edition, in six gigantic plates, included with the third edition of the *Atlas* edited by the Viscount of Santarém in 1852 and sponsored by the Portuguese government. Since the early 1800s, Zurla's work, the manuscript copy at the British Museum, and the published version edited by Santarém gave Europe's educated publics broader access to Fra Mauro's *mappa mundi*.[4]

In modern times, Fra Mauro's work has been the focus of more that forty different scholarly projects.[5] Among these, three are particularly outstanding. In 1956, Tullia Gasparrini Leporace, director of the Biblioteca Nazionale Marciana, in collaboration with Roberto Almagià, edited an extraordinary facsimile edition that entails 48 color plates and a diplomatic transcription of all the legends and toponyms, as well as indications that serve to identify the original passages and their locations on the map.[6] In 1987, Antonio Ratti compiled an inventory of documents and codices referring to Fra Mauro and his cartographic works.[7] In 2006, Piero Falchetta carried out for Brepols Publishers an English edition of Fra Mauro's world map, which includes a commentary on the toponyms and legends organized into a new and handy index. Moreover, Fra Mauro's world map was digitally reproduced and made available on a CD-Rom.[8] Between 2009 and 2010, the *mappa mundi* underwent an important restoration treatment and, to her great credit, Dr. Maria Letizia Sebastiani, current director of the Biblioteca Nazionale Marciana, led the very first material analysis of the work, resulting in important discoveries about its structure and composition.[9]

This book and the research it presents have benefited immeasurably from earlier studies (those just named, as well as many other works that will be cited in the pages that follow). And yet, despite all this research, there are still many open questions regarding Fra Mauro's world map's hermeneutic potential and interpretive challenges, both because of its encyclopedic dimensions and because of its deep roots in the culture and civilization of mid-fifteenth-century Venice – one of the richest and most complex civilizations of its time. In light of these issues, the present study of the *mappa mundi* interrogates the work's numerous sources ; its narrative

[4] For a complete bibliography of the cited works, see Part II, Chapter II below and its Documentary Appendix.

[5] See the Bibliography below. Also, P. Falchetta, "Carte veneziane dell'Asia da Fra Paolino a Giacomo Gastaldi," in G. Curatola (ed.), *Sciamani e dervisci dalle steppe del Prete Gianni : religiosità del Kazakhistan e percezione del fantastico a Venezia*, Venice : Multigraf, 2000, pp. 39-50 (pp. 49-50).

[6] T. Gasparrini Leporace (ed.), *Il Mappamondo di Fra Mauro*, Rome : Poligrafico della Zecca dello Stato, 1956 ; henceforth cited as *Mappamondo* (ed. 1956).

[7] A. Ratti, "A Lost Map of Fra Mauro Found in a Sixteenth Century Copy," *Imago Mundi* 40 (1988), pp. 77-85 (pp. 84-85).

[8] Piero Falchetta, *Fra Mauro's World Map*, Turnhout : Brepols, 2006 : henceforth cited as *Mappamondo* (ed. 2006).

[9] As this introduction is being written, the results still have not been disclosed. They will be publicized on the website of the Biblioteca Nazionale Marciana.

structure ; the various influences on Fra Mauro's intellectual outlook ; the factors that impacted the map's creation ; circulation and reception ; the map's material history ; and its documentary trail. With these investigations, it aims to help clarify central questions regarding the cultural history of the mid fourteenth century.

Testis unus, testis nullus ?[10]

Two questions emerge at this point : Why concentrate a research project on – but not limit it to – Fra Mauro's *mappa mundi* ? And can such narrowly defined project produce results of broad significance ? The first of these questions is actually a two-part query ; foremost it calls into doubt the decision to concentrate upon one single document, but it also requires a justification for choosing Fra Mauro's *mappa mundi* as the basis for an investigation of fifteenth-century culture. The answer to these questions lies in both the larger concerns revolving around cosmography's role in the knowledge-base of the *Quattrocento*, as well as in the more specific concerns sparked by current on the scholarship on Fra Mauro's *mappa mundi*.

In a debate organized in the early 1990s on microhistory, Carlo Ginzburg, Edoardo Grendi, Giovanni Levi and Jacques Revel considered the impact of various approaches to historical analysis that had held currency in the 1970s. Although speaking from diverse perspectives, they all largely agreed on the reason for the emergence in this period of a body of research, grouped under the term 'microhistory', which, despite the wide range of topics it treated, was bound by a shared enthusiasm for "a methodology centered upon the consciousness of the individual." With what amounted to an implicit declaration of mistrust about the potential applications of historical seriality and its assumptions of relative homogeneity, the producers of microhistory reacted against historical seriality by concentrating upon vastly more limited contexts, specific cases, and even anomalous ones, seeking to demonstrate that it is possible to gain an understanding of the past without resorting to the perspective of the macrohistory.[11] This is not the appropriate forum for a discussion of the validity and legitimacy of microhistory. Nevertheless, I would like to add a "codicil" to the conclusions reached in the course of the debate because it regards the origins of the present research project. There are entire phases of history for which the historian does not have the luxury of choosing between a macroscopic research project and a microhistoric one. There have been cultures, civilizations and historical events for which very few records exist and for which historical documentation is extremely rare, or even limited to one single documentary "witness". In these cases, the methodological scale of the historiographic project is predetermined and ineluctable. This seems to me to be a common situation for histo-

[10] "One witness [equals] no witness," meaning – according to a legal principle established in the Middle Ages – that testimony by any one person in court was to be disregarded unless corroborated by the testimony of at least one other. See A. Libois, "À propos des modes de preuve et plus spécialement de la preuve par témoins dans la jurisdiction de Léau au xve siècle," in *Hommage au Professeur Paul Bonenfant (1899-1965)*, pp. 332-546 (pp. 539-542). For a discussion on the implications of this principle in the framework of historical research, see C. Ginzburg, *Il filo e le tracce : vero, falso, finto*, Milan : Feltrinelli, 2006, pp. 205-225 (pp. 208-209).

[11] G. Levi, "On Microhistory," in P. Burke, *New Perspectives on Historical Writing*, London : Polity Press, 1991, pp. 93-113. C. Ginzburg, "Microstoria : due o tre cose che so di lei," *Quaderni storici* 86 (1994), pp. 511-539, E. Grendi, "Ripensare la microstoria ?," *ibid.*, pp. 539-549, J. Revel, "Microanalisi e costruzione del sociale," *ibid.*, pp. 549-575.

rians wishing to study cosmography and the representation of the *imago mundi* in the period under discussion here.

The widely held assumption that Medieval and Renaissance records of *imagines mundi* are very scarce has been disproved by many studies. Having said this, though, it should be underlined that records of ecumenical cosmography with the immense heuristic range that characterizes Fra Mauro's *mappa mundi* – in other words, the surviving traces of the diverse branches of fourteenth- and fifteenth-century ecumenical cosmography preceding the invention of the printing press – are of such a rarity that, in almost all cases, what we know of its various traditions often reaches us by means of single surviving documents, fragments of traditions that must have indeed been vast and myriad. The Ebstorf and Hereford *mappae mundi* (both from *c.* 1300) and the *mappa mundi* of Pietro Vesconte (*c.* 1320) are among the few complete surviving documents of this kind from the early fourteenth century; all the rest survive only in fragments. Representing the Catalan tradition of holistic representations of the *oikumene* are only two survivors: the so-called "Catalan Atlas" (*c.* 1375) at the Bibliothèque nationale de France and the "Catalan-Estense World Map" (*c.* 1460-1470) at the Biblioteca Universitaria Estense of Modena. In Florence, there is the early-fourteenth-century "Atlante Mediceo" at the Biblioteca Medicea Laurenziana, while a famous world map from 1457 is the so-called "Genoese World Map" at the Biblioteca Nazionale Centrale. In Rome, at the Biblioteca Apostolica Vaticana, there is a *mappa mundi* that was created in Konstanz by the Benedictine monk Andrea Walsperger in 1448. From the fifteenth-century Venetian tradition, there are three small *mappae mundi* by Giovanni Leardo (dated 1442, 1448 and 1452), an *Atlas* by Andrea Bianco (1436) which is as small as it is precious, and some fragments that are conserved at the Topkapı Sarayı Müzesi in Istanbul; apart from these, there remains nothing else but Fra Mauro's world map and a map that is almost identical to Fra Mauro's, but with a smaller territorial coverage, kept at the Biblioteca Apostolica Vaticana (Borgiano XVI, *c.* 1450-1460). All other "witnesses" to the Venetian tradition have been lost over time. The world maps of Albertin di Virga (*c.* 1411-1419) and that of don Antonio Leonardi (numerous copies of which, noted in documents beginning in 1462, were given to the Republic of Venice, Borso d'Este, Pope Pio II, and the pope's nephew, Cardinal Francesco Todeschini Piccolomini) are completely untraceable today.[12] Even the two copies of Fra Mauro's world map that were commissioned in the fifteenth century for the Portuguese court of Afonso V (1457) and for the Medici palace in Florence (*c.* 1480) have disappeared. In light of the singularity and disgregation of these documents, it is clear that surviving witnesses such as Fra Mauro's are single tesserae that must have once belonged to vast mosaics, flotsam of largely lost cultural traditions that must have once been of great breadth and profound richness.

The majority of the works mentioned above belong to the class of documents that Edoardo Grendi succinctly and brilliantly characterized with the oxymoron "normal exceptions" (*eccezionale normale*). This expression fittingly describes those unique witnesses, whether documents or phenomena, that, as Fra Mauro's world map does, gather together heterogeneous elements and transmit them in a more digested and systematic way, more fully and more explicitly.

[12] The bibliographic description and the analysis of the contents of the majority of the maps cited above is provided in the chapters that follow.

Notably, these elements, by virtue of their linguistic and narrative forms, as well as the specific history of their reception and their relationship with the cultural matrix to which they refer, furnish important interpretational keys for "decoding" historical moments that, because they are not documented in a serial way, would otherwise be little known and much misunderstood. Such documents, precisely due to their *exceptional* character (in the sense of being an exception to a rule) can be studied to reveal the practices, themes, and aspects that were both central and *normal* ('normal' meaning commonplace and frequent) in the cultures and civilization in which they were created.[13]

Grendi's methodological intuition encapsulates quite accurately the relationship between the history of mid-fifteenth-century Venice and Fra Mauro's world map. The *mappa mundi* shows itself to be a document of great value and importance for unpacking the three principal social and cultural processes that characterize the history of Venice in this period : the development of long-distance information networks ; the foundation of a global Capitalist economy in which Venice served as one of the capitals and leading protagonists ; and, finally, an expansion – both physically and epistemologically – into the spaces and seas that earlier were not believed to be accessible to man. These are processes in which Venice played a determining role, with consequences for all of Europe, and for the *oikumene* of the fifteenth century. The city of canals was a gravitational center for commercial networks and worldwide communication networks ; it was a place from which not only goods, but also ideas, knowledge and information arrived from far-flung places in the Orient, Northern Europe, and the West, flowing together only to be sent back again in a centrifugal information flow of enormous reach. Concentrating on one single document did not mean limiting research to that one document.[14] The *mappa mundi*, its documentary basis, and its heterogeneous, explicitly named sources, all served as points of departure for this study's selection, research, and comparison of a number of aspects regarding culture and knowledge in the mid fifteenth century. Fra Mauro's work, together with the documents and the sources connected to it, thus provided a kind of agenda of themes, subject matter, and debates, according to which precisely defined research problems have been determined, and these, in turn, have served as a basis for wider diachronic and comparative modes of investigation. Admittedly this study involves "microscopic" observation focused per force upon a single document ; however, this document's uniqueness is offset by its *forma encyclopediae*, and the "microscopic" observation is conducted with such an intensive procedure that I believe this project is capable of discerning processes, practices, and issues that more general studies are not equipped to perceive and analyze.

Fra Mauro's world no longer exists. Its remains are crystallizations of highly articulate and complex traditions of discourse that entail much more than strictly geographic knowledge. This circumstance called for the undertaking of an extremely meticulous comparison between the graphic and verbal aspects of the *mappa mundi*, as well as its sources, as a means of bring-

[13] E. Grendi, "Micro-analisi e storia sociale," *Quaderni storici* 35 (1977), pp. 506-520 (p. 512).

[14] Edoardo Grendi and Giovanni Levi have demonstrated in theoretical terms that the scale of research does not necessarily circumscribe the extension of the investigation. See E. Grendi, *Il Cervo e la repubblica. Il modello ligure di antico regime*, Turin : Einaudi, 1993, in particular, pp. vii-xvi, and G. Levi, "A proposito di microstoria," pp. 111-134.

ing out their links, discrepancies, and interactions with their contemporary cultural context. This analysis was conducted from the perspective of cultural history in order to recreate, with as much integrity as possible, the sense of the documents under examination, in light of the civilization and knowledge base from which they originated. This research employed analytical tools and bibliographies gleaned from a number of diverse disciplines : from the historical and philological analysis of texts, to the history of science ; from the history of ideas, to the history of European expansion ; from the history of art, to the history of language ; and from the study of the relationship between Latin and vulgate intellectual cultures. The goal was to engineer an analysis capable of teasing out the complexity of functions and meanings that converge in Fra Mauro's *mappa mundi*, whose heuristic application goes well beyond what is entailed by the modern notion of a geographic chart. The knowledge that can be extracted from these diverse contexts – keeping in mind that all of these contexts are complementary components of a single cultural enterprise – offers an array of view points whose fragmentation very effectively mirrors the diverse and heterogeneous discursive traditions that make up the epistemological and graphic unity of the *mappa mundi*.

This book is organized into eight chapters which tackle the study of the *mappa mundi* from eight different approaches, all necessary for producing the most precise image possible from a documentary perspective, and an image that most precisely conforms to its original cultural and social context. The chapters are divided into two sections titled "Encompassing the Fifteenth-Century World and Re-Creating the *Imago mundi*" and "Patronage. Marketplaces. The History of Cartography". The first part, which has six chapters, offers an in-depth study of Fra Mauro's *mappa mundi*, aimed at understanding the processes of cultural construction and ownership of intellectual space implemented by the cartographer, and also at deciphering how these processes were parsed in the fifteenth and sixteenth centuries. The second part shows how the study of the *mappa mundi* and its documentary basis offers the opportunity to observe and understand two distinct phenomena of which the *mappa mundi* was part : the appearance, already in the mid-fifteenth century, of a European market for maps that interacted with the system of patronage, and about three centuries later, the gradual recognition in the major cultural centers of Europe of maps as a cultural heritage deserving of specific study. In the remainder of this Introduction we present the individual chapters, highlighting the intellectual debates to which they relate and aim to contribute.

Chapter I traces the biography of Fra Mauro, discusses the dating of the *mappa mundi* in the Biblioteca Nazionale Marciana and draws the history of its early reception. The starting point of my research is the records published by Gasparrini Leporace in 1956 and by Ratti in 1987, which are here systematized and expanded with new documents. The collection of the testimonies of those who in the past commissioned copies of the *mappa mundi*, described it, or used it to draw new maps, which, though as detailed as possible, cannot be considered complete or final, has a particular hermeneutic goal : to grasp the significance of the work from the historical point of view, moving beyond contradictory, cumbersome, and unnecessary judgments about the map as medieval, Renaissance, doctrinal, experimental, *et alia*.

Chapter II is devoted to the study of the *mappa mundi* as a cosmographic image and as a treatise on natural philosophy. The research on the cosmographical legends written by Fra Mauro takes inspiration from the studies of Rita Librandi and Rosa Casapullo on the importance of translations and vernacular adaptations of medieval Latin scientific works in the formation of modern vernacular vocabularies.[15] These studies led me to investigate and try to understand how Fra Mauro adapted and transmitted in the vernacular ideas that had been developed within the framework of the Aristotelian natural philosophy of the universities. With regard to studies that have highlighted a cleavage between the texts located ouside the circular frame of the *mappa mundi* and those within it, this analysis shows a substantial unity and continuity in both the composition and in the type of reflections in these two parts of Fra Mauro's work. This unity can be appreciated through consideration of a central theme of his cosmography : the representation and description – cosmological, cosmographical, geographical, and as a mercantile space – of the seas and oceans.

Chapter III completes the study of the cosmographical contents of the *mappa mundi* through a study of the pictorial representation and written description of paradise. Both Fra Mauro's text and the image of paradise, painted by Leonardo Bellini, reveal the influence of *De Genesi ad litteram libri duodecim* (The Literal Meaning of Genesis in Twelve Books) by St. Augustine. Through a discussion involving cosmography, theology, and art, it is revealed how Fra Mauro tries to resolve, in his representation of Eden, the contradiction implicit in the Augustinian conception of paradise, a place which is both real and a "spiritual feeling."

Chapter IV, which is dedicated to the Venetian reception of Ptolemy's *Geography*, revises and substantially expands a study I published in *Geographia antiqua* in 2004. The research draws inspiration from an article by Marica Milanesi published in *Imago Mundi* in 1996, in which she presentes a forgotten manuscript of the *Geography* which shows "contamination" between Ptolemaic cartography and nautical charts, which took place in Venice in the first half the fifteenth century.[16] This study by Milanesi, along with research by Patrick Gautier Dalché on the reception of the work of Ptolemy,[17] led me to study in greater detail the Venetian reception of the *Geography*, investigating the complex relationship between the *mappa mundi* and the work of Ptolemy, and also considering a heretofore ignored manuscript in the Biblioteca Marciana which hands down a vernacular translation of the *Geography* in an innovative form.

Chapter V focuses on how Fra Mauro made use of accounts of travels to Asia and sailing in the Indian Ocean and represented them in the structure of his *mappa mundi*. The starting point of my analysis was the account of Giovanni Battista Ramusio, who was the first to recognize the influence of Marco Polo's *Milione* on the cosmographic writing of Fra Mauro. Ramusio's description, together with the insights of Ivar Hallberg and the Orientalist Paul Pelliot on the relationship between the Z version of the *Milione* and the *mappa mundi*, were the starting point for new studies of which versions of Polo's text were used by Fra Mauro ; my research here includes substantial new information not included in the study I published in *Itineraria* in

[15] R. Librandi, *La Metaura d'Aristotile. Volgarizzamento fiorentino anonimo del XIV secolo*, 2 vols., Naples : Liguori, 1995 ; R. Gualdo (ed.), *Le parole della scienza : scritture tecniche e scientifiche in volgare, secoli XIII-XV*, Galatina : Congedo, 2001 ; R. Casapullo and M. R. Policardo, "Tecniche della divulgazione scientifica nel volgarizzamento mantovano del 'De proprietatibus rerum' di Bartolomeo Anglico," *Lingua e stile* 38 (2003), pp. 139-176.

[16] M. Milanesi, "A Forgotten Ptolemy : Harley Codex 3686 in the British Library," *Imago Mundi* 48 (1996), pp. 43-64.

[17] See P. Gautier Dalché, "The Reception of Ptolemy's Geography (End of the Fourteenth to Beginning of the Sixteenth Century)," in D. Woodward (ed.), *The History of Cartography*, vol. III, 1, *Cartography in the European Renaissance*, pp. 285-364 ; Id., *La Géographie de Ptolémée en Occident (IVe-XVIe siècle)*, Turnhout : Brepols, 2009.

2004.[18] The research also includes a study of the presence in the *mappa mundi*, beside the *Milione*, of Book IV of the *De varietate fortunae*, where Poggio Bracciolini recounts the long journey of Niccolò de' Conti in Asia. The analysis addresses the influence of these works on both texts and images; the latter aspect has been neglected in earlier studies. The results of this research relate to two cultural areas generally considered separate: on the one hand, the inquiry focuses on the process of drawing and writing the map; on the other, the selective interpretation that Fra Mauro made of the *Milione* and of Book Four of the *De varietate fortunae*, considering them especially for their content related to commerce, illustrates one way in which the two narratives were read in the fifteenth century.

Chapter VI, which concludes the first section of the book, brings together the findings of the earlier chapters in order to address the overall narrative structure of Fra Mauro's work and its cultural matrix. Contrary to rigid and monolithic interpretations of the *mappa mundi* of Fra Mauro and its cultural context – whether as a "medieval map" (and therefore doctrinal, symbolic, moralizing, unscientific, and at best outdated), or as a Renaissance or "modern" map (and therefore realistic or even "experimental") – in fact there emerges a composite authorial image in which monastic culture, scholastic and encyclopedic culture, humanistic culture, and oral and practical knowledge interact in the process of design and arrangement of the *imago mundi*. Immersion in the work of Fra Mauro and in the numerous sources related to it result in an appreciation of the cartographer as a figure who tried to dominate the intellectual complexity of the *imago mundi* using ancient, medieval, and modern authors, encyclopedic anthologies, other maps, and also stories and projects, heard or merely imagined. Fra Mauro is fully aware that this kind of synthesis is always difficult, often contradictory, sometimes simply impossible and wants the reader to be aware of it too, and often remarks on this fact in the legends.[19] This, perhaps more than the geographic contents, is one of the really new reflections of Fra Mauro.

Part II, Chapter I is designed to identify the factors that determined the cost of fifteenth-century maps. The use of coeval Venetian, Florentine, and Catalan documentary accounting sources, particularly the records of San Michele di Murano on the production of a copy of the *mappa mundi*, and also commercial papers, records, contracts, legal documents, and recollections, has allowed us to verify that, beyond the cartographic genre, it was the material aspects of the maps – particularly their size and the number of decorations – that determined their price. Determining the relative fifteenth-century values of the currencies mentioned in the documents analyzed here, together with the description of the various cartographic works, have allowed us to show a significant consistency in the cost of buying a fifteenth-century map, at least in Venice, Florence, and Barcelona. The evidence suggests that the system of patronage

[18] I. Hallberg, *L'Extrême-Orient dans la littérature et la cartographie de l'Occident des* XIIIe, XIVe, XVe *siècles*, Göteborg: W. Zachrissons Boktryckeri, 1906; P. Pelliot, *Notes on Marco Polo*, Paris: Imprimerie Nationale, Librairie A. Maisonneuve, 1959-1963, 2 vols. and *Index*, 1973, pp. 569-570. A. Cattaneo, "Scritture di viaggio e scrittura cartografica. La *mappamundi* di Fra Mauro e i racconti di Marco Polo e Niccolò de' Conti," *Itineraria* 3-4 (2004-2005), pp. 157-202.

[19] See for example this eloquent legend: "Note that in this Asia Major there are many kingdoms and provinces which I have not put because of lack of space. Thus, I have decided to omit many things and take out those that seemed to be to be best known. And I have not even made mention of many rivers, mountains and deserts in various parts (....) Similarly, I have decided to say nothing about the novelties, customs and standing of the various peoples, about the magnificent and powerful seignories, about the great diversity of animals, or about an infinite number of other things." *Mappamondo* (ed. 2006) *793. See also *Mappamondo* (ed. 1956) Pl. XXI, 75.

through which most of the maps of the period were commissioned interacted with a market system.

Part II, Chapter II draws on research conducted for an exhibition on the cartographic work of the Second Viscount of Santarém held in 2006 at the Biblioteca Nacional de Portugal and coordinated by João Carlos Garcia and Maria Joaquina Feijão, in which I was asked to address the history of the reproduction of the *mappa mundi* of Fra Mauro in six huge plates in the *Atlas* published by Santarém in 1852.[20] This study led me to expand this research to analyze the several ways in which Fra Mauro's map began to acquire great weight in eighteenth- and nineteenth-century European historical imagination, becoming, in the course of about a hundred years, first a monument to the greatness of Venice ; then a proof – against Enlightenment anticlericalism – of the centrality of knowledge produced and broadcast by religious orders ; then a symbol of the Western ability to imagine the world in its entirety and to conquer it ; and finally a testimony of the alleged temporal priority (*prioridade*) of Portuguese discoveries in history and in geographical conquests in nineteenth-century colonial conflicts. From the confluence of these approaches the history of cartography was born in the mid-nineteenth century as a separate discipline. Fra Mauro's *mappa mundi* played a crucial role in the establishment of this composite cultural enterprise.

The *orbis terrarum* of Fra Mauro is the framework for a union of many fragmented geographies which combines knowledge that is heterogeneous, complex, and layered. Study of the *mappa mundi* reveals Fra Mauro's explicit awareness of the impossibility of composing a definitive work, and his efforts were rather to create a cosmography *in fieri*, in progress, influenced by new discoveries, literary and geographical, and by new historical contexts with which Fra Mauro and Venice came into contact, progressing not for the sake of mere accumulation, but rather for integration and evolution. It was, therefore, a process that assumed the coexistence of "different" cosmographies, but also a background cosmography that combined the fragmented geographies. Reading and analyzing the legends and images indicates that the *mappa mundi* is above all the representation of an interplay between cultural and social *milieux* which scholars have generally considered separate. By following this interplay, however, one perceives in the map a dialogue of an unexpected openness, intensity, and richness which transcends the disciplinary boundaries that historians have often imposed on works so layered and rich.

I have tried to select themes for research which accommodate this articulation of content and documentary perspective and which reveal connections and interdependencies. I am fully aware that these themes do not exhaust the study of the *mappa mundi* : this work is also a cosmography *in fieri*. Many issues important for a complete understanding of the work have not yet been processed by scholars. For example, the geography of Fra Mauro has not been the subject of a systematic comparative analysis with other contemporary Western *mappae mundi*, and even our knowledge of the representation of Asia, *Asia mediana* or the *Rossia*, in the *mappa mundi*, still depends to a large extent on studies made in the late nineteenth and early twentieth century (in particular those of Yule and Pelliot on the geographical content of the *Milione* of

[20] J. C. Garcia and M. J. Feijão, *A obra cartographica do 2º Visconde de Santarém*, Lisbon : Biblioteca Nacional de Portugal, 2006.

Marco Polo) that have not yet been fully updated, but only systematized.[21] There is no detailed comparison of Fra Mauro's geographical knowledge with contemporary Islamic maps or with the Sino-Korean *Honil kangni yŏktae kukto chi to* (Map of Integrated Lands and Regions of Historical Countries and Capitals, Korea, *c.* 1479-1485), the oldest surviving map which was designed in Asia and represents the western *oikumene*, which is contemporary with the *mappa mundi* and quite surprisingly shares some sources with it.[22] Even historians of language have not yet dealt with Fra Mauro's Venetian vernacular and only recently scholars have begun to study the material aspects of the *mappa mundi*. These issues are beyond the capabilities of a single researcher, and require the combined efforts of many scholars. I hope this study will be of help to those who in the future decide to pursue these and other studies.

*

The legends in the Venetian vernacular on the *mappa mundi* have been transcribed by comparing the high-resolution image on CD-ROM edited by Caterina Balletti and Francesco Guerra of the CIRCE laboratory at the IUAV University of Venice, which comes with the edition by Piero Falchetta (*Mappamondo* ed. 2006), with the admirable diplomatic transcription and forty-eight color plates of the edition edited by Tullia Gasparrini Leporace (*Mappamondo* ed. 1956).[23] The translation of the legends into English generally derives from Piero Falchetta's and Jeremy Scott's admirable English edition, unless otherwise indicated. With regard to the bibliography, full references are given the first time they are cited in each chapter and then in the Bibliography.

In the pages that follow the work of Fra Mauro is referred to as *mappamondo* or *mappa mundi*. I think it is a mistake to call it a "planisphere" for the following reasons. In Latin (and in translations into the vernaculars of the fifteenth century) *planispherium* referred to a geometric-mathematical representation of the celestial sphere and the visible stars on a plane circle, referring to the work of the same name by Ptolemy. Beginning in the late sixteenth century, the term was also used to refer to cartographic representations of the entire Earth using

[21] See the sources and bibliography cited in the comments on the legends regarding Asia in *Mappamondo* (ed. 2006).

[22] I refer here to a preliminary comparison I made between the *Kangnido* and the *mappa mundi* with the collaboration of two Korean scholars, Oh, Gil-sun, and Song, Sang-yong, and of an Italian sinologist, Paolo de Troia. See A. Cattaneo, "Il lungo viaggio della cartografia universale attraverso il mondo islamico, la Cina, la Corea e il Giappone (sec. XIV-XVI). Studio dell carta universale *Honil kangni yŏktae kukto chi to* (Corea, *c.* 1479-1485) in una prospettiva comparata," in M. Angelillo (ed.), *Culture, religioni e diritto nelle società dell'Asia Orientale*, Rome: Bulzoni, 2010, pp. 177-200 (*Asiatica Ambrosiana, Saggi e ricerche di cultura religioni e società dell'Asia* 2: Atti de convegno "Culture, religioni e diritto nelle società dell'Asia Orientale" organizzato dalla Accademia Ambrosiana, Classe di Studi sull'Estremo Oriente, Milano).

[23] With regard to the forty-eight tables that accompany the 1956 edition, it must be emphasized that, more than fifty years after their publication (edited by the printers of the Poligrafico e Zecca dello Stato), they prove to be extraordinarily precise and accurate. Comparison with the images on the CD-ROM shows the almost perfect agreement with the recent digital reproduction, which is a truly remarkable accomplishment for engravings made by hand.

one of several available projections.[24] Placido Zurla in his monograph of 1806 referred to the *mappamondo* as a "planisfero" in order to emphasize its scientific character, in a positivistic sense. In less scholarly contexts, the term "planisphere" has also come to be used as a synonym for "map." However, since the *mappa mundi* is not based on a mathematical projection and this is a scholarly study, I have decided not to use the term "planisphere."

[24] S. Battaglia and G. Bàrberi Squarotti (eds), *Il grande Dizionario della lingua italiana*, 21 vols., Turin: UTET, 1961-2002, vol. 13, p. 637.

I
ENCOMPASSING THE FIFTEENTH-CENTURY WORLD AND RE-CREATING THE *IMAGO MUNDI*

CHAPTER I

FRA MAURO'S LIFE AND WORK : A CRITICAL SURVEY OF SOURCES

WITH EMPHASIS ON HIS *MAPPA MUNDI* AND ITS RECEPTION TO 1600

We can never insist enough upon the necessity of rigorously examining, decade by decade, even year by year, that grand fact of civilization that was the Quattrocento.[1]

Eugenio Garin

I. – Life and works

The aim of this first chapter is to present Fra Mauro's biography and discuss the date of the *mappa mundi* in the Marciana Library ; later we shall consider the history of the reception of the work of Fra Mauro in the fifteenth and sixteenth centuries. In particular we shall try to clarify and define the reasons and context for the prestigious commissions for copies of Fra Mauro's work from the Portuguese Court of Afonso V (1457) and the Medici family (*c.* 1480). This section will permit a better understanding, beyond the useless and misleading labels that scholars continue to apply to the work of Fra Mauro – such as "medieval" or "late medieval" used in general in an inexplicably pejorative sense, or "Renaissance" and "early Renaissance" just as inexplicably "less" pejorative than the former – of the assessment and use of the *mappa mundi* in the one hundred and fifty years that followed its creation. These labels are simple generalizations, based on monolithic, linear and tautological visions of time and history – even obscuring the objects which they claim to enlighten – which do not help us to understand or explain the objects they pretend to identify.[2] In the best case scenario they merely label them from the chronological point of view, and even in that case in a very disputable if not uncertain way : it would be quite simple to identify for any extant fifteenth-century map (whether marine charts, or *mappae mundi*) the simultaneous use by different historians or historiographic traditions of all such labels, ranging from "medieval" to "late medieval," "Renaissance" or "early Renaissance" and even "modern."[3]

[1] "Non si insisterà mai abbastanza sulla necessità di scandire con rigore, decennio per decennio, e quasi anno per anno, quel grande fatto di civiltà che è il Quattrocento," E. Garin, *Umanisti, artisti, scienziati. Studi sul Rinascimento italiano*. Rome : Editori Riuniti, 1989, p. 165. The translation is mine.

[2] J.-M. Mandosio, "Alberti et l'esprit du caméléon," in F. Choay and M. Paoli (eds), *Alberti humaniste, architecte*. Paris : ENSBA, 2006, pp. 128-141 (p. 134) ; Id., *D'or et de sable*. Paris : Éditions de l'Encyclopédie des Nuisances, 2008, in particular "La Mesure de la réalité, ou la Grande Transformation racontée aux *golden boys*," pp. 107-141.

[3] "It has yet to be shown – argues Patrick Gautier Dalché convincingly – that there even was such a thing as a 'medieval way of describing and representing the world' ; a Weberian 'ideal-type' of this sort (the sole tenable, in

Following the methodological approach formulated by Carlo Ginzburg and Patrick Gautier Dalché – for the study of the *Utopia* of Sir Thomas More and the *Geography* of Ptolemy, respectively, two studies carried out based on the history of those two works' reception[4] – we will study the discursive testimony of the early spectators of the map, thus inserting the *mappa mundi* into a rich and partly forgotten literary as well as cartographic reception. This will prove to be an effective method to understand the historical and cultural significance of the cosmography of Fra Mauro, beyond useless value judgments that say very little, or even obscure the meaning of the work in the context of the knowledge and cultures of the fifteenth and sixteenth century. To collect and analyze the descriptions and, in general, the means by which men contemporary with the Camaldolese friar and those who succeeded him approached the *mappa mundi*, allow us to grasp the sense of the work in the context of knowledge and culture of the time (and the century immediately following). This also allows us to take an important step toward understanding the position and role of cosmography in the fifteenth and sixteenth century recognizing and distinguishing how maps worked in different *milieux* by using existing knowledge of those time.

1. A work in search of an author

The Camaldolese *converso* Mauro, celebrated by his contemporaries with the title *cosmographus incomparabilis* ("cosmographer without equal")[5] is remembered chiefly as the author of a monumental *mappa mundi* designed around 1450 and preserved for centuries in the monastery of San Michele di Murano in Venice. In May 1811, after the Napoleonic dissolution of all religious orders, it was acquired by the Biblioteca Marciana (Library of St. Mark, named after the patron saint of Venice; today Biblioteca Nazionale Marciana); a few months later it was deposited in the Ducal Palace of Venice, but finally and definitively returned in 1924 to the Marciana.[6]

principle) has never been produced by any historian". The same applies to an alleged "Renaissance way" of describing and representing the world. P. Gautier Dalché, "The Reception of Ptolemy's Geography (End of the Fourteenth to Beginning of the Sixteenth Century)," in D. Woodward (ed.), *The History of Cartography*, vol. III, 1, *Cartography in the European Renaissance*, Chicago: University of Chicago Press, pp. 285-364 (p. 287).

[4] For a well-conducted theoretical and methodological study of the relationship between the historical meaning of a work and the history of its literary reception see C. Ginzburg, "Il vecchio e nuovo mondo visti da Utopia," in *Nessuna isola è un'isola. Quattro sguardi sulla letteratura inglese*. Milan: Feltrinelli, 2002, pp. 17-44; and Id., *Rapporti di forza: storia, retorica, prova*. Milan: Feltrinelli, 2000. P. Gautier Dalché, "The Reception of Ptolemy's *Geography*," pp. 285-287; Id., *La Géographie de Ptolémée en Occident (IVe-XVIe siècle)*. Turnhout: Brepols, 2009, pp. 7-12.

[5] This commendation appears on a commemorative medal with a profile portrait of Fra Mauro: Venice, City Museums' Collection, Giovanni Boldù (attributed to), Frater Maurus, *c.* 1455-1459 or 1464, upon which the following words are inscribed "Frater Maurus S. Michaelis Moranensis de Venetiis ordinis Camaldulensis chosmographus incomparabilis." The medal is reproduced on the cover sheet of *Mappamondo* (ed. 1956) and in V. Meneghin, *San Michele in Isola di Venezia*, 2 vols., Venice: Stamp. di Venezia, 1962, vol. 1, p. 81. For a discussion on this medal and the attribution to Boldù, see S. Marcon, "Leonardo Bellini and Fra Mauro's World Map: the Earthly Paradise," in P. Falchetta, *Fra Mauro's World Map*, Turnhout: Brepols, 2006, p. 145; from here onwards, *Mappamondo* (ed. 2006).

[6] Fra Mauro's work was first hanged in a chapel in the left side of the church of San Michele and since 1655 in the monastic library where it was displayed on the wall above the entrance. Following the Napoleonic Edict of

Any attempt to reconstruct the biography of Fra Mauro must immediately confront the poverty of archival sources. There is a general lacuna in the documentary sources relating to the Camaldolese Order in the years between 1400 and 1453, and this problem is compounded by a lack of the registers of general priors and the dispersal of archives caused by the suppression of monasteries. In particular, the suppression of the monastery of San Michele di Murano, and the subsequent transfer of its archive to the Camaldolensian monastery of San Gregorio al Celio in Rome (*c.* 1806-1811), and then to the Archivio di Stato of Rome (1939), caused the loss of most of the documents pertaining to the first half of the fifteenth century.[7] Moreover, as we shall discuss later in this chapter, as early as 1480, when Pietro Dolfin was prior of San Michele, the Camaldolese Order appropriated the work of Fra Mauro and obliterated the memory of its creator, so as to cause even his name to be forgotten. The possibility, therefore, of sketching the life of Mauro at the time when he was a *converso* active as a cosmographer in the monastery of San Michele is vested in a few archival documents that will be presented systematically and in part transcribed in the Appendix to this chapter : some loose parchments, the register of a Venetian magistracy, and the register of the monastery of San Michele di Murano for the years 1453-1459 (with a note on Fra Mauro in 1464) edited by the prior Maffeo Gherardo. Working from the few documents that we have, we present the following biographical note.

Fra Mauro is registered as a lay brother in San Michele di Murano as early as 1409, and as some deeds and records demonstrate, he spent his whole adult life in the monastery.[8] In the four decades following his entry into the monastery, Fra Mauro distinguished himself not only by his *mappae mundi* and cosmographic knowledge, but also as both a cartographer of and expert about the lagoon of Venice and the Veneto region. In 1437 he drew a map of the possessions of the Camaldolese monastery of San Michele di Leme in Istria ; the cause of the commis-

Compiègne of April 25, 1810, which finally abolished ecclesiastic guilds and communities, the Camaldolesian monks had to leave the monastery of San Michele. Jacopo Morelli, director of the Biblioteca di San Marco, around May of 1811 managed to acquire the *mappa mundi* and had it transported to the Hall of Sansovino, on the upper floor of the Library. A few months later, however, this wing of the palace was confiscated by the Viceroy Eugène de Beauharnais and, on August 28, Morelli was ordered to transfer the *mappa mundi* to the Palazzo Ducale. In March 1812 the work was first carried to the Sala del Maggior Consiglio, and the following year to the Sala dello Scudo, where it was placed beside four large sixteenth-century canvases with painted maps (*teleri*), which had been restored between 1761 and 1763 by Francesco Grisellini. In 1882 the *mappa mundi* was assigned to the Archaeological Museum of Venice but, in the absence of adequate exhibition space, it remained in the Ducal Palace until 1924 when, following the jurisdictional reassignment of the building from the Italian state to the municipality of Venice, the Director of the Marciana Library at the time, Luigi Ferrari, managed to have the work transferred back to the Library. Registered with inventory number 106173, it has been exhibited since February 1924 in the Old Library of Sansovino "for contemplation by the most illustrious Lords of Venice." See the precise reconstruction by T. Gasparrini Leporace, "Introduzione," in Ead. (ed.), *Il Mappamondo di Fra Mauro camaldolese*, con la presentazione di R. Almagià. Rome : Poligrafico della Zecca dello Stato, 1956, pp. 11-16 ; from here onwards, *Mappamondo* (ed. 1956).

[7] See the chapter "Parcours historiographique et documentaire" in C. Caby, *De l'érémitisme rural au monachisme urbain. Les Camaldules en Italie à la fin du Moyen Âge*. Rome : École Française de Rome, 1999, pp. 9-56.

[8] Venice, Archivio di Stato, S. Michele di Murano, Busta 3, loose manuscript on vellum, proxy of the Chapter to the abbot Paolo Venerio and Nicolò da Modena, drafted by the notary, Prosdocimo Domenico da Padova, July 9, 1409. "Frater Maurus de Veneciis conversus" is recorded as being part of the chapter of the Monastery of San Michele of Murano as a lay brother (*converso*). See Document 1 in the Appendix of Chapter I. Venice, Archivio di Stato, S. Michele di Murano, Busta 1, fasc. I, f. 20r, April 15, 1434 : "Fra Maurus de Vinetiis" is listed among those who participated in the chapter of San Michele of Murano.

sion was a dispute between the monastery and an Istrian gentleman about property rights in the territory. The original manuscript on vellum, already heavily damaged in the mid-eighteenth century, is probably lost. It is preserved in a printed edition of 1762, published under the direction of Giovanni Battista Mittarelli and Anselmo Costadoni in the *Annales Camaldulenses*.[9] A few years later in 1444 Fra Mauro was retained by the *Savi ed Esecutori alle Acque* – a Venetian magistracy with executive powers to manage the lagoon – to provide technical expertise on a project that involved the diversion of the River Brenta.[10] These two episodes shed light on his skills, which included the ability to represent the world "on a small scale" and to draw maps of regions, provinces, and small territories for administrative purposes and as instruments to mediate relationships with the environment. Placido Zurla, in his monograph of 1806 devoted to the works of Fra Mauro, which is a veritable *summa* of over a century of Camaldolese historiography, shows, based on a register of San Michele which is now lost, that between 1448 and 1449 Fra Mauro was working to create "mappamondi."[11] In this context, and in the absence of the original, it is extremely difficult to indicate what lies behind that term ; probably Fra Mauro made not only *mappae mundi*, but also maps of various types, including perhaps nautical charts and chorographic maps. As we shall see in the next paragraphs, it is likely that Fra Mauro prepared during these years the *mappa mundi* now kept in the Biblioteca Marciana.

Between February 8, 1457 and April 24, 1459 the name of Fra Mauro appears often in the records of San Michele.[12] The Camaldolese was engaged in the preparation of a world map commissioned by Afonso V (1432-1481), King of Portugal and the Algarve. The ambassador João Fernandes Silveira was to request the map on behalf of the Portuguese Court ; he arrived in Milan and Venice in 1456 to negotiate the participation of Portuguese troops in the crusade which had been sponsored by Pope Nicholas V and renewed by Calicxtus III following the fall of Constantinople in 1453.[13] In the preparation of the new *mappa mundi* scribes and painters

[9] G. B. Mittarelli and A. Costadoni, *Annales camaldulenses ordinis Sancti Benedicti quibus plura interseruntur tum ceteras Italico-monasticas res, tum historiam ecclesiasticam remque diplomaticam illustrantia*, 9 vols., Venice : Giambattista Pasquali, 1755-1773, vol. 7, 1762, pp. 252-256. At this regard see Almagià, "Presentazione," in *Mappamondo* (ed. 1956), p. 5, 10. See doc. 3 in the Documentary Appendix of Chapter I.

[10] Venice, Archivio di Stato, Savi ed Esecutori alle Acque, Capitolare I, Busta 342, f. 47v, August 2, 1444 ; Venice, Archivio di Stato, MS 64, Bernardino Zendrini, *Memorie storiche dello stato antico e moderno delle lagune di Venezia*, 1726, vol. 2, pp. 172-180.

[11] Zurla quotes an account book of the monastery of San Michele from the year 1448 until July 1449 (Register 451 of the Monastery of San Michele di Murano, *Spese di D. Niccolò Monaco economo del Monastero dal 1448 fino al luglio 1449* (Expenses of Niccolò Monaco treasurer of the Monastery between 1448 until the end of July 1449). According to Zurla, many expenditures made by Fra Mauro for the acquisition of colors, parchment and gold for the preparation of maps were noted in this register. The register has been lost, most likely during the forced transfer of the archival documents of San Michele from Venice to Rome in 1811. See P. Zurla, *Il mappamondo di Fra Mauro camaldolese descritto ed illustrato da D. Placido Zurla dello stess'ordine*. Venice : s.n., 1806, pp. 79-80 ; and also D. Woodward, "Medieval *Mappae mundi*," in J. B. Harley and D. Woodward (eds), *The History of Cartography*, vol. 1, *Cartography in Prehistoric, Ancient and Medieval Europe and the Mediterranean*. Chicago : University of Chicago Press, 1987, p. 367.

[12] Rome, Archivio di Stato, San Gregorio al Celio, inv. 25/II, n. 9, n. 63, Gherardo Maffeo, *Entrata e uscita. San Michele di Murano* (1453-1460), ff. II, 41r, 42r, 44r, 123v-125r, 128v, 129r, 137v, 139r, 144r, 169r-171v, 172v, 173v, 174r, 175r, 180v, 181r. See the transcription in the Documentary Appendix of this chapter.

[13] *Monumenta Henricina*. Coimbra : Comissão Executiva das Comemorações do V Centenário da Morte do Infante D. Henrique, 1960-1978, 15 vols., vol. 12, doc. 109, pp. 202-204 ; doc. 175, pp. 350-352. On João Fernandes da

were employed, and Andrea Bianco, the Venetian cartographer and official of the galleys (*comitus*), also collaborated on the project. The map, for which a total of twenty-eight ducats were paid, was dispatched to Portugal on April 24, 1459, escorted by the Venetian patrician Stefano Trevisan. The *mappa mundi* made for the Portuguese Court is considered lost and it is likely that this map was a copy, perhaps revised on the basis of specific information received from the buyer, of that now housed in the Biblioteca Marciana in Venice.[14] The *mappa mundi* in the Marciana Library certainly served as a model for a copy of the work of Fra Mauro commissioned around 1480 by the Medici family of Florence. As we shall see below, this copy was made by Florentine painters especially appointed to San Michele of Murano, and instead of legends in the Venetian language, it had legends which had been translated into Latin by Piero Dolfin, the prior of San Michele.

The register that covers the years 1453-1460 mentions Fra Mauro not only for his work as a cosmographer but also as a collector of the monastery's benefices and rents. According to the register of Maffeo Gherardo, Fra Mauro was also responsible for the commercial dealings between the city of Venice and the monastery.[15] On October 20, 1459, Maffeo Gherardo (1405-1492), Abbot of San Michele di Murano, noted in his accounting book (*Entrata e uscita*) that all of the *mapamondi, desegni et scripture* (maps, drawings, and writings) of Fra Mauro – with the exception of the *mappa mundi* that was hanged in one of the chapels of the church of San Michele in August 1460 – had been transported to the Camaldolese monastery of San Giovanni Battista alla Giudecca,[16] sealed in a wooden box, and then were returned to the monks of San Michele in October 1464 :

> October 20, 1459. I note that I deposited the copies of Fra Mauro's maps of the world (*mapamondi*), drawings and writings in the [Camaldolese] monastery of *Zuano dela Zudecha* in the care of the prior don Andrea. [I also note] that these writings and drawings are held in a crate, or better a chest sealed with a lock whose key I keep here with me [in the monastery of San Michele of Murano]. I received everything back on October 25, 1464.[17]

This document suggested to Roberto Almagià the hypothesis that Fra Mauro died between May 1459, when the abbot Maffeo noted that he had received eight ducats from Fra Mauro, and

Silveira see *Portugal – Dicionário Histórico, Corográfico, Heráldico, Biográfico, Bibliográfico, Numismático e Artístico*. Lisbon : João Romano Torres, 1904-1915, vol 1, pp. 399-400.

[14] Rome, Archivio di Stato, San Gregorio al Celio, inv. 25/II, n. 9, n. 63, Gherardo Maffeo, *Entrata e uscita. San Michele di Murano* (1453-1460), ff. 125v and 169v ; Lisbon, Archivo da Torre do Tombo, Chanceleria de D. Afonso V, liv. 1, f. 2.

[15] See, for example, f. 42r of the *Entrata e uscita* : "L'officio conscripto dio auer adì sopradetto che ser Alvise Sandeli me mando contadi in monede ducati 10 [...] paga de marz 1446 in monede d. 186 per el pro de Ossoduro paga de setembre 1445 portali Frate Mauro converso."

[16] Giudecca is an island in the Venetian Lagoon that lies south of the central islands of Venice, from which it is separated by the Giudecca Canal. On the Camaldolese monastery of San Giovanni at Giudecca see F. Corner (1693-1778), *Notizie storiche delle chiese e monasteri di Venezia e di Torcello*, reprinted Bologna : Forni, 1990, pp. 457-458.

[17] *Entrata e uscita*, f. II : "1459 adì 20 octubre. Memoria fazo chomo le copie de mapamondi e desegni e scripture de Fra Mauro ho depositado al monastier de frari da Zuano dela Zudecha in man del prior del detto monastier zoè don Andrea, le qual scripture e desegni tuti sono posti in una chassa ouer bancho con uno luchetto, la chiave del qual ho qui apresso de mi. ho avuto | tutto in drietto e però adì 25 octubrio 1464."

October 1459.[18] But it is also possible that the Camaldolese friar did not die in 1459, but instead was transferred in that year, for unknown reasons, to the Camaldolese monastery of Giudecca – which was also home to a hospital – and that he died there before October 25, 1464, when his "*mappae mundi*, drawings, and writings" were finally brought back to San Michele. Whatever the case with Fra Mauro's biographical details, it is important to stress that the abbot Maffeo uses the plural in referring to "copies of *mapamondi*, drawings and writings," which constitutes undoubted proof that the overall work of Fra Mauro was conspicuous and also included writings among which it is safe to assume there were the cosmographical notes made in preparation for the extensive legends on the *mappa mundi*. If the life story of Fra Mauro can only be partially reconstructed, detailed analysis of the *mappa mundi* of the Biblioteca Marciana (the only work that can certainly be attributed to him[19]) its articulated reception, its complex cultural matrices, and composite narrative structure provide many important elements of his cultural and intellectual biography, and at the same time, important material for understanding the cultural processes of description and representation of the *imago mundi* in the fifteenth century.

2. The dating of the *mappa mundi* in the Biblioteca Marciana

Fra Mauro's *mappa mundi*, today held at the Biblioteca Nazionale Marciana in Venice, was originally framed and put on display in the sacristy of the church of San Michele on August 26, 1460.[20] While the date of the framing of the *mappa mundi* at the Biblioteca Marciana is certain, there is no concrete documentary evidence as to when it was actually drawn up. Two main theories have been put forward. The first, proposed by Placido Zurla, argues that Fra Mauro worked on the *mappa mundi* between 1457 and 1459, during roughly the same period in which he was working on a copy commissioned by the Portuguese court of Afonso V, produced between 8 February 1457 and 24 April 1459, and on this date dispatched to Portugal.[21] The second hypothesis, proposed by Roberto Almagià, suggests that Fra Mauro created the Marciana *mappa mundi* sometime between 1446 and 1453, some ten years before the copy for Afonso V.

In light of recent research concerning the depiction of the Earthly Paradise in the bottom left-hand corner of the map – an illustration now attributed to the Venetian miniaturist Leonardo

[18] R. Almagià, "Carta nautica con elementi corografici di Anonimo Veneziano (1450 circa)," in *Monumenta Cartographica Vaticana*, vol. I, Rome : Vatican City, 1944, pp. 32-39.

[19] The Vatican Library has a nautical chart which is very closely related to the *mappa mundi* of the Marciana Library, but which cannot be attributed directly to Fra Mauro. Vatican City, Biblioteca Apostolica Vaticana, Borgiano V, manuscript on vellum, 132.5 × 75 cm, oriented to the South, with 38 legends, of which at least 20 are almost identical to those of the Biblioteca Marciana *mappa mundi*. See Almagià, "Carta nautica con elementi corografici di Anonimo Veneziano," pp. 32-39 ; Id., "Presentazione," in *Mappamondo* (ed. 1956), pp. 7-8 ; for the transcription of the legends of the Borgia Map : H. Winter, "The Fra Mauro Portolan Chart in the Vatican," *Imago Mundi* 16 (1962), pp. 17-28. For a discussion and a comparison between the legends of the Borgia Map and those of Fra Mauro's *mappa mundi* see *Mappamondo* (ed. 2006), pp. 36-52 and pp. 739-753.

[20] The date of 26 August 1460 can be read on an inscription on the back of the wooden frame of the *mappa mundi* : "MCCCCLX a dì XXVI auosto fu complido questo lavor." See the image of this inscription in *Mappamondo* (ed. 1956), p. 11.

[21] For the date of the shipping of the *mappa mundi* to Portugal, see *Entrata e uscita. San Michele di Murano* (1453-1460), f. 125v.

Bellini (*c.* 1423/1425-*post* 1484) – a reinterpretation of Almagià's theory suggests two different periods of work on separate elements of the map. While the cosmological and geographical parts proper were probably worked upon from *circa* 1448 to 1453 (or are at least based upon geographical and historical material produced during those years), the portion depicting the Earthly Paradise was most likely added later, presumably after 1455.[22] It is important to remark that apart from the note regarding the framing on the back of the *mappa mundi*, which gives us 26 August 1460 as the latest possible date of production, we have no certain documentary evidence to settle the question. However, given the present state of research, the hypothesis of different stages of work upon the *mappa mundi* appears the most plausible. Although the difference proposed by these theories about the map's dates of completion appears minor these were, both politically and culturally, particularly important years. In the political sphere the Fall of Constantinople (May 29, 1453) and the subsequent, though ultimately unheeded, call for a new crusade from the three popes (Nicholas V, Calicxtus III, and Pius II) who reigned from 1453 to 1459 would have a tangible effect upon relations and borders between the Muslim and Christian worlds. Meanwhile, in the area of cartography, the period of 1448-1459 saw a substantial addition to the field of Western cosmographical knowledge – thanks, primarily, to the Portuguese voyages of exploration along the African coast, but also to the availability of new texts (in particular Strabo's *Geography*, of which Guarino Veronese and Gregorio Tifernate completed their Latin translation in 1458).[23]

a. *Zurla, Santarém, and Cortesão : 1459*

Cardinal Placido Zurla appeals to a particular feature of the *mappa mundi* as evidence of its creation between 1457 and 1459, pointing to the range of toponyms to be found on the Atlantic coast of Africa : "In that same year of 1456 [the Portuguese] sighted, and named *Capo rosso*,[24] the *Santa Anna* and *San Domenico River*, and finally reached as far as the *Rio Grande*, at around 11° degrees of latitude [...]. Therefore [...] I have no doubts in naming the date, at least for the completion of the map, as contemporary with that of the map made for king Alfonso – that is, from 1457 to 1459."[25] Zurla's argument has been very influential and, with rare

[22] On the attribution of the Earthly Paradise to Leonardo Bellini, see Marcon, "Leonardo Bellini and Fra Mauro's World Map : the Earthly Paradise," in *Mappamondo* (ed. 2006), pp. 135-161 (with an interesting discussion on the date of the *mappa mundi* at pp. 143-146).

[23] In 1453 Nicholas V commissioned Guarino Veronese and Gregorio Tifernate to translate Strabo's *Geography*, books I-X and XI-XVII respectively. The text was to be based on a Greek manuscript copied in Constantinople and brought to Italy by Ciriaco d'Ancona, where it was then divided into two parts for the translators (Florence, Biblioteca Mediceo Laurenziana, XXVIII 15, books I-X ; Eton, Eton College Library, ms. 141, books X-XVII). Veronese completed the translation of all the books of the *Geography* on 13 July 1458, dedicating the finished work not only to Nicholas V (who had died in 1455) but also to Iacopo Antonio Marcello, who presented the work to René of Anjou. On the history of the translation, see A. Diller, *The Textual Tradition of Strabo's Geography*, Amsterdam : Adolf M. Hakkert, 1975, pp. 119-129 ; A. Diller and P. O. Kristeller, "Strabo," in F. E. Cranz, P. O. Kristeller and V. Brown (eds), *Catalogus Translationum et Commentariorum. Medieval and Renaissance Latin Translations and Commentaries : Annotated Lists and Guides*, Washington, D.C. : Catholic University of America Press, 1960-1986, vol. 2, 1971, pp. 225-233 ; S. Gentile, *Firenze e la scoperta dell'America*, Florence : Olschki, 1992, entries 89-91, pp. 183-191.

[24] "Cavo rosso," *Mappamondo* (ed. 1956) Pl. XVIII, 11 ; *Mappamondo* (ed. 2006) *563.

[25] "Nello stesso anno 1456 [i portoghesi] videro, e denominarono il Capo Rosso, il Rio di S. Anna, e di S. Domenico, ed arrivarono fino al Rio Grande, all 11° di lat. all'incirca. [...] Dunque [...] che io non dubito stabilire, almeno

exceptions, the date still generally given for the *mappa mundi* is 1459.[26] One variant upon this theory dates back to the Portuguese ambassador and historian Manuel de Barros e Sousa, Second Viscount of Santarém (1791-1856), and the twentieth-century historian Armando Cortesão, who argued that in the years 1457-1459 Fra Mauro prepared two *mappae mundi*, copying the one that remained in the monastery of San Michele from that produced for the Portuguese court. The latter map, therefore, was the prototype, and its loss, according to Santarém and Cortesão, dated either from the unification of the Spanish and Portuguese crowns in 1581, when a significant amount of manuscripts and books was transferred to Spanish libraries, or from the terrible earthquake and fire which devastated Lisbon in 1755.[27] Though the difference between the dates proposed by Santarém and those suggested by Zurla is not very great, perhaps, it raises an issue that can best be answered by looking at the production methods of ancient cartography in general.[28] The idea that the *mappa mundi* produced for the Portuguese court was really some sort of prototype on which other copies were based is not tenable in light of what Maffeo Gherardo records in his accounts of the production of the map. Registering costs for *pentori* (illuminators) and *scriptori* (scribes), paid by the day, the accounts reveal that these artisans worked on the *mappa mundi* for about sixty days. The entry reads:

> On 8 February, Lio Roso, in the name of the magnificent King of Portugal, brought in cash 11 ducats (valued at 129 soldi per ducat) to pay the painters for their work on his [the king's] mappa mundi and for other expenses [...] On 21 October 1457, I note that I gave cash to Fra Mauro to pay a scribe who worked and wrote on the mappa mundi for 17 days, 12 soldi per day, for a total amount of lire 17, soldi 4 (valued at 124 soldi per ducat). On 7 October 1458, I gave 56 soldi (valued at 124 soldi per ducat) in cash to Don Francesco da Cherso to pay a scribe who worked on the aforementioned map for 4 days, 14 soldi per day. On 9 October [1458], I gave 14 soldi (valued at...) in cash to the same person [Francesco da Cherso] to pay a scribe who worked on the mappa mundi for one day.[29]

quanto al compimento di essa, contemporanea a quella fatta per il Re Afonso, cioè dal 1457 al 1459." P. Zurla, *Il mappamondo di Fra Mauro camaldolese descritto ed illustrato da D. Placido Zurla dello stess'ordine*. Venice, s.e., 1806, pp. 84-90 (the quote is from p. 90).

[26] To cite only the most important reference works for the history of cartography, Y. Kamal, *Monumenta Cartographica Africae et Aegypti*, 5 vols., Cairo, 1926-1951, vol. 5, p. 1495; M. Destombes, *Mappemondes A.D. 1200-1500. Catalogue préparé par la Commission des Cartes Anciennes de l'Union Géographique Internationale*. Amsterdam: N. Israel, 1964, file 52.14. Even the *History of Cartography*, the value and scholarly standing of which are beyond question, gives this date; see J. B. Harley and D. Woodward (eds), *The History of Cartography*, vol. 1: *Cartography in Prehistoric, Ancient and Medieval Europe and the Mediterranean*. Chicago and London: University of Chicago Press, 1987, p. 367.

[27] M. F. Santarém de Barros e Sousa, "Note sur le mappemonde de Fra Mauro," *Bulletin de la Société de Géographie de Paris*, 1846, pp. 251-252. A. Cortesão, *Cartografia e cartógrafos portugues dos séculos XV e XVI (Contribudo para um estudo completo)*, 2 vols., Lisbon: Seara Nova, 1935, vol. 1, pp. 121-125.

[28] The nationalism of the thesis put forward by Santarém and Cortesão reflects a tone that up until recently afflicted almost all traditions of history-writing in Europe and America, with a certain anxiety to establish national 'firsts'. See, for example, D. B. Durand, "Tradition and Innovation in Fifteenth Century Italy. "'Il primato dell'Italia' in the Field of Science," *Journal of the History of Ideas* 4 (1943), pp. 1-20, and the contributions on the subject by H. Baron, F. R. Johnson and P. O. Kristeller in the same issue of the review. See also R. Almagià, "Gli italiani primi esploratori dell'America," in *L'opera del genio italiano all'estero*, Rome: Libreria dello Stato, 1937, pp. 393-395.

[29] *Entrata e uscita. San Michele di Murano* (1453-1460), ff. 124r-125v: "Adì 8 febrier contadi pro Lio Roso per nome del magnifico Signor de Portogal per pagar pentori per lavurar el suo mapamundi e per altre spese in suma

Given the low pay (14 *soldi* – units of Venetian currency – a day), and the fact that the scribes remain unnamed, it seems likely that these were not specialists in cartographical drawing but, rather, simply copyists and painters skilled enough to be able to copy place names and the legends as well as to depict the magnificent ocean background and the small perspective views of castles, towers, cities, rivers, funeral monuments, and ships that adorn the *mappa mundi*. It is therefore to be supposed that these craftsmen were reproducing a *mappa mundi* already in existence (though not necessarily the one now held in the Marciana), adding the inscriptions and images to complete a cosmographical structure that had already been drawn up by Fra Mauro. The notion that these persons were copyists is further supported by the very fact that they were paid as journeymen, an arrangement which left little room for spontaneous creative contribution to the work.

During the fifteenth century the library of San Michele started to become a relevant cultural center whose importance subsequently expanded well beyond matters monastic. Thanks to the Prior General of the Order, Ambrogio Traversari (1386-1439), and to Paolo Venier, abbot of San Michele from 1392 to 1448, the library became a focal point for the circulation of books in the Venetian lagoon area, involved in both acquisitions and loans.[30] Instead, existing documents give us a much less clear picture of the function of the *scriptorium* of San Michele in the fifteenth century.[31] The evidence we have suggests that the works executed by the *scriptorium* did not involve the monks themselves so much as outside copyists and miniaturists, brought in to reproduce or adorn texts on a regular basis. This practice, then, within the monastery of San Michele, helps to contextualize Fra Mauro's decision to use outside journeymen to work on the copy of the *mappa mundi* for the Portuguese court.[32]

b. *Almagià : 1448-1453. A new binding proof*

Almagià's proposed dates for the composition of the *mappa mundi* (1446-1453), beyond considering the chronology suggested by the toponyms used to label the African coast, were based on still other important historical-cartographical features. Arguing for a beginning-of-production date that was a full decade before Zurla's, Almagià insisted that the toponyms *Cavo*

ducati 11 val [...] 129. Adì 21 hoctubre 1457 che io ho dado contadi a frar Mauro per pagar uno scriptor a laurado ouer scripto su il mapamundi zorni 17 a raxon da soldi 12 al zorno monta L. 17 S. 4 val a S. 124 per ducato. Adì 7 hoctubre 1458 contadi a don Francesco da Cherso per pagar quondam scriptor el qual scripse al dictto mapamundi zorni 4 a soldi 14 al giorno monta soldi 56 val 124 soldi per ducato. Adì 9 dictto contadi al dictto per pagar uno maistro scripse ala dicta opera zorno uno soldi 14 per val."

30 Meneghin, *San Michele in Isola di Venezia*, vol. 1, pp. 255-258.

31 Historians have proposed one of two basic theories : either there was no *scriptorium* as such and the copying of codices rested upon the initiative of individual monks ; or there was a *scriptorium* whose work and output would become increasingly important from the fifteenth century onwards ; see Caby, *De l'érémitisme rural au monachisme urbain*, p. 629 and L. Merolla, "Un corale di San Michele di Murano," *Nuovi Annali della Scuola speciale per archivisti e bibliotecari*, 9 (1997), pp. 120-122, rispectively.

32 Guglielmo Cavallo points how the Benedictines developed an aversion to the mechanical work of transcribing texts, probably because it was seen as occupying time that could be used for prayer and creative intellectual work ; see G. Cavallo, "Dallo *scriptorium* senza biblioteca alla biblioteca senza *scriptorium*," in *Dall'eremo al cenobio. La civiltà monastica in Italia dalle origini all'età di Dante*, Milan : Scheiwiller, 1987, pp. 331-421, esp. 400-412.

verde[33] and *Cavo rosso*,[34] found on Africa's Atlantic coast, began to figure in Western – and, in particular, Venetian – geography and cartography as early as the voyages of the Portuguese navigator Aires Gomes da Silva e Fernandes in 1444 and 1446.[35] This theory is confirmed by the fact that both names appear in the marine chart drawn in London in 1448 by the Venetian officer of the galleys and cartographer Andrea Bianco – a work that is the closest to the *mappa mundi*'s account of Africa and can plausibly be listed among its sources (plate IV).[36] That both Fra Mauro and Bianco knew each other and worked together is certainly beyond doubt, as Bianco worked on the preparation of the *mappa mundi* for the Portuguese court.[37]

As for the latest possible production date suggested by Almagià, the most cogent argument in support of 1453 is the total absence within the work of any reference to the Fall of Constantinople. The city was taken by the troops of Mehmed II (1430-1481) on May 29, 1453, and news of the capture of the ancient capital spread rapidly. By 30 June a gravely perturbed Venetian Senate had already informed Pope Nicholas V of the disaster[38] and, by August, the Ruthenian cardinal Isidor of Kiev, who managed to escape the ensuing slaughter, gave an account of the terrible event to Cardinal Basil Bessarion, king Alfonso V of Aragon, and Pope Nicholas V (in letters dating from August 6, 8 and 15, respectively).[39] Just a few days later Enea Silvio Piccolomini sent a letter from the imperial court to the pontiff, expressing sorrow and indignation, bemoaning the superb basilicas desecrated by the Turkish troops and lamenting the loss of the city's

[33] *Mappamondo* (ed. 1956) Pl. XVIII, 23 ; *Mappamondo* (ed. 2006) *568.

[34] *Mappamondo* (ed. 1956) Pl. XVIII, 11 ; *Mappamondo* (ed. 2006) *563.

[35] Almagià, "Presentazione," in *Mappamondo* (ed. 1956), pp. 5-9 ; Id., "Carta nautica con elementi corografici di Anonimo Veneziano," p. 35. The validity of the use of the chronology of toponyms in dating cartographical works has been demonstrated by Tony Campbell. See T. Campbell, "Portolan Charts from the Late Thirteenth Century to 1500," in J. B. Harley and D. Woodward (eds), *The History of Cartography*, vol. 1, *Cartography in Prehistoric, Ancient and Medieval Europe and the Mediterranean*. Chicago and London : University of Chicago Press, 1987, pp. 415-421.

[36] Milan, Biblioteca Ambrosiana, F. 260, inf. (1). A. Bianco, [marine chart], manuscript on vellum, dated and signed "Andrea Biancho venician comito de galia mi fexe a Londra M CCCC XXXX VIII" (Andrea Bianco from Venice, officer on galleys, made me in London, 1448), 63 × 85,5 cm. On the chronological limits implied by these toponyms, see Campbell, "Portolan Charts from the Late Thirteenth Century to 1500," p. 412 plate 19.2 ; A. Teixeira da Mota, *Topónimos de origem Portuguesa na costa ocidental de Africa desde o Cabo Bojador ao Cabo de Santa Caterina*, Centro de Estuds da Guiné Portuguesa, 14, Bissau, 1950. "Cetoschamar" (Pl. XVII, 19-*457) and "Dolcarmin" (Pl. XVII, 20 - *461 ; Pl. XVIII, 2 -*556), are the southernmost toponyms on the Atlantic coast of Africa and derive from medieval sources ; on these toponyms see *Mappamondo* (ed. 2006), pp. 100-108. On the first phase of Portuguese expansionism, see L. F. Thomaz, *Le Portugal et l'Afrique au XVe siècle : les débuts de l'expansion*, Lisbon : Inst. de Investigação Científica Tropical, Sep. Arquivos do Centro Cultural Português da Fundação Calouste Gulbenkian – Volume XXVI. Série Separatas 221, 1989.

[37] With regard to Andrea Bianco's collaboration on the *mappa mundi* for Alfonso V, see this entry in Maffeo Gherardo's *Entrata e uscita. San Michele di Murano* (1453-1460), f. 125v : "Adì 10 marzo 1459 per ducati 2 tolti in nui per messe per nome de ser Andrea Biancho che lui douea hauer per suo premio del lauorare lui fece al detto mapamundi val in chassa" (On 10 March 1459 I subtract from this account sheet 2 ducats that were due to Andrea Bianco for the work he did on the above mentioned *mapamundi* and I register them as a payment for two Masses celebrated on his behalf).

[38] A. Pertusi, *La caduta di Costantinopoli. Le testimonianze dei contemporanei*, Milan : Fondazione Lorenzo Valla and Arnoldo Mondadori Editore, 1997, p. lxxxviii (for the complete chronology, pp. lix-xci). See also L. von Pastor, *Storia dei papi dal Medio Evo*, 16 vols., Rome : Desclee, 1958-1964, vol. 1, pp. 603-620.

[39] See Isidorus of Kiev, "Lettere a Nicolò V, al Cardinal Bessarione, ai fedeli di Cristo, al Doge di Venezia, a Filippo di Brogogna," in A. Pertusi, *La caduta di Costantinopoli*, pp. 52-111 (Latin text and translation).

splendid libraries, declaring that Homer and Plato had died all over again (*extinctus est fons musarum*, Piccolomini wrote to the pope).[40]

According to Almagià, it would have been highly incongruous, then, for Fra Mauro, a Venetian cartographer, to have omitted mention of such a catastrophe ; even more so when one considers that Venice itself played a central role as a "clearing house" for news throughout Christendom. Justified in these very terms, Almagià's argument and explanation are not very convincing : there are other contemporary maps such as the so-called "Genoese *mappa mundi*" – the Portolano 1, drawn in 1457, today in the Biblioteca Nazionale Centrale in Florence (plate V) – or the Catalan-Estense *mappa mundi*, designed *circa* 1460, that display no hint of the Fall of Constantinople.[41] But there is a further argument ignored or not considered by Almagià that indirectly supports the hypothesis the map could not have been prepared after 1453 and, for sure, after 1457 : there is genuine documentary evidence that the *mappa mundi* produced for the Portuguese court was actually commissioned during the embassy to Italy led by Afonso V's Head Chancellor, João Fernandes da Silveira, who came to Venice in 1456 specifically to negotiate Portuguese participation in the crusade called by Calicxtus III (pope from 1455 to 1458) against the Turks. Therefore, if the map in the Biblioteca Marciana had been prepared at the same time as the map for the Portuguese Court – that is between 1457-1459 – an omission of reference to the Fall of Constantinople would be inexplicable indeed once the buyer visited several Italian cities to negotiate the possible participation of Portuguese soldiers in the crusade to take back Constantinople.

A third factor – this instead already considered by Almagià – that further strengthens the theory of a 1448-1453 finishing date occurs in a legend, near the city of Herat : "This [town] was under the seignory of Deli, but Tamerlane moved here that seat of power. Now Sia Roch Marzan, his son, is the lord of this, of Samargant and of all Persia as far as Bagadat and beyond."[42] One of the four sons of Tamerlane (1336-1405), *Sia roch Marzan* – Shah-Rokh – was *mirza* (literal Persian for "son of *Amīr*" or "son of the leader," that is, heir to the throne) from the death of his father until his own demise in 1447 when he was succeeded by his son, Ulugh-Beg (1394-1449).[43] Even allowing for the fact that such news might take a few years to reach Venice, it seems highly improbable that it was still not public knowledge eleven years later – that is, if one accepts the dating of the *mappa mundi* around 1457-1459, in 1457. Ignorance of Shah-Rokh's death, however, becomes much more explicable if Fra Mauro's map was produced between 1448-1453, leaving only a year or two for the news to reach Venice.[44] Zurla himself also provides

[40] R. Wolkan (ed.), "Der Briefwwhsel des Eneas Silvius Piccolomini," *Fontes Rerum Austriacarum* 2, 68 (1918), Epist. 109, p. 19.

[41] See their full trascriptions in A. Cattaneo, *Mappa Mundi 1457 (Biblioteca nazionale centrale di Firenze, Port. 1). Analisi, trascrizione e commentario*, Rome : Istituto dell'Enciclopedia Italiana, 2008, pp. 153-195 and in *Mapamundi Catalán Estense. Escuela Cartográfica Mallorquina*, Commentary by E. Milano, transcription of the original text by A. Battini. [Modena, Bibl. Estense, C.G.A.1], Barcelona : M. Moleiro, 1996, pp. 139-203, respectively.

[42] "Questa era za soto la signoria del deli | ma Tamberla(n) translatoe qui | quela sedia e ahora Sia roch Mar|zan el fiol suo ne è signor e de sa|margant e de tuta p(er)sia perfina | bagadat e oltra," *Mappamondo* (ed. 1956) Pl. XXVII, 87. The English translation derives from *Mappamondo* (ed. 2006) *1501.

[43] On the Tamerlane dynasty, see F. Adravanti, *Tamerlano*, Milan : Fabbri, 2001, pp. 286-287.

[44] On Venice as an international 'clearing-house' of news and information, see P. Burke, "Early Modern Venice as a Center of Information and Communication," in J. Martin and D. Romano (eds), *Venice Reconsidered : the*

documentary evidence against his own dating when he quotes from the (now lost) San Michele accounting book for the years of 1448-1449, which mention expenditures on blue and other pigments, gold, and parchment to be used to *formar mappamondi* (preparing world maps).[45] This, together with the fact that, from 1453 to1460, the only *mappa mundi* referred to in the account book of San Michele is that commissioned by the Portuguese court, thus further supports the dating suggested by Almagià.

c. A reinterpretation of Almagià's 1448-1453 dating

As already mentioned, the map's miniature of Earthly Paradise by Leonardo Bellini is thought to have been painted later and was clearly inspired by the drawings of the artist's uncle and teacher, Jacopo Bellini (1400-1470) – with traces of the influence of Jacopo's own son-in-law, Andrea Mantegna (1431-1506) – the style revealing the same naturalism to be found in the rocky landscapes of certain Bellini drawings and Mantegna paintings (especially those commissioned for private devotion).[46] Other features, such as the face of God the Father and the actual posing of the figures of Adam and Eve, also show the influence of works by Jacopo Bellini that have been dated between 1455-1470. These factors together suggest that the *mappa mundi*'s illustration of the Earthly Paradise was probably produced some time between 1455 and 1460, in which latter year the work was framed and hung in the church of San Michele. However, it cannot be proved with certainty that the scene was not added even later still when the *mappa mundi* was already on public display.

One important piece of documentary evidence that could be raised against dating the map around 1448-1453 merits more detailed analysis. One of the works cited in Fra Mauro's legends is Strabo's *Geography*, which was not translated into Latin until the years 1453-1458. Strabo is cited by Fra Mauro, together with Arrian, as an indispensable source for knowledge of India (the reference is clearly to Book XV of the *Geography*).[47] If we look at the history of the trans-

History and Civilization of an Italian City-State, 1297-1797, Baltimore: Johns Hopkins University Press, 2000, pp. 389-419.

[45] See the Appendix, doc. 6. Zurla writes: "ne' Libri d'entrata ed uscita di questo stesso nostro Monastero tra que' che spettano al secolo XV troviamo, io dissi, non di rado notato il nome di fra Mauro, e le spese per colori, per oro battuto, ecc per formar mappamondi, come nel registro num. 451 delle spese di D. Niccolò Monaco ed Economo del Monastero predetto nel 1448, fino Luglio 1449." Zurla *Il mappamondo di Fra Mauro camaldolese*, p. 79. Research in the Rome and Venice Archives and the Biblioteca del Monastero di Camaldoli to locate the *Libri di entrata e uscita di Niccolò Monaco* covering the years 1448-1449 has so far been in vain.

[46] On the Leonardo Bellini's depiction of the Earthly Paradise and its relations with Jacopo Bellini's and Andrea Mantegna's work, see S. Marcon, "Il Mappamondo di Fra Mauro e Leonardo Bellini," in *Mappamondo* (ed. 2006), pp. 103-108 and Chapter III in this volume.

[47] "San Hieronymo | nel prohemio de la Bibia dice che Apolonio phy|losopho passò q(ue)sto fiume phison | e uene ad alguni pop(o)li diti bragma|ni p(er) aldir Archan, el qual senta|ua in una sedia d'oro e i(n)signaua | de la natura, de i costumi e del | corso de le stelle. Q(ue)sto dico p(er)ché | q(ue)sti p(o)p(u)li, çoè bragmani, massa|geti, polibotri, pignei, pandi, | astiacani e de altri che de qui se | scriue esser, a nui no(n) sono noti mol|to, p(er)hò de li lor costumi et habiti | discrepatissimi chi ne uol hauer notitia | leça Ariano e Strabo, i qual de questa india | scriueno copiosamente." *Mappamondo* (ed. 1956) Pl. XXVI, 18. "In the Proem to his Bible, St. Jerome says that Apollonius the Philosopher crossed this river Phison to travel to certain peoples called Bragmani and hear Archan, who, seated on a gold chair, gave teachings on the nature, characteristics and course of the stars. I say this because these peoples – that is, the Bragmani, the Massageti, the Polibotri, the Pignei, the Pandi, and Astiacani and the others who are

lation of all seventeen books of Strabo's *Geography*, however, it becomes clear that Fra Mauro might well have had access to the work almost twenty years before the completion of the full Latin translation. And while the Latin translation was completed only in 1453-1458, we do know that the Greek manuscripts of the *Geography* arrived in Florence in 1438, brought over by the Metropolitan of Nicaea, Bessarion (1395-1472) who passed through Venice itself on his way to attend the Council held in Ferrara and later in Florence.[48]

One of the very first, if not the first, in the West to celebrate the arrival of the work of Strabo was the then Prior General of the Camaldolese, Ambrogio Traversari, a humanist learned in both Latin and Greek, who was Secretary to the ecumenical Council called by pope Eugenius IV (the Venetian Gabriele Condulmer, born 1383, pope 1431-1447).[49] The *Geography* was also known at that time through some excerpts from a Greek geographical treatise by the Byzantine philosopher Georgios Gemistos Plethon (*c.* 1355-1452), one of the laymen representing the Greek Church and a speaker at the Florence Council. A close associate of Bessarion – who would actually inherit the treatise upon the author's death[50] – Plethon is best known for having initiated the study in Florence of Plato (his treatise comparing said philosopher with Aristotle was very famous). But he also introduced Florentine humanists to Strabo's *Geography*, of which his own treatise offers a summary that is amended on the basis of Ptolemy's *Geography*.[51] Diller points out the similarity between a passage of Strabo, transcribed by Plethon, which deals with the *Permiani* who inhabited the north-east of *Rossia*, and what Fra Mauro says in an inscription in the *mappa mundi*, thus suggesting the possible influence of Plethon's short treatise on the map.[52] Though interesting, this idea has to be taken with some caution. In fact, it seems that Plethon himself derived his information from the suite accompanying Isidor of Kiev to the Florence Council (en route to which, the Metropolitan of all the Russias passed through Venice). Nevertheless, the fact remains that the name *Permiani* used by Fra Mauro for the peoples of eastern Russia does not figure in Marco Polo's book, his main source for the description of this part of the world. Thus the link between Bessarion, Traversari and Plethon ; the famous con-

written about here – are not very well known to us. Those who want information on their very different customs and practices should read Arrianus and Strabo, who write extensively about this India." *Mappamondo* (ed. 2006) *1380.

[48] See Diller, *The Textual Tradition of Strabo's Geography*, pp. 66-69 and 106.

[49] "Strabonem duo maxima volumina se [Bessarion] illic [Mothone] reliquisse professus est," *ibidem.* Diller identifies the *volumina* as the following two manuscripts of Strabo's *Geography* held today at the Biblioteca Marciana : Gr. Z. 377 (= 895) and Gr. XI, 6 (= 1142).

[50] Venice, Biblioteca Marciana, Gr. Z. 379 (= 520), a Greek manuscript in four different hands. The first of these is identified as that of Georgios Gemistos Plethon, who wrote ff. Ir-108v ; these pages include an extract from Book II of Strabo's *Geography* (ff. lr-12r), Plethon's own *Correctio quarundam Strabonis notitiarum* (ff. 12r-15v), other extracts from Books I, V, VI, VII, VIII, IX of Strabo (ff. 15v-18v) ; and the whole of Books XI-XVII. See Gentile, *Firenze e la scoperta dell'America*, entry 80, pp. 165-168.

[51] The excerpts from Strabo's Geography are transcribed and discussed in A. Diller, "A Geographical Treatise by Georgius Gemistus Pletho," *Isis* 27, 3 (1937), pp. 441-451. See also Gentile, *Firenze e la scoperta dell'America*, entry 80, p. 167. On Bessarion and Plethon, see E. Garin, *Umanisti artisti scienziati. Studi sul Rinascimento italiano*. Rome : Editori Riuniti, 1989, pp. 93-109. L. Labowsky, "Bessarione," in *Dizionario Biografico degli italiani*, Rome : Istituto della Enciclopedia italiana, 1967, vol. 11, pp. 686-696 ; Id., *Bessarion's Library and the Bibliotheca Marciana*, Rome : Edizioni Storia e Letteratura, 1979 ; D. J. Geanakoplos, *Greek Scholars in Venice*, Cambridge : Harvard University Press, 1962 and G. L. Coluccia, *Basilio Bessarione. Lo spirito greco e l'occidente*, Florence : Olschki, 2009.

[52] Diller, "A Geographical Treatise," pp. 459-460.

nection between Bessarion and the city of Venice; and the very fact that Traversari was a member of the Camaldolese Order all make it plausible that Fra Mauro may have had some access to Strabo's work, either through Bessarion's manuscripts or the excerpts in Plethon, before Guarino Veronese's and Giorgio Tifernate's translations. One can rule out that Fra Mauro, who did not know Greek, had direct access to Strabo's *Geography*. Still, the explicit reference to Strabo in an inscription within the *mappa mundi* would seem to indicate that his knowledge of that work was similar to the knowledge Patrick Gautier Dalché has shown there was of Ptolemy's *Geography* in the centuries before the Alexandrine's work was translated into Latin by Iacopo Angeli; for example, he may have known the title of all seventeen books or, indirectly, even some excerpts.[53]

All of these factors combine to support the notion that different parts of the *mappa mundi* were produced at different times. The geographical part – that is, the map within the circular frame – and the three cosmographical depictions – which in form, style, and content are very similar to the enclosed area – most probably date from 1448-1453, while the depiction of the Earthly Paradise was probably painted some time between 1455 and 1460 (the latter being the year in which the *mappa mundi* was framed and put on display in the church of San Michele). In addition to all this, while preparing the fundamental transcription of the map published in 1956, Tullia Gasparrini Leporace discovered that some of the legends – mostly those in the ocean that encircle the *oikumene* – were written on fragments of parchment that had been glued on the map in order to replace other legends or drawings. Though Almagià's paleographic analyses of these fragments revealed that they are contemporary to the rest of the legends, and written by the same main hands – two or three hands, according to my analysis – that compiled the map, their very existence implies that the map was corrected, and at least partially modified, in the years immediately after completion of the first version.[54]

II. – The reception of the *mappa mundi* until 1600

Most extant fourteenth- and fifteenth-century *mappae mundi* might be defined as "silent." Two typical examples of this silence are the famous Ebstorf and Hereford *mappae mundi* of *c.* 1239 and *c.* 1300, respectively, which both exist unaccompanied by any archival documentation – leaving the works' authors, origins, and purposes unknown, or the subjects of speculation at best. In the case of the Hereford map, we have only the name of its creator, Richard of Holdingham, without any details of his life, education, or when or where he lived. In the absence of documents that pointed to their existence, many *mappae mundi* have lain forgotten in monasteries and private collections. Indeed, their very survival may be due to their having been ignored for hundreds of years, as is suggested by the case of the Hereford map, of which Scott D. Westrem states, "the earliest independent witness to [its] existence and to its location in the

[53] See P. Gautier Dalché, "Le souvenir de la 'Géographie' de Ptolémée dans le monde latin médiéval (VIe-XIVe siècles)," *Euphrosyne. Revista de Filologia clássica* 24 (1999) pp. 80-106 and Id., *La Géographie de Ptolémée en Occident (IVe-XVIe siècle)*. Turnhout: Brepols, 2009, pp. 87-142.

[54] *Mappamondo* (ed. 1956), p. 6 and 17.

Hereford Cathedral dates from nearly four centuries after its production."[55] Frustratingly, this absence of source documentation of preserved works is counterbalanced by a relative abundance of material regarding some maps that have since disappeared. One example of this pattern is supplied by the work of a contemporary of Fra Mauro's, a Venetian cartographer by the name of Don Antonio Leonardi, known to have been active between 1462 and 1496. Although there is ample documentary proof of this priest's sizeable body of work and to the high regard in which it was held by his contemporaries, not a single map of his has survived.[56]

Fra Mauro and his *oeuvre* are, however, an important exception to this rule. While there remain significant documentary gaps regarding the genesis of the cartographer's works and his life, we possess a relatively abundant collection of material documenting how his work and in particular how one of his major achievements, the *mappa mundi* now in the Biblioteca Marciana in Venice, was viewed, copied and studied over the centuries. It is important to underline that for the work of Venetian contemporaries of Fra Mauro such as Andrea Bianco, Albertin di Virga, Antonio Leonardi, and Giovanni Leardo, we do not have any document attesting how their maps were received.[57] This documentary evidence does little to clarify questions about the genesis of Fra Mauro's *mappa mundi* – for this matter we will base the research on both the textual and visual analysis of the legends and the images of the map and their sources – but it sheds important light on the history of its reception since the fifteenth century, and on how the *mappa mundi* was observed and utilized over the centuries, and in this sense, is a key example of the cultural history of maps. The collection and study of the testimonies of those who, in the past, looked at, copied and used the *mappa mundi*, serves an important hermeneutic purpose. The meticulous collection of evidence regarding ancient readers of the *mappa mundi* is a critical step for understanding the historical significance of the work.

The history of the reception of the work leads us on the trail of its historical significance. Beyond what our eyes, at more than 550 years after its creation, can see and beyond what has been written about the *mappa mundi* the evidence that we collected and will analyze, which is quite close in chronological terms to the *mappa mundi* itself, indicate that fifteenth- and sixteenth-century readers such as the Portuguese Ambassador João Fernandes da Silveira, the Camaldolese general Pietro Dolfin, the prior of the monastery of San Michele Bernardino Gadolo, the Medici of Florence, the Venetian scholar Alessandro Zorzi, the humanist and Venetian secretary of state Giovanni Battista Ramusio, the cartographers Giorgio Sideri and Angelo Freducci,

[55] S. D. Westrem, *The Hereford Map. A Transcription and Translation of the Legends with Commentaries*, Turnhout: Brepols, 2001, p. xviii. This has not prevented scholars from advancing various theories regarding the circumstances of their production and their significance; the recent study of the Hereford Map by Westrem is a perfect example.

[56] Recent research of archival material has revealed that between 1462 and 1476 the cleric produced at least six *mappae mundi* and two large maps of Italy for such clients as Pius II, Borso d'Este, Francesco Todeschini Piccolomini, and the Venetian Republic itself (for the Doge's Palace); see R. Bianchi, "Notizie del cartografo veneziano Antonio Leonardi. Con una appendice su Daniele Emigli (o Emilei) e la sua laurea padovana," in V. Fera and G. Ferraù (eds), *Filologia umanistica. Per Gianvito Resta*, Padua: Antenore, 1997, pp. 165-211.

[57] On Albertin di Virga and Giovanni Leardo, see P. Falchetta, "Carte veneziane dell'Asia da Fra Paolino a Giacomo Gastaldi," in G. Curatola (ed), *Sciamani e dervisci dalle steppe del Prete Gianni: religiosità del Kazakhistan e percezione del fantastico a Venezia*, Venice: Multigraf, 2000, pp. 39-50; Id., "Maps ans Mapmakers in Venice in Cabot's Time," in R. Mamoli Zorzi (ed.), *Attraversare gli oceani. Da Giovanni Caboto al Canada multiculturale*, Venice: Marsilio, 1999, pp. 73-82.

the cosmographer Egnazio Danti, saw much more or at least something different than an "outdated" or "updated" cosmography, whether "medieval" or "Renaissance," "doctrinal" or "experimental." By tracing the history of the early reception of Fra Mauro's map, my aim is to identify *milieux*, to define modes of reading, understanding and using of cosmography and cartography based on the existing knowledge and testimonials of those time. By tracing the cultural history of the map, terms such as "innovation," "experience," "medieval" rather than "Renaissance" knowledge reveal their radical inconsistency as well as uselessness to understand Fra Mauro's effort and its legacy.

III. – The map for the Portuguese court of Afonso v (1457-1459)

1. An iconic "encounter"

Fra Mauro was commissioned to produce a *mappa mundi* for the Portuguese court in February 1457 and, as previously mentioned, the finished work was dispatched to Lisbon on April 24, 1459, in the safekeeping of the Venetian nobleman Stefano Trevisan.[58] The record of a payment of 30 and three-quarters ducats for "the painters who worked on the world map in Venice" (*aos pintores que pyntaram o mapa mundo em Veneza*) appears in the 1462 *Carta de quitação* (official release of obligation to confirm the payment of a service) from Afonso V to his ambassador João Fernandes da Silveira, thus confirming the *mappa mundi*'s arrival – authorization of payment would certainly not have been given otherwise.[59] The "encounter" between Fra Mauro and Afonso V is an iconic moment in both the history of European expansion and that of the cultural links between Portugal and Italian cities during the Renaissance. The *mappa mundi* not only exemplifies the commercial links which existed between Portugal and the cities of Italy, but it also illuminates the diplomatic and cultural policies pursued during the reign of Afonso V (1448-1481) – a period of significant Portuguese expansion in Africa and the south-western Atlantic. In the pages that follow I will show that the commissioning of Fra Mauro's *mappa mundi* is part of a broader set of economic, diplomatic and intellectual connections between Portugal and Italian cities, also involving the interchange of cosmographic and cartographic information.

This commission is to be seen in connection with Afonso V's attempt to provide both a diplomatic basis and dissemination to the project of exploring and conquering along the coast of Africa. In exchange for a formal commitment to participate in the crusades, proclaimed by Pius II and his successors Nicholas V and Calicxtus III, to take back Constantinople, which had fallen to the Turks in 1453, the Portuguese fleets were guaranteed by the bulls *Romanus Pontifex*

[58] "E adì 24 april 1459 que dom Nicholò nostro mea dictto che essendo io a capitolo a Camaldoli è stato saldà questa raxon a messer Stefano Trevisan per nome del dictto Segnor grando per el dectto Stefano li fo mandatto el suo mapamundi" ; *Entrata e uscita. San Michele di Murano* (1453-1460), f. 169v. (On April 24, 1459, [I record] that don Nicholò had told me that while I was attending the chapter [of the order] in Camaldoli, messer Stefano Trevisan was entrusted with concluding the project of the *mappa mundi* for the king of Portugal, to whom – through Stefano – he was sent his [the king's] *mappa mundi*). This is my translation.

[59] See the Documentary Appendix of Chapter I, doc. 9. Transcribed in *Monumenta Henricina* 1460-1469, vol. 14, doc. n. 73, pp. 199-204, esp. 202.

(from Nicolaus V on January 8, 1455) and *Inter caetera divinae* (from Calicxtus III in March 1456) the right to navigate along the southern coasts of Africa. These same bulls also prohibited all other nations, under threat of excommunication, to move in those waters, conceding the spiritual jurisdiction of the lands that would be conquered from Cape Bojador to the "Indies" to the Order of Christ led by Prince Henry the Navigator (*Infante Dom Enrique, duke of Viseu*, 1394-1460).[60] The documents of the Pontifical Curia do not, therefore, reflect the actual conquest of lands, but instead detail a pending expansion project for which the Pope guaranteed official political recognition. The reason for this was that the Portuguese expansion of the second half of the fifteenth century would have contributed to an overall strategy of resistance of Christianity against Islam, and specifically against the Turks. These two papal bulls thus express the ideological basis upon which Alfonso V predicated his policy of expansion, seeing it as an advance of Christianity "against the Turks" (*adversus Turcos*), a notion constantly reiterated in the letters exchanged between the Portuguese court, the Seignory of Florence, and the Republic of Venice in the years 1456-1457.[61]

Around 1450, African conquests began to offer European powers the sense of limitless opportunities in the trade of slaves, gold, and salt through a direct connection of sub-Saharan Africa to the Mediterranean and Atlantic. Africa was also believed to be home to the legendary Prester John – described in the *mappa mundi* as having "more than 120 kingdoms under his dominion" (*più de 120 regni soto el suo dominio*)[62] – who would obviously be a very attractive ally against the Moors. While the economic potential of Africa was still being explored, under Alfonso V, substantial quantities of goods were being shipped to Lagos and Lisbon via merchant companies in Florence, Venice, Bologna, Genoa and several other Italian cities – in part due to the fiscal privileges the Portuguese crown had granted to all fleets that put in to the nation's ports during their voyages northwards to England and Flanders.[63] Federico Melis has identified

[60] L. M. Jordão (ed.), *Bullarium Patronatus Portugalliae Regum in ecclesiis Africae, Asiae atque Oceanie : bullas, brevia, epistolas, decreta actaque sedis ab Alexandro III ad hoc usque tempus amplectens*, Lisbon : Typographia Nationali, 1868-1879, 5 vols., vol. 1, 1868, pp. 31-34 ; *Monumenta Henricina*, vol. 12, doc. 36, pp. 71-79 and 286-288. S. A. Gomes, *D. Afonso V o Africano*, Lisbon : Temas e Debates, 2009, pp. 221-245 ; J. P. Oliveira e Costa, *Henrique, O Infante*, Lisbon : Esfera dos Livros, 2009, pp. 112-135.

[61] This goal is evident in almost all the correspondence among the Portuguese Court, the Florentine government and the Venetian Republic in the years 1465 and 1457. See particularly the letters of the Florentine Seigniory and the King of Portugal Afonso V dated September 2, 1456, November 8, 1456, and March 5, 1457 in the Archivio di Stato of Florence, *Signori – Carteggio, Responsive*, Filza 39, f. 211r ; Filza 3, f. 41r ; *Signori – Carteggio, Missive*, Filza 40, f. 12v. These documents were published in *Serto di documenti attenenti alle Reali Case di Savoja e di Braganza per le auspicatissime nozze di Sua A. R. la Principessa Pia di Savoja con Sua Maestá Don Luigi I Re di Portogallo*, Florence : Stamperia Reale di Firenze, Francesco Cambiagi, 1862, docs XXXII-XXXIV, pp. 139-144.

[62] *Mappamondo* (ed. 1956) Pl. X, 37 ; *Mappamondo* (ed. 2006) *77.

[63] L. D'Arienzo, "La presenza italiana in Portogallo e nella Spagna meridionale all'epoca di Cristoforo Colombo," in G. Cavallo (ed.), *Cristoforo Colombo e l'apertura degli spazi*, 2 vols., Rome : Poligrafico e Zecca dello Stato, 1992, vol. 1, pp. 535-565 ; On the documents regarding the commercial privileges granted to Venetian and Florentine companies, see J. M. da Silva Marques and A. Iria (eds), *Descobrimentos portugueses : documentos para a sua história*, 3 vols., Lisbon : Instituto para a Alta Cultura, 1944-1971, vol. 1, 1944, Suplemento (1057-1460), pp. 263-264 and vol. 3, 1971, doc. 12, p. 20. See also G. Po, "La collaborazione italo-portoghese alle grandi esplorazioni geografiche e alla cartografia nautica," in *Relazioni storiche fra l'Italia e il Portogallo*, XVIII, Rome : Reale Accademia d'Italia, 1940, pp. 276-350 ; L. de Albuquerque, *Os descobrimentos portugueses*, Lisbon : Alfa, 1989, pp. 90-125 ; F. M. Sousa Viterbo, "A livraria real specialmente no reinado de D. Manuel," *Memoria da Academia Real das Scien-*

four main categories of goods traded with Portugal: textiles, primarily wool cloth (from Florence), silk (from Florence and Lucca), and brocade and taffeta (from Bologna); manufactured goods, such as weapons (from the valleys of the Brescia and Bergamo areas) and paper (from Fabriano and Colle Val d'Elsa); art products, such as illuminated manuscripts, paintings and quality furnishings; and, finally, manuscripts or books on all sorts of subjects (mainly from the *scriptoria* and later also the print shops of Florence, Venice and Bologna).[64] These trade relations also led to an interchange of knowledge and learning, with cosmography being one of the key subjects of interest. The information reported by Fra Mauro on the *mappa mundi* constitutes one of the first mentions of Portuguese navigations along African coasts and the first ever written reference to the existence of a Portuguese cartographic tradition (*nuoue carte de quel nauegar*), linked to the first phase of expansion:

> Many opinions and many texts claim that in the southern regions the water does not surround this whole inhabited and temperate area. But I have heard many opinions to the contrary, above all from those who were sent by his Majesty, King of Portugal, in caravels so that they might explore and see with their own eyes. These men say that they sailed around the coast from the south-west some 2.000 miles beyond the Strait of Gibraltar. And following that route, they then decided to sail south-southeast until they reach the meridian of Tunis, almost as far as that of Alexandria; and at each place they found good shores, with deep water and good navigation conditions without any hindrances. These men have drawn *new marine charts* and have given names to rivers, gulfs, capes and ports, of which I have had a copy [...].[65]

It is an issue that deserves detailed analysis. Without completely excluding the possibility that Fra Mauro had direct access to Portuguese marine charts, this legend could instead plausibly refer to a Venetian marine chart drawn in London after Portuguese sources by Andrea Bianco in 1448,[66] during the journey of the so called *muda di Fiandra* that regularly stopped at Lagos and Lisbon on the route towards England and Flanders. Bianco's 1448 marine chart is

cias de Lisboa, classe II, vol. 9, pp. 1, and 23, n. 95; F. Melis, "Gli italiani e l'apertura delle vie atlantiche," in *I trasporti e le comunicazioni nel Medioevo*, Florence: Le Monnier, 1984; and P. Dias, *Importaçao de esculturas de Itália no séculos XV e XVI*, Porto: Paisagem, 1982.

64 F. Melis, "Di alcune figure di operatori economici fiorentini attivi nel Portogallo nel XV secolo," in L. Frangioni (ed.) *I mercanti italiani nell'Europa medievale e rinascimentale*, intro. H. Kellenbenz, Florence: Le Monnier, 1990, pp. 1-18, esp. 15-16.

65 "Molte opinio(n) e leture se troua che i(n) le parte meridional l'aq(u)a | no(n) circunda questo n(ost)ro habitabile e temperado çona, ma | aldando molte testimonia(n)çe i(n) contrario e maxime q(ue)li iqual | la maiestà del Re de portogallo à mandato cu(m) le suo carauele | a çerchar e ueder ad ochio, i qual dice hauer circuito le spiaçe de garbin | più de 2000 mia oltra el streto de çibelter i(n) tanto che a uoler seguir quel camin hano (con)uenuto dar la p(ro)da quarta d'ostro i(n)uer sirocho | e p(er) suo çudisio hano passato l'i(n)dromo de tunisto e q(u)asi son ço(n)ti a q(ue)l d'alexa(n)dria, p(er) tuto troua(n)do bone spiaçe cu(m) puoco fondo e nauegar assai bo(n) e se(m)pre | se(n)ça fortuna. E i diti hano fato nuoue carte de quel nauegar e hano posto | nomi nuoui a fiumere, colfi, caui, porti, di q(u)al ne ho habuto copia..." (emphasis mine). See *Mappamondo* (ed. 1956), Pl. XI, 2. The English translation derives from *Mappamondo* (ed. 2006) *149.

66 For a biography of Andrea Bianco, see A. Conterio, "L'arte del navegar: cultura, formazione professionale ed esperienze dell'uomo di mare veneziano nel XV secolo," in *L'uomo e il mare nella civiltà occidentale: da Ulisse a Cristoforo Colombo*, "Atti della Società Ligure di Storia Patria", new series, 32 (1992), pp. 187-225. For an exhaustive bibliography about Bianco, see Falchetta, "Carte veneziane dell'Asia da Fra Paolino a Giacomo Gastaldi," pp. 39-50.

the most complete and up-to-date synthesis of Portuguese navigations along the African coast of the mid-fifteenth century before the Atlantic expedition narrated by Alvise Cadamosto and was surely based on Portuguese maps that are now lost (plate IV).

The legend on Fra Mauro's *mappa mundi* reflects the transfer of Portuguese cartographic information – probably by way of Andrea Bianco's chart – to Venice, as part of the active trade between Portugal and the *Serenissima* ; but in addition, Fra Mauro's account of the Portuguese navigations seems to be part of the same discursive tradition as the papal bulls which Portugal had obtained, which do not reflect the actual conquest of lands, but instead detail a pending expansion project. On Fra Mauro's map it is stated that the route followed by the Portuguese ships was to the south-southeast. The reference to the meridians of Tunisia and Alexandria would indicate that the Portuguese ships hugged the African coast, following the enormous indent of the Gulf of Guinea. This would entail that, at least in terms of the syntax and logic of representation, the navigations that Fra Mauro refers to in the legend extended beyond the *capo do mastos*, the furthest place name on Africa's western coast that is registered in surviving contemporary maps or texts. The cartographic design of the *mappa mundi* would thus express only part of the knowledge – probably through Portuguese maps, or more probably through Andrea Bianco's marine chart – to which Fra Mauro had access.[67] In other words, while the design of the western coasts of Africa on the *mappa mundi* corresponds to those on Andrea Bianco's contemporary Venetian map designed in London in 1448, what is transmitted in the *mappa mundi* in written form goes beyond the graphical and geographical limit of the cartography of the time. This is an important detail that shows that Fra Mauro's work, more than documenting voyages that already occurred, testifies to the diffusion, at least in the economic and cultural centers of the Mediterranean, of the projects of expansion planned by the Portuguese court of Afonso V.

While Fra Mauro's use of Portuguese charts – either direct, or mediated by Bianco – reflects the transfer of Portuguese cartographic information to Italian cities, there are also examples of such transfer in the other direction, from Italy to Portugal, which bear consideration in connection with the commissioning of Fra Mauro's map. Just two years after the commission of the *mappa mundi* certain "ambassadors of the king of Portugal" – probably some of the followers of Cardinal Jaime Aviz of Portugal (1434-1459), including Ambassador João Fernandes da Silveira, who spent the summer of 1459 in Florence on their way to a meeting with the Pope

[67] Scholars have often expressed doubts about the existence of these maps. The oldest Portugese nautical maps are conserved in Modena, Biblioteca Universitaria Estense, A.5c, [Map of Africa through the Gulf of Guinea, with the archipelagos of the Azores, Madeira, the Canary Islands and Capo Verde], manuscript on vellum, 617 × 732 mm, oriented North, place names in Portuguese ; see A. Cortesão and A. Teixeira da Mota, *Portugaliae Monumenta Cartographica*, Lisbon, 1960, 6 vols., vol. 1, pp. 3-5, pl. 2 ; a classical argument, based at least in part on nationalistic motivations, explained the absence from Portuguese archives of documents on the first years of the expansion by the so-called "politica do sigilo nos descobrimentos" (policy of secrecy), sanctioned by Manuel the Second in 1504 ; see J. Cortesão, *A politica de sigilo nos descobrimentos nos tempos do infante D. Enrique e de D. João II*, Lisbon : Comissão Executiva das Comemorações do V Centenário da Morte do Infante D. Henrique, 1960. For useful and appropriate criticism, see W. Diffie, "Foreigners in Portugal and the 'Policy of Silence'," *Terrae Incognitae* 1 (1969), pp. 23-34, and especially Thomaz, *Le Portugal et l'Afrique au XVe siècle*.

Pius II in Mantua[68] – had scholarly conversations in Florence with the physician and astrologer Paolo dal Pozzo Toscanelli (1397-1482) who also attended the cardinal.[69] The episode is described in the Florentine Francesco Castellani's *Ricordanze* :[70]

> I remember that on the day [blank space] of July I lent Andrea di Bochacino for master Paolo of master Domenico dal Pozzo Toscanelli my large fully annotated and decorated *mappamundo*, completely finished; his servant came to take it and was supposed to return it to me except when doing so he [Paolo dal Pozzo Toscanelli] had it for several days and showed it to some ambassadors of the king of Portugal; and so Andrea and master Paolo promised to return it to me.[71]

Though we have no other direct, contemporary evidence regarding the meeting between Toscanelli and the Portuguese ambassadors, they were undoubtedly discussing cosmographical matters, and referring to a "large fully annotated and decorated *mappamundo*." There is no need to spell out the similarity between the interests that lay behind this encounter and those that motivated the commission of Fra Mauro's *mappa mundi*.[72]

All of the aforementioned documents bear witness to a relatively swift circulation of cosmographical knowledge in the Early Modern period, while casting light on the cultural and political context in which Fra Mauro's map was commissioned – and in particular, on Portuguese

[68] G. Ciappelli, "Carte geografiche e politica nei rapporti tra Firenze e il Portogallo nel Quattrocento," *Annali dell'Istituto storico italo-germanico in Trento* 32 (2006), pp. 47-70, esp. 65-70.

[69] E. Garin, "Ritratto di Paolo dal Pozzo Toscanelli," in Id., *Ritratti di umanisti*, Florence: Sansoni, 1967, pp. 41-67.

[70] *Ricordanze*, mainly a Florentine genre, threaded laconic personal and family memoirs through account books. See G. Ciappelli, *Una famiglia e le sue ricordanze. I Castellani di Firenze nel Tre-Quattrocento*, Florence: Olschki, 1995, pp. 1-34.

[71] Florence, Archivio di Stato, Conventi soppressi XC 134, Francesco di Matteo Castellani, *Quaternuccio B, 1459-1485*, f. 4r: "Ricordo che insin a dì [...] di luglio prestai a Andrea di Bochacino per maestro Paolo di maestro Domenico dal Pozo Toscanelli el mio mappamundo grande storiato e compito di tucto; portò... suo famiglo et debbemelo restituire salvo aoperato l'arà per parechi dì e mostro a certi ambasciatori del rè di Portogallo; e così mi promisse decto Andrea e decto maestro Paolo di rendermelo." For the modern edition see F. Castellani, *Quaternuccio e giornale B, 1459-1485*, Florence: Olschki, 1995, pp. 14-17 and 33. See also C. Carnesecchi, "Paolo Toscanelli e gli ambasciatori del re di Portogallo nel 1459," *Archivio Storico Italiano* 5, 21 (1898), pp. 316-318; and Gentile, *Firenze e la scoperta dell'America*, entry 72, pp. 146-147.

[72] Further reference to this episode relating to Paolo Toscanelli and the Portuguese ambassadors occurs in a later manuscript compiled by the Florentine merchant Piero Vaglienti (1438-1514) sometime between the end of the fifteenth century and Vaglienti's death. In a panegyrical text addressed to King Manuel I of Portugal (1469-1521; reigned 1495-1521), Vaglienti credits Toscanelli with having identified the possibility of a route from Lisbon to the East that circumnavigated Africa – a possibility which he had outlined to Afonso V through an intermediary, Bartolomeo Marchionni, a Florentine merchant at the Portuguese king's court. Florence, Biblioteca Riccardiana, ms. 1910, P. Vaglienti, [*Compilazione di viaggi*], before 1514, ff. 83v-84v: "di tal chausa e opera n'è principal chagione uno dottore in medicina nostro fiorentino, il quale prima d'istologia [*sic*, astrology] e de segnali de' cieli avendo in ciò molto perduto tenpo, vide e chonobbe che non era sopra a la terra uomo che me' potesse in ciò travaglarsi chon più suo chomodità di tal viaggio fare e metere m opera, che la maestà de' rè di Portoghallo: e questo fu maestro Pagholo del Pozo Toschanelli, uomo singularissimo, el quale avisò a uno nostro fiorentino era in sua chorte, nominato Bartolomeo Marchiane, di tal trato: e lui ne fece avvisato a sua maestà, in modo dì d'oggi a fato un'opera di tal natura che è da eser lodato da tuto 'l mondo, e le spezierie dovevano o solevano andare al Chairo per la via del mare Rosso, ogi dì là a ridote a Lisbona." See also the critical edition: L. Formisano (ed.) *'Iddio ci dia buon viaggio e guadagno'. Biblioteca Riccardiana, Florence, Ms. 1910 (Codice Vaglienti)*, Florence: Edizioni Polistampa, 2006, pp. 159-160.

interest in both acquiring from and divulgating in the most important Italian commercial cities cosmographic and cartographic information. The commission from Fra Mauro of the *mappa mundi* by the Portuguese ambassador, as well as the ambassador's subsequent communications with Toscanelli in Florence, were part of the Portuguese diplomatic and cultural effort which, through papal bulls and diplomatic letters to major European powers, sought to promote, claim and defend Lusitanian rights to navigation and expansion along the African coast. Since at least 1455 this geopolitical goal was achieved through the divulgation of Portuguese plans of navigation to the major European courts. In the specific case of the commission of a map from Fra Mauro, the main goal was to frame them in and spread them through one of the most impressive representation of the mid-fifteenth-century *imago mundi*.

2. Traces of Fra Mauro's *mappa mundi* in Lisbon, *circa* 1490-1563

The copy of Fra Mauro's *mappa mundi*, which was dispatched to Lisbon on April 24, 1459, probably arrived at its destination in the late spring of 1459, without leaving any documentary trace of its receipt. This is not surprising, especially if the commission was intended to promote the expansion project in Venice by means of a famous cartographer. Once this project was incorporated into and described in the *mappa mundi*, there were not many other reasons for the Court of Afonso V to return to the matter. This may explain why after the *quitação* for the expenses incurred by João Fernandes Silveira in 1463, over the next hundred years no Portuguese document cites the work of the Camaldolese. A notable exception is the travelogue of Hieronymus Münzer (also known as Monetarius, *c.* 1437/1447-1508), a German humanist, physician, and cosmographer who between 1494 and 1495 visited the Iberian Peninsula, stopping in Lisbon at the court of D. João II.[73] In a diary page which refers to the last Sunday of November, 1494, Münzer says that he visited the Castle of São Jorge, which then served as the royal palace, which he describes thus :

> That same day we went to the Castle and saw two strong lions, the most beautiful I have ever seen. And then we also saw a *cosmographia*, painted with great care on a golden table whose diameter was 14 palms.[74]

Although the reference is very brief, the details supplied – in particular the fact that the *cosmographia* was circular (Münzer refers to the "diameter"), large (although it is not certain

[73] A. Classen, "Die iberische Halbinsel aus der Sicht eines humanistischen Nürnberger Gelehrten. Hieronymus Münzer, *Itinerarium Hispanicum* (1494-1495)," *Mitteilungen des Instituts für Österreichische Geschichtsforschung* III (2003), pp. 317-340.

[74] "Eodem die ascendimus Castellum et vidimus duos fortissimos leones adeo pulcros sicut nunquam vidi. Similiter cosmographiam in maxima et bene descripta tabula deaurata cuius dyameter erat 14 palmorum." B. de Vasconcelos, *'Itinerario' do Dr. Jerónimo Münzer (Excertos)*, Coimbra : Imprensa da Universidade, 1932, p. 21. The work survives in just one manuscript, which belonged to the famous author/editor of the *Nuremberg Chronicle*, Hartmann Schedel (1440-1514), and is now in Munich, Bayerische Staatsbibliothek, Clm 431, ff. 96-274v, *Itinerarium siue peregrinatio excellentissimi viri artium ac vtriusque medicine doctoris Hieronimi Monetarii de Feltkirchen ciuis Nurembergensis*. The manuscript has still not been published in its entirety in Latin, although there are two Spanish translations : Hieronymus Münzer, *Viaje por España y Portugal (1494-1495)*, introd. M. Gómez-Moreno, trnsl. J. López Toro, Madrid : Almenara, 1951 and Id. *Viaje por España y Portugal (1494-1495)*, Madrid : Polifemo, 1991.

what was the measure of the "palm" mentioned by Münzer), and "golden" – suggest a strong correlation with the physical characteristics of the only *mappa mundi* of Fra Mauro that has come down to us. Based on this correspondence of details, Alfredo Pinheiro Marques has recognized in the *cosmographia* described by Münzer a "clear reference" to the copy of world map of Fra Mauro commissioned by Afonso V.[75] The dimension given by Münzer – fourteen "palms" – would seem to indicate, in my opinion, that the *cosmographia* cannot be quite so easily attributed to Fra Mauro, but if viewed with caution, I think that this hypothesis is fully valid. So until at least 1494, the *mappa mundi* may have been exhibited in São Jorge Castle, the royal residence in Lisbon.

About sixty years later, the chronicler António Galvão in his *Tratado de todos os descobrimentos antigos & modernos, que são feitos até a era de mil & quinhentos & cincoenta*, published in 1563, wrote :

> In the year 1428, Dom Pedro, the king's eldest son, who was a great traveler, went into England, France and Germany, and thence into the Holy Land and other places, and came home by Italy through Rome and Venice. He is said to have brought [from Venice] a map of the world with him, in which all parts of the earth were described, by which the enterprises of Dom Henry for discovery were much assisted. In this map the Straits of Magellan are called The Dragon's tail and the Cape of Good Hope the Front of Africa and so of the rest. I was informed by Francisco de Sousa Tavares that in the year 1528, Dom Fernando, the king's eldest son, showed him a map which had been made 120 years before and was found in the Library of [the monastery of] Alcobaça, which exhibited all the navigations of the East Indies, with the cape of Good Hope, as in our latter maps ; by which it appears that there was as much discovered, or more, in ancient times as now.[76]

Galvão refers to two maps : first, a Venetian *mappa mundi* or a marine chart presumably brought to Lisbon by Dom Pedro around 1428, and a second, of unknown origin, which was mentioned by a certain Francisco de Sousa Tavares (probably *capitão-mor* of an *Armada da*

[75] A. Pinheiro Marques, *A maldição da memória do Infante Dom Pedro*, Figueira da Foz : Centro de Estudos do Mar e das Navegações, 1994, p. 186 : "E é um testemunho (...) claro e inequívoco [da utilização do mapamundo de Fra Mauro por D. João II] – pois pela descrição que faz não deixa qualquer dúvida (...). Referimo-nos à grande tabula deaurata, com catorze palmos de diâmetro, que foi vista em 1494 em Lisboa pelo viajante alemão Hieronimus Munzer, que na altura visitava Portugal e que chegou a entrevistar-se com o Rei D. João II."

[76] António Galvão, *Tratado que compôs o nobre & notauel capitão António Galuão, dos diuersos & desuayrados caminhos, por onde nos tempos passados a pimenta & especearia veyo da India às nossas partes, & assi de todos os descobrimentos antigos & modernos, que são feitos até a era de mil & quinhentos & cincoenta. Com os nomes particulares das pessoas que os fizeram : & em que tempos & as suas alturas, obra certo muy notauel & copiosa*, [Lisbon], Rua de Sã Mamede : em casa de Ioam da Barreira, 15 Dezembro 1563, f. 18r : "No anno de 1428. (i) diz que foy o Infante dom Pedro a Inglaterra, França, Alemanha à casa sancta, & a outras de aquella bãda, tornou por Itália, esteue em Roma, & Veneza, trouxe de lá hũ Mapamundo que tinha todo o âmbito da terra, & o estreito do Magalhães se chamaua, Cola do dragam, o cabo de Boa esperança, fronteira de África, & que deste pedram se ajudara o Infante dom Anrrique em seu descobrimento. Francisco de Sousa Tauares me disse que no anno de 1528 o Infante dom Fernando lhe amostrara hũa Mapa que se achara no cartório d'Alcobaça que auia mais de cento & vinte annos que era feito, o qual tinha toda nauegaçam da índia, com ho cabo de Boa esperança (...)." The English translation derives from Antonio Galvano, *Summary of the Discoveries of the World, from Their First Original to the Year 1555*, in R. Kerr (ed.), *A General History and Collections of Voyages and Travels*, 18 vols., London, 1811, vol. 4, pp. 23-116 (p. 45).

Índia during the reign of D. João III (1521-1557) which embarked in March 1540[77]) preserved in the Cistercian monastery of Alcobaça and designed around 1410. What is said about this second map by Galvão – that it showed the possibility of circumnavigating Africa – has often been interpreted, especially in Portuguese history, as a reference to the copy of Fra Mauro's *mappa mundi* commissioned by Afonso V. This identification, although perhaps appealing, is purely conjectural (not to mention that the discrepancy between the date of the alleged map which was preserved in Alcobaça and that of Fra Mauro's would be at least fifty years, according to what is reported by Galvão). It is therefore worth considering this identification, as this will allow us to clarify some points, still confused, that clutter the historiography of the works of the Camaldolese.

3. At the margin of a historical dispute

The commissioning by the Portuguese court of a copy of the *mappa mundi* of Fra Mauro has been the subject of some interest, especially among Portuguese historians. In particular, some scholars, on the basis of Galvão's *Tratado*, have suggested that the *mappa mundi* of 1457 had been commissioned by the Infante Dom Pedro of Portugal (1392-1449) in the course of his visit to Venice in 1427, during his trip to the European courts between 1425 and 1428. According to this hypothesis, initially advanced in 1955 by the historian Júlio Gonçalves and recently revived by Alfredo Marques Pinheiro, the *mappa mundi* of Fra Mauro of 1457 would have been promised to the Infante Dom Pedro about 20 years earlier, along with the gift of a (suspected) copy of Marco Polo's book. This question merits reconsideration and some clarifications.

The Venetian chronicler Antonio di Marco Morosini describes the arrival and residence of Dom Pedro in Venice with great detail in the *Cronica de Veniexia (1094-1433)*, which survives in two manuscripts.[78] Dom Pedro's residence in Venice lasted a week, from 5 to 12 April, 1428. The Portuguese Prince was received by the Doge at Mestre "with the highest honor," and was taken to San Marco via the Grand Canal, aboard the sumptuous ship *Bucintoro*, "armado," that is, decorated for a parade. Then Dom Pedro visited the Hall of Great Council in the Ducal Palace, the Arsenal, the merchants' stalls brimming with "infinite mountains of gold ducats" in the Rialto, and the most important Venetian churches and the relics that were kept in them. Before leaving for Rome to pay homage to Pope Martin XI, writes Morosini, Dom Pedro wanted to visit

[77] *Virtual Encyclopaedia of Portuguese Expansion* edited by the CHAM, the Centre for Overseas History, of the Faculty of Social and Human Sciences of the New University of Lisbon : http ://www.fcsh.unl.pt/cham/eve/index.php ?lang=en.

[78] The first manuscript, probably the autograph, is preserved in two volumes in Vienna, Österreichische Nationalbibliothek, 6586 and 6587 (for the account of Dom Pedro's visit, see cod. 6587, ff. 480-481) ; the second, a copy of the first made at the end of the nineteenth century, is preserved in Venice, Biblioteca Nazionale Marciana, Ms. It. VII, 2048 (= 8331) and Ms. It. VII, 2049 (= 8332), Antonio di Marco Morosini, *Cronicha de Veniexia*. For the account of Dom Pedro's visit, see Ms. It. VII, 2049 (= 8332[2]), pp. 859-65. The Venetian copy was made between 1887 and 1888 by Giuseppe Gallovich, archivist at the Archivio di Stato di Venezia on the order of the Director at the time, Bartolomeo Cecchetti, and is a very accurate and legible transcription of the original in Vienna. On the chronicle see A. Nanetti, *Antonio di Marco Morosini : Cronicha de Veniexia (1094-1434) : edizione critica*, Ph.D. dissertation, Bologna : Università degli Studi di Bologna, 2000, and also Id., *Il Codice Morosini. Il mondo visto da Venezia (1094-1433)*, 4 vols., Spoleto : Fondazione CISAM, 2009.

Murano and see the Mint of the gold ducats of Venice (*de veder Muran e bater dela Zecha di duchati doro de Veniexia*).[79] According to Gonçalves's interpretation, Dom Pedro went to Murano to meet Fra Mauro in the monastery of San Michele. Following this visit, deeply impressed by the work of the Camaldolese, Dom Pedro began an exchange of cartographic documents with Fra Mauro that would have persisted for more than twenty years and culminated in the official commission by the Portuguese King Afonso V of the *mappa mundi* from Fra Mauro.[80] All of this is false or at the least conjectural. First, the chronicle of Morosini does not report the giving of any manuscript of Marco Polo as a gift to Dom Pedro[81]; nor does it recount a visit to San Michele di Murano, much less the commission of a *mappa mundi*. Gonçalves's interpretation is based primarily on a conjectural and forced reading of the Venetian documents that describe the Infante Dom Pedro's short stay in Venice. Nothing precludes the possibility that Dom Pedro visited the Camaldolese monastery of San Michele; however, Morosini records with great precision the places visited by the Infante and his large following of Portuguese knights and Venetian nobles, and he makes no mention of San Michele. The misinterpretation of Gonçalves was probably caused by his having confused "Murano" with "San Michele di Murano." In fact it is well-known that the monastery is not on Murano, but rather on a small island in the lagoon between Venice and Murano. This confusion of the sites caused the misinterpretation of events. From Morosini we know that Dom Pedro visited the island of Murano to witness the minting of gold ducats and (probably) to see the glassmaking works, famous throughout Europe, but there is nothing about San Michele. Galvão's story – revived by Gonçalves and amplified by Pinheiro Marques – could be attributable to a cultural climate that sought to ennoble the figure of Dom Pedro, showing his virtues and intellectual curiosity and placing him at the origin of the Portuguese *descobrimentos*, and aimed to impart a "scientific" character to the Portuguese voyages and expansion right from the very beginning.[82] The further elaboration of this hypothesis by Pinheiro Marques, who goes so far as to identify the (alleged) *mappa mundi* brought to Lisbon by Dom Pedro with that held in Alcobaça – the two maps mentioned by Galvão – as a map authored by Fra Mauro seems to me totally unfounded.[83]

[79] Venice, Biblioteca Nazionale Marciana, Ms. Cl. VII 2049 (= 8332), pp. 859-865 (p. 865).

[80] J. Gonçalves, *O Infante D. Pedro as sete partidas e a génese dos descobrimentos*, Lisbon: Agência Geral do Ultramar, 1955, p. 247.

[81] C. Gadrat, "Le role de Venise dans la diffusion du livre de Marco Polo (XIVe-début XVIe siècle)," *Médiévales* 58, printemps 2010, pp. 63-78 (pp. 74-75): "La réalité historique du don ne repose en fait que sur l'introduction de Valentim Fernandes à son édition en portugais de Marco Polo en 1502, qui n'apporte pas de preuve ni n'indique de source. Mais Valentim Fernandes écrit plus de trois quarts de siècle après la visite de Dom Pedro à Venise. La tournure de ses propos laisse entendre, par ailleurs, qu'il n'a pas vu personnellement l'exemplaire conservé dans la bibliothèque royale. Il s'agit probablement d'une pure supposition, destinée à expliquer la présence du livre de Marco Polo dans la bibliothèque royale du Portugal et à donner du prestige à cet exemplaire en lui reconnaissant une origine vénitienne directe. C'est aussi, par contrecoup, la traduction éditée par Valentim Fernandes qui est ainsi valorisée. Cette traduction a en réalité été faite sur la version latine de Francesco Pipino." I am not quoting the footnotes.

[82] This paradigm of analysis has been discussed and discredited in many more recent works, including de Albuquerque, *Os descobrimentos portugueses*; Thomaz, *Le Portugal et l'Afrique au XVe siècle: les débuts de l'expansion*; F. Bethencourt and K. N. Chaudhuri (eds), *História da expansão portuguesa*, 5 vols., Lisbon: Temas e Debates, 1998-2000, vols. 1 and 2.

[83] A. Pinheiro Marques, *A maldição da memória do Infante Dom Pedro*, Figueira da Foz: Centro de Estudos do Mar e das Navegações, 1994, pp. 151-202 (p. 155 on the dating to 1448-1450, which otherwise we consider correct.

IV. – The *mappa mundi* in Latin for the Medici of Florence

On May 26, 1494 the Camaldolese Prior General Pietro Dolfin (1444-1525) replied to a letter from the prior of San Michele di Murano, Bernardino Gadolo (1463-1499). This was in response to a missive – which remains missing, presumably lost[84] – in which Gadolo had sought help to find the codex that contained Dolfin's Latin translation, compiled some years earlier, of the legends of the *mappa mundi*. Dolfin's letter to Gadolo offers crucial information about both the reception given to Fra Mauro's *oeuvre* and the circulation of the work during the fifteenth century.

Pietro Dolfin, born in Venice in 1444, entered the Camaldolese monastery of San Michele di Murano, and in 1479 was elected abbot of the same community. The following year he was made general of the order, an office which he held until 1513, when he resigned in favor of Paolo Giustianian.[85] Dolfin's letters, which number more than four thousand – addressed to different ecclesiastics of his own and other orders, as well as to various secular dignitaries – were in part published in Venice in 1524 by Giacomo da Brescia, acting under Bernardino Benalli ; and, later, several other pieces of correspondence which had been omitted from the Venetian edition were published by the French Benedictine historians Edmond Martène (1654-1739) and Ursin Durand in their *Veterum Scriptorum...amplissima collectio*.[86] The above-mentioned letter to Prior Gadolo, however, was not included in either of these printed editions. It is to be found instead in an early sixteenth-century manuscript held today in the Biblioteca Nazionale Marciana (Ms. Lat. XI, 92 (= 3828), pp. 502-503).[87] In this letter to Gadolo, besides expressing his satisfaction

of the "prototype" of the *mappa mundi*, provided that the author is referring to the *mappa mundi* now in the Marciana) : this section contains unsubstantiated and arbitrary statements such as "the origin of the 'plan of India' [and] the Venetian map of the Infante D. Pedro is the famous Fra Mauro *mappa mundi*... on the basis of which the King João II (grandson and heir of Don Pedro) conceived and executed the plan of the East," *ibid.*, p. 153. Id., "The Portuguese Prince Pedro's Purchase of the Fra Mauro Map from Venice," *The Globe* 48 (1999), pp. 1-34 ; and Id., *Vida e obra do Príncipe Perfeito Dom João II : Rei de Portugal, Senhor de Coimbra* etc., Figueira da Foz : Centro de Estudos do Mar e das Navegações : Mira : Câmara Municipal, 1997, pp. 96-101.

[84] The letter is not to be found in the handwritten collection by Gadolo, conserved today in Monastery library of Camaldoli (n. 945). I was not able to examine a second copy of this collection, held in the Camaldolese Hermitage of Frascati. See E. Massa (ed.), *Paolo Giustiniani. Trattati, lettere e frammenti dai manoscritti originali del beato Paolo Giustiniani dell'Archivio dei Camaldolesi di Monte Corona nell' Eremo di Frascati*, Rome : Edizioni di Storia e Letteratura, 1967.

[85] R. Zaccaria, "Dolfin Pietro," in *Dizionario biografico degli italiani*, Rome : Istituto della Enciclopedia italiana, vol. 40 (1991), pp. 565-571.

[86] E. Martène and U. Durand (eds), *Petri Delphini Epistolae, Veterum Scriptorum et monumentorum ecclesiasticorum et dogmaticorum amplissima collectio*, III, Paris, 1724.

[87] The first part of this manuscript contains correspondence from 11 March 1462 to 20 November 1480, followed by a section which groups together the letters from December 1480 to June 1505. Numbered 722 in the old San Michele catalogue system, the manuscript was briefly described in G. B. Mittarelli, *Bibliotheca codicum manuscriptorum S. Michaelis prope Murianum una cum Appendice librorum impressorum saeculi XV, opus posthumum Johannis-Benedicti Mittarelli Veneti, abbatis ex-generalis benedictino-Camaldulensis*, Venice : ex Typographia Fentiana, 1779, col. 360. Zurla transcribes the letter in *Il mappamondo di Fra Mauro camaldolese descritto ed illustrato da D. Placido Zurla dello stess'ordine*, Venice : s.e., 1806, pp. 150-151, though without making any reference to Dolfin's manuscript. A second codex, originally kept in Camaldoli and containing 3,000 letters, is today held in the Biblioteca Nazionale Centrale in Florence, with the marking of the Conventi Soppressi, E. III. 405 (in four volumes). Written

in acquiring a great many valuable books for a very modest sum.[88] Dolfin also asserts that, had Gadolo searched diligently among his writings still held at San Michele di Murano, he would certainly have found the former's Latin translation of Fra Mauro's legends:

> I thought that among my writings there should be the Latin translation that I made of the [legends] of [Fra Mauro's] map of the world. If you look for them more diligently, you will probably find all of them in your cell, written in my hand. Otherwise I do not think that I will easily be able to obtain a copy [of the Latin translation of the legends] from the Medici residence, as you requested. If it were the case of a manuscript, I do not doubt they would lend it to me for me to transcribe. In truth, what I would have to ask for is that someone, in my name, could be allowed in that house (the Medici house) to copy those annotations from the map of the world that the Florentine painters had once reproduced from our original ["archetype", that is Fra Mauro's *mappa mundi*, in San Michele]; I would not dare to do so, especially at this time, in which the city is not entirely immune to internal conflicts, and the two brothers of the Medici's family,[89] in such high offices, so completely loved by all the people, are now banned from the town.[90]

There are four main issues related to Fra Mauro's map that emerge from Dolfin's letter. First, Dolfin tells of his own Latin translation of the map's legends and confirms that in 1494 – at least in his mind – it was still among his writings (*inter cartulas nostras*; probably in a small notebook), in his cell in the San Michele monastery. The second issue raised by Dolfin is that his own translation had been used for a copy of the *mappa mundi*, produced from the original (*ex archeypo nostro*) by certain Florentine painters during the time in which Dolfin was still abbot of San Michele – that is, before 1480, when he moved to Camaldoli (close to Arezzo and Florence) in Tuscany, after he was appointed General of the Order. It follows, then, that the work sought by Gadolo was created before 1480, during a period in which a younger Dolfin, as yet unburdened by the responsibilities of greater power, was able to dedicate himself meticulously to transcribing and translating the legends directly from the original *mappa mundi*, or more probably from Fra Mauro's preparatory notes which are to be considered lost since from at least the mid-eighteenth century.

by a copyist working under the direction of Dolfin, these volumes contain letters gathered from the period of Dolfin's election as Prior General of the Order until his resignation.

[88] Venice, Biblioteca Nazionale Marciana, Ms. Lat. XI, 92 (= 3828), p. 502: "Bernardino Priori. Redierunt nostri Venetiis, cum sarcinulis, incolumes. Deo gratias. Miratus sum, tot, tantaque volumina, prae oculis exposita, tam parvo coempta fuisse pretio: ut tantidem profecto, aut pluris etiam, vel unum ex iis debuerit constitisse. Magna nostrorum temporum foelicitas, ut suppellex tam pretiosa tam vili veneat. Benedictus Deus, a quo omne datum optimum, et omne donum perfectum."

[89] Piero, called *Il Fatuo*, and Giuliano di Lorenzo de' Medici, Duke of Nemours; the third brother, cardinal Giovanni, the future pope Leo X, was already in Rome.

[90] Venice, Biblioteca Nazionale Marciana, Ms. Lat. XI, 92 (= 3828), p. 503: "Putabam esse inter alias cartulas nostras, quae olim in pictura orbis annotata, latina feceram. Si perquisieris illa diligentius, forte in cellula tua omnia invenies, conscripta manu mea: alias non facile speres me ex domo Medice, ut petisti, exemplar habiturum. Redacta in codicem, haud dubie commodarentur mihi transcribenda. Quod vero petere habeam, ut quis meo nomine in domo illa permittatur, annotationes ipsas ex eo orbe quem quondam Florentini pictores ex archetypo nostro effinxerunt, describere, non ego sane id auderem: hoc presertim tempore quo non penitus vacat intestinis seditionibus civitas, relegatis extra urbem duobus illis tantae autoritatis fratribus gentis Medicis, primariis utique civibus et praepotentibus, et universo populo quam carissimis. [...] Vale ex Fonte bono die XXVI maji 1494.".

At the same time – and this is the third issue raised in the letter – Dolfin clearly asserts that the copy of Fra Mauro's map containing the Latin translations has since been moved to the Medici Palace, in the very center of Florence. The *palazzo* mentioned in the letter is undoubtedly that which Michelozzo di Bartolomeo Michelozzi (1396-1472), under orders from Cosimo "il Vecchio" (the Elder) de' Medici, had begun to construct in 1444 on Via Larga (today, Via Cavour), and which was finished about ten years later.[91] The *palazzo* had been the Medici main residence for an entire century until 1537, when – at the founding of the Princedom of Cosimo I – the family took up residence in Palazzo Vecchio. Gadolo's implicit suggestion to Dolfin, to ask the Medici for a transcription of the Latin text from their copy of the *mappa mundi*, was regarded by Dolfin to be inappropriate in light of the recent revolt against the Medici and, in particular, Piero de' Medici, a close friend. Dolfin suggested instead that, if Gadolo could not find the original manuscript at the monastery of San Michele, someone would have to be granted permission on his (Dolfin's) behalf, to transcribe the Latin text directly from the *mappa mundi* in the Medici palace.

To this day we do not know for certain who commissioned this copy of the *mappa mundi*, but it was almost certainly a member of the Medici family. Lorenzo had assumed governorship of Florence immediately after the death of his father in 1469. Since, as mentioned, the transcription and translation of the map's legends were realized sometime before 1480 it is probable, then, that it was Lorenzo who commissioned this copy of the *mappa mundi*.

A clear evidence of Lorenzo de' Medici's interest in maps is found in the inventory of his possessions carried out after his death in 1492 – attributing to him ownership of no less than four world maps, as well as numerous others. The *Libro d'inventario dei beni*, copied in 1512 by the priest Simone di Stagio dalle Pozze, contained the following entries:

> "[...] A map depicting Italy / Another map depicting the castle of Milan / Another depicting a world map / Another, the Holy Land / Another, a world map / Two others, the [genealogical] tree of the emperor and king of France / A map depicting three realms of India / Another world map / Another illustration of the Holy Land / A long, narrow map of several countries / A map of Rome."[92]

Though the brevity of the descriptions makes it impossible to discern which of the four world maps mentioned in the inventory could be the copy of Fra Mauro's *mappa mundi*, the presence of notes in the list, such as "carta da navicare" (marine chart), "map of Italy," and "map of the Holy Land," suggests that in this specific case "world map" – otherwise a rather generic expression that in contemporary linguistic traditions could refer to any kind of map and even to textual descriptions of places – indicates ecumenical maps, whether Ptolemaic or non-

[91] Giorgio Vasari, *Le vite dei più eccellenti pittori, scultori e architetti*. Rome: Newton Compton, 2007, p. 362.

[92] Florence, Archivio di Stato, Mediceo avanti Principato, filza 165, *Libro d'inventario dei beni di Lorenzo il Magnifico, 1512. Questo libro d'inventarii è copiato da un altro inventario, el quale fu fatto alla morte del magnifico Lorenzo de' Medici, copiato per me, prete Simone di Stagio dalle Pozze, oggi, questo 23 di dicembre 1512, per commissione di Lorenzo di Piero de' Medici*, f. 29: "[...] Una carta dipintavi Italia | Un'altra carta dipintavi il castel di Milano | Una dipintavi el mappamondo | Una dipintovi Terrasanta | Una dipintavi l'apamondo | Du'altre dipintovi l'albero dello 'mperio e re di Francia | Una carta dipintavi tre reami d'India | Un'altra dipintovi l'apamondo | Un'altra dipintovi l'Italia | Una carta da navicare | Un'altra dentrovi l'apamondo | Un'altra dentrovi Terra Santa | Una carta stretta e lunga dentrovi più paesi | Una carta dentrovi Roma." There is also a printed edition: M. Spallanzani and G. Gaeta Bertelà (eds), *Libro d'inventario dei beni di Lorenzo il Magnifico*. Florence: SPES, 1992.

Ptolemaic. Nothing in the inventory excludes the possibility that, among Lorenzo's possessions, lays the copy of Fra Mauro's *mappa mundi* referred to by Dolfin and sought by Gadolo.

In the mid-eighteenth century, Giovanni Benedetto Mittarelli – abbot of San Michele di Murano since the second decade of the eighteenth century – searched in vain for any trace of the copy of Fra Mauro's *mappa mundi* in Florentine archives and libraries. Finding their search to be futile they concluded that the work was most likely among the material stolen, dispersed, and destroyed either during the sack of the Palazzo Medici at the end of 1494 or in the Bonfires of the Vanities (*Falò delle vanità*) that followed, when thousands of objects were deemed bearers of sin (precious clothes, cosmetics, mirrors, art, books, etc.), and were accordingly burned in Florence after the revolt against the Medici by the supporters of the Dominican Girolamo Savonarola (the most famous of them took place on 7 February 1497, on the Shrove Tuesday festival).[93] Nonetheless, the concrete reference to Fra Mauro's *mappa mundi* made in the legends of one of Egnazio Danti's maps – designed between 1563 and 1574 for the *Guardaroba nuova* (New Wardrobe) of Palazzo Vecchio (which will receive more attention at the end of this chapter) – indicates either that the copy held in Palazzo Medici was still in existence during the aforementioned period or at least had not yet been forgotten and went missing only subsequently. Another manuscript to go missing was that of Dolfin, containing his Latin translation of Fra Mauro's legends ; or, at least, it has not been successfully found.

Meanwhile, the map commissioned for Palazzo Medici was the second copy of Fra Mauro's world map executed in the space of twenty years, following the one commissioned between 1457 and 1459 for the Court of King Afonso V of Portugal. The very existence of a Florentine commission for a non-Ptolemaic Venetian cosmography around 1480 merits further discussion, especially with regard to two main aspects : the meaning and the function of such a commission in Florence, at that time one of the main centers for the diffusion of Ptolemy's *Geography* ; and the reasons for the translation into Latin of the vast text of Fra Mauro's *oeuvre* which was, after all, a huge and demanding task.

The Camaldolese monk's magnificent map, despite being conserved in a monastery, did not pass unobserved and Fra Mauro's fame spread quickly throughout the courts of Europe. In the three decades between 1460 and 1490 European expeditions geared toward geographic discoveries, and – even more significantly – the discoveries of long-forgotten classical literary works such as Strabo's *Geography*, were particularly focused on the Orient. Fra Mauro's Indian Ocean and *Cathay* were, and continued to be, the most breathtaking, ambitious representations of Asia until the early years of the sixteenth century when Portuguese maps such as the "Cantino Map" (bought by Alberto Cantino for the Dukes of Ferrara in 1502 for 12 ducats) began to make their appearance.[94] No Florentine Ptolemaic manuscript – not even the magnificent works of Henricus Martellus Germanus, drawn three decades later than Fra Mauro's map – could offer anything comparable. And in Florence, just like in the rest of Europe, all eyes were turned towards the Orient until the early sixteenth century – despite the pompous and often historically false celebrations of Columbus' or Vespucci's voyages that have been witnessed in recent years. Fra

[93] Mittarelli, *Bibliotheca codicum manuscriptorum*, col. 759-760.

[94] See M. F. Alegria, S. Daveau, J. C. Garcia *et al.*, "Portuguese Cartography in the Renaissance," in *The History of Cartography*, III, D. Woodward (ed.), *Cartography in the European Renaissance*, Chicago : Chicago University Press, 2007, pp. 975-1068.

Mauro provided the Florentines with a critical reading of Ptolemy's *Geography*, thanks to the unique interplay of influences within Venetian cosmography that brought together nautical cartography (in its description of the European and, in part, the African coastline), Ptolemaic maps, the accounts of modern travelers – such as Marco Polo, Niccolò de Conti, and certain Ethiopian monks – with new information regarding Africa and Asia. It is likely, therefore, that the commissioning of a copy of the *mappa mundi* could have also been inspired by the desire to update geographical knowledge of the time – a project that such Florentine figures as Nicolaus Germanus, Henricus Martellus, and Francesco Rosselli were all beginning to pursue during these years. Indeed, it was these very cartographers who were involved in the production of the *tabulae novellae* which, around the mid-fifteenth century, were added to the twenty-seven "classic" maps of the *Geography*.[95]

Commissioning a copy of the *mappa mundi* may also have been driven by a desire to impress. Displayed in a public area, or in a palace such as the Medici residence in Florence, it was undoubtedly an object that aroused awe and admiration : an appropriate adornment for a prince's abode. The choice of cosmography as the medium through which to celebrate the power of the Medici drew on well rooted historical models : that is, the study, possession, and commissioning of maps as closely related to powerful men (or governments), indeed, a theme with ancient roots. The particular idea meant to be transmitted through cartography is that of active sovereignty, played out through continuous acquisition of territory, aided by an ever-expanding geographical knowledge of the *orbis terrarum*.[96] Essentially, no matter how grand the manuscripts of Ptolemy's *Geography*, they could not compete with the "visual effect" of such a richly illuminated and extensively annotated map, with its wealth of details and impressive size. Moreover, most of the manuscripts of the *Geography* manufactured in Florence were copied in series, as happened in the case of the works by Piero Del Massaio and Niccolaus Germanus.[97] Several Florentine families, including the Medici and their rivals, the Strozzi, Gaddi, and Vitelli families, all owned lavishly illustrated manuscripts of the *Geography*.[98] Thus, Fra Mauro's *mappa mundi* was far more outstanding in comparison.

The question of the translation into Latin is rather more complex. The vernacular, in and of itself, did not constitute a problem in Florence, not even for prestigious patrons like the Medici, as is demonstrated by the *Septe giornate della Geographia* (Seven Days of Geography) by Francesco Berlinghieri (1440-1500). A pupil of Johannes Argyropoulos, a Greek scholar who

[95] The title page of the monumental codex, *Ptolemei Cosmographia* with maps drawn by Henricus Martellus (Florence, Biblioteca Nazionale Centrale, Magl. XIII, 16, *c.* 1490) clearly demonstrates the effort exerted by cartographers of German origin in Florence at the end of the fifteenth century to update and circulate (even in serial form) Ptolemy's *Geography*. The verso of the first sheet, displays, in gold letters on a blue background, the title of the work : "Cl. Ptolomei Cosmographia cum tabulis regionum n(ost)ri t(empor)is et universis portubus et locis maritimi tractus tam notis quam a rege Portus Galli nuper repertis hoc ornatissimo codice continentur." Or "In this most ornate codex lies the Cosmography of Claudio Ptolemy with modern tables [depicting] the regions of all the harbors and coasts, already known [by the ancients] and those recently discovered by the King of Portugal."

[96] See M. Milanesi, "Cartografia per un principe senza corte : Venezia nel Quattrocento," in *Micrologus. Natura, scienze e società Medievali – Nature, Sciences and Medieval Societies* 16 (2008), pp. 189-216.

[97] Gautier Dalché, *La Géographie de Ptolémée*, pp. 219-225.

[98] Gentile, *Firenze e la scoperta dell'America*, entries 101, 102, 103, pp. 202-215. G. Lazzi, "Mondi celesti e terrestri nei libri medicei," in A. Cecchi and P. Pacetti (eds), *La Sala delle Carte geografiche in Palazzo Vecchio : 'capriccio et invenzione nata dal Duca Cosimo'*, Florence : Polistampa, 2008, pp. 41-65.

taught in Florence from 1456 to 1471, and of Cristoforo Landino, who produced commentaries on Dante and Petrarch, Berlinghieri composed the *Septe giornate* (also know as *Cosmographia in terza rima*, Cosmography in Free Verse) as a free paraphrase in Tuscan vernacular of Ptolemy's *Geography*. The composition took him some fifteen years, from 1465 to 1478 or 1479. Working in the context of the *Accademia platonica* (Platonic Academy) he benefited from a relatively close relationship with the Medici family.[99] In fact, one of the four extant manuscripts of Berlinghieri's adaptation of the *Geography*, held today in the Biblioteca Nazionale Braidense in Milan, was identified with the "liber cosmografie quem composuit Franciscus de Berlinghieris et donavit Laurentio de Medicis ornatus minio et aliis," as described in Lorenzo de Medici's inventory.[100] Since no translation was deemed necessary before offering the work to Lorenzo 'il Magnifico', we can assume that the vernacular did not pose any serious issue : Lorenzo was a poet himself, writing in his native Tuscan vernacular.[101]

The question of translating Fra Mauro's legends into Latin was probably related to the version of the Venetian vernacular employed by Fra Mauro, heavily influenced by local dialects, which could consequently be considered inferior to the Tuscan vernacular and was therefore "not presentable" in the opinion of someone as well educated as Dolfin. More than simply a *desideratum* of its Florentine patrons, the translation of the *mappa mundi*'s legends from Venetian vernacular into Latin seems an initiative headed by Dolfin to increase the status and value of the work ; as well as to somehow connect Fra Mauro's work to a project begun by Dolfin and his student Bernardo Gadolo of writing (in Latin) the history of the Camaldolese order. Thus, by translating Fra Mauro's legends into Latin, Dolfin acquired a somewhat important role in a commission just as prestigious as if it had come straight from the Medici, without doubt one of the richest and most prominent families in Europe.[102] The version of the *mappa mundi* copied by Florentine painters with said legends – which were modified or corrected by Dolfin to what extent we know not – was transformed into proof of the cultural and literal riches of the Camaldolese ; and may have been the means through which Dolfin hoped to obtain the Medici's recognition and patronage. Put otherwise : from the second half of the fifteenth century Fra Mauro's *mappa mundi* was transformed into a Camaldolese *oeuvre* and Fra Mauro's name began to be forgotten. Moreover, it is likely that such motives were shared by Bernardino Gadolo in seeking out Dolfin's Latin translation of the legends – although Gadolo's initial written request to Dolfin stating his objective must remain a conjecture, as it missing from among the other letters of the *Epistolario*. What we do know for certain is that Gadolo studied at the University of Padua where he acquired a doctorate in canon law as well as a taste for classical literature.

[99] For the most recent study, finally re-evaluating the work of Berlinghieri, see Gautier Dalché, *La Géographie de Ptolémée*, pp. 252-255.

[100] Milan, Biblioteca Nazionale Braidense, AC XIV 44, manuscript on vellum, 438 × 355 mm, text in Tuscan vernacular, 31 maps, written before 1482. Signed : "Francesco Berlinghieri Florentino." In the period 1481-1482, Niccolò di Lorenzo della Magna printed the *Septe giornate della Geographia* in Florence, making it the first printed edition of the *Geography* in a vernacular language : *In questo uolume si contengono septe giornate della geographia di Francesco Berlingeri fiorentino allo illustrissimo Federigo duca durbino*. Florence : Niccolò di Lorenzo della Magna [before September 1482]. (IGI 1491). See Gentile, *Firenze e la scoperta dell'America*, entry 112, pp. 235-237.

[101] Lorenzo de' Medici, *Opere*, T. Zanato (ed.), Turin : Einaudi, 1992.

[102] R. De Roover, *The Rise and Decline of the Medici Bank, 1397-1494*, New York : W. W. Norton, 1966, pp. 194-356.

At the age of eighteen he entered the monastery of San Michele di Murano, and in June 1482 he made his profession there. The abbot of San Michele at that time, Pietro Dolfin, directed the young monk's energies to the study of scripture and the Church Fathers. Gadolo edited the Bible, comparing preceding editions with various manuscripts, and including in his new edition the *Glossa ordinaria* and the commentary of Nicholas of Lyra.[103] Two years later Gadolo edited St. Jerome's commentaries on Genesis and the prophets (Venice, G. de Gregoriis 1497), winning Erasmus' praise for his erudition and learning.[104]

More specifically relevant to us, in the context of Fra Mauro's *mappa mundi*, Gadolo also wrote the first historic memoir on the Camaldolese order, entitled *De vita sanctissimi patris nostri Romualdi, de initio item successu et nobilitate ordinis nostri* (The life of our most holy founder Romualdo, [and] of the founding, success, and nobility of our order),[105] a lengthy pamphlet which, on 30 August 1496, he addressed to the historiographer and Augustinian monk Jacopo Filippo Foresti di Bergamo (1434-1520), hoping the latter would include it in his widespread chronicle, the *Supplementum Chronicarum*.[106] Cécile Caby has further demonstrated that Gadolo probably also sent his work to the historian Marcantonio Sabellico (Marcantonio Coccio o Cocci, 1436-1506), a friend and regular correspondent, so that Sabellico might include it in his own work, the *Enneades*.[107] With few exceptions, until Gadolo, the Camaldolesi were relatively disinterested in the production and promotion of the order's history. A predilection for history and historiography began to emerge within the order around the end of the fifteenth century, a result of the Humanism movement which the order embraced after a significant cultural shift occurred internally – triggered by Ambrogio Traversari, head of the order until his death in 1439. Dolfin had encouraged the compilation of a written history of the order as a means of both recovering and transmitting the oral historical tradition of the Camaldolesi.[108] In this context, then, Bernardino Gadolo composed his *memorandum* on Camaldolese history, written for historians and contemporaries curious about monastic history. In addition to a historical biography of the order's founder, San Romualdo, Gadolo outlined the biographies of certain other illustrious Camaldolese monks from the order's beginnings in the eleventh to the fourteenth centuries, to his own contemporaries (among them, the Venetian Maffeo Gherardo, prior of San Michele during Fra Mauro's life, and Pietro Dolfin).

[103] *Biblia latina, cum glossa ordinaria Walafridi Strabonis; aliorumque et interlineari Anselmi Laudunensis; et cum postillis Nicolay de Lyra; expositionibusque Guillelmi Britonis in omnes prologos S. Hieronymi; Nicolaus de Lyra, Contra perfidiam Judaeorum*, ed. Bernardinus Gadolus, Eusebius Hispanus, et Secundus Contarenus. Parti I-III. Venice: Paganino de' Paganini, 1495.

[104] C. Caby, "Bernardino Gadolo ou les débuts de l'historiographie camaldule," *Mélanges de l'Ecole française de Rome. Moyen-Age* 109, 1 (1997) pp. 225-268.

[105] Camaldoli, Archivio di Camaldoli, cod. 735, manuscript on vellum, 22.5 × 16.7 cm, late fifteenth century, ff. 1-41. Published by C. Caby, "Lettre de Bernardino Gadolo à Jacopo Filippo da Bergamo (30 août 1496)," in Caby, "Bernardino Gadolo ou les débuts de l'historiographie camaldule," pp. 253-266.

[106] Jacopo Foresti published excerpts of the contents of *De vita sanctissimi patris nostri Romualdi* by Gadolo in the 1503 edition of the *Chronica* (*Supplementum Supplementi Chronicarum*, Venice, 1503, ff. 383v-384r). Transcription available in Caby, "Bernardino Gadolo ou les débuts de l'historiographie camaldule," pp. 268-269.

[107] M. A. Sabellico, *Enneades*, IX, lib. 2, in *Opera omnia*, C. S. Curio (ed.), Basel, 1560, t. II, col. 648-649. Transcription available in Caby, "Bernardino Gadolo," pp. 267-268.

[108] *Ibid.*, pp. 237-239.

Gadolo's search for the legends translated by Dolfin, then – two years before finishing *De vita sanctissimi patris nostri Romualdi* – could well constitute a critical part of his research on the composition of his *memorandum*, which dealt with, among other things, the successes of the order. Moreover, though Fra Mauro's work is not (directly) cited in the *memorandum*, we must not exclude the possibility that Gadolo was interested in the map's contents. At this regard, correspondence between Gadolo and the Aristotelian master Leone Michiel notably attests to the former's interests in astronomy and cosmography.[109] Whatever, then, were the true motives behind Dolfin and Gadolo's interest in the *mappa mundi* and in its diffusion, the commission by the Medici of a copy of Fra Mauro's work is particularly remarkable given that there are no clear (documentary) traces in Venice of a reaction to any of Fra Mauro's *oeuvre* until the second decade of the sixteenth century. As we will see in the following paragraphs, it is only in the period between 1519 and 1524 that Alessandro Zorzi, a Venetian erudite, included passages derived from the *mappa mundi* in a description of the Nile and its sources. Rather significantly, Zorzi attributes them to a "friar Nicola" (*frate Nicola da S. Michiel di Muran*) and not to Fra Mauro. The oblivion into which both Fra Mauro's name fell in Venice could have been also related to a specific strategy of the Camaldolese Order to take "intellectual" possession of his work by transforming it into a part of the monastic knowledge of the Camaldolesi. The presence of cosmographers such as Antonio Leonardi whose maps, regarded as more adherent to Ptolemy's *Geography* than that of the Camaldolese friar, were also likely written in Latin (making it more accessible to the educated in other cities) then contributed to reinforce the oblivion of Fra Mauro's name.[110] His maps were instead copied many times and anonymously in the course of the sixteenth century.

V. – Sixteenth-century Venetian literary reception

1. Alessandro Zorzi's *Aviso di Fra Nicola di San Michiel di Muran*

During the period from 1519 to 1524, Alessandro Zorzi, a scholarly collector of travel diaries, compiled four notebooks of manuscript sheets and annotated printed pages, containing letters, descriptions, and maps which guide the reader across Asia, Ethiopia, the Red Sea, Arabia, Persia, and even to the New World. In the last of the four notebooks, Zorzi traces four itineraries that lead either from Jerusalem or Alexandria to Ethiopia.[111] The third of these contains a discussion of the Nile and its sources that Zorzi attributes to a certain "frate Nicola da S. Michiel di Muran"

[109] Camaldoli, Archivio di Camaldoli, cod. 735, ff. 307-311. Published in *Annales Camaldulenses*, vol. VII, n. 50.

[110] P. Gautier Dalché, "Weltdarstellung und Selbsterfahrung : Der Kartograph Fra Mauro" in H.-D. Heimann and P. Monnet (eds), *Kommunikation mit dem Ich*, Bochum : 2004, pp. 39-51.

[111] Florence, Biblioteca Nazionale Centrale, Banco rari 236, A. Zorzi, [*Miscellanea di cose geografiche*], manuscript on paper, *c.* 1530, ff. II, 192, I' ; modern pagination in pencil : 1-192. The description of Ethiopia by fra Nicola of San Michele di Murano, entitled *Aviso di Fra Nicola inscripto fra di S. Michiel di Muran*, is in Venetian vernacular and has hand-written comments by Alessandro Zorzi ; it occupies ff. 38r-43v (old numering : ff. 25r-29v). Cf. the modern edition, with transcription and English translation in O. G. S. Crawford (ed), *Ethiopian Itineraries circa 1400-1524. Including those collected by Alessandro Zorzi at Venice in the years 1519-1524*, Cambridge : Hakluyt Society at the Cambridge University Press, 1958, pp. 108-125. See also P. F. Gomes, "Alessandro Zorzi e l'invenzione

(brother Nicola of San Michele of Murano). In fact, the brother's text appears to have been drawn directly from the previously-mentioned *scripture*, Fra Mauro's preparatory notes for the *mappa mundi*, and can thus be considered the first example of Fra Mauro's work being referred to as a credible source of geographical knowledge.[112] The second part of the *Aviso*, on f. 43v, is almost a literal transcription of these four legends from the *mappa mundi*: "It is said that Prester John has more than 120 kingdoms under his dominion…"[113]; "Benichileb. This name is interpreted as meaning "son of a dog" because these people have dog-like faces"[114]; "Around 1420 a ship, or junk of the Indian Ocean …[115]; "Note that the Abyssinians say their territory is more extensive…".[116]

It is significant, however, that when (around 1519-1524) Zorzi was transcribing the *mappa mundi*'s notes in his description of Ethiopia, the former does not attribute them to Fra Mauro, but to another monk from San Michele who, Zorzi tells us, visited the regions of Ethiopia in 1470. The *Aviso* thus provides clear evidence that seventy-five years after the completion of the planisphere, the description of Ethiopia – a description, according to Fra Mauro, based on maps provided by certain Coptic monks visiting San Michele[117] – was still considered up-to-date. Moreover, modern scholarship has shown that most of the Ethiopian toponyms used in the *mappa mundi* – and subsequently in Zorzi's *Itinerarium ad Ethyopiam* – correspond to those confirmed by archaeological investigation of the monasteries, roads and cities of the ancient region of Ethiopia.[118]

dei Tropici," in A. Cattaneo, D. Ramada Curto and A. Ferrand Almeida (eds), *La cartografia europea tra primo Rinascimento e fine dell'Illuminismo*, Florence: Olschki, 2003, pp. 109-132.

112 For the textual comparison of of Zorzi's *Aviso* with Fra Mauro's legends, see the Appendix of this chapter. On the hypothesis that Fra Mauro's lost notebooks could have been Zorzi's source, see A. Cattaneo, "Fra Mauro Cosmographus Incomparabilis and His Mappamundi: Documents, Sources, and Protocols for Mapping," in *La cartografia europea tra primo Rinascimento e fine dell'Illuminismo*, pp. 19-48 (p. 23).

113 *Mappamondo* (ed. 2006) *77: "Se dice che presto Jane ha più de 120 regni …." *Mappamondo* (ed. 1956) Pl. X, 21.

114 *Mappamondo* (ed. 2006) *155: "Benichileb. Questo nome è interpretado fiol de can …," *Mappamondo* (ed. 1956) Pl. XI, 11.

115 This is my translation of the beginning of the legend. For a different interpretation see *Mappamondo* (ed. 2006) *19: "Circa hi ani del Signor 1420 una nave over çoncho da India…," *Mappamondo* (ed. 1956) Pl. IV, 6.

116 *Mappamondo* (ed. 2006) *454. "Nota che abassini dicono haver più teritorio …." *Mappamondo* (ed. 1956) Pl. XVII, 1.

117 "Perché ad alguni par da nuouo che io parli | de questa parte meridional, la qual quasi està | incognita a li antichi, perhò io respondo ch(e) | tuto questo desegno da sayto in suso io l'ò ha|buto da queli proprij che sono nasudi qui, | che son sta' religiosi, i qual cu(m) le suo man | me hano desegnato tute q(ue)ste p(ro)ui(n)cie | e citade e fiumi e mo(n)ti cum li | suo nomi, le qual tu|te cosse no(n) le ho possu|do meter cu(m) el debi|to ordine per | no(n) esserui logo." *Mappamondo* (ed. 1956) Pl. X, 30; "Because to some it will appear as a novelty that I should speak of these southern parts, which were almost unknown to the Ancients, I will reply that this entire drawing, from Sayto upwards, I have had from those who were born there. These people are clerics who, with their own hands, drew for me these provinces and cities and rivers and mountains with their names; all these things I have not been able to put in due order for lack of space," *Mappamondo* (ed. 2006) *98.

118 Crawford, *Ethiopian Itineraries*, pp. 16-26 and 194-204.

2. Ramusio's interpretation: "one amongst the various miracles of Venice"

In 1559, the Giunti, a Venetian family of printers, published the second, posthumous volume of *Navigationi et viaggi* by Giovanni Battista Ramusio (1485-1557), a notable humanist and high-ranking functionary in the Venetian government. In that volume Ramusio publishes Marco Polo's book in Italian with the title of *Historia delle cose de Tartari, et diuersi fatti de loro Imperatori, descritta da M. Marco Polo Gentilhuomo Venetiano, et da Haiton Armeno* (History of things of the Tartars, and various facts of their emperors, described by M. Marco Polo, Venetian gentleman, and by Haiton the Armenian).[119] In the dedication of the proeme to Marco Polo's text to the doctor and humanist Girolamo Fracastoro (1478-1553) Ramusio underlines the originality and importance of the new edition, corrected of errors and imprecision through "most ancient manuscripts, written one hundred and fifty years ago" (*antichissimi [codici] e scritti già centocinquanta anni*). It was the first attempt at a critical edition of Marco Polo's book, based on comparative examinations of both the manuscripts and printed editions then available in Venice.[120] In the final part of an explicatory text (the *Dichiarazione d'alcuni luoghi ne' libri di messer Marco Polo con l'istoria del reubarbaro* or Declaration of several places in the books by messer Marco Polo with the history of rhubarb) that precedes Marco Polo's book, Ramusio states what is to be considered the first description and analysis of Fra Mauro's *mappa mundi*.[121] Ramusio asserted that he had used the testimony of Don Paolo Orlandini, then abbot of San Michele, "excellent cosmographer" and his friend, who told him that the map in the monastery of San Michele – Ramusio never credits the map to Fra Mauro, but quite generically to a "a lay brother of the monastery who had pleasure in the study of cosmography" (*uno loro converso del monasterio, quale si dilettava della cognizione di cosmografia*) – had been

> diligently drawn and copied from a most beautiful and very old marine chart and a world map that had been brought from Cathay by the most honorable Messer Marco Polo and his father. This Marco Polo as he travelled around the provinces upon the orders of the great Khan would make additions and notes on his maps regarding the cities and places where he was, as is described above.[122]

[119] On the *Navigazioni e viaggi* by Ramusio, see the introductions by Marica Milanesi to the six volumes of the Einaudi edition, G. B. Ramusio, *Navigazioni e viaggi*. See also Ead., "G. B. Ramusio e le Navigazioni e viaggi (1550-59), in R. Zorzi (ed.), *L'Europa delle scoperte*, Florence: Olschki, 1994, pp. 75-101. For the modern edition of the *Historia delle cose de Tartari (...) descritta da M. Marco Polo Gentilhuomo Venetiano*, G. B. Ramusio, *Navigazioni e viaggi*, vol. III, 1980, pp. 21-299.

[120] For Ramusio's sources, see Milanesi's introduction the *Historia delle cose de Tartari* as well as C. Gadrat, "Le rôle de Venise dans la diffusion du livre de Marco Polo," *Médiévales* 58 (2010), pp. 63-78. See also Chapter V in this book.

[121] G. B. Ramusio, *Dichiarazione d'alcuni luoghi ne' libri di messer Marco Polo, con l'istoria del reubarbaro*, in *Navigationi et viaggi*, in Venetia, appresso i Giunti, 1559, vol. II, f. 17r. Modern edition, Ramusio, *Historia delle cose de Tartari*, pp. 68-71.

[122] "[...] diligentemente tratto e copiato da una belissima e molto vecchia carta marina e da un mappamondo, che già furono portati dal Cataio per il magnifico messer Marco Polo e suo padre: il quale, cosí come andava per le provincie d'ordine del gran Can, cosí aggiugneva e notava sopra le sue carte le città e luoghi che egli ritrovava, come vi è sopra descritto." *ibidem*.

The original maps brought to Venice by Marco Polo and his father had become corrupted "due to the ignorance of another man who painted and reproduced them after him, adding the description of men and animals of all sorts and many other nonsenses [...] so that for many years it [the book of Marco Polo] almost lost all authority with men of any judgment.[123] Ramusio proceeds, stating that as

> When [people of judgment] began to read and consider rather more carefully than they had previously done what is written in this book by Messer Marco Polo, and to compare what he writes with the picture drawn after him, it immediately became clear that the said world map was undoubtedly drawn from that of Messer Marco Polo, [whose contents were there] displayed in good order and with accurate measurements. Right up to the present day it has been held in great veneration and prestige in all of this city ; and every day at least some of those who take most pleasure in the things of cosmography come to see and consider this map with great delight ; and amongst the *various miracles* of this divine city, it is shown as a rare and beautiful thing to those foreigners who come to see the works of glass produced on Murano.[124]

Therefore Fra Mauro's map, defined as "one amongst the various miracles of Venice," was derived from a copy – scarcely faithful, if not corrupt, with added decorative and imaginative details – of a world map and a marine chart designed by, or belonging to, Marco Polo. Ramusio believed that Marco Polo's (presumed) original cartographic documents had been already lost around the mid-fifteenth century. According to him, in fact, Fra Mauro had therefore critically and carefully corrected – and with the additional aid of Marco Polo's text – the erroneous derivatives copies of the maps of the Polo's, to achieve an accurate result.

After this fitting celebration of the beauty and fame of the *mappa mundi* in fifteenth-century Venice Ramusio begins the comparison with Marco Polo's narration, indicating the correspondence between what was written in the *Milione* and what was displayed by Fra Mauro in the *mappa mundi*, then compares it all with the new knowledge that men of Ramusio's generation disseminated and that no-one again could possibly verify after Marco Polo's return to Venice at the end of the thirteenth century. For Ramusio the errors and the disorder which at times are found in Fra Mauro's map were mostly due to copyists who had corrupted Marco Polo's original maps :

> And here again one sees many things rendered in a confused way without order, degree or measure which [faults] must be attributed to the person who painted and produced it [not Fra Mauro's copy, but that used by Fra Mauro].

For the rest, the *mappa mundi* was a faithful representation of Marco Polo's text, compared to which it indicated some examples of perfect correspondence and the capacity of second sight :

[123] "[...] ignoranzia d'un altro che dopo lui lo dipinse e forní, aggiugnendovi la descrizione d'uomini e animali di piú sorti e altre sciocchezze (...) che appresso gli uomini di giudicio quasi per molti anni perse tutta la sua auttorità." *ibidem.*

[124] "[...] di confrontare quello ch'egli scrive con la pittura di lui, immediate si è venuto a conoscere che 'l detto mappamondo fu senza alcuno dubbio cavato da quello di messer Marco Polo, e incominciato secondo quello con molto giuste misure e bellissimo ordine : onde fin al presente giorno è dapoi continuamente stato in tanta venerazione e precio appresso tutta questa città, e coloro massime che si dilettano delle cose di cosmografia, che non è mai giorno che d'alcuno non sia con molto piacere veduto e considerato, e fra gli altri miracoli di questa divina città, nell'andare dei forestieri a vedere i lavori di vetro a Murano, non sia per bella e rara cosa mostrato." *ibidem.*

> Beautiful and worthy things can be admired there, which had been unknown to or unrecognized by the ancients: for example, such as in the direction of the Antarctic, where Ptolemy and all the other cosmographers place *terra incognita* without sea, in this [map] of San Michele of Murano made already many years ago, can be seen that the sea surrounds Africa and that it is possible to navigate toward the West, which was already known in Messer Marco's time, although that cape is un-named [on Fra Mauro's map], but was called Cape of Good Hope by the Portuguese in our own time, the year 1500. In the map it can be seen next to the island of Magastar, now called [island of] San Lorenzo, and the island of Zinzibar, of which Messer Marco speaks in Chapters 35 and 36 of the third book, and many other particularities are seen in the names of the eastern islands, which have since been discovered by the Portuguese in our times.[125]

For Giovanni Battista Ramusio the *mappa mundi* represented a first attempt at a faithful reading and criticism of the *Milione*. Fra Mauro would have used the copies of Polo's maps critically in conjunction with the travelogue, thus correcting the errors contained in the maps.[126] Ramusio's analysis clearly illuminated several issues: the relationship between what was designed in the *mappa mundi* and – as Ramusio thought – what Marco Polo had designed in a *mappamondo* and in a marine chart, other than the text of the *Milione*. It is unknown if these cartographic works, and their more or less corrupt copies, really did exist or not. Ramusio was absolutely convinced that Marco Polo, as a merchant, a traveler, and an officer for the Gran Khan, must have been map conscious: surely he had to use maps, even if only in simple skeletal or itinerary form, and to draw others to bring back to Venice. At the same time, Ramusio was also convinced that Polo's original maps as well as all derivative copies were all lost probably by the mid-fifteenth century and therefore the only extant evidence of their existence was Fra Mauro's map. We can sympathize with Ramusio. Nevertheless, we may also recognize his conviction of an anachronistic projection back to the thirteenth and fourteenth centuries of his own sixteenth-century *modus operandi*, in terms of map consciousness. Closely connected to Ramusio's work of gathering the textual sources on European expansion was in fact the extraordinary and updated cartographic work of the cartographer Jacopo Gastaldi (*c.* 1500-1566) who, among other things, designed the maps in the first edition of the *Navigazioni e viaggi*.[127] Ramusio takes for granted that Marco Polo must have prepared the geographic maps and that these

[125] This is my translation of the following sentence: "... vi si comprendono per ciò di molto belle e degne particularità, non sapute ancora né conosciute meno dagli antichi: come che verso l'antartico, ove Tolomeo e tutti gli altri cosmografi mettono terra incognita senza mare, in questo di San Michele di Murano già tanti anni fatto si vede che 'l mare circonda l'Africa e che vi si può navicare verso ponente, il che al tempo di messer Marco si sapeva, ancor che a quel capo non vi sia posto nome alcuno, qual fu per Portughesi poi a' nostri tempi l'anno 1500 chiamato di Buona Speranza. Vi si vede appresso l'isola di Magastar, ora detta di San Lorenzo, e quella di Zinzibar, delle quali messer Marco parla nè capitoli 35 e 36 del terzo libro, e molte altre particularità nelli nomi dell'isole orientali, che dapoi per Portughesi a' tempi nostri sono state scoperte." *ibidem*.

[126] Piero Falchetta's interpretation differs regarding Ramusio's comment on the *mappa mundi*. This is Falchetta's opinion: "In effect, for him [Ramusio], the San Michele planisphere must have been nothing more than a poor copy of the Marco Polo original." As we have seen, Ramusio's narrative states otherwise. Cf. *Mappamondo* (ed. 2006), p. 29.

[127] D. Perocco, "Giacomo Gastaldi e la Universale Descrittione del mondo," in S. Ballo Alagna (ed.), *Esplorazioni geografiche e immagine del mondo nei secoli XV e XVI*, Messina: Grafo, 1994, pp. 211-222; M. Milanesi, "El mundo visto desde Venecia en el siglo XVI. Ramusio y Gastaldi," in *El mundo de los mapas*, *Quadernos de la Fundaciòn M. Botin* 7 (2005), pp. 131-160.

might have been transmitted, more or less corrupt, until Fra Mauro's times. The fact that Ramusio omits to mention them seems to imply that these works might have already been lost from the first half of that century. An important example of a sixteenth-century discussion on cartography, Ramusio's account would subsequently be taken up and summarized by the historian of the Camaldolese Order, Agostino Fortunio, in 1579.[128]

VI. – The legacy of Fra Mauro's *Œuvre* in sixteenth-century maps

1. A 1541 marine chart of Giorgio Sideri and a1556 atlas of Angelo Freducci

Antonio Ratti around 1984 discovered that a nautical chart of the Mediterranean, dated 1541 and signed by the cartographer Giorgio Sideri, also known as Callapoda (originally from Crete, who worked in Venice from about 1537 to 1562) reproduced most of the content of the Borgiana V map – this being contemporary to, and very much probably derived from, the prototype also used to create Fra Mauro's *mappa mundi* (plate II).[129] The Sideri map, which Ratti was able to consult only in a photographic reproduction of low quality, was sold at auction by Gallery Augustoni-Salomon-Agrati of Milan, probably to a French buyer, on October 24, 1984, and since then its whereabouts are unknown. The comparative analysis of Ratti, which he conducted on graphic details, on the cartographic design, and some textual elements of Sideri's map (the few that the photograph allowed him to read) confirms the derivation (direct or indirect, it is hard to tell) from the work of Fra Mauro.[130] The accidental loss of the archive of the Gallery Salamon-Augustoni-Agrati (which has since gone out of business) makes it impossible to conduct further research until the map resurfaces. However, the information already acquired and analyzed by Ratti constitutes a clear proof of the dissemination and reuse of the knowledge and maps of Fra Mauro until the middle of the sixteenth century. In the face of alleged epistemological ruptures often advanced by historians between allegedly "symbolic" medieval cartography and "scientific" Renaissance mapping, and also in the face of interpretations which declare maps of the period obsolescent by evaluating them according to current parameters, Sideri's map testifies to the longevity not only of the cartography and cartographic vision of Fra Mauro, but also to that of the procedures and results in the processes of constructing space, between the fifteenth and sixteenth centuries.

An important documentary addition that matches the case of Sideri's marine chart, enabling this analysis to develop further, is provided by an atlas designed by Angelo Freducci "Anconitanus" in 1556. Around the same time that Ramusio was preparing his description of the world map at San Michele, a map similar in both content and design to the Biblioteca Apostolica Vaticana's Carta Borgiana V was the model for some maps of those contained in the nautical atlas prepared and signed by Freducci, currently held at the Biblioteca Teresiana in

[128] See the last document in the Documentary Appendix to this chapter.

[129] R. Almagià, "Carta nautica con elementi corografici di Anonimo Veneziano (1450 circa)," in *Monumenta Cartographica Vaticana*, vol. I. Rome : Vatican City, 1944, pp. 32-39 ; H. Winter, "The Fra Mauro Portolan Chart in the Vatican," *Imago Mundi* 16 (1962), pp. 17-28 ; Falchetta, *Fra Mauro's World Map*, pp. 36-52.

[130] A. Ratti, "A Lost Map of Fra Mauro Found in a Sixteenth-Century Copy," *Imago Mundi* 40 (1988), pp. 77-85.

Mantua (plate XI).[131] This atlas, then, provides an important insight into the history of how the geographic knowledge – and the works created from, and communicating, such knowledge – illustrated by Fra Mauro was viewed and utilized in the sixteenth century. In 1953 Giuseppe Caraci relied on a careful philological study of the Freducci's atlas to accurately reveal that, of its nine maps the first two – depicting Florida and South America – looked to one, today in the Archivio di Stato of Florence and attributed to Freducci's father, of the New World. He then demonstrated that the third to the seventh maps were based on one similar to the Borgiana V.[132] In addition to this, Caraci argued that the last two maps of the atlas (maps n. 8 and 9), which depict regions of Asia only partially included in the Carta Borgiana V, show cartographic influence of the geographic principals found in Fra Mauro's *mappa mundi*.[133] Not only do Freducci's maps re-use, nearly to the letter, numerous toponyms and legends from Borgiana V (and to a lesser extent, Fra Mauro's *mappa mundi*), but their design, too, mirrors the graphic style of Fra Mauro's cartography.[134]

But what is most important to note is the fact that Fra Mauro's *mappa mundi*, over one hundred years after its realization, was still considered scientifically valid, and worth copying. Through channels which, at the present moment, remain uncertain, the maps authored by Fra Mauro (or their reproductions) had a wide circulation throughout Italian cities and Europe, subsequently influencing geographical knowledge, ideas, and graphic representations. Freducci's copying and adapting of Fra Mauro's maps is, moreover, a testament of yet another means of reproduction and diffusion of the Camaldolese friar's work – together with the copy ordered for the Medicean Palace in Florence at the end of the fifteenth century. Presuming that, in Florence, an accurate copy of the *mappa mundi* was available around 1480-1490 – as attested by the letter, already analyzed, sent in 1494 by the Camaldolese Pietro Dolfin to Bernardino Gadolo – then one can reasonably surmise that, in Ancona, it was another work of Fra Mauro's that was adapted to a multi-tablet atlas format. In short, during the sixteenth century, too, Fra Mauro's *oeuvre* was appreciated for what it truly was : a valuable cartographic work. As such, it shared the fate of every work of widespread success, being used and adapted – dismembered, as it were – according to new demands.

131 Mantua, Biblioteca Comunale Teresiana, ms. 646, Angelo Freducci, [Nautical Atlas], manuscript on vellum, ff. 5 with 9 maps, 1556, signed and dated "Angelus Eufredutius Anconitan(us) composuit Anconae in anno d(omi)ni MDLVI."

132 The map of the Atlantic coast from Morocco to the British Isles (map n. 3) ; of the western Mediterranean from the Italian Peninsula to Crete (map n. 4) ; of the eastern Mediterranean with Black Sea and surrounding regions (map n. 5) ; the Caspian Sea and surrounding regions (map n. 6) ; and finally of the Red Sea, Arabian Peninsula, and Mesopotamia (map n. 7). For an extremely detailed description of these maps, see G. Caraci, "The Italian Cartographers of the Benincasa and Freducci families and the So-Called Borgiana Map of the Vatican Library," *Imago Mundi* 10 (1953), pp. 23-45 (pp. 31-33).

133 *Ibidem*.

134 Caraci's conclusion that these maps of Freducci's *Atlas* are indeed derived from the Borgiana V or from another, highly similar, map, renders no longer feasible some of his other hypotheses concerning the Borgiana map's origins. He believes this map to have been drawn much later than Fra Mauro's *mappa mundi* and to have been designed in Ancona, by a cartographer of the Freducci family. The studies carried out by Roberto Almagià, Henry Winter, and, most recently, Piero Falchetta have, however, convincingly demonstrated that the Borgiana map is contemporary to, if not a few years older than, Fra Mauro's *mappa mundi* : and was prepared in Venice on the basis of a (now lost) prototype, traceable to the hand of Fra Mauro. See *Mappamondo* (ed. 2006) p. 52.

Without necessarily excluding other means of contact, a probable link between Fra Mauro and Angelo Freducci was likely another "anconitanus" cartographer: Grazioso Benincasa who lived in Venice – apart from periods of absence – from 1436, when Fra Mauro was still alive, to 1474 and was in communication with various cartographers and travelers who clearly had an influence on his work. Indeed, it was in Venice that Benincasa prepared more than half (13 out of 25) of the works traceable to him today. It is more than probable that Benincasa, during his stay in Venice, could have seen and studied the famous world map made in San Michele di Murano, or other maps, drawings and notes of Fra Mauro that were held in the monastery until 1459 (and then again after 1464, as attested by Maffeo Gherardo, the Abbot of San Michele, in his *Entrata e uscita*). The interest with which Benincasa followed Portuguese explorations along the west coast of Africa – for example, the voyages of Alvise Cadamosto – constitutes another important connection between Benincasa and the work of Fra Mauro who, between 1457 and 1459, prepared a copy of his world map for the Portuguese court of Alfonso V – evidence of the awareness, already in 1450, in Venice, of Portugal's plans to circumnavigate Africa.[135]

2. Reminiscences of Fra Mauro's *mappa mundi* at the Florentine Court of Cosimo I

At the conclusion of this chapter we will briefly touch on a particular form of reminiscence inspired by Fra Mauro's work: this is attested to by the acknowledgment of its existence by Egnazio Danti (1536-1586), the cosmographer of the Medicean court of Cosimo I (1519-1574). Proof of such recognition lies in the explanatory legend on a map entitled *Costa della China, e Isola del Giapan ovvero Cipangu* (Coast of China, and Island of Japan or Cipangu), designed between 1563 and 1574 by Danti for the *Guardaroba nuova* (New Wardrobe) of Florence's Palazzo Vecchio. A thorough understanding of the significance of Danti's citation of the *mappa mundi* merits a brief explanation of the cosmographic project prepared around 1560 to decorate a new wing of the municipal Florentine *palazzo*, which was transformed by Cosimo I into a ducal palace.

The architectural and cosmographical plans for the Palazzo Vecchio's *Sala della Guardaroba nuova* (New Wardrobe Room), known today as the *Sala delle Carte geografiche* (Map Room), were conceived between 1561 and 1563 by Cosimo I de' Medici and Giorgio Vasari, with the contribution of the court cosmographer, the Olivetian monk Don Miniato Pitti.[136] The plans envisioned the realization of a complex *theatrum mundi*, displaying the depictions of the Earth and heavens, with 57 maps encompassing the whole world, portraits of ancient and modern illustrious figures, marble busts of emperors and rulers, an astronomical clock by Lorenzo

[135] See M. Emiliani, "Le carte nautiche dei Benincasa cartografi anconitani," *Bollettino della Reale Società Geografica Italiana* 7, 1 (8-9), 1936, pp. 485-510; also as an offprint paginated 1-27. Refer also to A. Cattaneo, "From the Mediterranean to the Atlantic. Venice, Florence and Lisbon. Commercial Routes and Networks of Knowledge, 1300-1550," in J. A. Levenson (ed.), *Encompassing the Globe: Portugal and the World in the 16th & 17th Centuries*. Essays [Vol. 3]. Washington, D.C.: Arthur M. Sackler Gallery, 2007, pp. 13-21.

[136] Vasari described the entire project in an emphatically page in the second edition of *Le Vite*. See Giorgio Vasari, "Degl'accademici del disegno pittori, scultor et architetti e dell'opere loro e prima del Bronzino," in *Le vite de' più eccellenti pittori, scultori, e architettori... di nuovo dal medesimo riviste et ampliate con i ritratti loro et con l'aggiunta delle vite de' vivi, & de' morti dall'anno 1550 infino al 1567...*. 3 vols., Florence: Giunti, 1568, vol. II, p. 877.

della Volpaia, and two globes (terrestrial and celestial) to be hung from the ceiling.[137] The attainment of the Ptolemaic maps – the simplest and certainly the least expensive part of a very ambitious, if not visionary, project – was entrusted to the Dominican Egnazio Danti (1536-1586), not yet thirty years old, but already an accomplished mathematician, who would later become an even more well-known cosmographer.[138] In 1575, following Cosimo I's death, Danti left Florence – to escape the Inquisition, it seems[139] – and the project was taken over by the Olivetian brother Stefano Buonsignori († 1589). The two cosmographers had been given the task of transforming the codices and printed editions of Ptolemy's *Geography*, 26 ancient regional maps, and a greater number of more modern maps, into a "mural" cycle; all the while maintaining the formal characteristics of the *Geography*'s manuscripts: notably, the spacing of the grid of latitudinal and longitudinal lines, and the division of the *oikumene* into separate maps. However, the ambitious cosmographic project for the Palazzo Vecchio was never finished. Of the 57 planned map panels only 53 were completed and this only after a space of twenty years. The *tavole di cosmografia*, mounted on the cupboards of the *Guardaroba nuova* at different times, were finally put on display in 1587. Cosimo's sons, Francesco I and Ferdinando I, who succeeded him to the title of Grand Duke, favored other projects which were carried out in the adjoining *Palazzo degli Uffizi*.[140]

The explanatory legend of the map entitled "Coast of China, and Island of Japan or Cipangu" – one of the 23 maps drawn by Danti between 1563 and 1574 – states:

> There are many differences among geographers regarding the location of this coast of China and the island of Japan, which in this map are situated according to the opinion of the most modern

[137] P. Barocchi and G. Gaeta Bertelà, *Collezionismo mediceo e storia artistica 1545-1743*, 2 vols. in 5 books: vol. I: *Da Cosimo I a Cosimo II (1540-1621)*, Florence: S.P.E.S, 2002, pp. 91-143. F. Fiorani, *The Marvel of Maps. Art, Cartography and Politics in Renaissance Italy*, New Haven CT and London: Yale University Press, 2005, pp. 27-183; A. Cattaneo, "La cosmografia di Cosimo," in F. Camerota and M. Miniati, *I Medici e le scienze. Strumenti e macchine nelle collezioni granducali*. Catalogo della mostra (Firenze, maggio 2008-gennaio 2009), Florence: Giunti, pp. 147-151; A. Cecchi and P. Pacetti (eds), *La Sala delle Carte geografiche in Palazzo Vecchio. 'Capriccio et invenzione nata dal Duca Cosimo'*, Florence: Polistampa, 2009.

[138] The Dominican friar Egnazio Danti, cosmographer at the Court of Cosimo I de' Medici (1519-1574), made notable contributions to the improvement of scientific instruments. He was the author of widely circulated texts on the construction and use of the astrolabe and on various astronomical topics. Danti mounted astronomical instruments – a quadrant and two equinoctial *armillae* – on the façade of the church of Santa Maria Novella, still visible today. In the same church he began the construction of two gnomons, but was unable to finish them as, after his patron Cosimo I's death, he was forced to leave Florence. He then moved to Bologna to teach mathematics at the university and contributed to the reform of the calendar. In 1576 he went to Rome and was appointed cosmographer to Pope Gregory XIII (1502-1585, pope from 1572), for whom he oversaw the preparation of the maps displayed in the Vatican palaces. He became bishop of Alatri (Frosinone), where he died in 1586. See J. del Badia, *Egnazio Danti cosmografo e matematico e le sue opere in Firenze*, Florence: M. Cellini, 1881; P. Dubourg-Glatigny,"Egnazio Danti O.P. (1536-1586). Itinéraire d'un mathematicien parmi les artistes," *Mèlanges de l'École Française de Rome* 114, 2 (2002), pp. 543-605. G. R. Levi-Donati, *Le trentacinque cartelle della Guardaroba medicea di Palazzo Vecchio in Firenze*, Perugia: Benucci, 2002 (first edition 1995).

[139] See M. Marcolin, "I cartigli delle tavole della sala della Guardaroba," in A. Cecchi and P. Pacetti (eds), *La Sala delle Carte geografiche in Palazzo Vecchio*, pp. 107-133 (pp. 116-119).

[140] See A. Cattaneo, "Inventare musei per ordinare e rappresentare il mondo. La Guardaroba nuova di Palazzo Vecchio e le Sale della cosmografia e delle matematiche agli Uffizi a Firenze," in *Revista de História da Arte* 8 (2010), pp. 24-47.

[geographers], though it seems that this coast from the cape Liambo to Pangiu should have been instead oriented South-North and not West-East as it is depicted here [in this map], and as for the Island of Japan, it should also have been oriented South-North, and not East-West. But due to the fact that in the great globe (following the wishes of the Grand Duke of Tuscany Cosimo I) they were situated as described by Marco Polo and João de Barros, the latter of whom states explicitly that the easternmost cape in the Indies is cape Liampo, which is close to the city of Nimbo, and which is also known as cape Ninfù, [we have therefore mantained the same orientation]. One cannot see this South-North oriented coast of China in an old *mappamondo* that was depicted on the basis of an even older one designed in Venice by Marco Polo, who wrote many things about this part of the world, and although no one at the time believed them, they were all later discovered to be true.[141]

The "old *mappamondo*" referred to by Danti is doubtless the copy of Fra Mauro's *mappa mundi* commissioned by the Medici family around 1480 and authored – with the legends translated into Latin by Pietro Dolfin – by Florentine painters sent to this purpose to San Michele. Danti's descriptions of the connections between the Camaldolese friar's work and a Venetian world map by Marco Polo were evidently drawn from Ramusio's writings, published in the preface to the edition of Polo's *Milione* included in *Navigationi et viaggi* (to which we have already given much attention). The reference to the *mappa mundi* in this particular context – that of the illustration of the western coast of China and Japan – is not accidental. Fra Mauro's world map is one of the first cartographic documents designed in the West to depict the island of *Cimpagu* (*Ixola de Cimpagu*[142]) – found, according to Marco Polo, on the eastern margin of the known world opposite the "magnificent Port of Zaiton,"[143] and running north to south. Danti's written description appears to indicate that the depiction of China's southwestern coasts was a novelty: it was knowledge gleaned first from a written description by Marco Polo, and then updated by Portuguese exploratory voyages recorded by João de Barros. While aware of the erroneous orientation of Japan on the map – consciously left incorrect in order to conform to the depiction of Japan on the globe completed in 1569 under Cosimo I's direction – Danti nevertheless demanded recognition for completing the first comprehensive cartographic render-

[141] Florence, Palazzo Vecchio, Guardaroba nuova, Egnazio Danti, [Costa della China, e Isola del Giapan ovvero Cipangu], oil on wood. The original legend on the left side of the map reads "Gran diversità è fra i Geografi nel situare questa costa della China, e l'isola del Giapan le quali nella presente tavola si sono poste seco[n]do l'oppinione de più moderni se bene par che si dovesse far correre questa costa dal capo Liampo fino a Pangiu per Scirocco et Maestro et non per Greco et Libeccio come sta qui et l'isola del Giapan dovria correre per Ostro e Tramontana et no[n] per Levante et Ponente. Ma perché nel gran globo si son posti (co[n] la determinazione dello Seren[issi]mo Cos[i]mo Gran D[uca] di Toscana) nel modo che li descrive Marco Polo et nel modo che la pone il S[ignor] Giovan di Baros il quale [e]spressamente dice che il più orientale capo dell'Indie scoperte sia capo Liampo il quale capo è vicino alla città di Nimbo dal quale è chiamato anche capo Ninfù. Senza che si vede questa costa della China posta per Scirocco e Maestro in un antico Mappamondo fatto a simigl[i]anza d'uno che già fece in Venetia Marco Polo il quale scrisse molte cose di questa provincia le quali se bene alora non se li credevano si è poi trovato non di meno tutto esser vero [...]." See M. Marcolin, "Atlante iconografico: le mappe e i cartigli," in A. Cecchi and P. Pacetti (eds), *La Sala delle Carte geografiche in Palazzo Vecchio*, pp. 151-262 (pp. 213-214). The *hidalgo* João de Barros (1496-1570) wrote three *Decades de Asia* (1552, 1553, & 1563), his most important work, which celebrates Portuguese expansion in Asia. Danti cited the *Decades* in Ramusio's partial Italian edition published in *Navigationi e viaggi*.

[142] *Mappamondo* (ed. 1956) Pl. XXV, 17; *Mappamondo* (ed. 2006) *1334.

[143] "Magnifico porto de zaiton." *Mappamondo* (ed. 1956) Pl. XXV, 8; *Mappamondo* (ed. 2006) *1335.

ing of the Far East. According to Ramusio and Danti, even Fra Mauro, though possessing the (alleged) world map of Marco Polo had not succeeded in such a comprehensive illustration of all the territories described in writing, first in Polo's *Milione* and later, around the middle of the sixteenth century, by the Portuguese. Instead, Danti – while using as a model the map entitled *Il disegno della terza parte dell'Asia* designed by the Venetian cartographer Jacopo Gastaldi († 1569) and published by Antonio Lafrery (1512-1577) in 1562[144] – claimed the credit for the cartographic representation of areas previously described only in writing.

In keeping with the historical views of Fra Mauro's work, Danti's note inserted in the description of Japan seems to indicate that the latter cosmographer had not merely heard of it indirectly but had seen the *mappa mundi* with his own eyes. Danti implicitly writes of having verified the correspondence between Marco Polo's writings and Fra Mauro's cartographic renderings. Since, however, Danti does not seem to have ever visited the monastery of San Michele di Murano, the legend of the *Carta del Giapan* might, instead, testify that the copy of the map attained by "Florentine painters" around 1480 for the Medici *palazzo* was still consultable between 1563 and 1574 when Danti worked on the maps for Palazzo Vecchio – and that the said map only went missing subsequently.

[144] Giacomo Gastaldi (cartographer) and Antonio Lafrery (engraver? and publisher), *Il disegno della terza parte dell'Asia. All' illmo. Sigr. il Sr. Marcho Fucharo Barone di Kirchberg e' d' Waisenhoren ; Giacomo di Castaldi Piamotese Cosmographo in Venetia* ; engraved and printed on paper, 70 × 71.8 cm, 1562. Plate 75 of Antonio Lafrery, [Tavole moderne di Geografia de la maggior parte del mondo di diversi autori raccolte…], Rome : A. Lafrery, 1545-1566. I used for the comparison Gastaldi's *Terza parte dell'Asia* held in a copy of Lafrery's [Tavole moderne di Geografia…] held at the Biblioteca Nazionale Centrale in Florence. For a detailed description of this map see M. Milanesi, *s.v.*, in I. Massabò Ricci, (ed.), *Architettura militare. Luoghi, città, fortezze, territori in età moderna*. Rome, Ministero per i beni e le attività culturali. Direzione generale per gli archivi, 2003, pp. 245-246.

CHAPTER II

THE COSMOS OF A MID-FIFTEENTH CENTURY MONK

The very order of things created by God shows the unity of the world. For this, world is called one by the unity of order, whereby some things are ordered to others. But whatever things come from God, have relation of order to each other, and to God Himself, as shown above.[1]

Thomas Aquinas, *Summa theologiae*

I have said that, in my opinion, all was chaos, that is, earth, air, water, and fire were mixed together; and out of that bulk a mass formed – just as cheese is made out of milk – and worms appeared in it, and these were the angels. The most holy majesty decreed that these should be God and the angels, and among that number of angels, there was also God, he too having been created out of that mass at the same time.[2]

Domenico Scandella, called Menocchio, a miller at Monreale Valcellina, first half of the sixteenth century

I. – Natural Philosophy

1. Cosmography

This chapter focuses on the images and legends on the *mappa mundi* that describe the Earth as part of the celestial and sublunar worlds. It first considers the *mappa mundi* as a cosmographic image and then proceeds to a reading and analysis of the *mappa mundi*'s numerous legends that describe the *oikumene* as a whole, thus contextualizing Fra Mauro's cosmo-

[1] *The Summa Theologica of St. Thomas Aquinas.* Second and Revised Edition, 1920, literally translated by Fathers of the English Dominican Province, p. I, q. 47, a. 3. The original quotation in Latin runs: "Ipse ordo in rebus sic a Deo creatis existens, unitatem mundi manifestat. Mundus enim iste unus dicitur unitate ordinis, secundum quod quaedam ad alia ordinantur." See *S. Thomae Aquinatis Summa theologiae*, cura et studio sac. Petri Caramello cum textu ex recensione leonina, pars prima et prima secundae, Turin: Marietti, 1952, p. 334.

[2] See the original quotation in Italian: "Io ho detto che, quanto al mio pensier et creder, tutto era un caos, cioè terra, aere, acqua et foco insieme; et quel volume andando così fece una massa, aponto come si fa il formazo nel latte, et in quel deventorno vermi, et quelli furno li angeli; et la santissima maestà volse che quel fosse Dio et li angeli; et tra quel numero de angeli ve era anche Dio." in C. Ginzburg, *Il formaggio e i vermi: il cosmo di un mugnaio del '500*, Turin: Einaudi, 1976, p. 8. The quotation in English derives from C. Ginzburg, *The Cheese and the Worms. The Cosmos of a Sixteenth-Century Miller*, trans. J. and A. Ledeschi, London and Henley: Routledge & Kegan Paul, 1980, pp. 5-6. This passage comes from the account of the first interrogation of Menocchio, held in the Tribunal of the Inquisition in Udine on February 7, 1584.

graphic culture within the wider landscape of medieval and early modern natural philosophy. At the time of Fra Mauro interest in natural philosophy and cosmography and their cosmological implications were all but confined to the learned academic circles of masters and their students at both secular and conventual schools and universities (*studia*). The semantics of the heavens and the earth were part of medieval biblical exegesis and preaching, as well as of philosophy, the courtly novel, visionary literature, and hagiography.[3] Therefore, a study of the cosmographic contents of Fra Mauro's *mappa mundi* sheds light on a central aspect of the knowledge and beliefs of fifteenth-century men in terms of both learned and popular culture.

Natural philosophy (*philosophia naturalis*), which in the period from the thirteenth century to the end of the seventeenth century has been described as the "study of the material world in so far as it is carried in the stream of change (*motus*),[4] rarely excites curiosity in contemporary historiography. Fra Mauro's *mappa mundi* is no exception. Scholars who have studied it have either ignored or undervalued this aspect of the work.[5] All too easily relegated to the genre of *mirabilia* or that of dusty fossilized knowledge, medieval and early modern natural philosophy, in the form of *quaestiones*, *disputationes*, *tractati*, and *mappae mundi*, represents instead a fundamental part of medieval and early modern culture and civilization, and is characterized by rigor of argumentation and logic of exposition. At the end of the nineteenth century Pierre Duhem began to emphasize the relationship between medieval thinkers and medieval cosmology and their first modern successors in his encyclopedic work.[6] In contrast to the grand "hero" narratives of older histories of science during this period, commonly known as the "Scientific Revolution," more recent cultural histories of science identify the original cultural *humus* of sixteenth- and seventeenth-century European investigations into nature represented in the physical theories and vocabulary of the Aristotelian masters of the thirteenth and four-

[3] See *Cieli e terre nei secoli XI-XII. Orizzonti, percezioni, rapporti*. Atti della tredicesima Settimana internazionale di studio, Mendola, 22-26 agosto 1995, Milan : Vita e pensiero, 1998, in particular the essays of R. Grégoire, "Semantica del cielo e della terra nell'esegesi biblica medievale," pp. 3-29 ; A.-D. von den Brincken, "Mappe del Medio Evo : mappe del cielo e della terra," pp. 51-73 ; J. Verger, "Ciel et terre dans l'enseignement des 'artes'," pp. 111-133 ; M. L. Meneghetti, "Cielo e terra nel romanzo cortese," pp. 179-195 ; P. Dinzelbacher, "Voli celesti e contemplazione del mondo nella letteratura estatico-visionaria del Medioevo," pp. 215-259 ; and P. Tomea, "Rappresentazioni e funzioni del cielo della terra nelle fonti agiografiche del Medioevo occidentale," pp. 311-349. See also A. Cattaneo, "Cosmographie et prédication médiévale et renaissante," *Cahiers V.-L. Saulnier* 26 (Paris, Editions Rue d'Ulm, 2008) pp. 57-70 (Proceedings of the conference "Les méditations cosmographiques à la Renaissance," organized by J.-M. Besse, M.-D. Couzinet, F. Lestringant).

[4] M. Clagett, "Some General Aspects of Physics in the Middle Ages," *Isis* 39, 1-2 (1948), pp. 29-44.

[5] R. Almagià, who briefly discussed Fra Mauro's theory of tides, is an exception ; see R. Almagià, "La dottrina della marea nell'antichità classica e nel Medio Evo," *Atti della R. Accademia dei Lincei* 301, *Memorie della classe di scienze fisiche, matematiche e naturali*, s. V, 5 (1904), pp. 377-513.

[6] P. Duhem, *Le système du monde : histoire des doctrines cosmologiques de Platon à Copernic*, Paris : Libr. scientifique A. Hermann, 1913-1917, 5 vols. (reprinted Paris : Hermann, 1954-1965, 10 vols.). Though criticized for its anachronistic philosophical interpretation of the history of medieval science, this work remains a fundamental and seminal contribution to the field. See E. Grant, *Planets, Stars, and Orbs : the Medieval Cosmos, 1200-1687*, Cambridge and New York : Cambridge University Press, 1994, pp. 3-9 ; and A. C. Crombie, "Intuizioni storiche della scienza medievale," in P. Tourbert and A. Paravicini Bagliani (eds), *Federico II e le scienze*, Palermo : Sellerio, 1994, pp. 15-24, in particular p. 22.

teenth centuries.[7] Several works of William Clagett, Edward Grant, Emmanuel Poulle, Stefano Caroti, Patrick Gautier Dalché and Anthony Grafton demonstrate that while medieval natural philosophy unveiled a world with a different structure than that perceived and affirmed by modern science, the scientific revolution of the beginning of the seventeenth century originated in the discussion, critique, and renovation of the concepts space, movement, speed and time that emerged in the physical theories of the Aristotelian masters of the thirteenth and fourteenth centuries.[8]

Cosmography was a relevant part of natural philosophy. According to a tradition that persisted until the beginning of the modern age,[9] this composite field of knowledge included elements of Christian cosmology (the Earth situated within the creation story), astronomy and Ptolemaic astrology (the Earth placed in relation to the heavenly world of planets and stars), Aristotelian natural philosophy (the Earth placed in relation to other elements of the sublunar world, i.e. water, air and fire), and universal geography.[10] Such an eclectic body of knowledge emerged from a miscellaneous set of material and written cultures, including written description of the *imago mundi*, treatises on astronomy, universal chronologies, and encyclopedias. Cosmographic contents were also discussed in, and disseminated by, travel narratives, merchants' letters and handbooks, diplomatic correspondence, astrological treatises, medical handbooks, *herbaria*, and chronicles. Between the late medieval period and the beginning of the modern age, also major pictorial representations of the *imago mundi*, such as, for example, the

[7] E. Grant, "Late Medieval Thought, Copernicus, and the Scientific Revolution," *Journal of the History of Ideas* 23, 2 (1962), pp. 197-220; Id., "A New Look at Medieval Cosmology, 1200-1687," *Proceedings of the American Philosophical Society*, 129, 4 (1985), pp. 417-432.

[8] P. O. Kristeller, *Renaissance Thought and its Sources*, M. Mooney (ed.), New York: Columbia University Press, 1979 (especially chapter 5); A. Blair and A. Grafton, "Reassessing Humanism and Science," *Journal of the History of Ideas* 53, 4 (1992), pp. 535-540; D. C. Lindberg, *The Beginnings of Western Science: the European Scientific Tradition in Philosophical, Religious, and Institutional Context, 600 B.C. to A.D. 1450*, Chicago: University of Chicago Press, 1992, pp. 355-367; A. C. Crombie, *Styles of Scientific Thinking in the European Tradition*, London: G. Duckworth, 1994; E. Grant, *The Foundations of Modern Science in the Middle Ages: Their Religious, Institutional, and Intellectual Contexts*, Cambridge and New York: Cambridge University Press, 1996, pp. 168-205; M. L. Colish, *Medieval Foundations of the Western Intellectual Tradition, 400-1400*, New Haven: Yale University Press, 1997, pp. 319-326.

[9] See at this regard Gerard Mercator's emblematic *Atlas siue Cosmographicae meditationes de fabrica mundi et fabricati figura*. Dusseldorf: Albertus Busius, 1595 and *Chronologia. Hoc est temporum demonstratio exactissima, ab initio mundi, vsque ad annum Domini 1568. Ex eclipsibus et obseruationibus astronomicis omnium temporum, sacris quoque Biblijs, ... summa fide concinnata*. Koln: apud haeredes Arnoldi Birckmanni, 1569. See the proceedings of the conference "Les méditations cosmographiques à la Renaissance," organized by J.-M. Besse, M.-D. Couzinet, F. Lestringant, published in *Cahiers V.-L. Saulnier* 26 (2008).

[10] A.-D. von den Brincken, "Mappa mundi und Chronographie," *Deutsches Archiv für Erforschung des Mittelalters* 24 (1968), pp. 118-186; A. de Smet, "Les géographes de la Renaissance et la cosmographie," in *Album Antoine de Smet*, Brussels: Centre national d'Histoire des Sciences, 1974 (Publ. du Centre national d'Historie des Sciences, IV), pp. 149-160; M. Milanesi, "La Cartografia italiana nel Medio Evo e nel Rinascimento," in M. Milanesi, V. Valerio, E. Casti Moreschi, and L. Rombai, *La Cartografia italiana. Cicle de conferències sobre Història de la Cartografia. Tercer curs, 1991*, Barcelona: Institut Cartogràfic de Catalunya, 1993, pp. 15-80; P. Gautier Dalché, "Saperi geografici del Mediterraneo cristiano," in P. Tourbert and A. Paravicini Bagliani (eds), *Federico II e le scienze*, Palermo: Sellerio, 1994, pp. 162-184; Id., "Pour une histoire du regard géographique. Conception et usage de la carte au XVe siècle," *Micrologus* 4 (1996), pp. 77-103; N. Bouloux, *Culture et savoirs géographiques en Italie au XIVe siècle*, Turnhout: Brepols, 2002.

Hereford and Ebstorf *mappae mundi* (*c.* 1300), but also the so-called 'modern' planispheres of Martin Waldseemüller (Saint-Dié des Vosges, *c.* 1475-1522), Sebastian Münster (Basel, 1488-1552), Jacopo Gastaldi (active in Venice, † around 1565), Gerard Mercator (Louvain and Duisburg, 1512-1594) and Abraham Ortelius (Antwerp, 1527-1598), conveyed relevant cosmographic notions – generally, in long cartouches placed in the corners or the margins of the maps – along with their geographic contents. In a similar way, Fra Mauro works outlines most of the (written) cosmographic knowledge in seven long and complex legends placed in the four corners between the square and circular frames of his work, which are closely linked to several, shorter, and less complex legends located inside the circular *mappa mundi*. In doing so, he developed a complex scientific lexicon in the Venetian vernacular that translated and creatively adapted notions and concepts that were originally expressed in Latin in treatises of natural philosophy. Recently, historians of language have emphasized the importance of these medieval scientific texts in the vernacular, considering them a valuable reservoir for the reconstruction of early vernacular vocabularies.[11] Thanks to the studies of Rita Librandi, Rosa Casapullo, and Riccardo Gualdo on the methods of translation of scientific texts, it is now clear that the scientific vernaculars – including that of Fra Mauro – forged new ways of scientific expression and vocabularies, and, through the need to describe a difficult subject matter, found ways to avoid approximations and created new linguistic formulas and constructions. These considerations constitute the framework in which this research has been developed.

2. The cosmographic structure of the *mappa mundi*

Fra Mauro's world map is magnificently painted in sumptuous colors, with a profusion of gold and *azzurro*, the latter a precious powdered blue pigment obtained by crushing lapis lazuli, which in this case was placed on assembled sheets of parchment glued and nailed onto wooden tables reinforced by horizontal lozenges. The *mappa mundi* does not have a scale or rhumb lines, equator or tropics. Four winds (Auster, North wind, Gregale, East wind) are positioned at the margins of the map indicating the four cardinal points. The work is placed in a square wooden frame that measures 223 cm a side containing a circular frame with a diameter of 196 cm on the horizontal axis and 193 cm on the vertical axis. The two frames delimit two spaces : the *oikumene* within the circular space is represented chorographically with numerous designs and legends ; and the space in the four corners delimited by the two frames that I define as cosmographical represents the sublunar and celestial worlds, and describes their fundamental properties in seven long legends, three diagrams and an image of the earthly paradise. In these seven

[11] R. Casapullo and M. R. Policardo, "Tecniche della divulgazione scientifica nel volgarizzamento mantovano del 'De proprietatibus rerum' di Bartolomeo Anglico," *Lingua e stile* 38 (2003), pp. 139-176 ; M. Barbato, *Il libro VIII del Plinio napoletano di Giovanni Brancati*, Naples : Liguori, 2001 ; R. Gualdo (ed.), *Le parole della scienza : scritture tecniche e scientifiche in volgare, secoli XIII-XV*, Galatina : Congedo, 2001 (in particular, the chapters by R. Librandi, "Auctoritas e testualità nella descrizione dei fenomeni fisici," pp. 99-126 and R. Casapullo, "Segmentazione del testo e modalità d'uso delle enciclopedie tra latino e volgare, " pp. 153-185) ; R. Librandi, "Tratti sintattico-testuali e tipologia di testi : la trattatistica scientifica," in M. Dardano and G. Frenguelli (eds), *La sintassi dell'italiano antico*, Atti del Convegno internazionale (Roma, 18-21 settembre 2002), Rome : Aracne, 2004, pp. 271-292.

legends Fra Mauro presents the macrostructure, or the "grand picture," in which the *oikumene* is situated.

In the upper left corner the celestial world composed of ten spheres is painted and described. The legend in the cartouche to the right of the spheres indicates the distances of the celestial spheres from the center of the world and the planets' diameters which were taken, directly or indirectly, from the *Theory of the Planets* (*Theorica planetarum*) of Campanus of Novara.[12] In the cartouche positioned under the diagram Fra Mauro discusses the number of heavens according to the authority "of the holy theologians,"[13] and the text consists of a literal translation into Venetian of Article 4, "Whether there is only one heaven," of *Quaestio* 68, "On the work of the second day," of the first part of the *Summa theologica* of Thomas Aquinas.[14] Proceeding clockwise around the map, the sublunar world is painted in the upper right corner with the spheres of the four elements and the sphere of the moon. Five concentric circles of different colors are represented: the innermost is brown representing the element of the earth; outside it a pale blue circle represents the sphere of water, which is surrounded by a more intense blue sphere of air, represented by clouds and fog, followed by a red sphere of fire with flames that rise and expand toward the first sphere of the celestial world, that of the moon. A legend entitled "How waters rise and subside due to the attractive properties of the moon"[15] is positioned to the right of this image. Here Fra Mauro expounds the luni-solar theory of the tides, which, as we shall see, was of great interest in Venetian culture; the text adapt into the vernacular ideas that derives from *Book I* of the *Liber de causis proprietatum elementorum* of Pseudo-Aristotle in the commentary of Albert the Great.[16] I am not at all saying that Fra Mauro directly read or consulted this work, but simply that the ideas he discusses in the legend, as well as the way he frames them, are traceable to Albert the Great's *Liber de causis*. Underneath the figure the contents of the legend "How by Divine Providence the earth is raised above the water"[17] can be traced to Quaestio 7, "Whether the entire earth is inhabitable" (*Utrum tota*

[12] "On the Distance of the Heavens. Rubric" (*De la distantia de hi | cieli. Rubrica*), *Mappamondo*, upper left corner (right): *Mappamondo* (ed. 1956), pp. 19-20; transcribed as Doc. 2 in the Appendix. See also *Mappamondo* (ed. 2006), pp. 722-723.

[13] "Del numero de hi cieli secondo l'auctorità de hi sacri theologi," *Mappamondo*, upper left corner (left): *Mappamondo* (ed. 1956), p. 19; transcribed as Doc. 1 in the Appendix. See also *Mappamondo* (ed. 2006), pp. 720-722.

[14] *S. Thomae Aquinatis Summa theologiae*, Pars I, q. 68, a. IV; see *S. Thomae Aquinatis Summa theologiae*, cura et studio sac. Petri Caramello cum textu ex recensione leonina, pars prima et prima secundae, Turin: Marietti, 1952, p. 334; see Doc. 1a in the Appendix of this chapter.

[15] "Questa rubrica de sopra dechiara come per la uirtù atractiua de la luna le aque cresce e descresce," *Mappamondo*, upper right corner (left): *Mappamondo* (ed. 1956), p. 21; transcribed as Doc. 5 in the Appendix. See also *Mappamondo* (ed. 2006), pp. 724-726.

[16] Albertus Magnus, *Liber de causis proprietatum elementorum*, in *Alberti Magni ... Opera Omnia*, Tome V, Part II, edited by P. Hossfeld, Monasterii Westfalorum in aedibus Aschendorff, 1980, Lib. I, Tract. 2, chapter 5, *Et est digressio ostendens veram causam accessionis maris in communi et excludens errores, qui sunt contra hanc*, pp. 68b-70b; chapter 6, *Et est digressio declarans, quot et quae confortant accessionem marium*, pp. 70b-72b.

[17] "Come per diuina prouidentia la terra habitabile è subleuada da l'aqua," *Mappamondo*, upper right corner (right): *Mappamondo* (ed. 1956), p. 21; transcribed as Doc. 4 in the Appendix. See also *Mappamondo* (ed. 2006), pp. 726-729.

terra sit habitabilis) of Book 2 of the *Quaestiones in Aristotelis De caelo* of John Buridan.[18] In the lower right corner the astronomical circles are depicted within a circle where, unfortunately, half of the colors are vanished. In contrast to the *mappa mundi*, which is oriented with South at the top, the astronomical diagram is oriented with North at the top. The equator and the tropics are recognizable, as well as the ecliptic with the constellations and the "southern parallel" (*paralelo meridionalis*), a line of latitude. To the right of the image is the legend "How the Earth which is under the [celestial] Equator and the torrid zone may be habitable,"[19] which transmits knowledge that seems to have been adapted – either directly or indirectly is difficult to be determined – from Chapters VI and VII of Albert the Great's *De natura loci*, as well as the already cited Chapter VII of the *De caelo* of Buridan.[20] Above the image is placed the legend "Of the Elements. That is, how much one element exceeds another in quantity."[21] Finally, in the lower left-hand corner is an illumination that represents the earthly paradise, recently attributed to the Venetian miniaturist Leonardo Bellini (*c.* 1420-*post* 1484), nephew of Jacopo Bellini and cousin of Giovanni and Gentile Bellini.[22] Above the earthly paradise the legend explains various theological arguments about paradise, as well as its iconography.[23] In contrast to almost all of the medieval *mappae mundi* in which paradise is placed within the chorographic space, Fra Mauro's *mappa mundi* paradise is placed outside it : a choice that will be analyzed in Chapter III. Taken together, these legends and diagrams demonstrate that Fra Mauro set about to describe and analyze not only the *oikumene*, but also the structure and functioning of the cosmos.

[18] B. Patar, *Ioannis Buridani Expositio et questiones in Aristotelis De Caelo*, édition, étude critique et doctrinale, Louvain-la-Neuve : Éditions de l'Institut supérieur de philosophie ; Louvain and Paris : Éditions Peeters, 1996, Lib. II, q. VII, *Utrum tota terra sit habitabilis*, pp. 410-417. See also E. A. Moody, "John Buridan on the Habitability of the Earth," *Speculum* 16, 4 (1941), pp. 415-425, in particular pp. 421-425 ; for the Italian translation see Giovanni Buridano, *In de caelo, Liber II, Quaestio VII*, "Se tutta la terra sia abitabile" in *Il cielo e il mondo. Commento al trattato 'Del cielo' di Aristotele*, A. Ghisalberti (ed.), Milan : Rusconi, 1983, pp. 286-297. As in the case of Albert the Great's *Liber de causis*, *we* are not suggesting that Fra Mauro made direct use of the *Quaestiones in Aristotelis De caelo* of John Buridan, but simply that the ideas put forth by Camaldolese can be traced to those developed by Buridan.

[19] "Come la terra supposita a l'equinoctial e a | la torrida çona è abitabile," *Mappamondo*, upper right corner (left) ; *Mappamondo* (ed. 1956), pp. 23-24 ; transcribed as Doc. 6 in the Appendix. See also *Mappamondo* (ed. 2006), pp. 733-735.

[20] As observed above in connection with Buridan, we are not suggesting that the *De natura loci* of Albert the Great was the direct source utilized by Fra Mauro, but that the ideas maintained by Camaldolese can be traced to those developed by Albert Magnus in the chapters indicated in the *De natura loci*. For a study of the way that Fra Mauro draws on his sources and for the identification of the codices that he utilizes, please see Chapter VI.

[21] "De hi elementi, cioè quanto uno elemento | excieda l'altro in quantità," *Mappamondo*, right lower corner (right) ; *Mappamondo* (ed. 1956), pp. 22-23 ; transcribed as Doc. 3 in the Appendix. See also *Mappamondo* (ed. 2006), pp. 731-732.

[22] S. Marcon, "Il Mappamondo di Fra Mauro e Leonardo Bellini," in M. Piantoni and L. de Rossi (eds), *Per l'arte da Venezia all'Europa. Studi in onore di Giuseppe Maria Pilo*, Venice : Edizioni della Laguna, 2001, pp. 103-108 ; A. Cattaneo, "God in His World. Leonardo Bellini Illuminator of the Earthly Paradise in Fra Mauro's Mappa mundi," *Imago Mundi* 40, (2003), pp. 97-102.

[23] "The location of the earthly paradise" (*Del sito del paradiso terrestro*), *Mappamondo*, left lower corner ; *Mappamondo* (ed. 1956), p. 22 ; transcribed in the Appendix to Chapter III. See also *Mappamondo* (ed. 2006), pp. 729-731.

The graphic and textual modalities with which Fra Mauro describes the earth as part of the sublunar and celestial world will be analyzed in the epistemological context of natural philosophy of the Aristotelian-Thomistic tradition. In the first part of this analysis we will consider the cosmographic knowledge contained in the *mappa mundi*: first, as an image which is of much greater importance than the text of the legends that cover it. It is an image whose contents will be grouped in three principle divisions: its orientation to the South; the "fullness" of the world as a metaphor for its inhabitability, which also includes the seas; and finally, the graphic function of the text as part of this fullness, which we hypothesize precedes its function as legible text. Thus we will consider the long legends in the four corners of the all-encompassing square frame, as well as several others written inside the circular *mappa mundi*, situating them within fifteenth-century cosmographic debate and identifying their sources when possible. At least three of the legends are close adaptations of their source texts, which has allowed us to identify the works from which Fra Mauro drew them. For the other legends where this was not possible we have tried, nonetheless, to trace his ideas to specific *auctores* and suggest works that could be possible sources.

II. – Cosmography and images

1. Looking at Fra Mauro's *mappa mundi*

When we study or contemplate a *mappa mundi* on the wall of a museum or library, or through a facsimile reproduction, we frequently forget that the work has undergone a process of radical estrangement, not only from the point of view of meaning and function – in the case of a *mappa mundi*, from a scientific, didactic instrument with a celebratory function to an object of collection and museum exposition – but also from the visual point of view, in as much as the original conditions of perception have been unmistakably altered. The *mappa mundi* of Fra Mauro was first exhibited within "un gran armaro" (or, a large armoire; one can imagine a structure with doors, like that formerly used to display the Hereford *mappa mundi*) in a room adjacent to the sacristy of the church of San Michele di Murano called *sala del mappamondo*, and then in the library of San Michele.[24] It is very unlikely that spectators could read the legends. Ramusio, in fact, in the first analytical description of a *mappa mundi*, is very clear in this respect:

> Right up to the present day it [Fra Mauro's map] has been held in great veneration and prestige in all of this city; in particular by those who delight themselves with cosmography, and every day at least some visitors come to see this map with great delight, regarding it with great consideration; and amongst the various miracles of this divine city, it is shown as a rare and beautiful thing to those foreigners who come to see the works of glass produced on Murano. And still today it is possible to see [in the *mappa mundi*] many [valuable] things.[25]

[24] On the history of the exposition of the *mappa mundi* see the indispensable essay by T. Gasparrini Leporace, "Presentazione," in *Mappamondo* (ed. 1956), pp. 11-16.

[25] "[...] onde fin al presente giorno è dapoi continuamente stato [il mappamondo] in tanta venerazione e precio appresso tutta questa città, e coloro massime che si dilettano delle cose di cosmografia, che non è mai giorno

The *mappa mundi* was, first and foremost, an object to be looked at rather than read. The lack of appropriate lighting and the fact that the map was hanging on a wall made reading or textual analysis of the legends, which required the use of a ladder, platform and candles, quite difficult. It was much easier to look at it : the *mappa mundi* was therefore likely perceived primarily as an image.

During my frequent visits to the first floor of the Biblioteca Marciana, where the *mappa mundi* now hangs in a showcase on a wall illuminated by a window with views of Piazza San Marco, I noticed that the experience of a modern spectator is probably not much different from that of a fifteenth-century spectator : once the curtain that protects the *mappa mundi* from the sun is drawn back it is only possible to read the legends up to the middle of the map, and even these require the aid of a magnification lens. The minutely ornamented design is illuminated in vibrant colors, predominantly blue, white, red and green. Among the designs, the ships that surround the *oikumene*, earthly paradise, cities and the funerary monuments of Asia stand out. Every attempt to read the text is made difficult by the overabundance of colors, forms and designs that confuse the vision : the *mappa mundi* is teeming with streets, cities, mountains and ships that enchant the eye and invite one to focus on the pictorial details in the monumental luminosity of the whole, rather than reading the texts. If one considers Medieval and Renaissance maps as images, their beauty and elegance can be interpreted as purely ornamental. However, this would be an error : the *mappa mundi* of Fra Mauro helps demonstrate that the role of images in ancient cartography is more than decorative. The individual images on the *mappa mundi*, and the *mappa mundi* as a total image, constitute a significant part of Fra Mauro's cosmographic discourse. Even though it is not always easy to determine the symbolism intended by these representations it is still important to emphasize the richness of meanings that fifteenth-century spectators could take away from the *mappa mundi*, which structured cartographic and cosmographic semantics even while functioning both as an image and through its images. In the next paragraph we will analyze the specific visual cosmographic elements of the *mappa mundi* including its orientation to the South and the function of its legends.

2. The center and confines of the world

The *mappa mundi* is first and foremost a circle. This structural element recalls, on a much larger scale, the form and function of the so-called T-O *mappae mundi*. It is important to note that the so-called T-O *mappae mundi* of the *Etymologies* (*Etymologiae*) – in particular in Books XIII and XIV.2.3, that is "On the Universe" (*De mundo*) and "On the Earth" (*De terra*) – of the *On the Nature of Things* (*De natura rerum*) of Isidore of Seville, the *Bellum Iugurthinum* of

che d'alcuno non sia con molto piacere veduto e considerato, e fra gli altri miracoli di questa divina città, nell'andare de' forestieri a vedere i lavori di vetro a Murano, non sia per bella e rara cosa mostrato. E ancor che quivi si vegghino molte cose." Giovanni Battista Ramusio, "Dichiarazione d'alcuni luoghi ne' libri di messer Marco Polo, con l'istoria del reubarbaro," in *Di messer Giovanbattista Ramusio prefazione sopra il principio del libro del magnifico messer Marco Polo. All'eccellente messer Ieronimo Fracastoro*, in *Navigationi et viaggi*, in Venetia, appresso i Giunti, 1559, vol. II, f. 17r. Modern edition ed. by M. Milanesi, Giovanni Battista Ramusio, "Dichiarazione d'alcuni luoghi nè libri di messer Marco Polo, con l'istoria del reubarbaro," in Giovanni Battista Ramusio, *Navigazioni e viaggi*, Turin : Einaudi, vol. III, 1980, pp. 69-71.

Sallust (traditionally in Chapter XVII, corresponding to the beginning of the description of Africa), the *Commentarius in somnium Scipionis* of Macrobius, and of the *De nuptiis Philologiae et Mercurii* of Martianus Capella (usually in the geographical section of Book VI, the *De geometria*), are not the fruits, as is often written, of a medieval barbarization of geographical knowledge.[26] Placed at the beginning or end of the chapters of the aforementioned books they are not a form of cartographic naivety, but rather mnemonic figures that described the 'scheme' of the world so as better to fix that scheme in the reader's mind.[27] Notwithstanding the temporal distance and the unbridgeable gap in terms of content between the T-O *mappae mundi* and Fra Mauro's *mappa mundi*, the circular form of the latter serves the function of a 'scheme' of the terrestrial world. Almost 1000 years after T-O *mappae mundi* were created, Fra Mauro in his *mappa mundi* expressed his complex cosmographic vision based on this simplified cosmographic scheme.[28] As a circle the *mappa mundi* has a geometric center (a metal pin around which Fra Mauro traced a red circle that surrounds the *oikumene* close to the wooden frame, at a regular distance of about one and one half centimeters), and a symbolic center, Jerusalem, placed towards the West with respect to the geometric center. The red circle traced with a compass that surrounds the *oikumene* is an *unicum* in the history of the Western representation of cosmographic space. As we lack any explanation of this circle from Fra Mauro it is not easy to understand its meaning. The red circle may represent the confine between the inhabitable space that is explorable and knowable by man and the uninhabitable and unknowable space of darkness. This is an interpretation that we derive from certain legends placed in the Indian Ocean (*Mare indicum*) concerning the confines of human space. In the mid-fifteenth century Fra Mauro relied on an ancient cosmographic conception that was destined to survive for a long time after him, according to which the borders of the *oikumene* are marked and surrounded by thousands of islands. The physical and metaphysical confines of the knowable and the known assume the geographic morphology of an insular reality, a reality evidently suspended between the two elements of water and earth.[29] Fra Mauro writes:

> Shown here all around its circumference, this uninterrupted chain of islands surrounds the Sea of India. They are inhabited by different species of birds and in this sea one can find ambergris, which the people who live here say originates in these islands and is then carried into the Indian Sea by the currents that wash their shores; others believe it is the sperm of whales. Note that when navi-

[26] For an example of a reading of medieval geography in terms of barbarization, see G. H. T. Kimble, *Geography in the Middle Ages*, London: Methuen, 1938 (particularly chapter 2, entitled "The Dark Ages of Geography"). For a useful criticism of this approach, see *La 'Descriptio mappe mundi' de Hugues de Saint-Victor*, with introduction and commentary by P. Gautier Dalché, Paris: Études augustiniennes, 1988, pp. 117-121.

[27] P. Gautier Dalché, "De la glose à la contemplation. Place et fonction de la carte dans les manuscrits du Haut Moyen Age," in *Testo e immagine nell'Alto medioevo, Settimane di studio del Centro italiano di studi sull'Alto Medioevo* 51, Spoleto: 1994, pp. 693-771.

[28] For an elaborate, but not exhaustive, catalogue of T-O world maps, see M. Destombes, *Mappemondes A.D. 1200-1500*, Amsterdam: N. Israel, 1964; D. Woodward, "Medieval *Mappae mundi*," in J. B. Harley and D. Woodward (eds), *The History of Cartography*, vol. 1, Chicago & London: Chicago University Press, 1987, pp. 286-370, in particular pp. 294-304 and pp. 359-367.

[29] M. Donattini, *Spazio e modernità: libri, carte, isolari nell'età delle scoperte*, Bologna: CLUEB, 2000, pp. 250-256.

> gators see the birds of these islands, they think that they have come too close to them and they pull away – because beyond these islands are the Shadows, which are so dense that if a ship ventured into them it would not be able to go either backwards or forwards. And this is known from the experience of those who did venture into them, and who perished.[30]

The region of darkness brushes up against the *oikumene* near southern Africa: "Just beyond these off-shore islands the Shadows begin to appear, but within this cape they do not hinder navigators."[31] Surrounding the entire *oikumene* the darkness prevents navigation, and constitutes a grave danger for navigators in that the air and water there are dense and sticky (*tegnente*), trapping ships that would venture near it without any hope of escape, to such an extent that Fra Mauro writes "it is better to die" (*conuien perir*) than be stuck there: "The ships which, sailing south, pass too close to the Lost Islands are carried by the currents into the Shadows, which – like the water below them – are so dense that the ships cannot go forward, and they inevitably perish."[32] The red circle, then, may represent and sanction, in graphic form, the boundary between the explorable part of the earth and the part precluded from man and not represented in the *mappa mundi*.

Concerning Jerusalem as the "shifted," or as we shall see, human and historical center of the *oikumene*, Fra Mauro deems it necessary to explain this dislocation in a complex legend:

> By latitude, Jerusalem is at the center of the inhabited world; but by longitude, it is further west. However, as Europe is more [densely] populated, Jerusalem does appear to be at the center [of the inhabited world] by longitude as well, if one takes into account not [geographical] space but the number of inhabitants.[33]

Fra Mauro holds that Jerusalem is situated at the center of the inhabited earth with respect to latitude, but with respect to longitude, it is located more to the West. As the Western part of the *oikumene* is more densely populated than the East, it follows that if one takes into consideration the number of inhabitants, Jerusalem is also at the human center with respect to longitude. Fra Mauro's idea of two distinct centers of the earth, a geometric center and what could be called an "anthropological" one, which is determined by population density, can be traced

[30] "Quela tirada de isole che cençe el [*mar d'india come è designato a la uia circular sono habitade* | per deuerse qualità de oseli [...]. *E nota che qu(a)ndo i naueganti uedeno li oseli de le iso* | le predite, pare(n)doli esser acost[*adi tropo a quele, se delongano da esse p(er)ché oltra quele* | sono le tenebre, le qual son [*tanto dense che le naue che se abatesse intrar i(n) quele* | non poria nì andar nì tornar [*in driedo. E questo se sa per experientia che quele che* | se hano abatude sono peride." *Mappamondo* (ed. 1956) Pl. VIII, 2. The English translation derives from *Mappamondo* (ed. 2006) *25.

[31] "Pocho lonçi da queste isole foreane | co(m)ença a parer le tenebre, le qual q(u)i olt(r)a q(ue)sto cauo no(n) impaçano i nauega(n)ti." *Mappamondo* (ed. 1956) Pl. IV, 1. The English translation derives from *Mappamondo* (ed. 2006) *17.

[32] "Nota che le naue le qual nauegando per ostro | se lassa acostar a le isole perse, le corentie le por | tano a le tenebre et intrade in quele per le sue | densità e anchor de quela de le aque, le qual son molto tegnente, conuien perir." *Mappamondo* (ed. 1956) Pl. VIII, 1. The English translation derives from *Mappamondo* (ed. 2006) *24.

[33] "HIERVSA|LEN è i(n) mezo de la te(r)ra | habitabile seco(n)do la latitudine | de la terra habitabile, be(n)|ch(é) seco(n)do la lo(n)getu|dine la sia più oc|cide(n)tal, ma p(er)ch(è) la | p(ar)te ch'è più oc|cide(n)tal è più habi|tada p(er) l'europa | p(er)hò l'è i(n) mezo an|cora seco(n)do la lon|gitudine, no(n) co(n)|sidera(n)do el spatio de la terra | ma la moltitudi|ne di ha|bita(n)ti." *Mappamondo* (ed. 1956) Pl. XXII, 201. The English translation derives from *Mappamondo* (ed. 2006) *1011.

back to concepts originally developed by Ali Ibn Ridwan (*c.* 998-*c.* 1061) in his commentary on Ptolemy's *Tetrabiblos*.[34] A Muslim physician, astrologer, and astronomer, born in Giza, generally cited by European authors as Haly, or Haly Abenrudian, he became known through the translations of his works by Gerard of Cremona in the thirteenth century, and became highly respected for his astronomical-astrological and medical knowledge. He wrote a commentary on the *Ars parva* of Galen and also a commentary on the *Tetrabiblos* which was later quoted by prominent mathematician-astrologers such as Paulo Dagomari (also known as Paolo dell'Abbaco, 1282-1374) who were able to follow his complex reasoning and mathematical descriptions.[35] Fra Mauro's case shows, however, that some of Ali Ibn Ridwan's ideas, simplified and emptied of their mathematical content, and removed from the studies of the most erudite and learned, became a knowledge also acquired and discussed in the vernacular.

3. Why the *mappa mundi* is oriented towards the south

The orientation of the *mappa mundi* towards the south is perhaps the first aspect that surprises and intrigues the modern spectator who is used to north-oriented maps, and who is therefore disoriented by the effort required to identify landmasses which not only have 500-year-old outlines, but which are also turned "upside down," thus losing their familiar shapes. Most of the diagrammatic manuscript *mappae mundi* of the period 1150-1500 are oriented to the east. But among the maps contemporary with that of Fra Mauro the 1448 world map of Andreas Walsperger,[36] the so-called "Borgia world map"[37] of the first half of the fifteenth century, and the "Zeitz Mappa mundi"[38] of the last quarter of that century are all oriented to the south. Thus it is not surprising that, around the mid-fifteenth century, a *mappa mundi* was oriented to the south. Fra Mauro, in fact, did not even think it necessary to address the issue, probably

[34] For example, in the Latin translation of the Arabic version of the *Tetrabiblos* with Ali ibn Ridwan's commentary, published together with other treatises in Claudius Ptolemaeus. *Liber quadripartitus*, Venice: heirs of Octavianus Scotus, 1519, II, 3 (f. 27rb). I owe this important reference to Patrick Gautier Dalché who presented his own transcriptions of Ali ibn Ridwan's work during his research seminar at the École Pratique des Hautes Études between 2006 and 2008.

[35] P. Gautier Dalché, "Usages critiques et scientifiques de la carte marine au XIVe siècle : Pétrarque, Boccace, Paolo dell'Abbaco," in T. Michalsky, F. Schmieder and G. Engel (eds), *Aufsicht – Ansicht – Einsicht. Neue Perspektiven auf die Kartographie an der Schwelle zur Frühen Neuzeit*, Frankfurt : 2009, pp. 81-92.

[36] Vatican City, Biblioteca Apostolica Vaticana, MS Palat. Lat. 1362B. Andreas Walsperger, [Mappa mundi sive descriptio orbis geometrica facta ex cosmographyca Ptholomey... Et cum vera et in-|tegra cartha navigationis marium...], vellum, 589 × 746 mm, south at the top, 1448. In the lower margin the last line of a six-line note in black ink states : "Facta est hec mappa per manus fratris Andreas Walsperger ordinis Sancti Benedicti de Salisburga Anno Domini 1448 in Constantia." See K. Kretschmer, "Eine neue mittelalterliche Weltkarte der vatikanischen Bibliothek," *Zeitschrift der Gesellschaft für Erdkunde* 26 (1891), pp. 371-406 ; Almagià, *Monumenta cartographica vaticana*, vol. 1, pp. 30-31 and plate 12 ; and Destombes, *Mappemondes*, entry 52.10, pp. 212-214.

[37] Vatican City, Biblioteca Apostolica Vaticana, MS Borgiano XVI, [Borgia world map], engraved on two semicircular plates of bronze held together by nails, first half of the fifteenth century, diam. 630 mm. See Almagià, *Monumenta cartographica vaticana*, vol. 1, pp. 27-28 and plate 8 ; Destombes, *Mappemondes*, entry 52.10, pp. 212-214.

[38] Zeitz, Stiftsbibliothek, MS Hist. Fol. 497. For a reproduction see D. B. Durand, *The Vienna-Klosterneuburg Map Corpus of the Fifteenth Century : A Study in the Transition from Medieval to Modern Science*, Leiden : Brill, 1952, plate 16.

considering it congruent with the wider culture of his readers and patrons. But what was the reason for this cosmographic choice ? In the historiography of the *mappa mundi* this aspect has often been attributed to the influence of Islamic cartography through the manuscripts of the *Nuzhat al-Mushtāq* of al-Sharif al-Idrisi (1099-† between 1166 and 1180), in which the world maps were oriented to the south.[39] The fact of the matter is that no trace of ancient Arabic geographical manuscripts, either by Idrisi or by any other author, can be found today in Venice nor in old monastic library registers or other records (in the *Navigazioni et viaggi* Ramusio recalls two maps that Marco Polo carried back to Venice which possibly served as a model for the *mappa mundi* of Fra Mauro – without referring to their orientation – but he mentions no Arab maps).[40] Furthermore, there are other south-oriented Venetian maps, such as the extraordinary early fifteenth-century manuscript map of Italy, entitled *Italie provincie modernus*, in the British Library, "probably the earliest surviving 'modern' map of the region" that was presented as a gift to Henry VII or Henry VIII, presumably by a Venetian ambassador.[41] As a result, the hypothesis of a direct Arab influence does not at all seem convincing to me. Indeed, the suggestion of Arabic influence, aside from the fact that it is impossible to verify by documentary evidence, relies on elements that are "exogenous" to the cosmographic culture of Fra Mauro : it seems very unlikely that the Camaldolese would have adopted the approach of another culture, Arab in this case, without deeming it necessary to give an explanation.

Within the general narrative structure of the *mappa mundi* this would represent an anomaly. When he did not follow Ptolemy's *Geography* "either in the form of the world or in the measurement of latitude and longitude" (*Unde se algun contradirà a questa p(er)ché no(n) ho seguito | Claudio Tolomeo, sì ne la forma come etia(m) ne le sue mesure p(er) longeça e p(er) | largeça*) the Camaldolese friar justified his decision in two long legends.[42] It would have been highly unusual for a cosmographer like Fra Mauro, who was involved in an intense dialogue with his readers, to adopt a cosmographic choice so foreign to the intended readership of the *mappa mundi* without giving the slightest explanation. Ramusio, in his long description of the *mappa mundi* in the second volume of the *Navigazioni e viaggi*, does not even mention the orientation of the map, which indicates that such a choice was part of the symbolic universe and habits of graphical and cosmographical representation of Venice at that time. Surely this culture included knowledge of the maps made by other civilizations, for example, the Muslim civilization. The frequent imports from Alexandria by Venetian merchants and navigators could have very well been the means by which Arab as well as Persian manuscript maps, or at least second-hand knowledge of them, arrived in Venice. Arab, Persian and Syrian, the three languages

[39] E. Cerulli, "Fonti arabe del mappamondo di Fra Mauro," *Orientalia : Commentarii periodici Pontifici Istituti Biblici* 4 (1935) pp. 336-338 ; S. Maqbul Ahmad, "Cartography of al-Sharif al-Idrisi," in *The History of Cartography*, Vol. 2, Book 1, *Cartography in the Traditional Islamic and South Asian Societies*, pp. 156-173. See also the other essays in the volume for an analysis of Islamic cartography and geometry.

[40] On this subject see Chapter VI, "The narrative structure of Fra Mauro's *mappa mundi*."

[41] London, The British Library, Cotton Roll xiii.44, *Italie provincie modernus*, manuscript on vellum, 137 × 65 cm, Venice, *c.* 1425-50. See M. Milanesi, "Antico e moderno nella cartografia umanistica. Le grandi carte d'Italia nel Quattrocento," *Geographia Antiqua* 26 (2007-2008), pp 153-175 (Atti del IV Seminario di *Geographia Antiqua* "La cartografia degli antichi e dei moderni," Perugia 2006) ; P. Barber and T. Harper, *Magnificent Maps. Power, Propaganda and Art*. London : British Library, 2010, pp. 28-29 ("Henry VIII's map of Italy").

[42] *Mappamondo* (ed. 1956) Pl. XL, 49 and XLI, 21. *Mappamondo* (ed. 2006) *2834 and *2892, respectively.

of the different Islamic cosmographic schools, were familiar to Venetian merchant navigators, mariners and ambassadors, not to mention professional interpreters.[43]

In any case other explanations are possible: and they have the merit of being part of the culture to which Fra Mauro belonged, in particular, part of his cosmographical culture. One such possible source of the southern orientation of the *mappa mundi* is one of the most important treatises of medieval natural philosophy: the *De caelo* of Aristotle and its commentaries. In this work Aristotle associates the Antarctic pole of the universe, and by extension, the southern hemisphere, with the "uppermost part":

> Of the poles, the one which we see above us is the lowest part, and the one which is invisible to us the uppermost. For we give the name of right-hand to that side of a thing whence its motion though space starts. Now the beginning of the heaven's revolution is the side from which the stars rise, so that must be its right, and where they set must be its left. If this is true, that it begins from the right and moves round to the right again, its upper pole must be the invisible one, since if it were visible, would be leftward, which we deny. Clearly therefore the invisible pole is the upper, and those who live in the region of it are in the upper hemisphere and to the right, whereas, we are in the lower and to the left.[44]

It is important to emphasize that the four fundamental spatial directions for Aristotle and his commentators – defined as *sursum, deorsum, dextrum* and *sinistrum* in middle-Latin translations of the *De caelo* and its commentaries – were absolute: "above" and "below," "high" and "low," were intrinsic properties of matter and the universe, comparable with the opposed couples "head" and "feet" in the animal world. To understand Aristotle's reasoning it is necessary to examine his theory concerning the concept of place, which is founded on four postulates. Among these is the principle that each place has a high and a low and that each body has a natural tendency to move towards its proper place and to remain in it. A passage from the *Physics* of Aristotle, in which he explains the dynamics of movement, is very important for our argument:

> Moreover the trends of the physical elements (fire, earth, and the rest) show not only that locality or position is a reality but also that it exerts an active influence; for fire and earth are borne, the one upwards and the other downwards, if unimpeded, each towards its own "position," and these terms – "up" and "down" I mean, and the rest of the six dimensional directions – indicate subdivisions or distinct classes of positions or localities in general. Now these terms – such as up and down and right and left, I mean – when thus applied to the trends of the elements are not merely relative to ourselves. For in this relative sense the terms have no constancy, but change their meaning according to our own position, as we turn this way or that...: whereas in Nature each of these directions is distinct and stable independently of us. "Up" or "above" always indicates the "whither" to which things buoyant tend; and to "down" or "below" always indicates the "whither" to which weighty and earthy matters tend, and does not change with circumstance.[45]

[43] U. Tucci, "Mercanti veneziani in Asia lungo l'itinerario poliano," in L. Lanciotti (ed.), *Venezia e l'Oriente*, Florence: Olschki, 1987, pp. 307-321.

[44] Aristotle, *De caelo*, II.2.15-25. See Aristotle, *On the Heavens*, trans. W. K. C. Guthrie, London: Heinemann; Cambridge, MA: Harvard University Press, 1939, pp. 145-146.

[45] Aristotle, *Physica*, IV.1, 208b. See Aristotle, *The Physics*, with an English translation by P. H. Wicksteed and F. M. Cornford, London and Cambridge, MA: Harvard University Press, 1970, pp. 281-283.

This last assumption makes space a subject of absolute qualitative difference and provides the metaphysical foundation not only of natural movement, but also of spatial representation.[46] High and low are linked, in an absolute way, to the concepts of place and space. These conditions are also applied to "things" : for example, to the elements of the sublunar world, as entities which "occupy" places. From the *De caelo* and the *Physics* of Aristotle we can clearly deduce that the right, left, and high and low of the world are respectively the east, west, south and north. All of this has been demonstrated graphically in the traditional iconography of Atlas, whose head, the noblest part, faces the Antarctic pole, his feet face the Arctic pole, his right hand is turned toward the east and his left hand toward the west.[47] The cosmographic structure of the *mappa mundi* of Fra Mauro is of an Aristotelian type, based on a theory of the elements derived from the medieval treatises and comments of the *philosophia naturalis* of Aristotle. Its southern orientation could also be an integral part of this cosmographic vision.

There is, however, another consideration that justifies or, better yet, contextualizes, this cosmographic choice. The *mappa mundi* is a Venetian map in which Fra Mauro crystallizes the experience and also the dreams, hopes and limits to expansion and commerce contemporaneously perceived in Venice. A reading of the legends of the *mappa mundi* clearly indicates that the Venetian vision of commercial and maritime expansion was above all turned towards the Indian Ocean, and thus towards the south. As we will see in the last part of this chapter, navigation routes and commercial circuits towards and in the "Sea of India", which at the time were still in the form of projects, take form and are minutely traced and described in various legends. Fra Mauro would be the interpreter of a vision according to which Venice benefited from the riches generated by her merchants, navigators and mariners in the southern and eastern parts of the world.[48] The orientation of the *mappa mundi* towards the south may also reflect the vision of the world that Fra Mauro had come to have as a participant in, and interpreter of, Venetian culture. Fra Mauro conceptualized and gave graphic form to the possibility of combining the circumnavigation of Africa with Arab, and possibly even Chinese, navigations in the "Sea of India" into a single immense route : from his cell in San Michele di Murano, Fra Mauro clearly oriented his intellectual and cultural interest in cosmography towards the south and south-east.

4. A bounded full world

In the cosmographic space of the *mappa mundi* hundreds of coloured and vibrant images, along with hundreds of rectangular cartouches, give form to a single unified narrative. Fra Mauro seems to insist on the aspects of visual cosmography that attract the imagination of the

[46] The other fundamental postulates by Aristotle on the concept of place are : 1) the place where an object lies is not part of the object, nor a component, but that which contains it ; 2) the exact place of an object is neither smaller nor greater than the object itself ; 3) an object can separate itself from the place where it lies, and therefore does not coincide with the place where it lies. See M. Jammer, *Storia del concetto di spazio*, Milan : Feltrinelli, 1963.

[47] B. Nardi, *La caduta di Lucifero e l'autenticità della 'Quaestio de aqua et terra'*, Turin-Genoa-Milan : SEI, 1959, pp. 9-12.

[48] These considerations on the sea and the navigational routes and expansion will be addressed below in a section entitled "The idea of the ocean and the navigational directions at the dawn of the European expansion."

spectator and present the universe as an immense *liber figurarum*, a minutely "annotated" body, full of signs and images that include all aspects of the world. In the following section I will try to isolate the different parts of this unity in order to determine their roles and functions. The images that fill and give form to the *mappamondo* can be divided into thirteen groups: cities, temples (churches and mosques, and also *hedifici di idoli*, or pagan temples), bridges, sepulchral monuments, ships, animals, mountains, rivers, seas, frontiers (of provinces and of the world itself), streets, and text. Fra Mauro and the *pictori* who painted the *mappa mundi* constantly used and reused these models, varying their dimensions to create a diverse and sumptuous representation. Omitting for now the iconographical analysis of these images – in Chapter V we will investigate the relationship between some of the images in the *mappa mundi* and the *Milione* of Marco Polo and Book IV of Poggio Bracciolini's *De varietate fortunae* – we address here the question of what was the primary function of such a composite creation. According to our interpretation these signs, which compose the surface of the map like the stones of a mosaic, suggest and recreate the cosmographic concept of a full world, entirely knowable, recognizable and explorable.

Alexander Koyré saw the pre-Galileo world as a "finite and closed world."[49] The world of Fra Mauro is, even more than finite and closed, a full world: full from a cosmological point of view, which means according to the refined and complex fusion of, and convergence between, Christian theology and Aristotelian cosmology created by scholastic philosophers. It is full from a cosmographical point of view and from the point of view of natural philosophy according to the physical principles of Aristotle and his Arab and Christian commentators, who consider emptiness an aberration.[50] It is also full from a chorographic and anthropological point of view. If the cosmos cannot contain empty spaces, the created world and, in particular, the earth, cannot admit *terrae incognitae* or unknown spaces. Cities, sepulchral monuments, streets, borders of provinces, ships that cross the ocean and seas, and cartouches are placed in, and fill every part of, the *mappa mundi*, even the farthest eastern reaches of Asia, the extreme north of "Rossia," and the southern latitudes of Africa. Fra Mauro assigns the task of conveying the human and known fullness of the world to the hundreds of images and texts that fill the *mappa mundi*.

5. The text of the *mappa mundi* as image

In this context the function of the texts on the map merits specific attention. The textual richness of the *mappa mundi* is imposing. The *mappa mundi* appears, in fact, to suffocate under the legends that Fra Mauro interposes everywhere between the representations of cities, rivers, mountains, streets and bridges. The map contains three thousand textual units, of which around two hundred – the rest are single place names – constitute a grand treatise of cosmography including descriptions of places and peoples, commercial geography, history, navigation and expansion, and, finally, questions that would today be defined as methodological. Without

[49] A. Koyré, *From the Closed World to the Infinite Universe*, Baltimore: Johns Hopkins University Press, 1957.

[50] Aristotle, *De caelo et mundo*, III.6, 305 a 21; J. Hamesse, *Les auctoritates Aristotelis: un florilège médiéval. Étude historique et édition critique*, Louvain: Publications Universitaires; Paris: Béatrice-Nauwelaerts, 1974, p. 166, n. 89.

negating the importance of the text as legible content in the *liber figurarum* of the *mappa mundi* it does play a primary role as image. From a cosmographical point of view, the way in which Fra Mauro places his text on the *mappa mundi*, in synergy with hundreds of drawings, reinforces and alludes to a representation of space in which the world is seen as completely knowable and inhabitable. The text, written all over the *mappa mundi*, contributes to the creation of a concept of terrestrial space that is completely knowable and human. For Fra Mauro the Earth, or better, all of the earth, is *oikumene*. Without necessarily reading the cartouches, but just looking at their placement, the ancient and modern spectator could plausibly perceive the fullness of the world. Those among them who knew of the form and extension of the *oikumene* of Ptolemaic planispheres could also appreciate the difference of dimensions between the ancient and modern *oikumene*.[51]

III. – Cosmography and writing

1. Remnants of a lost cosmographic anthology

Settling our gaze on the *mappa mundi*, we see the macrostructure give way to a world that is minutely detailed with numerous designs and legends. Fra Mauro gives form to a proper cosmographic treatise by completing the *mappa mundi* with nearly three thousand inscriptions. The legends of the *mappa mundi*, particularly the most lengthy, constitute a work unto themselves almost separate from the world map. As we will see in the last chapter, there was a tradition which was thought to have been lost, but which in fact was only forgotten in the archive of the Camaldolese monastery: the text alone of the *mappa mundi* was transmitted, with an increasingly complete re-copying of the legends, and was passed down in manuscripts until the printed publications of 1806.

The text of the *mappa mundi* was likely based on a hand-written cosmographic anthology prepared by Fra Mauro. We can reasonably deduce that at least more than two hundred extended legends on the map were based on, and copied from, an anthology that was written in a format well suited to this task, such as a notebook. It would have been a collection of quotations rather than a complete and discursive work, an anthology of cosmographic excerpts that were derived from direct and mediated readings of the principal authors who wrote on the sublunar world, including works such as the *De sphaera* of Sacrobosco and its numerous commentaries in Latin, the *Summa theologica* of Thomas Aquinas, the *De natura loci* and the commentary on the *Liber de causis proprietatum elementorum* by Albert the Great and Ptolemy's *Geography*, among others. This anthology made it possible to copy the legends onto different maps, and also as separate essays, as in the case of Pietro Dolfin's Latin translation (prepared for a copy of the *mappa mundi* commissioned by the Medici family around 1480), or the *Aviso di fra Nicola inscripto fra di s. Michele di Muran*, compiled by Alessandro Zorzi around 1520.[52]

[51] They were definitely able to appreciate the similarities evident in the design of the Asiatic coasts that face the Indian Ocean. On the complex relationship between Fra Mauro and Ptolemy see Chapter IV.

[52] See Chapter I and its Documentary Appendix.

This argument seems very likely to me, especially in connection with the seven long cosmographic legends placed in the four corners of the *mappa mundi* between the square and circular frames. In contrast to the legends placed within the *oikumene*, which are directly related to the images, the content of these seven legends is more articulated than that of the accompanying diagrams and is independent of them. Their complex language and narrative structure, similar to those found in scholastic *disputationes*, distinguish them from the other legends which are, in general, shorter, less cogent and much more specific. This clearly implies the existence of a collection of texts, which was perhaps more systematic than the legends of the *mappa mundi*, in which Fra Mauro had collected his cosmographic thoughts and knowledge in written form.

2. *Quaestiones philosophiae naturalis*

Fra Mauro presents most of his cosmographic knowledge in the seven long legends placed in the four corners of the map along with seven other cartouches placed in the chorographic space of the work. Often considered "external" and marginal to the *mappa mundi*, and therefore often neglected, they are actually crucial to understanding the general cosmographic meaning of Mauro's total *oeuvre*. It is chiefly in the seven long legends that he sets forth the vocabulary that he uses to trace and define his cosmography. In these legends, and in other extensive cartouches inside the *mappa mundi*, the dimensions, shape, qualities and confines of the celestial sphere, the sublunar world with the four spheres of the elements and, especially, the earth and water, literally take form. The following un-numbered list indicates the main subjects related to the physics of the celestial and sublunar world that Fra Mauro addresses in his *mappa mundi* (the titles are placed in quotation marks when they are lacking in the original, and are supplied here for clarity of exposition, deriving them either from the *incipit* or from the content of the legends) : On the number of the heavens according to the authority of the holy theologians[53] ; On the distance of the heavens[54] ; How waters rise and subside due to the attractive properties of the Moon[55] ; How by Divine Providence the earth is raised above the water[56] ; Of the elements. That is, how much one element exceeds another in quantity[57] ; How the earth which is under the [celestial] equator and the torrid zone may be habitable[58] ; The location of the earthly paradise[59] ; On whether in the southern regions the water does not surround our inhabited and

[53] "Del numero de hi cieli secondo l'auctorità de hi sacri theologi." *Mappamondo* (ed. 1956) p. 19 ; *Mappamondo* (ed. 2006), p. 720.

[54] "De la dista(n)tia de hi | cieli." *Mappamondo* (ed. 1956), p. 19 ; *Mappamondo* (ed. 2006), p. 722.

[55] "Come per la uirtù atractiua de la luna le acque cresce e decresce." *Mappamondo* (ed. 1956), p. 21 ; *Mappamondo* (ed. 2006), p. 726.

[56] "Come per diuina prouidentia la terra habitabile è subleuada da l'aqua." *Mappamondo* (ed. 1956), p. 21 ; *Mappamondo* (ed. 2006), p. 726.

[57] "De hi elementi, cioè quanto uno elemento | excieda l'altro in quantità." *Mappamondo* (ed. 1956), p. 22 ; *Mappamondo* (ed. 2006), p. 731.

[58] "Come la terra supposita a l'equinoctial e a | la torrida çona è abitabile." *Mappamondo* (ed. 1956), p. 23 ; *Mappamondo* (ed. 2006), p. 733.

[59] "Del sito del paradiso terrestro." *Mappamondo* (ed. 1956), p. 22 ; *Mappamondo* (ed. 2006), p. 729.

temperate zone[60]; "On the size of the *orbis terrarum*" (*circunferentia*)[61]; "On the center of the world"[62]; "Regarding the divisions of the world – that is, of Asia from Africa and of Europe from Asia"[63]; "On the number of provinces in the world"[64]; "If the Sea of India is enclosed like a pond and does not communicate with the ocean"[65]; "On the length of the sea route that links Europe and India via Africa"[66]; "On the regions of darkness."[67]

Fra Mauro's legends constitute one of the most ample treatises on cosmography in the Venetian vernacular of the first half of the fifteenth century. Fra Mauro collects and, in part, gives original form to the cosmographic knowledge transmitted in Latin by many medieval and Renaissance authors on cosmology and cosmography. The questions of natural philosophy treated in the legends of the *mappa mundi* can be divided into four principle thematic units: the structure and dimensions of the celestial world; the dimensions and structural properties of the sublunar world; the form, extension and navigability of the seas; and the earthly paradise as a part of cosmographic space. In this chapter we will address the first three themes. The earthly paradise will be discussed in the following chapter.

IV. – The structure and dimensions of the celestial world

1. Saint Thomas and the number of heavens

The legend dealing with the number of heavens recalls a central dispute in the Christian theology of creation, and posed two problems: one concerning the number of heavens according to the Bible, and another concerning the consistency of the Bible with scientific descriptions of the celestial world, such as, for example, the concentric spheres described by Plato in the *Timaeus* or Aristotle in the *De caelo*, derived from the theories of Eudoxus of Cnidos (fourth century BC).[68] The first problem was born of a distinction that was already present in the bibli-

60 "Che i(n) le parte meridional l'aq(u)a | no(n) circunda questo n(ost)ro habitabile e temperado çona." *Mappamondo* (ed. 1956) Pl. XI, 2; *Mappamondo* (ed. 2006) *149.

61 "Delle dimensioni della circunferentia." *Mappamondo* (ed. 1956) Pl. XXXIX, 88; *Mappamondo* (ed. 2006) *2828.

62 ["Del centro del mondo."] *Mappamondo* (ed. 1956) Pl. XXVIII, 24; *Mappamondo* (ed. 2006) *1011.

63 "Circa la diuision de la terra, çoè de la affrica da l'a|sia e similiter de la europa da l'asia." *Mappamondo* (ed. 1956) Pl. XXXIV, 79; *Mappamondo* (ed. 2006) *2489.

64 "Del nu(mer)o de le p(ro)vi(n)cie del mo(n)do." *Mappamondo* (ed. 1956) Pl. XXXIX, 88; *Mappamondo* (ed. 2006) *2828.

65 "Del mar d'i(n)dia ch'el sia serado come un stagnon | e che'l mar occea(n) no(n) li entri." *Mappamondo* (ed. 1956) Pl. IX, 26; *Mappamondo* (ed. 2006) *53.

66 ["Della lunghezza della rotta di circumnavigazione dell'Africa verso l'India."] *Mappamondo* (ed. 1956) Pl. XI, 2; *Mappamondo* (ed. 2006) *149.

67 *Mappamondo* (ed. 1956) Pl. IV, 6; IV, 1; VIII, 1; *Mappamondo* (ed. 2006) *2, *19, *24.

68 For the theory of the homocentric spheres in Plato, see *Timaeus* 31a, 36d; 38cd and 39a; in Aristotle, see for example *De caelo* I.9, 277 b 27-29. Eudoxus of Cnidos, using a group of three or four concentric spheres for every planet, was able to explain the motion of Mercury, Jupiter and Saturn along the zodiac. Calippus of Cizicus (fourth century BC) introduced some extra spheres for the Moon, the Sun, Venus and Mars. The resulting models were worked out by Plato (427-347 BC) and Aristotle (384-322 BC) in order to develop their physical conceptions of the universe:

cal account of creation between the heaven created the first day (Gen. 1:1) and the firmament created by God the second day (Gen. 1:6-7), to which may be added Saint Paul's affirmation in the *Second Letter to the Corinthians* (2 Cor. 12:2) implying the existence of a third heaven. So, how many heavens existed? And, moreover, how is the biblical account compatible with the Platonic and Aristotelian description of the concentric spheres of the planets and the fixed stars, or with the more complex astronomical geometry of the epicycles, equants and deferents described by Ptolemy? Fra Mauro did not evade such questions, but he addresses them by entrusting himself entirely to the authority of Saint Thomas.

His lengthy legend "On the number of the heavens according to the authority of the holy theologians," is in fact a literal translation into the Venetian vernacular of Article 4, "Utrum sit unum caelum tantum," of *Quaestio lxviii*, 'De opere secundae diei in quatuor articulos divisa,' of the first part of the *Summa theologica* of Saint Thomas.[69] This fact, which has escaped both scholars of the *mappa mundi* and commentators on Thomas, has an importance that goes far beyond the history of cosmography. The long legend concerning the number of heavens represents a very rare, if not singular, translation into Venetian of an article of Thomas' *Summa theologica*. It also unequivocally demonstrates how the depiction of the *imago mundi*, or the universal description of land and seas, plants and animals, and above all, people, their history, customs, and trade – within the framework of the creation of the universe – was founded on a synthesis of knowledge derived from diverse and eloquent sources from the late medieval period to the beginning of the modern age. Article 4 is based on the authority of four Fathers of the Church and two *auctores* (Basil of Caesarea, John of Damascus, John Chrisostom, Augustine, Rabanus Maurus and Bede the Venerable), and is divided in two thematic units. In the first, Thomas evokes the cosmologic doctrine of John Chrisostom (347-407), according to which only one heaven existed, beginning at the sphere of the Moon, embracing the seven planets and extending to the fixed stars. Thomas Aquinas contrasted this doctrine with the theses expressed by Basil of Caesarea (330-379) in the third of his *Homilies on the Hexaëmeron* (III.1) and by John of Damascus (*c.* 675-749) in his *De fide orthodoxa* (II.4, in *PG* XCIV, 879 ff.; and II.11, in *PG* XCIV, 910 ff.). Following the exegeses of the Bishops of Caesarea and Damascus, and citing the *De universo* of Rabanus Maurus, Thomas affirms that despite their formal diversity, a sub-

the immobile Earth was surrounded by eight groups of crystalline spheres that transported the Moon, the Sun, the planets and the static stars. From the twelfth century, following the tradition of the *De caelo* of Aristotle, the eight groups of spheres were enclosed in a ninth sphere, the *primum mobile* and later in a tenth sphere, the Empyrean, God's abode. Finally, the Empyrean was divided into nine metaphysical spheres corresponding to the nine levels of the angelic hierarchy. For the formulation of the theory of the homocentric spheres of Eudoxus of Cnidus, see E. Maula, *Studies in Eudoxus' Homocentric Spheres*, Helsinki: Societas Scientiarum Fennica, 1974. For a general study of the doctrine of the homocentric spheres see G. Schiapparelli, "Le sfere omocentriche di Eudosso, di Callippo e di Aristotele," *Memorie del Reale Istituto Lombardo, Classe di Scienze matematiche e naturali* 13 (1875) (reprinted in *Scritti sulla storia dell'astronomia antica*, II, Bologna, 1926, pp. 126-141). For an analysis of the theory of the eight spheres as suggested by Plato and that of 55 homocentric spheres by Aristotle and their medieval interpretations, see Duhem, *Le système du monde*, vol. 1, pp. 126-150.

[69] *S. Thomae Aquinatis Summa theologiae*, Pars I, q. 68, a. IV (see *S. Thomae Aquinatis Summa theologiae*, edited and study by Petri Caramello cum textu ex recensione leonina, pars prima et prima secundae, Turin: Marietti, 1952, p. 334. In the *Quaestio* 68 Thomas discusses one of the *quaestiones* of the scholastic tradition: "On the number of spheres, whether there are eight or nine, or more or less." See Grant, *Planets, Stars, and Orbs*, p. 697. The same *quaestio* is also discussed by Roger Bacon, Albert of Saxony, and Pierre d'Ailly.

stantial unity of the heavens can be found.[70] In the second part of the Article, through a celebration of the mystical light derived from the *Homilies on the Hexaëmeron* of Basil, Thomas presents the thesis of the existence of ten heavens, beginning with that of the moon and, beyond it, the seven planets (the Moon, Mercury, Venus, the Sun, Mars, Jupiter and Saturn), the heaven of the starry firmament, and the Prime Mover or Empyrean. The *Homilies on the Hexaëmeron* represent an interpretation of Genesis that also pays attention to the cosmological and cosmographic aspects of the Bible.[71] According to Basil, the expression *caeli caelorum* indicated that the spheres of planets and stars are reassembled in the unity of the heavens.[72] In the last part of the legend, again citing Thomas (*Summa theologica*, I, q. 66, a. 3; I, q. 102, a. 2.1m), Fra Mauro affirms: "To conclude, current opinion is that there are ten heavens, from the heaven of the moon to the empyrean, as this present drawing shows."[73]

The diagram of the heavens in the upper left corner of the *mappa mundi* is a schematic representation of the celestial spheres, but only of the so-called "total spheres". Between 1200 and 1600 most diagrammatic representations of the cosmos included only the total and homocentric spheres, ignoring the partial and eccentric spheres and, as a consequence, deferents, epicycles, and equants. They are represented as circular orbits that form a geometric system with the centers of revolution placed one on top of the other.[74] The diagram of the *mappa mundi* includes the four elements (only three of which are labeled, *aqua* | *aere* | *fuoco*) and the ten celestial spheres: the Empyrean heaven, the ninth sphere, the sphere of the fixed stars, Saturn, Jupiter, Mars, Sun, Venus, Mercury, and the Moon.[75] The ninth orb was that of the Prime Mover. In contrast to the other spheres the Empyrean did not have an astronomical function, but was rather a theological creation introduced as an autonomous immobile celestial sphere in the cosmological debate of the twelfth century, when theologians like Peter Lombard (cited by Fra Mauro as *maistro de le sentenzie*, master of the *Sentences*), Hugh of Saint Victor, and astronomer-theologians like Campanus of Novara described it as a place of dazzling light where God, the angels and the blessed resided for all eternity.[76] Fra Mauro therefore followed a well-established model and iconographic tradition.

[70] Rabanus Maurus, *De universo*, in *Patrologia Latina*, vol. 111, col. 9-614, esp. col. 258-279.

[71] V. Novembri, *La geografia dei Padri della Chiesa: il caso di Basilio di Cesarea*; Ph.D. dissertation, Florence: University of Florence, 2005.

[72] Basil the Great, *Omelia III.3* in *Sulla Genesi*, M. Naldini (ed.), Milan: Fondazione Lorenzo Valla – Arnoldo Mondadori Editore, 1990, pp. 73-81.

[73] "Ma cho(n)cluda(n)do dei cieli, | sego(n)do la chomuna opinion, cho(m)me(n)çando dal cielo de la luna so|no diece cieli computando el cielo empyreo, chome apar | qui ne la presente pictura." *Mappamondo* (ed. 1956) p. 19; *Mappamondo* (ed. 2006), upper left corner, p. 722.

[74] For a fascinating look at the representation of the heavens in medieval miniatures, see *Il cielo: intorno alla Cosmologia di Plinio. Capolavori delle rappresentazioni miniate nei codici delle principali biblioteche del mondo*. Turin: Allemandi, 1994.

[75] "Cielo empireo | Sfera nona | Stelle fixe | Satvrno | Iove | Marte | Sol | Venvs | Mercvrio | Lvna." *Mappamondo* upper left corner.

[76] Grant, *The Foundations of Modern Science in the Middle Ages*, p. 164.

2. The dimensions of the universe: echoes from Campanus of Novara's *Theorica planetarum*

The legend titled "On the Distance of the Heavens. Rubric"[77] placed in the upper left corner of the *mappa mundi* lists the distances from the center of the universe (the term *mondo*, derived from the Latin word *mundus*, literally "world," refers to the universe) to the internal and external surfaces of the four sublunar and the eight heavenly spheres. Even though the *Theorica planetarum* (*Theory of the Planets*) of Campanus of Novara (*c.* 1220-1296) is never explicitly cited, all of these distances derive, no doubt indirectly, from this work. The *Theorica* was commissioned by Pope Urban IV and written between 1261 and 1264 while Campanus was residing at the Pontifical Curia in Orvieto.[78] In this work Campanus corrected and updated the dimensions of the universe and planets indicated by Ptolemy in his *Planispherium* (*Planetary Hypotheses*). Its calculations became widely accepted until the first half of the sixteenth century and were still cited even in the seventeenth century.[79]

The comparison between the distances and measurements reported by Fra Mauro and those in the *Theorica* makes it clear that the *mappa mundi* summarizes only a small part of Campanus' data. While the *Theorica* reports the dimensions of each planet and its spheres (the radius, the diameter, and the circumference of the planet; the internal and external circumferences of the sphere that carries the planet; the circumference of the deferent and of the equant, the distance between the center of the epicycle and the equant; the circumference of the epicycle of the planet; and finally the distance between the moon and its epicycle),[80] the legend on Fra Mauro's *mappa mundi* indicates just two sets of measurements: the diameters of the seven planets and the distances of all of the spheres from the center of the universe.[81] Even if it is not possible to exclude the possibility that the Camaldolese friar read and directly employed the *Theorica*, this detail strongly suggests that Fra Mauro derived these magnitudes from cosmographic *florilegia* that include excerpts of the *Theorica*, rather than directly from the very difficult and technical work of Campanus. As for the diameters of the seven planets, the following are the distances reported on the *mappa mundi*:

[77] "De la dista(n)tia de hi | cieli. Rubrica." *Mappamondo* upper left corner.

[78] Iohannes Campanus of Novara was one of the most important scientists and mathematicians of the thirteenth century. His fame is above all linked to the commentary on the *Elementa* of Euclid (*c.* 1250), of the *Theorica planetarum* (1261-64), and of the *Tractatus de sphaera* in which he commented on the work of Sacrobosco. Campanus was affiliated with the Roman Curia as both doctor and astronomer of Pope Urban IV; see Campanus Novariensis, *Campanus of Novara and Medieval Planetary Theory. Theorica Planetarum*, edition with an introduction and English translation and commentary by F. S. Benjamin and G. J. Toomer, Madison: University of Wisconsin Press, 1971, pp. 3-24; see also Duhem, *Le système du monde*, vol. 3 (1915), pp. 317-326; and G. J. Toomer, "Campanus of Novara," in *Dictionary of Scientific Biography* 3 (1981), pp. 23-29.

[79] See "Influence of the *Theorica*," in *Campanus of Novara and Medieval Planetary Theory*, pp. 36-41.

[80] These are the magnitudes calculated by Campanus for each planet and its spheres as reported in the *Theorica* in Latin: "semidiameter corporis, tota diameter corporis, totus ambitus corporis, ambitus superficiei convexe [sphere]; ambitus superficiei concave [sphere]; ambitus deferentis aut equantis; dieta centri epicicli de partibus equantis; ambitus epicicli; dieta [corporis] in suo epiciclo." See *Campanus of Novara and Medieval Planetary Theory*, pp. 356-361.

[81] *Campanus of Novara and Medieval Planetary Theory*, pp. 356-361; A. van Helden, *Measuring the Universe: Cosmic Dimensions from Aristarchus to Halley*, Chicago and London: University of Chicago Press, 1985, pp. 28-52.

> The moon is of diameter 1,896 and 26/33. | […] | Mercury is of diameter 230 and 26/33 miles. | […] | Venus is of diameter 2,884 and 560/660 miles. | […] | The Sun is of diameter 35,700 miles. | […] | Mars is of diameter 7,572 and 480/660 miles. | […] | Jupiter is of diameter 29,641 and 540/660 miles. | […] | Saturn is of diameter 29,202 miles.[82]

Each of these figures is followed by the measurement of the radius of the sphere on which each planet is carried, calculated from the center of the universe. The legends begin with the magnitude of the radius of the earth and that of the whole sublunary world below the sphere of the moon :

> From the center of the world to the surface of the earth there are 3,245 and 5/11 miles. | From the center of the world to the lower surface of the heavens of the moon there are 107,936 and 20/33 miles. | From the center of the world to the lower surface of the heaven of Mercury there are 209,198 and 26/33 miles.| From the center of the world to the lower surface of the heaven of Venus there are 579,320 and 560/660 miles. | From the center of the world to the lower surface of the heaven of the Sun there are 3,892,866 and 560/660 miles. | From the center of the world to the lower surface of the heaven of Mars there are 4,268,629 miles. | From the center of the world to the lower surface of the heaven of Jupiter there are 323,520 and 420/660 miles. | From the center of the world to the lower surface of the heaven of Saturn there are 52,544,702 and 280/660 miles. | From the center of the world to the lower surface of the eighth sphere – that is, the heavens of the fixed stars – there are 73,387,747 and 180/660 miles. Each of these miles measures 400 cubits.[83]

The last specification referring to the mile as equivalent to "400 cubits" is clearly a simple mistake of transcription for "4000 cubits," an equivalence originally established by the Persian Muslim philosopher and astronomer al-Farghani (ninth century, known in the West as Alfraganus) in his *Elements of Astronomy on the Celestial Motions*.[84] Written in about 833, this descriptive summary of Ptolemy's *Almagest* was widely used by Western cosmographers after it was translated into Latin in the twelfth century. It is important to emphasize that these measurements on Fra Mauro's map likely derived from an intermediate source that transmitted some of the data of Campanus' *Theory of Planets* with great precision.[85] There are very few discrepancies between the magnitudes on the map from those of Campanus. Fra Mauro con-

[82] "Tuto el diametro de la luna è miglia 1896 26/33 […] | Tuto el diametro de Mercurio sono miglia. 230 26/33 […] | Tuto el diametro de Uenere sono miglia 2884 560/660 […] | Tuto el diametro del sol è miglia 35700 […] | Tuto el diametro de Marte è miglia 7572 480/660 […] | Tuto el diametro de Joue è miglia 29641 540/660 […] | Tuto el diametro de Saturno è miglia 29202." *Mappamondo*, upper left corner (right), lines 3, 5, 7, 9, 11, 13, and 15.

[83] Dal centro del mondo a la superficie de la terra sono miglia 3245 5/11 | Dal dicto centro a la superficie inferior del cielo de la luna sono miglia 107936 20/33 | Tuto el diametro de la luna è miglia 1896 26/33 | Dal dicto centro del mondo a la superficie i(n)ferior del cielo de Mercurio sono miglia 209198 26/33 […] | Dal centro del mondo a la superficie inferior del cielo de Uenere sono miglia 579320 560/660 […] | Dal centro del mondo a la superficie inferior del cielo del sol sono miglia 3892866 520/660 […] | Dal centro del mondo a la superficie inferior del cielo de Marte sono miglia 4268629 […] | Dal centro del mondo a la superficie inferior del cielo de Joue son miglia 323520 420/660 […] | Dal centro del mondo a la superficie inferior del cielo de Saturno sono miglia 52544702 280/660 […] | Dal centro del mondo a la superficie inferior de la octaua spiera, çoè del cielo | de le stelle fixe, sono miglia. 73387747 180/660 | Questi miglia sono cadauno de lor de mesura de cubiti .400." *Mappamondo*, upper left corner (right), lines 1-2, 4, 6, 8, 10, 12, 14, 16-18.

[84] See Abdelhamid I. Sabra, "Farghānī, al-," *Dictionary of Scientific Biography*, vol. IV, New York : Charles Scribner's Sons, 1971, pp. 541-545.

[85] Piero Falchetta is of the opposite opinion.

cludes the legend by declaring the impossibility of calculating the size of the eighth, ninth and tenth spheres. Since they rotate around the center of the universe without epicycles and deferents the lack of a geometrical model makes it impossible, Fra Mauro argues, to determine their circumferences (*circuli*) and, consequently, their distances from the center of the world:

> It is not possible in any way to deduce the depth of the eighth sphere in which there are the fixed stars, because these move around the center of the world in a different manner to the seven planets. In fact, those latter move around other centers so it is possible to calculate the depth of their circles on the basis of the distance of these centers from the center of the world. For this reason, the depth of the ninth sphere and also of the crystalline heavens is unknown to us.[86]

Aside from its technical content, which was the subject of extended disputes in medieval and early modern natural philosophy,[87] this conclusion of the legend, together with another quotation explaining the phenomenon of the tides which mentions the epicycle of the moon,[88] suggests that Fra Mauro was referring to the Ptolemaic astronomical system: planets are assumed to move in a small circle called an epicycle (literally "on the circle" in Greek) which, in turn, moves on a larger circle called a deferent. In Ptolemaic astronomy the epicycle, deferent and equant (a point that is equidistant from the earth and the centre of the deferent) comprise a geometric model designed to explain the variations of speed and direction in the apparent motion of the seven planets. In particular, it explained their retrogade motion and the changes in their distances from the Earth.[89] Thus there is no doubt that Fra Mauro was aware of the complex astronomical geometry of Ptolemy's *Almagest* and *Planetary Hypotheses*, quite probably by way of Albert the Great.

[86] "Quanto sia la grosseça de questa octaua spiera ne la qual sono le stelle fixe | non è possibile p(er) alguna demo(n)stratiua raxon provar p(er)chè le predicte | stelle fixe se moueno sopra el centro del mondo e non come li sette | planeti i qual se muoueno sopra altri ce(n)tri p(er) la diuersità de li qual | ce(n)tri, dal centro del mo(n)do se mesura la qua(n)tità de hi suo | circuli. E p(er) questa medema raxo(n) rima(n) ignota | a nui la qua(n)tità de la spera nona | e quela etia(m)dio de cielo cristalin," *Mappamondo*, upper left corner (right), lines 19-26.

[87] C. Trifogli, "The Place of the Last Sphere in Late-Ancient and Medieval Commentaries," in S. Knuttila, R. Työrinoja, S. Ebbesen (eds), *Knowledge and the Sciences in Medieval Philosophy. Proceedings of the Eighth International Congress of Medieval Philosophy (SIEPM)*, II, Helsinki, pp. 342-350.

[88] "La | segonda caxon è la propinquità de la luna a le aque, zoè qua(n)do la luna è in la parte inferior del suo epiciclo, questa propi(n)quità fa trar più le aque" ("The second cause is the closeness of the moon to the waters – that is, when the moon is in the lower part of its epicycle"). The quotation is taken from a complex legend placed in the upper right corner of the *mappa mundi* with the following *explicit*: "Questa rubrica de sopra dechiara come per la uirtù atractiua de la luna le aque cresce e descresce" ("The above rubric describes how waters rise and subside due to the attractive properties of the moon"). The passage derives from the comment on the *De causis proprietatum elementorum* of Albert the Great. See the section below titled *The Seas and Tides, a Venetian Debate* and also Doc. 5 in the Appendix, transcribed from the upper right corner (left) of the *mappa mundi*. For the citation of Albert the Great, see Albertus Magnus, *De causis proprietatum elementorum*, Book I, Tract. II, chapter 6, in *Alberti Magni... Opera Omnia*, Tomus V, Pars II, P. Hossfeld (ed.), Monasterii Westfalorum in aedibus Aschendorff, 1980, p. 71, 21-35.

[89] The equant was a point near the center of a planet's orbit from which the center of the planet's epicycle would always appear to move at the same speed. Since planets actually moved at different speeds when the epicycle was at different points on its deferent Ptolemy, by using an equant, could keep the motion of heavenly spheres uniform and circular as had been assumed by Plato and Aristotle. See A. Murschel, "The Structure and Function of Ptolemy's Physical Hypotheses of Planetary Motion," *Journal for the History of Astronomy* 26 (1995), pp. 33-61.

3. The absolute and relative dimensions of the sublunar and celestial world. At the frontiers of natural philosophy and theology

The distances reported on the *mappa mundi*, which are not particularly eloquent in themselves, acquired a more concrete meaning when compared to the diameter of the earth (by this term I refer to the Aristotelian element), which was calculated to be 6.490 10/22 miles, a figure obtained by multiplying the distance from the centre of the universe to the surface of the earth by two. Through simple operations of subtraction and division, using the distances supplied by Fra Mauro, it is possible to reconstruct both the absolute and relative dimensions of the planets and sublunar elements and the relative dimensions of the sublunar world and Earth with respect to the entire universe. This operation allows us to visualize the universe as imagined and perceived in mid-fifteenth century Europe with great precision.

The radius of the entire universe, including the sublunar and heavenly worlds, is 73.387.747 and 18/66 miles. This figure represents the distance from the center of the universe to the internal surface of the sphere of fixed stars, or the external surface of the seventh sphere. The radius of the sublunar world, including the four elementary spheres of earth, water, air and fire, is 107.936 20/33 miles. Dividing the radius of the universe (73.387.747 and 18/66 miles) by the radius of the sublunar world (107.936 20/33 miles – in the formulation of Fra Mauro, the distance from the world's center to the interior surface of the moon's sphere), we see that the sublunar world is approximately 680 times smaller than the celestial world (precisely 679.919). Dividing the radius of the eighth sphere, that of the fixed stars, by the Earth's radius (3245 5/11 miles), the relative dimensions of the Earth and universe are 1 to 22.612.47 ; that is, the radius of the Earth is 22.612.47 times smaller than that of the entire universe. The relative dimensions of the Earth with respect to the whole sublunar world are approximately 1 to 33 (precisely 33.25, a result obtained by dividing the radius of the inner surface of the moon's sphere of the moon by the Earth's radius). If we designate the diameter of the Earth as "T," the following are the relative dimensions of the seven planets : Earth = T ; Moon = 0.29 T (the diameter of the Moon is 1/3 that of the Earth) ; Mercury = 0.03 T (the diameter of Mercury is 3/100 that of the Earth) ; Venus = 0.44 T (the diameter of Venus is nearly half that of the Earth) ; Sun = 5.50 T (the diameter of the Sun is 5½ times that of the Earth) ; Mars = 1.16 T (the diameter of Mars is about 17% larger than that of the Earth) ; Jupiter = 4.56 T (the diameter of Jupiter is about 4½ times that of the Earth) ; Saturn = 4.56 T (the diameter of Saturn is 4½ times that of the Earth).

Using the same distances and dimensions of the heavens written in the left margin of the *mappa mundi*, and proceding to calculate the volumes of the sublunar and heavenly spheres, we find that the sublunar world, including earth, water, air and fire, represents 0.147 percent of the universe. Similarly, the earth represents just 3 percent of the sublunar world ; in other words, 97% of the sublunar space is occupied by the three lightest elements (water, air and fire), and the remaining approximately 3% by the heavier and more compact elements. Given that the distance from the center of the universe to the surface of the earth is 3245 5/11 miles, and that the earth is spherical, we can determine a circumference equal to 20.387 5/11 miles. Fra Mauro's *mondo*, that is, the universe in its entirety, has a circumference of 460.875.052.872 miles.

Beyond the clear astronomical significance of determining the relative sizes of the sublunar and heavenly parts of the universe, the image of the heavens and the description of heavenly distances are located at the frontiers of astronomy, religion, philosophy and aesthetics. Within medieval and early modern Christian cosmography the design and the calculation of distances are also, if not mainly, a way to approach the God-Creator of the Bible, the supreme artist who sketched the diagram of the world during the Creation and molded all of nature's surfaces. In designing an image of the heavens and specifying their numerical distances, the cosmographer assumes the role of interpreter of the biblical creation story, re-read from the point of view of natural philosophy. In Christian theology, beginning with the exegesis of Philo of Alexandria (20 BC-AD 50), God is, first and foremost, the architect of the existing.[90] This form of meditation also finds expression outside Christian theology. Beginning in the mid-fourteenth century, Hebrew cabbalists perfected a refined technique of visualization of heavenly forces, a topography of the divine based on the binome measurements/diagrams. Kabbalists fixed the proportions of the divine architecture of heavens on maps and diagrams, seeking to capture the mystery of creation through the harmony of geometrical forms. At the same time as Campanus of Novara was correcting the *Planispherium* of Ptolemy for the Roman Curia, during the second half of the fourteenth century, the Cabbalists' works were widely diffused, even in Rome itself. Giulio Busi, who has recently studied this corpus of manuscripts, identified them as belonging to the "first phase of visual qabbalah" before an even wider diffusion took place among the Hebrews of Spain.[91]

V. – The dimensions and structural properties of the sublunar world

1. The *quaestio de aqua et terra*

The two legends "Of the elements. That is, how much one element exceeds another in quantity,"[92] and "How by Divine Providence the earth is raised above the water"[93] placed in the lower and upper right corners respectively of the *mappa mundi*, are to be understood as part of a centuries-old dispute, the so-called *Quaestio de aqua et terra* (Dispute on [the elements] water and earth)[94] This *quaestio* was one of the most debated problems of natural philosophy in universities and in commentaries on the *De caelo*, *De generatione et corruptione*, and *Meteorologica* of Aristotle concerning the sublunar world from the thirteenth through the

[90] D. T. Runia, "Philo of Alexandria and the Beginnings of Christian Thought. Alexandrian and Jew," *Studia Philonica Annual* 7 (1995), pp. 143-160.

[91] G. Busi, *Qabbalah visiva*, Turin : Einaudi, 2005, pp. 25-56.

[92] "De hi elementi, cioè quanto uno elemento excieda l'altro in quantità" : see Doc. 3 in the Appendix.

[93] "Come per diuina prouidentia la terra habitabile è subleuada da l'aqua" : see Doc. 4 in the Appendix.

[94] The medieval *quaestio* normally consists of five parts : the posing of the question ; objections to the position that the author, in case of a written *quaestio* like the articles of Thomas' *Summa theologiae*, or the master (*magister*), in case of an oral dispute, will ultimately defend : the *sed contra*, most often a brief scriptural passage or passage from a Father of the Church, or an *auctor*, like Aristotle, suggesting the position that the master will defend : the *responsio*, or *corpus* of the article : the master's determination of the question ; and finally, the reply to each of the objections posed at the beginning. On the structure of the *quaestio* and of the *disputatio* as rhetorical genres and scientific procedures, consult B. C. Bazàn, J. W. Wippel, G. Fransen and D. Jacquart, *Les questions disputées et les questions quodlibétiques dans les facultés de théologie, de droit et de médecine*, Turnhout : Brepols, 1985 and S. Caroti, "Filosofia e scienza della natura nel Medioevo e nel Rinascimento," in W. R. Shea (ed.), *Storia delle scienze. Le scienze fisiche e astronomiche*, Turin : Einaudi, 1992, pp. 110-147.

sixteenth centuries. Most of the major scholastic philosophers were involved in it.[95] Much more than simply an argument among university professors, the *quaestio de aqua et terra* was also addressed in debates held in urban churches where participation did not require any university education, as it has been clearly shown by the investigations around the attribution (albeit controversial) to Dante Alighieri of a work entitled *Quaestio de aqua et terra*. Analysis of the validity of this attribution has prompted archival digging into the social context in which the *quaestio* was debated, showing its wide dissemination outside the *studia*.[96] It was, therefore, a debate that interested a vast public that ranged from professors and their students to the educated urban classes.[97] Closely related to the *quaestio de aqua et terra* was the one over the relative size of the elements of the sublunar world.

This scholastic dispute originates in the meeting of Aristotelian natural philosophy with Patristic exegesis of the Bible, and in particular, of the story of the creation contained in Genesis. If the four elements were arranged in concentric spheres placed one within the other, and water must enclose within itself the earth, how was it that the inhabitable part of the earth was not immersed in water? The problem was never considered in these very terms by Aristotle: in the *De caelo* he does not formulate a descriptive discourse, but rather reflects on the theoretical organization of the elements in the sublunar world. Even though the Stagirite philosophers had distinguished the spheres of earth and water in his theory of the elements, he held that the two together, from an astronomical and geographic point of view, formed a single sphere.[98]

When translations from Greek and Arabic introduced the numerous works of the rich philosophical and scientific heritage of ancient Greece and the Islamic civilization to Christian Europe in the second half of the twelfth century, the *Quaestio de aqua et terra* became one of the most burning disputes in natural philosophy. Works such as the *Canon of Medicine* of the Persian philosopher Ibn Sīnā, known in the West as Avicenna (980-1037), one of the most important and thoroughly-discussed texts used for teaching medicine in medieval universities,

[95] N. Bouloux, *Culture et savoirs géographiques en Italie au XIVe siècle*, Turnhout: Brepols, 2002, pp. 19-44.

[96] Over the last two centuries the major philologists of Dante have engaged in a lively debate over this attribution, which may have been definitively concluded in 1979 when the *Quaestio de aqua et terra* was published among Dante's minor works in the Ricciardi edition of Italian literary classics. See Dante Alighieri, *Quaestio de aqua et terra*, in *Opere minori*, tomo II, F. Mazzoni (ed.), Naples: Ricciardi, 1979, pp. 693-880. Among those disagreeing with the attribution of the *Quaestio* to Dante, the position of Bruno Nardi is paradigmatic: see Id., *La caduta di Lucifero e l'autenticità della 'Quaestio de aqua et terra'*, Turin, Genoa and Milan: SEI, 1959; however, the opinion of those favoring the attribution to Dante has become dominant. For a review of the *querelle* see M. Pastore Stocchi, "Quaestio de aqua et terra," in *Enciclopedia Dantesca*, Rome: Istituto della enciclopedia italiana, vol. IV (1973), pp. 761-765.

[97] Nardi, *La caduta di Lucifero*, pp. 28-31.

[98] While attributing a spherical form to the four elements, Aristotle recognized that their spheres were not perfect and geometrical like the heavenly spheres; the earth in particular presented depressions and profound cavities (*De caelo* II.32) into which water had poured to form seas and the ocean (*De caelo* II.31). By its innate property to tend towards the center of the world, water takes its place next to the element of the earth, collecting in the earthly cavities and thus creating seas, lakes, and the ocean, the natural place of water (*Meteorologica* II.2). As the waters collected in the cavities of the earth, certain parts of the earth emerged and can be inhabited by earthly animals and by man (*Meteorologica* II.5). The inflence of celestial bodies on the elements of the sublunar world provoked continuous changes in the distribution of the emerged land, so that new land could emerge while other parts were submerged in a continuous process of "generation and corruption" (*Meteorologica* I.14).

or Johannes de Sacrobosco's *The Sphere*, furnished material relevant to the *quaestio*.[99] It is in comparing Aristotelian notions transmitted by works like the *Canon of Medicine* or *The Sphere* with the text of Genesis 1:9, on the third day of creation, that the Christian reader perceives a contradiction. The literal interpretation of the six days of creation as described in Genesis 1:9-10 ("And God said, 'Let the waters under the heavens be gathered together into one place, and let the dry land appear'") conflicted with literal interpretations of passages such as "The earth is a simple body, whose natural location is at the center of the universe...,"[100] and "Also the water is a simple body, whose natural location is around the earth, while being surrounded by air, all of which are meant to remain in their natural locations,"[101] made by Avicenna and many other *auctores*, such as Sacrobosco, Campanus of Novara, and Michael Scotus. The contradiction between the biblical text and Aristotelian definitions of the four elements resulted in a dispute that raged for centuries over how the *oikumene*, the inhabited part of the world, had emerged from the waters. Authors grappling with the *Canon of Medicine* or *The Sphere* put forward various solutions to overcome this apparent contradiction, mostly as a result of Avicenna's and Sacrobosco's lack of precise explanation as to why part of the earth emerged from the water to allow plants and other air-breathing animals to live, as so eloquently argued by Prosdocimo de' Beldomandi, professor at the University of Padua († 1428).[102] The analysis of the diverse interpretations of the passage that addresses the question "What is the form of the universe" (*Quae forma sit mundi*) in the numerous commentaries on *The Sphere* of Sacrobosco proves that medieval physics did coalesce around three main positions on this crucial question of natural philosophy.

An initial theory proposed by various temporally distant authors such as Isidore of Seville (sixth century), Bartholomaeus Anglicus (mid-thirteenth century), Thomas Aquinas (thirteenth century) and John Buridan (fourteenth century), and reintroduced to thirteenth- and fourteenth-century vernacular culture in the writings of Brunetto Latini (*c.* 1220-1295) and Ristoro d'Arezzo (the second half of the thirteenth century), is built upon the idea of the eccentricity

[99] The *Qanun* was translated into Latin as *Canon medicinae* by Gerard of Cremona in the twelfth century and into Hebrew in 1279. Its encyclopedic content, systematic arrangement and philosophical plan made it the reference work in the medical literature of Europe, even displacing the works of Galen. The text was read and commented in most of the universities until the Renaissance. See Andrea Borruso's introduction to Avicenna, "Degli elementi," in A. Borruso (ed.), *Il poema della medicina*, Turin: Zamorani, 1996, pp. 1-38 and N. G. Siraisi, *Avicenna in Renaissance Italy: the Canon and Medical Teaching in Italian Universities after 1500*, Princeton: Princeton University Press, 1987.

[100] "Terra est corpus simplex, cuius locus naturalis in medio totius existit, in quo naturaliter manet quieta, et ad ipsum, cum ab eo separata fuerit, naturaliter movetur." Avicenna, "Degli elementi," in Borruso, *Il poema della medicina*, p. 1.

[101] Aqua autem est corpus simplex, cuius naturalis locus est, ut sit circundans terram, et circundata ab aere, quum quidem in suis sitibus naturalibus permanserint," *ibidem*.

[102] Prosdocimo de' Beldomandi, who held the chair of Biagio Pelacani (*c.* 1347-1416), wrote in his comments on the chapter "Quae forma sit mundi" of the *De Sphaera*: "Notandum quod causa quare pars terrae sit taliter aquis discoperta, a diversis solet alia et alia assignari causa." See Prosdocimus de Beldemandis, "Super tractatu sphaerico commentaria," in Johannes de Sacro Bosco, *Sphaere tractatus Joannis de Sacro Busto. Gerardi Cremonensis theoricæ planetarum veteres. G. Purbachii theoricae planetarum novae. Prosdocimi de Beldomando super tractatu sphaerico commentaria, nup in lucem diducta per L. Ga[uricum] nunquam amplius impressa. Joannis Baptistae capuani expositio in sphaera & theoricis...*, Venetiis: in aedibus Luceantonii Iunte Florentini, 1531 mense Ianuario, f. 7v.

of the two spheres, albeit with different shadings and implications. A second theory mainly supported by Campanus of Novara and theologians like Giles of Rome – but which maintained adherents up to the late fifteenth century, as evidenced by Christopher Columbus' account of his Third Voyage – posited the concentricity of the two elements and held that the earth was not spherical, but rather a bump that rose out of the waters, making the *oikumene* an enormous island that emerged from the ocean. A third theory, closer to the explanation of Aristotle, suggested by Antonio Pelacani of Parma (fourteenth century), professor of physics and mathematics in Bologna and Verona, proposed that the water and earth were concentric, but that the latter contained most of the marine water in its hollows.[103]

Among these solutions to the *quaestio de aqua et terra*, Fra Mauro opts for the theory of eccentricity of the spheres of earth and water, similar to the interpretation of John Buridan. Fra Mauro introduces the legend "How by Divine Providence the earth is raised above the water"[104] with a definition of the earth: it is "the place which generates animal life and all the things necessary to human life, such as trees, plants and herbs."[105] The earth is not a homogenous body, but is characterized by porosity and concavity:

> This earth has to be of diversified parts. That is, in some places the earth must be less dense: it must be light and of clear porosity and concavity, as one sees in the mines of metals and of precious and non-precious stones. So that, if one weighs two parts of earth that were equal in quantity or measure but different in density and compactness, the non-porous part will weigh more.[106]

Concavity and porosity were especially present in the parts not covered by water:

> If, therefore, such differences appear manifest and visible in some parts of the earth – and the part that is not submerged by water is the place that generates animals and other things – without any doubt, this part is less dense and more porous than that which is covered by water, which does not need such qualities.[107]

[103] For a well-documented and comprehensive review of the diverse interpretations outlined, see Vogel, *Sphaera terrae: Das mittelalterliche Bild der Erde und die kosmographische Revolution*, chapter III b, 6-12, "Sphaera terrae. Konzeptionen der Erdgestalt bis zum 14. Jahrhundert," pp. 153-239; and chapter IV.1, "Sphärische Kosmographie und der Beginn der europäischen Expansion nach Übersee," pp. 246-306. F. Mazzoni, *Contributi di filologia dantesca. Prima serie*, Florence: Sansoni, 1966, pp. 75, 83 and 87-88; Dante Alighieri, "Quaestio de aqua et terra," in *Opere minori*, pp. 703-705; Nardi, *La caduta di Lucifero*, p. 40 and pp. 145-150; Duhem, *Le système du monde*, vol. 9, pp. 129-132 and 142-145.

[104] "Come per diuina prouidentia la terra habitabile è subleuada da l'aqua," *Mappamondo*, upper right corner (right).

[105] "Essendo la t(er)ra luogo de la generatio(n) di a(n)i(m)ali e de le altre cosse a la vita humana neccessarie, come sono li arbori e piante over herbe…"; è un concetto che viene ribatito poco oltre, aggiundo la specifica " […] la terra e quellarte ch'è descoperta da le aque [è] luogo de la generatio(n) di a(n)i(m)ali…. " *Mappamondo*, upper right corner (right), lines 1, 5-6.

[106] "[…] bixogna | q(ue)lla t(er)ra auer diuersità ne le sue parte, zoè che i(n) alguni luogi de quella sia più rara et habia manifeste et sensibile porosità e co(n)cauità, come | uedemo esser ne li luogi de le minere de li metali et etia(m) de le pietre preciose e no(n) p(re)ciose. Onde, se'l se pesasse do' parte de terra equal de qua(n)tità de | mole ouer de grandeza ma desp(ar) i(n) esser dense e rare, ouer cu(m) ma(n)ifeste porosità e (con)cauità e ssenza alcuna porosità e co(n)cauità, quella parte no(n) porosa più | peseria de l'altra, " *Mappamondo*, upper right corner (right), lines 1-5.

[107] "Se adoncha tal differencia se troua ma(n)ifesta e sensibelme(n)te ne le parte de la terra e quella parte ch'è descoperta da le aque sia luogo | de la generatio(n) de li a(n)i(m)ali e de le cosse p(re)dicte, senza alguna dubietà

This entailed that the part not covered by water was more "rare," and one therefore reads :

> Natural reason concludes, therefore, that if by hypothesis one divided the earth into two parts of equal measure and quantity, so that one half comprised the earth not covered by the waters, which is where we are and which is less dense, and the other half comprised that covered by the waters, which is more dense, then the first, less dense, first half would weigh less than the dense.[108]

The difference in density and weight would, therefore, provoke an asymmetric disposition of the earthly sphere around the center of the world, so that one half emerges from the waters :

> In fact, if the two halves of the earth were of equal measure and weight, the center of the [element] earth would coincide with the center of the world [the universe]. But as the heavier things tend naturally towards the center of the world, even with violence, and drive from that center the less heavy things, therefore it cannot but be the case that this heavier half of the earth [sphere] is closer to the center, I mean the center of the world [the universe]. Thence it follows that there is a greater distance from the center of the world to the surface on which we live than from the same center to the surface of the opposite part. These things have been predisposed by God with wonderful providence. If all the surfaces of the earth were of an equal distance from the center of the world, then water would enclose and cover all parts of the earth, as it is an element of less density. And if the earth was entirely of equal weight and thus of equal distance from the center [of the universe], that would mean that, in covering one part of it, water would cover all of it – to the great detriment of the life of terrestrial animals.[109]

It is worth studying the syntactical and logical means by which Fra Mauro presents his argument in this passage. The logical presupposition for the immersion of part of the earthly sphere in the sphere of water is a lack of homogeneity in the earthly mass. The purpose of this lack of homogeneity is to allow the divine creation to be accomplished. It is evident – and not

q(ue)sta tal parte è più rara e porosa che l'altra parte de la t(er)ra coperta da l'aque, priuada de nec|cessità de le cosse predicte." *Mappamondo*, upper right corner (right), lines 5-7.

[108] "Co(n)clude adoncha la natural raxo(n), che se p(er) imaginaria diuisio(n) fosse partita la t(er)ra i(n) do' parte equal qua(n)to a la mole ouer qua(n)tità, i(n) | tal modo che una mità e parte co(m)prehe(n)desse q(ue)lla portio(n) de la terra scoperta da le aque che è uerso nui e rara e l'altra mitade q(ue)lla che è cop(er)ta et | densa, la prima mità rara men peseria de la densa." *Mappamondo*, upper right corner (right), lines 7-9.

[109] "E più oltra seguita che quella parte più densa sia dep(re)sa zoè più bassa et propi(n)qua al mezo del mo(n)do | de l'altra, perché essendo la mità de la terra q(uan)to a la mole cum gra(n)deça in peso equal, el mezo tuta la terra seria pu(n)ctalmen(n)te nel centro del | mondo, ma perché le cosse più graue naturalme(n)te uano al dito centro etiamdio cu(m) uiolentia, chazando dal dito centro le cosse me(n) graue, no(n) può ess(er) de meno che tal p(ar)te e mità de la terra non sia più uicina al centro, dico zoè l'è mezo el mondo et e più spatio da tal | centro a la sup(er)fitie de la p(ar)te de la tera da nui habitada che dal dicto centro a la sup(er)fitie da l'altra p(ar)te a questa nostra opposita, la qual cossa il guber | natore sumo Idio cu(m) mirabel prouide(n)tia ha p(er) tal forma ordinado, p(er)chè se tuta la sup(er)fitie de la tera fosse equalme(n)te distante dal cent(r)o ouer | mezo del mondo, l'aqua da ogni p(ar)te circu(m)daria e copriria equalmente tuta la terra no(n) auendo lo eleme(n)to de la qual tanta dinsità ne le suo parte | qua(n)ta se truoua ne la p(ar)te de la terra la qual equalmente graue e così equame(n)te remota dal dito mezo q(uan)to de la p(ar)te sup(er)ior zoè sup(er)fitie | e p(er) (con)seque(n)te copra(n)do una p(ar)te de la t(er)ra tute le coure(r)ia, la qual cossa seria i(n) detrime(n)to de la uita de li animali terestri i qual sopra l'aqua no(n) se poso | no sustentar p(er) la lor grauità nè sotto l'aqua posono uiuer p(er) la respiration et i(n)spiratio(n) de l'aere p(er) lo anelito ouer fiador, le qual sono | necessarie a la uita animal p(er) refrigerio del cor e mu(n)dification sua." *Mappamondo*, upper right corner (right), lines 9-19.

a surprise – that Fra Mauro reasoned using the logical categories of scholasticism. A close analysis of Fra Mauro's argumentation shows that he proceeds in the classical manner of the scholastics, considering the final cause (creation), the material cause (the porosity of the earth), the formal cause (the emergence of the earth in water), and the efficient cause (the lightness of the emerged hemisphere with respect to that submerged in water) in their traditional order.[110] The rubric proceeds to explain the natural tendency of the elements of the sublunar world to place themselves in such an order as to allow the emergence of the earthly sphere from that of the water, caused by the natural inclination of the elements. This conclusive remarks by Fra Mauro recalls a passage of Albert the Great's commentary on Aristotle's *Meteorologica*, in which the Dominican friar asserted that the final goal of the natural properties of water and earth – properties that prevented the two elements from perfectly occupying the concave surface of the spheres of air and fire – was precisely to enable the creation of animals and plants.[111]

The legend "How by Divine Providence the earth is raised above the water", though containing elements that may be ascribed to the theories of Albert the Great and Antonio Pelacani, nonetheless seems to be derived indirectly from the *Quaestiones in De caelo* of Buridan. One of the most important and influential commentators on Aristotle, John Buridan (1290-† *ante* 1358; professor at the Faculty of Arts of the University of Paris from 1328 to 1340) formulated the theory of the eccentricity of the two spheres from his observations of erosion, and thus his argument reflected his recognition of the importance of firsthand experience in understanding natural philosophy. In the second part of *Quaestio VII* of his *Quaestiones in De caelo*, a commentary on Aristotle's *On the Heavens*, Buridan discusses the inhabitability of the earth in relation to the distribution of water and formulates the hypothesis that the spheres of water and earth share the same center, i.e. the center of the universe, even though the distribution of water and land is necessarily uneven around this center.[112] According to this theory, the mostly dry part of the sphere of earth was filled with air, and thus was less heavy than the more dense submerged part, which was impregnated with water. From this observation, Buridan concluded that it is necessary to distinguish between the earth's center of gravity and its geometric center, both of which cannot coincide given the unequal distribution of weight within the sphere of the earth. Because the earth's center of gravity tends towards the center of the universe, it follows that this center is only the earth's center of gravity, and not its geometric center. With this conclusion Buridan further explains the variations of the surface of the earth, specifically the formation and erosion of mountains: these phenomena are all based on movements of the earth's center of gravity. According to Buridan, differentiating between the geometric center and the center of gravity could reconcile the structural concentricity of the elements and their geometric eccentricity: the earth is not at the center of the universe because of its location, but because its

[110] Caroti, "Filosofia e scienza della natura nel Medioevo e nel Rinascimento," pp. 110-147.

[111] Albert the Great, *Meteora*, P. Hossfeld (ed.), Monasterii Westfalorum in aedibus Aschendorff, 2003, *Tractatus tertius secundi libri Meteororum de natura maris*, cap. 1, "Quae sit tractatus intentio, et de duabus opinionibus antiquorum de esse et generatione maris"; cap. 2 "Et est degressio declarans, an aqua aliquando totam terra operuerit et an siccabilis sit per totum procedente tempore," pp. 82-85 (*Alberti Magni... Opera omnia*, vol. VI.1); see at this regard also Duhem, *Le sistème du monde*, vol. 9, pp. 121-122.

[112] *Ioannis Buridani Quaestiones in Aristotelis De caelo*, Liber II, Quaestio VII, *Utrum tota terra sit habitabilis*, pp. 410-417. See also Giovanni Buridano, *Il cielo e il mondo. Commento al trattato 'Del cielo' di Aristotele*, A. Ghisalberti (ed.), Milan: Rusconi, 1983, pp. 286-297.

center of gravity coincides with the center of the universe. Aristotelian natural philosophy, in this way, could be harmonized with Christian theology; and in fact Buridan's analytical techniques came to be identified with the very method of philosophy, understood as a secular practice distinct from theology. Fra Mauro's legend in the vernacular simplifies the technical terminology in Latin developed by Buridan and reframes it in a new synthesis. It is very unlikely that Fra Mauro consulted directly Buridan's commentary on Aristotle's On the Heavens; rather, he probably used an intermediary work. Nevertheless, even if this were the case, it is important to note how complex theories originating in the university circles permeated common knowledge and were restated in its terms.

VI. – The habitability of the Earth: the contradiction between zones and climates

The rubric "How the earth which is under the [celestial] Equator and the torrid zone may be habitable"[113] is one of the most elaborate and dense of Fra Mauro's cosmographic treatises. Concerned with the causes that determine and limit human habitation of the earth, it argues that the whole earth is inhabitable by invoking astronomical arguments and discussing the natural properties of the elements (*per raxon naturale*). The following thoughts have no claim to completeness. The medieval debate on the antipodes and the habitability of the Earth (and hence on the climatic zones and *klimata*) is very articulate and far less linear than has been suggested in frequently cited works, such as the book by Gustavo Uzielli and Giovanni Celoria[114] and the recent synthesis of Gabriella Moretti.[115] Research seminars led by Patrick Gautier Dalché at the École Pratique des Hautes Études between 2006 and 2008 revealed many medieval texts, both published and unpublished, which deconstruct the linearity of the debate and call for further research.[116] In this context, I will recall just a few important and well-established points regarding the medieval scientific debate on the climatic zones, which will help to contextualize and better understand the knowledge of Fra Mauro.

The habitability of the Earth was one of the crucial problems of medieval natural philosophy, and debate mostly focused on whether or not human habitation was possible in the equatorial and southern temperate zones.[117] Medieval cosmography did not provide a single consistent system for dividing the Earth into climatic zones or knowing the habitability of each

[113] "Come la terra supposita a l'equinoctial e a | la torrida çona è abitabile. Rubrica." *Mappamondo*, lower right corner (left). See Doc. 6 in the Appendix.

[114] G. Uzielli, *La vita e i tempi di Paolo Dal Pozzo: ricerche e studi con un capitolo sui lavori astronomici del Toscanelli di Giovanni Celoria*, Rome: Auspice il Ministero della Pubblica Istruzione, 1894.

[115] G. Moretti, *Gli antipodi: avventure letterarie di un mito scientifico*, Parma: Pratiche, 1994.

[116] P. Gautier Dalché, "Les répresentations de l'espace de l'antiquité tardive au XVIe siècle," in *École pratique des hautes études. Sciences historiques et philologiques. Annuaire. Résumé des conférences et travaux*, 139e année, 2006-2007, pp. 119-122; 140e année, 2007-2008, pp. 144-146. Some aspects of this research have already been published: see P. Gautier Dalché, "Le paradis aux antipodes? Une *Distinctio divisionis terre et paradisi delitiarum* (XIVe siècle)," in D. Barthélemy and J.-M. Martin (eds), *Liber largitorius. Études d'histoire médiévale offertes à Pierre Toubert par ses élèves*, Geneva: Droz, 2003, pp. 615-637.

[117] E. A. Moody, "John Buridan on the Habitability of the Earth," *Speculum* 16. 4 (1941), pp. 415-425; A. Hiatt, *Terra Incognita. Mapping the Antipodes before 1600*, Chicago: University of Chicago Press, 2008.

of them. Two main systems, those of zones and of *klimata*, offered two solutions which were partially contradictory.[118] Apparently invented by Parmenides (fifth century BC), the theory of zonal division was delineated by Aristotle in the *Meteorologica* (II.5) and his account appears to be a systematization of earlier Greek knowledge.[119] The theory received new impulses in Alexandria, where geographical speculation on the sphericity of the earth and the theory of zones met with particular success.[120] Later, the theory became traditional in medieval geographical and cosmographical works through the influence of treatises composed during Late Antiquity like the *Commentarium in somnium Scipionis* (Commentary on the Dream of Scipio) by Macrobius (fifth century) and *De Nuptiis Philologiae et Mercurii et de septem Artibus liberalibus libri novem* (On the Wedding of Philology and Mercury and of the Seven Liberal Arts in Nine Books) by the pagan encyclopedist Martianus Capella (fifth century), though with different interpretations and nuances. According to this theory the earth was divided into five latitudinal bands, differentiated by their climatic conditions. The poles, the Arctic and Antarctic Circles, and the Tropics served as a basis for the definition of the five astronomical and climatic zones, whose main function was to delimit the parts of the earth that were inhabitable from those that were uninhabitable. The traditional vision affirmed that the central zone (*perusta zona*), delimited by the Tropics, with the equator at its center, was uninhabitable because of the unbearable heat of the sun's rays, which in this area were almost always perpendicular to the surface of the earth. The Arctic and Antarctic zones (*septentrionalis frigida inhabitabilis* and *frigida australis inhabitabilis zona*), symmetric one to other with respect to the equator, and located between the Arctic and Antarctic circles and the North and South Poles, were also uninhabitable, but for the opposite reason. Two symmetric temperate and inhabitable zones (*temperata nostra* and *temperata Antipodum zona*) were placed between the Tropic of Cancer and the Arctic Circle, and between the Tropic of Capricorn and the Antarctic Circle. The *oikumene*, the known and inhabitable part of the earth, was positioned in the northern temperate zone.

[118] P. Gautier Dalché, "Saperi geografici del Mediterraneo cristiano," in P. Toubert and A. Paravicini Bagliani (eds), *Federico II e le scienze*, Palermo : Sellerio, 1994, pp. 162-184, in particular pp. 162-164.

[119] Regarding the introduction to the theory of the zones into Greek science, Strabo writes : "Poseidonius, then, says that Parmenides was the originator of the division into five zones, but that Parmenides represents the torrid zone as almost double its real breadth, in as much as it falls beyond both the tropics and extends into the two temperate zones, while Aristotle calls 'torrid' the region between the tropics, and 'temperate' the regions between the tropics and the 'arctic circles.' But Poseidonius censures both systems, and with justice, for by 'torrid,' he says, is meant only the region that is uninhabitable on account of heat : and, of the zone between the tropics, more than half is uninhabitable if we may base a conjecture upon the Ethiopians who live south of Egypt." See Strabo, *Geographia* 2.2.2, trans. H. L. Jones, Cambridge, MA : Harvard University Press, 1917, pp. 362-363. For a general discussion of the positioning of the *oikumene* on the terrestrial sphere in classical geography, see F. Prontera, "Prima di Strabone : materiali per uno studio della geografia come genere letterario," in F. Prontera (ed.), *Strabone : contributi allo studio della personalità e dell'opera*, vol. 1, Ist. di Storia Antica, Med. e Mod., Facoltà di Lettere e Filosofia, Università di Perugia, 1984, pp. 187-256 ; see also "The Foundation of Theoretical Cartography in Archaic and Classical Greece" (prepared by the editors from materials supplied by G. Aujac), in *History of Cartography*, vol. I, pp. 130-147, esp. pp. 144-147 ; and "Greek Cartography in the Early Roman World" (prepared by the editors from materials supplied by G. Aujac), pp. 148-175.

[120] O. A. W. Dilke (with additional material supplied by the editors), "The Culmination of Greek Cartography in Ptolemy," in *History of Cartography*, vol. I, pp. 177-200.

The second system, that of climes (*klima*, plural *klimata*, meaning "inclination," referring to the angle between the axis of the celestial sphere and the horizon) originated in Greek astronomy and geography. The list of *klimata* numbered seven, counted from south to north. This number was taken up by Ptolemy who in his *Geography* divided the northern temperate zone into seven climes.[121] Each *klima* differed from the others by the length of the summer solstice, the longest day of the year.[122] This system, diffused in the West beginning with the first Latin translations of cosmographical treatises from the Arabic during the tenth century, held that the inhabitable zone, divided into seven *klimata*, extended from 12° to 50° of northern latitude.

The two systems were contradictory in that they proposed two different extensions of the part of the earth thought to be inhabitable : according to the theory of zones the temperate zone extended from the Tropic of Cancer to the Arctic Circle ; while according to the theory of *klimata*, the inhabitable part was more limited with respect to both the polar circle and the Tropics with a difference of about 12° from both latitudes. The first *clime*, inhabitable according to the theory of *klimata*, was found in the Torrid Zone, which according to certain interpreters of Aristotle like Sacrobosco was considered uninhabitable. By the same token the regions above 50° north, inhabitable according to the theory of zones, were regarded as uninhabitable by the theory of *klimata*. The most widely diffused medieval cosmographical treatise, *The Sphere* of Johannes Sacrobosco, superimposed these two systems without offering a synthesis.[123] In the second chapter of *The Sphere*, Sacrobosco adopted the traditional division of the earth into five zones, with the Torrid Zone and the two frigid zones considered uninhabitable :

> That zone which lies between the tropics is said to be uninhabitable because of the heat of the sun, which ever courses between the tropics. Similarly, the zone of earth directly beneath it is said to be uninhabitable because of the fervor of the sun, which ever courses above it. But those two zones which are described by the Arctic circle and the Antarctic circle about the poles of the world are uninhabitable because of too great cold, since the sun is far removed from them. The same is to be understood of the zones of earth directly beneath them. But those two zones of which one is between the summer tropic and the Arctic circle and the other between the winter tropic and the Antarctic circle are habitable and tempered from the heat of the torrid zone between the tropics and from the cold of the extreme zones which lie about the poles. The same is to be understood of the stretches of earth directly beneath them.[124]

[121] *Ptolemy's Geography: An Annotated Translation of the Theoretical Chapters*, A. Jones and J. L. Berggren (eds), Princeton : Princeton University Press, 2000, pp. 177-199.

[122] Isidore of Seville had already given a 'non-astronomical' definition in the *Etymologiae* : "Sunt et alia septem climata caeli, quasi septem lineae ab oriente in occidentem, sub quibus et mores hominum dispares atque animalia specialiter diversa nascuntur, quae vocata sunt a locis quibusdam famosis : quorum primum est Merois, secundum Syene, tertium Catachoras, id est Africa, quartum Rhodus, quintum Hellespontus, sextum Mesopontum, septimum Borusthenes." *Isidori Hispalensis episcopi Etymologiarum sive originum*, III.4.

[123] Gautier Dalché, "Saperi geografici del Mediterraneo cristiano," pp. 162-164.

[124] "Illa igitur zona que est inter duos tropicos dicitur inhabitabilis propter calorem solis discurrentis semper inter tropicos. Similiter plaga terre illi directe supposita dicitur inhabitabilis propter fervorem solis discurrentis semper super illam. Ille vero due zone que circumscribuntur a circulo artico et circulo antartico circa polos mundi inhabitabiles sunt propter nimiam frigiditatem, quia sol ab eis maxime removetur. Similiter intelligendum est de plagis terre illis directe suppositis. Ille autem due zone quarum una est inter tropicum estivalem et circulum articum et reliqua inter tropicum hyemalem et circulum antarticum habitabiles sunt et temperate a caliditate torride zone

Later, at the end of the third chapter, the division of the inhabitable zones into *klimata* was introduced. The first climate was located in the area that was identified as uninhabitable in the zonal division of the earth:

> Between these two extreme lines suppose six lines parallel to the equator, which, with the two former, divide the whole habitable quarter into seven parts which are called the "seven climes." First Clime. The middle of the first clime is where the length of the longest day is 13 hours and the pole is elevated above the horizon 16 degrees, and it is called the "clime of Meroe." It begins where the length of the longest day is 12¾ hours and the pole is elevated above the horizon 12¾ degrees. And its breadth extends to the place where the length of the longest day is 13¼ and the pole is elevated above the horizon 20½ degrees, which distance is 440 miles. [...] Seventh Clime. The middle of the seventh clime is where the longest day is 16 hours and the elevation of the pole above the horizon is 48 2/3 degrees and it is called "the clime of Ripheon." Its breadth is from the end of the sixth clime to where the maximum day is 16¼ hours and the pole is elevated above the horizon 50½ degrees, which space on earth is 185 miles.[125]

It is important to emphasize that Fra Mauro not only was aware of the contradictions between the systems of zones and climates, but his lengthy reasoning aimed to resolve the contradictions inherent in Sacrobosco's description of the systems. Fra Mauro begins his analysis by listing the Aristotelian causes of overheating of the Torrid Zone:

> [...] the fact that below the Equator it is less hot than in the first climate is proved by an examination of the causes for the excess of heat in the air and on the earth. First of all, when the sun passes over a place in such a way that its rays are perpendicular – so, as a result, the reflected rays are added to the direct rays – then in that place the rays of the sun cause more heat because their power is greater if they are combined rather than when they are dispersed. The second cause is the duration of the day. In fact, the longer the day lasts, the more the sun is above the earth, and hence the more heat there is. The third is the length of summer in relation to the length of winter – that is, if summer lasts longer than winter. The fourth is the continuity of the warm season – that is, if the hot weather is not interrupted by cold weather.[126]

existentis inter tropicos et frigiditate zonarum extremarum que sunt circa polos. Idem intellige de plagis terre illis directe suppositis." Johannes de Sacrobosco, *Tractatus de sphaera*, II. Cf. Thorndike, *The Sphere of Sacrobosco*, p. 94.

125 "Inter istas etiam duas lineas extremas intelligantur sex linee paralelle equinoctiali, que cum duobus prioribus dividunt partem totalem quarte habitabilem in septem portiones, que dicuntur septem climata. Medium igitur primi climatis est ubi maioris diei prolixitas est 13 horarum, et elevatur polus mundi supra circulum emisperii 16 gradibus, et dicitur clima Diameroes. Initium vero eius est ubi maioris diei prolixitas est 12 horarum et dimidie et quarte unius hore, et elevatur polus supra orizontem 12 gradibus et dimidio et quarta unius, et extenditur eius latitudo usque ad locum ubi longitudo prolixioris diei est 13 horarum et quarte unius, et elevatur polus 20 gradibus et dimidio, quod spatium terre est 440 miliariorum. [...] Medium septimi climatis est ubi prolixitas diei sit 16 horarum, et, elevatio poli supra orizontem 48 graduum et duarum tertiarum et dicitur clima Diaripheos. Eius vero latitudo est ex termino sexti climatis usque ubi maxima dies est 16 horarum et quarte unius, et elevatur polus supra orizontem 50 gradibus et dimidio, quod spatium terre est 185 miliariorum." Thorndike, *The Sphere of Sacrobosco*, pp. 110-112.

126 "Ma che sotto el dicto equinotial sia men caldo cha nel p(r)imo clima se proua p(er)ch(è) | le caxo(n) ch(e) fano excesso de caldo ne l'aiere e la terra sono: p(r)ima, q(ua)n(do) el sol passa de directo sopra quel luogo, onde i razi suo sono perpe(n)dicular e p(er) co(n)se | que(n)te essendo reflexi se unisse cu(m) i primi directi e perché la uirtù unida è più forte che la uirtù dispersa, p(er)tanto in quel luogo i razi del sol sono più ef | ficaci a scaldar. La sego(n)da caxo(n) si è la longeza di zorni, perch(è) qua(n)to el zorno è più lo(n)go tanto el sol sta più sopra la terra

This passage is a citation of the *Meteorologica* of Aristotle,[127] a work of great importance in discussions of the habitability of the Earth because it affirms that for every zone of the boreal hemisphere there is a corresponding symmetrical zone in the austral hemisphere.[128] In light of this principle it follows that, once the inhabitability of the regions placed under the celestial equator is demonstrated, the entire northern hemisphere (except for the polar zone) and the regions placed immediately south of the equator, could be considered inhabitable. This is precisely the reasoning that Fra Mauro pursued in referring to the authority of the *De natura loci* of Albert the Great, the *Canon medicinae* of Avicenna, the commentary on the *De caelo et mundo* of Averroes, and Ptolemy's *Geography*.[129] The first three philosophers held that the equatorial regions were habitable, even if intense heat made life there unpleasant (*indelectabilis*).[130] This theory was also accepted by John Buridan who, like Albert the Great, referred to the theory of Avicenna in Chapter VII of his commentary on the *De caelo*.[131] Fra Mauro did not cite Avicenna and Averroes directly, but rather by way of the *De natura loci* of Albert the Great[132] or Buridan's commentary on *De caelo*, in particular Chapter VII, whose theory on the relation between the earth and water seems to have been known to Fra Mauro. Even though he simplifies the long disquisitions of Albert the Great, and there are no literal translations from Albert, the authors cited by Fra Mauro in his legend on the habitability of the Earth ("Socrates"[133]; the *De dispositione sphaerae* of Ptolemy "qui est introductorius ad Almagesti," according to

e p(er) co(n)seque(n)te più scalda. La terza | si è la lo(n)geza del te(m)po de l'instate sopra q(ue)llo de l'i(n)uerno, zoè che'l te(m)po de l'instate sia mazor cha q(ue)llo de l'inuerno. La quarta caxo(n) è la co(n)tinuatio(n) del caldo, zoè | q(ua)n(do) el te(m)po del caldo no(n) è interoto dal fredo," *Mappamondo*, lower right corner (left), lines 6-11. 733.

[127] Aristotle, *Meteorologica*, 348b 2-3; 349a 7-9; see also *Les auctoritates Aristotelis: un florilège médiéval*, J. Hamesse (ed.), Louvain: Publications Universitaires; Paris: Beatrice-Nauwelaerts, 1974, p. 172, n. 10.

[128] Aristotle, *Meteorologica*, 2.5 362b.

[129] "Et a questa raxo(n) co(n)sona l'auctorità del phylosopho i(n) diu(er)si luogi ma più exp(re)ssa nel libro de le proprietà e n(atur)a di eleme(n)ti e Alberto Mag(n)o nel suo simel libro et Auerois come(n)tator nel come(n) to del libro del cielo e del mondo et Auicena nel suo libro di canoni de medicina," *Mappamondo*, lower right corner (left), lines 22-23.

[130] Albert the Great writes: "Omnibus autem huiusmodi considerationibus habitis consentiendum videtur Ptolemaeo et Avicennae, ut dicamus torridam non omnino esse torridam, sed esse habitatam tam in litoribus maris, quod ibi est et mare Indicum vocatur, quod multos habet adamantes in fundo, quam etiam in insulis maris multis, quae ibidem a Philosophis diversis esse describuntur. Sed distinguendum est, quoniam sub tropico aestivo propter rationes superius inductas (see at this regard *De natura loci*, Tract. I, chapter 6, p. 9, lines 47-86, p. 10, lines 1-86, p. 11, lines 1-22) non potest esse continua et delectabilis habitatio, sed aliquando delectabilis et aliquando laboriosa et indelectabilis"; see Albertus Magnus, *De natura loci*, Tract. I, chapter 6, p. 11, lines 68-78.

[131] It seems, nonetheless, that the origin of the citation of Avicenna both in the *De natura loci* of Albert the Great and in the commentary on *De caelo* of Buridan could be Pars IV of the *Opus maius* of Roger Bacon. See Albertus Magnus, *De natura loci ad fidem autographi*, p. 11, note to line 23, and p. 12, note to lines 28-30. On Buridan and the *quaestio* of habitability, see E. A. Moody, "John Buridan on the Habitability of the Earth," *Speculum* 16, 4 (1941), pp. 415-425, esp. p. 419. See also Roger Bacon, *Opus maius*, Pars IV, d. IV, c. 4 (edited, with introduction and analytical table, by J. H. Bridges, Oxford: Clarendon Press, 1897-1900), in which the sources indicated are the *Canon of Medicine* by Avicenna (Book I, Fen. II, Doctrine II, chapter 8), and the first book of his *Treatise on Animals*.

[132] In particular chapters VI, *De distinctione terrae per loca habitabilia et non-habitabilia*, and VII, *Utrum habitabilis sit quarta terrae, quae est ab equinoctiali usque in polum australem*, of Tractatus I.

[133] *De natura loci*, Tract. I, chapter 7, p. 13, line 44. "Socrates" is not better specified in the *mappa mundi* nor in the *De natura loci*.

Albert the Great[134]; Avicenna; and Averroes[135]) coincide with those cited in the two chapters on the inhabitability of the earth in the *De natura loci*.

Fra Mauro does not limit himself to an argument *ex auctoritate*. Through indirect reasoning (a procedure he also uses in the legend "Of the elements. That is, how much one element exceeds another in quantity") he aims to show that if it is possible to live in the first climate, which is just above the equator, by extension it is also possible for humans to live in the Torrid Zone between the two Tropics and divided by the Equator, which was uninhabitable according to the zonal theory. Fra Mauro argues that the general astronomical conditions of the equatorial regions are not that different from those of the first *klima*: the inclination of the sun's rays, the length of summer, and, more generally, the quantity of heat that accumulates in the first *klima* are very similar to those registered at the Equator. So if these conditions do not prevent man from living in the first *klima*, it follows that the zones immediately south and north of the equator could also be inhabited:

> For these reasons, when the sun passes perpendicularly over the head of those who inhabit the regions of the Equator, it does not create as much heat as when it passes over the head of those who live in the first climate. If, therefore, with all of this it is still possible to live in the first climate, as all philosophers agree, one can also live below the Equator, because there is no impediment due to excessive heat. One can, therefore, conclude that all climates are inhabitable. In fact, we know that the northern climate is inhabited like ours, which is situated between the Tropic of Cancer and the northern parallel. And the same is true of the hot zone, as we have just seen above. And as the southern zone, which lies between the Tropic of Capricorn and the austral parallel, and the zones that lie below that parallel, correspond to the northern zones that are inhabited, it follows that the southern zones are inhabited too.[136]

Fra Mauro affirms a conception already clearly expressed by the French Cardinal Guillaume Fillastre[137] († 1428) in his *Introductio in Pomponium Melam*.[138] Grounding himself on the author-

[134] *De natura loci*, Tract. I, chapter 7, p. 13, line 52.

[135] "Averroes enim in Commento super quartum librum caeli et mundi dicit Aristotelis et suam fuisse opinionem, quod loca illa sint habitabilia, et rationem adducit id necessario probantem, ut mihi videtur; dicit enim inter calidissimum et frigidissimum esse necessario temperatum." see Albertus Magnus, *De natura loci*, Tract. I, chapter 7, p. 13, lines 1-7.

[136] "E pertanto se nel primo clima el transito del sol de directo sopra el capo no(n) remuoue la habitation p(er) exces | so de caldo, neanche el dicto transito sopra la t(er)ra supposita a l'equinotial remuove la habitatio(n) p(er) excesso de caldo e maxime essendo prima i zorni dei predicto | transito nel primo clima mazor cha i zorni proporcional sotto l'equinotial, et etia(m) sta(n)do più tempo el sol propi(n)quo al zenith ouer al capo de li habitatori del | p(r)imo clima di quello el sta quelli habita sotto lo equinotial. Onde p(er) le do'p(re)dicte raxo(n) el sol nel circulo equinotial, passando de directo sopra el capo de li ha | bitatori de quelle parte, no(n) scalda tanto qua(n)to lui fa q(ua)n(do) el passa sopra el capo de li habitatori del p(r)imo clima Se adoncha cu(m) tuto questo nel p(r)imo clima se | può habitar come concorda tuti i phylosophi, p(er) co(n)seque(n)te etia(m)soto l'equinotial se può habitar, zoè no(n) è impazada la habitatio(n) humana p(er) excesso de | caldo." *Mappamondo*, lower right corner (left), lines 27-33. For the English translation see *Mappamondo* (ed. 2006), pp. 734-735.

[137] For the bibliography on Cardinal Fillastre see H. Millet, "Guillaume Fillastre: esquisse biographique," in D. Marcotte (ed.), *Humanisme et culture géographique à l'époque du Concile de Constance: autour de Guillaume Fillastre*, Turnhout: Brepols, 2002, pp. 7-35.

[138] Guillaume Fillastre, *Introductio in Pomponii Melae Cosmographiam*, 8, 8a, 12-13; see P. Gautier Dalché, "L'œuvre géographique du Cardinal Fillastre († 1428). Représentation du monde et perception de la carte à l'aube

ity of Ptolemy's *Geography*, as well as on reports of voyages south towards the Torrid Zone, Fillastre supported the idea that the equatorial regions were inhabitable :

> de illa media zona dico quod non credo verum quod iste asserit illam propter calorem inhabitabilem primo auctoritate Tholomei in sua *Cosmographia* (capitulo ultimo *versio A*) de qua supra. [...] Verum est autem quod sunt plures terre inhabitabiles non propter estum solis sed naturam terre sterilis et bestias incompatibiles cum hominibus. Preterea actor iste probat per experienciam terram peragratam ac navigatam a mari occeano meridionali usque ad mare occeanum septentrionale, ab Ethiopia ultra Egiptum usque ad Gades Herculis.[139]

By this argument, Fillastre affirmed an important theological principle : "nec Deum mundum quem propter hominem creauit in tam magnas partes fecisse inhabitabilem neque solem quem ad generationem creauit tanti – caloris ut tantam terre partem inhabitabilem reddat."[140] God would not have created large stretches of the Earth so that they were uninhabitable and, therefore, in a certain sense, wasted.[141]

VII. – The ocean and the seas

1. Natural philosophy, navigation and trade

In contrast to the monumental thirteenth-century *mappae mundi* of Ebstorf (*c.* 1246) and Hereford (*c.* 1300), to the cosmographic representations of Andrea Bianco (1436), Giovanni Leardo (1442, 1448, 1452), and Andreas Walsperger (1448), and also to the numerous Latin and Greek manuscript copies of Ptolemy's *Geography*, Fra Mauro's *mappa mundi* presents a complex cosmographic analysis of the ocean and the seas, including a detailed analysis of tides, their navigability and an in depth description of the main maritime and commercial routes linking Asia and the Indian Ocean to the Mediterranean via the circumnavigation of Africa. In the pages that follow I will investigate how Fra Mauro combined principles of natural philosophy with mercantile considerations and seafarers' knowledge to create one of the most fully articulated reflections on the natural properties, size, shape and "habitability" of the Ocean in the mid-fifteenth century. This part of Fra Mauro's cosmographic culture is developed in several legends and images that are of significant documentary value : the *mappa mundi* provides a spatial and visual dimension to contemporary written descriptions of navigation and commerce in Asia and the Indian Ocean as narrated in merchant's books (*pratiche della mercatura*) and letters, as well as in merchants', diplomats' and missionaries' travel accounts, which constituted the cor-

des découvertes," in Marcotte, *Humanisme et culture géographique à l'époque du Concile de Constance*, pp. 293-355, in particular pp. 331-332 (originally published in *Archives d'histoire doctrinale et littéraire du Moyen Âge* 59, 1992, pp. 319-383).

[139] Guillaume Fillastre, *Introductio in Pomponii Melae Cosmographiam*, pp. 331-332.

[140] *Ibidem*.

[141] At this renard see also N. Wey-Gómez, *The Tropics of Empire. Why Columbus Sailed South to the Indies*, Cambridge, MA and London : MIT Press, 2008.

nerstone of representations of both maritime and terrestrial travels from Europe to the Orient and vice versa from the thirteenth century onwards.[142]

Fra Mauro's legends and the images describe the natural properties of the seas, analyze the possibilities and the limits of navigation, detail the principal routes and trading cities and ports frequented by European, Arab and Asian merchants (from *Zaiton* in the eastern Sea of China to Hormuz, Alexandria and Baghdad, towards the western part of the *oikumene*), and specify the merchandise and business customs prevalent in each of these regions. Thus, we have a pioneering cosmographic representation of navigation and the spice trade as perceived in Venice in the mid-fifteenth century. It is also important to state that Mauro's map openly asserted the possibility of circumnavigating Africa by joining the Atlantic and Indian Oceans in a single immense sea route, the so-called *carreira da Índia*, that would actually be travelled by Portuguese ships guided by Vasco da Gama between 1497 and 1499.

2. The seas and tides : echoes from Albert the Great's commentary on the *De causis proprietatum elementorum*

During antiquity the periodic rising and falling of the seas constituted a constant source of wonder, and the opening sentence of Fra Mauro's legend "How waters rise and subside due to the attractive properties of the moon"[143] evokes this feeling : "Amongst all the things that we observe, one that appears of no small account is the fact that in twenty-four hours the waters of the sea rise and fall twice, over the same interval of time."[144] The rise and fall can happen differently and can occur at different times, one, two, or four times a day, depending on the

[142] F. Balducci Pegolotti, *La pratica della mercatura*, A. Evans (ed.), Cambridge, MA : Mediaeval Academy of America, 1936 ; A. Stussi (ed.), *Zibaldone da Canal : manoscritto mercantile del sec. XIV*, Venice : Committee for the publications of sources relative to the history of Venice, 1967 ; *Merchant culture in fourteenth century Venice. The Zibaldone da Canal*, translated with an introduction and notes by J. E. Dotson, Binghamton, NY : Medieval & Renaissance texts and studies, 1994 ; B. Cotrugli, *Il libro dell'arte di mercatura*, U. Tucci (ed.), Venice : Arsenale, 1990. The literature on the Franciscan mission in the Orient is vast ; see *Sinica franciscana*, Vol. I : *Itinera et relationes fratrum minorum saeculi XIII et XIV*, Collegit, ad fidem codicum redegit e adnotavit p. Anastasius Van Denert O.F.M., Florence, Apud Collegium S. Bonaventurae, 1929 (Barbera, Alfani e Venturi) ; and U. Tucci, "Mercanti, viaggiatori, pellegrini nel Quattrocento," in *Storia della cultura veneta*, vol. 3.2, Vicenza : Neri Pozza, 1981, pp. 317-353. For Ruy González de Clavijo : *Embajada a Tamorlán. Estudio y edición de un manuscrito del siglo XV*, F. López Estrada (ed.), Madrid : C.S.I.C., 1943 ; Ruy González de Clavijo, *Viaggio a Samarcanda 1403-1406. Un ambasciatore spagnolo alla corte di Tamerlano*, P. Boccardi Storoni (ed.), Rome : Viella, 1999 ; L. Kehren, *La route de Samarkand au temps de Tamerlan. Relation du voyage de l'ambassade de Castille à la cour de Timour Beg par Ruy González de Clavijo, 1403-1406*, Paris : Imprimerie Nationale, 2002. For Marco Polo and Niccolò de' Conti see Chapter V, which is dedicated specifically to the relationship between the *mappa mundi* and the narration of the Venetian travellers, two fundamental sources for Fra Mauro.

[143] The title is derived from the *explicit* of the rubric : "Questa rubrica de sopra dechiara come per la uirtù atractiua de la luna le aque cresce e descresce." *Mappamondo*, upper right corner (left) ; *Mappamondo* (ed. 1956), p. 21. Full transcription in the Appendix (Doc. 5). For the English translation see *Mappamondo* (ed. 2006), p. 726.

[144] "Non picola cossa par tra quelle che nui uedemo, che le aque marine do' fiade i(n) uinti quatro hore cresca et in quel medemo spacio de tempo do' fiade descresca ouer minuisse," *Mappamondo*, upper right corner (left), line 1.

place. The tides and currents in the Euripus Strait, a narrow channel of water separating the Greek island of Evia (Euboea) in the Aegean Sea from Viotia (Boeotia) in mainland Greece, change direction between four and fourteen times a day depending on the lunar cycle, and were recognized in antiquity as an example of the complexity of the phenomenon (Pliny, *Natural History* 2.99.219). The geographic variability of tides did not, however, prevent the uniformity of some of their characteristics from being noted.[145] Fra Mauro offers an interesting testimony to the wonder and interest raised by tides in the past while, at the same time, elaborating a complex theory to explain their cause, derived principally from Book I, Tract. 2, Chapters V and VI of the *Liber de causis proprietatum elementorum* of Pseudo-Aristotle in the commentary of Albert the Great.[146]

The comparison between Fra Mauro's text and specific passages of the two cited chapters of the *De causis* demonstrates the Camaldolese friar's dependence on those passages as his source for a theory usually called the "lunar-solar theory." Briefly hinted at by Pliny in the *Natural History* (2.99-105.212-224) and Ptolemy in Book I of the *Tetrabiblos*, the theory was discussed at length by Albert the Great who dedicated to the subject a total of five chapters of the *De causis*.[147] In his legend on the tides Fra Mauro translates into the vernacular various *excerpta* from the two chapters of the *De causis* cited above and explains the periodicity of the tides by the conjoined action of the sun and moon on the sphere of water. The lunar influence alone was not enough for the Camaldolese to explain the waters' daily rise and fall. It was also the sun that caused the rising (*elevation*), by inducing the formation of vapors both on the surface and in the depth of the waters:

> Thus, on the basis of certain special knowledge, I will say that the Sun, which has the property of absorbing, consuming and dispersing humidity, draws many vapors out of the waters – that is, transforms them in part and dissolves them as vapors. The vapors are not generated only from the visible surface of the water but also from the depths. And if the waters in the depths of which these vapors are generated are subtle, then these vapors are not held back by such subtlety and emerge a little at a time, as they are generated. Hence in such emergence there is no movement in the water. But where the waters are dense and heavy – and above all, when they are deep – the vapors gener-

[145] L. Russo, *Flussi e riflussi. Indagine sull'origine di una teoria scientifica*, Milan: Feltrinelli, 2003, p. 21.

[146] This would be a work originally written in Arabic, composed around 830-875 in Basra, translated into Latin by Gerard of Cremona. The work by Aristotle was also believed to be the work of Albert the Great. It was his student Thomas Aquinas who proposed that it was not a work by Aristotle, but a later composition. See S. L. Vodraska, *Pseudo-Aristotle, De causis proprietatum et elementorum, critical edition and study*, Ph.D. dissertation, London: University of London, 1969, p. 7; P. Hossfeld, "Prolegomena," in *Alberti Magni ... Opera Omnia*, Tomus V Pars II, P. Hossfeld (ed.), Monasterii Westfalorum in aedibus Aschendorff, 1980, p. ix. Albertus Magnus, *Liber de causis proprietatum elementorum*, Book I, Tract. 2, chapter 5, "Et est digressio ostendens veram causam accessionis maris in communi et excludens errores, qui sunt contra hanc," pp. 68b-70b; chapter 6, "Et est digressio declarans, quot et quae confortant accessionem marium," pp. 70b-72b; the subject of *accessio maris* is dealt with by Albert Magnus also in *Meteora* I.3, Tr. 1, chapters 20-21.

[147] In addition to the works already cited, see Albertus Magnus, *Liber de causis proprietatum elementorum*, Lib. I, Tract. 2, chapter 4, "Et est digressio declarans octo, quae praenoscenda sunt ad sciendum accessum et recessum maris" (pp. 67-68); Book I, Tract. 2, chapter 7, "Et est digressio destruens duas sectas errantes circa causam accessus maris" (pp. 72-74) and also Book I, Tract. 2, chapter 8, "Et est digressio declarans diversitatem marium in accedendo et recedendo" (pp. 74-76).

ated within those depths in the way we have said cannot emerge without some clear help, because they are resisted by the depth, heaviness, thickness and quantity of the waters.[148]

While the vapors of the surface waters evaporated into the air, those of the depths, because they originated in "thick and heavy waters" (*aque grosse et ponderose*), were held back, leading to a swelling of the waters. This passage adapts and summarizes a passage of Chapter V of the *Tractatus I* of Albert the Great's *De causis proprietatum elementorum* (Tract. I, Chap. V, p. 68, 95 and p. 69, 9). At this point Fra Mauro, relying on Ptolemy's *Tetrabiblos*, introduced the moon into his argument, saying that it naturally attracts humidity, pulling heavy and "corrupt" vapors toward it, withholding the "thick waters." This explanation was based on the Aristotelian principle that the four elements tend toward their own natural place. Because the moon is humid it attracts watery bodies which by their own nature are subject to its dominion, thus causing them to swell. Fra Mauro, citing Ptolemy *Tetrabiblos* I.1, writes : "Given that the moon has the property of attracting all humidity, it consequently attracts to itself also the vapors that are generated from that humidity, above all if they are impure (as is stated in the first book of the *Tetrabiblos*)."[149]

Rising vapors provoke the elevation of the waters in accordance with the cyclical movement of the moon around the earth. Therefore, the rising of the waters (*bulimento*, literally, "the boiling of waters") results from the conjoined daily action of the sun with the monthly periodicity of the moon. For Fra Mauro, who follows Albert the Great's *De causis*, the inclination of the lunar rays to the surface of the waters is essential : the more directly the lunar rays fall, the stronger the attraction of the waters would be. In this same vein the Camaldolese writes that when the moon is in the other hemisphere, "and the attractive force of the moon's rays begins to decrease" (*declinando la virtù de li razi de la luna*), its attraction also decreases and, consequently, so does the swelling of the waters in the western hemisphere. Fra Mauro continues specifying the multiple and more specific causes (*molte caxon*) that affect the rise and fall of the waters : first, "the waters rise most when the moon is in Cancer" (*le aque più cresce qua(n) do la luna è in cancro*), that is to say, when it is closer to the zenith of the northern part of the celestial hemisphere, or, in the language of Fra Mauro, "our part." The second cause is "the closeness of the moon to the waters – that is, when the moon is in the lower part of its epicycle" (*La segonda caxon è la propinquità de la luna a le aque*). This proximity implies that the moon exerts more attraction over the waters, which increases the amplitude of the tides. A third cause

[148] "Donde per alguna particular informatio(n) nui diremo che el sol, auendo efficacia de un caldo digestiuo et co(n)sumatiuo de humidità et de quelle resolutiuo, molti uapori de le aque lieua, quelle | in parte conuertendo ouer resoluendo in uapori. I qual uapori no(n) solamente se genera ne la exterior et apparente sup(er)ficie de le aque, ma etia(m) nel profondo de quelle. Et se le aque, nel fondo de le quale sono tai uapori geneti, ano in sì subtilità, quelli uapori, | no li facendo resistentia la subtilità de le aque, esceno a pocho a pocho, secondo chomo quelli uapori sono geniti. Et pertanto in questo exito no(n) se fa apparente movimento i(n) quelle aque. Ma doue le aque sono grosse et ponderose e maxime essendo profonde, li uapori per el dicto | modo geniti nel fondo de quelle aque no(n) possono insire senza notabile adiutorio, facendoli ressistentia a la dicta eleuatio(n) la grosseza, ponderosità et quantità de le aque." *Mappamondo*, upper right corner (left), lines 2-5. For the English translation see *Mappamondo* (ed. 2006), p. 724.

[149] "Et perchè la luna ha in sì uirtù attractiua de ogni humidità e per consequente tira ad sì etia(m) li ua|pori de queste humidità geniti e maxime se quelli sono corrupti, como se dichiara nel primo libro del quadripartito." *Mappamondo*, upper right corner (left), lines 5-6. For the English translation see *Mappamondo* (ed. 2006), p. 724.

is linked to the different rates at which the moon increases and decreases : there is a larger flux or reflux of the waters in the phase of increase, which is longer than that of decrease.[150] Finally, the fourth cause is "the wind, which can blow the waters towards the coast. The rising of the waters can also be helped by the sun. This happens when the moon is in conjunction with the sun – that is, when the moon is either full or in its fifteenth day. Due to the power that the moon receives from the sun in these two situations, it has more force to attract the vapors. In this period, the tides are always at their highest ; sometimes more so than others because of the presence of one or more other cause that can help or hinder the tide."[151] Translating literally the end of the fifth chapter of the *De causis*,[152] Fra Mauro concludes his legend by explaining why river waters are not subject to tides. In contrast to the waters of seas that "are very thick – because many terrestrial substances are dissolved in them – and also deep" (*permixte cu(m) molte parte terrestre, et etia(m)dio profonde*), the waters of rivers are "thin" (*sottili*), in the double sense of not being deep and in remaining unmixed with soil. This quality makes them unable to contain the vapors, and they allow the vapors to leak out in such a way that "the rise of these latter does not cause ebullition" (*tal aque no(n) cresceno nì discresceno come le aque del mar*).

A specific interest linked Venice and its culture to the explanation of the tides. Venice is one of the main places where the delicate balance between land and water and the phenomenon of the "boiling of the waters," in the language of Fra Mauro, acquired a special significance : the lagunar city is one place in the Mediterranean where the tides are most evident, having there a variation of approximately half a meter. Since the fourteenth century the lunar-solar theory of the tides was particularly prominent in the Venetian Republic and was taught at the University of Padua. Among the Venetian works on this subject, the *De fluxu et refluxu maris* (On the Flux and Reflux of the Sea), composed by Jacopo Dondi before 1360, is the oldest documented evi-

[150] "Terza qua(n)do el tempo che la luna se muoue da l'oriçonte oriental a la linea meridiana è mazor | de te(m)po che quela luna se muoue da la predicta linea meridiana a l'orizonte occidentali le | aque cresceno più perché auanti che l'aqua sia partida da le riue descresce(n)do, la qual cresce(n)do era mossa da le riue, un'altra fiada crescendo torna a quelle riue." *Mappamondo*, upper right corner (left), lines 17-18. English translation : "Thirdly : when the period of time in which the moon travels from the eastern horizon to the meridian line is greater than that it takes to pass from the meridian line to the western horizon, the waters rise more because before they can subside entirely from the coast, they already begin to rise again at that coast." For the English translation see *Mappamondo* (ed. 2006), p. 726.

[151] "Quarta se sono uenti i qual muouano le aque uerso le riue. Item el crescer de le aque | aiuta el lume del sol, zoè qua(n)do la luna è cu(m) el sol conzonta, ouer quando lei è piena de lume zoè nel qui(n)todecimo, alora per el uigor el qual la luna receue dal sol in queste due situation, lei ha più forza de trar i sopraditi uapori, e però | che questa ultima caxon è più potente de le altre, semp(re) a questi do' tempi le aque sono colme e una fiada più che l'altra, segondo como una ouer più de le altre caxon dicte aiuta oue le contrarie minuisse. E nel tempo che la luna | è meza piena del lume del sol, lei no(n) ha pien uigor de mouer i uapor del profondo e per tanto i(n) questo te(m)po le aque no(n) cresceno molto, nè ancor molto discresceno, saluo se una ouer più de le predicte caxon no(n) aiuta el crescer | ouer el discrescer." *Mappamondo*, upper right corner (left), lines 18-22. For the English translation see *Mappamondo* (ed. 2006), p. 726.

[152] *De causis proprietatum elementorum*, Lib. I, Tract. II, chapter 5, p. 70, lines 40-44 : "Nos autem his respondentes dicimus, quod aquae dulces subtiles sunt et vaporem venti non retinent in seipsis et ideo nec accedunt nec recedunt, praecipue cum fluunt, sicut faciunt flumina, quia tunc motus facit eas subtiliores."

dence of the teaching of this theory.[153] Jacopo Dondi (*c.* 1293-1359) entered the medical profession very young, practicing in Chioggia and then Padua. At the University of Padua he became a professor of medicine and astrology. His most famous work is the *Aggregator medicamentorum, seu de medicinis simplicibus*, a treatise on drugs composed in 1355, published for the first time in Strasbourg in 1470 and reprinted several times in the fifteenth and sixteenth centuries.[154] In the *De fluxu et refluxu maris* Dondi explains the tides by the conjoined action of the moon and sun, whose effects add one to the other in conjugation and opposition, and subtract one from the other during the *quadratura* when the moon appears half-illuminated. The lunar-solar theory (definitively accepted in 1726 when Isaac Newton published his *Philosophiae naturalis principia mathematica*) remained peculiar to Venetian physics and, in particular, the Paduan teaching of Jacopo Dondi and his successors to the Chair of Astronomy, Federico Crisogono (first half of the sixteenth century), Federico Delfino (1477-1547) and Marc'Antonio de Dominis (1560-1624) up to the end of the seventeenth century.[155]

3. The cosmography of the seas : beyond natural philosophy to a geo-political vision

In his *mappa mundi* Fra Mauro interprets and combines fragments from different cultures knowledge, creating a unique synthesis in which the cosmography of the seas goes beyond natural philosophy to become a geo-political and economic vision of the future world. At least three distinct aspects of this vision, all of which demonstrate Fra Mauro's mediation of diverse areas of knowledge, can be recognized in the *mappa mundi* : the foundation of a world capitalist economy ; the creation of a network of relations over great distances ; and, finally, the development of geographic exploration and a mental opening to spaces and seas that had previously been considered inaccessible to man.[156] Venice enjoyed a particular centrality with respect to these three processes : trading voyages on galleys achieved their maximum development in the first half of the fifteenth century, expanding to include distant geographical areas.[157]

[153] Milan, Biblioteca Ambrosiana, MS. N. 334 sup., Jacopo Dondi, *De fluxu et refluxu maris*, ff. 1-17, a copy of the sixteenth century. Jacopo Dondi, *Il trattato della marea*, introduction, Latin text, translation and appendix by P. Revelli, Florence : Ricci, 1911. Russo, *Flussi e riflussi*, pp. 52-53.

[154] Jacopo Dondi is often confused with his son Giovanni who succeeded him in his Chair of Astronomy and Medicine at the University of Padua. Giovanni Dondi is the famous designer and maker of astronomical clocks who brought international fame to the Dondi family later referred to as "Dondi dell'Orologio" (*Dondi of the clock*).

[155] In addition to Jacopo Dondi's *Trattato della marea*, these are the works that constitute the "Venetian history" of the theory of tides : Federico Delfino, *De fluxu et refluxu aquae maris*, Venice : Accademia Veneta, 1528 ; Federico Crisogono, *Tractatus de occulta causa fluxux et refluxus maris* in *Federici Chrisogoni nobilis Jadertini Artium et Medicine doctoris Subtilissimi et Astrologi excellentissimi de modo Collegiandi Prognosticandi et curandi Febris Necnon de humana Felicitate ac denique de Fluxu et Refluxu Maris* …, Venetiis : impressum a Joan. Anto. De Sabbio et fratribus, 1528 ; and Marc'Antonio de Dominis, *Euripus, seu de fluxu et refluxu maris sententia*, Rome : Andreas Phaeus, 1624.

[156] For a recent study of these three processes and a discussion of the vast bibliography on the subject, see D. Curto Ramada and A. Molho, "Les réseaux marchands à l'époque moderne," *Annales* 58, 3 (2003), pp. 569-579.

[157] F. Braudel, *Civilisation matérielle, économie et capitalisme XVe-XVIIIe siècle*, 3 vols., vol. 3. *Le temps du monde*, Paris : Armand Colin, 1979, chapter 2, "En Europe, les économies anciennes à domination urbaine : avant et après Venise," pp. 99-200 ; A. Tenenti and U. Tucci, "Nota preliminare," in *Storia di Venezia*, vol. IV, *Il Rinascimento. Politica e cultura* cit., p. 4. See also I. Wallerstein, *The Modern World-System I : Capitalist Agriculture and the*

Located at the center of the Mediterranean basin, Venice was an important node in world commercial and communication networks; a node not only for merchandise, but also for knowledge and information that arrived from the East (Romania, Russia, Central and Eastern Asia), Northern Europe and the West, and which was combined and re-diffused throughout Europe and the Orient.[158] The Venetian postal network is exemplary in this respect, reliably connecting the Serenissima to Rome, Milan, Lyon and Northern Europe (the so-called 'Post of Flanders' and the 'Post of Vienna'), while the Venetian post of Constantinople played a fundamental role in the communication of all of Europe with the East.[159]

As demonstrated in the preceding paragraphs, the Ocean is considered from various points of view in the *mappa mundi*. As an element of the sublunar world, within the context of a critical revision of the Aristotelian theory of the elements, the Ocean is considered quantitatively equal to the earth, or at least its "circumference" in macro-cosmological terms, and marks the physical and metaphysical boundary of human space.[160] Another cosmographic characterization of the seas that is fundamental in the *mappa mundi* is Fra Mauro's insistence on "openness" of the Sea of India through the circumnavigation of Africa, so that it communicates with the western sea, the Sea of Darkness:

> Some authors write that the Sea of India is enclosed like a pond and does not communicate with the ocean. However, Solinus claims that it is itself part of the ocean and that it is navigable in the southern and south-western parts. And I myself say that some ships have sailed it along that route. This is confirmed by Pliny when he says that in his day two ships loaded with spices coming from the Sea of Arabia sailed around these regions to Spain and unloaded their cargo at Gibraltar (he gives the reason for this choice of route, but I omit it here). Fazio [degli Uberti] says the same; and those who have taken this route, men of great prudence, agree with these writers.[161]

Origins of the European World-Economy in the Sixteenth Century, New York: Academic Press, 1974 and J. Richard, "Les navigations des occidentaux sur l'océan Indien et la mer Caspienne (XIIe-XVe siècles)," in M. Mollat (ed.), *Sociétés et compagnies de commerce dans l'Océan Indien*, Paris: S.E.V.P.E.N, 1970, pp. 353-363; reprinted in *Orient et Occident au Moyen Âge. Contacts et relations*, Aldershot: Ashgate, 1976, n. XXI.

158 P. Sardella, *Nouvelles et spéculations à Venise au début du XVIe siècle*, Paris: Armand Colin, 1948, pp. 10-14; F. Braudel, *La Méditerranée et le monde méditerranéen à l'époque de Philippe II*, Paris: Armand Colin, 1949; F. Melis, "La diffusione dell'informazione economica nel Mediterraneo," in E. Le Roy Ladurie (ed.), *Histoire économique du monde Méditerranéen 1450-1650, Mélanges en l'honneur de Fernand Braudel*, 2 vols., Toulouse: Privat, 1973, vol. 1, pp. 389-424; P. Burke, "Early Modern Venice as a Center of Information and Communication," in J. Martin and D. Romano (eds), *Venice Reconsidered*, Baltimore and London: Johns Hopkins University Press, 2000, pp. 389-418.

159 A. Cattani, *Da Venezia in viaggio con la Posta. Pagine e documenti di storia veneta*, Padua: Editrice Elzeviro, 2002, in particular the chapter "Le comunicazioni postali con l'Oriente," pp. 97-134, and the Documentary Appendix.

160 This topic is discussed in the legend "De hi elementi, cioè quanto uno elemento | excieda l'altro in quantità." (Of the Elements. That is, how much one element exceeds another in quantity) *Mappamondo*, lower right corner (right); *Mappamondo* (ed. 1956); transcribed as Doc. 3 in the Appendix.

161 "Alguni autori sc(r)iue del mar d'i(n)dia che'l sia serado come un stagnon | e che'l mar occea(n) no(n) li entri, ma Solin uol che'l sia occea(n) e ch(e) q(ue)la p(ar)te austral | e del garbi(n) sia nauigabile, et io affermo che algune naue açira e uol|ta quel cami(n). E questo a(n)chor (con)ferma Plinio qua(n)do el dice che al te(m)po | suo do' naue se mosse del mar de arabia, e dice la cason la qual q(ui) lasso, | ma charge de spetie ziroe q(ue)le p(ar)te perfin in spagna e a çibelter descar|gò. Ancora Facio afferma q(ue)sto et simelme(n)te hi exp(er)imentadori de | quel cami(n), ho(min)i de gra(n) prudentia i qual co(n)corda co(n) queli autori," *Mappamondo* (ed. 1956) Pl. IX, 26. The English translation derives from *Mappamondo* (ed. 2006) *54.

As we will see in Chapter IV, which is dedicated to the study of the Venetian reception of Ptolemy's *Geography*, in this legend Fra Mauro argues against Ptolemy who imagined that the Indian Ocean was surrounded by land, with Africa and Asia uniting in the South.[162] The Camaldolese first confutes Ptolemy by citing contrary passages from Pliny's *Natural History*, Gaius Julius Solinus' *Collectanea rerum memorabilium* (also known as *Polyhistor*), and Pomponius Mela's *De chorographia*.[163] He then refers to Book 1.9 of the *Dittamondo* of Fazio degli Uberti (1305-*post* 1367). The *Dittamondo* (*Dieta mundi*, that is, "Measure of the world"), written in Florence by Farinata degli Uberti's great nephew, is a geographical encyclopedia in Tuscan rhyming verse. In this work Fazio degli Uberti describes the *orbis terrarum* under the guidance of the Roman geographer Caius Iulius Solinus, following the model of the *Commedia* in which Dante was guided by Virgil. In Fazio's poetry a "long and thin" Africa (*lunga e stretta*) takes shape, encircled by the same ocean that laps the coasts of India "from East to West," the poet argues.[164] This is the image of Africa that Fra Mauro depicts in his *mappa mundi*. If, therefore, the Indian Ocean is not a "huge pond" (*gran stagnon*) as Ptolemy thought, and the Sea of Darkness is navigable, then Africa becomes potentially circumnavigable; the only remaining task is to determine, or try to imagine, the length of the circumnavigation.

4. Sailing to and from the Indian Ocean

The *mappa mundi* describes vast marine spaces. The seas represented in the *mappa mundi* are part of the *oikumene*, in the sense that they are navigable and their thousands of islands are either inhabited or inhabitable. The only limit of navigation is the "gripping waters" (*aque tegnenti*) and the obscurity that marks the physical and metaphysical border of the *oikumene*, highlighted in the *mappa mundi* by a red circle, close to the circular wooden frame. Within this cosmographic context Fra Mauro sets out to demonstrate the possibility of navigating the Indian Ocean and circumnavigating Africa all the way to Gibraltar and the Mediterranean basin. Two legends are extremely important in this regard. The first narrates the voyage of a ship or "junk from India" (*naue ouer çoncho de india*) which was thrust by a storm into the Sea of Darkness,

[162] *Geography* 7.5. See *Ptolemy's Geography: An Annotated Translation of the Theoretical Chapters*, A. Jones and J. L. Berggren (eds), Princeton: Princeton University Press, 2000, pp. 108-110.

[163] Caius Iulius Solinus, *C. Iulii Solini collectanea rerum memorabilium iterum*, T. Mommsen (ed.), Berlin: Weidmann, 1895, Par. 55, 4-5, pp. 227-228 (indeed, paragraphs 55 and 56 work out a "circular" description of Africa and of the nearby regions and islands; *Ibid.*, pp. 226-231); Gaius Plinius Secundus, *Storia naturale. Cosmologia e geografia: Libri 1-6*, introductory essay by G. B. Conte; bibliographic note by A. Barchiesi, C. Frugoni, and G. Ranucci; translation and notes by A. Barchiesi [et al.], Turin: Einaudi, 1982, pp. 560-561 and 760-763.

[164] "Etiopia di sopra, in oriente, | con le selve d'Egitto s'accompagna | e, di verso aquilone, il Nilo sente. | Dal mezzogiorno l'Ocean la bagna | e 'n vèr zeffiro tanto si distende, | che porge ad Atalante le calcagna: Book 1.9.49-52: Poi, dove il sole al vespro par che caggia, | è Tingitana e questa con la coda | perde la terra e l'Oceano assaggia; Book 1.9.56-72 | ... or qui puoi, dunque, | l'Africa imaginar ch'è lunga e schietta. | E pensa l'Etiopia, con qualunque | provincia nomo, ch'io la truovo sempre | dal mezzogiorno: e questo non falla unque. | Poi dietro a l'Etiopia par che stempre | tanto il calore la giacente rena, | che natura vi perde le sue tempre. | Qui sono i gran deserti e la Carena | e, dietro a tutto, l'Oceano è poi, | che da levante a ponente incatena. | Di vèr settentrion, dove siam noi, | d'Africa il nostro mar le piaggia immolla | con quanto tien di Libia i liti suoi. | Or, perché veggi in fino a la merolla, | le Sirti, ch'io nomai, son acqua e terra | che sempre tira e ciò che prende ingolla." See Fazio degli Uberti, *Il Dittamondo e le Rime*, G. Corsi (ed.), Bari: Laterza, 1952, 2 vols., pp. 28-29.

traveling approximately 2,000 miles westward in 40 days in the area that today would be identified as the South Atlantic:

> Around 1420 a ship, or junk, from India crossed the Sea of India towards the Island of Men and the Island of Women, off Cape Diab, between the Green Islands and the shadows. It sailed for 40 days in a south-westerly direction without ever finding anything other than wind and water. According to these people themselves, the ship went some 2,000 miles ahead until – once favourable conditions came to an end – it turned round and sailed back to Cape Diab in 70 days […].[165]

This legend may refer to the navigations of the Chinese junks of the Muslim admiral Zheng He during the first quarter of the fifteenth century, as already noted by von Humboldt[166] and subsequently Braudel.[167] Between 1405 and 1433 the Ming Dynasty sponsored a series of seven maritime expeditions to the Indian Ocean which aimed to expand the Chinese sphere of influence and ensure its hegemony through tangible manifestations of Chinese power and wealth. The Chinese maritime expeditions led by Zheng He (1371-1433), whose father was a Muslim who had made a pilgrimage to Mecca, established Chinese maritime power from Japan to the east coast of Africa, each time employing an armada of approximately sixty-three ships and more than twenty thousand men. Each expedition lasted two years; the early voyages reached Java, Malacca, Ceylon and Calicut, and the later voyages reached Hormuz and African city-states such as Malindi. On the whole, the expeditions not only contributed to the establishment of Chinese hegemony – relying in general on ceremony and, only when necessary, on force – but also effectively stimulated commerce. Another result of these voyages was the collection of information about other territories. Ma Huan, who participated in some of the voyages, published in 1451 an account of the expeditions entitled *Ying-yai sheng-lan.* Around 1434-1436 this work was circulating in manuscript copies, and contributed to the dissemination of the complementary or identical travel writings by King Chen and Fei Xin. The *Ming Shi* of the latter author draws more attention to political matters in contrast to Ma Huan's emphasis on customs and habits. After the death of Fei Xin in 1460 his book seems to have had little impact, but it was rediscovered in the Imperial Library in 1773. Two dedicatory inscriptions by Zheng He reveal the extent to which the voyages to the distant "barbarian" lands were made under the protection of the gods, as well as the importance attached to gifts and embassies.[168] Fra Mauro's legend may be the first European echo of Chinese expansion in the Indian Ocean.

[165] "Circa hi ani del Signor 1420 una naue ouer çoncho de india discorse per una | trauersa per el mar de india a la uia de le isole de hi homeni e de le done de fuo|ra dal cauo de diab e tra le isole uerde e le oscuritade a la uia de ponente e de | garbin per 40 çornade, no(n) trouando mai altro che aiere e aqua, e p(er) suo arbitrio | iscorse 2000 mia e declinata la fortuna i fece suo | retorno in çorni 70 fina al sopradito cauo de diab. […]," *Mappamondo*, (ed. 1956) Pl. IV, 6. The English translation derives from *Mappamondo* (ed. 2006) *19.

[166] A. von Humboldt, *Examen critique de l'histoire de la géographie du nouveau continent et des progrès de l'astronomie nautique aux quinzième et seizième siècles*, Paris: Librairie de Gide, 5 vols., 1836-1839, vol. 1 (1836), p. 337.

[167] F. Braudel, *Civilisation matérielle, économie et capitalisme XVe-XVIIIe siècle*, 3 vols., Paris: Armand Colin, 1979, vol. 1, *Les structures du quotidien*, pp. 456-458. For a complete view of fifteenth-century Chinese navigations, see P. Pelliot, "Les grands voyages maritimes chinois au début du xve siècle," *T'oung Pao* 30 (1933), pp. 237-452.

[168] J. V. G. Mills, "Introduction," in Ma Huan, *Ying-yai sheng-lan 'The Overall Survey of the Ocean's Shores' (1433)*, Cambridge: Hakluyt Society, 1970, pp. 1-66; J. J. L. Duyvendak, "The True Dates of the Chinese Maritime Expeditions in the Early Fifteenth Century," *T'oung Pao – Archives concernant l'histoire, les langues, la géographie,*

Elsewhere the cartographer sets out to determine the length of the African circumnavigation. In one of the most interesting legends on the *mappa mundi* Fra Mauro links an account of Portuguese navigations along the coast of Africa with a story that the he claims to have heard in person from a person "worthy of trust" who was caught in a storm that blew his ship southwest down the eastern coast of Africa. In doing so Fra Mauro implicitly defines the dimensions of a single immense route that could unite the Mediterranean and the western coasts of Europe to the Indian Ocean:

> Many opinions and many texts claim that in the southern regions the water does not surround this whole inhabited and temperate area. But I have heard many opinions to the contrary, above all from those who were sent by his Majesty, King of Portugal, in caravels so that they might explore and see with their own eyes. These men say that they sailed around the coast from the south-west some 2,000 miles beyond the Strait of Gibraltar. And following that route, they then decided to sail south-southeast until they reach the meridian of Tunis, almost as far as that of Alexandria; and at each place they found good shores, with deep water and good navigation conditions without any hindrances. These men have drawn *new marine charts* and have given names to rivers, gulfs, capes and ports, of which I have had a copy. So, if one wants to disagree with these men, who have seen with their own eyes, then there is all the more reason to dissent from and not believe those who have left writings on things they did not see with their own eyes but only believed to be the case. What is more, I have spoken with a person worthy of trust, who says that he sailed in an Indian ship caught in the fury of a tempest for 40 days out in the Sea of India, beyond the Cape of Soffala and the Green Islands towards west-southwest; and according to the astrologers who act as their guides, they had advanced almost 2,000 miles. Thus one can believe and confirm what is said by both these and those, and that they had therefore sailed 4,000 miles [...].[169]

The reasoning of Fra Mauro is clear: adding together the distances covered by the Portuguese ships, that is, 2,000 miles along the coasts of western African (from north to south-southeast, passing the meridian of Tunis and almost reaching that of Alexandria), with that travelled by an anonymous source on a "ship of India" that was carried by a storm toward the

l'ethnographie et les arts de l'Asie orientale 34 (1939), pp. 341-412; J. Gernet, *A History of Chinese Civilization*, J. R. Foster trans., Cambridge: Cambridge University Press, 1982 (trans. of *Le Monde Chinois*, Paris: Armand Colin, 1972), pp. 398-405; E. L. Farmer, *Early Ming Government: The Evolution of Dual Capitals*, Cambridge, MA: Harvard University Press, 1976, pp. 104-114.

[169] "Molte opinio(n) e leture se troua che i(n) le parte meridional l'aq(u)a | no(n) circunda questo n(ost)ro habitabile e temperado çona, ma | aldando molte testimonia(n)çe i(n) contrario e maxime q(ue)li iqual | la maiestà del Re de portogallo à mandato cu(m) le suo carauele | a çerchar e ueder ad ochio, i qual dice hauer circuito le spiaçe de garbin | più de 2000 mia oltra el streto de çibelter i(n) tanto che a uoler seguir quel camin hano (con)uenuto dar la p(ro)da quarta d'ostro i(n)uer sirocho | e p(er) suo çudisio hano passato l'i(n)dromo de tunisto e q(u)asi son ço(n)ti a q(ue)l d'alexa(n)dria, p(er) tuto troua(n)do bone spiaçe cu(m) puoco fondo e nauegar assai bo(n) e se(m)pre | se(n)ça fortuna. E i diti hano fato nuoue carte de quel nauegar e hano posto | nomi nuoui a fiumere, colfi, caui, porti, di q(u)al ne ho habuto copia. Unde se'l se uorà | (con)tradir a q(ue)sti i qual hano uisto ad ochio, maçorme(n)te se porà no(n) asse(n)tir nè creder a q(ue)li | che hano lassato i(n) scriptis q(ue)lo hi no(n) uete mai ad ochio, ma cusì hano opinado esser. | Anchora io ho p(ar)lato *cu(m) p(er)sona dig(n)a de fede*, che afferma hauer scorso cu(m) una naue de i(n)dia | p(er) rabia de fortuna de trauersà p(er) zorni 40 fuora del mar d'india oltra el cauo de soffala | e de le i(n)sule uerde e qui pur al garbi(n) e al pone(n)te e *p(er) lo arbitrar de i suo astrologi* i q(u)al son | lor guida i scorse circa 2000 mia. *Unde certamente el se può affermar e cre|der cussì a questi come a queli i qual uien hauer scorso mia 4000...*" (emphasis mine). See *Mappamondo* (ed. 1956) Pl. XI, 2. *Mappamondo* (ed. 2006) *149.

west beyond Soffala in the Sea of Darkness (the southern point of Africa in the *mappa mundi*), that is, another 2,000 miles, he deduces that the length of the African circumnavigation would have to be at least 4,000 miles. To support this opinion he also refers, at the end of the legend to Eudoxus of Cyzicus' circumnavigation of Africa from east to west, departing from Arabia and arriving at Gibraltar, which is reported in Book III of the *De chorographia* of the Latin geographer Pomponius Mela:

> In the third book of his *Cosmography*, Pomponius Mela says that a certain *Eudoxo*, fleeing from Lathmin, king of Alexandria, sailed out of the Arabian Gulf and navigated though those southern regions until he arrived at Gades, which is the Strait of Gibraltar. One can therefore claim without any doubt that this southern and south-western part is navigable, and that the Sea of India is an ocean and not an inland sea. This is what is said by all those who sail this sea and live in those islands.[170]

What is narrated in written form on Fra Mauro's map regarding the commercial routes that crossed the Indian Ocean goes well beyond geographical limits of mid-fifteenth-century European expansion, and is structured more as a project of future expansion, a reflection on the possibility of navigation in seas and at latitudes not yet explored, rather than a descriptive report of navigations already completed. Therefore, in contrast with several histories of the European expansion and of cartography in general, in which Fra Mauro's effort has either been celebrated or denigrated for transmitting or failing to record the discovery of this or that promontory,[171] I will not emphasize here the specific geographical data reported by the *mappa mundi*. Instead I will focus on something more interesting, namely, the evidence the *mappa mundi* supplies of the existence of detailed knowledge in mid-fifteenth-century Venice of Portuguese voyages along the African coasts, as well as of Arab or "Indian" navigations in the

[170] "Dice | ancora Po(m)ponio Mela nel terço libro de la sua Cosmographia | che uno hauea nome Eudoxo, el qual sca(m)pando Lathmin, | Re de alexandria, usì del colfo arabico e nauegò quela | parte austral e uene fin a gades ch'è al streto de çi|belterra. | Adoncha sença alguna dubitatio(n) se può affermar | che questa p(ar)te austral e de garbin sia nauigabile e ch(e) | quel mar indiano sia occeano e no(n) stagnon, e cusì | affermano tuti queli che nauegano quel mar | e che habitano quele i(n)sule." *Mappamondo* (ed. 1956) Pl. XI, 2; *Mappamondo* (ed. 2006) *149. Fra Mauro refers to the following passage of the *De chorographia*: "verum et [si] Hanno Carthaginiensis exploratum missus a suis, cum per oceani ostium exisset magnam partem eius circumvectus, non se mari sed commeatu defecisse memoratu rettulerat, et Eudoxus quidam avorum nostrorum temporibus cum Lathyrum regem Alexandriae profugeret, Arabico sinu egressus per hoc pelagus, ut Nepos adfirmat, Gades usque pervectus est: ideo eius orae notae sunt aliqua" (3.90). Cf. Pomponii Melae, *De chorographia. Libri tres*, P. Parroni (ed.), Rome: Edizioni di Storia e Letteratura, 1984, p. 170. See also Pomponius Mela, *Chorographie*: Latin text, translation and notes by A. Silberman, Paris: Les Belles Lettres, 1988, 3.9.89-91, pp. 90-92, and L. Beltramo, *Un antico volgarizzamento veneziano della Chorographia di Pomponio Mela*, Alessandria: Edizioni dell'Orso, 2002 (printed in 2003), pp. 70-71.

[171] The literature on these subjects is vast. See, for example, W. G. L. Randles, *L'image du Sud-Est Africain dans la littérature européenne du XVIe siècle*; pref. A. da Silva Rego, Lisbon: Centro de Estudos Históricos Ultramarinos, 1959; Id., *La signification cosmographique du passage du cap Bojador*, Lisbon: Centro de Estudos Históricos Ultramarinos, 1961; Id., *La représentation de l'Atlantique dans la conscience européenne au Moyen Age et à la Renaissance*, Funchal: Secretaria Regional do Turismo, Cultura e Emigração, Centro de Estudos de História do Atlântico, 1989. For a recent study on the European expansion of Africa, which at least in part goes beyond the "geographic" determinism evident in studies like those by Randles, see F. Relaño, *The Shaping of Africa. Cosmographic Discourse and Cartographic Science in Late Medieval and Early Modern Europe*, Aldershot: Burlington-Ashgate, 2002.

Indian Ocean. This knowledge was probably acquired from Venetian merchants who regularly anchored at ports in Lagos and Lisbon with the galleys of Flanders (*muda di Fiandra*), and in port-cities in Persia and India as well.[172]

5. Sailing in the Indian Ocean at the dawn of the European Expansion

The *mappa mundi* of Fra Mauro is one of the first European documents that explicitly testifies to the growing importance of navigation beyond the Mediterranean and Western commercial routes. Relating cosmography to expansion and trade projects in the Indian Ocean, Fra Mauro minutely describes the principle navigation routes and commercial circuits of the Indian Ocean (*Mare indicum*) in two legends near Java Minor and the Island of Hormuz. They are worth studying:

> [...] on this said Java grow ginger and other noble spices in great quantity, and at the time of harvest, all that grows on this and the other islands is taken to Java Major and there is divided into three parts: one part [is sent] to Caiton and Cathay, another to Hormus, Cide and Mecca, by the Sea of India, and the third is sent northwards across the Sea of Cathay. And according to the testimony of those who sail this sea, from this island one sees the Southern Cross a yard above the horizon.[173]

And:

> The province of Mogolistan, shown above to the right, lies opposite the island called Hormus... Here come some of the ships from India with their merchandise of pearls, pepper, ginger and other spices in great quantity. These then travel by way of Balsera and Bagadat – that is, the Babilonia of the Chaldeans – to the river Tigris and Euphrates and then to Mesopotamia, Armenia, Cappadocia and even the Sea of Pontus. The main city of the island has the same name – that is, Hormus – and in ancient times it was built by the philosophers.[174]

[172] On the organization of the lines of Venetian navigation see B. Doumerc, "Il dominio del mare," in *Storia di Venezia*, vol. IV, p. 113. See also U. Tucci, "I servizi marittimi veneziani per il pellegrinaggio in Terrasanta nel medioevo," *Studi veneziani* 9, 1985, pp. 43-55.

[173] "E ne la dita Giaua nasce çe(n)çero e altre specie nobile gra(n) | qua(n)tità e tute q(ue)le che nasce i(n) questa e ne le altre al te(m)po | de recolte uie(n) portade a giaua maçor e de lì se despe(n)sa | in tre p(ar)te: una p(er) çaito(n) e chataio, l'altra p(er) el mar de i(n)|dia a hormus, çide e a la mecha, e la terça per el mar | del chataio da la parte de tramontana. (Et) in questa iso|la, sego(n)do el testimonio de queli che nauegano q(ue)sto | mar, el se uede el polo antartico leuado uno braço," *Mappamondo* (ed. 1956) Pl. XIII, 34*. *Mappamondo* (ed. 2006), * 0589.

[174] "La prouincia dita mogolistan, posta qui de sopra | a man dextra, è in dromo de la isola dita hormus (...). *E qui capitano parte | de le naue de india cu(m) le sue mar|chantie, che sono perle, piper,| çençero e altre specie in gra(n) | quantità, le qual poi sono | condute de lì per la uia | de la balsera e da bagadat | çoè babilonia de caldea, per | el fiume tygris et eufrates et | p(er) mesopotamia, armenia, capa|docia e perfina al mar de po(n)to. La cità principal de la dita | isola tien el nome de l'iso|la, çoè hormus, la qual an|tichame(n)te fo edifi|cada per phylosophi,"* *Mappamondo* (ed. 1956) Pl. XV, 96. *Mappamondo* (ed. 2006) *389. The texts derives from a passage of Book 4 of *De varietate fortunae* in which Poggio Bracciolini writes in Latin the so-called "Voyage to India" of Niccolò de' Conti. See Chapter VI below. The reference "Hormus, la qual antichamente fo edificada per phylosophi" refers to the belief brought by Odorico da Pordenone and John Mandeville that Hormuz might have been founded by Hermes, the God of Commerce, from whom the name derives. See P. Gautier Dalché, "Représentations antiques et médiévales (monde latin) du Golfe Persique," soon to be published in the proceedings, Actes du Premier Colloque international sur la

In these two legends five major routes of commercial navigation in the Indian Ocean and from the Indian Ocean to Persia, Central Asia and Europe are identified and traced: first from Zaiton, on the north-eastern coast of China, towards the southern Chinese coasts; second from Zaiton towards the Indian coasts; the third and fourth from the western coast of India to Hormuz and Mecca respectively; and a fifth route, part marine and part caravan, from Hormuz to Persia and Europe, via the Tigris and Euphrates and ending at the Black Sea. These two legends on the spice trade routes on the *mappa mundi* of Fra Mauro can, in my opinion, be considered the first detailed account of the main commercial routes on a *mappa mundi*, and derive, like most of Fra Mauro's legends that concern Asia and the Indian Ocean, from the narrations of Marco Polo and Niccolò de' Conti.[175] In fact these legends contain an embryonic representation of what modern scholars have defined as the three main commercial circuits of the Indian Ocean; I refer to the studies of S. D. Gotein on commercial relations between the Mediterranean and India using documents of the *Geniza* of old Cairo, and more recently the studies of Kirti N. Chaudhuri and Janet L. Abu-Lughod, Michael Pearson and Ashin Das Gupta on commerce in the Indian Ocean.[176]

6. Iconography of the sea

While the depictions of the sea in the *mappa mundi* express enthusiasm for the potential of navigation and commerce, the written and figurative narration of the Camaldolese also reveal a consciousness of the sea's tragic nature. It is important to emphasize that this is part of Fra Mauro's cosmographic analysis of the sea. Next to marvelous scenes of galleys, round ships, and even "Indian junks" – which, as we will see in Chapter V, derive directly from the description of Marco Polo (plate I.1) – confidently cutting the waves are scenes of terrible shipwrecks, whirlpools, and dangerous currents, and legends that warn of getting too close to the Lost Islands, the Sea of Darkness and the "gripping dense waters" – in short, the physical and meta-

cartographie historique du golfe Persique, "Bibliothèque iranienne" (Institut Français de Recherche en Iran – Institut d'Étude Iranienne de l'Université de Paris III).

[175] On the commercial and spatial knowledge of Asia and the Indian Ocean linked to the mercantile presence of the Venetians and the Genoese, with references to the Pisans and Piacentini in the fourteenth and fifteenth centuries, see *Genova, Venezia, il Levante nei secoli XII-XIV: atti del Convegno internazionale di studi, Genova-Venezia, 10-14 marzo 2000*, G. Ortalli and D. Puncuh (eds), Genoa: Società ligure di storia patria; Venice: Istituto veneto di scienze, lettere ed arti, 2001; also see M. Balard, "Precursori di Cristoforo Colombo. I genovesi in Estremo Oriente nel XIV secolo," in *Atti del convegno internazionale di studi colombani, 13-14 ottobre 1973*, Genoa, 1974, pp. 149-164; Id., "Les Génois en Asie centrale et en Extrême-Orient au XIVe siècle: un cas exceptionnel?" in *Mélanges offerts à Edouard Perroy*, Paris, 1973, pp. 681-689; Id., "Génois et Pisans en Orient (fin du XIIIe - début du XIVe siècle)," in *Genova, Pisa e il Mediterraneo tra Due e Trecento. Atti della Società ligure di Storia Patria*, Genoa, 1984, pp. 179-209.

[176] See S. D. Goitein, "From the Mediterranean to India: Documents on the Trade to India, South Arabia, and East Africa from the Eleventh and Twelfth Centuries," *Speculum* 29, 2 (1954), pp. 181-197; J. L. Abu-Lughod, *Before European Hegemony. The World System A.D. 1250-1350*, New York and Oxford: Oxford University Press, 1989; K. N. Chaudhuri, *Trade and Civilisation in the Indian Ocean: an Economic History from the Rise of Islam to 1750*, Cambridge: Cambridge University Press, 1985; A. Das Gupta, *The World of the Indian Ocean Merchant, 1500-1800*, New Delhi: Oxford University Press, 2001; and M. N. Pearson, *The Indian Ocean*, New York: Routledge, 2003.

physical limits of human habitation. The *mappa mundi* contains a wide variety of images that focus on the most dramatic elements of navigation, and which are able to provoke the intense emotional participation of the spectator (plate I.5). Fra Mauro wants to tell of the sea, the history of navigations and commerce, and of the immense wealth that circulated on ships from Zaiton to Calicut and from Aden to Alexandria, but these were only fragments of a more articulated and, to some extent, tragic story. Another aspect of Fra Mauro's account of the sea that links text and images is the rich visual chronicle of the many shipwrecks that provided a dramatic undercurrent and distressing background, especially in the Indian Ocean, to the great commercial fervor that the stories of merchants like Marco Polo, and closer to Fra Mauro, Niccolò de Conti, gave rise to in the West.

In the southern Indian Ocean and off the Northern coast of Africa between the Canary Islands and the Strait of Gibraltar the Camaldolese created a kind of shipwreck narrative in the form of a "tragic history of the sea" (*história trágico-marítima*) before the birth of this literary genre.[177] In 1735-1736 the Portuguese sailor Bernardo Gomes de Brito collected in the libraries and archives of Porto, Coimbra, Lisbon and Évora numerous manuscript stories written in the sixteenth and seventeenth centuries that recorded the tragic shipwrecks that befell the Portuguese fleet as it sailed the route to India (*carreira da Índia*). Brito transcribed and published them with titles that became famous and paradigmatic of the *História trágico-marítima*. His anthology is considered today the most complete corpus of a literary genre that would grow around the Portuguese shipwrecks of that time. This is not to argue that the *mappa mundi* develops the literary model of the shipwreck narrative identified by Giulia Lanciani in the sequence "background, sailing, tempest, shipwreck, corsair attack, surrender, capture, wandering, cruelty of enemies and return."[178] Still, in Fra Mauro's figurative account, a taste for the sea's tragic element can be clearly distinguished in the ships with upturned keel, dismasted galleys at the mercy of the waves, the boots, ladders, decks, and masts among the billows; in short, scenes of the poor remains of fleets destroyed by storms and encircled by enormous fish. Through these depictions of storms and shipwrecks the spectator can recognize, and maybe even vicariously participate in, the dangers and misfortunes of the sea route traced by Fra Mauro in the Indian Ocean and the Sea of Darkness, imagining himself in the scenes of panic and fury on the overturned ships off the coast of Gibraltar and the Indian Ocean, and feeling the agony

[177] Bernardo Gomes de Brito, *História trágico-marítima, em que se escrevem chronologicamente os naufragios que tiverão as naos de Portugal, depois que se pozem exercicio a navegação da India*, Tom. 1-2, Lisboa Occidental, Na Officina da Congregação do Oratorio, 1735-1736. In the same years reports of shipwrecks, already published separately in the seventeenth century, were republished together and in fact considered the "terceiro volume" of the *História Trágico-Marítima* (it is thus described, for example, in the British Library's catalog). Id., *História Trágico-Marítima*, A. Bueno (ed.), Rio de Janeiro: Lacerda Editores/Contraponto, 1998. The bibliography on the *história trágico-marítima* as a literary genre is extensive; here we limit the citations to several essential titles. See J. Duffy, *Shipwreck & Empire: Being an Account of Portuguese Disasters in a Century of Decline*, Cambridge, MA: Harvard University Press, 1955; C. R. Boxer, "An Introduction to the História Trágico-Marítima," in *Miscelânea de Estudos em Honra do Prof. Hernâni Cidade*, Lisbon: Publicações da Faculdade de Letras, 1957, pp. 48-99; Id., "An Introduction to the História Trágico-Marítima. Some Corrections and Clarifications", *Quaderni portoghesi* 5 (1979), pp. 99-112; J. Blackmore, *Manifest Perdition: Shipwreck Narrative and the Disruption of Empire*, Minneapolis: Minnesota University Press, 2002; and G. Lanciani (ed.), *Viaggi e naufragi portoghesi sulla via delle Indie*, Naples: Liguori, 2002.

[178] "Antecedentes, partida, tempestade, naufrágio, ataque corsário, arribada e captura, peregrinação e impiedade dos inimigos, retorno": see Lanciani, "Introduzione," in *Viaggi e naufragi portoghesi*, pp. 81-82.

of shipwrecks and ships as they are swallowed by the waves (plate I.5). The dangerous gripping waters enveloped in darkness at the edges of the *mappa mundi* mirror the fears of fifteenth-century men in the face of the unknown. This particular cosmographic conception is reminiscent of biblical cosmology, which posits the sea as a primordial aquatic abyss, understood as the physical and metaphysical limit of creation, or perhaps an unfinished creation. With regard to the sources that furnished the inspiration, language and intellectual and cultural instruments for describing the borders of the *orbis terrarum* in the Middle Ages and the early modern period, research has focused on the Bible and works derived from it.[179] If, in the Bible, the locution "towards the sea" means "west," and the "sea" par excellence is the Mediterranean basin, Fra Mauro shifts the terror and negativity that in the Bible are associated with the Mediterranean to the Indian Ocean and the Sea of Darkness. In the sacred text the sea, far from evoking contemplation or peace, is above all a grand negative symbol, a mythical category that evokes the "big waters," *majim rabbim*, that brought flooding and death with them.[180] Many passages in the Bible describe the sea with terror, considering it almost an omen of death and annihilation. Only God can control it. Théophile Briant in his anthology *Les plus beaux textes sur la Mer* emphasized this aspect of the Biblical cosmology of the sea, citing two stories taken from the Bible that parallel classic descriptions of sea storms from Homer and Virgil to Ovid.[181] The universal moral fable of Jonah swallowed by an enormous fish for three days and nights evokes this feeling of horror and fear. From the stomach of the sea monster Jonah proclaims : "You hurled me into the deep, into the very heart of the seas, and the currents swirled about me ; all your waves and breakers swept over me. [...] The engulfing waters threatened me, the deep surrounded me ; seaweed was wrapped around my head" (Jonah 2 :4-6). In Psalm 107 four people appear who are fulfilling their vows in the temple of Jerusalem. There is a caravan guide who had lost his way in the desert and had found it again, a freed prisoner, and a gravely sick person who is cured. At the end the most touching story is of a sailor who stands to keep his vow. The devout sailor says (verses 23-30) :

> Others went out on the sea in ships ; they were merchants on the mighty waters. They saw the works of the Lord, his wonderful deeds in the deep. For he spoke and stirred up a tempest that lifted high the waves. They mounted up to the heavens and went down to the depths ; in their peril their courage melted away. They reeled and staggered like drunken men ; they were at their wits' end. Then they cried out to the Lord in their trouble, and he brought them out of their distress. He stilled the storm to a whisper ; the waves of the sea were hushed. They were glad when it grew calm, and he guided them to their desired haven.

[179] For the philological analysis of these sources and of their spread into thought from Antiquity to the Middle Ages, see P. Gautier Dalché, "Modes de connaissance des *fines orbis terrarum* du Nord-Ouest (de l'Antiquité au XIIIe siècle)," in *L'Europe et l'Océan au Moyen Age. Contribution à l'Histoire de la Navigation*, S.H.M.E.S.-Cid édition, 1988, pp. 217-233 and the bibliography cited there.

[180] For the sea in biblical cosmology I am referring to a book by O. Kaiser, *Die mythische Bedeutung des Meeres in Ägypten*, Berlin : Ugarit und Israel, 1959 (second edition 1962), and A. Manodori (ed.), *La preghiera del marinaio : la fede e il mare nei secoli della Chiesa e nelle tradizioni marinare*, Rome : Istituto Poligrafico e Zecca dello Stato, 1992, in particular the section "Le sacre scritture e il mare" (essays by M. Guarducci, A. Amicarelli Scalisi, and M. d'Ottone), pp. 153-237 ; C. Militello, "La teologia e il mare," pp. 319-333 ; and the section "Metafore religiose del mare" (several essays written by M. Scollo, A. Amicarelli Scalisi, and A. Manodori), pp. 343-381.

[181] T. Briant, *Les plus beaux textes sur la Mer*, Paris : La Colombre, 1950, pp. 3-15.

But the Bible does not only tell of the primordial fear of man in the face of unleashed natural forces and in the face of the unknown. It also tells of the metaphysical emotions implicit in man's confrontation with the void, the sensation of running up against the metaphysical limit of thought and human action. In the legends and in certain images of the Christian cosmography of Fra Mauro that describe the sea and the limits of human navigation at the distant Lost Islands, one detects a strong echo of the biblical cosmology of the sea.

VIII. – At the origins of a vernacular scientific vocabulary

The most recent edition of Fra Mauro's work introduced a cleavage, if not a fracture, between the map's geographical texts (placed mainly inside the circular frame), which are praised as "scientific" and "experimental," and its cosmographic texts (displayed, according to Falchetta, mainly in the four corners between the square and circular frames), which are instead regarded as "traditional," and "doctrinal."[182] Even the fact that the earthly paradise – a central part of the cosmography and of the *cosmographicae meditationes* in the medieval and early modern periods[183] – is represented outside the world, is considered by Falchetta as evidence of "a sharp distinction between the epistemological status of geographical knowledge, which can be subjected to critical evaluation and judgment, and religious doctrine, which is simply to be observed and followed."[184]

[182] See for example this passage : "The first thing to note is that the critical standard adopted in this material [the legends placed in the four corners of the map] is very different to that one finds in the cartography of the world map. There, Fra Mauro follows what one might describe as an open, composite method, which is rationalistic and 'scientific' *avant la lettre* in that it places great importance on the information gleaned from the eye-witness accounts of those who have travelled to the four corners of the globe. Furthermore, no theological considerations interfere with the geographical theory [...] This cosmographical material, which discusses important cosmological and astronomical themes, is itself characterised by a certain open-mindedness and eschews resort to dogmatic positions. However, the rationalism demonstrated here is undoubtedly channelled within the ideas of the great Classical and medieval auctoritates ; the limits of speculative thought are determined by Aristotle, Ptolemy and various Fathers of the Church. In short, one might say that Fra Mauro is as 'experimental' as a geographer and cartographer, as he is traditionalist as a cosmographer." P. Falchetta, "The Cosmological Notes," in *Mappamondo* (ed. 2006), p. 123.

[183] See the recent work of A. Scafi, *Mapping Paradise : A History of Heaven on Earth*, London : British Library ; Chicago : University of Chicago Press, 2006, in particular, with regard to medieval notions of earthly paradise and its representation in cartography, see chapters 2-8 (pp. 32-253) ; the depiction of paradise in Fra Mauro's map is discussed at the end of chapter 8, pp. 235-240. For a detailed analysis of the representation of paradise in the *mappa mundi* see the next chapter here. On the *cosmographicae meditationes*, see *Les méditations cosmographiques à la Renaissance* (Actes du colloque organisé par J.-M. Besse, M.-D. Couzinet, F. Lestringant), *Cahiers V.-L. Saulnier* 26, Paris : Editions Rue d'Ulm, 2008.

[184] "The presence here [left lower corner of the map] of this note and the miniature that illustrates it is interesting for various reasons. Though, as we have seen, Fra Mauro does not contradict the commentators, the lack of any clear link between the geographical-cartographical contents of the world map and the representation of Eden would seem to suggest that he did not accept the traditional topographical indication regarding the location of Paradise ; he repeats it simply as a question of doctrine, which has little to do with the geography of the oecumene. So, once again, one has a sharp distinction between the epistemological status of geographical knowledge, which can be subjected to critical evaluation and judgement, and religious doctrine, which is simply to be observed and followed and offers no space for critical comments that contradict it. The fact that such a separation should exist in a work by a cleric – and that that same work could be the boast of a religious institution – without anyone ever pointing

In the light on the analysis developed in this chapter, I do not share these interpretations. The division between the inside of the map, which is held to be "geographical and experimental," and the external part, in the margins, which is though to be "doctrinal" and "traditional," is the result of a forced reading, of a cliché. The rigid schematicism which some see in the work is not confirmed by a close analysis of the text and image. Reading and analysis of the inscriptions and images that make up Fra Mauro's complex work reveal an intimate connection between individual legends and pictures, united both in narrative structure and in vocabulary. In both the "external" and "internal" legends, Fra Mauro used the best and most respected authorities available, and the ideas he proposed in the "external" legends continued to be debated until the end of the sixteenth century. One has only to read the following legend about Gog and Magog, located in the interior part of the *mappa mundi* to grasp clearly how interpretations that suggest "a clear separation between geographical content and doctrinal matters" are not useful in understanding either the work of Fra Mauro or the culture from which it originated :

> Some write that on the slopes of Mount Caspian, or not far from there, live those peoples who, as one reads, were shut in by Alexander the Macedonian. But this opinion is certainly and clearly mistaken and cannot be upheld in any way because the diversity of the peoples who live around that mountain would certainly have been noticed ; [...] Hence, if those peoples had been enclosed there, I think that others would have had notice of it, and the fact would be known to us. But given that those peoples are at the limit of the earth – something of which I have information that is certain – this explains why all the peoples I listed above know no more about them than we do. Hence I conclude that these peoples are very far from Mount Caspian and are, as I said, at the extreme limit of the world, between the north-east and the north, and they are enclosed by craggy mountains and ocean on three sides. They are to the north of the kingdom of Tenduch and are called the Ung and the Mongul, which people know as Gog and Magog and believe that they will emerge at the time of the Antichrist. *But certainly this mistake is due to the way some force the Sacred Scriptures to mean what they want them to mean.* So, I am not differing here from the authority of St. Augustine, who in his *De Civitate Dei* rejects all the opinions of those who claim that Gog and Magog are the peoples that will support the Antichrist. And Nicholas of Lyra agrees with this claim, explaining the two names by their Hebrew origin. But elsewhere one can speak more fully of this matter and of how it is to be understood.[185]

out this clear contradiction between science/knowledge and doctrine, can probably be taken as indicating that by the middle of the fifteenth century the Middle Ages were definitely part of the past (at least as far as Venice was concerned). One should also point out that the note on the Earthly Paradise seems to have been planned as part of the world map from the very beginning : style, syntax and lexicon are identical to those one finds in the inscriptions within the map. In this case, one can see that Fra Mauro's intention was for there to be a clear separation between geographical content and doctrinal matters – a distinction that must have seemed entirely legitimate to him." Cf. *Mappamondo* (ed. 2006), pp. 131-132.

[185] "Alguni scrive che a le radice del monte caspio over pocho lutan sono queli populi i qual co|me se leçe sono seradi per Alexandro Macedo. Ma certo questa opinio(n) manifestamente è erronea | e da non esser sostenuta p(er) algun modo, p(er)ché certo le si notò la diversità de le natio(n) che habitano circa | q(ue)l monte che'l no(n) è possibile che ta(n)ta numerosità de p(o)p(u)li ne fosse ignoti cum sit che tute q(ue)le p(ar)te sono assai | domestege p(er) esser freque(n)tade sì da nostri come da altre natio(n) che sono çorçani, mingresi, armini, | cercassi e tartari, e molte altre g(e)n(er)ation de populi, i q(ua)l fano (con)tinuam(en)te quel cami(n). Unde se q(ue)sti po|puli fosse de li rechiusi credo che se q(ue)li ne havesse notiçia anchora | seriano a nui noti. [...] P(er)hò (con)cludo che q(ue) sti p(o)p(u)li siano molto lu|tani dal mo(n)te caspio e siano come ho dito ne la extremità de la terra tra grieco e tramo(n)tana e sono | ci(r)cu(n)dadi da mo(n)ti asp(er)imi e dal mar occea(n) q(u)asi da tre ba(n)de, e q(ue)sti

The problem, then, is not at all in the opposition between "geographical content" and "doctrinal matters" – a problem that did not exist for Fra Mauro – but in being able to read the Scriptures without exploiting them, without using them to make them say "what one wants them to mean," to paraphrase what is written very clearly and profoundly by Fra Mauro. The long legends placed in the outer corners of the map and those inside the *oikumene* do not disagree with, or contradict, each other; rather they are contiguous and complementary. And as such they must be analyzed if we are to understand the overall cosmographical representation and analysis presented in the map. This is demonstrated by examining how Fra Mauro develops one of the crucial issues of the *mappa mundi*: his meditation on the sea and the ocean, which without a doubt is a central theme not only in the work of the Camaldolese, but also in fifteenth-century cosmography and cartography generally. It is only by jointly analyzing the "external" and "internal" legends together with the images on the *mappa mundi* that it becomes possible to understand the complexity of Fra Mauro's thought, and the five semantic levels that compose his cosmography of sea and ocean.

Contrary to Falchetta's argument, there is no contrast or contradiction between the style and linguistic form of the internal and external legends: in fact, a close reading of them reveals a substantial continuity of style, language and grammar. It is true that in the circular portion of the *mappa mundi* there are around thirty legends in which Fra Mauro, through the repetitive use of the pronoun "io" ("I"), enters in the first person into the cosmographic discourse, creating a genuine dialogue with his readers, which he does not do in the "external" legends. But it is also true that in the circular portion of the map, that which Falchetta calls "experimental," as they are based on sources such as Marco Polo and Niccolò de' Conti (who are improperly categorized, even within quotation marks, as "experimental geographers"), there are hundreds of legends in which Fra Mauro summarizes or paraphrases passages of *Il Milione*, the *De varietate fortunae*, and other authors both identified and unidentified,[186] in a style and language quite similar to those of the legends in the outer margins.[187] In other words, the only real difference is that the latter are more extensive, and therefore use a vocabulary and syntax more complex than the "internal" legends that were necessarily synthetic for "lack of space" (as Fra Mauro himself says).

This gives us a further element for reflection. Setting aside judgments such as "doctrinal," "traditional" and "experimental," which are misleading and anachronistic, the external legends

sono soto el regno de te(n)duch e sono chia|mati vng e mo(n)gul, i q(u)al el uulgo dice gog e magog extimando che q(ue)sti sia q(ue)li che diebano usir de lì al | te(m)po de antichristo. Ma certo q(ue)sto error è adeuenuto p(er) alguni che tirano la | sacra scriptura al suo sentime(n)to. P(er)hò io me acosto a la auctorità de sanc|to Augustino, el q(u) al nel suo De Civitate Dei reproua la opini|on de queli che dicono che gog e magog | significa q(ue)li populi che darano | fauor ad antichristo et i(n) q(ue)sta | sente(n)cia etia(m) co(n)corda Nicolò de Lira, el q(u)al | expone questi do' nomi secondo la he|braica uerità, ma de questa materia | i(n) altro luogo se ne par|lerà cu(m) mior modo | quelo se douerà | tignir." *Mappamondo* (ed. 1956) Pl. XXXIII, 77. The English translation derives from *Mappamondo* (ed. 2006) *2403.

[186] On this subject see the following chapters here, but also Leporace's 1956 edition of the legends of the *mappa mundi*, and that by Falchetta as well, particularly his commentary on the legends. See *Mappamondo* (ed. 2006), pp. 174-719.

[187] On Fra Mauro's use of Marco Polo and Niccolò de' Conti see Chapter V below; for an analysis of the overall narrative structure of the work, see Chapter VI.

are of vital importance not only in the study of Fra Mauro's work, but also with regard to the history of so-called "vernacular science." These legends, together with those within the circular *mappa mundi*, employ one of the most complex scientific vocabularies in the vernacular, not only in Venice, but anywhere in Europe in the middle of the fifteenth century. The long cosmographical legends discuss a wide range of ideas from medieval cosmography and cosmology, and either draw inspiration from, paraphrase, or partly translate into vernacular Venetian excerpts from works such as the commentary on the *De causis proprietatum elementorum* and *De natura loci* of Albertus Magnus and the *Summa theologica* of Thomas Aquinas. The legend on the number of heavens in the upper left corner, for example – and this fact had previously eluded scholars – is the only known translation (probably made by Fra Mauro himself) into the Venetian vernacular of an article of the *Summa theologica* of St. Thomas. This makes the legends of Fra Mauro one of the most significant scientific popularizations of the time, whose importance for understanding the formation of the vernacular scientific lexicon is equal to that of the translations of treaties such as the encyclopedic *De rerum proprietatibus* of Bartholomeus Anglicus, translated by the Mantuan notary Vivaldo Belcalzer in the early fourteenth century, and Aristotle's *Meteorologica* (among the less theoretical contributions of the *corpus aristotelicum* and the only one to be popularized before the fifteenth century), translated into the Florentine dialect during the fourteenth century.[188] In fact, its value in this regard is perhaps equal to that of the first original scientific works in the vernacular, such as the *Trésor* of Brunetto Latini.

In the work of Fra Mauro the intention of the author is clear: he aims to guide the readers of his treatise on cosmology and cosmography through a carefully focused selection of terms. A rich variety of adjectives (to define the waters, the heavens, the elements, etc.) is clear evidence of the exceptional nature of the text that we are discussing: many of the contents were taken from Latin and vernacular sources, but others were deduced from sensory perceptions (even if they were seen or experienced by others).[189] The text does not consist of exemplifications, then, but rather of linguistic choices that are focused and meditated. In the Appendix of this chapter a complete list of contextualized and analyzed cosmographic lemmas employed by Fra Mauro is compiled. This analysis shows that the legends of the *mappa mundi* are located at the birth and earliest affirmation of the scientific lexicon in the vernacular.

[188] R. Librandi, *La Metaura d'Aristotile. Volgarizzamento fiorentino anonimo del XIV secolo*, 2 vols., Naples: Liguori, 1995.

[189] These topics will be at the center of Chapter VI that focuses on the narrative structure of Fra Mauro's work.

CHAPTER III

THEOLOGY, COSMOGRAPHY, ART
THE EARTHLY PARADISE IN FRA MAURO'S *MAPPA MUNDI*

> I am aware that many authors have written at great length on paradise, but their theories on the subject in general can be reduced to three. There is, first, the opinion of those who interpreted the word 'paradise' in an exclusively corporeal sense. Then there are those who prefer to give an exclusively spiritual meaning to the word. Finally, there are those who accept the word 'paradise' in both senses, sometimes corporeally and at other times spiritually. Briefly, then, I admit that the third interpretation appeals to me.[1]
>
> Augustine

I. – Representing St. Augustine's exegesis of the earthly paradise

1. The reasons for this research

The exegesis of the Bible has given rise to an age-old debate about the earthly paradise that involves theologians, ecclesiastics, scholars, humanists, poets and writers.[2] Representing

[1] "Non ignoro de paradiso multos multa dixisse ; tres tamen de hac re quasi generales sunt sententiae. Una eorum qui tantummodo corporaliter paradisum intellegi volunt : alia eorum qui spiritaliter tantum ; tertia eorum qui utroque modo paradisum accipiunt ; alias corporaliter, alias autem spiritualiter. Breviter ergo ut dicam, tertiam mihi fateor placere sententiam." Augustine, *La Genesi alla lettera : testo latino dell'edizione maurina confrontato con il Corpus scriptorum ecclesiasticorum Latinorum* ; translation, notes and indices by L. Carrozzi, Rome : Nuova biblioteca agostiniana, Città nuova, 1989, Book VIII.1.1, p. 386. The English translation derives from Augustine, *The Literal Meaning of Genesis*, transl. and annotated by J. H. Taylor, 2 vols., New York-Ramsey : Newman Press, 1982, pp. 32-33.

[2] Significant studies include : A. Scafi, *Mapping Paradise : A History of Heaven on Earth*, London : British Library ; Chicago : University of Chicago Press, 2006 (in particular, with regard to medieval notions of earthly paradise and its representation in cartography prior to Fra Mauro, see chapters 2-8 (pp. 32-253) ; the depiction of paradise in Fra Mauro's map is discussed at the end of chapter 8, pp. 235-240 ; Id., *The Notion of the Earthly Paradise from the Patristic Era to the Fifteenth Century*, Ph.D. dissertation, London : University of London, 1999 ; Id, "Mapping Eden," in D. Cosgrove (ed.), *Mappings*, London : 1999, pp. 51-69. See also J. Delumeau, *Une histoire du paradis. Le jardine des délices*, Paris : Fayard, 1992, pp. 11-127 ; Id., *Que reste-t-il du paradis ?*, Paris : Fayard, 2000 ("Première partie" and "Deuxième partie", pp. 1-194). Also see E. Testa, "Il paradiso dell'Eden secondo i SS. Padri (Contributo alla storia dell'esegesi)," *Studi Biblici Franciscani*, Liber Annus 18 (1968), pp. 94-102. For a study of paradise at the antipodes and, at the same time, a review of the debate in Scholastic circles about the location of paradise, see P. Gautier Dalché, "Le paradis aux antipodes ? Une *Distinctio divisionis terre et paradisi delitiarum* (xive siècle)," in D. Barthélemy and J.-M. Martin (eds), *Liber largitorius. Études d'histoire médiévale offertes à Pierre Toubert par ses élèves*, Geneva : 2003, pp. 615-637. Some erudite studies dating from the nineteenth century and early twentieth century are still fundamental texts, some of which have recently been republished : A. Graf, *Miti*,

the Garden of Eden has always been a topos in Medieval and Renaissance painting and illumination.[3] Similarly, the description and depiction of paradise on earth is one of the significant *loci* of Medieval and Renaissance *mappae mundi*.[4] This chapter analyses the pictorial representation of, as well as the theological and cosmographic discussion on, the earthly paradise placed in the lower left corner of Fra Mauro's *mappa mundi*. It will also discuss the identification of the painter who, in the late 1450s, depicted one of the most artistically relevant depictions of a Garden of Eden in the history of *mappae mundi* : Leonardo Bellini (Venice, *c.* 1423/1435-*c.* 1490), nephew and pupil of Jacopo Bellini.[5] As is habitual in Fra Mauro's work the iconography, theological, and cosmographic meanings of the pictorial representation are explained by an extensive and complex inscription placed above the illumination of the paradise. Text and image complete, and complement, each other : in methodological terms it is therefore important that any interpretation simultaneously addresses both these idioms, emphasizing the relationships and semantic convergence between the text and the image.

Fra Mauro's theological and cosmographical reflections were based on an exegesis derived, above all, from the *De Genesi ad litteram libri duodecim* (or *The Literal Meaning of Genesis in Twelve Books*) by Augustine, a fundamental work amongst medieval biblical exegeses. The analysis of the link between Fra Mauro's description of the paradise and the *De Genesi ad litteram* is crucial especially in order to understand the reasons that induced him to place the earthly paradise outside the *oikumene*, a theme that is developed with a new emphasis when compared to other preceding and coeval *mappae mundi*. As for the attribution, in conformity with the results of Susi Marcon's independant research, I will only briefly reconstruct the process that, under the guidance of Giordana Mariani Canova, an authoritative expert on the history of Venetian miniatures, led me to attribute the representation of paradise to the Venetian painter Leonardo Bellini. Since I am not an art historian, I will not delve greatly into the aesthetic-artistic aspects of the representation and on Leonardo Bellini's iconographic sources – only as much as is necessary to illustrate the reasons for this attribution.[6] Instead, I will analyze the

leggende e superstizioni del Medio Evo, C. Allasia and W. Meliga (eds) ; introduction by M. Guglielminetti ; critical essays by E. Artifoni and C. Allasia, Milan : Mondadori, 2002, pp. 1-116 ; Id., *La leggenda del Paradiso terrestre : lettura fatta nella R. università di Torino, addì 11 novembre 1878*, Turin : Loescher, 1878 (reprinted Rome : Edizioni del Graal, 1982) ; G. Marinelli, *Scritti minori. Metodo e storia della geografia*, vol. 1, Florence : Le Monnier, 1878, pp. 304-308 ; E. Coli, *Il Paradiso terrestre dantesco*, Florence : Carnesecchi, 1897.

[3] *La Genesi. Capitoli 1-11*, translated from Hebrew by Gianfranco Ravasi, with an iconographic project by M. G. Ciardi Duprè Dal Progetto and M. Bernabò, Turin : Allemandi, 1998 ; M. Levi d'Ancona, *The Garden of the Renaissance. Botanical Symbolism in Italian Painting*, Florence : Olschki, 1977.

[4] Scafi, *Mapping Paradise*, pp. 191-240 ; Delumeau, *Storia del paradiso*, pp. 76-94 (an analysis that is not always up-to-date).

[5] This part of the study further develops Susi Marcon's as well as my research published respectively in : S. Marcon, "Il Mappamondo di Fra Mauro e Leonardo Bellini," in M. Piantoni and L. de Rossi (eds), *Per l'arte da Venezia all'Europa. Studi in onore di Giuseppe Maria Pilo*, Venice : Edizioni della Laguna, 2001, pp. 103-108, Ead., "Leonardo Bellini and Fra Mauro's World Map : The Earthly Paradise," in P. Falchetta, *Fra Mauro's World Map*, Turnhout : Brepols, 2006, pp. 135-161 ; A. Cattaneo, "God in His World. Leonardo Bellini Illuminator of the Earthly Paradise in Fra Mauro's Mappa mundi," *Imago Mundi* 55 (2003), pp. 97-102.

[6] For a detailed analysis of the artistic aspects of his work, see Marcon, "Il Mappamondo di Fra Mauro e Leonardo Bellini," pp. 103-108, Ead., "Leonardo Bellini and Fra Mauro's World Map : The Earthly Paradise, " in *Mappamondo* (ed. 2006), pp. 135-161.

significance and the documental value of the fact that one of the indisputable masters of Venetian miniature assisted in the completion of this *mappa mundi*.

2. The earthly paradise in Fra Mauro's cosmography

As has been definitively demonstrated by Alessandro Scafi's research, earthly paradise was a central part of the Western cosmographic discourse, embracing theological, philosophical and cosmographical traditions that are crucial to Medieval and Renaissance episteme, even well after Fra Mauro's age.[7] Fra Mauro's analysis of the Garden of Eden deals mainly with the history of paradise in the Bible and its location in space and time, and includes theological, cosmological and geographical elements. With Jerusalem, the site where Christ lived, placed at the centre of the map (as has been stressed in Chapter II, it is not the geometric centre, but rather an "anthropological" one, which thus refers to the density of population[8]) and Noah's Ark,[9] which echoes the rebirth of man and all living creatures on Earth after the great flood, the placing of this earthly paradise amongst the spaces represented in the *mappa mundi* signified modeling a cosmography that sprang forth mainly from the Christian doctrine of Creation and Re-Creation.

The detailed narration of the earthly paradise in the Book of Genesis (Gen. 2, 4-3, 24) is contained in the so-called "second telling of the Creation," the older version, with a more descriptive and corporeal character.[10] The Garden of Delights, in which God "placed the man to whom he had given shape" (Gen. 1, 26), unequivocally expresses the tangible bond between the Creator and His creatures, establishing the starting point for the history of mankind. The immediate world in the Middle Ages and the Renaissance is one of faith, and the universe appears as a creation of God. Thus, in this age, reality (and therefore even cosmographical and geographical reality) was experienced and viewed as a coherent whole in terms of the facts of the revelation. What needed to be understood and taught, in terms of cosmology, also depended on the decrees of councils and their definitions with regard to the origins of the world and

7 Scafi, *Mapping Paradise*, pp. 16-31.

8 "HIERUSA|LEM. è i(n) mezo de la terra | habitabile, seco(n)do la latitudine | de la terra habitabile, be|(n)ché seco(n)do la lo(n)getu|dine la sia più oc|cidental. Ma p(er)ché la p(ar)te ch'è più occi|dental è più habi|tada p(er) l'Europa | p(er)hò l'è i(n) mezo an|cora seco(n)do la lon|gitudine, no(n) co(n)|siderando el spa|tio de la terra | ma la moltitudi|ne di ha|bita(n)ti." *Mappamondo* (ed. 1956) Pl. XXVIII, 24. This is the English translation quoted from *Mappamondo* (ed. 2006) *1011 : "By latitude, Jerusalem is at the centre of the inhabited world ; but by longitude, it is further west. However, as Europe is more [densely] populated, Jerusalem does appear to be at the centre [of the inhabited world] by longitude as well, if one takes into account not [geographical] space but the number of inhabitants."

9 "Archa Noè," *Mappamondo* (ed. 1956) Pl. XXVIII, 99 ; *Mappamondo* (ed. 2006) *1679. About the symbolic significance of the ark in Christian theology, see, for example, Augustine, *De civitate Dei*, Book XV, Chap. 26, 'De mensura arcae'.

10 The first description (Gen. 1, 1 and 2, 4a), considered to be a later version in terms of composition, proves to be the most systematic and theological of all ; it resorts to the term *Elohim* to designate God. In the opinion of biblical scholars, the second telling (Gen. 2, 4 and 3, 24), which uses the name *Jahvè* to address God, the story of Adam's placement in the Garden of Eden, the creation of Eve and their expulsion from the *hortus deliciarum* is narrated in a more figurative and concrete way. It is generally reputated to be an older version of the story of the Creation. See J. Auer, *Il mondo come creazione*, Assisi : Cittadella Editrice, 1977, pp. 35-73 and 214-241.

mankind, as also to their destiny. Divine will created the world, and the Creation has been revealed in the Bible (Gen. 1, 1). It then follows that everything that cosmographers cannot directly experience can nevertheless be represented by reading the scriptures, the exegesis of the Fathers of the Church, the Christian interpretation of Aristotle, Plato's natural philosophy, and their Muslim and Christian commentators. It is essential to grasp this vitally important fact if one wishes to understand medieval and early modern Christian civilizations with their belief and thought systems, of which Fra Mauro's *mappa mundi* was an integral part.[11] Fra Mauro's work, in which he discusses cosmography, trade, spices, journeys and navigation, is based on this system of beliefs, and not just in a "rhetorical" sense.

3. The image of the earthly paradise in Fra Mauro's *mappa mundi*

In this *mappa mundi*, the circular earthly paradise is described and portrayed as the peak of a rocky landscape, entirely surrounded by water, almost as though it were an island.[12] Its enclosing walls are reinforced by three towers and a crenellated door surrounding the luxuriant garden, within which God and Adam and Eve are depicted standing beside the tree of good and evil (Gen. 2, 10) ; around this tree is entwined the serpent, while an angel armed with a sword stands guard over the door "to protect the path to the tree of life" (Gen. 3, 24). At the foot of the tree a watercourse rises from a spring that, after passing under the walls, separates into four streams that immediately flow downwards. They re-emerge in the *oikumene* and become the four rivers that were called the Pishon, the Gihon, the Tigris and the Euphrates in the Book of Genesis (Gen. 2, 10-14) and that Fra Mauro identified – in accordance with a tradition that began with the historian Josephus Flavius in the first century AD – as the Ganges, the Nile, and, obviously, the Tigris and the Euphrates.[13]

This iconography essentially interprets and echoes the description of paradise given in Genesis 2, 4-3, 25, according to stylistic features that art historians have deemed to be quite typical for the fourteenth and fifteenth centuries.[14] The only exception is the rocky outcrop that extends in the right half of the image – a peculiarity that I shall discuss in the course of this chapter in terms of its artistic relevance. However, this representation is quite singular in the context of *mappae mundi* and the history of cosmography for other reasons : above all, in the fact that it is placed explicitly outside the area of the *oikumene* of the *mappa mundi*, in conjunction with a legend that was evidently inspired by Augustine's *De Genesi ad litteram libri duodecim* and the *De civitate Dei*. Fra Mauro interpreted paradise as coterminous : both as an

[11] É. Gilson, *La filosofia nel Medioevo*, Bari : Laterza, 1998, p. 907.

[12] Scafi proposes a different interpretation : more than the summit of a mountain-island, the representation in the *mappa mundi* could portray paradise within the circle, while the rocky cave would represent the Earth, both of them being surrounded by the ocean. See *Mapping Paradise*, pp. 239-240.

[13] Iosephus Flavius, *Ant. Iud.* I, 37-38 ; see L. Moraldi (ed.), *Antichità giudaiche*, 2 vols., Turin : UTET, 1998, vol. 1 *Libri I-X*, p. 53.

[14] Specifically about the iconographic contents of Fra Mauro's paradise : A. Scafi, "Il Paradiso Terrestre di Fra Mauro," *Storia dell'Arte* 93/94 (1998), pp. 411-419 (which nonetheless does not identify Bellini's hallmark) ; for an overview of the iconography of representations of earthly paradise, see Delumeau, *Storia del paradiso*, pp. 81-97. The discourse about the pictorial quality of paradise is quite different, as will be analysed in the second part of this chapter.

immanent corporeal space, placed within the story of Creation; and as a spiritual space, that is, a theological allegory.

It is not so much the fact that Fra Mauro followed Augustine's exegesis that is so significant (it was perhaps the most influential interpretation of paradise in exegetic history, echoed by many theologians such as Bede, Rabanus Maurus, Peter Lombard, Thomas Aquinas and Nicolaus of Lyra[15]). Instead, it is the dialogue that emerges between the theological-cosmographical reflection that is developed in the written description and the figurative representation of paradise that makes so interesting the manner in which Fra Mauro presents it. This is framed within an epistemological context in which the Garden of Eden is neither a decorative element nor an obligatory doctrinal note, perhaps out of respect for the ecclesiastical hierarchy or monastic culture. In contrast, this is a central part of the narrative structure of the *mappa mundi*. The presumed "marginality" of paradise, which could be deduced from the fact that it is not placed within the *oikumene*, is instead a central feature in terms of the cosmographic structure of the work as a whole, a theme that has been developed and interpreted by Fra Mauro with a remarkable emphasis and originality.

4. Paradise in coeval *mappae mundi*

It is useful to compare the earthly paradise depicted in Fra Mauro's work with those depicted in other fifteenth-century *mappae mundi* in order to understand the singularity of his representation. In the coeval world maps of Albertin di Virga (drawn between 1411 and 1419), Andrea Bianco (1436), Andreas Walsperger (1448) and Giovanni Leardo (1442, 1448 and 1452), the so-called "Borgia world map" (*c.* 1430), and in the "Catalan Estense world map" (*c.* 1460-1470), paradise is placed at the edges of the *oikumene*. There is a kind of pendulum affecting the location of paradise in these mid-fifteenth-century maps: the Garden of Eden, from the eastern edges of the *oikumene*, swings towards Africa and *vice versa*.

The anonymous engraver of the Borgia world map,[16] Bianco,[17] Walsperger[18] and Leardo all place paradise in the east, in (or beyond) India. In the maps drawn by Leardo paradise is

[15] The literature about the transmission of the Augustinian exegesis of the *Genesis* is exceedingly vast; with regard to paradise, see Scafi, "The Notion of the Earthly Paradise from the Patristic Era to the Fifteenth Century," in *Mapping Paradise*, pp. 37-184.

[16] Vatican City, Biblioteca Apostolica Vaticana, Borgiano XVI, [Borgia World Map], engraved on two semicircular copper plates joined by nails; Austria, southern Germany or Bohemia, diameter *c.* 63 cm, South on the top, legends written in Latin, *c.* 1430. See Almagià, *Monumenta cartographica vaticana*, vol. 1, pp. 27-28, Plate 8; Destombes, *Mappemondes*, entry 52.10, pp. 212-214; A. H. Heeren, "Explicatio Planiglobi Orbis terrarum faciem exhibentis ante medium saeculum XV summa arte confecti Musei Borgiani Velitrensi," *Commentarius Soc. Regiae Scientiarum Gottingensis*, 16 (1808), pp. 250-284; Scafi, *Mapping Paradise*, pp. 210-212.

[17] Venice, Biblioteca Nazionale Marciana, Ms. It. Z. 76 (= 4783), Andrea Bianco, [Nautical Atlas] manuscript on vellum, ff. 10, 260 × 380 mm, 1436. On ff. 9v-10r, the *mappa mundi*, with a diameter of 246 mm, is the last of the ten maps that make up the atlas, preceded by eight nautical charts that cover the Mediterranean Sea, the Black Sea and the Atalntic coasts of Europe, followed by a Ptolemaic planisphere. A facsmile copy of the codex exists: Andrea Bianco, *Atlante nautico del 1436*, P. Falchetta (ed.), Venice: Arsenale Editrice, 1993. For a more in-depth analysis of this manuscript, see Chapter IV.

[18] Vatican City, Biblioteca Apostolica Vaticana, MS Palat. Lat. 1362B, Andreas Walsperger, [Mappa mundi sive descriptio orbis geometrica facta ex cosmographyca Pthlomey (...). Et cum vera et in-|tegra cartha navigatio-

depicted in the form of a fortified city[19]; in that by Walsperger it is instead represented as a fortified castle out of which flow the four rivers; Bianco and the engraver of the Borgiana map portray paradise in the form of an unadorned *hortus conclusus* from which four rivers spring.

One can, however, observe a different placement of the Garden of Eden in the Catalan Estense world map (*c.* 1460-1470) preserved at the Biblioteca Estense Universitaria in Modena. In this map paradise is situated in Africa instead of in the East and, particularly in the "Third India," not far from the realm of Priest John, at the same latitude as the equator.[20] As for the rest, the representation of paradise follows traditional iconography: the luxuriant garden with the tree of life placed between Adam and Eve, the spring from which the four rivers emerge, and the entire scene surrounded by six encircling and protecting mountains.[21] A legend which describes the image, placed beside paradise, also reveals that Isidore of Seville was the textual source that was used for the iconography.[22] Even the now lost *mappa mundi* that was drawn sometime between 1411 and 1419 by the Venetian contemporary of Fra Mauro (and *patronus*

nis marium (...)], manuscript on vellum, 589 × 746 mm, South at the top, 1448. In the lower margin, the last line of a note written in black ink on six lines reads thus: "Facta est hec mappa per manus fratris Andreas Walsperger ordinis Sancti Benedicti de Salisburga Anno Domini 1448 in Constantia". See Almagià, *Monumenta cartographica vaticana*, vol. 1, pp. 30-31, Plate 12; Destombes, *Mappemondes*, entry 52.10, pp. 212-214.

[19] Verona, Biblioteca Civica, Giovanni Leardo, *Mapamundi. Figura mundi*, manuscript on vellum, East at the top, 28 × 35 cm, diam. of the world map 27.8 cm, 1442; Vicenza, Biblioteca Civica Bertoliana, ms. 598a, Giovanni Leardo, *Mapamundi. Figura Mundi*, manuscript on vellum, 35 × 31 cm, 1448, signed and dated below the scroll on the bottom right; Milwaukee, University of Wisconsin, American Geographical Society Library, Rare (3, East), 50 A, Giovanni Leardo, [World map], ms., vellum, 60 × 73 cm, 1452. See G. Crivellari, *Alcuni cimeli della cartografia medievale esistenti a Verona*, Florence: Bernardo Seeber, 1903, pp. 1-33; G. Berchet, *Il planisfero di Giovanni Leardo dell'anno 1452. Fac-simile nella grandezza originale*, Venice: Ferdinando Ongania Editore, 1880.

[20] Modena, Biblioteca Estense Universitaria, C.G.A.I, [Catalan Estense World Map], manuscript on vellum, North at the top, legends in Catalan, 113 × 100 cm, *c.* 1460-1470. See the facsimile edition with a translation of the legends in Italian and German: *Il mappamondo catalano estense*; commentary: E Milano; transcription of the original text: A. Battini, Zürich: Dietikon, 1995. For the tradition that placed paradise in Africa, see F. Relaño, "Paradise in Africa. The History of a Geographical Myth from Its Origins in Medieval Thought to Its Gradual Demise in Early Modern Europe," *Terrae Incognitae* 36 (2004), pp. 1-11; Id., *The Shaping of Africa*, Aldershot; Burlington: Ashgate, 2002, pp. 197-204 and Scafi, *Mapping Paradise*, pp. 16-18 and pp. 207-234.

[21] "Monts de diamats losquals son anguardia dep(ar)adis teranall."

[22] "Aquesta regio es p(ar)adis teranall | loquall es molt delitable loch en | loqual p(ar)adis as murat tot defoch | los quals puien trofins alcell | en lo quall es labre deuida (...) dit paradis hix una ffont la qual separatix en qual parts la hu eufrates laltra triges laltra trageon laltra frixon... *de aquest lac parla ysodorus* dient que paradis teranall es miga de laquinocsiall lo qual es molt delitable loch sens comparacio an lo quall es paradis teranall," see Battini's transcription in *Il mappamondo catalano estense*, pp. 184-185. For the reference to Isidore of Seville, see *Etymologiae*, XIV.3.2-4. Another source for this part of the Catalan Estense Map could have been the *Mirabilia descripta* (*c.* 1330) written by the Franciscan Jordanus Catalani, who wrote: "Between this India Tertia and Ethiopia is said to be, towards the East, the terrestrial Paradise; far from those parts come down the four rivers of Paradise, which abounds exceedingly in gold and in gemstones" (Jordanus Catalani, *Mirabilia descripta*, VII.6). I have quoted from the 1863 English translation of the Latin original, *Mirabilia descripta: The Wonders of the East, by Friar Jordanus, of the Order of Preachers, and Bishop of Columbum in India the Greater*, translated from the Latin original, as published at Paris in 1839, in the *Recueil de voyages et de mémoires*, of the Society of Geography, with the addition of a commentary, by Colonel Henry Yule, London: Hakluyt Society, 1863, p. 43. The translation was based on the *Mirabilia descripta per fratrem Jordanum, Ordinis Praedicatorum, oriendum de Severaco, in India Majori episcopum columbrensem*, in *Recueil des voyages et mémoires, publié par la Société de Géographie*, Tome quatrième, Paris: Arthus-Bertrand, 1839, pp. 1-68, which I have been unable to consult.

navis or merchant and ship owner). Albertin di Virga (*c.* 1350-1428/29), seems to place paradise in the West: two concentric circles in the southern part of 'Africa', could, in fact, refer to an earthly paradise.[23]

A significative change with respect to both the conceptualization and representation of paradise appears in the anonymous so-called "Genoese World Map" drawn in 1457, one of the most important and up-to-date map of the *oikumene* in the mid-fifteenth century (plate V).[24] A new transcription of the text (including all the legends and toponyms) would indicate this map as the only known example of a *mappa mundi* of possible Castilian origin before 1500. The majority of the legends are written in Latin, but most toponyms are in a *patois* with traces of Castilian flavour, though one legend in Asia seems to be in clear Castilian. The Genoese coat of arms placed in the left top corner of the *mappa mundi* and several Genoese flags in the Black Sea, rather than witnessing a Genoese origin, suggest instead that this map once had a Genoese dedicatee, or owner.[25]

[23] Albertin di Virga, [Mappa Mundi], manuscript on vellum, 696 × 440 mm, diam. of the *mappa mundi* 410 mm, signed and dated: *A. 141*[...] *Albertin diuirga me fecit in vinexia*, *c.* 1411-1419. Albertin di Virga's map, which has been lost from 1932, has been reproduced with its original dimensions in F. von Wieser, *Die Weltkarte des Albertin de Virga aus dem Anfange des XV. Jahrhunderts in der Sammlung Figdor in Wien*, Innsbruck: Heinrich Schurick, 1912. Also see R. Almagià, "Il Mappamondo di Albertin de Virga," *Rivista Geografica Italiana* 21 (1914), pp. 92-96; Destombes, *Mappemondes*, entry 52.6, pp. 205-207; A. Dürst, "Die Weltkarte von Albertin de Virga von 1411 oder 1415," *Cartographia Helvetica* 13 (1996), pp. 18-21; K. Seaver, "Albertin de Virga and the Far North," *Mercator's World* 2.6 (1997), pp. 58-62. For a biography of Virga, see P. Falchetta, "Marinai, mercanti, cartografi, pittori. Ricerche sulla cartografia nautica a Venezia (sec. XIV-XV)," *Ateneo Veneto* 182 (1995), pp. 40-45; Id., "Carte veneziane dell'Asia da Fra Paolino a Giacomo Gastaldi," in G. Curatola (ed.), *Sciamani e dervisci dalle steppe del Prete Gianni: religiosità del Kazakhistan e percezione del fantastico a Venezia*, Venice: Multigraf, 2000, pp. 39-50.

[24] Florence, Biblioteca Nazionale Centrale, Portolano 1, [*Hec est vera cosmographor(um) cum marino accordata* [*descri*]*cio, quoru(n)d(am) i(n) frivolis narracionibus reiectis, 1457*], manuscript on vellum, ellipse-shaped, North at the top, legends in Latin, 39.5 × 82 cm, 1457. The title is placed in the left margin of the parchment. There are two main translations, whose difference regards the interpretation of "marino." The first was by proposed by Joaquim Lelewel in 1850: "This is the true description of the cosmographers drawn in accordance with *marine charts*, from which all frivolous stories have been rejecgted". See J. Lelewel, *Géographie du Moyen Âge*, 7 vols., Bruxelles: V. Pilliet and J. Pilliet, 1852-1857, vol. 1, pp. 82-84. The other interpretation was proposed by German geographer Theobald Fischer: "This is the true description of the cosmographers drawn in accordance with *Marinus of Tyrus*, from which all frivolous stories have been rejecgted"; see T. Fischer, *Sammlung mittelalterlicher Welt und Seekarten italienischen Ursprungs und aus italienischen Bibliotheken und Archiven*, Venice: Ongania 1886, pp. 155-206, in part. p. 156 (English edition, E. L. Stevenson, *Genoese world map 1457, fac-simile and critical text incorporating, in free translation, the studies of professor Theobald Fisher*, New York: American Geographical Society and Hispanic Society of America, 1912). Outside the ellipse there are two scales of reduction: the first is placed to the North-East ("pro C miliaribus"); the second is placed to the South-East ("pro L miliaribus"). Also see Gentile, *Firenze e la scoperta dell'America*, entry 83, pp. 173-175; A. Capacci, "Planisfero detto 'genovese'," in G. Cavallo (ed.), *Cristoforo Colombo e l'apertura degli spazi*, Rome: Istituto Poligrafico e Zecca dello Stato, 1992, 2 vols., vol. 1, entry III.18, pp. 491-494.

[25] In 2007, I was commissioned by the Biblioteca Nazionale Centrale of Florence to study this map, to transcribe (together with Sara Bonechi) and translate all the toponyms and legends, and to analyze the text and the rich iconography. Previous transcriptions of the legends resulted incomplete and unreliable, partly because much of the gold pigment had flaked off the parchment. Archival research into the provenance of this map helps confirm the hypothesis that this *mappa mundi* was part of the Medici ducal collections and in all likelihood belonged to the *corpus* of maps on display in the Stanza delle matematiche in the Uffizi Palace probably since the beginning of the seventeenth century. See A. Cattaneo, *Mappa mundi 1457 (Biblioteca nazionale centrale di Firenze, Port. 1). Analisi, trascrizione e commentario*, Rome: Istituto dell'Enciclopedia Italiana, 2008; Id., "Découvertes littéraires et géogra-

In this *mappa mundi*, paradise is voluntarily not represented, as clearly emphasized in the following legend placed at the southernmost part of Africa :

> There were those who depicted the paradise of delights in this region [in Ethiopia], although others say that in reality it is in the East beyond the Indies. But since this is the true description of the cosmographers, who did not mention paradise at all, I abstain from accounting for it.[26]

The unidentified cartographer of this *mappa mundi* deliberately abstained from indicating the location of paradise, without either denying or confirming its location. On the one hand, the author was fully aware of the discrepancies between sources on its location (*alii in Aethiopia… alii ultra Indias in orientem…*) ; on the other hand, he justified his choice to exclude the *paradisus deliciarum* from the cartographic space of the *mappa mundi* in keeping with the main scientific tradition on which his work was based : that is, the work of the "cosmographers" (Ptolemy, Pomponius Mela and Solinus) all of whom were pagans and therefore did not refer to this topos of Christian cosmography either for reasons of not giving credit to it, or for ignoring it.

With respect to coeval representations, Fra Mauro's choice of placement of the paradise not only proves to be a rather eccentric, but also an innovative, reflection even in terms of the syntax of the cartographical representation itself. The anonymous cartographer of the 1457 *mappa mundi* is perhaps the only one who expresses a similar sensibility, albeit *sans* the theological dimensions that Fra Mauro included in his cosmography. Fra Mauro did not remain aloof from prevailing theological and doctrinal traditions : the originality of his interpretation, however, is evident in the way in which he interprets these traditions. He was clearly aware that the issue of the location of paradise, in the form of a cosmographic problem, implied dealing with the issue of how and where this topical moment in the history of Creation could be represented in a coherent cartographic way. In comparison to preceding and coeval *mappae mundi* Fra Mauro proffered a new interpretation of the challenging task of locating the historical site of the Garden of Eden, a place that was at the same time corporeal and spiritual.

5. Fra Mauro's written description of paradise

The opening words of the long legend placed above the image of the Garden of Eden make it clear that, for Fra Mauro, paradise was, above all, a real place on Earth and far away from regions that were known and inhabited by mankind. At the same time it was also a "spiritual sentiment" (*sentimento spiritual*) :

phiques au xve siècle. Le 'Portolano 1' de la Bibliothèque Nationale Centrale de Florence," *Médiévales* 58 (2010), pp. 79-98. For previous (partial) transcriptions of the legends of the Portolano 1 see T. Fischer's book quoted in the previous footnote and S. Crinò, *Come fu scoperta l'America*, Milan : Hoepli, 1943. See also Destombes, *Mappemondes*, entry 52.13, pp. 222-223.

[26] "In hac regione [Aethiopia] depinxerunt quidam paradisum deliciarum, alii vero ultra Indias ad orientem eum esse dixerunt. Sed quoniam hec est cosmographorum descripcio, qui nullam de eo fecerunt mentionem, adeo omittitur hic de eo narratio." Cattaneo, *Mappa mundi 1457*, p. 190 (legend E3).

> The Paradise of Delights does not only have a spiritual meaning ; it is also a real place on the Earth, as St. Augustine says in his *De Genesis* and in his book *De civitate Dei*. This place is very far from all human settlements and knowledge …[27]

Fra Mauro was explicitly referring to Augustine's Book VIII.1 (*Paradisus in Eden, et proprie et figurate accipiendus*) of the *De Genesi ad litteram libri duodecim* and to Book XIII.21 (*Quidam as intellegibilia paradisum referunt*) of the *De civitate Dei*. In comparison to the three main patristic forms of exegesis about paradise – reality, allegory, and reality with strong symbolic overtones[28] – in these two works Augustine firmly chose the third interpretation, thus implying that everything in earthly paradise was deemed to be real, but was also spiritual and symbolic :

> I am aware that many authors have written at great length on paradise, but their theories on the subject in general can be reduced to three. There is, first, the opinion of those who interpreted the word 'paradise' in an exclusively corporeal sense. Then there are those who prefer to give an exclusively spiritual meaning to the word. Finally, there are those who accept the word 'paradise' in both senses, sometimes corporeally and at other times spiritually. Briefly, then, I admit that the third interpretation appeals to me.[29]

Even though Fra Mauro cited diverse authoritative sources (*auctoritates*), his main and constant source for the legend on paradise was Augustine, whose reading of the Book of Genesis

[27] "El paradiso de le delicie no(n) solamente ha sentimento spiritual ma etia(m) quello esser uno luogo ne la terra si|tuado mette s(an)c(t)o Augustino sopra el Genesis et ancora nel libro De ciuitate Dei el qual luogo è molto remoto da la habi/tatio(n) e cog(ni)tion humana…," *Mappamondo*, lower left corner, lines 1-3 ; *Mappamondo* (ed. 1956), p. 22 ; for the English translation : *Mappamondo* (ed. 2006), p. 729.

[28] With regard to the literal interpretation of paradise, S. Hieronymus, *Epistula* LI, 5 : see S. Hieronymus, *Epistulae I-LXX*, edidit I. Hilberg. Editio altera supplementis aucta. Vindobonae : Osterreichischen Akademie der Wissenschaften, 1996 (*Corpus Scriptorum Ecclesiasticorum Latinorum* 54), pp. 403-405. The origins of the allegorical interpretation of paradise can be traced back to Philo Alexandrinus (13 BC-45 AD) and Origenes Alexandrinus (185-*c.* 253 AD), who were amongst the most important theologians of the Greek Church. See Philo Alexandrinus, *Le Allegorie delle leggi*, I, XIV 'La simbologia del giardino dell'Eden' ; I, XV 'La virtù, di cui è simbolo il giardino, è grazia di Dio e non conquista dell'uomo' ; I, XVII 'Gli alberi del giardino rappresentano le virtù particolari' ; I, XVIII 'Il significato dell'albero della vita e dell'albero della conoscenza del bene e del male e la loro collocazione nel giardino' ; I, XIX 'L'allegoria dei fiumi dell'Eden', in R. Radice (ed.), *Tutti i trattati del Commentario allegorico alla Bibbia*, in collaboration with G. Reale, C. Kraus and C. Mazzarelli, Milan : Rusconi, 1994, pp. 84-90 ; Origenes Alexandrinus, *De principiis*, IV.3.1, Manlio Simonetti (ed.), Turin : UTET, 1968, pp. 353-356, 514. Ambrose (339-397 AD) played a fundamental role in transmitting the interpretations of Philo and Origenes Alexandrinus in Latin. It is important to keep in mind that Augustine had heard the Sunday sermons proffered by Ambrose, the bishop of Milan, during his sojourn in the capital of the Western Roman Empire between 384 and 387. For Ambrose's spiritual or allegorical interpretation, see Ambrosius, *De paradiso*, I, 5-6 (Ambrosius, *Il giardino piantato a Oriente : De paradiso* ; introduction by U. Mattioli ; translation and notes by C. Mazza, Rome : Edizioni Paoline, 1981, pp. 61-63). For a study that is still highly relevant about the historical and cultural origins of the allegorical interpretation of the Bible in Philo and Origenes Alexandrinus' work, see W. Jaeger, *Early Christianity and Greek Paideia*, Cambridge, MA : Harvard University Press, 1961, pp. 1-16, 49-109. For a comprehensive discussion of the literal and allegorical interpretation of the Book of Genesis, see L. Carrozzi, "Introduzione," in Augustine, *La Genesi alla lettera : testo latino dell'edizione maurina*, pp. i-cx.

[29] Augustine, *De Genesi ad litteram libri duodecim*, Book VIII.1.1. See Augustine, *The Literal Meaning of Genesis*, transl. and annotated by J. H. Taylor, 2 vols., NewYork-Ramsey : Newman Press, 1982, pp. 32-33. For the Latin text, see note 1.

established an interpretation of the Garden of Eden as a real historical place.[30] Both the illumination and the written description of paradise explicitly took form according to the style and theological views of the *De Genesi ad litteram* and the *De civitate Dei*, as did the cosmographic contents attached to them. The history of Creation was one of Augustine's main theological interests and he placed it at the origin of Christian beliefs.[31] His invitation to ascertain the literal truth of the biblical exposition along with its spiritual significance makes the *De Genesi ad litteram libri duodecim* one of the most relevant contributions of the hexameronal exegetic tradition.[32] The Augustinian interpretation of paradise is a paradigmatic example of this approach to the biblical exegesis.[33]

As for its location, Fra Mauro followed a tradition established by Bede (*c.* 673-735), the scholar monk of Tyneside, Northumbria, and the commentator Peter Lombard (*c.* 1100-1160) who, on the basis of Isidore of Seville's *Etymologiae* (*c.* 635), affirmed that the Garden of Eden was to be found in the *eastern part* of the earth.[34] As clearly analyzed by Scafi, "the grounds for identifying the site of the earthly paradise rested on the *Vetus latina* (the old Latin translation of the Hebrew text of Genesis) which had explained that paradise had been planted not 'from the beginning' – as Jerome rendered in the Vulgate – but 'in the east'.[35] Fra Mauro then specified, referring to Albert the Great's *De natura loci*, that paradise could be found more precisely beyond the circle of the equinox.[36]

In truth, in his *De natura loci*, Albert the Great does not pronounce upon the location of earthly paradise. Instead he affirms the principle of the habitability of the equatorial areas and

[30] Scafi, *Mapping Paradise*, chapters 3 and 4, pp. 44-83.

[31] In *De Genesi ad litteram*, Augustine once again returned to the Genesis for the fourth and final time. He had done so previously in *De Genesi ad litteram libri xii* (*Patrologia Latina*, vol. 34, col. 245-286 ; *Corpus Scriptorum Ecclesiasticorum Latinorum* 28/1, pp. 3-435) and in AD 388-389 with the *De Genesi contra Manicheos libri duo* (*Patrologia Latina*, vol. 34, col. 173-219). This was followed by an attempt to expound upon its literal significance in 393 CE, in the *De Genesi ad litteram liber imperfectus* (*Patrologia Latina*, vol. 34, col. 219-246 ; *Corpus Scriptorum Ecclesiasticorum Latinorum* 28/1, pp. 457-503), which was, however, left unfinished and then again, in 398 CE, with the last three books (XI-XIII) of the *Confessions* (398-400 CE), which view it in an allegorical manner. For an excellent introduction to Augustine's works on the Book of Genesis, see L. Carrozzi, "Introduzione" and "Bibliografia" in Augustine, *La Genesi alla lettera : testo latino dell'edizione maurina*, pp. i-cxi.

[32] This was the view expressed by Augustine about the *De Genesi ad litteram* during a revision of all his works ; see Augustine, *Retractationes*, 2, 24.1, in Augustine, *La Genesi alla lettera : testo latino dell'edizione maurina...* p. 8 (*Patrologia Latina*, vol. 32, col. 640).

[33] G. Ricciotti, "St. Agostino come interprete del Genesi," *Didaskaleion* 10 (1931), pp. 23-52.

[34] "[...] el qual luogo è molto remoto da la habi|tatio(n) e cog(ni)tion humana : posto ne le parte oriental, sego(n)do la doctrina del sacro doctor Beda p(er) la cui auctorità el mai|stro da le sente(n)cie tal oppinio(n) aferma, " *Mappamondo*, lower left corner, lines 3-5. "This place is very far from all human settlements and knowledge ; and according to the teachings of the holy doctor Bede, whose authority is also followed by the Master of the Sentences [Peter Lombard], it is to be found in the East." The English translation derives from *Mappamondo* (ed. 2006), p. 729. See Isidorus Hispalensis, *Etymologiae*, XIV, 3.2-4. Isidorus Hispalensis, *Etimologie, o Origini di Isidoro di Siviglia* ; A. Valastro Canale (ed.), 2 vols., Turin : UTET, 2004, vol. 2, pp. 166-169.

[35] Scafi, *Mapping Paradise*, p. 35 and p. 47. Jerome's translation "from the beginning" could have been rooted in that the east, in one factual sense, is the "beginning of a day," because the sun rises there.

[36] "Auegna ch'el come(n)tator Alberto Magno nel libro de la n(atur)a di luogi metta quello oltra el circlo | equinotial, pur ne la regio(n) oriental," *Mappamondo*, lower left corner, lines 4-5 ; "In his book *De natura loci*, Albert the Great also puts it in the East, beyond the circle of the equinox," *Mappamondo* (ed. 2006), p. 729.

refutes traditional notions (of Aristotelian origin) that affirmed that the equatorial regions were uninhabitable.[37] Albert dealt with earthly paradise and its possible location in the East, either within the equator or beyond it in the southern hemisphere, in two other works : the *Summa de creaturis* and, especially, in his *De homine*.[38] While discussing the climatic characteristics of the equatorial region in the light of Ptolemy's *Almagest*, and confirming the reflections contained in the *De natura loci* about the inhabitability of the region, in the *De homine* Albert concluded that paradise could be found in the East, beyond the equator and thus in the southern hemisphere. This speculation should be considered in conjunction with his commentary on Peter Lombard's *Sententiae*, in which he affirms that paradise was to be found on the *orbis terrarum* but in a place that God had kept secret from mankind.[39] The analogy between a temperate equatorial zone and an earthly paradise that was held to be similarly temperate, together with the idea that the precise whereabouts of the Garden of Eden were unknowable, were at the origin of the interpretation expressed by Albert the Great.[40] On the basis of his exegesis, this notion that earthly paradise could possibly be located in the equatorial region had become diffused extensively at the time : for example, it is expressly mentioned, at least as a possibility, in widely circulated texts such as in the *Summa theologica* by Thomas Aquinas (who reflected upon two alternatives, a location on, or south of, the equator), or in Bonaventura's commentary on Peter Lombard's *Sententiae*, in which he formulated the hypothesis that paradise could be found slightly south of the equator (*vergens ad meridiem*).[41]

In the midst of these diverse indications it is important to emphasize that the geographical location of paradise on the *oikumene* was not the central issue of Fra Mauro's legend. Similar to the case of Augustine's *De Genesi as litteram* which, while maintaining that paradise was indeed a real place, regarded the issue of its whereabouts of secondary importance, Fra Mauro

[37] Albert the Great, *De natura loci ad fidem autographi*, edidit Paulus Hossfeld, Monasterii Wesfalorum in aedibus Aschendorff, 1980, Tract. 1, cap. 6, p. 12, lines 13-15. Albert the Great also contested the opposing view of authors such as Robertus Anglicus – who lived in the second half of the thirteenth century and authored a commentary of Sacrobosco's *De sphaera*, amongst many other works – who were of the opinion that the equatorial regions were perhaps the most temperate of all areas on account of the fact that day and night were of an equal duration. See Robertus Anglicus, *Commentus in Tractatum de spera Iohannis de Sacrobosco*, in L. Thorndike, *The Sphere of Sacrobosco and Its Commentators*, Chicago : University of Chicago Press, 1949, p. 190.

[38] Albert the Great, *Summa de creaturis* II, quaest. LXXIX, art. I (*Opera omnia*, E. Borgnet (ed.), t. 35, Paris, 1896, p. 640). See Scafi, *Mapping Paradise*, pp. 179-181 and P. Gautier Dalché, "Le paradis aux antipodes," pp. 615-637 (pp. 628-629). Albert the Great changed his mind about the location of the earthly paradise on several occasions, but in the *De homine* he finally placed paradise south of the equator.

[39] "Ad aliud dicendum, quod ille locus ab habitabili nostra secretus est : non quod non sit in habitabili nostra, sed quod Dominus clausit eum a nobis," see *In II sententiarum*, dist. XVII, E, art. 4, E. Borgnet (ed.), *Opera omnia*, Paris : 1890-1899, t. 27, 1894, p. 304 and Gautier Dalché, "Le paradis aux antipodes," p. 628.

[40] Scafi, *Mapping Paradise*, pp. 179-181.

[41] Thomas Aquinas, *Summa theologica*, Pars I, quaest. 102 ; Pars II.2, quaest. 155, art. 4 ; Bonaventura da Bagnorea, *Liber II sententiarum*. Lib. II, dist. XVII, dub. 3 : "paradisi corporalis situs est valde eminens et altus et iuxta aequinoctialem in oriente, quodam modo vergens ad meridiem" : see *Liber II sententiarum*. Editio minor. Ad Claras aquas, Florentiæ (Quaracchi, Florence) : Ex typographia Collegii S. Bonaventuræ, 1938, pp. 440-441 (this is part of the *Doctoris seraphici S. Bonaventurae s.r.e. episc. cardinalis opera theologica selecta*, 5 vols., iussu et auctoritate R.mi P. Leonardi Bello, P. Pacifici M. Perantoni, P. Augustini Sepinski, cura pp. Collegii S. Bonaventurae edita, Ad Claras aquas, Florentiae (Florence, Quaracchi) : Ex typographia collegii s. Bonaventurae, 1934-1964, vol. 2). For a discussion about the location of paradise at or beyond the equator, see Scafi, *Mapping Paradise*, pp. 172-179.

did not specify where paradise was located : he limited himself to reporting the most influential among various early medieval interpretations that distilled Augustine's exegesis. Rather, in the *mappa mundi*, paradise was interpreted as a fundamental theme in the context of theology and cosmography concerning the Creation, instead of in the context of the geography of the *oikumene*.

As will be discussed in Chapter VI, dedicated to the "Narrative structure of the *mappa mundi*", Fra Mauro, by using the pronoun "I", personally intervened about 80 times in the legends while presenting his own authoritative and learned opinions on several cosmographic, geographic and historical matters to readers whom he addressed directly. The fact that he avoided expressing his own opinion about the uncertain location of paradise is an important indicator that he was not interested in the strictly geographical aspects concerning the "Paradise of Delights."[42] In order to sustain the historical, tangible, and earthly existence of a terrestrial paradise Fra Mauro thus turned to the patristic exegesis of some passages of the so-called "historical and prophetical books" of the Old Testament :

> It is believed that Enoch was placed in this Paradise and transported to heaven and, in the fourth chapter of Genesis and also in Ecclesiastes, one can read of his ascent. This is similarly believed to have happened to the prophet Elijah after he was carried up to heaven ; this is recorded in the fourth Book of Kings. It is also believed that here the souls of the Holy Fathers were visited by the Redeemer after his death on Good Friday ; that, with them, the soul of our Redeemer, united with God, descended into Limbo until the day of his Resurrection ; and that the Fathers were freed from the bonds of Original Sin and placed in this Paradise until the day of the Ascension, as St. Augustine says in his sermon *De passione Domini*. This was so that Christ should be the first to enter into the heavens and paradise of the blessed : "And on the day of his Ascension, he showed men the way to get there, as had been predicted by the prophet Micaiah.[43]

The "fourth chapter of Genesis" (in reality, the fifth)[44] and the Book of Ecclesiastes for the story of Enoch ; and the "fourth Book of Kings" (actually, the second, in current numbering

[42] For a study of the way in which Fra Mauro personally enters the narrative structure of the *mappa mundi*, in addition to Chapter VI of this book, also see P. Gautier Dalché, "Weltdarstellung und Selbsterfahrung : Der Kartograph Fra Mauro," in H.-D. Heimann, P. Monnet (eds), *Kommunication mit dem Ich*, Bochum : 2004, pp. 39-51 (pp. 45-47).

[43] "Qui ancora pietosame(n)te se crede fosseno le anime de li s(an)c(t)i padri uisitade dal n(ost)ro Rede(m)ptor | el uenerdì de la sua passio(n) da poi la morte sua, cu(m) li qual se crede l'anima del nostro Rede(m)ptor unita a la diuinità esser stata nel li(m)bo fina | al zorno de la sua resurrectio(n), dal qual luogo q(ue)lli padri liberadi dal uinculo del peccato original, come parla s(an)c(t)o Augusti(n)o nel sermo(n) de la pas|sion, fono posti i(n) questo paradiso, nel qual fin al zorno de la asce(n)sion lor steteno, cu(m)ciosiaché Christo fosse el p(r)imo che nel cielo e paradisi de li | b(ea)ti intrasse mo(n)stra(n)do a li homini la uia de andar a quello nel dì de la sua asce(n)sio(n), come predisse el proph(et)a Michea. Et auegna che'l Rede(m)ptor n(ost)ro | per la passio(n) e morte sua meritasse a la humana n(atur)a lo i(n)troito del paradiso de li b(ea)ti, ampuo li piaque per la asce(n)sion indur le anime predicte | in possessio(n) del dicto luogo de beatitudine za a li ho(min)i aperto per li meriti de la sua passio(n). E p(er)ò da questo paradiso de le delicie se crede le anime | sancte predicte ascendesse in cielo el zorno de la ascensio(n) seguendo Christo ampuo uisibilme(n)te dal monte Oliueto ascese in cielo, ne la costa del | qual monte era situado el castello ouer uila de Betania, come narra el sacro doctor Beda." *Mappamondo*, lower left corner. *Mappamondo* (ed. 1956), p. 22 ; *Mappamondo* (ed. 1956), p. 730.

[44] These must have been the passages of the Holy Scriptures to which Fra Mauro referred. For "Genesis al quarto capitulo", see Genesis 5, 21-24 : [21]When Enoch had lived 65 years, he became the father of Methuselah. [22]And after he became the father of Methuselah, Enoch walked with God 300 years and had other sons and daughters.

of the Old Testament[45]) for the story of the translation of the prophet Elijah, are the historical books *par excellence* in the Biblical tradition. Fra Mauro's passage needs a clarification: the books that he cited from the Holy Scriptures do not, in fact, speak of paradise. The identification of the divine abode where God took Enoch and Elijah, before the Passion of Christ, with the Garden of Eden is a belief that was expressed as early as the first centuries of Christianity, both in the Greek as well as Latin patristic traditions, which then firmly took hold in medieval doctrinal tradition. This tradition further affirmed that Jesus, while on the cross at Golgotha, by promising the good thief that "Today you will be with me in Paradise," (Luke 23, 43), reopened paradise, which had been closed after Adam and Eve had been expelled, to the souls of the just. Thus, earthly paradise became a "place of waiting" for the souls of the chosen ones before they ascended to the heavenly paradise, after Judgment Day.[46] In Augustine's exegesis the admission of the fleshly Enoch and Elijah into paradise, well after the expulsion of Adam and Eve, "was a way of saying that, in allowing the two Old Testament patriarchs into Eden, God was showing mankind what Adam and Eve would have enjoyed had they not sinned; namely, a period of fleshy immortality, on earth (granted by the Tree of Life) before being transferred to heaven," Scafi argues.[47] It would seem that it was this kind of belief, the subject of heated doctrinal disputes, which had even been confirmed by the Council of Florence during Fra Mauro's age, that the legend on paradise sought to transmit.

Once the "historical" description was concluded there then followed a cosmography of paradise, in which Fra Mauro introduced the fundamental theme of the fountain from which then emerge the "four large and main rivers"; he next lingers over "the tree that distinguishes good and evil" and the "tree of life", interpreted by means of an explicit reference to Augustine's

[23]Altogether, Enoch lived 365 years. [24]Enoch walked with God; then he was no more, because God took him away." The words suggest that Enoch did not die, but was transported by God to some divine abode. See Scafi, *Mapping Paradise*, pp. 54-55. In this regard, it is curious to note how Fra Mauro gets confused while indicating the "fourth chapter of the book of Genesis" for the story of Enoch, which is instead to be found in Genesis 5, 24. As for the "Book of Ecclesiastes," see Ecclesiastes 44, 16: "Enoch piacque al Signore e fu rapito, esempio istruttivo per tutte le generazioni."

[45] For the "Fourth Book of Kings, see Second Books of Kings 2, 1-12: "[1]... Elijah and Elisha were on their way from Gilgal.... [8]Elijah said to Elisha, "Tell me, what can I do for you before I am taken from you?" "Let me inherit a double portion of your spirit," Elisha replied. [10]"You have asked a difficult thing," Elijah said, "yet if you see me when I am taken from you, it will be yours – otherwise not." [11]As they were walking along and talking together, suddenly a chariot of fire and horses of fire appeared and separated the two of them, and Elijah went up to heaven in a whirlwind. [12]Elisha saw this and cried out, "My father! My father! The chariots and horsemen of Israel!" And Elisha saw him no more. Then he took hold of his own clothes and tore them apart." While indicating the Fouth Book of Kings for this passage referring to Elijah, Fra Mauro follows a consolidated tradition of medieval numbering for the Old Testament.

[46] See Delumeau, *Storia del paradiso*, pp. 35-55 (pp. 42-44), who cites Jerome (*Epistula XXII*, *Patrologia Latina*, vol. 22, col. 405), John Crisostomo (*De cruce et latrone*, *Patrologia Graeca*, vol. 49, col. 401), later transmitted via Isidore of Seville (*De ordine creaturarum*, *Patrologia Latina*, vol. 83, col. 927, 928, 940) and Bede (*Homilia IV* and *Homilia XXII*, *Patrologia Latina*, vol. 94, respectively col. 30, 96).

[47] Augustine, *De civitate Dei*, XV.19. See Scafi, *Mapping Paradise*, pp. 55-56 and p. 61 note 71. Fra Mauro's reference to Augustine's *De passione Domini* seems to be misleading.

exegesis.[48] Fra Mauro's description ended with a discussion about the fountain of Eden and the four rivers that spring out of it, setting out from Genesis 2, 10-14:

> 10 A river watering the garden flowed from Eden; from there it was separated into four headwaters.
> 11 The name of the first is the Pishon; it winds through the entire land of Havilah, where there is
> gold. 12 (The gold of that land is good; aromatic resin and onyx are also there.) 13 The name of the
> second river is the Gihon; it winds through the entire land of Cush. 14 The name of the third river
> is the Tigris; it runs along the east side of Asshur. And the fourth river is the Euphrates.

This is Fra Mauro's version:

> As the Holy Scriptures speak of the spring in Paradise and the four rivers that arise from it, this drawing depicts that. As there are still many who wonder how it is possible that these four rivers, arising from that most remote place, should yet have sources that are very far from one another, I will answer with the words of St. Augustine's *De Genesi*: these rivers, whose sources are known to us, are linked by underground routes, running through many regions and then coming to the surface in different places. One – that is, the Ganges – arises in India; the Tigris in Armenia, at Mount Charabach; the Euphrates also in Armenia, near the city of Erzurum; and the Gihon – or Nile – in Ethiopia, in the province of Meroe in Abassia.[49]

48 "Nel qual paradiso fo posto el primo nostro progenitor Adam nel stato de la sua innocentia, e | nel mezo de quello è uno fonte p(er) irrigatio(n) del dicto luogo, dal qual poi procede i quatro grandi e principal fiumi, et etia(m)dio | in questo paradiso Idio su(m)mo auea piantado nel mezo do' arbori, uno di qualli era l'arboro de la scientia del ben e del mal, zoè | quello del fructo del qual Dio fece comandame(n)to ha Adam che no(n) ne ma(n)zasse, dal qual coma(n)dame(n)to parte(n)dosse Adam per inobe|[16]die(n)tia, zoè ma(n)zando del fructo, oltra el bene che prima lui auea ueduto e sentito, lui p(er) esperimento cognoscete el male, cadendo | ne li diffecti che seguitano el peccato, dal qual aduenime(n)to quello arboro è stato nominato de scientia de bene e de male. L'altro fo l'ar|boro de uita, el qual ma(n)zato hauea proprietà de co(n)seruar l'omo in uno stato per alqua(n)to tempo. Onde essendo sta' continuato el man|zar de questo arboro l'omo seria co(n)tinuame(n)te uiuesto, come dechiara s(an)c(t)o Augustino nel libro De Ciuitate Dei e anchor nel libro de le q(ue)|stio(n) del Nuouo et Vechio Testame(n)to." *Mappamondo* lower right corner. *Mappamondo* (ed. 1956) p. 22. *Mappamondo* (ed. 2006) p. 730. See Augustine, *De civitate Dei contra paganos*, Lib XIII, 20 'Quod caro sanctorum, quae nunc requiescit in spe, in meliorem reparanda sit qualitatem, quam fuit primorum hominum ante peccatum'; XIII, 21 'Quidam ad intellegibilia paradisum referunt'; Lib. XIV, 17 'De nuditate primorum hominum, quam post peccatum turpem pudendamque viderunt'; also see Augustine, *De Genesi ad litteram*, VIII, 5.9 'Lignum vitae et rem veram et figuram fuisse' e VIII, 6.12. 'Lignum scientiae boni et mali arbor vera et innoxia'.

49 "Perchè la sacra scriptura fa mention del fonte del paradiso et ancora de queli quatro fiumi che nasceno da q(ue)llo, | però ne la presente pictura se demostra q(ue)sto, e p(er)chè sono molti che se merauegliano come sia possibile che q(ue)sti quatro | fiumi nassando da questo luogo remotissimo habi li suo fonti distantissimi l'uno da l'altro, respondo seco(n)do sancto | Augustino sopra el Genesi, che q(ue)sti fiumi, hi fo(n)ti di qual ne sono noti, habino luogi subterranei e passa(n)do molte | regio(n) nassano i(n) diuerse p(ar)te e che uno cioè Ganges nassa i(n) India, Tygris in Armenia dal mo(n)te Charabach, Eufrates | similiter i(n) Armenia apresso la cità Arçeron, (et) Gen ouer Nilo i(n) Ethyopia i(n) Abassia, in la prouincia de Meroa." *Mappamondo*, lower left corner; *Mappamondo* (ed. 1956), p. 22. *Mappamondo* (ed. 2006), p. 730. With regard to the sources of the Nile, according to Fra Mauro: "Io ho notado de sopra | che'l nilo nasce i(n) abassia tra | do' proui(n)cie, zoè marora ouer meroa e salgu. Ma j libri punici dicono che | nasce i(n) Mauritania, la qual cossa | io no(n) credo tuta esser uera che'l Nilo | habi qui el suo origine p(er) le i(n)formatio(n) | ho habuto, ma che questo sia uno ramo | del Nilo io affermo, p(er)ché se truoua q(ue)li | simili animali che se truoua nel | Nilo." *Mappamondo* (ed. 1956) Pl. XVII, 41: "I have noted above that the Nile arises in Abassia between two provinces – that is, Marora or Meroa and Salgu. But the Punic books say that it arises in Mauritania, which I do not think can be entirely true – that the Nile has its origins here – because of the information that I have had. But I say that this is a branch of the Nile, because within it one finds animals similar to those in the Nile." *Mappamondo* (ed. 2006) *480.

In order to resolve the contradiction between biblical dictates and geographical knowledge Fra Mauro further referred to Augustine's *De Genesi ad litteram*, especially to two passages extracted from Book VIII, 7.13 (*Paradisi flumina vera fuisse*)[50] and Book VIII, 7.14 (*De fonte et decursu fluminum illorum*).[51] Strictly with regard to the cosmographic facet of the four rivers, following an interpretation by Iosephus Flavius, Augustine thought that they were a concrete reality – and identified them as the Nile, the Ganges, the Euphrates, and the Tigris – but they were also symbols of the four cardinal virtues.[52] He explained their common origin, envisaging long subterranean routes and placing their ultimate source in an unknown location. This is the version to the letter of Fra Mauro, who thus followed a tradition that was deeply ingrained throughout medieval Christianity.

Although explicit references were made to Augustine's *De Genesi ad litteram* and *De civitate Dei* the legend on paradise conceals intermediary texts. The discussion on the terrestrial sources of the four rivers clearly corroborates this view. The fact that Fra Mauro resorted to intermediate sources is evident when one keeps in mind that although Augustine identified the biblical rivers as the Nile, the Ganges, the Tigris and the Euphrates, and affirmed the principle that these rivers all had subterranean courses, he did not, in fact, describe the locations of their earthly sources. Instead, it was Bede who, in his *Libri in principium Genesis*,[53] a didactic commentary on the *Genesis*, focused on this particular aspect of the geography of paradise, indicating the terrestrial sources of the four rivers – the Caucasus mountains for the Ganges, Armenia for the Tigris and the Euphrates, the region of the Atlas Mountains for the Nile – and explained their common origin from the fountain of Eden by means of subterranean courses.

Already mentioned by Fra Mauro in the first part of the legend, Bede's work was a didactic commentary that summed up and, above all, simplified Augustine's complex exegesis of Genesis, thus facilitating the reception of Augustine's works during the Middle Ages.[54] Bede's

[50] "De his fluminibus quid amplius satagam confirmare, quod vera sint flumina nec figurate dicta, quae non sint, quasi tantummodo aliquid nomina ipsa significent, cum et regionibus, per quas fluunt, notissima sint et omnibus fere gentibus diffamata? Quin immo ex his, quoniam constat ea prorsus esse – nam duobus eorum nomina vetustas mutavit, sicut Tiberis dicitur fluvius, qui prius Albula vocabatur; *Geon quippe ipse est, qui nunc dicitur Nilus*; *Phison autem ille dicebatur, quem nunc Gangem appellant; duo vero cetera, Tigris et Euphrates, antiqua etiam nomina tenuerunt* – nos admoneri oportet cetera quoque primitus ad proprietatem litterae accipere, non in eis figuratam locutionem putare, sed res ipsas, quae ita narrantur, et esse et aliquid etiam figurare." See Augustine, *La Genesi alla lettera*, VIII, 7.13, pp. 404-407.

[51] "An eo movebimur, quod de his fluminibus dicitur aliorum – esse fontes notos, aliorum autem prorsus incognitos, et ideo non posse accipi ad litteram, quod ex uno paradisi flumine dividuntur? Cum potius credendum sit, quoniam locus ipse paradisi a cognitione hominum est remotissimus, inde quatuor aquarum partes dividi, sicut fidelissima Scriptura testatur, sed ea flumina, quorum fontes noti esse dicuntur, alicubi isse sub terras, et post tractus prolixarum regionum locis aliis erupisse, ubi tamquam in suis fontibus nota esse perhibentur. Nam hoc solere nonnullas aquas facere quis ignorat? Sed ibi hoc scitur, ubi non diu sub terris currunt. Exibat ergo flumen de Eden, id est de loco deliciarum, et irrigabat paradisum, id est ligna omnia pulchra atque fructuosa, quae omnem terram regionis illius opacabant." See Augustine, *La Genesi alla lettera*, VIII, 7.14, pp. 406-407.

[52] Iosephus Flavius, *Ant. Iud.*, I, 37-38. Turin: UTET, vol. 1, book I-X, p. 53.

[53] Bede, *Libri quatuor in principium Genesis usque ad nativitatum Isaac et electionem Ismahelis adnotationum*, C. W. Jones (ed. and studies), Turnhout: Brepols, 1967 (Corpus Christianorum. Series Latina; 118A), II.10, rr. 1520-1539, pp. 48-49.

[54] With regard to the relationship between Augustine's and Bede's exegesis of Genesis, see P. Siniscalco, "Due opere a confronto sulla creazione dell' uomo: il De Genesi ad litteram libri XII di Agostino e i Libri in principium

commentary did, in fact, have a large following: it was cited by Peter Lombard, who, in the *Sententiae* (*c.* 1155-1158), while upholding the dual interpretation of the spiritual and material nature of paradise, indicated that Bede's work was one of the main sources of his discussion of paradise.[55] However, I would not exclude other possibilities such as the *Glossa ordinaria*, a twelfth-century anonymous collective and systematic commentary on the Bible, compiled in the monastic school at Laon, providing synoptic explanatory comments. The *Glossa* was widely used in the Middle Ages as a standard reference work, "systematically quarried by later medieval clerics eager to take advantage of the *Glossa* as a single source for quotations from the Church Fathers for any verse of Scripture."[56] In conclusion, the analysis of the legend on paradise shows that Fra Mauro interpreted a nucleus of notions which, rooted in Augustine's *De Genesi ad litteram libri duodecim* and *De civitate Dei*, were transmitted to the Quattrocento *via* Isidore of Seville, Bede, Peter Lombard and the *Glossa ordinaria*, becoming a common patrimony in medieval exegesis.

6. Fra Mauro and St. Augustine's exegesis of the earthly paradise

The clear link between Fra Mauro's description of the Garden of Eden and Augustine's *De civitate Dei* and, above all, *De Genesi ad litteram libri duodecim* (albeit, probably an indirect one, *via* Bede's comments on the Book of Genesis and the *Glossa ordinaria*), could also provide one of the keys to reading, interpreting and explaining Fra Mauro's choice of representing paradise outside the limits of the *oikumene*, in continuance with the cosmographic space of the heavens, the sub-lunar world, the astronomical structure of the universe, drawn and described in the other three corners between the square and circular frame of the *mappa mundi*. There could also be other reasons for this. For example, simple but crucial practical aspects of composition: Fra Mauro often lamented in his *mappa mundi* that he "did not have enough space" for everything that he could, or wished to, represent. This was, of course, a habitual and recurrent rhetorical formula aimed at guarding against possible criticism and *lacunae*, thus thwarting any eventual reproof *a priori*.[57] Nevertheless, shifting paradise from the edges of the map where it was generally placed did, in fact, leave room for other representations. Even aesthetic and artistic choices could have played a role: an illumination of the quality of the paradise, by the celebrated and masterful hands such as those of Leonardo Bellini, who worked for doges, patrician families and the most important Venetian religious confraternities (*scuole*), would have undoubtedly exalted the *mappa mundi*, conferring immense prestige upon the work. For

Genesis di Beda," *Augustinianum* 25 (1985), pp. 435-452 (*Miscellanea di studi in onore di P. Agostino Trapé*).

[55] For a description of earthly paradise in the *Sententiae*, see Peter Lombard, *Sententiarum libri quatuor*, II. dist. XVII chap. 5, 'Quod paradisus terrenus fuit et localis', in Peter Lombard, *Sententiae in IV libris distictae*, t. 1.2, Grottaferrata: Editiones Collegii S. Bonaventurae ad Claras Aquas, 1971, pp. 413-414 (also see *Patrologia Latina*, vol. 192, col. 686). There is a vast bibliography about Peter Lombard. For a study of his doctrine about an overall interpretation of Creation, especially with regard to the complex relationship with St. Augustine's works, see M. Colish, *Peter Lombard*, 2 vols., Leiden: Brill, 1994, vol. 1, pp. 303-397.

[56] Scafi, *Mappimg Paradise*, p. 49.

[57] See the final considerations in Gautier Dalché, *Weltdarstellung und Selbsterfahrung: Der Kartograph Fra Mauro*, pp. 39-51 (pp. 45-47).

this to happen it was necessary to find an adequate location for paradise, which was impossible within the representation of the *oikumene* given the already choked spaces on the maps that were suffocated by all kinds of drawings and scrolls. Thus, there was no option but to choose one of the external corners, which were perfect for accommodating both the textual as well as the pictorial representation.

These kinds of reasons of a broad aesthetic and compositional nature are not wholly incompatible with cultural explanations. As a matter of fact it would seem that one would support the other and vice versa. The fact of a lack of space, the impossibility of providing a "topographical" representation for paradise because of the uncertainty surrounding its location, combined with an awareness of the theological and cosmological value and pertinence of the *hortus deliciarum*, a site that was simultaneously a real and spiritual place, implied the need to be able to resort to an artistic hand that knew how to draw anthropomorphic figures in an appropriate manner (the rough sketches of paradise in Bianco's *mappa mundi*, acceptable in terms of the small size of the map, would have proved unsustainable on the scale of Fra Mauro's map). Thus, many different reasons could have combined to influence the location of paradise.

The fact that the image of paradise could have been defined while also keeping in mind Augustine's exegesis is also worthy of note. If, in fact, in terms of written arguments, tradition provided numerous and refined examples for interpreting Augustine's exegesis of a paradise that was at the same time a real place and a "spiritual sentiment" (*sentimento spirituale*), actually giving it a pictorial and cartographical visibility was a far more problematic proposition. The choice of explicitly placing paradise outside the boundaries of the *oikumene* can possibly be interpreted as an effective means – and not an accidental outcome – of concretely depicting St. Augustine's fondness for "those who accept the word 'paradise' in both senses, sometimes corporeally and at other times spiritually."

The act of placing it outside the *oikumene* represented in the circular *mappa mundi*, like the nine heavens (upper left corner), and the sub-lunar world formed by the four elements (upper right corner), could be compatible with the dictates of preparing a coherent visual representation corresponding to Augustine's description in the *incipit* of Book VIII of the *De Genesi ad litteram*, as also in Book XIII of the *De civitate Dei*. This is a choice that fits into, and confirms the distinction between, the space of the *oikumene* accessible to man, and a cosmographic and cosmologic space that was undoubtedly discernable but inaccessible, at least during an earthly existence. The two spaces are separated but nonetheless still communicate. The water of the fountain of Eden and the four rivers that emerge from it unite these two areas : *via* this water the cosmographical space communicates directly with the *oikumene*. According to St. Augustine the fountain of Eden and the rivers that sprang from it also had a spiritual significance : the water of paradise flowing towards the *oikumene* renewed for mankind the grace and mystery of God's gesture of creation. Fra Mauro arrived at an effective solution for the theological subtlety of a paradise *on* earth that was simultaneously not part of the *oikumene*, but which was an integral part of the story of creation and was thus a part of the macro-space of cosmography and cosmology.[58]

Therefore, in this regard for Fra Mauro, the depiction of paradise was not a purely symbolic element, nor was it a "due" doctrinal element or even the exclusion of an earthly paradise

[58] Augustine, *La Genesi alla lettera*, p. 386.

from the ambit of what could be mapped. On the contrary, Fra Mauro removed paradise from the margins of the *oikumene*, where it had generally been relegated in preceding and coeval *mappae mundi* (being situated close to Asia or Africa, as has been seen), and placed it at the centre of the cosmographic discourse. The fact that the geographical reality of paradise was not placed within, and included in, the *oikumene* in this *mappa mundi* does not really indicate that the *quaestio* about the position of paradise was a "doctrinal element," extraneous to cosmographical knowledge.

It is important to stress and highlight that there was no separation or "epistemological rupture" between geographical knowledge, increasingly based on experience and knowledge transmitted *via* books, and "doctrinal" cosmographical knowledge. The legend and representation of paradise is proof of this : the compositional style, the syntax, and lexicon of the written description of paradise correspond to those of the inscriptions that can be read within the map. Paradise was thus part of the overall programme of Fra Mauro's *liber figurarum*. The fact that he did not execute it *manu propria* was a result of the apparent awareness that the representation of human figures required very different, and more sophisticated, pictorial skills than those that were necessary in cartographic drawings of small figures used for ships, buildings and cities, essential points that filled the spaces of the *mappa mundi*. In fact, the figures of God, Adam and Eve are the only anthropomorphic images to be represented in the chart. A final note in the legend on paradise specifies what could have been the relationship between the cosmographer and the illuminator : "As the Holy Scriptures speak of the spring in Paradise and the four rivers that arise from it, this drawing depicts that."[59] This last sentence of the long legend describing paradise denotes that the iconography was defined by Fra Mauro and that, at least for the strictly cosmographical aspects, the miniaturist followed his indications so that the details of the rivers, which were fundamental elements in the context of the cosmography, depicted the Augustinian interpretation of the biblical narration of a terrestrial paradise.

II. – Leonardo Bellini and the earthly paradise in Fra Mauro's *mappa mundi*

1. A brief history of an attribution

It is patently evident to anyone who admires the *mappa mundi* preserved at the Biblioteca Marciana that the cosmographical drawing and its details were not painted by the same hand that drew the earthly paradise in the lower right-hand corner. God, Adam and Eve, the angel Gabriel and the trees and flowers that adorn the *hortus deliciarum* were all drawn by a more refined and artistically erudite hand, familiar with the new Renaissance techniques of three-dimensional representation of space and a humanistic pictorial sensibility that, especially in the representation of faces and bodies, proposed a fresh vision and a confident plasticity that distinguishes these figures both from the Gothic calligraphy that characterizes the numerous drawings of cities, temples, bridges, and ships within the space of the *oikumene* in the *mappa mundi*

[59] "Perché la sacra scriptura fa mention del fonte del Paradiso et ancora de queli quatro fiumi che nasceno da quello, però ne la presente pictura se demostra questo…, " *Mappamondo*, left lower corner. The English translation derives from *Mappamondo* (ed. 2006), p. 730.

as well as the Gothic style that characterized Venetian miniatures until the 1450s.[60] Furthermore, the figures of God, Adam and Eve and the angel guarding paradise are the only human figures portrayed in the *mappa mundi*.

A research project I carried out on miniaturists and painters who operated in Venice between 1450 and 1465, and a critical comparison of works attributed to this period in the history of Venetian miniatures while searching for iconographic traces that could facilitate the attribution of works, enabled me to ascribe the authorship of the paradise depicted in the *mappa mundi* to the Venetian, Leonardo Bellini (Venice, *c.* 1423/1435-*c.* 1490), who was a nephew of Jacopo Bellini (*c.* 1400-1470 or 1471) and cousin of the celebrated Giovanni (1430-1516) and Gentile Bellini (1429-1507).[61] This attribution is in conformance with the views of the art historian Susi Marcon who suggested a similar authorship in an article published at around the same time when I presented my research at the 19th International Conference on the History of Cartography held in Madrid in July 2001.[62] Both these research projects were carried out independently and considered only partially coincidental aspects while reaching the same conclusions. Marcon's research was particularly interested in the longstanding debate in the history of Venetian art about the possible involvement of the Bellini family and Andrea Mantegna in miniature art.[63] As will be analyzed shortly, the extraordinary congruity of styles, sceneries, colors and postures of the human figures, in other words the same modelling and physiognomic and naturalistic characterization, between the earthly paradise of Fra Mauro's *mappa mundi* and many other drawings unanimously attributed by historians of miniature art to Leonardo Bellini, were decisive elements in this attribution, which was not linked to documentary evidence but rather to iconological procedures.

The characteristics that critics have identified and emphasized as being drawn by the hand of Leonardo Bellini are : the clearly three-dimensional conception of space ; a quest for effects reflecting regularity and geometrical simplicity combined with the acquisition and development of experiences derived from the Ferrara school of miniature painting, especially with regard to the figure of Franco dei Russi ; as well as the stimuli of Tuscan influences in Venice and Padua that are evident in the delicacy of the contours ; the soft and joyous interplay of colors combined with a clear and airy representation of the landscape. These traits, so typical of Leonardo

[60] For an overview of the history of Venetian miniature painting during the fifteenth century, G. Mariani Canova, "Miniatura e pittura in età tardogotica (1400-1440)," in *La pittura nel Veneto. Il Quattrocento*, vol. 1, Milan : Electa, 1989, pp. 193-222 ; Ead., "La miniatura a Venezia dal Medioevo al Rinascimento," in R. Pallucchini (ed.), *Storia di Venezia*, vol. 2, *Temi. L'arte*, Rome : Istituto della Enciclopedia italiana, 1995, pp. 769-843 ; J. J. G. Alexander, "Notes on Some Veneto-Paduan Illuminated Books of the Renaissance," *Arte Veneta* 23 (1969), pp. 9-20.

[61] This research was carried out between 1999 and 2000 and was submitted in October 2000 and presented in July 2001 in Madrid, at the 19th International Conference on the History of Cartography. It was published in *Imago Mundi* 55 (2003), pp. 97-102. I am extremely grateful to Prof. Giordana Mariani Canova, a renowned expert on Venetian miniatures and especially on Leonardo Bellini, for having guided and familiarised me with the stylistic elements that enabled the attribution of this depiction of paradise. I am also indebted to Prof. Lilian Armstrong for her revision of the text.

[62] Marcon, "Il Mappamondo di Fra Mauro e Leonardo Bellini," pp. 103-108. See note 5 (p. 132).

[63] "The unmistakeable Bellini hallmark visible in at least one of the depictions annexed to Fra Mauro's celebrated *Mappamondo* leads us to an argument that has long generated rivers of ink, that is, the fact that the Bellini family possibly worked in artistic genres similar to paintings carried out on canvas and boards ; and particularly the activity of the Bellini studio in the context of miniature art or paintings on parchment". See Marcon, "Il Mappamondo di Fra Mauro e Leonardo Bellini," p. 103.

Bellini's distinctive style, also characterize the earthly paradise depicted in Fra Mauro's *mappa mundi*. In particular, an iconographical analysis of this representation of paradise confirms, according to Giordana Mariani Canova, a common visual culture shared by Leonardo Bellini and the Florentine masters Silvestro Barbetta and Antonio de Giacomo, who, in 1458, signed and dated four figures of saints rendered in mosaic in St. Mark's Basilica.[64]

2. Leonardo Bellini

Leonardo Bellini's existence has been known only since 1894 thanks to Pietro Pauletti's rediscovery of a contract between one of the founders of the Venetian artistic Renaissance, the painter Jacopo Bellini, and Leonardo, the son of Paolo dai Remi and Elena Bellini, Jacopo's sister.[65] The contract, signed on 23 August 1443, concerned the conditions for Leonardo's work in the studio of Jacopo, with whom he had already worked since the age of twelve after the death of his father, and whose surname he also adopted.[66] In this agreement, emphasizing Leonardo's talent for painting, Jacopo Bellini hired his nephew to work at the studio for two years, along with his own sons Gentile and Giovanni, for an overall sum of 25 gold ducats. Leonardo's date of birth can also be deduced from this document since, having affirmed that he was "free", he must have already attained the age of 18-20 years and would have therefore been born between 1423 and 1425.[67] Leonardo was thus able to grow up in the most extraordinary and productive studio in Venice and even of all Northern Italy – along with Francesco Squarcione's studio in Padua where, it would be opportune to recall, Nicolaus Germanus also developed his skills before moving to Florence around 1460 and preparing several manuscripts of Ptolemy's *Geography*.[68] In addition to Jacopo Bellini's sons, Giovanni and Gentile, another distinguished visitor to the studio was in fact the most celebrated of Squarcione's students, Andrea Mantegna (1431-1506) who, by marrying Nicolasia Bellini, one of Jacopo's daughters, was now part of the family and artistic milieu of the Bellini clan. This is amply testified by the intense pictorial "dialogue" between the works painted by Mantegna and Giovanni Bellini between the sixth and seventh decade of the fifteenth century. As we shall see Leonardo was undoubtedly inspired by the innovations and pictorial styles of his uncle Jacopo, his cousin Giovanni, and Andrea

[64] G. Mariani Canova, "Per Leonardo Bellini, *Arte veneta* 22 (1968), pp. 9-20 (p. 12).

[65] P. Paoletti, *Raccolta di documenti indediti per servire alla storia della pittura veneziana nei secoli XV e XVI*. Fascicolo I. *I Bellini*, Padua : R. Stabilimento P. Prosperini, 1894, pp. 7-8.

[66] For a general overview of the figures, style and works of Leonardo Bellini see Mariani Canova, "Per Leonardo Bellini," pp. 9-20 ; Ead., *La miniatura veneta del Rinascimento*, Venice : Alfieri, 1969, pp. 303-304, figures 10-19 ; U. Bauer-Eberhardt, "Die Rothschild-Miscellanea in Jerusalem Haupwerk des Leonardo Bellini," *Pantheon* 42 (1984), pp. 229-237 ; S. Marcon, "Bellini, Leonardo," in Milvia Bollati (ed.), *Dizionario biografico dei miniatori italiani*, Milan : Sylvestre Bonnard, 2004, pp. 76-78.

[67] For a collection of documents on Leonardo Bellini see L. Moretti, "Di Leonardo Bellini, pittore e miniatore," *Paragone. Arte* 99 (1958) pp. 58-66 ; Id., "Bellini, Leonardo," in *Dizionario biografico degli italiani*, vol. VII, Rome : Istituto della Enciclopedia italiana, 1965, pp. 712-713 ; E. Billanovich, "Per la storia della pittura nel Veneto," *Italia medioevale e umanistica* 16 (1973), pp. 362-367.

[68] A. De Nicolò Salmazo, "Padova," in M. Lucco (ed.), *La pittura nel Veneto. Il Quattrocento*, vol. 2, Milan : Electa, 1990, pp. 481-540 ; Ead., "Presentazione," in A. De Nicolò Salmazo (ed.), *Francesco Squarcione 'pictorum gymnasiarcha singularis'*, Padua : Il Poligrafo, 1999, pp. 14-15.

Mantegna in terms of the landscapes and environment portrayed in the paradise contained in the *mappa mundi*. As has been mentioned in Chapter II, a watershed breakthrough in the process of identifying Leonardo Bellini's works occurred when Lino Moretti recognized a work mentioned in a receipt dated 7 December 1463 : for a payment made to Leonardo of four gold ducats in the *Promissione dogale di Cristoforo Moro* (Official Oath of Doge Cristoforo Moro) preserved at the British Library (Add. MS 15816, f. 5r).[69] After this identification the style of the Venetian master became instantly recognisable and unmistakable in the history of Venetian fifteenth-century miniature art, thus allowing art historians to attribute the decorations of many codices to Leonardo, despite a situation of documental *lacunae*. Research carried out by Lino Moretti, Giordana Mariani Canova, J. J. G. Alexander, Ulrike Bauer-Eberhardt, and more recently by Susi Marcon, has even indicated that Leonardo Bellini was responsible for reforming Venetian miniature art. Although he maintained some of the distinctive traits of Gothic miniatures from the lagoons of Venice – the merging of colors, a predilection for shades of blue and dusky rose, the marked contours of human figures – he introduced motives in book illustrations that were clearly inspired by the Renaissance, especially in terms of the perception and representation of volumes and the depth of space, which recreated on a miniature scale the artistic style that had been developed in the drawings and paintings of his uncle Jacopo, his cousins Giovanni and Gentile, and Andrea Mantegna. This is evident above all in the realistic corporeal depiction of human figures, in the scenic environment that was evidently inspired by naturalism, and the quest for effects of perspective. In addition to being in conformance with Bellini's pictorial habits these aspects are all consistent with the tastes and styles that were represented by miniature masters in Ferrara around the mid-fifteenth century for the adornment of the celebrated Bible prepared for Borso d'Este, especially those of Taddeo Crivelli, Franco dei Russi, and Maso Finiguerra.[70]

3. Recognizing styles : the details of attribution

The affinities between some particulars of the earthly paradise depicted in the *mappa mundi* and those of numerous other works by Leonardo Bellini proved to be decisive elements in the process of attribution. These other works include the decoration of the *Promissione* [*dogale*] *di Cristoforo Moro* (1463),[71] the Bible of the Archdiocesan Seminary in Padua (1461),[72]

[69] For the document see Moretti, "Di Leonardo Bellini, pittore e miniatore," pp. 58-96.

[70] Modena, Biblioteca Estense Universitaria, Ms Lat. 422 and Ms Lat. 423, manuscripts on vellum, 280 × 400 mm. Facsimile edition : *La Bibbia di Borso d'Este : ms. Lat. 422-423, Biblioteca estense e universitaria, Modena – Ferrara, seconda meta del XV secolo*, 3 vols., Rome : Istituto della Enciclopedia italiana, 1997. With regard to the influence of masters from Ferrara on Leonardo Bellini, Mariani Canova, "Per Leonardo Bellini," p. 9. There is a vast bibliography on this celebrated Bible in two codices as well as on this stage in the history of Ferrara miniatures. See vol 3, *Commentario*, of the facsimile edition.

[71] London, British Library, Add. Ms. 15816, f. 5r ; miniature on vellum, 340 × 235 mm. ; for reproductions, see : Mariani Canova, *Per Leonardo Bellini*, p. 10 ; Ead., "Leonardo Bellini," in *Storia della miniatura veneta*, fig. 10 and fig. 11.

[72] Padua, Biblioteca del Seminario, Ms. XXI ; for reproductions of the decoration see : A. Donatello, G. M. Florio, N. Giovè, L. Granata, G. Mariani Canova, P. Massalin, A. Mazzon, F. Toniolo, and S. Zamponi (eds), *I mano-*

the miniatures of the *Rothschild Miscellanea*,[73] perhaps his largest work, the illustrations of the 1473 *Promissione dogale di Niccolò Marcello* (plate VII))[74] and the *Commissione di Niccolò Marcello e Andrea Lion*[75] and the codex of Ptolemy's *Cosmography* preserved in Oxford.[76] An analysis of the figure of God in this earthly paradise reveals the extremely strong stylistic and somatic affinities with the figure of God in the initial letter of the *Promissione di Niccolò Marcello*, especially the drawing of the beard, the expression and the shape of the face.[77] This kind of representation can also be linked to another miniature that can be found in the *Mariegola* of the San Girolamo school, which was initially drawn by Leonardo in 1457 with the work entitled "San Girolamo in cattedra", which shows the same modelling and a similar physiognomic characterization.[78] Another noteworthy fact is the stylistic coherence with the figure of God in the 1461 Bible prepared for the Archdiocesan Seminary in Padua (ms. XXI, f. 4r), copied from Martinus de Galliis and reduced to a miniature scale in the studio of Leonardo Bellini. As a matter of fact a strong congruity can be observed in the figure of God depicted in the earthly paradise with one of the stylistic elements that best identifies the hand of Leonardo : the representation of old men characterized by rather long white beards and hair, standing in a dominant position with respect to the rest of the representation. Moreover, the beard is divided into two parts, just like in other miniatures in which men are drawn from a frontal angle with a pictorial modality that is not to be found in works by other master miniature painters from Venice and Padua. This stylistic characteristic is also visible, in addition to the works already cited, in a couple of illuminations in the *Miscellanea Rothschild*, f. 83v, which are referred to as "The Morning Prayer" and "Abraham with the Angels". The posture proves to be quite significant – this is a very particular and characteristic element in the history of Venetian miniatures during the period under study – and the pictorial style in which the lower sections of the figures of Adam and Eve have been drawn, similar in every aspect to those of the little angels in the *Promissione di Cristoforo Moro* (f. 5r), the *Promissione di Niccolò Marcello* (f. Ir), the *Commissione di Niccolò Marcello e Andrea Lion* (f. 5r), and the Oxford manuscript of Ptolemy's *Cosmography* (f. I). On the whole this is a set of congruities and affinities that would be difficult to explain without attributing all these works to the same hand. Thus, the depiction of earthly paradise must have been drawn by Leonardo between the years 1455-1457, the period when his first dated works appear, and August 1460, when the *mappa mundi* was framed and displayed. It would

scritti della Biblioteca del Seminario vescovile di Padova, Florence : Sismel, Edizioni del Galluzzo, 1998, p. 11, entry 19, and plate XXII.

73 Jerusalem, Israel Museum, Ms. 180\51 ; for a reproduction of a good part of the miniatures and an exhaustive study of the codex, see U. Bauer-Eberhardt, "Die Rothschild-Miscellanea in Jerusalem Haupwerk des Leonardo Bellini," *Pantheon* 42 (1984), pp. 229-237.

74 Venice, Museo Correr, Ms. I. III, 322, f. Ia ; for a reproduction of the decoration, see Mariani Canova, "Leonardo Bellini," in *Storia della miniatura veneta*, fig. 18.

75 Venice, Biblioteca Nazionale Marciana, Ms. lat. X, 358 (= 3517), f. 5r ; for a reproduction see Bauer-Eberhardt, "Die Rothschild-Miscellanea in Jerusalem Haupwerk des Leonardo Bellini," p. 234, fig. 5.

76 Oxford, Eton College, Ms. 140. It was the art historian J. J. G. Alexander who attributed the decoration of folio I to Leonardo Bellini ; for the attribution and a reproduction of the miniature, see J. J. G. Alexander, "Notes on Some Veneto-Paduan Illuminated Books of the Renaissance," *Arte veneta* 23 (1969), pp. 9-11.

77 *I manoscritti della Biblioteca del Seminario vescovile di Padova*, p. 11.

78 Venice, Museo Correr, Ms. Cl. IV, 133. Reproduced in Mariani Canova, "Leonardo Bellini," in *Storia della miniatura veneta*, p. 12, fig. 17, p. 18.

therefore belong to the proto-Renaissance phase of this master's *œuvre*.[79] In fact, the paradise drawn in Fra Mauro's map would thus constitute one of the most important works by Leonardo Bellini: it was precisely during these years, the sixth decade of the fifteenth century, that the Venetian master, who had already been working since 1443 when he had begun to receive a stipend in his uncle's studio, began to develop a characteristic style that art critics have unanimously described as clearly recognisable in the history of fifteenth-century miniature painting.[80]

4. Echoes of Jacopo Bellini and Andrea Mantegna

If the circular form of paradise with its walls and towers – Marcon argues – seems to have been derived from a French iconographic model, which was widely diffused in Northern Italian miniature art, the figure of God and the rocky landscape on the right-hand side of the drawing reveal a clear stylistic affinity with, and were possibly inspired by, the work of Jacopo Bellini and Andrea Mantegna.[81] Jacopo Bellini, a Venetian mediator between the international Gothic and the new pictorial style that had been developed in Florence during the early decades of the fifteenth century, laid the groundwork for the maturing of a truly extraordinary pictorial renaissance in Venice in which his two sons, Giovanni and Gentile, would be the two most important protagonists. The figure of God in the *mappa mundi*, as well as that in the Bible in the library of the Archdiocesan Seminary in Padua, shows great similarities with the figure of God in the lunette of the *Trittico di San Sebastiano* attributed to Jacopo Bellini and his school, preserved at the Galleria dell'Accademia in Venice,[82] as also with the figure in the depiction of San Bernardino of Siena that Gentile Bellini painted after a drawing by Jacopo Bellini on the left panel of the Altar of Gattamelata, which has been entirely lost.[83] However, the most striking echo of Leonardo Bellini's paradise can be seen in the drawing books of Jacopo Bellini that are today preserved at the Louvre and the British Museum and the work of Andrea Mantegna.

[79] With regard to the documental aspects of the dating of the *mappa mundi*, see Chapter I.

[80] Mariani Canova, *Leonardo Bellini*, pp. 103-104; Ead., "Per Leonardo Bellini," pp. 9-20; Bauer-Eberhardt, "Die Rothschild-Miscellanea in Jerusalem Haupwerk des Leonardo Bellini," pp. 229-237; Marcon, "Bellini, Leonardo," pp. 76-78.

[81] "It is likely that the representation of the walls with the cylindrical towers is derived from a French matrix, which appeared in images that were soon widely diffused in Northern Italy, such as those by the Limbourg brothers. Examples of an Earthly Paradise surrounded by circular walls, with biblical figures inside, can be found in representations such as the work on the same subject painted by Jean Limbourg in Jean de Berry's *Très riches heures* [Chantilly, Musée Condé, ms 65, f. 25v]", see Marcon, "Leonardo Bellini and Fra Mauro's World Map: the *Earthly Paradise*," pp. 160-161. There is a vast bibliography on the *Très riches heures* by Jean de Berry and the Limbourg brothers. About the influence of the Limbourg brothers on miniature art in Northern Italy, see J. J. G. Alexander, "The Limbourg Brothers and Italian Art: A New York Source," *Zeitschrift für Kunstgeschichte* 46 (1983), pp. 425-435.

[82] Venice, Galleria dell'Accademia, *Trittico di San Sebastiano*, *c.* 1460, tempera on panels; main panels 127 × 45 cm; lunette 57 × 151 cm. For a good reproduction see C. Eisler, *The Genius of Jacopo Bellini. The Complete Paintings and Drawings*, New York: Abrams, p. 73. Also see M. Boskovits, "Per Jacopo Bellini pittore (postilla ad un colloquio)," in *Paragone. Arte* 36, (1985), pp. 113-123.

[83] New York (private collection), *Sant'Antonio abate e San Bernardino da Siena*, left panel of the Altar of Gattamelata, 1459/60, tempera on panels, 110 × 57 cm. For a reproduction see Eisler, *The Genius of Jacopo Bellini*, p. 63.

Bellini's drawing books belong to a precocious genre that had already been diffused in the Middle Ages by collections of sketches, in the form of figurative notes, as also in finished drawings of diverse subjects which were generally prepared by the master of the studio, in which references to antiquity are a constant presence, with portraits, building profiles, statues, reliefs and epigraphs. These were thus books, both for personal study as well as for studio use, from which apprentices such as Leonardo Bellini and Jacopo's sons Gentile and Giovanni, learnt the art of drawing and composition.[84] Jacopo's notebooks, just like the coeval notebooks of Gentile da Fabriano (*c.* 1370-1427), are thus characterized by a collective usage, especially in the works of apprentices. However, they were also used by artists outside the *studio* whereby the drawings, both originals and copies of paintings, were collected together with the sole scope of providing a repertoire. Leonardo Bellini's paradise presents some notable affinities with some of Jacopo Bellini's drawings in the notebooks, especially with those that have rocky, bare, landscapes, dominated by rough and powerful massifs, such as, for example, in f. 19r (fig. 173), f. 44r (fig. 193), f. 60r (fig. 139) in the drawing book preserved at the British Museum, f. 25r (fig. 185), or ff. 70v-71r (figs 255-256) in the *Livre des dessins* preserved at the Louvre, as also with an altar-step entitled *Christ in the Limbo* painted by Jacopo with the assistance of his sons around 1459-1460.[85] In particular, the *Agony in the Garden* in the British Museum notebook seems to have been, in its turn, the source of a painting on the same subject by Andrea Mantegna that was perhaps commissioned by the Venetian *condottiere* Jacopo Antonio Marcello around 1457, and can be dated to no later than 1459-1460 when Giovanni Bellini also painted a work that depicted the *Agony in the Garden*.[86] In this *Agony in the Garden* by Mantegna an austere landscape, dominated by stratified and arid rocks, provides a backdrop for a praying Christ while the apostles Peter, John, and James sleep at the foot of the relief (plate VIII). A comparison with Jacopo's drawings and the aforementioned works of Andrea Mantegna and Giovanni Bellini

[84] London, British Museum, Inv. 1855-8-11, Jacopo Bellini, [*Book of Drawings*], the following words are written on f. 2r *De mano de ms. Iacobo Bellino veneto, 1430, In Venetia*, ms., paper, ff. 101 (134 drawings), original size of the leather cover 360 × 275 mm; Paris, Musée du Louvre, Inv. R. 1475-1556: Jacopo Bellini, [*Livre des dessins*], manuscript on vellum, ff. 93, with a paper folio inserted, 427 × 290 mm. For an in-depth codicological study, see A. J. Elen, "A Codicological Analysis and Reconstruction of Jacopo Bellini's Drawing Books," in Eisler, *The Genius of Jacopo Bellini*, pp. 455-474. The two manuscripts have been published in a facsimile edition in Eisler, *The Genius of Jacopo Bellini*. The drawings have also been reproduced in B. Degenhart, A. Schmitt, *Corpus der Italienischen Zeichnungen 1300-1450*, Teil II: *Venedig. Jacopo Bellini*, Berlin: Mann, 1990: V: *Text*; VI: *Katalog*; VII, figs 1-119 (Drawing book at the Louvre); VIII, Tab. 120-319 (Drawing book at the British Museum). Also see G. Mariani Canova, "Riflessioni su Jacopo Bellini e sul libro dei disegni del Louvre," *Arte Veneta* 26 (1972), pp. 9-30.

[85] The numbers of the illustrations refer to C. Eisler, *The Genius of Jacopo Bellini* cited above. Padua, Museo Civico, Jacopo Bellini e figli, *Discesa di Cristo al Limbo*, 28,5 × 58, *c.* 1459-1460 (predella of an altar-step). Reproduced in R. Goffen, *Giovanni Bellini*, New Haven: Yale University Press, 1989, p. 9.

[86] London, National Gallery, Andrea Mantegna, *The Agony in the Garden*, oil on wood, 63 × 80 cm, *c.* 1455-1460. See R. Lightbown, *Mantegna: With a Complete Catalogue of the Paintings, Drawings and Prints*, Oxford: Phaidon-Christie's, 1986, pp. 60-61, 404; J. Martineau, *Andrea Mantegna*, London: Thames and Hudson, 1992, p. 106. London, National Gallery, Giovanni Bellini, *Prayer in the Garden*, oil on wood, 81 × 127 cm, 1459. See R. Goffen, *Giovanni Bellini*, pp. 105-111 (the painting has been reproduced on p. 105). The rocky landscape is characteristic of another painting depicted by Giovanni Bellini around 1460, a *Trasfiguration of Christ*, preserved in Venice at the Museo Civico Correr (134 × 68 cm); for a reproduction see Goffen, *Giovanni Bellini*, pp. 14-15. For Giovanni Bellini: in addition to Goffen, also see G. Robertson, "The Earlier Work of Giovanni Bellini" *Journal of the Warburg and Courtauld Institutes* 23 (1960), pp. 45-59; Id., *Giovanni Bellini*, Oxford: Clarendon Press, 1968.

shows how Leonardo Bellini's paradise also once again proposes the same structure : an extended spur of vivid rocks, with a few trees, surrounded by a watercourse at the foot of the rock, just like the path that abuts the rocky cave in Mantegna's painting. Critics have acknowledged that Andrea Mantegna and the Bellinis were the inventors of this kind of "rocky" landscape setting, employed in paintings commissioned for private devotion. Along with the painting on the same theme by Giovanni Bellini, Mantegna's *Agony in the Garden*, in its turn, can be closely linked with an analogous composition in the Louvre's *Livre des dessins* by Jacopo Bellini. If Leonardo thus belonged to the family milieu of the Bellini clan, which included Andrea Mantegna, an iconographic analysis of this paradise reveals that this identification was also, and supremely, an artistic one. On the basis of these observations it would perhaps be appropriate to suggest that, for this depiction of paradise, Leonardo Bellini's primary sources of inspiration, both for the human figures and, above all, for the landscape setting, were the drawings and paintings of Jacopo Bellini and Andrea Mantegna. The affinities with Andrea Mantegna's *Agony in the Garden* could prove to be a decisive element for dating this representation of paradise that, if it was inspired by this work by Mantegna, must have been drawn around 1459-1460.

5. Beyond art : interpreting an attribution

This representation of earthly paradise clearly reflects a convergence of two skills : Fra Mauro's cosmographic abilities and the artistic-symbolic skills of the miniaturist, where the former guided the hand of the latter who, in his turn, used his own awareness of the symbolic language of the art of painting to complete the cosmography. This collaboration between Fra Mauro and Leonardo Bellini, that is, between a cosmographer and a painter, was not the only example of such an association in Italian courts and cities during the fifteenth century. In Ferrara, at the d'Este ducal court, there was an important collaboration between the Florentine miniaturist Giovanni Falconi and the Latinist Guglielmo Capello, a student of Pietro d'Abano. Guglielmo Capello availed himself of Falconi's art to illustrate and decorate the *Commento al Dittamondo* (1435-37), a rare and valuable commentary aimed at teaching geography and cosmography *via* Fazio degli Uberti's *Dittamondo*, the most important and most widely circulated geographic work in Italy during the fourteenth century. For the copy dedicated to the lords of Ferrara, Falconi drew 77 miniatures with astronomical and geographical subjects under the direct guidance of Guglielmo.[87] It is also important to stress that Giovanni Falconi had also executed, again for the lords of Ferrara, a Ptolemaic map and drawings on astronomical subjects[88] and that Nicolaus Germanus, one of the most extraordinary cartographers of the fifteenth

[87] Four manuscripts have survived of the *Commento al Dittamondo* : Modena, Biblioteca Universitaria Estense, Ms. Ital. 483 ; Venice, Biblioteca Marciana Ms. It. Cl. IX. 40 (= 6901) ; Paris, Bibliothèque Nationale, Ms. It. 81 ; Turin, Biblioteca Nazionale Universitaria, N.1.5. Unfortunately, the manuscript with an illustrated dedication by Giovanni Falconi has been lost. See M. Milanesi, "Il Commento al Dittamondo di Guglielmo Capello (1435-37)," in *Alla corte degli Estensi. Filosofia, arte e cultura a Ferrara nei secoli XV e XVI, Atti del Convegno di studi, Ferrara, 5-6-7 marzo 1992*, Ferrara : Edizioni dell'Università di Ferrara, 1994, pp. 365-388 ; for a biography of Capello, see F. R. Haussmann, "Guglielmo Capello," in *Dizionario Biografico degli Italiani*, vol. 28, Rome, 1975, pp. 494-495.

[88] F. Toniolo, "Marco dell'Avogaro e la decorazione all'antica," in *Le Muse e il Principe. Arte di corte nel Rinascimento pagano*, 2 vols., Panini : Modena, vol. II, pp. 133-140.

century, had been one amongst the apprentices in Francesco Squarcione's studio in Padua before moving to Florence and updating Ptolemaic cartography with the codices dedicated to Borso d'Este and Pope Paul II. Another illustrious example of collaboration between cartographers and miniaturists is that of Berlinghieri's version of Ptolemy's *Geography*, which was illustrated by two celebrated miniaturists, Marco Attavante and Antonio del Cherico.[89]

The first dated works by Leonardo Bellini, well-known today, date back to 1457, when he decorated the *Institutiones* by Lattanzio for the Venetian patrician Leonardo Sanudo (who had written the codex and noted that the payment for the miniature was to be made to Leonardo)[90]; a Bible realized in the same year for the Carthusian church of Sant'Andrea del Lido[91]; and a second Bible that was decorated for the Venetian patrician Stefano Trevisan in 1461 – the same individual who was entrusted with sending a copy of the *mappa mundi* to Portugal, commissioned by the Portuguese court of Afonso V.[92]

These works, together with the portrait of the doge Francesco Foscari, painted in 1457 on the first folio of the official oath (*promissione*) at the end of his long tenure as doge, and the commissioned works for the decoration of the doges' *promissioni* in 1463 (for Cristoforo Moro) and 1473 (for Niccolò Marcello) indicate that, from 1460 onwards, the artistic qualities of Leonardo Bellini were appreciated both in official Venetian circles as well as in religious confraternities, which also played an extremely important role in civil life in Venice.[93] The fact that the painting of earthly paradise in the *mappa mundi* had been entrusted to Leonardo Bellini, a master patronised by high Venetian society, especially by patricians and political circles of the institution of the doge, indirectly contributes towards shedding light upon the "placement" of a work such as the *mappa mundi*. As has been mentioned, in a legend that was very important in the overall scheme of the work, and in which Fra Mauro dealt with the central issue of the relationship between his work and Ptolemy's *Geography*, he stated that his work had been "realized to be contemplated by these Illustrious Lords". This is a rhetorical formula that can be interpreted in diverse ways and does not allow one to precisely identify the target readership of this *mappa mundi*, especially in terms of relations with official circles of the Venetian state. My interpretation "translates" the words of Fra Mauro in this manner: "This work, made to be contemplated by Illustrious Lords", is a reading that excludes a direct or indirect commissioning on the part of the Venetian state.

[89] P. d'Ancona, *La miniatura florentina (Secoli XI-XVI)*, Florence: Olschki, 1914, vol. II, pp. 777-784 and index; A. Garzelli, *Miniatura fiorentina del Rinascimento, 1440-1525: un primo censimento*, Florence: Giunta regionale toscana; Scandicci: La Nuova Italia, 1985.

[90] Venice, Biblioteca Nazionale Marciana, Lat. II, 75 (= 2198), Lattanzio, *Institutiones*, manuscript on vellum, 1457. See Mariani Canova, *La miniatura a Venezia dal Medioevo al Rinascimento*, pp. 807-808. Also see the reproductions in Marcon, "Il Mappamondo di fra' Mauro e il miniatore Leonardo Bellini," figures 4-5, pp. 165-166.

[91] Venice, Biblioteca Nazionale Marciana, Ms. Lat. I, 16 (= 2280) *Biblia*, manuscript on vellum, 1456-1457; See Mariani Canova, *Per Leonardo Bellini*, pp. 9-11, 14.

[92] Padua, Biblioteca del Seminario vescovile, Cod. 21, ms., paper, 1461, f. 4r. See *I manoscritti della Biblioteca del Seminario vescovile di Padova*, p. 11. For Stefano Trevisan's involvement in dispatching the *mappa mundi* to Lisbon, see Chapter I and its Documentary Appendix.

[93] For the portrait of Foscari: Venice, Biblioteca Nazionale Marciana, Ms. Lat. X, 190 (= 3555), *Promissione di Francesco Foscari*, manuscript on vellum, 1457 or 1458, f. I.

There is documental evidence that shows how, in these same years between 1450 and 1460, the Republic commissioned cartographical works from Don Antonio Leonardi that, at least going by their documental description, were apparently closer to Ptolemy's work than Fra Mauro's map (as has been mentioned in Chapter I, in stark contrast to thorough documental records relating to Leonardi's commission, it seems that all the maps drawn by the Venetian priest have been lost). Perhaps this was the precise reason why these works were commissioned from him since they were more similar to the humanistic tastes of the time.[94] However, the participation in the preparation of the *mappa mundi* by Leonardo Bellini, a master who in precisely those years was illuminating codices commissioned by the highest levels of Venetian official and political circles, together with the medal dedicated to Fra Mauro at the end of the 1450s – although the work's individual commissioner remains unknown – are indications that allow one to deduce that Fra Mauro's work enjoyed an elevated level of prestige, even without official commissions by the Venetian Republic. The intervention of such a highly esteemed master (and moreover an expensive one, as we will see in Chapter I of the second part of this book) would, in fact, be hard to justify if the *mappa mundi* had been perceived as being "extraneous" to this cultural and, in a broader sense, political environment that was universally acknowledged to be one of great excellence.

[94] Gautier Dalché, "Weltdarstellung und Selbsterfahrung : Der Kartograph Fra Mauro," pp. 39-51.

CHAPTER IV

READERS AND READINGS OF PTOLEMY'S *GEOGRAPHY* IN VENICE IN THE MID-FIFTEENTH CENTURY

> I do not think that I have belittled Ptolemy by not having followed his Cosmography, because if I had wanted to observe his meridians, parallels and degrees, I would have had to omit from this depiction of the world [circumferentia] many provinces that are known today that Ptolemy does not mention, especially in the extreme latitudes, for to the north and south, he labels these [regions] terra incognita, because in his age they were not known.[1]
>
> Fra Mauro

I. – A heterodox and forgotten reception

While the history of the Florentine reception of Ptolemy's *Geography* during the fifteenth century (as also its reception by the d'Este ducal court in Ferrara and the Pontifical Curia) has been well documented,[2] Venice's reception of this work has been little studied, albeit with some significant exceptions.[3] According to Marica Milanesi, this has been due to two main reasons. First, there is a paucity of Venetian manuscripts preserved in Venetian libraries that directly or indirectly transmit Ptolemy's *Geography*. Apart from two Greek copies of the *Geography* that

[1] This is my translation of the following Venetian text which appears on Fra Mauro's map and in *Mappamondo* (ed. 1956) Pl. XLI, 21*: "Io no(n) credo derogar a Tolomeo | se io non seguito la sua cosmographia, | perchè se hauesse uoluto obs(er)var i sui | meridiani ouer paralleli ouer gradi | era necessario quanto a la demo(n)st(r)a|tion de le parte note de questa circu(m)|ferentia lassar molte proui(n)cie de le | qual Tolomeo no(n) ne fa me(n)tio(n), ma | per tuto maxime i(n) latitudine çoè tra | ostro e tramo(n)tana dice terra i(n)cognita, | e q(ue)sto p(er)chè al suo te(m)po no(n) li era nota." A translation into English also appears in *Mappamondo* (ed. 2006) *2892.

[2] The bibliography is vast. Here I just refer to the most relevant and recent works, which are quoted in chronological order: S. Gentile, "L'ambiente umanistico fiorentino e lo studio della geografia nel secolo XV," in L. Formisano et al. (eds), *Amerigo Vespucci. La vita e i viaggi*, Florence: Banca Toscana, 1991, pp. 9-63; Id., *Firenze e la scoperta dell'America. Umanesimo e geografia nel '400 fiorentino*, Florence: Olschki, 1992; M. Milanesi, "Presentazione" of *La cultura geografica e cartografica fiorentina del Quattrocento*, in *Firenze e il mondo nuovo. Geografia e scoperte fra XV e XVI secolo*, Lucca: Pacini, 1993, pp. 15-32 (*Rivista geografica italiana*, Centenary Issue, fasc. 1, 1993); P. Gautier Dalché, "The Reception of Ptolemy's Geography (End of the Fourteenth to Beginning of the Sixteenth Century)," in D. Woodward (ed.), *The History of Cartography*, vol. III, 1, *Cartography in the European Renaissance*, pp. 285-364; Id., *La Géographie de Ptolémée en Occident (IVe-XVIe siècle)*, Turnhout: Brepols, 2009. Also see A. Jones and J. L. Berggren (eds), *Ptolemy's Geography: An Annotated Translation of the Theoretical Chapters*, Princeton: Princeton University Press, 2000.

[3] For an emblematic example that demonstrates how the reception of the *Geography* in Venice has been overlooked, see A. Bevilacqua, "Geografi e cosmografi," in *Storia della cultura veneta*, vol. 3, part 2, *Dal primo Quattrocento al concilio di Trento*, Venice: Neri Pozza, 1980, pp. 355-374.

the Byzantine Cardinal Bessarion (*c.* 1395-1472) donated to the Senate of Venice along with the rest of his valuable library in 1468 and which are now held in the Biblioteca Nazionale Marciana (Venice, Biblioteca Marciana, Graecus 388 and Graecus 516),[4] there are no traces of manuscripts of the *Geography* in Venetian libraries and archives that can be compared to those preserved in Florence, Modena, Paris, London, Naples and Rome, including the Vatican.[5] This dearth of Venetian documents, which contrasts the wealth of archival and library sources of Florentine provenance, has been incorrectly interpreted as proof of a limited or non-existent reception of scientific works attributed to a humanism that has been deemed to be "more rhetorical" in nature.[6] It has even been erroneously ascribed to a "certain disinterest – or better still, a certain passivity – in the transmission of traditional technical-scientific information" in Venetian culture.[7]

The second reason, which perhaps better explains this lack of studies on the subject – Milanesi argues effectively – is linked to the cataloguing of the manuscripts of the *Geography* carried out by Father Joseph Fischer and its reception by historians of cartography and geography. Fischer concentrated his research on manuscripts that contained Ptolemaic maps and identified about fifty extant copies. Most are monumental manuscripts, commissioned by popes, cardinals, and leading families of the age, crafted and copied in the Florentine workshops (*botteghe*) such as that of Vespasiano da Bisticci, by cartographers and "painters" (*dipintori*) like Nicolaus Germanus and Piero del Massaio.[8] In addition to the fact that he did not consider manuscripts of the *Geography* without maps, Fischer completely overlooked the more modest manuscripts prepared for scholars who were interested in the work as part of their education or study. The fact that Fischer almost exclusively considered "orthodox" luxurious manuscripts (that is, editions that were structured into the eight books that comprise the text of the *Geography*, followed by a planisphere, twenty-six regional maps and, eventually, the *tabulae novellae*) had an unexpected result: whenever Ptolemy's "Guide to Cartography" did not conform to the

[4] For a codicological description of these two manuscripts see E. Mioni, *Bibliothecae Divi Marci Venetiarum codices Graeci manuscripti. Thesaurus antiquus*, vol. 2, *codices 300-625*, Rome, 1985, pp. 381-384; S. Bernardinello, "Le carte dell'Africa nella Geographia di Tolomeo. Lettura dal codice Marciano gr. Z. 516," *Atti e memorie dell'Accademia Patavina di Scienze, Lettere ed Arti. Parte III* 109 (1996-97), pp. 48-95; S. Marcon, "Cardinal Bessarion's Ptolemy Codex: A Book of Inestimable Worth (Il prezioso Tolomeo di Bessarione)," in Klaudios Ptolemaios, *Geographike yphegesis. O ellenikos kodikas 388 tes Markianes Bibliothekes tes Benetias*, Athens: Militos, 1999, pp. 11-32.

[5] M. Milanesi, "A Forgotten Ptolemy: Harley Codex 3686 in the British Library," *Imago Mundi* 48 (1996), pp. 43-64.

[6] For an interpretation of Venetian humanism that sheds light upon the rhetorical aspects, see the nearly eighty biographic profiles in M. L. King, *Umanesimo e patriziato a Venezia nel Quattrocento*, Rome: Il Veltro, 2 vols., 1989. See also Ead., *Venetian Ideology and Reconstruction of Knowledge: Giovanni Caldiera (c. 1400-1474)*, Ph.D. dissertation, Palo Alto: Stanford University, 1972; F. Gilbert, "Humanism in Venice," in S. Bertelli, N. Rubinstein and C. H. Smyth (eds), *Florence and Venice, Comparisons and Relations*, 2 vols., Florence: La Nuova Italia, 1979, vol. 1, pp. 13-26; C. Vasoli, "A proposito di umanisti fiorentini e umanisti veneziani," in *Continuità e discontinuità nella storia politica, economica e religiosa. Studi in onore di Aldo Stella*, Vicenza: Neri Pozza, 1993, pp. 89-116.

[7] *Mappamondo* (ed. 2006), p. 127.

[8] *Claudii Ptolemaei Geographiae codex Urbinas graecus 82*, phototypice depictus consilio et opera curatorum Bibliothecae Vaticanae, [Vatican City], Biblioteca Apostolica Vaticana, 1932: Tomus prodromus: J. Fischer SJ, *De Cl. Ptolemaei vita operibus Geographia praesertim eiusque fatis*. Pars prior. *Commentatio*, Vatican City: Biblioteca Apostolica Vaticana, 1932.

Florentine model, these "heterodox" manuscripts have been almost completely ignored.[9] In this context, the points that Marica Milanesi has raised in her introductory note to the study of the British Library's Harley MS 3686, which is still thought to be the only manuscript of the *Geography* produced in Venice, and which was ignored by Fischer, are effective arguments :

> More humble codices, with or without maps, belonging to students of geography... have been overlooked. Yet in late medieval and early Renaissance Europe only very few individuals could have afforded codices with maps. The luxurious and costly Florentine codices were largely 'mass produced' and, apart from prototypes, are repetitive, and full of uncorrected errors in both text and diagrams. Their maps are copied from other maps of old design, occasionally with additions.[10]

These words clearly reflect the real state of the Venetian – as opposed to Florentine – documentation upon which this chapter is based. The analysis of the correlation between Fra Mauro's *mappa mundi* and Ptolemy's *Geography* will be introduced by the brief presentation of the results of the research carried out by Marica Milanesi and Patrick Gautier Dalché on other contemporary Venetian readings and interpretations of the work. I'm referring to Andrea Bianco's Ptolemaic planisphere, the anonymous Harley MS 3686 of the British Library and Giovanni Fontana's *Liber de omnibus rebus naturalibus*. These testimonies enables us to observe the diffusion and use of, and the response to, the *Geography* in different *milieux*, avoiding the rigidity of interpretations that associate Ptolemy's "Guide to Cartography" to humanistic circles exclusively. This framework provides the background to better understand Fra Mauro's approach to the *Geography* and also to study a further crucial document attesting to the complex reception of Ptolemy's work in Venice : a "forgotten" manuscript by an anonymous Venetian cosmographer who, around 1430-1450, prepared the first known vernacular translation and commentary on the *Geography*, while reinventing Ptolemy's cartographic syntax in a manuscript that has been overlooked until now.[11]

II. – The *Geography* in mid-fifteenth-century Venice

1. Venetian readers of the *Geography* contemporaries to Fra Mauro

There are at least four main testimonies that document knowledge of the *Geography* in Venice during the first half of the fifteenth century. A few years after the Latin translation of

[9] At this regard, Milanesi highlighted that the *Catalogus translationum* has not yet inventorised the Latin manuscripts of the *Geography*. Douglas Marshall has, however, published a survey that, although incomplete, would indicate that the number of manuscripts of the *Geography*, especially editions without the plates, would have been quite large. See D. W. Marshall, "A List of Manuscript Editions of Ptolemy's Geography," *Bulletin of the Geography and Maps Division. Special Libraries Association* 87 (1982), pp. 17-38.

[10] Milanesi, "A Forgotten Ptolemy," p. 44.

[11] Venice, Biblioteca Nazionale Marciana, Ms. It. VI, 24 (= 6111), paper, ff. 97, II', 205 × 281 mm, second quarter of fifteenth century. See note 39 below for a more detailed codicological description. The manuscript has been forgotten, in the same sense in which Milanesi explained the concept while referring to the British Library's Harley 3686 : "Forgotten in the sense of unstudied, not unknown to librarians" (Milanesi, "A Forgotten Ptolemy," p. 43). In the case of the Venetian manuscript, it has also not been duly noted that it is a translation and critical revision of Ptolemy's *Geography*.

the *Geography* by Iacopo Angeli da Scarperia, which was concluded in 1409, Ambrogio Traversari wrote to Niccolò Niccoli in December of 1423 about the visit of somebody called *Petrus* who had told him that he was engaged in a revision of Ptolemy's text.[12] Traversari maintained that this *Petrus* did not have the necessary preparation to overcome the linguistic and mathematical difficulties of the text of the *Geography* and had gently tried to dissuade him from the task. What interests us in this context is that this letter by Traversari indicates that Iacopo Angeli's Latin translation and the geographical and cosmographic contents of Ptolemy's work were available in Venice even before 1434, when Palla di Nofri Strozzi (*c.* 1373-1462) went into exile in Padua and took with him the Greek manuscript of the *Geography* that the Byzantine humanist Emmanuel Chrysoloras (*c.* 1355-1415) – who had been invited to teach Greek in Florence in 1397 by Coluccio Salutati (1331-1406), bringing with him a selection of works in Greek that were essential aids for teaching the language, or had them sent over later with the financial assistance of Palla Strozzi – had bequeathed to him (plausibly the manuscript Urb. gr. 82 held at the Biblioteca Apostolica Vaticana).[13]

The second testimony is that of the Venetian cartographer Andrea Bianco[14] who, in 1436, drew a Ptolemaic planisphere, the last map in a magnificent autographed atlas that also included eight marine charts of the Mediterranean basin, North Africa, and northern Europe, and a

[12] Giovanni Mercati once identified this *Petrus* as the Venetian physician and humanist Pietro Tomasi or Tommasi (*c.* 1375-1458. See J. Fischer SJ, *De Cl. Ptolemaei vita operibus Geographia praesertim eiusque fatis.* Pars prior. *Commentatio*, Vatican City : Biblioteca Apostolica Vaticana, 1932, pp. 543-544 (first volume of *Claudii Ptolemaei Geographiae codex Urbinas graecus 82*, phototypice depictus consilio et opera curatorum Bibliothecae Vaticanae, [Vatican City], Biblioteca Apostolica Vaticana, 1932). This identification has been recently critized and rejected with suitable arguments by Patrick Gautier Dalché : *La Géographie de Ptolémée en Occident*, pp. 204-205. For a recent and valuable critical edition of Traversari's letter see A. Favi, *Ambrogio Traversari. Lettere a Niccolò Niccoli. Edizione critica e commento*, Ph.D. dissertation, Florence : Università degli Studi di Firenze, 2003-2004 : "Letter 4" (December 27th, 1423), pp. 11-13 (pp. 111-113 for the commentary). Consider also the "Letter 46" (May 30th, 1433), pp. 86-87 (pp. 192-194 for the commentary). See also S. Gentile in "Toscanelli, Traversari, Niccoli e la geografia," *Rivista geografica italiana. Annata del Centenario* 100 (1993), p. 115 ; Id., *Firenze e la scoperta dell'America*, entry 52, pp. 102-103. For a biography of Pietro Tommasi see King, *Umanesimo e patriziato a Venezia nel Quattrocento*, vol. 2, pp. 642-644.

[13] Fischer, *De Cl. Ptolemaei vita operibus Geographia*, pp. 180-182 and 192-201 ; A. Diller, *De Ptolemaei Geographiae codicibus editionibusque*, pp. v-xv (republished in Id., *Studies in Greek Manuscript Tradition*, Amsterdam : Hakkert, 1983, pp. 125-135). For a critical discussion of the complex documentary questions concerning this identification, see S. Gentile, "Umanesimo e cartografia : Tolomeo nel secolo XV," in D. Ramada Curto, A. Cattaneo, and A. Ferrand Almeida (eds), *La cartografia europea tra primo Rinascimento e fine dell'Illuminismo*, Florence : Olschki, 2003, pp. 4-18. On Palla Strozzi's crucial agency in the Florentine reception of Ptolemy's *Geography*, see Gautier Dalché : *La Géographie de Ptolémée en Occident*, pp. 145-154. With regard to Chrysoloras, see at least S. Gentile, "Emanuele Crisolora e la 'Geographia' di Tolomeo," in M. Cortesi and E. V. Maltese (eds), *Dotti bizantini e libri greci nell'Italia del secolo XV*, Naples : D'Auria, 1992, pp. 291-308.

[14] For a biography of Andrea Bianco, see A. Conterio, "L'arte del navegar : cultura, formazione professionale ed esperienze dell'uomo di mare veneziano nel XV secolo," in *L'uomo e il mare nella civiltà occidentale : da Ulisse a Cristoforo Colombo*, *Atti della Società Ligure di Storia Patria*, new series, 32 (1992), pp. 187-225. For an exhaustive bibliography on Bianco see P. Falchetta, "Carte veneziane dell'Asia da Fra Paolino a Giacomo Gastaldi," in G. Curatola (ed.), *Sciamani e dervisci dalle steppe del Prete Gianni : religiosità del Kazakhistan e percezione del fantastico a Venezia*, Venice : Multigraf, 2000, pp. 39-50.

circular *mappa mundi* (plate III).[15] The atlas fluidly combined all the known techniques for the graphic representation of the *oikumene* that were available in the first half of the fifteenth century. Bianco's planisphere is the first known Venetian map using Ptolemaic techniques that has survived to modern times. Even though its dimensions are much smaller than those of Urb. gr. 82 and its Greek and Latin copies – Bianco's planisphere measures 26 × 38 cm, while the Ptolemaic one measures 57.5 × 83 cm – and although it has toponyms and inscriptions in Latin, Bianco's planisphere was designed using the first Ptolemaic projection and has graphic and toponymic characteristics that are very similar to those of the planisphere in Urb. gr. 82. These similarities are sufficiently strong for Sebastiano Gentile to hypothesize that Bianco's map derived directly or indirectly from this manuscript, with partial corrections in the representation of the Ethiopian regions and the addition of some rivers that are missing in the Greek manuscript's planisphere, but easily derivable from the regional maps of the same manuscript.[16] Bianco's planisphere constitutes the first documentary evidence of the presence of manuscripts of Ptolemy's *Geography* and of the Ptolemaic maps in Venice, and attests to the circulation of the text and maps even in cultural *milieux* not directly linked to learned humanistic circles: Bianco was an official of the Venetian galleys who undoubtedly had no knowledge of Greek and only a rudimentary grasp of Latin.[17] Bianco's *Atlas* represents an important moment of synthesis between diverse geographical and cartographical traditions without manifesting contaminations or superimpositions between the various techniques of representing space. For example, the circular *mappa mundi* (ff. 8r-9v) – an "erroneous" form of representation according to Book I of the *Geography* – displays a cosmographical conception that is different from the one expressed in the Ptolemaic planisphere in the following map. By the same token, also the coastal outline of the Mediterranean and the Black Sea in the eight marine charts that comprise the manuscript is very different – and is more precise – compared to that of the planisphere. The marine charts, the circular *mappa mundi* and the Ptolemaic planisphere are presented in succession without any critical analysis that evaluates or compares the geographical contents or graphic concepts of the maps.

A third testimony about the Venetian reception of Ptolemy's *Geography* can be found in Harley 3686 of the British Library, which dates probably to sometime between 1430 and 1450

[15] Venice, Biblioteca Nazionale Marciana, Ms. It. Z. 76 (= 4783), manuscript on vellum, ff. I, 10, 260 × 395 mm. The Ptolemaic planisphere is the last of the ten maps that comprise the atlas. For a detail codicological description and a high resolution reproduction of the atlas, consult GeoWeb ‹http ://geoweb.venezia.sbn.it/›, edited by Piero Falchetta. There is also a facsimile copy of Bianco's atlas: P. Falchetta (ed.), *Andrea Bianco. Atlante nautico del 1436*, Venice: Arsenale Editrice, 1993.

[16] "Exercising all possible caution in this case, one could still conclude that Bianco's Ptolemaic planisphere derives from the Urbinate manuscript, either directly or indirectly"; for the quotation and for a cartographic and toponymic analysis of Bianco's planisphere and the Urbinate manuscript, see Gentile, "Umanesimo e cartografia," pp. 8-9.

[17] His culture could possibly be identified with what Carlo Maccagni has defined as being the knowledge of "technicians," skills that were learned and transmitted in artisans' workshops, in warehouses and on board the galleys, almost always without Latin, but with some notions of the abacus. This culture was mostly based on practical experience and was summarized in fragmentary precepts that were not framed within a structured system of knowledge. C. Maccagni, "Leggere, scrivere e disegnare la 'scienza volgare' nel Rinascimento," *Annali della Scuola Normale Superiore di Pisa. Classe di lettere e filosofia* 23 (1993), pp. 631-676; Id., "Sapere e cultura dei tecnici nel Rinascimento," in M. Dalai Emiliani and V. Curzi (eds), *Piero della Francesca tra arte e scienza*, Venice: Marsilio, 1996.

and has been carefully studied by Marica Milanesi.[18] The numerous errors in the Latin text, which reveals elements of the vernacular; the fact that the author did not understand some Greek words, indicate that this manuscript was not prepared by a learned humanist.[19] Instead of the planisphere and the twenty-six ancient maps of the Ptolemaic manuscripts an anonymous Venetian cartographer added to the eight books of the *Geography* eighteen modern maps, in which one can find an unprecedented level of contamination between nautical charts and Ptolemaic maps: while the coastal profiles have been drawn in the standard manner of nautical cartography, both in terms of the graphic forms as well as the geographical contents, the hinterlands have been drawn according to Ptolemaic precepts. In contrast to Bianco's *Atlas*, in Harley 3686, in addition to the contamination between nautical and Ptolemaic cartography, one can also find a mixture of humanistic culture (in which the Latin translation of the *Geography* was rooted) and less erudite elements derived from agents who were not fully familiar with Latin grammar and linguistics and who had no knowledge of Greek although, albeit without any artistic training, they were very competent cartographers.

To these testimonies we must add a fourth: that of the Aristotelian physician, astronomer, astrologer and engineer Giovanni Fontana, who studies and makes extensive use of the Ptolemy's *Geography*, especially in his *Liber de omnibus rebus naturalibus*, or *Book of all natural things*, whose date of composition must be placed between the Jubilee Year of 1450, which Fontana mentions, and 1454-1455 when he probably died.[20] Giovanni Fontana, a contemporary of Fra Mauro and Andrea Bianco, an eclectic scientist, was probably understood only with difficulty by his contemporaries, and this could explain the lack of dissemination of his works, mostly still unpublished or published only in part, and preserved in unique manuscripts distributed in various European libraries.[21] The presence of Ptolemy's *Geography* in the *Liber de omnibus rebus naturalibus* has been the subject of a recent and detailed analysis by Patrick Gautier Dalché who has shown that while being conscious of the extraordinary strength and theoretical

[18] Milanesi, "A Forgotten Ptolemy."

[19] *Ibid.*, pp. 42-43.

[20] There are no known manuscripts of the *Liber de omnibus rebus naturalibus*, but the *Liber* was published in Venice in 1544 under the name of Pompilio Azali, a plagiarist: Liber Pompilij Azali Placentini De omnibus rebus naturalibus quae continentur in mundo videlicet. Coelestibus et terrestribus necnon mathematicis. Et de angelis motoribus quae coelorum. Venetijs: apud Octauianum Scotum D. Amadei, 1544. The definitive attribution to Fontana was made by L. Thorndike, "An unidentified work by Giovanni Fontana: 'Liber de omnibus rebus naturalibus'," *Isis* 15 (1931), pp. 31-46.

[21] Even a sample listing of the works that are generally attributed to him can give a sense of the vastness and eccentricity of Fontana's interests and the complexity of his culture: the treatises *Nova compositio Horologii*, *Horologium aqueum*; and *Tractatus (Metrologum) de pisce, cane et volucre*, composed during Fontana's years in Padua, as a student and later teacher and rector, and related to the construction of clocks; the treatise *Bellicorum instrumentorum liber cum et figuris fictitiis literis conscriptus*, written in cipher and enriched by an extraordinary series of illustrations that deals with the construction of machines and *automata*; the *Secretum de thesaurus experimentorum ymaginibus hominum*, in which Fontana deals with the different types of memory and illustrates the functions of artificial memory; the *De trigono balistario* that describes the sextant, an instrument designed and developed for making optical trigonometric measurements, transcribing cartographic and perspective data graphically. See M. Clagett, "The Life and Works of Giovanni Fontana," *Annali dell'Istituto e Museo di Storia della scienza di Firenze* 1 (1976), pp. 5-28; and the biography of Fontana written by M. Muccillo in *Dizionario biografico degli Italiani* 48 (1997), pp. 672-675 that reports also the manuscritps transmitting Fontana's works.

potential of Ptolemy's work, Fontana's reading unifies Ptolemy's *Geography* with medieval Aristotelian astronomical geography. Fontana used data from the *Geography* of Ptolemy to "confirm definitively some of the results of the astronomical geography of the universities, and some of its astrological corollaries."[22] While aware of the superiority of Ptolemy's mathematical and geometrical approach to cartography, Fontana knew that *Geography* described the *oikumene* of the ancients, which was more limited than that known in the fifteenth century. At the same time, new astronomical data became available, as a result of the astronomical and astrological research developed by natural philosophers, mostly in the universities. For all these reasons, in Fontana's interpretation, Ptolemy's *Geography* was not in contrast to, but rather complemented *mappae mundi*, nautical charts, as well as the astronomical geography of the universities, and the *Alfonsine Tables*, in the representation of the *imago mundi*.[23]

The cases briefly described above constitute the evidence we have about the context in which Fra Mauro and a hitherto unidentified Venetian author approached Ptolemy's *Geography* sometime in the mid-fifteenth century, when they prepared detailed critical readings of the work. Their criticisms were based partly on the fact that their objectives were quite different from Ptolemy's.

2. Fra Mauro : apparent contradictions and an instrumental reading

Although he was well acquainted with Ptolemy's *Geography* Fra Mauro, as is well-known, did not follow its cartographical approach.[24] In fact, Fra Mauro drew his *mappa mundi* without resorting to any of the four methods of projection that were described in the *Geography*. However, a study of his *oeuvre* reveals that he maintained a continuous dialogue with Ptolemy's "Guide to Cartography." An analysis of the legends in which the Camaldolese friar criticizes, corrects and generally reflects upon the *Geography* clearly demonstrates the profound relationship between the two works, thus thwarting the initial impression that the graphic syntax of Fra Mauro's *mappa mundi* was bereft of Ptolemaic elements :

> This work, created to be contemplated by this most illustrious Seignory, is not as complete as it should be because it is not possible for the human intellect, without the help of some higher demonstration, to verify completely this cosmography or mappa mundi ; from these, one gets what is more like a sample of information rather than a full satisfaction of one's desire. So if someone contests this work because I have not followed Claudius Ptolemy either in the form of the world or in the measurement of latitude and longitude, I do not want to defend this map in any other

[22] Gautier Dalché, *La Géographie de Ptolémée*, pp. 198-203 (pp. 200-201).

[23] As Gautier Dalché writes, Fontana's *Liber de omnibus rebus naturalibus* helps us understand that "la *Géographie* de Ptolémée confirme et couronne l'*imago mundi* médiévale : ce n'est pas [...] la révélation d'un espace nouveau."*Ibid.*, p. 203.

[24] With regard to the relationship between Fra Mauro and Ptolemy, see W. Iwanczak, "Entre l'espace ptolémaïque et l'empirie : les cartes de Fra Mauro," *Médiévales* 18 (1990), pp. 53-68 ; P. Gautier Dalché, "Weltdarstellung und Selbsterfahrung : Der Kartograph Fra Mauro," in H. D. Heimann and P. Monnet (eds), *Kommunication mit dem Ich*, Bochum, 2004, pp. 39-51, esp. 45-47. See also the recent analysis in Gautier Dalché, *La Géographie de Ptolémée en Occident*, pp. 195-198. Fra Mauro refers to Ptolemy's *Geography* as the "cosmographia," following the translation by Iacopo Angeli.

> way than that in which Ptolemy defends himself when, in the first chapter of the second book [of his Geography], he says that one can only speak correctly of regions that are visited continually: of those which are less frequented no-one should think himself capable of speaking with equal accuracy. But if he meant by this that he had not been able to verify his cosmography thoroughly, because it is a long and difficult thing, and life is short and experience fallible, he is actually admitting that with the passage of time the work could be improved – that is, one could have more certain information than that available to him. So I say that in my own day I have been careful to verify the written accounts by practical experience, investigating for many years and frequenting trustworthy individuals, who have seen with their own eyes what I faithfully report above.[25]

It is precisely in light of Ptolemy's writings about the practices that shaped the cosmographical knowledge contained in Book II of the *Geography* that Fra Mauro explains the impossibility, and perhaps the futility, of drawing a *mappa mundi* according to Ptolemy's cartographic guidelines. Fra Mauro writes: "One can speak accurately about those parts that are continuously frequented, but about those parts that are not frequented thus, I do not feel that anyone can speak about them so accurately," literally translating a passage from the *Geography*, II.1.[26] One must also stress that this legend is placed in a part of the *mappa mundi* that is easy to read and situated at a spot designed to catch the reader's eye. Fra Mauro thus places his work, drawn to be contemplated by the "illustrious Seignory" of Venice, on the same heuristic level as the *Geography*. He possibly even framed it as a continuation of this work in the guise of a critical updating that extended to parts of the *oikumene* that were unknown to the ancients because, "With the passage of time, such a work can benefit from the more precise information that has come to light since Ptolemy's age". A second legend better explains Fra Mauro's approach:

> I do not think that I have belittled Ptolemy by not having followed his *Cosmography*, because if I had wanted to observe his meridians, parallels and degrees, I would have had to omit from this depiction of the world [*circumferentia*] many provinces that are known today that Ptolemy does not mention, especially in the extreme latitudes, that is to the north and south, he labels these [regions] *terra incognita*, because in his age they were not known.[27]

[25] This is my translation of the following legend: "Questa opera, fata a conte(m)platio(n) de questa illustrissima signoria, no(n) ha i(n) sì | quel co(m)pimento che la doueria, p(er)chè certo no(n) è possibile a l'intellecto huma(n) se(n)za | qualche superna demostratio(n) uerificar in tuto questa cosmographia over ma|pamu(n)di, de la qual se può haver qualche noticia più a degustation cha a suppli|[mento del desiderio. Unde se algun contradirà a questa p(er)chè no(n) ho seguito | Claudio Tolomeo, sì ne la forma come etia(m) ne le sue mesure p(er) longeça e p(er) largeça, no(n) uogli più curiosamente defenderlo de quel che lui | proprio no(n) se defende, el qual nel seco(n)do libro capitulo primo | dice che quele parte de le qual se ne ha co(n)tinua pratica se | ne può p(ar)lar corretamente, ma de q(ue)le che no(n) sono cussì fre|que(n)tade no(n) pe(n)si algun se ne possi p(ar)lar cussì correctam(en)te. Però | inte(n)dando lui no(n) haver possudo in tuto verificar la sua cosmogra|phia, sì p(er) la cossa longa e difficile e p(er) la vita brieue e l'experime(n)to fa|llace, resta che'l co(n)ciede che cu(m) longença de te(m)po tal opera se possi meglio | descriuer over hauerne più certa noticia de | quel habuto lui. Perta(n)to dico che io nel te(m)|po mio ho solicitado verificar la scriptura | cu(m) la experie(n)tia, i(n)vestiga(n)do p(er) molti a(n)ni | e pratica(n)do cu(m) p(er)sone degne de fede, | le qual hano ueduto ad ochio q(ue)lo che | qui suso fedelme(n)te demostro." *Mappamondo* (ed. 1956) Pl. XL, 49. For a different version, see *Mappamondo* (ed. 2006) *2834.

[26] *Ptolemy's Geography*, Book II.1,"Preface to the detailed guide,"pp. 94-95.

[27] *Mappamondo* (ed. 1956) Pl. XLI, 21* "Io no(n) credo derogar a Tolomeo | se io non seguito la sua cosmographia, | perchè se hauesse uoluto obs(er)var i sui | meridiani ouer paralleli ouer gradi | era necessario quanto a la demo(n)st(r)a|tion de le parte note de questa circu(m)|ferentia lassar molte proui(n)cie de le | qual Tolomeo no(n)

Fra Mauro was well aware, and almost apologetic about the fact, that he was not following the "forms" and "measurements" of Ptolemy's *Geography*. "Forms" in this context refer to the cartographical graticule of the first and second Ptolemaic projections, i.e. the general "form" of representation of the Ptolemaic *oikumene* ; the "measurements" refer rather to the geographic contents of Ptolemy's work, in the specific sense of the coordinates of almost eight thousand toponyms that are listed in Books II-VII of the *Geography* and thus, implicitly, to the overall design of the *oikumene*. Following the Ptolemaic definition of longitude and latitude proposed in Book II.1 of the *Almagest* (repeated in *Geography* I.6) Fra Mauro defined the co-ordinates of latitude and longitude as "height" and "width."[28] It was above all in terms of the "height," i.e. from South (*ostro*) to North (*tramontana*) beyond Agysimba at 16°S and Thule at 63°N, areas that Ptolemy had dubbed *terra incognita*, that the *Geography* revealed its inadequacies in representing the world as it was known to mankind around the mid-fifteenth century. These inadequacies, according to Fra Mauro, also extended to remote regions of the Orient as well as to areas closer to home in Europe.[29] In addition to its incompleteness in its coverage of land, the *oikumene* described by Ptolemy was also inadequate in terms of its overall cosmographic structure, especially with regard to the shapes and sizes of the seas. To briefly return to my analysis at the end of Chapter II Fra Mauro diverged from Ptolemaic cosmography precisely in his conception and representation of the seas. Unlike the *Geography* Fra Mauro's *mappa mundi* describes vast marine spaces : he wished to demonstrate the concrete possibility of sailing from the Indian Ocean to Gibraltar by circumnavigating Africa to reach the Mediterranean basin. The seas he represented beyond the confines of the Ptolemaic *oikumene* not only included regions that Fra Mauro felt were inhabited and inhabitable, but also vast and navigable expanses of ocean. In addition to the geographical details linked to toponymic changes, it was thus the cosmographic structure of the Earth as described in the *Geography* that Fra Mauro found so discordant with modern knowledge, and it was this that made it impossible for him even to attempt to correct and update the individual Ptolemaic maps.

Notwithstanding his criticism of the *Geography*, an analysis of Fra Mauro's *mappa mundi* reveals that he was well aware of the richness of Ptolemy's geographic knowledge, even with regard to regions that were still not very well known in the mid-fifteenth century : "It seems to me that in the fourth plate of Europe Ptolemy mentions this Scandinavia and says that there

ne fa me(n)tio(n), ma | per tuto maxime i(n) latitudine çoè tra | ostro e tramo(n)tana dice terra i(n)cognita, | e q(ue)sto p(er)chè al suo te(m)po no(n) li era nota." A translation into English also appears in *Mappamondo* (ed. 2006) *2892.

[28] "La largeza se intende da ostro in tramontana," *Mappamondo* (ed. 1956) Pl. XXXIX, Q 31, "By height I mean from the South to the North" *Mappamondo* (ed. 2006) *2918 ; "La longeza se i(n)ten|de leua(n)te a pone(n)te" *Mappamondo* (ed. 1956) Pl. XXIV, 141 "By width I mean from the East to the West." For a different version, *Mappamondo* (ed. 2006) *1298. See Claudius Ptolemaeus, *The Almagest*, II.1, trans. C. Tagliaferro, in *Great Books of the Western World*, XVI, Chicago : Encyclopedia Britannica, 1952, p. 34.

[29] "Da questa proui(n)cia seri|ca in çoso Tholomeo fa ter|ra ignota" *Mappamondo* (ed. 1956) Pl. XXXI, 9. "Southwards from this province of Serica, Ptolemy labels the land as *terra ignota*" *Mappamondo* (ed. 2006) *2243. Also with regard to Northern Europe : "Questo colfo, el qual | Tolomeo no(n) mete, ha piusor | nomi, e fi dito lubech, pru|sico, sarmatico, germani|co, e p(er)ché questo ultimo | no[me è più chiaro, p(er)ò | ho notado colfo ger|manico." *Mappamondo* (ed. 1956) Pl. XL, 23. "This gulf, which Ptolemy has not shown, has various names, and is called the Gulf of Lübeck, the Prussian Gulf, the Sarmatic Gulf, the Germanic Gulf, and because this last name is the clearest of them all, I have therefore noted it as the Germanic Gulf," *Mappamondo* (ed. 2006) *2862.

are 18 hours of daylight, whereas I am amazed that he knew about this and this entire part of Norway and Sweden because was unknown to him."[30] Moreover, in at least one case, Fra Mauro fully agrees with Ptolemy, and does not fail to say so openly:

> Because there was not enough space in Europe to give the full verdict of Ptolemy with regard to the division of Africa and Asia, I will observe here that he practically makes two divisions: the first begins from the tip of Ethiopia and runs down the coasts of the Gulf of Arabia. [But then] he says that, in order not to divide Egypt, he places the limit of Africa at the Nile – and I agree with him here.[31]

Fra Mauro also cites the *Geography* to correct what he felt was a widespread error: "Those who write that Ninive and Babilonia are the same city make a great mistake; I will let Ptolemy answer them for me."[32] This was not all. The *Geography* was the main source that Fra Mauro used to draw the Asian coasts. In fact, he adapted and transformed the coastline of Asia in the Ninth to Twelfth *tabulae* and in the Asian part of the *Geography*'s planisphere, in a style reminiscent of nautical cartography, i.e. without the geographic grid of meridians and parallels. A comparison of the Asian coast in Bianco's planisphere and that of Fra Mauro's *mappa mundi* shows several similarities. Furthermore, almost all the Ptolemaic toponyms that had "changed" or had been "corrupted" were replaced with names that had been extracted, above all, from the travel accounts of Marco Polo and Niccolò de' Conti.[33] The contradiction is only apparent because, around the mid-fifteenth century, despite all the imperfections that Ptolemy himself had admitted (*Geography* I.12-14; II.1), the Ptolemaic maps provided the only systematic graphical representation of Asia and the Indian Ocean that was available in the West. Fra Mauro, the Dominican Andreas Walsperger (who prepared a *mappa mundi* dated 1448),[34] the anonymous

[30] "Parme che Tolomeo ne la q(u)ar|ta taola de Europa nomina q(ue)sta | sca(n)dinaria e dice che'l à 18 hore | de zorno, unde me meraueio ch(e) | q(ue)sta li sia sta' nota e tuta | questa p(ar)te de norue|gia e suetia li sia | sta' ig(n)ota" *Mappamondo* (ed. 1956) Pl. XLI, 27. The English translation is mine. For a slightly different version, see *Mappamondo* (ed. 2006) *2904.

[31] "Perchè io no ho habuto loco | in europa de dir la uera sente(n)tia | de Tolomeo circa la diuisio(n) de | l'affrica da la asia, p(er)hò qui dico ch(e) | lui fa quasi do' diuisio(n): la p(r)ima el co(m)me(n)ça | da la po(n)ta de ethiopia e uie(n) zoso p(er) la costiera del | colfo de Arabia, poi el dice che per non | diuider l'egypto el mete | el fin de l'affrica nel nilo e | cu(m) el modo suo io me | acordo," *Mappamondo* (ed. 1956) Pl. XXIII, 51. The English version derives from *Mappamondo* (ed. 2006) *1077.

[32] "Queli che | scriueno | che niniue | e babilonia | sia una me|dema cita|de pre(n)deno | grande er|ror, ai q(u) al | bastame che | Tolomeo | li respo(n)da," *Mappamondo* (ed. 1956) Pl. XXVII, 68. The English version derives from *Mappamondo* (ed. 2006) *1466.

[33] See the *Descriptio none tabule Asie* [Aria, Paropamsadas, Drangiana, Arcosiam, Gedrosiam]; *Descriptio decime tabule Asie* [India intra Gangem]; *Descriptio undecime tabule Asie* [Sinus Magnus, Sinus Gangeticus, Indicum Pelagus]; *Descriptio duodecime tabule Asie* [Taprobana insula], in the manuscripts with plates and in the printed editions of Ptolemy's *Geography* (the main regions described in these plates are mentioned in the square brackets).

[34] Vatican City, Biblioteca Apostolica Vaticana, Palat. Lat. 1362B, Andreas Walsperger, *Mappa mundi sive descriptio orbis geometrica facta ex cosmographyca Ptholomey (...). Et cum vera et in|tegra cartha navigationis marium* (...), manuscript on vellum, 589 × 746 mm, oriented towards the south, 1448. In the lower margin the last line of a note written in black ink on six lines, states: "Facta est hec mappa per manus fratris Andreas Walsperger ordinis Sancti Benedicti de Salisburga Anno Domini 1448 in Constantia." See Almagià, *Monumenta cartographica vaticana*, vol. 1, pp. 30-31, plate 12; and Destombes, *Mappemondes*, entry 52.10, pp. 212-214.

dipintori of the so-called "Genoese" *mappa mundi* that dates from 1457 (plate V),[35] and the Catalan-Estense *mappa mundi* from around 1460-70[36] all adopted Ptolemy's outline of the Asian coast since there were no available systematic alternatives.

Fra Mauro's reading did not consider the aspects that Ptolemy had indicated as the fundamental part of the *Geography*, i.e. the codification of the procedures of representing the *oikumene* that occupied about a quarter of the terrestrial sphere on the flat surface of a map. Ptolemy held that these techniques were the only ones that were capable of making cosmographic knowledge transmittable and correctable, and he criticized Marinus of Tyre precisely for his lack of such codification (*Geography*, I.6-7, I.11, I.24.3, I.24.9 and VII.6-7 ; *Almagest* II.1). Fra Mauro was not interested in, and really did not explore, the theoretical aspects of Ptolemy's "Guide to Cartography," which he treats purely as an instrument. The cosmographical language used by Fra Mauro is quite different from that of the *Geography*. As has been seen in Chapter II, his "theoretical" interest in cosmography concerned, above all, themes pertaining to terrestrial physics and the structure of the sub-lunar and celestial world within the framework of Aristotelian cosmology.[37] The *mappamondo* of the Biblioteca Marciana does not contain any element that can be linked to Ptolemy's theoretical considerations. Like Guglielmo Capello in his interpretation of the *Geography* in his *Commento al Dittamondo* (*c.* 1437), in which he used Ptolemy to correct Fazio degli Uberti, Fra Mauro shows no interest in Ptolemy's mathematical calculations.[38] He limited himself to producing a general, geographic image of Ptolemy's *oikumene*, updating it wherever possible or using it when he did not have better sources. A reading of the legends that refer to the *Geography* indicates that not only was Fra Mauro not interested in Ptolemy's cartographical methods but, further, he had not understood the methodological lessons and the cartographical instructions contained therein. This was in stark contrast to individuals such as Emmanuel Chrysoloras, Pierre d'Ailly, Guillaume Fillastre and Francesco Berlinghieri who, despite their lack of sophisticated mathematical skills, had nonetheless grasped that one of the added values of Ptolemaic cartography compared to other forms of mapping lied precisely in the geometrical and mathematical approach to the description of the *imago mundi*. A critical and radically different reading of the *Geography*, centered on the theoretical aspects of this work, would instead be written in Venice during the same period by an anony-

[35] Florence, Biblioteca Nazionale Centrale, Portolano 1, [*Hec est vera cosmographor(um) cum marino accordata* [*descri*]*cio, quoru(n)d(am) i(n) frivolis narracionibus reiectis, 1457*], manuscript on vellum, 395 × 820 mm. For the description of the map, see Chapter III, notes 25 and 26.

[36] Modena, Biblioteca Estense Universitaria, C.G.A.I. [Catalan-Estense World Map], manuscript on vellum, 113 × 113 cm, *c.* 1460-1470 ; see F. L. Pullè, "Illustrazione del Mappamondo catalano della Biblioteca Estense di Modena" *Atti del Sesto Congresso Geografico Italiano*, Venice, 1908, pp. 341-397 and E. Milano (ed.), *Commentario al facsimile Mappamondo Catalano Estense*, supplemented by a transcription by A. Battini, Dietikon-Zürich, 1995.

[37] For the analysis of the main themes of natural philosophy examined by Fra Mauro see Chapter II.

[38] Guglielmo Capello (d. sometime after 1449) was a Latinist and preceptor at the d'Este court from 1421 until his death. He wrote a work in vernacular entitled *Commento al Dittamondo*, which was concluded in 1437. There are four manuscripts of the *Commento* : Turin, Biblioteca Nazionale Universitaria, N.I.5 ; Modena, Biblioteca Estense, Ital. 483 ; Venice, Biblioteca Marciana, It. IX, 40 (= 6901) ; Paris, Bibliothèque Nationale de France, It. 81). Capello commented upon, and corrected, the geographical part of Fazio degli Uberti's *Dittamondo* on the basis, above all, of Ptolemy's *Geography*, Pomponius Mela's *Cosmography*, and Pliny the Elder's *Natural History*. See the in-depth analysis by Milanesi, "Il commento al Dittamondo di Guglielmo Capello (1435-1437)," pp. 365-388.

mous author-cartographer, revealed in a splendid manuscript that has so far been forgotten by history and that merits further study.

3. A forgotten document: Venice, Biblioteca Nazionale Marciana, Ms. It. VI, 24 (= 6111)

MS. It. VI, 24 (= 6111) preserved at the Biblioteca Nazionale Marciana in Venice currently represents the sole testimony of confirmed Venetian origin that transmits Ptolemy's *Geography*.[39] It transmits translations of extensive excerpts from all the eight Books, inserted amongst a larger and more complex treatise, which contains a synthesis that includes elements of Ptolemaic geography and cartography, medieval *mappae mundi* and marine charts, *auctores veteres* such as Pliny the Elder and Gaius Julius Solinus, Julius Caesar and Tacitus, and *auctores novi*, such as Marco Polo.[40] Even though the author did not transcribe the entire text of the *Geography*, this Venetian manuscript contains what could prove to be one of the first translations and critical revisions in the vernacular of the so-called theoretical books (*Geography* I.22-24; VII.6). Furthermore, in the more than 60 maps that illustrate the work, the author transformed the Ptolemaic cartographic syntax to produce an innovative graphic and cartographic synthesis that has no equivalent in the history of ancient or medieval cartography. The planispheres and cosmographical grids propound an extension of Ptolemy's projections beyond the confines of the Ptolemaic *oikumene* to includes the more expansive world known to fifteenth-century cartography.

The manuscript is undated but calligraphy, watermark analysis, and contents allow one to date it to the second quarter of the fifteenth century. The entire text and all the maps appear to be in the same hand and are drawn in black ink and pencil on paper. The writing is a so-called "bastard" script, with humanistic influences that are apparent in the rounding of some characters such as the letters "r" and "d."[41] On f. 97, the last page of the manuscript, there is a recognizable watermark with a "five petal flower," a circle in the centre and a stem, 85 mm

[39] Venice, Biblioteca Nazionale Marciana, Ms. It. VI, 24 (= 6111), paper, 205 × 281 mm, mid-fifteenth century; ff. 97, II', contemporary numbering on the top right, ff. 55 and 56 inverted; f. 67 missing; ff. 68-70 are blank. The manuscript is made of seven quires, of which five consist of 16 double pages (ff. 1-16, ff. 17-32, ff. 33-48, ff. 49-66, ff. 76-91), one has nine double pages (ff. 67-91) and the six final folios (ff. 92-97); ff. 1-2v have one column, the others have two; the sections of the pages meant for the cartographical designs are blank on ff. 6rv, 7rv, 8rv, 9v, 10rv, 11rv, 12r, 14rv, 15v, 19r, 53rv, 57r, 59r, 60rv, 88r; ff. 12v, 13rv, 14r, 47v, 48rv, 71rv, 72rv are struck through; ff. 6rv, 81v have toponyms, but no cartographical designs; written in the Venetian vernacular in a single hand. There are 62 maps and geometric designs drawn in pen and in pencil. The original binding, consisting of leather over wooden boards with traces of claps on the back cover, is intact, albeit very worn.

[40] Marica Milanesi has defined *mappae mundi* such as those prepared by Pietro Vesconte (*c.* 1320) and by Fra Mauro, the so-called "Genoese World Map," dating from 1457, and the "Catalan-Estense World Map" dating to about 1460-1470, as a "cartography of synthesis," in which several graphic and technical syntaxes mingle. See M. Milanesi, "La Cartografia italiana nel Medio Evo e nel Rinascimento," in *La Cartografia italiana. Cicle de conferències sobre Història de la Cartografia. Tercer curs*, 1991, Barcelona: Institut Cartogràfic de Catalunya, 1993, pp. 15-80, esp. 23.

[41] A comparison with other mid-fifteenth century Venetian manuscripts corroborates the hypothesis that this manuscript could also have been written at this time. See C. Cassandro, N. Giovè Marchioli, P. Massalin, and S. Zamponi (eds), *I manoscritti datati della provincia di Vicenza e della Biblioteca Antoniana di Padova*, Tavarnelle (Florence): Sismel-Edizioni del Galluzzo, 2000, pl. 38 (Vicenza, Bib. Civ. 297, f. 3r, 1448); pl. 44 (Padua, Bibl. Ant. 378,

high and 35 mm wide, which corresponds to Briquet 6384. This is a watermark that was probably of Venetian origin and could be found in Udine in 1425 and Pisa between 1433 and 1437.[42] The writing on f. 97 is completely homogeneous in terms of its *ductus*, format and ink to that of the manuscript's other pages. This watermark is thus one more documentary element corroborating the hypothesis that MS. It. VI, 24 (= 6111) dates to the second quarter of the fifteenth century, with a *post quem* around the 1430s. This hypothesis is further reinforced by the fact that neither the numerous maps nor the text contain any hints of Atlantic and African discoveries. This dating thus coincides with the date proposed (1430-1450) by Marica Milanesi for the British Library Harley MS 3686, with the 1423 attempt of revising Iacopo Angeli's Latin translation of the *Geography* by a Venetian called *Petrus*, and the Ptolemaic planisphere that Andrea Bianco drew in 1436. All this evidence indicates that sometime around the third decade of the fifteenth century, Venice was the setting for an important reflection upon Ptolemy's *Geography*, which involved various cultural *milieux*.

This manuscript was not entirely unknown to scholars. Giuseppe Caraci reproduced the cartographical outline of the world map drawn on f. 47r in the entry "Cartografia" in volume XI of the *Enciclopedia italiana* without, however, making any reference either to the manuscript or its contents.[43] The manuscript was then briefly described in volume LXXVII of the *Inventari dei manoscritti delle biblioteche d'Italia*, as a "fifteenth-century geographical treatise."[44] Subsequently, in 1964, Marcel Destombes included one of the two *mappae mundi* in a survey

f. 37v, 31 October 1454); pl. 42 (Vicenza, Bib. Civ. 137, f. 1r, 8 June 1453); pl. 46 (Montagnana, Padua, Bib. Ant. 45, f. 111r, 1455); pl. 40 (Padua, Bib. Ant. 432, f. 27r, 14 November 1464); pl. 64 (Vicenza, Bib. Civ. 225, f. 105r, 1465).

[42] The "five petal flower" watermarks (Briquet 6369-6391) are described as follows: "Les 6362 à 6368 et 6369 à 6391 (ces derniers avec un petit rond au centre de la fleur) se rapprochent, comme dessin, de la pervenche. Cette marque est italienne probablement vénitienne. Les Mocenigo portaient une fleur similaire dans leurs armoiries. Un membre de cette famille, Leonardo, était en 1407 capitaine à Vicence. (C'est lui qui fit construire la *Porta Lupia*, et qui y fit placer son écu. Lors de la démolition de cette porte, ces armoiries furent reportées sur une tourelle de la nouvelle Barrière de Porta Lupia. Les Mocenigo jouèrent un role important dans l'histoire de Venise. Trois membres de cette famille, Thomas, Pierre et Jean, ont été doges dans le courant du xve s., et un autre, Louis, dans la seconde moitié du xvie. En se rappelant que Vicence a possédé de très bonne heure des papeteries dans son voisinage et que l'une d'elles, à Negrisola, dans le district de Cenada [e.g. Vittorio Veneto, some 60 kilometers north of Venice], est mentionnée en 1768 (dans un règlement approuvant les marques adoptées par les papeteries) comme appartenant à la sérénissime maison des Mocenigo, nous sommes portés à attribuer à Negrisola et à la famille Mocenigo, la plupart des figures du groupe 6362 à 6391, auxquels se rattachent les 6439 à 6449. " See C. M. Briquet, *Les filigranes: dictionnaire historique des marques du papier dès leur apparition vers 1282 jusqu'en 1600: a facsimile of the 1907 edition with supplementary material contributed by a number of scholars*, A. Stevenson (ed.), 4 vols., Amsterdam: Paper Publication Society, 1968, vol. 1, pp. 364-365. This was thus a paper that came from Northern Italy, made in the second quarter of the fifteenth century, probably in the vicinity of Venice. I am grateful to Marica Milanesi for having shared the results of her research on the watermark of the manuscript with me.

[43] *Enciclopedia italiana*, vol. XI, pl. LXXI, fig. 13, "Mappamondo anonimo del secolo XV," Rome: Istituto Giovanni Treccani, 1931, pp. 230-250.

[44] This is the description provided by P. Zorzanello: (f. 97a) 'here I will put a figure in which I will draw them with their ancient and modern names' (compass rose). The author states he based his work on Ptolemy, Pliny the Elder, Agrippa, Ugo Floriacense and Marco Polo. Fifteenth century, paper. In. fol. (298 × 220), ff. 97. Gothic writing, many geographic and geometric designs in ink and in pencil. Wooden binding, leather spine. Origin Farsetti No. CXXV." See G. Mazzatinti, *Inventari dei manoscritti delle biblioteche d'Italia*, vol. LXXVII, Venice-Marciana Mss. Italiani – Classe VI (Storia e Geografia), P. Zorzanello (ed.), Florence: Olschki, 1950, p. 7.

on medieval world maps that was sponsored by the International Geographical Union.[45] With the exception of these sporadic citations the Marciana manuscript has remained largely forgotten: these modern authors did not realize, let alone mention, that the manuscript contained the first commentated vernacular translation of the theoretical chapters of Ptolemy's *Geography* and that its fifteenth-century author had reinvented its cartographical protocols.[46]

The Marciana manuscript consists of six main parts: ff. 1r-2v contain a general description of all the provinces "indicated by Ptolemy" followed by their geographical coordinates, and these pages serve as a general presentation of the work; ff. 3v-46v contain a written and cartographical description of Europe ("And I will describe all of Europe, and the first part of this will be Spain...," ff. 3vb-26v),[47] Africa ("From here on I will begin with Africa, divided into XII provinces...," ff. 27va-31v),[48] and Asia ("From here on, I will talk of Asia, the third part of the *oikumene*, divided into XLVIII provinces...," ff. 32ra-46v),[49] using an innovative graphic syntax that is unique in the history of pre-modern cartography and will be examined on the following pages. Ff. 30v-31v contain a map of Europe and Africa, and a map of Africa only in the graphic form of a *mappa mundi*; ff. 47v-72v contain what I consider to be the first vernacular translation and transformation of the 27 Ptolemaic maps, introduced by a written description of the "round Earth" (on f. 46v, e.g., the author refers to a round graduated *mappa mundi*)[50] and by a very peculiar *mappa mundi* on f. 47r, surrounded by the twelve winds, on which two graduated axes intersect at Jerusalem, the center of the map; ff. 73r-82r contain a geographic description of the *oikumene* (*universale*) based on ancient and modern authors, where the main sources are Pliny the Elder, Gaius Julius Solinus, Ptolemy and Marco Polo; ff. 82va-91v include the first translation and critical revision in the vernacular of Books I.22-24 and VII.6 of the *Geography*, applying the Ptolemaic cartographical method to the larger *oikumene* known in the fifteenth century by inventing the diagram of a "double" first Ptolemaic projection; and ff. 95v-97 (ff. 93r-95r are blank) contain a description of Asian

[45] The description provided by Destombes reads thus: "97. ff. Anonyme, Trattato di Geografia, en langue italienne, pour lequel l'auteur dit avoir utilisé Agrippa, Pline, Ptolémée, Marco Polo, etc. f. 67 [sic], mappemonde circulaire, type D, nomenclature 41 noms (Asie 23, Afrique 9, Europe 9)" ("97. ff. Anonymous, geographical treatise, in Italian, whose author states that he used Agrippa, Pliny the Elder, Ptolemy, Marco Polo, etc.; f. 67, circular world map, type D, contains 41 place names – Asia 23, Africa 9, Europe 9." His description is followed by a reference to Caraci's article in the *Enciclopedia Italiana*. Destombes, *Mappemondes*, entry 51.36, p. 188.

[46] In 1985 the manuscript was not listed amongst works of a scientific nature preserved at the Biblioteca Marciana; see C. Maccagni (ed.), *La scienza a Venezia tra Quattrocento e Cinquecento. Opere manoscritte e a stampa*, Venice: Comune di Venezia and Biblioteca Nazionale Marciana, 1985. The manuscript was introduced to the scientific community in the course of the *20th International Conference on the History of Cartography*, held at Cambridge MA and Portland ME in June 2003. See A. Cattaneo, "A Forgotten Mid-Fifteenth Century Venetian Manuscript Codex with Sixty Maps and also the First Vernacular Adaptation of Ptolemy's Geography," in *ICHC 2003 Program, Abstracts, and List of Participants*, Cambridge and Portland, p. 110; Id., "Letture e lettori della Geografia di Tolomeo a Venezia intorno alla metà del Quattrocento," *Geographia Antiqua* 4-5 (2005), pp. 41-65.

[47] "E dirò a | pieno de Europa, e la prima parte | d'essa serà la Spagna...," ff. 3vb-26v.

[48] "De qui in avanti accomincerò de la Affrica divisa per XII provintie...," ff. 27va-31v.

[49] "De qui in avanti seguirò a dire de Asia, la | terza provintia de l'universo, divisa per XLVIII | provintie...," ff. 32ra-46v.

[50] "tondo de l'universo," f. 46v.

regions "according to the tradition of modern writers," which cites parts of Marco Polo's book.[51]

The manuscript seems to have been written in the author's own hand, rather than being the work of a copyist. The numerous cancellations and textual corrections, the addition of glosses and cross-references (e.g. on ff. 58v, 59r, 60v-62rv), and the changes in compositional intentions (e.g. f. 16r, in which the expected ninth, tenth, eleventh and twelfth *tabulae* of Asia are replaced by a hemispheric globe) suggest that it is more likely a draft reflecting the process of the author's preparation and revision rather than an example of a copyist's transcription, which is perforce more mechanical and sequential. It is thus a preparatory manuscript, a draft for a definitive later version "in a neat hand," which was perhaps never even written or at least has not yet been found. As in the case of British Library Harley 3686 the form of some specific technical terms, such as "paralellus" used instead of the humanistic "parallelus,"[52] along with the text's syntax and linguistic structure, which is frequently rather convoluted and generally not very linear in its narrative flow, could indicate that the author did not belong to the humanistic tradition. The work in the manuscript is anonymous but we should nonetheless formulate a hypothesis about its possible author. Giovanni Fontana, as we have seen, was among those in Venice in the second quarter of the fifteenth century who mentioned the work of Ptolemy and understood its theoretical aspects; in addition to his interest in cartography and geography, Fontana also had a remarkable skill in drawing as shown by his autographed manuscripts. Given this evidence, it is clear that if he were informed that a translation of the work of Ptolemy was available, the Venetian physician would have attempted to obtain a copy. It also seems possible that Biblioteca Marciana MS. It. VI, 24 (= 6111) could be the work of Fontana himself, which he wrote in order to make himself intimately familiar with Ptolemy's work. Fontana was ignorant of Arabic and Greek, but had studied both Arabic and Greek authors through translations; he certainly knew Latin, though his knowledge was not comparable to that of the Humanists. The partial translation of Ptolemy into the Venetian language in Ms It. Cl. VI, 24 (= 6111) could have been made for personal edification rather than with the goal of possible publication and diffusion. To be able to fully appreciate the value and scope of Ptolemy, in short, for a full understanding of the text, Fontana found it necessary to translate it into a language that for him could be easier to use: the Venetian language. Sadly, there is still no study of Fontana's calligraphy; and it would only be through such a study, as well through analysis of the text, that confirmation or refutation of our hypothesis could be achieved.

[51] "Mostratto de sopra in pratica et in theori|ca simul et semel la descriptione de l'univer|so, sì per Ptholomeo, per Plinio e diversi al|tri scriptori... intendo de qui in avanti scrivere, secondo la | traditione dei moderni scriptori, pervi|gnerò a la descriptione dei ‹universi› fiu|mi de l'universo notabili, dei monti, de le | fonte, dei lachi e de le selve, nelle | cui fini termenerà l'opera mia," f. 95va; in English: "I have shown above the practice and theory of the description of the Earth, by Ptolemy, Pliny [the Elder] and various other writers... from here on, I intend to write according to the tradition of the modern writers, rendering a description of the rivers of the known world, the mountains, springs, lakes and forests, at the end of which my work will conclude."

[52] For example: "Paralelli septentrionales," f. 16rb; "Affrica che da la parte occidentale si appressa | dal paralello extivale," f. 50vb; "Sitia che è for | del monte Imao, iuxta el paralello de l'isola | de Tille," f. 63vb; "dicen|do la linea b esser paralello occidentale," f. 82vb. For an analysis of the use of "paralellus" and "parallelus" see Milanesi, "A Forgotten Ptolemy," pp. 63-64.

The author of the Venetian manuscript carried out his vernacular translation with critical comments of lengthy excerpts of the *Geography*, at least twenty or thirty years before Francesco Berlinghieri's *Septe giornate della Geographia in terza rima* (1469-1479 or 1482), so far considered to be the first vernacular translation of Ptolemy's work. Unlike Berlinghieri's *Septe giornate*, the translation and critical comments by the anonymous Venetian author focus on an interpretation of the geographical aspects and cartographical theory of Ptolemy's *Geography*. Together with Andrea Bianco's 1436 atlas, Harley 3686 and Fra Mauro's *mappa mundi* this manuscript is a crucial document, if not the most important that has come to light till now, for its insights into the Venetian reception of the *Geography* in the fifteenth century and also for the wider context of science in vernacular languages. The author explains his intentions at the beginning of the text:

> So, with this intense study, to the best of my abilities, I am willing to show you, for your pleasure, the description of the world and its main parts. As a first step I will divide this [i.e. the description of the world] into the habitual parts (e.g. Europe, Africa and Asia) and then I will divide each part into its regions and I will draw them, so that it will be clearer for you, by rendering on a flat surface the curvature of the sky, from which all the measurements of the Earth are derived, including the latitudes and the longitudes calculated from the equator, from which all measurements are taken.[53]

"Inducing in a flat plane the flexion of the sky… you can clearly see the division of the entire universe," he declares on f. 1v, before listing the regions and provinces of Europe, Africa and Asia, indicating their geographical coordinates, derived from Ptolemy's *Geography*: "These are all the provinces indicated by Ptolemy with the distances and their heights. Henceforth, I will show each part of the Earth, depicting each part by itself, so that each thing that we wish to see is clearer; then I will join them together, showing the universal description of the world (f. 3ra)."[54] Only a few aspects of this invaluable manuscript will be addressed in this chapter: in particular, an analysis of the vernacular translation and adaptation of the geometric procedures for the design of the first Ptolemaic projection (ff. 90va-91v) and of the "Circularis spherae cum habitabili terrae descriptio" (*Geography* VII.6), the so-called third Ptolemaic projection (f. 92vb), and finally the unique graphic scheme by which the anonymous author "reinvented" the Ptolemaic cartographical protocols to draw regional charts (ff. 17vb-46ra).

a. *A new Ptolemaic projection: the "double figure"*

On f. 90va, in the context of a long analysis and critical review of the procedures used to draw the chart of the world in circular form using a Ptolemaic-style graduated cartographic grid, the author begins with a description of the geometric techniques employed to build the

[53] "Per tanto, con quello fermo studio a me fia possibile, la descriptione del mondo con le | sue partisone, vogliendo a voi per vostra complacentia redure, partirò prima quella | iuxta le sue consuette parte e cadauna de quelle puoi dividerò per le sue regione, | designando quelle, a ciò che a voi più perspicace se renda inducendo in figura plana | el flexione del cielo, dal qual se proceder si può ogni mesura de le terre, cossì de | altezza a l'equinoziale come de long[h]eza ad ponto de occidente, dal qual si prende | ogni misura," f. 1r.

[54] "Queste sono tutte le provincie poste | per Pthολομeo con le distantie e loro alte|ze. De qui in avanti vegnerò a la | partisone de cadauna parte de le terre | figurando cadauna parte per se istessa, | afinché meglio sia chiara ogni | cossa noi voremo vedere; po' quelle giun|gerò insieme, mostrando la universale | descriptione del mondo" f. 3ra.

"triangular figure," the diagram of the so-called First Ptolemaic Projection (*Geography* I.24.3).[55] This is followed by a detailed description (written in an appendix) of the algorithm used to draw the diagram of the first Ptolemaic projection, paraphrasing Book I.24 of the *Geography* step by step. The first method of projection, the simplest of all the techniques described in the *Geography*, had straight meridians that converged at a theoretical point, placed to the North at a distance equivalent to 34° with respect to the dimensions of the grid. The parallels are concentric arcs drawn around this geometric centre. In this projection the *oikumene* is represented in a conical grid which was called a "small cloak" (*mantellino*) in early sixteenth-century Florentine vernacular, and in the manuscript under study was termed a "triangular figure."[56] As was the author's habit he indicated that he also described the procedure via diagrams ("And so that it is even clearer, via a figure, whose position I will shortly indicate") very similar to those of many other manuscripts that transmitted Book I of the *Geography*. Thus, once the "triangle" was drawn, the author stated:

> This figure does not fully show, however, all the inhabitable parts of the world. In order to see the true habitable areas, it is better to draw another figure, quadrangular like this, and make the triangle inside it ‹which should begin from the equinox›, if for any other reason one does not wish to proceed, as I will touch upon below; and to better illustrate this, I will draw the figure (f. 90va).[57]

Like Fra Mauro the author appears to be aware of the fact that the classic diagram using the first Ptolemaic projection did not allow a cartographer to include the known inhabited world of the mid-fifteenth century which was larger, especially in terms of latitude, than the one described by Ptolemy. Unlike Fra Mauro who, based on this same awareness, decided not to follow Ptolemy "in his shape and measurements," the author of the Venetian manuscript suggested the construction of a "double figure" (*doppia figura*, f. 91ra), that is a "double projection," with which to include the parts of the *oikumene* that had been omitted by Ptolemy (plate VI.1). The diagram of the first Ptolemaic projection was thus "redoubled" so that even the southern part of the hemisphere to 63° – well beyond the 16° below the equator that Ptolemy had represented – would be included in the geographic grid. The "double figure" is then followed by a detailed explanation that illustrates the technique used to draw it (f. 91ra), which is transcribed in the Documentary Appendix to this chapter at the end of this book. Both in its technical details and over all the manuscript is a critical reflection upon Ptolemy's work, and is exceedingly

[55] "From here on, I will deal with the description of the universe, according to various writers, first demonstrating the said description in a round figure, then in a triangular figure, preparing these for many regions, as will be seen in their figures" (f. 82va).

[56] In the original version of the inventory of the properties of Alessandro Rosselli, the son of the Florentine illuminator, etcher and cartographer Francesco Rosselli (*c.* 1445-*post* 1513, *ante* 1527), the notary Giovan Domenico di Filippo Spadini noted five Ptolemaic maps of the known inhabited world, calling them "five Ptolemaic small cloaks on painted canvas" ("5 *mantellini* di Tolomeo in tela dipinti"). See Florence, Archivio di Stato, Magistrato dei Pupilli avanti il Principato 189, Inventario di maserizie di chasa e di bottegha apartenette alla rede di Lesandro di Francesco Rosselli merciaio, 20 February 1527, ff. 735r-743v; See Gentile *Firenze e la scoperta dell'America*, entry 116, pp. 247-250 (p. 250).

[57] "Né però questa figura sia tutta habitabile, | onde per vedere la vera parte de lie habitabile | se convien formar una altra figura | quatrangula como questa, e far in quella el | triangulo ‹che comenci da l'equinociale›, se per | altra ‹p› rasone non se volesse procedere, | como tocherò de sotto; e per meglio dechi[a]|rare porrò la figura" (f. 90va).

innovative in its reworking of the *Geography*. To the best of my knowledge there are no known precedents for such an innovation in the history of cartographical design.

b. *The "Description of the circular sphere on a flat plane"*

After his translation and comments on the first and second Ptolemaic projections the author then turned his attention to the third projection, introduced thus : "As an epilogue to this pre-made extension according to Ptolemy, there will appear a description of the circular sphere on a flat surface with the Earth comprised in it" (f. 92vb).[58] Paraphrasing Book VII.6 of the *Geography*, the description – transcribed in the Appendix to this chapter – deals, above all, with the position of the beholder who, when placed at a correct distance, should be able to take in at a glance the entire *terra*, i.e. the *oikumene*, laid out on a terrestrial globe and viewed in between the astronomical circles of an armillary sphere. At the end of his explanation, the author lamented the difficulties involved in understanding Ptolemy's explanation in the "tradition" (i.e. in Iacopo Angeli's translation) and announced that he would depict it in a figure : "And for greater clarity, I will draw this image in the form of a diagram, so that the obscure tradition transmitting the *Geography* can be understood, and by this design the description of the *oikumene* will be concluded" (f. 92vb).[59] But the figure he mentions is missing from f. 93r. The page contains a circle but there is no trace of a drawing of the "description of the circular sphere on a flat plane" (*descriptione de la spera circulare in plano*). The absence of the diagram of the so-called "third Ptolemaic projection" is not surprising in itself. While the four diagrams illustrating the procedures to draw the first and second projections were drawn in almost all the ancient Greek and Latin manuscripts, and also in the printed editions of the *Geography*, the diagram pertaining to the "Circularis spherae cum habitabili terrae descriptio" is generally missing in fifteenth-century manuscripts and their Greek prototypes.[60] In 1477 a woodcut of the flat diagram of Ptolemy's third projection was printed in the Bologna edition of the *Geography*, the first in which the maps were published.[61] It was only in the early sixteenth century, sometime between 1503 and 1516, that an anonymous illuminator added a clear and coherent representation to a fifteenth-century manuscript of the *Geography*, which depicted the globe with the rings inserted within an armillary sphere, thus reproducing for the first time the third projection as described by Ptolemy.[62] In the Rome editions of 1507 and 1508, which contain

[58] "Epilogo de questa prefata extensione secondo Ptholomeo | apparerà la descriptione de la spera circula|re in plano con la terra compresa in quello avendose" (f. 92vb).

[59] "E per più chiarezza redu|rò questa figura in forma, a ziò che la | sua obscura traditione per la figura | sia compresa, et in quella se fia fine quanto | a la descriptione de | l'universo. (f. 92vb).

[60] The only known exceptions are two Greek manuscripts of the *Geography* : London, British Library, Burney 111, *Γεωγραφική ὑφήγησις*, manuscript on vellum, in Greek, 440 × 310 mm, 65 maps, fourteenth century, f. 104v, and Chicago, Newberry Library, Ayer 41, *Γεωγραφική ὑφήγησις*, paper, in Greek, no maps, fourteenth century, ownership mark of the Camaldolese monastery of San Michele di Murano (i.e. the monastery where Fra Mauro lived), 430 × 224 mm, f. 101r. However, they both provide very simple representations of this diagram.

[61] Hain, *13538, IGI 8181. Facsimile : Claudius Ptolemaeus, *Cosmographia : Bologna, 1477*, introduction R. A. Skelton, Amsterdam : N. Israel, 1963.

[62] Paris, Bibliothèque Nationale, Lat. 4801, Claudius Ptolemaeus, *Cosmographia*, manuscript on vellum, *ante* 1471, 28 maps. Ff. 73v-74r and 76v-77v, with the magnificent three-dimensional representation of the "Circularis spherae cum habitabili terrae descriptio," were added to the manuscript in the early sixteenth century.

a reprint of the copperplate engraved maps of the earlier Rome editions (1478 and 1490), along with two astronomical treatises by Ptolemy (the *Planisphaerium* and the *De circulis aequinoctialis*). Marcus Beneventanus discussed Ptolemy's projections, illustrating Book VII.6 with a large diagram entitled the "Sphaera in plano," in which a section of the globe with a diagram of the *oikumene* was depicted on a flat surface.[63] Finally, it was in the 1525 Strasbourg edition of the *Geography* that Albrecht Dürer created the most famous representation of the "Circularis spherae cum habitabili terrae descriptio," in which the globe is seen in perspective within the armillary sphere.[64]

Even though the diagram of the "Circularis spherae cum habitabili terrae descriptio" is missing even in the Marciana manuscript, f. 16r has the representation of a three-dimensional hemisphere completely covered with a grid of meridians and parallels clearly depicting the terrestrial sphere in such a way that the three dimensions are immediately perceptible in the drawing (plate VI.2). It is important to stress that this is not a representation of the third Ptolemaic projection : the armillary sphere with the astronomical circles is absent and the parallels are all curved which contradicts not only the original Ptolemaic instructions, but also the vernacular translation of the description of the third projection made by the author himself, who was aware that the parallel that passed through Syene would appear as a "straight line" (*una diritta linea*).

Nonetheless, the diagram constitutes one of the first endeavors in the West, or at least one of the earliest attempts of which proof has survived to modern times, of creating a three-dimensional representation of the terrestrial sphere with a geographic grid derived from Ptolemy's work. Even though it is not a representation of the third projection it reflects some of its characteristics, albeit in an unorthodox manner : for example the three-dimensional aspect (the "spherical shape" in the author's words) and the fact that the design presupposes that the beholder is located sufficiently far away from the sphere to take in the entire terrestrial globe at a single glance.[65] The curvature of the parallels and the meridians creates the illusion of a three-dimensional representation and shows how the completely graduated globe would appear from a distance that allowed the viewer to see the entire hemisphere. The drawing is followed by a note stating that "This figure will be placed below, after the universal description of the world". It is not clear to which part of the manuscript this observation refers. In my opinion there are two plausible hypotheses. First, the note could refer to f. 47v, struck through, which follows a *mappa mundi* drawn on f. 47r in pencil with very faint geographical contours, within a circle drawn in ink with a compass, surrounded by twelve winds (which, quite curiously, are in Latin), with two main graduated axes, which the author describes thus on f. 46v :

[63] *In hoc opere haec continentur Geographia Cl. Ptolomaei a plurimis uiris… emendata : et cum archetypo graeco ab ipsis collata. … Figura de proiectione sphaerae in plano, quae in libro octavo desiderabatur ab ipsis… instaurata…*, Rome, 1508 (Nouiter impressum per Bernardinum Venetum de Vitalibus expensis Euangelista Tosino Brixiano… die 8. Septembr. 1507). The diagram that illustrates Book VII.6 is on f. 92r (printer's signature m 8).

[64] Albrecht Dürer, [Descriptio spherae armillaris una cum tota habitabili in plano], in *Claudii Ptolomaei Geographicae enarrationis libri octo Bilibaldo Pirckeymhero interprete. Annotationes Joannis De Regio Monte, in errores commissos a Jacobo Angelo in translatione sua*, Argentorati, Johannes Grieningerus, communibus Johannis Koberger Impensis excudebat. Anno Christi nativitate MDXXV. Tertio Kal. Apriles, f. 69v.

[65] For the complete text of the "description of the circular sphere on a flat surface ‹with› the Earth within it," see the documentary Appendix of this chapter.

And so as to keep with the order announced earlier,[66] before starting the description of the maps according to Ptolemy, I will draw a circular map of the world, to show in the clearest way possible each part of the world on its own; and on each part of the circle [e.g. the circular world map] I will put the smaller parts [e.g. provinces], so that everybody interested [in them] for his own curiosity could easily describe, i.e. draw them [the single details] on a map.[67]

A second, and perhaps more probable, hypothesis is that the three-dimensional hemisphere on f. 16r should have followed what was defined in the manuscript as the "description of the universe in a spherical form," i.e. the representation of the Ptolemaic *oikumene* on the terrestrial sphere, drawn in such a way so as to create the optical effect of three-dimensions (f. 87r), which, according to the manuscript's author, was simply a variant of the first Ptolemaic projection: "An easy way of describing the universe (...) and this ‹measure› figure[68] allows one to see that there is no difference between this one and the one given above in a triangular shape, just that this is in a spherical form, and that one is in a flat form" (f. 87va). With respect to this other diagram, the representation on f. 16r would seem to extend the grid of meridians and parallels beyond the limits of the Ptolemaic *oikumene*, covering the entire hemisphere, in a way that seems prefigured on f. 87r in the eighth meridian on the left side of the figure, which extends to the two poles. The caption pertaining to the drawing on f. 16r – "This figure will be placed below, after the universal description of the world" – coupled with the author's awareness that the modern *oikumene* extended beyond the limits indicated by Ptolemy both towards the North and to the South thus confirm that, just like the "double figure," the three-dimensional hemisphere with the cartographical grid was also to be interpreted as one of the first attempts to apply Ptolemaic cartographical protocols to the vaster known inhabited world of the mid-fifteenth century and, possibly, to the entire terrestrial surface.[69]

c. *The cartographical description of the provinces*

As has been mentioned above, the author developed an innovative graphic syntax, setting out from a profound reformulation of the Ptolemaic maps, in the part dedicated to the description of the 89 provinces mentioned in the manuscript (29 in Europe, 12 in Africa, 48 in Asia;

[66] Correction in the margin: *l'ordine acominciatto*.

[67] "E per servare l'ordine acominc[i]atto | ‹ordine accominc[i]ato avanti che vegna | a la positione de le tavole in figura | secondo Ptholomeo›, formarò el tondo | de l'universo, a demostrar per evidentia | manifesta cadauna de le parte da | per sé instessa, et in questa positione del | tondo tocherò ‹c› le general particule, | a ciò che più lievemente a tutte le | particularità pervenire possa qualunche | vorrà per suo vag[h]ezza descriver quello, | over designarlo in figura piana," f. 46va.

[68] Correction between lines: *figura* instead of ‹*mesura*›.

[69] The geometric design on f. 16r has distinct similarities with a representation of the *oikumene* on a terrestrial globe contained in an incunabulum of Ptolemy's *Geography* that was printed in Nuremberg in 1495, the so-called "Deutsche Ptolemäus" (Hain *13542). See *Der 'Deutsche Ptolemäus' aus dem Ende des XV. Jahrhunderts (um 1490)*, in Faksimile-druck herausgegeben mit einer Einleitung von J. Fischer SJ, Strassburg, J. H. Ed. Heitz, 1910, ff. 2v-3r (before the chart marked "a2," pp. 52-53, unnumbered, in the facsimile edition). The *mappa mundi* has also been reproduced in M. Herkenhoff, *Die Darstellung aussereuropäischer Welten in Drucken deutscher Offizinen des XV. Jahrhunderts*, Berlin: Akademie, 1996, p. 134. On the "Deutsche Ptolemäus," apart from Fischer's introduction to the facsimile edition, also see E. Rosenthal, "The German Ptolemy and Its World Map," *Bulletin of the New York Public Library*, 1944, pp. 1-15.

ff. 17v-46r).[70] For the regional maps the author created a cartographical protocol in which they are inserted amongst and placed alongside the text, in a manner similar to the graphic model and pagination of texts and images that can be found in the so-called "Group B" of Greek manuscripts of Ptolemy's *Geography* with 65 maps instead of the more common corpus of 27. He then proceeds to describe the graphic structure of the work. Each of the two columns into which the page is divided is separated into three main fields. There are two squares in the upper margin. The first square contains a brief text in one column, with four or five lines on average which, in addition to a list of the provinces described in the map shown below the text, also contains the Ptolemaic geographical coordinates of the center of the map. This center is identified by a cross ("X") in a miniature planisphere drawn in pencil within a circle whose average diameter is 50 mm, placed immediately below or, in some cases, beside the text. The province described in the text is then drawn in detail in another map, placed below the small planisphere. In the regional maps a second cross, always shown as an "X," indicates the correspondence with the miniature planisphere. These maps have the graphic form of cartographical sketches which are sometimes very confused and at other times drawn with great precision, which allow the beholder to clearly distinguish cities, rivers and mountains. They look like preparatory designs, as they are rather rough and appear to have been sketched in a hurry, but nonetheless they were drawn by a hand that reveals a certain familiarity with artistic techniques. Almost all the maps follow this two-column scheme, with the exception of ff. 21v, 25r, 42rv, 43r, 44v, 45rv and 46r which occupy the whole page (plate VI.3). In these exceptions the text is placed alongside the miniature planisphere in the upper half of the page, while the regional map is placed in the lower half.[71]

Neither the miniature planispheres nor the chorographies placed below have graduated grids. The references to the geographical coordinates of the places depicted are instead mentioned in the brief texts that introduce the cartographical drawings, which are mainly derived from Book VIII of the *Geography*. The following notes are examples of some of these descriptions. In Europe : "England, the island of Britain is distant from the Western point by XX degrees and by XXXXXIIII [degrees] from the equator, as shown in the figure below" (f. 17vb).[72] In Africa : "From here onwards I will begin to talk of Africa and I will deal with this part according to its XII provinces, of which the first is Mauritania Tingitana, in front of Sibilia Alergarbia in the part where the Inhabited world ends, that is beyond the pillars of Hercules ; and this is distant from the Western point by VIII degrees and [from the equator] XXXIII, as shown in its figure" (f. 27va).[73]

This singular pagination of texts, planispheres and regional maps, in the form of cartographical sketches, represents a particularly innovative graphic syntax which is unparalleled

[70] For a list of these provinces see the Appendix at the end of this chapter.

[71] For reasons of visual clarity, I have chosen to reproduce one of these full-page representations (f. 25r depicting the region of Macedonia) instead of a page with two cartographical representations in the two-column format.

[72] "Ing[h]elterra isola de Britania dista dal | puncto occidentale gradi XX et alta | XXXXXIIII, como nella figura infrasscripta" (f. 17vb).

[73] "De qui in avanti accomincerò de la Affrica | e quella porrò secondo le suo XII provintie, | de le quale la prima fia Mauritania Tingi|tana, posta a l'opposito de Sibilia Alergar|bia da la parte ove è Habilis termene, over | altra colona de Hercule ; e questa dista dal punto | occidentale gradi VIII alta XXXIII, como | nella figura sua" (f. 27va).

in any other manuscript dating from the fifteenth century or even later. Another exceedingly important fact, which has great heuristic value, is that on f. 2r the author himself defines this syntax and provides an accurate explanation of his technique:

> These are all the provinces given by Ptolemy with the distances and their heights. From here onwards I will proceed to deal with each part of the Earth, drawing each part by itself, so that everything that we wish to see is clear; then I will put them together, showing the universal description of the world. And in each of the parts where there is an X, that will be the centre of the province; then around this cross I will put the more prominent lands and places and rivers in the said province. And I will first begin with Europe, victor over every nation, because of that one part that is in it, i.e. Italy (f. 3ra).[74]

The anonymous Venetian author thus revealed that he was aware of the graphic and visual limitations of such small maps for the reader, and sought to remedy this difficulty by dedicating specific chapters to each province in the latter part of the work: "Because the figures are small, they will not be able to describe the lands, but at the end I will make chapters for each province with their lands, especially for Europe; and these will have the modern names as also the ancient names" (f. 19rb).[75]

The entire cartographical apparatus reformulates and reinvents the cartographical protocol of Ptolemy's *Geography*, in the sense that it is based on and simultaneously extends and applies the conceptual relationship between universal and specific elements (i.e. between the planisphere and the regional charts) described in the *Geography*. The singular graphic syntax of this manuscript clearly illustrates the theoretical-conceptual dimension of the *Geography* in each individual cartographical description. This is a procedure that the anonymous author developed during the course of his work: in fact, some maps still have the titles of the *Geography*'s twenty-seven e.g. f. 16r. This scheme was later abandoned in favor of a subdivision of the world into provinces drawn according to the graphic syntax that has been analyzed above.

III. – Venice and Florence: a few considerations

The manuscripts that have been analyzed in the course of this study attest to the fact that Ptolemy's *Geography* was not received in a uniform manner. Before the invention of the printing press allowed a greater uniformity of texts profoundly different readings, interpretations and revisions of the *Geography* were documented, especially in mid-fifteenth-century

[74] "Queste sono tutte le provincie poste | per Ptholomeo con le distantie e loro alte|ze. De qui ‹in Av› in avanti vegnerò a la | partisone de cadauna parte de le terre | figurando cadauna parte per se istessa, | afinché meglio sia chiara ogni | cossa noi voremo vedere; po' quelle giun|gerò insieme, mostrando la universale | descriptione del mondo. Et in cadauna | de le parte dove fia la *X* lì serà | el mezzo de la provintia; poi atorno | quella croce porrò le terre e' luochi | e' fiumi nella ditta provintia più | notabili. E prima comincerò de Europa, vincitrice de ogni natione, per | quella sola parte che è in essa, c[i]oè Italia" (f. 3ra).

[75] "Perché le figure sono picole non seriano | capace de le descriptione de le terre, ma ne l'ul-|timo farò capitoli de cadauna provintia, | maximamente in Europa, de le terre; e quella prende|rà con li nomi moderni ‹me›, ponendo | anchora gli nomi antichi" (f. 19rb).

Venice.[76] These documents clearly demonstrate that Ptolemaic geography and cartography underwent important innovations in Venice – and not in Florence – at least until the age of Francesco Berlinghieri's *Septe giornate della geographia*, a free paraphrase in Tuscan verse (*terza rima*) of the *Geography* composed from 1465 to 1478 or 1482,[77] and Henricus Martellus Germanus' maps illustrating two manuscripts of the *Geography* and five manuscripts of the *Insularium illustratum*, i.e. around the last two decades of the fifteenth century. Mid-fifteenth-century Florentine documents that transmitted the *Geography* tended to be lavish manuscripts, often produced serially. Although Venetian documents are fewer in number and more "heterodox" in nature, they allow one to deduce the existence of a vast and diffused interest in the *Geography* that has yet to be studied fully, especially in terms of its diffusion in contemporary Latin scientific circles as well as in less erudite *milieux*, that is, in the less erudite environments in which marine charts were made.

In Florence around the mid-fifteenth century Nicolaus Germanus,[78] Piero del Massaio[79] and, subsequently, Henricus Martellus updated the maps of Ptolemy's *Geography* by adding the so-called *tabulae novellae* to Ptolemy's twenty-seven 'original' maps. In addition, a trapezoidal graduated grid was introduced for regional charts. The first *tabula novella* to be added was the Ptolemaic map of Northern Europe prepared by Claudius Claussøn Swart. This chart was

[76] See the final considerations in P. Gautier Dalché, "Jean Fusoris et la géographie. Un astronome, auteur d'un globe terrestre, à la découverte de Ptolémée," in Marcotte, *Humanisme et culture géographique à l'époque du Concile de Constance*, pp. 161-176 and Id., *La Géographie de Ptolémée en Occident*, pp. 203-207.

[77] The *Septe giornate della geographia* is Berlinghieri's only major literary contribution and survives in four manuscripts and a printed edition: Milan, Biblioteca Nazionale Braidense, AC XIV 44, manuscript on vellum, 438 × 355 mm, 31 maps, written before 1482, signed "Francesco Berlinghieri Florentino"; Città del Vaticano, Biblioteca Apostolica Vaticana, Urb. lat. 273, manuscript on vellum, 458 × 312 mm, 31 maps, written *c.* 1482, signed "Francesco Berlinghieri"; Paris, Bibliothèque Nationale, Lat. 8834, manuscript on vellum, 690 × 435 mm, 26 maps (without the map of Germany), no signature or date; and Florence, Biblioteca Medicea Laurenziana, Pl. XXX.1, manuscript on vellum, 604 × 435 mm, 31 maps, late fifteenth century. Printed edition IGI 1491: *In questo uolume si contengono septe giornate della geographia di Francesco Berlingeri fiorentino allo illustrissimo Federigo duca durbino*. Florence: Niccolò di Lorenzo della Magna, s.d. (but printed before September 1482 when Federigo died). Long considered to be nothing more than an eccentric exercise – Roberto Almagià and Robert Skelton wrongly thought that Berlinghieri was equipped with very modest geographical and cartographical skills – the *Septe giornate della geographia* is clearly a crucial testimony to the way in which the representation of the *imago mundi* and, in particular, Ptolemy's "Guide to Cartography" were perceived in the neo-Platonic culture of late fifteenth-century Florence. Fo an innovative analysis of Berlinghieri's work: Gautier Dalché, *La Géographie de Ptolémée en Occident*, pp. 252-255 ("Francesco Berlinghieri, ou Ptolemée platonicien"). Cf. with R. Almagià, "Osservazione sull'opera geografica di Francesco Berlinghieri," *Archivio della R. Deputazione Romana di Storia Patria* 68 (1945), pp. 211-255 and R. A. Skelton, "Bibliographical Note," in Francesco Berlinghieri, *Geographia. Florence 1482*, Amsterdam: Theatrum Orbis Terrarum, 1966, pp. v-xiii.

[78] For a critical analysis of Nicolaus Germanus' reception of the *Geography*, see B. Maracchi Bigiarelli, "Niccolò Tedesco e le carte della Geografia di Francesco Berlinghieri autore-editore," in B. Maracchi Bigiarelli and D. E. Rhodes (eds), *Studi offerti a Roberto Ridolfi direttore de 'La bibliofilia'*, Florence: Olschki, 1973, pp. 377-397; Gentile, *Firenze e la scoperta dell'America*, entries 102-103, pp. 207-215 and Gautier Dalché, *La Géographie de Ptolémée en Occident*, pp. 219-224 ("Nicolaus Germanus, ou la fausse exactitude").

[79] G. Aujac, "Le peintre florentin Piero del Massaio et la 'Cosmographia' de Ptolémée," *Geographia Antiqua* 3-4 (1994-1995), pp. 187-204 and figs 1-10; Gautier Dalché, *La Géographie de Ptolémée en Occident*, pp. 224-225 ("Piero del Massaio, ou la production en série"). For a discussion of the material aspects of the Ptolemaic maps depicted by Piero del Massaio see Part II, Chapter I.

already available in 1427, when Cardinal Guillaume Fillastre ordered a replica of the map to be included in one of his manuscript copies of the *Geography* (Nancy, Bibliothèque municipale, 441).[80] This was followed by a new Ptolemaic chart of Germany that has traditionally (and fully arbitrarily) been attributed to Nicolaus of Cusa.[81] The gradual process culminated in Henricus Martellus Germanus' twelve *tabulae novae* in the *Ptolomei Cosmographia* (Florence, Biblioteca Nazionale Centrale, Magl. XIII, 16), which were prepared around 1490.[82] As already mentioned, the only significant intervention in the text of the *Geography* that occurred in Florence was Francesco Berlinghieri's effort in his *Septe giornate della geographia*. Instead in Venice, between 1430 and 1460, Andrea Bianco, Giovanni Fontana, Fra Mauro and two unknown Venetian author-cartographers prepared innovative interpretations, translations and revisions of the *Geography*. These differed considerably from both Greek and the more common Latin manuscripts of Ptolemy's celebrated *magnum opus*. The works we have analyzed show that Ptolemaic maps were received and updated in Venice with the integration of marine charts, regional maps, and even new forms of mapping based on the innovative application of the instructions in Ptolemy's text (I am referring, in particular, to the regional maps in the Ms. It. VI, 24 (= 6111). The text of the *Geography* was critically examined and revised with data provided by other *auctores veteres*, as well as through the integration of information derived from Medieval natural philosophers and travelers. Ptolemy's *Geography* was compared to works by Solinus, Pliny, Julius Caesar, Albert the Great, Marco Polo, and new data was integrated and oversights were corrected, as shown in different shades in Fra Mauro's' map, the Ms. Cl. VI, 24 (= 6111) and Fontana's *Liber de omnibus rebus naturalibus*.

The marked differences in the interpretation, understanding and transmission of Ptolemy's *Geography* in fifteenth-century Venetian and Florentine culture lead one back to a classic theme of Renaissance historiography: a comparison of these two cultures that, despite being pro-

[80] On Claudius Clavus see Gautier Dalché, *La Géographie de Ptolémée en Occident*, pp. 174-175.

[81] Gentile, *Firenze e la scoperta dell'America*, entry 80, pp. 165-168.

[82] See Martellus' manuscripts of the *Geography* (Florence, Biblioteca Nazionale Centrale, Magl. XIII, 16 and Vatican City, Biblioteca Apostolica Vaticana, Vat. Lat. 7289) and the *Insularium illustratum* (Florence, Biblioteca Medicea Laurenziana, Plut. XXIX, 25, considered to be Martellus' working copy and the basis for the other four manuscripts: London, British Library, Add. MS 15760; Leiden, Bibliotheek der Rijksuniversiteit, Vossianus lat. F. 23, Chantilly, Musée Condé, Ms. 698-0483; Minneapolis, The James Ford Bell Library, Ms. 4). The maps in these manuscripts intended update the *corpus* of maps contained in Ptolemy's *Geographiké Uphégesis* through the combination of Ptolemaic and nautical cartography and the *insularii*. See Gautier Dalché, *La Géographie de Ptolémée en Occident*, pp. 245-252. Martellus was also the first to integrate Portuguese discoveries along the African coast in the framework of Ptolemaic cartography, as shown by the frontispiece of the manuscript Magl. XIII, 16 of the Biblioteca Nazionale Centrale of Florence ("Cl. Ptolomei Cosmographia cum tabulis regionum n(ost)ri t(empor)is et universis portubus et locis maritimi tractus tam notis quam a rege Portus Galli nuper repertis: hoc ornatissimo codice continentur") and by the extraordinary planispheres displayed in two copies of the *Insularium* (London, British Library, Add. MS 15760, ff. 68v-69r and Florence, Biblioteca Medicea Laurenziana, Plut. XXIX, 25, ff. 66v-67r) with the representation of the African coast as far south as the Cape of Good Hope (to include Diogo Cão's 1482-1483 and 1485-1486 as well as Bartolomeu Dias' 1487-1488 navigations) and the widening up of the Ptolemaic *oikumene* to 220 degrees of longitude, an enlargement of over 40 degrees to the east, depending most of all on the cartographic inclusion of Marco Polo's description of Asia. See A. Cattaneo, "La 'Ptolomei Cosmographia' di Henricus Martellus Germanus," in *Ptolomei Cosmographia*, pp. 23-55; "Orb and Sceptre: Cosmography and World Cartography in Portugal and Italian Cities in the Fifteenth Century," *Archives Internationales d'Histoire des Science* 59 (2009), pp. 531-555.

foundly linked via uninterrupted exchanges, nonetheless manifested distinct peculiarities.[83] What were the underlying causes of this diverse reception of a classic author? In what way could these different attitudes shed light upon the cultural *milieux* of the two cities and thus highlight general practices and dynamics in the formation and diffusion of knowledge? While examining these issues Marica Milanesi has opportunely recalled a valuable lesson by Eugenio Garin about the reception of Plato's works in Florence and Venice during the fifteenth century: while Florence was open to Plato, there was considerable Aristotelian resistance in Venice.[84] Garin, following in Giuseppe Billanovich's footsteps,[85] partially attributed the favorable Florentine reception of Plato to the fact that between the fourteenth and fifteenth centuries Florence did not have a consolidated university tradition. The absence of a strong pre-existing cultural configuration ensured that Florence was far more receptive and open to cultural novelties than Venice.[86] In the same way it could have been precisely the absence of a consolidated tradition in Florence of depicting and representing the *imago mundi*, in both the graphic syntaxes of *mappae mundi* and nautical cartography, that could have facilitated a favorable and relatively quick reception of Ptolemy's *Geography*.[87] Just like its neighbor Padua, the seat of a prestigious university, Venice had deep-rooted traditions of studying geography (via *auctores veteres* and *novi*). The city also had a thriving interest in cartography (names such as the fourteenth-century Venetian cartographer Pietro Vesconte, followed by Andrea Bianco and Fra Mauro, come instantly to mind). Works of great prestige and international renown (Pietro Vesconte's world maps and nautical charts were widely circulated, and the Portuguese court even commissioned a copy of Fra Mauro's celebrated *mappa mundi*) represented the world on both a large and a small scale. Thus, as had been the case with Plato (who was immediately compared with the Venetian tradition of Aristotelian studies), Ptolemy's work was critically appraised and, via complex processes, immediately compared with pre-existing systems of representing the earth. This was, in short, the common approach that explains the heterodox interpretations of the *Geography* by Andrea Bianco, Fra Mauro and the anonymous authors of the Harley 3686 of the British Library and the MS. Cl. VI, 24 (= 6111) of the Biblioteca Marciana.

Eugenio Garin's study, particularly his reflections about humanistic education – as Milanesi has indicated – is especially useful in ascertaining the value of Venice's interpretation of Ptolemy's *Geography*. For Garin, humanism did not emerge from the contents of ancient books, but instead from the attitude with which the humanists read these works. They read the works by the Greek and Latin *auctores* to position themselves as interlocutors of great authors from the past. This, according to Garin, is the essence of the humanistic approach to the Ancients: the classics were not read merely for erudition, but also to create awareness of man, as a historic being. Humanism

[83] S. Bertelli, N. Rubinstein and C. H. Smyth (eds), *Florence and Venice, Comparisons and Relations. Acts of Two Conferences at Villa I Tatti in 1976-1977*, Florence: La Nuova Italia, 1979.

[84] M. Milanesi, "Cartografia per un principe senza corte: Venezia nel Quattrocento." *Micrologus. Natura, scienze e società Medievali / Nature, Sciences and Medieval Societies* 16 (2008).

[85] G. Billanovich, "Dall'antica Ravenna alle biblioteche umanistiche," *Aevum* 30 (1956), pp. 319-353, revised version in *Annuario dell'Università Cattolica del Sacro Cuore*, 1955-1957, Milan, 1958, pp. 73-107, then reprinted in *Italia medioevale e umanistica* 36 (1993), pp. 107-174.

[86] E. Garin, "Cultura filosofica toscana e veneta nel Quattrocento," in V. Branca (ed.), *Umanesimo europeo e Umanesimo veneziano*, Florence: Sansoni, 1963, pp. 11-33.

[87] As Marica Milanesi has noted. See "Cartografia per un principe senza corte," pp. 215-216.

reads the classics as a legacy, a Greek or Latin heritage, restores the historical dimension of these works, and then analyzes this historical dimension. These reflections, generally mentioned in the context of Florentine humanism, would also seem to precisely capture the spirit with which Venice, from the 1430s onwards, read and pondered the *Geography*. As a commercial and maritime city, Venice offered multiplicity in lieu of centralization. It offered rich opportunities for a dynamic interchange between the wide assortment of social and professional classes that constantly thronged this emporium – patricians, merchants, *gente di mare*, prelates, *popolani*, students, exiles and diplomats.[88] In mid-fifteenth-century Venice, Ptolemy's *Geography* was read, criticized and transmitted at the crossroads of these social and cultural *milieux*, as well as those of the composite networks of contemporary knowledge, including scholasticism, humanism and monastic culture coupled with more technical skills such as marine cartography and mercantile practices.

[88] See D. Romano, *Patricians and 'Popolani': The Social Foundations of the Venetian Renaissance State*, Baltimore: Johns Hopkins University Press, 1987, pp. 120-148, esp. p. 123; E. Muir and R. F. E. Weissman, "Social and Symbolic Places in Renaissance Venice and Florence," in J. A. Agnew and J. S. Duncan (eds), *The Power of Place: Bringing Together Geographical and Sociological Imaginations*, Boston: Unwin Hyman, 1989, pp. 81-103, esp. 85 and 87.

CHAPTER V

TRAVEL ACCOUNTS AND CARTOGRAPHY FRA MAURO, MARCO POLO AND NICCOLÒ DE' CONTI

I. – The representation of Asia

1. Research agenda

The objective of this chapter is to itemize and interpret the methods of transmission of knowledge passed down from travel narratives to cartography. In particular I will concentrate on how, around the mid-fifteenth century and at the dawn of the European expansion, Fra Mauro perceived the experience of journeys in Asia and navigations in the Indian Ocean and inserted them into the graphic and epistemological unity of his map of the world. Text and image are inseparable in Fra Mauro's work ; therefore the analysis is conducted not on his text alone but rather as a unified whole constituted of both text and the images of the *mappa mundi* in relation to its sources, in particular to travel accounts. This approach differentiates my research from those carried out hitherto.

Other than Asia and the basin of the Indian Ocean (*Mare Indicum*) there are four main areas of direct and explicit derivation of travel accounts : Northern Europe, particularly the Scandinavian peninsula that Roberto Almagià already identified as one of the most innovative representations of the fifteenth century ; the western coasts of Africa ; the Ethiopian region, or "from the province of Sayto (Asyut, in Egypt) upwards" (*da Sayto in suso*, in Fra Mauro's words) ; lastly, the Caspian Sea region. The representation of each of these areas is based on the reading and re-elaboration of specific travel narratives.

Chapters I has already discussed the representation of the Ethiopian regions, particularly with reference to their sources. For this part of the *mappa mundi*, and for ancient geography of Africa in general, O. G. S. Crawford's unsurpassed study has brought evidence that Fra Mauro did in fact draw on first-hand sources. Place names and, in general, the design of the Ethiopian regions in the *mappa mundi* assume an archeological value.[1] What Fra Mauro wrote in one of the most cited of the map's cartouches is therefore credible, and reminds us of the fundamental contribution to the design of the southern part of Africa by a group of clerics born in Ethiopia :

> Because to some it will appear as a novelty that I should speak of these southern parts, which were almost unknown to the Ancients, I will reply that this entire drawing, from Sayto upwards, I have

[1] O. G. S. Crawford (ed.), *Ethiopian Itineraries* circa *1400-1524. Including Those Collected by Alessandro Zorzi at Venice in the Years 1519-24*, Cambridge, MA : Hakluyt Society at the Cambridge University Press, 1958. See also G. Barbieri and G. Fiaccadori (eds), *Nigra sum sed Formosa. Sacro e bellezza dell'Etiopia cristiana*, Treviso : Terra-Ferma, 2009.

> had from those who were born there. These people are clerics who, with their own hands, drew for me these provinces and cities and rivers and mountains with their names; all these things I have not been able to put in due order for lack of space.[2]

This could refer to several clerics of the Ethiopian and Coptic Churches who, on their way to Rome on the occasion of the Council of Florence, had reached Venice around 1440-41. Pope Eugene IV had invited them to Rome in order to make an agreement on the ecumenical union by sending the Franciscan legate Albert of Sarteano to Egypt.[3]

In the design of Northern Europe, for Fra Mauro's explicit citation, it is important to remember the account of the shipwreck off the Norwegian coast of a Venetian patrician, noble of the *Maggior Consiglio*, and merchant, Piero Querini.[4] The account – preserved in several manuscripts in Venice, in Florence (rewritings attributable to Matteo di Corrado Cardini's compilation), and in the Vatican City (attributable to Cristofalo Fioravante and Nicolò de Michiele)[5] – narrates the expedition from its departure from Crete to its shipwreck off the Lofoten Islands, the long and generous hospitality of the fishermen who saved the castaways on the island of Rustene, and ultimately the long terrestrial return voyage that was concluded in Venice on October 12, 1432.

Regarding instead the representation of the regions near the Caspian Sea, Ivar Hallberg has accurately demonstrated that, other than the knowledge ascribable to Marco Polo,[6] the *mappa mundi* transmits some important geographical innovations that were diffused in the

[2] "Perchè ad alguni par da nuovo che io parli | de questa parte meridional, la qual quasi està | incognita a li antichi, perhò io respondo ch(e) | tuto questo desegno da sayto in suso io l'ò ha|buto da queli proprij che sono nasudi qui, | che sono sta' religiosi, i qual cu(m) le suo man | me hano desegnato tute q(ue)ste p(ro)ui(n)cie | e citade e fiumi e mo(n)ti cum li | suo nomi, le qual tu|te cosse no(n) le ho possu|do meter cu(m) el debi|to ordine per | no(n) esserui logo," *Mappamondo* (ed. 1956) Pl. X, 30. The English translation derives from *Mappamondo* (ed. 2006) *98.

[3] S. Tedeschi, "Etiopi e Copti al concilio di Firenze," *Annuarium historiae Conciliorum* 21, 2 (1989), pp. 380-407. The cultural and artistic relationships between Venice and Ethiopia since the fifteenth century were very intense: the presence at the Ethiopian Court of the Venetian painter Nicolò Brancaleone who deeply influenced Ethiopian art – particularly icons – is evidence of this. See G. Fiaccadori, "L'Etiopia, Venezia e l'Europa," in Barbieri and Fiaccadori (eds), *Nigra sum sed Formosa*, pp. 27-48.

[4] "In questa proui(n)cia de norue|gia scorse misier Piero Queri|ni, come è noto," *Mappamondo* (ed. 1956) Pl. XXXV, 161; "In this province of Norway came ashore Piero Querini, as is well known." The English translation derives from *Mappamondo* (ed. 2006) *2674.

[5] See the manuscripts Venice, Biblioteca nazionale marciana, Ms. It. VII, 368 (= 7936), and Vatican City, Biblioteca Apostolica Vaticana, Vat. lat. 5256, ff. 42-55v, edited in Pietro Querini, Nicolò de Michiele, Cristofalo Fioravante, *Il naufragio della Querina: Veneziani nel circolo polare artico*, P. Nelli (ed.), Rome: Nutrimenti, 2007. See also Florence, Biblioteca Nazionale Centrale, Panciatichiano 20, ff. 97r-108r, "Compilatione per me Antonio di Matteo di Churrado de' Chardhini da Firenze questo dì xiii di dicembre 1432" (f. 97r); see S. Morpurgo, P. Papa and B. Maracchi Bigiarelli (eds), *Catalogo dei manoscritti Panciatichiani della Biblioteca Nazionale Centrale di Firenze*, I, Rome, 1887-1962, pp. 17-18; and S. Gentile, *Firenze e la scoperta della America. Umanesimo e geografia nel '400 fiorentino*, Florence: Olschki, 1992, pp. 163-164, n. 79. Modern edition in G. B. Ramusio, *Navigazioni e viaggi*, M. Milanesi (ed.), Turin: Einaudi, 1983, vol. IV, pp. 51-77 and 79-98.

[6] P. Pelliot, *Notes on Marco Polo*, 2 vols. and *Index*, Paris: Imprimerie Nationale, Librairie A. Maisonneuve, 1959-1963, 1973, t. I, 1959, *s.v.* "Badascian", p. 63; "Balc", p. 71; "Belor", p. 91; "Bucara", p. 108; "Caracorom", p. 165; "Catai", p. 216; "Cotan", p. 408.

West following the travels in Central Asia of León Rui González de Clavijo († 1412).[7] Chamberlain of King Henry III of Castile, in 1403 Clavijo was invited as ambassador of Samarqand and Bukhara to the court of Tamerlane, finally returning to Spain in 1406. Clavijo described his long voyage and the history and customs of the populations that lived in the Timurid Empire in a work entitled *Embajada a Tamorlán* (Embassy to Tamerlane).[8] In the *mappa mundi* the representation of the kingdoms of *Çagataï, Organça* e *Samargante*, to the east of the Caspian Sea (therefore Central Asia, near the present-day border between Afghanistan, Turkmenistan and Uzbekistan[9]), displays some of the geographical innovation transmitted by *Embajada a Tamorlán*. One detail proves decisive. Following the voyage towards Bukhara between Termiz to Samarqand, Clavijo had crossed and described the "Iron Gates", the two fortresses that guard the valleys near the river *Termit* and the city of Termiz.[10] In the *mappa mundi* the "Iron Gates" of Termiz,[11] followed by the explanation "This valley is held closed by two fortresses, which are called Iron Gates,"[12] are represented clearly and quite distinctly from the more celebrated ones of *Derbent* (Darband) in today's Daghestan, that Fra Mauro places near the Caspian Sea's western shore.[13] This detail involves an important cartographic advance that reveals renewed cognition of Central Asia surpassing Pliny's description in the *Naturalis historia*[14] whose echoes

[7] For the relationship between Fra Mauro and Clavijo, see I. Hallberg, *L'Extrême-Orient dans la litérature et la cartographie de l'Occident des* XIII*e*, XIV*e*, XV*e* *siècles*, Göteborg: W. Zachrissons Boktryckeri, 1906, p. 291, 370, 414 (in particular "Portes de fer", p. 414). Fra Mauro's *mappa mundi* includes the cities of *Mesiet chorasian* (XXVII, 51; *1473), *Nassabor* (XXVII, 66; *1491), *Bastan* (XXVII, 108; *1506), the uplands of Karabakh (*Monte Charabach*, XXVII, 82; XXVIII, 262; *1498) that are described by Clavijo.

[8] Ruy González de Clavijo, *Embajada a Tamorlán. Estudio y edición de un manuscrito del siglo XV*, F. López Estrada (ed.), Madrid: C.S.I.C., 1943; Id., *Viaggio a Samarcanda 1403-1406. Un ambasciatore spagnolo alla corte di Tamerlano*. P. Boccardi Storoni (Italian ed.), Rome: Viella, 2002.

[9] It is important to remember that the notion of "Central Asia" was elaborated at the beginning of the nineteenth century. Marco Polo writes of an *Asia mediana*: a notion, however, that was not assimilated into contemporary cartography. See Pelliot, *Notes on Marco Polo*, t. I, "Asya Media," p. 55 and *s.v.* "Ciagatai," p. 254.

[10] Clavijo, *Embajada a Tamorlán*, pp. 204-206.

[11] *Termici* (XXXIII, 73; *2399).

[12] "Questa uale è serada da do' forteze, | le qual se chiama porte de fero," *Mappamondo* (ed. 1956) Pl. XXXIII, 63. The English translation derives from *Mappamondo* (ed. 2006) *2388.

[13] "Queste porte, le qual ne la lengua de | questi sono dite derbent çoè porte de | ferro e questo p(er) esser inexpugnabile, so|no anchor dite porte chaspie p(er)chè a uo|ler passar q(ue)sto mo(n)te chaspio è necessità | passar per q(ue)ste porte, le q(u)al p(er) tuto se dice chaspie," *Mappamondo* (ed. 1956) Pl. XXXIII, 43. "These gates in the language of the people of this place are called Derbent – Iron Gates – because they are impregnable. They are also called the Caspian Gates because, if you want to pass these Caspian mountains, you have to go by these gates, which are therefore called Caspian"; the English translation derives from *Mappamondo* (ed. 2006) *2360. The Gates of Alexander, or Caspian Gates of Derbent, with their thirty north-looking towers at one time stretched for forty kilometres between the Caspian Sea and the Caucasus Mountains, effectively blocking the passage across the Caucasus. The historical Caspian Gates were most likely built long after Alexander's time, probably during the reign of Khosrau I in the sixth century AD; yet they came to be credited to him throughout the centuries. See S. M. Gorshenina, *De la Tartarie à l'Asie centrale. Le cœur d'un continent dans l'histoire des idées entre la cartographie et la géopolitique*, Ph.D. dissertation, Paris: Université Panthéon-Sorbonne, and Lausanne: Université de Lausanne, 2007, pp. 53-79. See also A. Runni Anderson, "Caspian Gates," *Transactions and Proceedings of the American Philological Association*, 59 (1928), pp. 130-163; J. F. Standish, "The Caspian Gates," *Greece & Rome*, 2nd Ser., 17, 1 (1970), pp. 17-24.

[14] Pliny, *Naturalis Historia*, VI.40: "Corrigendus est in hoc loco error multorum, etiam qui in Armenia res proxime cum Corbulone gessere. namque hi Caspias appellavere Portas Hiberiae, quas Caucasias diximus vocari,

were transmitted by many medieval authors and cartographers such as John de Mandeville[15] and Giovanni Leardo, as well as the *Romance of Alexander*, a rewriting of the history of the voyage and conquests of Alexander the Great in Asia, widely diffused in the Middle Ages.[16] Furthermore, in the *mappa mundi* the Iron Gates of Termiz and Darband are clearly disassociated from the tradition of Gog and Magog, a story openly rejected by Fra Mauro.[17] The toponomastic coincidences between Fra Mauro's *mappa mundi* and the *Embajada a Tamorlán* therefore suggest a diffusion of the information gathered and transmitted by Clavijo independent of the manuscript circulation of his work. Clavijo completed the long return voyage from Tana on the Black Sea to Spain aboard the Genoese galleys, which were efficient intermediary vessels for the transmission and diffusion of geographical knowledge in the Mediterranean, besides the circulation of written texts.

In the light of these considerations, my choice in studying Asia rather than the other regions indicated was dictated by the observation of a greater quantity and density of information recorded by the Camaldolese friar in this area of the *mappa mundi*, joined by the consideration that, for Asia, sources are better identifiable and their textual traditions is richer and better understood. It is in fact noted that the *mappa mundi*'s Central and Southern Asia and the enormous Indian Ocean basin were designed, on the basis of a cartographic plan substantially of Ptolemaic derivation, to offer a graphic and spatial representation of Marco Polo's book in addition to the *Liber IV* of *De varietate fortunae* (The Vicissitudes of Fortune) in which the humanist Poggio Bracciolini recounted the long voyage of the Venetian merchant Niccolò de' Conti in Asia. Nevertheless, most historians have not investigated these aspects in detail, limiting themselves to brief mentions. Important exceptions, mostly regarding the identification of the toponomastic correspondence, are the studies by Henry Yule, Ivar Hallberg, Paul Pelliot and recently, Piero Falchetta. Their contributions to the textual comparisons between Fra Mauro's *mappa mundi* and the complex textual narrative tradition of the travels of Marco Polo, Niccolò de' Conti, and Odorico da Pordenone, remain fundamental.[18] The comparison between the texts

situsque depicti et inde missi hoc nomen inscriptum habent. Et Neronis principis comminatio ad Caspias Portas tendere dicebatur, cum peteret illas, quae per Hiberiam in Sarmatas tendunt, vix ullo propter oppositos montes aditu ad Caspium mare. Sunt autem aliae Caspiis gentibus iunctae, quod dinosci non potest nisi comitatu rerum Alexandri Magni."

[15] Jean de Mandeville, *Le livre des merveilles du monde*, C. Deluz (ed.), Paris : CNRS, 2000, p. 428.

[16] P. Boitani, C. Bologna, A. Cipolla, and M. Liborio (eds), *Alessandro nel Medioevo occidentale*, introduction by P. Dronke, Rome : Fondazione Lorenzo Valla, 1997.

[17] The tribes of Gog and Magog are described in Genesis 10 :2, Ezekiel 38 : 1-3 ; and Revelation 20 : 7-8. According to the complex traditions of the *Alexander Romance* – the text was recast into various versions as well as languages between the fourth and the sixteenth centuries, including Middle Greek, Latin, Armenian, Syriac, and most medieval European vernaculars – Alexander came to a northern land devastated by incursions from barbarian peoples, including Gog and Magog. Alexander defended the land by constructing the Gates of Alexander, an immense wall between two mountains that will stop the invaders until the end of times. In the *Romance*, these gates are built between two mountains in the Caucasus called the "Breasts of the World" ; this has been often taken as a reference to the historical "Caspian Gates" in Derbent, Russia. See Boitani, *Alessandro nel Medioevo occidentale*.

[18] I. Hallberg, *L'Extrême-Orient dans la litérature et la cartographie de l'Occident des* XIIIe, XIVe, XVe *siècles*, Göteborg : W. Zachrissons Boktryckeri, 1906 ; H. Yule, *The Book of Ser Marco Polo the Venetian Concerning the Kingdoms and Marvels of the East*. Translated and edited, with notes, by Colonel Sir H. Yule. Third edition, revised throughout in light of recent discoveries by Henry Cordier, 2 vols., London : Murray, 1921 ; Pelliot, *Notes on Marco*

of the *mappa mundi*, the *Milione* and the *Liber IV* of the *De varietate fortunae*, through the most relevant philological studies on the various manuscript traditions of these books, on the basis of the toponomastic studies by Ivar Hallberg and Paul Pelliot, lead to the conclusion that Fra Mauro drew on at least 150 toponyms and legends from Marco Polo and Niccolò de' Conti. These are distributed in a non-uniform way over two geographical areas derived from these two main sources respectively: Marco Polo for Northern and Central Asia; and Niccolò de' Conti for Southern Asia and the Indian Ocean.

In the following pages these textual and visual relations between travel writings and the *mappa mundi* will be analyzed beginning with the investigation of which textual recension(s) of Marco Polo's book the Camaldolese friar referred to; than it will be evaluated how information derived from Marco Polo's and Niccolò de' Conti's travelogues was integrated into the cartographic layout of Asia. We will seek to understand the different details of information that Fra Mauro considered and which aspects were instead neglected in the reading that he made of his sources. This analysis shows that there are four methods of transmission and re-elaboration of knowledge in the passage from the travel narratives of Marco Polo and Niccolò de' Conti to the cosmography of Fra Mauro. The first method is the citing of toponyms that identifies cities, rivers, mountains, seas, provinces, kingdoms, and historical events; a second one is the summarizing, or paraphrasing, of entire passages, and mostly those in which the commercial routes are described as well as the places of production of spices and various commodities, with a predilection for gold, pearls, and precious stones.[19] For Marco Polo's book there are, lastly, two further ways in which the Venetian traveler's narrative enters into Fra Mauro's cosmography: through the visual, in some cases literary, in others creative and highly imaginative, citation of entire excerpts of the *Milione*; then, ultimately, through the *Milione*'s influence in its narrative and syntactic form on Fra Mauro's writing style. While the first three methods will be investigated in this chapter, the relationship between the several parts of the *Milione*'s narrative structure and the *mappa mundi* – particularly of those parts in which the story proceeds through the employment of the first person narrative voice – will be instead analyzed in Chapter VI.

Against established historiographic interpretations that reduce the narration and depiction of the East to the genres of the "marvels" and the "monstrosities" this textual and visual analysis and comparison will contribute to the highlighting of the complex intellectual and pictorial strategies that medieval authors, illuminators and cartographers developed to describe a world that in most cases they could not directly experience. The results of this research regard two general cultural areas often left divorced from one another. On one hand the research examines the cartographic writing and design procedures; on the other it confronts the history of the reception by cosmographers and cartographers of two central works in western culture – the *Milione* and the *De varietate fortunae* – a debate substantially neglected in the rich his-

Polo. On the identification of the toponyms quoted by Marco Polo see also H. Cordier, *Ser Marco Polo. Notes and Addenda to Sir Henry Yule's Edition, Containing the Results of Recent Research Discovery*, London: Murray, 1920; F. Pullè, *Cartografia antica dell'India*, Florence: Tip. Carnesecchi, 1932, pp. 43-51; L. Olschki, *L'Asia di Marco Polo. Introduzione alla lettura e allo studio del Milione*, Venice: Fondazione Giorgio Cini, 1957 (rep. 1978); G. R. Cardona, "Indice ragionato," in Marco Polo, *Milione: versione toscana del Trecento*, V. Bertolucci Pizzorusso (ed.), Milan: Adelphi, 2003, pp. 489-761.

[19] In the *mappa mundi* there are 31 goods listed, most of them being spices and the rest pearls, gold and precious stones.

toriographic studies of these works. The selective interpretation that Fra Mauro makes of the *Milione* and the *Liber IV* of *De varietate fortunae*, considered in the *mappa mundi* above all for their commercial contents, documents one of the main ways in which the two travel narratives were read in the fifteenth century.

The comparison between text and images in the *mappa mundi* and the *Milione* will be based on two main textual recensions of Marco Polo's work : the so-called Z tradition, in Latin,[20] and a particular Venetian version witnessed by the Ms. Ital. Cl. VI. 56 (= 6140) of the Biblioteca Nazionale Marciana, a manuscript undoubtedly available in Venice since the first half of the fifteenth century, which so far has never been worked on. This analysis shows that, if on the one hand, it is impossible to identify which manuscript(s) Fra Mauro employed directly, it is conversely possible to detect unequivocal traces of both the Z and the Venetian textual traditions of the *Milione* in the *mappa mundi*'s legends. For the English translation of passages of the *Milione* that will be cited I will use the edition by Sir Henry Yule revised by Henry Cordier in 1903.[21] Regarding the comparison with *Liber IV* of the *De varietate fortunae* in which Poggio Bracciolini, between 1441 and 1448, transposed the story of twenty-five years of travels in Asia by the merchant Niccolò de' Conti (1395-1469), I will use the Latin textual tradition, in the two critical editions established by Outi Merisalo and Michèle Guéret-Laferté.[22] For the translation I will use the English version by Winter Jones, edited by Richard Henry Major and Lincoln Davis Hammond in 1857.[23]

It is furthermore important to underline how, regarding the textual sources, Marco Polo's and Niccolò de' Conti's narration and, for the cartographic ones, Ptolemy's *Geography*, are not the sole, but just the principle, sources used by Fra Mauro for the design of Asia. The still valid

[20] Marco Polo, *Milione. Redazione latina del manoscritto Z*, A. Barbieri (ed.), Parma : Fondazione Pietro Bembo, Guanda, 1998.

[21] *The Book of Ser Marco Polo the Venetian Concerning the Kingdoms and Marvels of the East*, translated and edited, with notes, by Sir Henry Yule, Third edition revised by Henry Cordier, 2 vols., London : John Murray, 1903.

[22] Poggio Bracciolini, *De varietate fortunae*, critical edition with an introduction and commentary edited by O. Merisalo, Helsinki : Suomalainen Tiedeakatemia, 1993 (Annales Academiae Scientiarum Fennicae, Ser. B) ; Poggio Bracciolini *De l'Inde : les voyages en Asie de Niccolò de' Conti* ; text established, translated and commented by M. Guéret-Laferté, Turnhout : Brepols, 2004. Piero Falchetta instead uses the Italian edition of *Liber IV* of the *De varietate* edited by Ramusio around the middle of the sixteenth century : *Viaggio di Nicolò di Conti*, in Giambattista Ramusio, *Navigationi et viaggi*, M. Milanesi (ed.), Turin : Einaudi, 1978-1988, vol. 2, pp. 781-820. As demonstrated by Merisalo and Guéret-Laferté this edition was strongly revised and updated by Ramusio in light of new discoveries of the mid-sixteenth century. For this reason the Latin editions seems to me preferable.

[23] *The Travels of Nicolò Conti in the East in the Early Part of the Fifteenth Century*, translated from the original of Poggio Bracciolini, with notes, by J. Winter Jones. [Esq. FSA, Keeper of Printed Books, British Museum], in Richard Henry Major, *India in the Fifteenth Century. Being a Collection of Narratives of Voyages to India in the Century Preceding the... Discovery of the Cape of Good Hope ; from Latin, Persian, Russian and Italian sources. Now First translated into English*. Edited, with an introduction, by R. H. M., London : Hakluyt Society, 1857, Part 2, pp. 1-39. The translation by John Frampton is less adequate in that it is based on the Italian text by Ramusio. See J. Frampton (ed.), *The Most Noble and Famous Travels of Marco Polo together with the Travels of Nicolò de Conti*. Edited from the Elisabethan Translation [1579] of John Frampton with introduction, notes and appendixes by N. M. Penzer, London : Argonaut Press, 1929 (repr. Amsterdam and New York 1971).

studies by Ivar Hallberg and the recent contribution of Piero Falchetta clearly demonstrate the existence of textual convergences, above all regarding the toponyms, with other travel narratives: for example the already cited *Embajada a Tamorlán* by León Rui González de Clavijo, but also the *Relatio* of the Franciscan Odorico da Pordenone (1265-1331, mission in Asia from 1318 to 1330).[24] It is certainly not surprising that this happens. The philological research of Luigi Foscolo Benedetto, one of the most noted experts of the entire manuscript tradition of the *Milione*, on the extensive and complex *corpus* of manuscripts that passed down the work of Marco Polo demonstrates that, out of around one hundred and forty manuscripts, twenty-five included the *Milione* together with the works of the Franciscan and Dominican friars who travelled to Asia between the thirteenth and the fourteenth century; and seven, in particular, of the *Relatio* by Odorico da Pordenone. This implies that to fourteenth- and fifteenth-century readers these works were known and perceived as belonging to the same narrative corpus.[25]

2. Fra Mauro and the textual traditions of the *Milione*

It was already pointed out in Chapter I that the relationship between Marco Polo's *Milione* and Fra Mauro's *mappa mundi* had been revealed at least since the mid-sixteenth century when Giovanni Battista Ramusio wrote that Fra Mauro had used copies of maps originally designed by Marco (a world map and a marine chart) in conjunction with the travelogue.[26] Ramusio was absolutely convinced that Marco Polo, as a merchant, a traveler, and an officer for the Gran Khan, must have had to use maps and to draw others to bring back to Venice. It is unknown if these cartographic works, and their copies, really did exist or not; whatever the truth, we can legitimately deduce from Ramusio's account that by the mid-sixteenth century (alleged) Polo's original maps as well as all derivative copies were already all lost and therefore the only extant evidence of their existence was Fra Mauro's *mappa mundi*. What instead about the text of Marco Polo used by Fra Mauro? What was the textual tradition(s) it did belong to?

Marco Polo's book had ample multi-lingual transmission, articulated in different textual recensions, with different titles: *Livre des merveilles*; *Devisement dou monde*; *Del Gran Khan*, *Historia Tartarorum*; *De Condicionibus et consuetudinibus orientalium regionum*, and *Milione* are some of the most diffused.[27] The oldest draft of the work, in *langue d'oïl*, with evident

[24] See the numerous references in Hallberg, *L'Extrême-Orient dans la litérature et la cartographie de l'Occident des XIIIe, XIVe, XVe siècles* and in *Mappamondo* (ed. 2006).

[25] A. Andreose, "Tra ricezione e riscrittura: la fortuna romanza della 'Relatio' di Odorico da Pordenone," in G. Carbonaro, M. Cassarino, E. Creazzo and G. Lalomia (eds), *Medioevo romanzo e orientale. Il viaggio nelle letterature romanze e orientali*, Soveria Mannelli: Rubbettino, 2007, pp. 5-21; A. García Espada, "Marco Polo, Odorico of Pordenone, the Crusades, and the Role of the Vernacular in the First Descriptions of the Indies," *Viator* 40 (2009), pp. 201-222.

[26] G. B. Ramusio, *Dichiarazione d'alcuni luoghi ne' libri di messer Marco Polo, con l'istoria del reubarbaro* (Declaration – "elucidation" – of several places in the books by messer Marco Polo with the history of rhubarb), in *Navigationi et viaggi*, in Venetia, appresso i Giunti, 1559, vol. II, f. 17r. Modern edition, Ramusio, *Historia delle cose de Tartari*, pp. 68-71. The description of Fra Mauro's map is in the final part of the *Dichiarazione* and precedes Ramusio'edition of Marco Polo's text.

[27] V. Bertolucci Pizzorusso. "Enunciazione e produzione del testo nel *Milione*," *Studi mediolatini e volgari* 25 (1977), pp. 5-43 (p. 8); reprinted in Ead., *Morfologie del testo medievale*, Bologna: Il Mulino, 1989, pp. 209-241. The

morphological and lexical Italianisms, was composed at the end of the thirteenth century in Genoese prisons by Marco Polo, *auctor-dictator*, and by Rustichello da Pisa, in role of *auctor-scriptor* and over all compiler of the narrative.[28] The holograph has been lost. The oldest extant manuscript of the *Milione*, which is Ms. 1116 of the Bibliothèque nationale de France at Paris (referred to by Luigi Foscolo Benedetto as "Draft F") was copied by an Italian at the very beginning of the fourteenth century.[29] French copyists using the *langue d'oïl* draft as a basis, prepared a more orthodox French version (Benedetto calls it "Draft FG") that is divided into four subgroups.[30] In the Tuscan and Veneto regions, early in the fourteenth century, two versions that translated the *langue d'oïl* work into the Tuscan and Venetan vernaculars were drafted; which were divided into many branches.[31] From the Venetan draft, VA – this also widely disseminated[32] – around 1315-1320, the Dominican friar Pipino da Bologna composed one of the six known Latin translations of the work, generally known under the title *De condicionibus et consuetudinibus orientalium regionum*. Pipino's translation is probably the most diffused textual tradition: passed down *via* several manuscripts (more than sixty copies are recorded) and one fifteenth-century printed edition, it was addressed to a public of clerics and scholars.[33] For the

first effective attempt to draw the complete *stemma* of the complex testual traditions of the *Milione* should be credited to Luigi Foscolo Benedetto, "Introduzione," in M. Polo, *Il Milione*. First edition by L. F. Benedetto, Florence: Olschki, 1928, pp. i-ccxxi. (English edition: *The travels of Marco Polo*; translated into English from the text of L. F. Benedetto by A. Ricci; with an introduction and index by E. Denison Ross, London: Routledge, 1931). Benedetto's edition, printed in only 600 copies, is now very rare. Therefore, see also Marco Polo, *Milione. Le divisament dou monde. Milione nelle redazioni toscana e franco-italiana*, G. Ronchi (ed.), introduction by C. Segre, Milan: Mondadori, 1982. For the most recent and updated *stemma*, which carefully and critically revises the philological research of the last century, see C. Gadrat, "Le rôle de Venise dans la diffusion du livre de Marco Polo," *Médiévales* 58 (2010), pp. 63-78 (p. 65 for a chart of the textual derivation).

[28] Rustichello da Pisa was a writer of chivalric stories, the author of a compilation of Arthurian legends, known as the *Roman de Meliadus*. See F. Cigni (ed.), *Il romanzo arturiano di Rustichello da Pisa*, foreword by V. Bertolucci Pizzorusso, Pisa: Pacini and Cassa di Risparmio di Pisa, 1994. On the relationships between Rustichello da Pisa and Marco Polo, with an emphasis on the composition of the *Devisement dou monde*, see A. Barbieri, "Marco, Rustichello, il "patto", il libro: genesi e statuto testuale del *Devisement dou monde* alla luce degli studi recenti," in G. Carbonaro *et al.* (eds), *Medioevo romanzo e orientale. Il viaggio nelle letterature romanze e orientali*, Genoa: Rubbettino, 2006, pp. 23-42.

[29] For an analysis of the manuscript, see Benedetto, "Introduzione," pp. xii-xxxiii.

[30] Marco Polo, *Le Devisement du monde*, P. Ménard (ed.), 6 vols., Genève: Droz, 2001-2009, vol. 1, p. 28-31. See also Benedetto, "Introduzione," pp. xxxiv-lxxix.

[31] The Tuscan translation from the beginning of the fourteenth century is passed down by five apograph manuscripts; see Benedetto, "Introduzione," pp. lxxx-xcix. Marco Polo, *Il Milione*, introduction, edition of the Tuscan text (Ottimo), illustrative notes, exegetics, linguistics, onomastic repertoires and lexicons edited by R. M. Ruggieri, Florence: Olschki, 1986; see also Bertolucci Pizzorusso (ed.), *Milione: versione toscana del Trecento*.

[32] The different branches of the composite Venetan tradition of Marco Polo's book are witnessed by at least twenty manuscripts. The VA version is represented by six principal manuscripts and three additional later copies. See Benedetto, "Introduzione," pp. c-cxxxii and Barbieri, "Introduzione," in Marco Polo, *Il "Milione" veneto. Ms. CM 211 della Biblioteca Civica di Padova*, A. Barbieri and A. Andreose (eds), Venice: Marsilio, 1999, pp. 23-65.

[33] There are around sixty manuscripts that transmit the Latin translation of friar Pipino; for the study of this textual tradition of the *Milione*, see C. W. Dutschke, *Francesco Pipino and the Manuscripts of Marco Polo's "Travels"*, Ph.D. dissertation, Los Angeles: University of California, 1993, unpublished) and Gadrat "Le rôle de Venise dans la diffusion du livre de Marco Polo," who added further manuscripts to Dutschke's survey. *Editio princeps*: [Marco Polo], *De consuetudinibus et condicionibus orientalium regionum*, Gouda: Gerard Leeu, between 1483 and 11 June

scope of this research, it is important to recall that among the six Latin translations, there is a key recension, called by Benedetto "Z draft" that survives in a single manuscript of around the mid-fifteenth century held at the Archivo y Biblioteca Capitulares in Toledo.[34] In the Biblioteca Ambrosiana in Milan there is an eighteenth-century manuscript copy commissioned by Giuseppe Toaldo (1719-1798), a professor of geography and astronomy at the University of Padua.[35] Both manuscripts have proved to be extremely important for the reconstruction of the textual tradition of the *Milione*.[36]

Already in the nineteenth century Ivar Hallberg and of Paul Pelliot established that Fra Mauro used at least one particular manuscript of the *Milione* belonging to this Latin textual tradition.[37] The latest research – which begins with Hallberg and Pelliot's work, extends to

1484 (IGI 7976). Modern edition: Marka Pavlova z Benátek, *Milion. Dle jediného rukopisu spolu s příslušným základem latinským*, Prague: Justin V. Prášek, 1902. Fra Pipino's text was also printed in German (1477 and 1481), in the Venetan vernacular (1496) in Portuguese (1502) and in Castillian (1503). See Gadrat "Le rôle de Venise dans la diffusion du livre de Marco Polo," pp. 66-67. On friar Pipino: L. Manzoni, "Frate Francesco Pipino di Bologna dei PP. Predicatori, geografo, storico e viaggiatore," *Atti e memorie della R. Deputazione di Storia Patria per le Provincie di Romagna*, s. 3, 13 (1895), pp. 257-334; G. Zaccagnini, "Francesco Pipino traduttore del *Milione*, cronista e viaggiatore in Oriente nel secolo XIV," *Atti e memorie della R. Deputazione di Storia Patria per le Provincie di Romagna*, s. 5, 1 (1936), pp. 70-95.

[34] Toledo, Archivo y Biblioteca Capitulares, ms. 49.20 Zelada, manuscript on paper, ff. 135. Benedetto called it Z in honour of the Spanish cardinal Francisco Xavier De Zelada (1717-1801), the last owner of the manuscritp before it entered the Toledan Biblioteca Capitular.

[35] Toaldo attempted a critical edition of the *Milione* that remains unfinished. The *corpus* of copies drawn on manuscripts of the *Milione* made by Toaldo, initially destined for the Biblioteca Marciana in Venice, is conserved in the Biblioteca Ambrosiana in Milan; see Benedetto, "Introduzione," p. clxi. Until 1937 it was thought that the Z textual tradition was only passed down by the eighteenth-century copy commisioned by Toaldo and that the original might have been lost. The original fifteenth-century manuscript was then re-discovered by John H. Herriott in the Archivo Capitular at Toledo in the 1930's. See J. H. Herriott, "The 'Lost' Toledo Manuscript of Marco Polo," *Speculum* 12, 4 (1937), pp. 456-463. It was then transcribed and translated, with some significant errors and gaps, by P. Pelliot and A. C. Moule. See Marco Polo, *The Description of the World*, edited, translated, and with an introduction by A. C. Moule and P. Pelliot, 2 vols., London: Routledge, 1938 (facsimile New York: AMS Press, 1976). The manuscript was again transcribed, translated into Italian and published: Marco Polo, *Milione. Redazione latina del manoscritto Z*, A. Barbieri (ed.), Parma: Fondazione Pietro Bembo, Guanda, 1998.

[36] Benedetto writes: "Z is the literal version, in a Latin that lends easily to the original language, of a French-Italian exemplar clearly better than F": that is the manuscript in the *langue d'oïl* on the basis of which the fourteenth-century Tuscan and Venetian vernacular translations of Marco Polo's book were made. Z hands down two hundred passages more compared to the Tuscan translation as well as to Fra Pipino's Latin translation that derives from a Venetian version of the *Milione*. Moreover, Z conserves a better rendition for the names of people and for the toponyms and provides "significant contributions to the geographic material. See Benedetto, "Introduzione," pp. clxvi-clxvii. Nevertheless, Barbieri appropriately points out, "this advantage of Z over F does not apply to the entire text. The Latin version taken from the Toledan manuscript is strongly abbreviated in the initial part: the long proem is reduced to a few lines, entire chapters are skipped (...) others suffer from considerable cuts" ("Questo vantaggio di Z su F non riguarda l'intero testo. La versione latina tràdita dal codice toledano è fortemente abbreviata nella parte iniziale: il lungo proemio è ridotto a poche righe, interi capitoli sono saltati (...) altri sono colpiti da tagli notevoli"). See Barbieri, "Nota al testo," pp. 576-7. The English translation is mine. For the study and transcription of the chapters of F suppressed in Z, see Barbieri, "Appendice," in Marco Polo, *Milione. Redazione latina del manoscritto Z*, pp. 611-686.

[37] Se for example Hallberg, *L'Extrême-Orient dans la littérature et la cartographie de l'Occident*, pp. 415-416 and p. 487; Pelliot, *Notes on Marco Polo*, pp. 766-770 and p. 812; Barbieri, "Introduzione" and "Nota al testo e

my own investigations published since 2004, and includes Piero Falchetta's 2006 study and Eugenio Burgio's 2009 research – has demonstrated that some thirty of the *mappa mundi*'s toponyms were drawn from the textual tradition Z of Marco Polo's book. This fact indisputably attests to a relationship between the two works. More precisely the toponyms *Balch, Charaçan, Chauzuzu, Çiansan, Chuçu, Choiganzu, Cinzu, Ianzu, Ghindu, Linzinfu, Pinzu, Sindinfu, Sugzu,* and *Xandu,* and also the name of a ruler, *Çagatai,*[38] can be traced back only to Z, with only minimum variations in form (such as the substitution of *z* with *ç*, or the addition of *h* after *c*). This circumstance was termed by Pelliot as "one more striking parallel [of Fra Mauro's *mappa mundi*] with Z and with Z only."[39] It should be emphasized that, in contradiction to what Paul Pelliot has argued, the manuscript held in Toledo cannot have been used by Fra Mauro. The very existence of the Toledan manuscript simply and only enables us to recognize that Fra Mauro's work was based on the Z textual recension of Marco Polo's book. Nevertheless, although the form of a number of the *mappa mundi*'s toponyms and a few of its legends derive from this textual tradition, deviations in the transcription of the toponyms and a notable infidelity in Fra Mauro's longest legends prove that his *mappa mundi* did not derive directly from the Toledan manuscript itself. What is more, pursuing this vein of research, an analysis of the *mappa mundi*'s legends confirms that the Z textual tradition was not the only one through which Fra Mauro read the *Milione*. This is most evident with regard to the description of Persia : "Persia contie(n) viij Reami de i qual | el primo uie(n) dito *chassu*, s(econdo) *lor*, terço | *cu(r)dista(n)*, quarto *thymochain*, qui(n)to | *celsta(n)*, sexto *istaruch*, septimo *ce|raci*, octauo *sonçara*. Ma nel regno | de *thymochai(n)* nasce nobilissimi caua|li e aseni de gra(n) priesio e sono pote(n)ti | e ueloci, di qual se ne fa gra(n) marcha|da(n) tie i(n) india."[40] A comparison between Fra Mauro's designation of the eight kingdoms of Persia with the forms in which they appear in the principle versions of the *Milione* reveals a clear deviational trend :

apparati," in Marco Polo, *Milione. Redazione latina del manoscritto Z*, pp. ix-xxxv and pp. 565-582 (pp. 575-580).

[38] *Balch* (Pl. XXVII, 98 ; *2536), *Charaçan* (Pl. XIX, 59 ; *653), *Chauzuzu* (Pl. XX, 6 ; *652), *Çiansan* (Pl. XXV, 54 ; *1365), *Chuçu* (Pl. XXV, 48 ; *1364), *Choiganzu* (Pl. XXXII, 61 ; *2300), *Cinzu* (Pl. XXXI, 19 ; *2248), *Ianzu* (Pl. XXVI, 23 ; *1391) il nome del principe *Çagatai* (Pl. XXXIII, 24 ; *2371), *Ghindu* (Pl. XXV, 15 ; *741), *Linzinfu* ((Pl. XXXII, 66, 68 ; *2303), *Pinzu* (Pl. XXXII, 65 ; *2302), *Sindinfu*, (Pl. XXV, 39 ; XXVI, 34 ; *1349 ; *1396), *Sugzu* (Pl. XX, 48, *677), *Xandu* (Pl. XXXVIII, 13 ; *2716). See A. Cattaneo, "Scritture di viaggio e scrittura cartografica. La *mappamundi* di Fra Mauro e i racconti di Marco Polo e Niccolò de' Conti," *Itineraria* 3-4 (2004-2005), pp. 157-202 (pp. 173-174 and pp. 181-182) ; *Mappamondo* (ed. 2006), pp. 64-66 and E. Burgio, "'Cartografie' di viaggio. Sulle relazioni fra la *Mappamundi* di Fra Mauro e il *Milione*," *Critica del testo* 12.1 (2009), pp. 59-106 (pp. 87-90).

[39] Pelliot, *Notes on Marco Polo*, I, pp. 569-570.

[40] *Mappamondo* (ed. 1956) Pl. XXI, 101. "Persia contains eight realms. The first is called Chassu, the second Lor, the third Curdistan, the fourth Thymochain, the fifth Celstan, the sixth Istaruch, the seventh Ceraci and the eighth Sonçara. But in the realm of Thymochain are raised most noble horses and highly-prized mules, for which there is a great market in India." The English translation derives from *Mappamondo* (ed. 2006) *815.

Fra Mauro	*Z*[41]	*T*[42]	*R*[43]	*V*[44]
Chassu	Chasuin	Causom	Casibin	Chasom
Lor	Curdistan	Distan	Curdistan	Ciudistan
Curdistan	Lor	Lor	Lor	Lori
Thymochain	Suolistan	Cielstan	Suolistan	Cielstan
Celstan	Yspaan	Istain	Spaan	Lorstanie
Istaruch	Çirac	Zerazi	Siras	Ceraci
Ceraci	Soticara	Soncara	Soncara	Socchara
Sonçara	Tunocain	Tunocain	Timocaim	Tenochain

By contrast, this same legend of the *mappa mundi* exhibits a high degree of proximity to a vernacular version that derives from the Milione's textual tradition in the Venetian vernacular, namely, Ms. It. VI. 56 (= 6140) at the Biblioteca Nazionale Marciana in Venice[45]:

Fra Mauro	*Ms. It. VI. 56 (= 6140), f. 8r*
Chassu	Cassu
Lor	Lor
Curdistan	Curdistam
Thymochain	Timochaim
Celstan	Cielstam
Istaruch	Istainch
Ceraci	Ceraci
Sonçara	Sonçara

Moreover, the legend that describes the Persian city of *Cobinam* ("In q(ue)sta cità de Cobinam è gra(n)de | habundantia de ferro e | açali e endego e bele|tissimi spechi de a|çal,"[46]) which is positioned near Persia's legend, would appear to depend directly on this same Venetian manuscript: "‹C›obina è una grande citade in la quale è grande habondantia de ferro e de azal

[41] Marco Polo, *Milione. Redazione latina del manoscritto Z*, ch. X, p. 38.

[42] Marco Polo, *Milione. Versione toscana del Trecento*, pp. 44, 693. For Polo's description of Persia, S. Franchi, *L'itinerario di Marco Polo in Persia*. Turin: Tipografia Collegio Artigianelli, 1941.

[43] I consulted the edition by Giovan Battista Ramusio, *Secondo volume delle nauigationi et viaggi, raccolto gia da m. Gio. Ramusio, et hora in questa nuoua editione accresciuto: nel quale si contengono l'Historia delle cose de Tartari, & diuersi fatti de' loro imperatori, descritta da m. Marco Polo gentil'huomo venetiano & da Hayton Armeno. Varie descrittioni di diuersi auttori,... et il viaggio della Tana. Con la descrittione de' nomi de' popoli,... Aggiuntoui in questa vltima editione la Descrittione dell'vna e dell'altra Sarmatia,... Con l'indice...*, In Venetia appresso i Giunti, 1583, ch. XI, "Del nome di otto regni, che sono nella provincia di Persia, e della forte di cavalli e asini, che ivi si trovano," f. 6v.

[44] Marco Polo, *Il 'Milione' veneto. Ms. CM 211 della Biblioteca Civica di Padova*, A. Barbieri and A. Andreose (eds), Venice: Marsilio, 1999, ch. XX.

[45] Venice, Biblioteca Nazionale Marciana, Ms. It. VI, 56 (= 6140), *Libro de lo savio et honorevole homo e fidele signore miser Marcopolo de Venesia*, first half of the fifteenth century, ff. 1r-74r; the lack of an apposite study can be attributed perhaps to Foscolo Benedetto's dismissal of it as a "mediocre work of the first half of the fifteenth century" (Benedetto, "Introduzione," p. cxlvii).

[46] *Mappamondo* (ed. 1956) Pl. XXI, 31. "In this city of Cobinam there is a great abundance of iron and steel and indigo; and very fine steel mirrors." The English translation derives from *Mappamondo* (ed. 2006) *773.

e de andarrico [*sic*] quello logo se fano spechi de azal molto belli e grandi" (f. 10va). Ms. It. VI, 56 (= 6140) at the Biblioteca Marciana, which was undoubtably available in Venice since the first half of the fifteenth century, has never been worked on, despite being known well enough to have been cited in both bibliographic and historiographic studies. It transmits a unique Venetian tradition that was described by Benedetto as "una curiosità" because of its borrowings from several distinct textual traditions of Marco Polo's book.[47] At this point it should be clarified that the Biblioteca Marciana manuscript was not among the sources used by Fra Mauro. Except for the two passages cited above regarding Persia no other correspondences can be found. Nonetheless, these two instances provide important evidence about the precise nature of the relationship between Fra Mauro's *mappa mundi* and the writings of Marco Polo.

Thus far the comparison between the toponyms and cartouches of the *mappa mundi* and the different traditions of Marco Polo's text do not allow us to point unequivocally to which specific textual tradition of the *Milione* Fra Mauro referred, nor to which manuscript or manuscripts, among those reviewed or surviving, would have been the *direct* source of the *mappa mundi*. Still, the analysis of similarities and differences between the *mappa mundi* and at least two versions of the *Milione* (the Z tradition represented by the Toledo manuscript, and the Venetian vernacular tradition represented by the Biblioteca Marciana manuscript) suggests two primary hypotheses regarding the origin of Fra Mauro's work and, at the same time, the Venetian circulation of manuscript copies of Marco Polo's *Milione*. According to the first hypothesis, Fra Mauro would have had access to two manuscripts representing two distinct textual traditions originating from Marco Polo's writings. One of these manuscripts must have either descended from manuscript Z, written in Latin, or otherwise been strongly influenced by this branch of the textual tradition. The second must have belonged to the branch descended from the composite Venetan tradition.

The second hypothesis is that Fra Mauro would have had at his disposition a single manuscript, now lost, in which elements of the Z tradition and the Venetan tradition were intermixed. The existence of manuscripts fitting this definition, which Benedetto termed "bizzarri," is demonstrated by manuscript Hamilton 424 in Berlin's Staatsbibliothek which, although written in the Venetan vernacular, possesses strong similarities to the Z textual tradition.[48] At the present moment, it is not possible to draw more conclusive results from the current state of research. As will be shown in greater detail in Chapter VI, the library of the monastery of San

[47] I have also compared Fra Mauro's legends with another important Venetian textual tradition of Marco Polo's *Milione*, identified as VB, whose most important representative is held in Venice, Museo Correr, Ms Donà delle Rose 224, manuscript on paper, II. 266. II, before 1450, [Meser Marco Polo. La so lezenda], ff. 229r-279v. Though this manuscript was copied in Venice, or in the Venetan area, in the same years in which Fra Mauro was preparing the *mappa mundi*, the textual comparison shows that Fra Mauro's legends and toponyms do depend on other recensions of the *Milione*. Cf. the diversity in the form of the eight kingdom of Persia (*Causum*, *Cardistan*, *Lor*, *Cielstan*, *Ilstaint*, *Ceratin*, *Somcarra*, *Tinochan*, f. 233r) and of the description of *Cobinam* ("Coobinam è una citade grandissima et i abitanti de quella adorano Machometo. In questa cità lavorano finissimo açal e fassene bellissimi spechi de açale; éne andanicho," f. 234r). See P. Gennari, "*Milione*", *redazione VB. Edizione critica commentata*. Ph.D. dissertation, Venice: Universita Ca' Foscari, 2010, pp. iii-iv, 29, 35. The dissertation is available on-line at the following address: ‹http://dspace.unive.it/bitstream/10579/937/1/Gennari_955311.pdf›.

[48] For this description, Benedetto, "Introduzione," pp. clcciii-clcciv. Giuseppe Toaldo in 1793 prepared a copy based on the Berlin manuscript (Milan, Biblioteca Ambrosiana, Y 162 sup.). I have examined this manuscript and determined that it does not bear similarities with Fra Mauro's *mappa mundi*.

Michele in Murano never owned a copy of Marco Polo's book. The *Milione* therefore must have numbered among the texts that Fra Mauro had access to through loans; this is all the more plausible since, at this time, books in Venice circulated among monasteries and private individuals with remarkable frequency.

This comparative analysis between Fra Mauro's Asian toponyms and those of Marco Polo allows us to highlight a further element regarding the manuscript circulation of the *Milione* in Venice, for herein are the links that connect Fra Mauro and Giambattista Ramusio, a interesting aspect in the cultural history of the *Serenissima*. Luigi Foscolo Benedetto has argued correctly that, among the manuscripts Ramusio consulted in order to produce his "critical edition" in Italian of Marco Polo's book published in 1559 there must have been a manuscript, now lost to us, that belonged to the Z textual tradition.[49] To summarize broadly the results of Benedetto's research, in the light of both Alvaro Barbieri's and Christine Gadrat's more recent and updated investigation on the Z textual recension, it can be posited that a Franco-Italian manuscript which adhered closer to the original draft than any other version was the source of all manuscripts in the Z textual tradition. Of the diverse manuscripts that were based on this Latin prototype only two are known today. One of them is the previously mentioned manuscript Z at the Archivo Capitular in Toledo; the other, Z[1], is presumed to have been the so-called "Ca' Ghisi" manuscript which was quite possibly that also used in the early fourteenth century by the Dominican Pietro Calo di Chioggia († 1348) – a coastal town on the Southern side of the Venetian lagoon – whose *Legendae Sanctorum* includes a hagiography of Saint Thomas containing a transcription of the chapter that Marco Polo wrote about the saint. It has been firmly demonstrated that Pietro Calo, who even specified Marco Polo's chapter number from which he copied, drew his transcription from a manuscript belonging to the Z textual tradition.[50] Yet in the single surviving manuscript of the Z branch, namely the one in Toledo, the chapters are unnumbered; by contrast, both Calo's short transcription and Ramusio's published edition of the *Milione* refer to enumerated chapters. From this circumstance, it can be deduced that Pietro Calo's *Legendae*, Fra Mauro's *mappa mundi* and Ramusio's *Historia delle cose de Tartari* depended instead on the same textual tradition of the *Milione*. This is not to claim that the Dominican and Camaldolese friars and the Venetian Humanist consulted the same manuscript at a century's distance from each other; rather, it is merely being proposed here that the Z tradition from which one of Ramusio's sources for the *Milione* undoubtedly springs was also the tradition that gave rise – either through direct transmission or indirectly through a translation – to the texts consulted by Pietro Calo and Fra Mauro. Still, the manuscripts used by the three authors presumably beared some mutual similarity and all stemmed from the Z recension

[49] Ramusio's edition was based on several manuscript sources, including a testimony of the Z textual tradition, a Latin manuscript with Pipino's translation; two manuscripts belonging to the Venetian textual recencions, VA and VB; another in the Tuscan vulgate; as well as the edition published in Portuguese in 1502. See Benedetto, "Introduzione," pp. clviii-clxxxii; Milanesi, "Introduzione," in G. B. Ramusio, *Navigazioni e viaggi*, p. 60; Renzi, "Premessa," in Marco Polo, *Il 'Milione' veneto*, p. 18.

[50] A. Barbieri, "Nota al testo e apparati," in Marco Polo, *Milione. Redazione latina del manoscritto Z*, pp. 574-575. For Pietro Calo see A. Poncelet, "Le légendier de Pier Calo," in *Analecta Bollandiana* 29 (1910), pp. 30-34 and 44-116. For the chapter on Saint Thomas, P. Devos, "Le miracle postume de S. Thomas l'Apôtre," *Analecta Bollandiana* 66 (1948), pp. 270-272. For the most complete survey on the dissemination of the Z textual recension in Venice, see Gadrat, "Le rôle de Venise dans la diffusion du livre de Marco Polo," pp. 69-71.

of Marco Polo's book. This explanation would account for the fact that a comparison of Ramusio's *Historia delle cose de Tartari*, Fra Mauro's *mappa mundi* and Calo's *Legendae* reveals variations in the forms of the toponyms and in the enumeration of the chapters derived from Polo. In the absence of evidence regarding the circulation of the textual tradition Z beyond Venetian circles this Latin textual tradition seems therefore to pertain exclusively to the culture of the Venetian region being on the principal, if not the key, recension that circulated in the Veneto area between *circa* 1350 to 1550.[51]

3. The *mappa mundi* and the *Liber IV* of *De varietate fortunae*

The legends of the *mappa mundi* that describe the Indian Ocean basin and the southern regions of Asia attest that Fra Mauro not only utilized Marco Polo's *Milione* as a source, but that he was also among the first readers of the *Liber IV* of the *De varietate fortunae* in which the humanist and Apostolic Secretary Poggio Bracciolini recounts the long voyage to Asia East of the Venetian merchant Niccolò de' Conti. Other than the knowledge of Bracciolini's work, some evidence leaves room for the hypothesis that Fra Mauro personally knew the Venetian traveler : this is a relevant detail that allows us to understand how bookish knowledge and experience not only non contradicted but, on the contrary, complemented each other by interacting in both the epistemic unity of the representation of the *imago mundi* and the culture of the time.

Niccolò di Giovanni Conti (*c.* 1395-1469), originally from Chioggia, like many other Venetian merchants of the time, traveled and traded in the Near East, where he learned Arabic. In the year 1414, departing from Damascus, Conti began a long voyage in which, in the course of twenty-five years, he reached the Euphrates, near the city of Baghdad ; crossed Mesopotamia to the city of Basra ; passed through the Persian Gulf to Khalat, and finally arrived in the port city of Cambay, in Gujarat, on the western coast of India. He visited the innermost regions of the Indian sub-continent of which, as opposed to Marco Polo, he had a profound knowledge, and entered the kingdom of Vijayanagar. He then reached the Kingdom of Ava and the Mon kingdom of Hanthawady Pegu in Burma (Myanmar), the Kingdom of Champa in today's Vietnam, and the island of Borneo, thus completing the circumnavigation of the Indian Ocean. From the Kingdom of Champa he returned by sea to India, sailed up the western coast and, from Cambay, crossed the Indian Ocean westwards, finally reaching the island of Socotra (Suqut□rá) beyond the Horn of Africa, and the cities of Berbera, Aden, and Jidda in the Arabian Peninsula. Returning along the Red Sea, Conti ultimately reached Egypt, from whence he returned to the Mediterranean. After returning to Venice in 1439 he went to Florence where, sometime between then and 1443 in the context of the Ferrara-Florence Ecumenical Council, he encountered the humanist Francesco Poggio Bracciolini (1380-1459, from 1431 Apostolic Secretary of Pope Eugenius IV), who interrogated him on the long voyage he had undertaken.[52]

According to Bracciolini, Conti went to Florence to ask Pope Eugenius IV to be re-admitted to the Catholic faith after having been forced to recant, in order to save his wife and four

[51] This manuscript, although now at the Archivo Capitular in Toledo, did not reach Spain until the early nineteenth century following the death of its previous owner, Cardinal Francisco Xavier de Zelada, a native of Toledo.

[52] For the biography of Niccolò de' Conti, see the article by F. Surdich, *Dizionario biografico degli italiani*, vol. XIII (1971), pp. 640-646.

children, while *en route* on their return voyage in Muslim hands. Bracciolini, in what seems more literary invention than historical evidence, explained that Eugenius IV invited Conti to provide details about his voyage as a penance for his (forced) renunciation of Christianity.[53] Niccolò de' Conti's phenomenal voyage, well beyond the limits of the *oikumene* of ancient geographers, inspired Poggio to elaborate the story and to include it as the fourth, and last, part of a work that he was completing, the *De varietate fortunae*.[54] Although Book IV is attributed to Conti, it is doubtless the product of much editing and organization by Poggio Bracciolini. As the work's title states, at the centre of the *De varietate fortunae* is a theme highly debated by both classical and medieval authors: the concept of Fortune as mover of the great theatre of the world. The world, for Poggio, is subject to an endemic and irresoluble disorder: Divine Providence, and the cardinal virtues of both Platonic and Aristotelic philosophy, are not necessarily enough to protect from *fortuna*, *fatum* and *casus* even those who, stoically, know how to measure their own limits and to cultivate wisdom and virtue.

The *De varietate fortunae* was made public in 1448; however, Books II (dedicated to the history of the popes from Urban VI) and III (on the affairs of Eugenius IV) had already been prepared during 1442-43. Between 1444 and 1448 Poggio completed Books I and IV (on the ruins of Rome, and on the voyages of Niccolò de' Conti, respectively): narratives that, throughout the centuries, received the highest praise.[55] *Circa* forty-four manuscripts of the *De varietate fortunae* are recorded. Of these, twenty-one transmit all four books; twenty-three others leave us only Book IV. Before the combined diffusion of all four books in 1448[56] Book IV seems to

[53] "Nicolaus quidam Venetus, qui ad ultima Indiæ penetrauit, ad Eugenium ponteficem (is tum secundo Florentiæ erat) accessit, ueniæ impetrandæ gratia, quoniam, cum ab Indis rediens ad Ægypti fines mari Rubro peuenisset, fidem abnegare, neque suo tantum, quantum uxoris liberorumque, quos secum aduexerat, mortis metu, coactus est" *De varietate fortuna* IV.11-16 (ed. Guéret-Laferté, pp. 76-78). The translator of the 1502 Portuguese-language edition included it in his introduction and from there it has been handed down in the literature up to the present.

[54] For the history of the text, see Merisalo, *Introduzione, Appendici, Commento*, in Poggio Bracciolini, *De varietate fortunae*, pp. 9-83. The critical edition of the *Liber iv* and the commentary to the text are available in two editions: the one by Outi Merisalo to pp. 153-177 and pp. 225-248 of the edition previously mentioned and that of M. Guéret-Laferté, *Poggio Bracciolini (Le Pogge). De l'Inde. Les voyages en Asie de Niccolò de' Conti. De varietate fortunae livre IV*, Brepols: Turnhout, 2004. See also A. Grossato (ed.) *L'India di Nicolò de' Conti: un manoscritto del Libro IV del De varietate fortunae di Francesco Poggio Bracciolini da Terranova (Marc. 2560)*, Padua: Editoriale Programma, 1994. The bibliography of *De varietate fortunæ*, even if only for the aspects regarding the *Liber iv*, is very vast; see the edition by Merisalo (pp. 255-275) and by Grossato (pp. 97-98). Also very useful is R. Fubini and S. Caroti (eds), *Poggio Bracciolini nel VI Centenario della nascita. Codici e documenti fiorentini*, Florence: Biemme, 1980; R. Fubini, *Il "teatro del mondo" nelle prospettive morali e storico politiche di Poggio Bracciolini*, Florence: INSR, 1982. See also *Viaggio di Niccolò di Conti*, in G. B. Ramusio, *Navigazioni e viaggi*, pp. 781-819. Still valid are M. Longhena, *I manoscritti del iv libro del De varietate fortunae di Poggio Bracciolini contenenti il racconto dei viaggi di Niccolò de' Conti*, Rome: Reale Società geografica italiana, 1925 (extracted from Bollettino della R. Società geografica italiana, 1925); M. Longhena (ed.), *Viaggi in Persia, India e Giava di Niccolò de' Conti, Girolamo Adorno e Girolamo da Santo Stefano*, Milan: Alpes, 1929.

[55] As noted in R. Fubini, *Storiografia dell'umanesimo in Italia da Leonardo Bruni ad Annio da Viterbo*, Rome: Edizioni di storia e letteratura, 2003 pp. 148-149.

[56] Merisalo, *I codici*, pp. 25-82 (p. 81). The oldest copy of Book IV most likely dates to the year 1444 and is conserved in the manuscript Pl. XC sup. 55, ff. 27-37 of the Biblioteca Medicea Laurenziana in Florence. See Merisalo, *I codici*, n. 45, pp. 64-65. Among the oldest manuscripts Merisalo and Guéret-Laferté note Paris, Bibliothèque Nation-

have had a separate and striking success. This Book, particularly, amongst Bracciolini's work, saw a considerable circulation in the Veneto-Friuli area, as at least eight extant manuscripts bear witness.[57]

The tale of the voyage of Niccolò de' Conti is extremely rich. It contains descriptions of countries visited, customs of the peoples, information about plants and animals, personal anecdotes that happened to the traveler, as well as being a veritable mine of geographical knowledge of the Orient. The travel memoir of Niccolò de' Conti thereby constitutes a fundamental source of information, above all on India and the many islands of the Indian Ocean of which he describes rituals and traditions that until that time remained unknown. Bracciolini's recounting of Conti's voyage is appended to provide completion to, and verifications of, Polo's claims about southern Asia, as it is attested by the case of Fra Mauro's *mappa mundi*, but also of the *editio princeps* of Book IV, published by Cristoforo da Bollate in 1492 in Milan and entitled *India recognita* (The Indies Rediscovered)[58] as well as by the 1502 Portuguese edition in which Bracciolini's work was meant to complete the account by Marco Polo, the central work of the publication.[59]

The comparison between several passages taken from Bracciolini's recounting of the voyage of di' Conti and the cartouches of the mappa mundi supports the hypothesis that Fra Mauro plausibly read the original in Poggio Bracciolini's Latin and, other than the toponomastic citation, he summarized its contents, as was the case for the *Milione*. The following toponyms – *Macin, Regno de Scerno, Scierno, flumen Scierno, Bisenegal, Anagundi,* and *Urguer* – would all seem to have been derived exclusively from the *De varietate fortunae*.[60] As has already been emphasized, the study of their morphology further allows us to capture an important aspect of Fra Mauro's knowledge: the presence of oral knowledge and its interaction with the written tradition, a topic that will be confronted next in Chapter VI dedicated to the narrative structure of the *mappa mundi*.

ale, lat. 7854, ff. 73-138; Berlin, Staatsbibliothek, Hamilton 523; Oxford, Bodleian Library, Canon. Misc. 557, ff. 3-77v; London, British Library, MS Add. 8799, ff. 2-51v.

[57] The ample Venetian circulation appears to be attested by three manuscriptes: Padua, Biblioteca Civica, C.M. 207, ff. 1-52; Venice, Biblioteca Nazionale Marciana, lat. IV. 151 (= 2560), ff. 1-85v and lat. VI. 157 (= 3488); and San Daniele del Friuli, Guarneriana 121, ff. 31-51v.

[58] India Recognita. Christoforus bullatus ducis Isubrium senator Petro Carae ducis alobrogum Senatoris. Poggii Florentini de uarietate Fortunae [The Indies Rediscovered. Printed by Cristoforo da Bollate, Senator of the Duke of Milan, for Petro Caro, Senator of the Duke of Savoy. Poggio of Florence's *De Varietate Fortunae*], Milan: Uldericus Scinzenzeler, 1492. This edition is extremely rare: the Incunabula Short Title Catalogue of the British Library lists five extant copies (one of which at the British Library, IA.26738). Bollate appears to have edited Bracciolini's *Liber IV* to serve as a guide for Petro Caro, who was preparing to travel to India.

[59] Marco | Paulo. | Ho liuro de Nycolao veneto. | O trallado da carta de huu genoues das ditas terras. Cõ priuilegio del Rey mosso senhor... [Marco Polo. The Book of Nicolò the Venetian. A Translation of a Letter of a Genoese on the Said Lands]. Imprimido per Valentym fernandez alemao. Em a muy nobre çidade de Lyxboa, 1502. See *Livros antigos portuguezes da bibliotheca de sua Majestade fidelissima – Early Portuguese Books in the Library of His Majesty the King of Portugal: 1489-1600*, 3 vols., London: Maggs Bros, 1929-1935, vol. 1, pp. 111-157.

[60] *Macin* (Pl. XX, 27; *255, *258, *651, *656, *657, *663, *688, *706, *709, *728, *788); *Regno de Scerno* (Pl. XX, 51; *683); *Scierno* (Pl. XX, 56; *684,*685, *692, *722); *Flumen Scierno* (Pl. XX, 61; *684); *Bisenegal* (Pl. XV, 46; *312; *317, *329); *Anagundi* (Pl. XV, 84; *317; *330); *Urguer* (Pl. XVI, 23; *371).

II. – Patterns of textual derivation

1. Summarize, paraphrase, and annotate

In addition to citing their toponyms Fra Mauro made use of the *Milione* and *De varietate fortunae* by summarizing, paraphrasing and annotating them in the briefest of paragraphs. Indicative of Fra Mauro's method for synthesizing the text of Marco Polo's book is the following eloquent passage, a legend describing "The Old Man of the Mountain" :

> In ancient days here ruled a lord known as The Old Man of the Mountain, who through his cunning had created a place full of every delight and pleasure. Here he brought men and fooled them into thinking it was paradise. Out of devotion to this lord, these men then committed great robbery and murder, which was the reason why a Tartar lord seized control of this place.[61]

Among all the many versions of the *Milione*, the amount of space dedicated to "The Old Man of the Mountain" ranges from a minimum of a single long chapter (as in manuscript Z) to two long chapters (as in the Venetian manuscript tradition) ; in the Tuscan version, the story takes up as many as three chapters. Fra Mauro condensed this long narrative into just a few lines, retaining only a couple of details while omitting all the rest.[62] Also characteristic of this technique are the following brief legends : "Province of Çardandam. These people gild their teeth"[63] ; "In this mountain [in the province of Balasian] one finds a lot of balas rubies"[64] and, "In this gulf [the gulf of Malabar] pearls are fished."[65]

Drawing from the textual tradition Z Fra Mauro summed up in a single, lapidary line three of the longest chapters of the *Milione*, namely chapter lviii 'Quando vero itum est versus ponentem .v. dietis a província Caraiam discedendo…', chapter xxi 'Balaxia est provincia quedam…', and chapter cvii 'Hic narratur de provincia Maabar'.[66] In contrast to Marco Polo's

[61] "Qui antichamente dominaua uno signor dito el uechio | da la montagna, el qual per sua sagacità hauea fato | uno logo amenissimo de tute le delicie, piaceri e | deleti, e lì conduceua alguni homeni e aliena|uai p(er) modo li credeua q(ue)lo fosse el paradiso | e per questi piaceri a co(n)te(m)platio(n) de q(ue)sto signo(r) feuano gra(n) robarie | et occisio(n), p(er) la qual cossa | uno signor tartaro | prese questo luogo." *Mappamondo* (ed. 1956) Pl. XXVII, 43. The English translation derives from *Mappamondo* (ed. 2006) *1467.

[62] *Il "Milione" veneto. Ms. CM 211*, chs. xxviii-xxix, pp. 138-140 ; *Milione. Redazione latina del manoscritto Z*, ch. xvii, pp. 56-61 ; *Milione ; versione toscana del Trecento*, chs. xl-xlii, pp. 56-58.

[63] "P(rouincia) çardandam. | Questi po|puli se do|ra i de(n)ti." *Mappamondo* (ed. 1956) Pl. XXVI, 7 ; The English translation derives from *Mappamondo* (ed. 2006) *1379

[64] "In questo | monte se | troua ba|lassi assai." *Mappamondo* (ed. 1956) Pl. XXXIII, 10. The English translation derives from *Mappamondo* (ed. 2006) *2376.

[65] "In questo colfo | se pesca | perle." *Mappamondo* (ed. 1956) Pl. XV, 6 ; The English translation derives from *Mappamondo* (ed. 2006) *271.

[66] For the first citation, compare the passage in the Z manuscript in Toledo (here indicated with italics) : "Quando vero itum est versus ponentem .v. dietis a província Caraiam discedendo, invenitur quedam província nomine Çardandan, cuius gentes adorant ydola et sunt subdite Magno Can. Magistra civitas huius provincie Vocian appelatur. *Gentes ipsius dentes habent deauratos, videlicet quod omnes dentes sunt coperti auro* : nam fieri faciunt in similitudinem dentium unam formam de auro (…)." *Milione. Redazione latina del manoscritto Z*, ch. lviii, pp. 148-149. Due to the extreme length of the texts regarding Balasian and the Gulf of Malabar these passages have not been cited here (*Milione. Redazione latina del manoscritto Z*, ch. xxi, pp. 64-69 ; ch. ccvii, pp. 308-339). More wieldy are the following passages from *Historia delle cose de Tartari* (…) *descritta da M. Marco Polo Gentilhuomo Venetiano*,

remarkable descriptive prolixity, Fra Mauro gathered and selected just a few succinct elements – and sometimes one element alone. This tendency suggests that Fra Mauro worked much as an annotator does, using the shortest notes possible to encapsulate a text's long and complex contents. Guided by this analogy with the annotator's mode of highlighting a text's most significant aspects, we can deduce what Fra Mauro considered to be the most important aspects of his sources. It will be demonstrated below that when Fra Mauro consulted his textual sources, he usually focused on details regarding commerce, including three commodities in particular: spices, jewels and gold.

This procedure of both summarizing and evaluation was applied some thirty times to passages from Marco Polo's book, as well as to Book IV of *De varietate fortunae*.[67] Take, for example, the following two legends: "Bandan, a small island close to the shadows, on which grow a lot of cloves. Also one there finds parrots that are totally red, except for their beaks and claws, which are yellow"[68]; and "The Island of Sondai is near Bandan. Here grows nutmeg and other spices in great quantities. And here you can find parrots of seven different colors that are as big as pigeons; and again there is another sort, as big as seagulls, which are entirely white except for the beak and claws, which are red."[69] Both derive from (and rework) this passage in Book IV of *De vari-*

in G. B. Ramusio, *Navigazioni e viaggi*, Turin: Einaudi, vol. III, 1980: bk. II ch. 41 'Della provincia di Cardandan…', pp. 209-211; bk. I ch. 25 'Della provincia di Balaxiam…', pp. 116-118; bk. III ch. 20 'Della provincia di Malabar', pp. 267-274.

[67] While hardly an exhaustive list, the following locations pertain to a few of the *mappa mundi*'s passages with firm basis in the texts of the *Milione* and *De varietate fortunae*, arranged according to order of plates in the facsimile edition of 1956, and with reference to passages in Ramusio's edition of Marco Polo's book, curated by Marica Milanesi (MPR), and to Book IV in Outi Merisalo's edition of *De varietate fortunae* (DV). The study concerns plates III, IV, VIII, X, XIV, XV, XIX, XXI, and XXV of the facsimile. In plate VIII: "Nota | de la | nobilis|sima iso|la de say|lan e de | le sue mi|raueie" (VIII, 5): MPR III.19, *Dell'isola di Zeilan*, pp. 266-267 and DV IV.85. "Isola colombo. Questa è abondante | d'oro (…)" (VIII, 3*): MPR III.25, *Del regno di Coulam*, pp. 278-279. In plate IX: "Le naue ouer çonchi che nauegano questo mar…" (IX, 16): MPR III.1, *Dell'India maggiore, minore e mezzana…*, pp. 250-251. In plate III: "Mahal isola habita[da] per christiani…" (III, 6): MPR III.34, *Dell'isola di Soccotera*, p. 285. In plate X: "… e acostandose la naue (…) i marinari uedeno uno ouo de uno oselo nominato chrocho…" (X, A 13): MPR III.35, *Della grand'isola di Magastar*, pp. 286-287. In plate XV: "In questo cauo | de chomari se perde | la tramontana" (XV, 5): MPR III.25, *De Cumari*, p. 280. "Insula andama(n) in dromo de taprobana…" (XIV, 39): MPR III.18, *Dell'isola di Angaman*, p. 265. "Iso[*la de* | Sayla[*m* (XIV, 11): DV IV.85. "Isole lamuri, nauagari e arij…" (XIV, 8): MPR III.15, *Del quinto regno di Lambri*, pp. 263-264. "Nota che queli che nauegano questo ma(r) | de india …" (XV, 9): MPR III.25, *De Cumari*, p. 280; note here that Fra Mauro summarizes and cites twice the same information. In plate XV: "Isola hormus…" (XV, 69): MPR III.43, pp. 294-296 (note that Fra Mauro reduces three pages of the *Milione* to three lines) and DV IV, 44-45. "Isola sochotra posta tra el mar de persia…" (XVI, 26): MPR III.34, *Dell'isola di Soccotera*, p. 285. In plate XIX: "Giaua maçor, isola nobillissima posta i(n) leua(n)te…" (XIX, 16*): MPR III.7, *Dell'isola detta Giava*, pp. 257-258 and DV IV.111-19. In plate XXI: "In questa proui(n)cia | se troua turchese…" (XXI, 63): MPR I.13, *Del regno di Chiermain*, pp. 103-104. In plate XXV: "Questa nobillissima città dita chansay…" (XXV, M 30): MPR II.68, *Della nobile e magnifica città di Quinsai*, pp. 232-243. For the transcription and translation of both Fra Mauro's legends and Polo's excerpts, see *Mappamondo* (ed. 2006) *28, *27, *48, *4, *19, *264, *220, *32, *195, *282, *298, *385, *602, *792, *2240.

[68] "Bandan, isola picola propi(n)qua a le tene|bre, ne la qual nasce garofali assai. Item lì | se troua papaga' tuti rossi, saluo i piedi | e'l becho che son çali." *Mappamondo* (ed. 1956) Pl. XIII, 13. The English translation derives from *Mappamondo* (ed. 2006) *178.

[69] "Sondai insula propinqua a banda(n). In questa na|sce nose muscade et altre specie i(n) qua(n)tità. E qui se troua | papaga' de vij colori, grossi come uno | colombo, et ancora se ne troua un'altra sorte, | grossi chome uno

etate fortunae: "(...) altera Badan, nomine in qua sola gariofoli producentur deferunturque ad Iauas insulas. Badan triplices fert psitacos rubeis pennis croceoque rostro et uersicolores, quos noros hoc est lucidos ambos magnitudine palumborum et item albos gallinis pares."[70] In a similar instance, the brief legend, "This great city of Bisenegal can furnish 900,000 armed men,"[71] summarizes this passage from *De varietate fortunae*: "Profectus hinc est procul a mari miliaribus trecentis ad ciuitatem ingentem nomine Biçenegaliam ambitu miliarium sexaginta circa preruptos montes sitam. Conualles obductu ad montes muro claudunt ab ea parte urbem; ita eius circuitu ingentior. In ea censentur hominum qui arma ferre possunt milia nongenta."[72]

Fra Mauro's approach to the work by Poggio Bracciolini does not differ, therefore, from his approach to the *Milione*. Nonetheless, the legends drawn from Niccolò de' Conti's account tend to transcribe a greater quantity of the source text compared to those deriving from Marco Polo's writings. This circumstance can be explained by the fact that the citations from *De varietate fortunae* refer above all to Southeastern Asia and the Indian Ocean; because these two areas have only a sparse sprinkling of toponyms and drawings in contrast to continental Asia Fra Mauro had more space at his disposition for the elaboration of legends, as can be seen in the case of Taprobana's legend, whose central part reports a passage in *De varietate fortunae* almost *verbatim*:

Fra Mauro

Taprobana is a most noble island which is said to be more than 4,000 miles around. It is divided into four kingdoms, in which you find a lot of gold, pepper, cloves, aloe wood and a wood called galambech, which has the finest scent in the world and is sold for its own weight in gold. Here trees bear a fruit called the durian, which is the size of a reasonably big watermelon and has a green, knobbly skin, rather like a pine cone. The fruit inside is each one the size of a reasonably big pine

Bracciolini / Niccolò de' Conti

Tapobranem sexdecies centenis milibus passuum patere ambitu affirmat, uiri crudeles et moribus asperi (...) pipere reliquo maiore et item longo pipere, canfora et auro plurimo abundant. Piperis arbor persimilis edere; grana eius uiridia ad formam grani iuniperi, que modico cinere aspersa torrentur ad solem. Fructum uiridem habent nomine 'durianum' magnitudine cucumeris, in quo sunt quinque ueluti mala arancia oblonga uarii saporis instar butiri coagulati.[73]

cocal e tuti bianchi saluo | i piedi e'l becho che sono rossi," *Mappamondo* (ed. 1956) Pl. XIX, 14*. The English translation derives from *Mappamondo* (ed. 2006) *601.

[70] *De varietate fortunae*, IV.234-238, p. 162.

[71] "Questa | cità bise|negal gran|dissima pò | far 900000 | homeni d'ar|me," *Mappamondo* (ed. 1956) Pl. XV, 52. The English translation derives from *Mappamondo* (ed. 2006) *312.

[72] *De varietate fortunae*, IV.60-65. Merisalo, *Il Commento*, p. 228: "Vijayānagara, Città delle Vittorie, sul fiume Tungābhadrā, visitata da Conti all'epoca dei sovrani Devayagiri I (1406-24) e Devayagiri II (m. 1446), della prima dinastia di Vijayānagara, la Sangama".

[73] *De varietate fortunae*, IV.105-110, Merisalo pp. 156-157. Compare the sixteenth-century English translation by John Frampton: "It is estimated that Taprobana it is sixteenth passu in compass ... They are Idolatours, and have muche Pepper named the greatest, and of the long Pepper, and greate plentie of Camphore and golde ... there is a greene fruite named Duriano, of the bignesse of Cucumbers. And there be some of them lyke long Orengies or Lemans, of diverse favours and taste, as like butter, lyke milke, and like curdes," in *The Most Noble and Famous Travels of Marco Polo together with the Travels of Nicolò de Conti. Edited from the Elisabethan Translation* [1579] *of John Frampton with Introduction, Notes and Appendixes by N. M. Panzer*, London: Argonaut Press, 1929; repr. Amsterdam and New York, N. Israel – Da Capo Press, 1971, pp. 128-129.

cone and each one of them has its own pleasant taste; inside they are reddish-purple in colour and they are very warm. Here there are also a lot of elephants, and the people of this island live in better condition than those of the nearby islands; they are well-formed, strong, polite and good astrologers. They are also taller than those who are born in India, just as their elephants are bigger than those in India, and those in India are bigger than those in the Mauritanias. This is due to its fine location and air.[5]

The comparison between Fra Mauro's legend and *De varietate* results in another insight that is fundamental for understanding Fra Mauro's operational strategy, particularly in regard to the way he handles and reworks his sources. Even in the cases in which the source appears at first to be obvious – as in this specific case, where there is no doubt that the source is the passage cited from *De varietate* since the durian, a fruit that was unknown in the West, had never been cited in a literary source before Poggio's narrative of the travels of Niccolò de' Conti – the information supplied does not necessarily all derive from that same source. For instance, careful analysis reveals that the first part of the legend stands in contradiction to what is written in *De varietate*: according to Fra Mauro the outer edge of Trapobana is about four thousand miles, while for Bracciolini/de' Conti it is about sixteen-hundred miles. Also, in both the initial and the final parts of the legend, there is information that can be traced back, at least indirectly, to Pliny's *Natural History*.[75] It should be emphasized that none of this information derives from the *Milione*, Fra Mauro's principal source for the design of Asia, as this work does not even contain a description of Taprobana (it contains instead a description of *Samara*, or Sumatra, albeit in a form that diverges significantly from the one found in the *mappa mundi*). Thus, it can be concluded that Fra Mauro's reworking of his sources is always a vigorous operation, whether it consists of compiling information of diverse origins, or transmuting information found in intermediary sources.

In analyzing the representation of Asia in Fra Mauro's work, we can grasp the complexity of his methodology, one which pertains to his representation of other regions and the writing of the cosmographic legends placed in the outer corners as well. Studies carried out since the nineteenth century to determine the toponyms and identify the correspondences in the works of Marco Polo and Bracciolini/de' Conti furnish exegetical models that, in most cases, should

[74] "Taprobana isola nobilissima, la qual se dice uolta mia più de 4000 et è diui|sa in quatro regni, ne la qual setroua oro assai, piper, ganfora, legno aloe, el | qual se chiama gala(m)bach e ha odor più nobel sia al mondo e uendesse a peso d'oro. | E qui nasce i(n) arbori uno fruto chiamado duria(n), è de grandeça de una raso|neuel anguria e ha el scorço uerde e gropoloso come la pigna et ha de(n)|tro v fructi, chadauno de gra(n)deça de una rasoneuel pigna e chadauno de | questi v fructi hano differente suauità de sapor, e sono de(n)tro de colo(r) | paonaço e sono molto calidi. Qui se trova elefanti assai, e hi ho|meni de questa isola sono de mior condicio(n) cha queli de le altre iso|le vicine e sono formosi homeni, forti e çentil e boni astrologi | ma idolatri. Sono anchora maçor de statura che queli che na|sceno in india e li lor elefanti maçor de queli de le indie, e queli de le indie maçor de queli che nasce ne le maurita|nie, e questo è per el suo optimo sito e bontà d'aiere," *Mappamondo* (ed. 1956) Pl. VII, 1*. The English translation derives from *Mappamondo* (ed. 2006) *23.

[75] Gaius Plinius Secundus, *Naturalis historia*, VI, 24.

not be applied *sic et simpliciter* to Fra Mauro's legends, unless with extreme caution. The legend for Taprobana examined above offers unequivocable demonstration of this. While it is true that *De varietate* serves as the source for part of what Fra Mauro recounts, the rest of the legend supplies information that does not derive from *De varietate* and which cannot be ascribed directly, with absolute certainty, to any specific source.[76]

2. Selective reading: the *Milione* and *De varietate fortunae* as handbooks for merchants

The preceding discussion raises a new question: how did Fra Mauro view Marco Polo's and Niccolò de' Conti's travel accounts? In other words: what concerns guided his reading of the *Milione* and *De varietate fortunae*? Which of their aspects did he integrate into the *mappa mundi* and which did he omit or pass over? A study of the legends indicates that Fra Mauro read the *Milione* and Book IV of *De varietate fortunae* as if they were both handbooks for merchants (*pratiche di mercatura*) and ship pilots' guides. Studies regarding the *Milione*'s literary form have already demonstrated the intertextual presence of merchants' handbooks and pilots' guides – serving as a kind of palimpsest – in the narrative structure of Marco Polo's work.[5] From the vast repertoire of topics treated by his sources Fra Mauro selected passages describing commerce in the Levant, Asia, and the Indian Ocean, and he gave greatest attention to any passages indicating the origins of spices. This information, as has been demonstrated in the conclusion of Chapter II above, is compiled in the representation of five principle commercial routes that commence in the Indian Ocean, intersect in the Middle East, and continue to Europe. As a result of a selective reading of his sources Fra Mauro is able to furnish details about the great markets of Europe as well as of Asia; a wealth of information about the Levant; and an extraordinary description of the principle commercial activities, trade routes, and products exported from the Indian Ocean.

Above all in regard to the *Milione*, Fra Mauro took scant interest in his source's lengthy accounts of battles, its anecdotes, its parables, its hagiographies, and its anthropological and ethnographic descriptions (such as Polo's detailed reports about the religion, the sexual practices and the kinship structures of the various peoples he encountered). As an exception to this general rule there are occasional descriptions in the *mappa mundi* of the peoples of Asia, the opulence of various Asiatic cities, the court of the Khan, and even certain religious practices – namely, funerary rites of the "idolaters."[77] Fra Mauro ignored completely those parts of Marco Polo's book in which Rustichello da Pisa's prose and stylistic traits inflect it with the characteristics

[76] Cf. the recension of the several classical interpretations and descriptions of Taprobana that the Middle Ages inherited from Antiquity. See A. Herrmann, 'Taprobane', in *Paulys Real-Encyclopädie der Classischen Altertumswissenschaft. Neue Bearbeitung begonnen von Georg Wissowa* [...], 2nd ser., vol. IV [Stuttgart, 1932], col. 2260-2272; D. P. M. Weerakoddy, "Greek and Roman Notices of Sri Lanka and their Historical Context," *Sri Lanka Journal of Humanities* 20 (1994), pp. 65-86; J. André and J. Filliozat, *L'Inde vue de Rome. Textes latins de l'Antiquité relatifs à l'Inde*, Paris: Les Belles Lettres, 1986. For a different interpretation, see the commentary of the legend on Taprobana in *Mappamondo* (ed. 2006) *23.

[77] These pratices of the "idolaters" play a relevant part in the narrative structure of the *Milione*. See E. Burgio, "Marco Polo e gli 'idolatri'," in N. Pasero and S. Barillari (eds), *Le voci del Medioevo. Testi, immagini, tradizioni*, Alessandria: Edizioni dell'Orso, 2006, pp. 31-62.

of a courtly romance ; much in the same way as Fra Mauro discounted the figure of Marco Polo himself, who is depicted in the narrative as an errant medieval knight wandering through the immense territories of Asia, which in turn serve as a backdrop to the plot.[78] Fra Mauro treated the *Milione* and Bracciolini's account of de Conti's voyage as if they were the sort of merchants' handbooks sold in the market squares of Asia and the Levant, whose selected facts and tales he then attempted to distribute across the geographic framework of the *mappa mundi*.

The nature of Fra Mauro's approach to his sources, especially the *Milione*, supports the hypothesis formulated by Franco Borlandi, Enrico Vicentini and Ugo Tucci. Taking their cue from one of Luigi Foscolo Benedetto's intuitions, Borlandi, Vicentini and Tucci theorized that at the origin of Marco Polo's book there must have a merchants' handbook containing such information as the distances between cities and provinces, the transportation possibilities and the time they required, the origins and trade routes of commodities (particularly spices), measure and weight systems, and monetary exchange rates – all of which could have been compiled by Marco Polo during his long journey through Asia, most probably in the Venetian dialect.[79] According to these three scholars, this hypothesis is demonstrated by the fact that in the oldest surviving manuscript of the *Milione*'s textual tradition – namely, ms. fr. 1116 at the Bibliothèque nationale de France, dating from the first decade of the thirteenth century – Asian toponyms seem to have been transliterated according to an Italianate phonetic system (this is not the case in the other parts of the text, which were edited by Rustichello da Pisa in *langue d'oïl*) ; this fact perhaps indicates that at the origin of Marco Polo's work there existed some writings in an Italian dialect, which, in all probability, would have been Venetian.[80] The longest of the *mappa mundi*'s legends regarding merchandise reveals a structural affinity with precisely those parts of the *Milione* where Borlandi found the greatest similarity between its narrative structure and that of merchants' handbooks, such as the *Pratica di mercatura* by Francesco Balducci Pegolotti (1301-1350), which was compiled around 1340 (roughly forty years after Polo's work), yet which was still based on information gathered at the beginning of the previous century. Noteworthy in this context is the fact that Pegolotti's *Pratica di mercatura*, despite being devoted primarily to commerce in the Mediterranean, the Levant, and Northern Europe, begins with a description of the immense commercial route from Europe to Cathay – the route that is also the framework of Marco Polo's book.[81]

[78] Again, Luigi Foscolo Benedetto is to be credited with yet another valuable observation regarding the narrative form of Marco Polo's book : "l'arrivo dei Polo alla corte di Kubilai, il loro soggiorno, il loro ritorno a Venezia diventano un'avventura di cavaliere errante, ospitato per qualche tempo in qualche castello e poi rilasciato ai proprio destino". Cf. Benedetto, "Introduzione," p. xx ; see also, Bertolucci Pizzorusso, "Introduzione," in Polo, *Milione ; versione toscana del Trecento*, pp. 61-62.

[79] Benedetto, "Introduzione," pp. xxvi-xxvii ; F. Borlandi, "All'origine del libro di Marco Polo," in *Studi in onore di Amintore Fanfani*, vol. I, Milan : Giuffrè, 1962, pp. 107-147 ; U. Tucci, "I primi viaggiatori e l'opera di Marco Polo," in G. Arnaldi e M. Pastore Stocchi (eds), *Storia della cultura veneta*, I, *Dalle origini al Trecento*, Vicenza : Neri Pozza, 1976, pp. 633-670 (pp. 642-646) ; E. Vicentini, "Il Milione di Marco Polo come portolano," *Italica* 71, 2 (1994), pp. 145-152.

[80] Benedetto, "Introduzione," p. xi ; Borlandi, "All'origine del libro di Marco Polo," pp. 108-110. Barbieri, "Introduzione," in Polo, *Milione. Redazione latina del manoscritto Z*, pp. xiii-xvii.

[81] Borlandi, "All'origine del libro di Marco Polo," pp. 111-115.

3. Mapping the global spice trade

The *mappa mundi* mentions around thirty commodities and spices, mostly based on information gathered from the *Milione* and *De varietate fortunae*. These commodities and spices are listed below in alphabetical order:

aloes (*aloe*); aloes wood, called 'Galembech' (*legno aloe, el qual se chiama galambach*); amber (*ambra*); ambergris (*ambracan*); azur indigo (*açuro*); balas rubies (*balasi*); brocades (*pani d'oro e de seda*); camphor (*ganfora*); China porcelane (*porcelane*); cinnamon (*cinamomo*); cloves (*garofali*); diamonds (*diamanti*) durian (*duriano*); gemstones (*çoie, çoie nobilissime, çoie pretiose*; *piere pretiose*); ginger (*çençero*); gold (*oro*); incense (*incenso*); indigo (*endego*); iron (*ferro*); lapislazuli (*piera d'azuro*); manna (*manna, mana*); myrrh (*mirra*); naphtha (*nephto*); nutmeg (*nose muscade*); pearls (*perle, perle oriental*); pepper (*pevere*); pepper cubeb; rhubarb (*reobarbaro*); rubies (*rubini*); silk (*sede*; *sede thalai, sede chanarvi, sede siechi*); silver (*arçento*); steel (açalo); sugar (*zucharo*); turquoise (*turchese*).

Of all the varieties of merchandise mentioned in the *mappa mundi*, spices are the most prevalent. In the Middle Ages, 'spice' was a generic term that embraced a vast number of vegetable, animal and mineral substances. Florentine merchant Francesco Balducci Pegolotti compiled a list of 286 spices, organized according to a highly complex taxonomy; by comparison, of the roughly two hundred commodities cited in the *Milione*, some forty of these concern spices, and, besides these, there is a highly detailed description of metals and precious stones.[82] In addition to the spices listed in the table above, Fra Mauro indicates another sub-category of these substances: *specie nobile* (noble spices), which he also refers to as *specie sotil* (fine spices).[83] There is no consensus in the medieval sources on the definition of these two terms. According to Pegolotti, cardamom, Java pepper, and nutmeg were *spezierie minute* (fine spices), while ginger and pepper were *grosse* (bulky).[84] However, according to Marin Sanudo the Elder (*fl.* fourteenth century), fine spices were constituted by "omnia mercimonia modici ponderis & magni praetii, sive valoris: ut cubebe, spicum, gariofili, nuces moscatae, maci et hiis simila"; by contrast, ginger was classified "ut piper, cannella & simila hiis" among "mercimonia gravioris ponderis & minorisi praetii" as a *spezia grossa*.[85] In the *Pratica di mercatura* composed by the Datini, a merchant family of Prato, "zaferano, chubebe, gharofani, macie, spigho, noci moscade, scamonea, pepe lungho, carsamone, seta" all were categorized under the umbrella of *ispezierie sottili* [fine spices], the most costly category, and they were sold "in small amounts and paid for in real silver" (*libbre sottile e ariento vivo*).[86] But aside from the issue of these small discrepancies, it can be seen that Fra Mauro's conflation of these two commercial categories, "fine spices"

[82] F. Balducci Pegalotti, *Nomi di spezierie*, in *La pratica della mercatura*; A. Evans (ed.), Cambridge, MA: Harvard University Press, 1936, pp. 293-297 (reprint, New York, 1970); F. Brunello, *Marco Polo e le merci dell'Oriente*, Vicenza: Neri Pozza, 1986, pp. 35-61, 81-105.

[83] See *Mappamondo* (ed. 1956) XIII, 34*; *Mappamondo* (ed. 2006) *589.

[84] Balducci Pegolotti, *La pratica della mercatura*, pp. 293-297.

[85] Marin Sanudo il Vecchio, *Liber secretorum fidelium Crucis super Terrae Sanctae recuperatione et conservatione, quo et Terrae Sanctae historia ab origine et eiusdem vicinarumque provinciarum geographica descriptio continetur*..., Jerusalem: Massada Press, 1972, p. 23 (Partial reprint of the edition from Hannover, Typis Wechelianis apud heredes Ioannis Aubrii, 1611).

[86] C. Ciano, *La pratica di mercatura datiniana secolo XIV*, introd. F. Melis, Milan: Giuffré, 1964, p. 75.

and "noble spices," was based on an equivalency between the units of measure by which certain spices were priced and the inherent values of those spices that is found in Venetian Marin Sanudo's definition. According to Sanudo, 'sottili' described spices that were priced very highly and measured out in small units, such as saffron, mace, and cloves, while "grosse" applied to sugar and pepper, since these were sold in larger quantities and were relatively less expensive due to their greater availability.[87] The importance Fra Mauro attributed to spices and precious stones among the many commercial goods produced in the Far East reflects the central place these particular commodities had in Western trade and consumerism, until at least the end of the eighteenth century, based on four primary categories of usage: pharmaceutical ingredients (*medicamenta*), perfumes (*aromata*), pigments (*pigmenta*) and food condiments (*condimenta*).[88] As is well known, Venice was then the most important entrepôt in the Western world: here, spices coming from the Egyptian port of Alexandria were shipped out again by land and by sea towards northern Europe, the Iberian peninsula, and Flanders, from which points they penetrated profoundly into European commerce.[89] Venice maintained its primacy in this role until the seventeenth century, by which time the gravitational center of European commerce had already shifted some time previously to Antwerp, as a consequence of the new route around the African cape navigated by Vasco da Gama's Portuguese fleet in 1498.[90] Of course, a good portion of these imported spices remained in Venice, where they were transformed in the local *speciarie* (spice shops) into *medicamenta* and *condimenta* for retail sale. Both pharmaceutical and culinary recipe books of this period abound with references to *spezie alla veneziana* (Venetian spices), a term that indicates a mixture of ground spices consisting usually of pepper, cinnamon, ginger and saffron. The importance of the spice trade and the spice industry is witnessed by the exceptionally severe laws promulgated by Venice punishing not only smugglers but also the vendors who adulterated their products: to monitor this latter group, special officials called *stimatori* (assay-masters) inspected the spices to insure that they were high quality and unadulterated. They ensured, for example, that costly saffron was not fraudulently adulterated with almond oil, and that spice mixtures purporting to contain the highly valuable and aromatic "pevere tondo" (pepper corns) did not secretly substitute this ingredient with "pevere longo" (long pepper).

87 On the distinction between "sotil" and "grosse" spices in the context of the diverse classifications of spices in merchants' handbooks, encyclopedias, cookbooks, and medieval travel diaries, see A. Durel, *L'imaginaire des épices. Italie médiévale, Orient lointain* XIIe-XVIe *siècle*, Paris: Harmattan, 2006, pp. 32-34.

88 The literature on this topic is vast. Above all see the pioneristic W. von Heyd, *Histoire du commerce du Levant au Moyen-Age*, 2 vols., Leipzig: O. Harrassowitz, 1885-1886 (reprint Amsterdam: Hakkert, 1967); E. Ashtor, *Levant Trade in the Late Middle Ages*, Princeton: Princeton University Press, 1983; Brunello, *Marco Polo e le merci dell'Oriente*; A. J. Grieco, O. Redon, L. Tongiorgi-Tomasi (eds), *Le monde végétal (*XII*e-*XVII*e siècles), savoirs et usages sociaux*, Saint-Denis: Presses Universitaires de Vincennes, 1993. S. Halikowski Smith, *Portugal and the European Spice Trade, 1480-1580*, Ph.D. dissertation, Florence: European University Institute, 2001; Durel, *L'imaginaire des épices*.

89 H. G. Beck, M. Manoussacas and A. Pertusi (eds), *Venezia centro di mediazione fra Oriente e Occidente (secoli XV-XVI). Aspetti e problemi*, Florence: Olschki, 1977, vol. 1; F. C. Lane, "The Spice Trade," in Id., *Venice a Maritime Republic*, Baltimore and London: Johns Hopkins University Press, 1973, pp. 285-294.

90 A. Tenenti and C. Vivanti, "Le film d'un grand système de navigation: les galères vénitiennes, XIVe-XVIe siècles," *Annales: histoire, sciences sociales* 16, 1 (1961), pp. 83-86.

By selecting certain commercial topics and information from the travel accounts Fra Mauro perpetuated a perspective and an interest that were already strongly apparent in fourteenth-century documents such as the chart drawn by the Pizzigano brothers in Genoa in 1367, or the *Catalan Atlas* drawn around 1375.[91] With regard to this long-standing phenomenon, however, it is argued here that Fra Mauro's innovation was not his interest in the commercial exploitation of the East (this being a common denominator among many documents of this period). Rather, his innovation consisted in furnishing a vastly increased amount of detailed information – in other words, a more systematic and complete overview of the spice trade on a global level. This can be accounted for by two circumstances : on the one hand, he had access to Venice's extensive mercantile information network, one which Fra Mauro compared against, and expanded with, information drawn from classical sources, above all, Pliny's *Naturalis Historia* ; on the other hand, Fra Mauro had at his disposition more graphic space in comparison to other cartographic documents of this time, and this factor allowed him to describe with greater specificity and precision the products, markets, ports, and commercial routes that crisscrossed the Indian Ocean and Central Asia.

Once the *mappa mundi* is seen within the context of a continuous tradition of depicting the East according to mercantile interests and mercantile knowledge, then it becomes possible to recognize the significant documentation value that is inherent in Fra Mauro's transmission of mercantile knowledge from travel accounts to the *mappa mundi*. Through this operation Fra Mauro was able to perceive, and to enhance with a portrayal in spatial dimensions, the representation of commerce that had been passed down in written form in the so-called *Libri della mercatura*, such as those of Pegolotti and Canal, and in the merchants' writings that had served as the cornerstone of European trade since the twelfth century. Upon sifting out from the complexity of Fra Mauro's narrative those passages in which spices are mentioned there emerges what is, in fact, one of the first thematic cartographic representations of the production and trade of spices in the Indian Ocean, as understood from a mid-fifteenth-century Venetian perspective ; or, at the very least, as transcribed by Fra Mauro in the *mappa mundi*. Here is the map that results from this process :

Tinçu (Eastern China) : China porcelane.[92]
Fucui, *Valle fausta*, *Mar Breunto*(China)[93] : ginger, gold dust, and rhubarb.[94]
Paugin (Baoying in the county of Jiangsu, in China) : brocades.[95]
Isola di Saylam, *Isola di Colombo* (Sri Lanka), *Giava* (Java)[96] : ambergris, silver, rubies, gold, pepper, indigo, gemstones, and rubies.[97]

[91] A useful overview of the classical origins of commerical instructions regarding the Persian Gulf within the larger context of the Indian Ocean is P. Gautier Dalché, "Représentations antiques et médiévales (monde latin) du golfe Persique," in M. Taleghani, D. Silva Couto and J.-L. Bacqué-Grammont, *Cartographie historique du golfe Persique*. Actes du colloque organisé les 21 et 22 avril 2004 à Téhéran par l'EPHE, l'Université de Téhéran et le Centre de Documentation et de Recherche d'Iran réunis Téhéran, 2006 (Bibliothèque iranienne, 64), p. 35-60.

[92] Porcellane (XXV, 36).

[93] These place names are to be found in *Mappamondo* (ed. 1956) Pl. XXVI.

[94] Çençero and reobarbaro (XXVI, 80), oro de paiuola (XXVI, 91).

[95] Pani d'oro e de seda (XXXI, 31 : XXXII, 59).

[96] These place names are to be found in *Mappamondo* (ed. 1956) Pl. VIII.

[97] Ambracan (VIII, 2), arçento, rubini, oro, pevere, endego (VIII, 3*), oro, arçento, piere pretiose, rubini (VIII, 5).

Isola Sondai, *Giaua maçor*, *Sinus Gangeticus* (eastern side of the Indian Ocean basin)[98] : aloe, gold, finest gemstones, silver, nutmeg, gold, and "precious spices."[99]
Giava minor (Java the Less), *Isola Bandan* (Banda Island) *Isola Siamotra* (Taprobana)[100] : ginger, "noble spices," "fine spices," and cloves.[101]
Isola Siamotra (Taprobana) : aloes wood, called 'Galembech', durian, camphor, gold.[102]
Isola Abapaten (middle of the Indian Ocean basin)[103] : pearls and "oriental pearls."[104]
Isola Mahal (Maldives) : Amber.[105]
Arabia, *Arabia Sabea*, *Golfo de Chomari*, *Golfo di Milibar*, *Guçirat*, *Mare persicum*, *Mogolistan* (western side of the Indian Ocean basin)[106] : pearls, ginger, pepper, diamonds, myrrh, cinnamon, incense, precious stones and metals.[107]
India seconda, *Macin*, *Pandon*, *Sefar* (the regions between India and South-West China, including Bengala and Birmania)[108] : gold and fine rubins.[109]
India prima : diamonds, lichnites, beryls, chrysoberyls, chrysoprases, jacinths.[110]
Cremania, *città di Chobinam* (India, province of Kerman, city of Kuhbonan, in Southern Iran)[111] : turquoise, azur indigo, feather works, iron, stell, indigo.[112]
Provincia del Pamir, *Balch* (Central Asia, between Tajikistan and the province of Badakshan, Afghanistan)[113] : silver, lapislazuli, balas rubies, and pearls.[114]
Provincia de Siroan, *Lago Insical* (Central Asia, region of Shirvan in modern day Azerbaijan)[115] : naphtha and pearls.[116]
Sacara (tip of Africa) : manna.[117]

The way in which Fra Mauro read and interpreted the *Milione* and Book IV of *De varietate fortunae* allows us to recognize and understand the attraction held by these two accounts (according to legend, Marco Polo's book was "chained down" in the Rialto, Venice's mercantile center). In working out a synthesis of these two works, focusing particularly on information

98 These place names are to be found in *Mappamondo* (ed. 1956) Pl. XIX.

99 Legno aloe, oro, çoie nobilissime (XIX, 16*), arçento, oro, çoie, spetie (XIX, 12), nose muscade, specie (XIX, 14*).

100 These place names are to be found in *Mappamondo* (ed. 1956) Pl. XIII.

101 Çençero, specie nobile, specie sotil (XIII, 34*), garofali (XIII, 13).

102 Aloe, detto Gambalach, durian, ganfora, oro (VII, 1*).

103 These place names are to be found in *Mappamondo* (ed. 1956) Pl. XIV.

104 Perle, perle oriental (XIV, 20).

105 Ambra (III, 6).

106 These place names are to be found in *Mappamondo* (ed. 1956) Pl. XV and Pl. XVI.

107 Manna (XVI, 77) ; perle (XV, 6), çençero, pevere (XV, 36), diamanti (XV, 82), mira, cinamomo, incenso, piere pretiose, metali (XVI, 33).

108 These place names are to be found in *Mappamondo* (ed. 1956) Pl XX. On the geographic meaning of *Macin* see *Mappamondo* (ed. 2006) *258.

109 Oro, (XX, h 4), rubini fini (XX, 121), rubini (XX, 70)

110 Diamanti, lichiniti, berrilli, chrisoberilli, chrisopassi, iacincti (XXI, 57)

111 These place names are to be found in *Mappamondo* (ed. 1956) Pl. XXI.

112 Turchese, açuro endego, lavori de piuma nobilissimi (XXI, 63), ferro, açali, endego (XXI, 31).

113 These place names are to be found in *Mappamondo* (ed. 1956) Pl. XXXII and Pl. XXXIII.

114 Arzento (XXXII, 85), piera d'azuro (XXXII, 49 ; XXXIII, 9), balasi (XXXIII, 12 ; XXXIII, 15).

115 This place name is to be found in *Mappamondo* (ed. 1956) Pl. XXVII.

116 Nephto (XXXIII, 42) ; perle (XXXIII, 56).

117 Manna (III, 17).

regarding commerce, Fra Mauro may very well have taken the point of view of the public for whom the *mappa mundi* was intended : Venetian, Portuguese, and Florentine merchants, those same social groups who would lead voyages of discovery and conquest with sponsorship from the Portuguese and, later, the Spanish crowns.[118]

III. – Fra Mauro's visual citation of Marco Polo's book

Spread across the two-dimensional space of the *mappa mundi* are a seemingly infinite number of cities, castles, bridges, temples, and ships, all portrayed with a striking degree of stylistic uniformity in terms of both form and color. As discussed above in Chapter II, the range of imagery found in this map can be divided into eleven categories : cities, temples, sepulchral monuments, bridges, ships, mountains, rivers, seas, borderlines (whether of provinces or inhabited areas), roads, and text legends in the form of cartouches. These eleven image-types are repeated frequently throughout the map, with no changes except in terms of their relative dimensions. The result is a sumptuous, uniform and well-ordered representation. The majority of these miniaturized depictions are stylized, conventional representations whose function is to indicate symbolically the outstanding characteristics and aspects of a territory – not to provide a realistic pictorial depiction of them – using an approach that, in large part, resembles that of modern cartography.

This last aspect indicates that the pictorial ability of Fra Mauro and the copyists who assisted him did not exceed the realm of technical drawing : they were incapable of representing the human figure, and even animal and vegetable forms. The only exception to this are the fish that abundantly populate the Indian Ocean, most of which are depicted large. They recall the depictions of fish in San Marco's mosaics, particularly those in the Zen Chapel, as well as the depictions of fish in fourteenth-century paintings such as Guido da Graziano's *The Calling of St. Peter* from around 1290.[119] There is a second important aspect that deserves attention in regard to the *mappa mundi*'s iconography. Fra Mauro completely ignored the *mirabilia*, or natural wonders, described in Solinus's and Pliny's works, the *Romance of Alexander*, and Jean de Mandeville's *Livre des merveilles du monde*, as well as those mentioned in Marco Polo's book. He explicitly addressed his opinion on this matter when writing about the monsters that supposedly inhabited Africa :

> Because there are many cosmographers and most learned men who write that in this Africa – and, above all, in the Mauritanias – there are human and animal monsters, I think it necessary to give my opinion. Not because I want to contradict the authority of these men but because of the care

[118] F. Melis, "Di alcune figure di operatori economici fiorentini attivi nel Portogallo nel XV secolo," in L. Frangioni (ed.), *I mercanti italiani nell'Europa medievale e rinascimentale*, intro. H. Kellenbenz, Florence : Le Monnier, 1990, pp. 1-18 (pp. 15-16) ; L. D'Arienzo, "La presenza italiana in Portogallo e nella Spagna meridionale all'epoca di Cristoforo Colombo," in G. Cavallo (ed.), *Cristoforo Colombo e l'apertura degli spazi*, 2 vols., Rome : Istituto Poligrafico e Zecca dello Stato, 1992, vol. I, pp. 535-565.

[119] See Guido da Graziano's *San Pietro in trono. Annunciazione, Natività di Cristo e Storie di san Pietro*, gold and tempera on panel, 100,5 × 141 cm, Siena, Pinacoteca Nazionale, inv. no. 15 ; see also A. Bagnoli (ed.), *Duccio : alle origini della pittura senese*, Cinisello Balsamo : Silvana, 2003, pp. 87-91 and the illustration on p. 91.

I have taken in all these years in studying all possible information concerning Africa [...] *I have never found anyone who could give me information on what those men have written.* Thus, not knowing anything, I cannot bear witness to anything; and I leave research in the matter to those who are curious about such thing[s].[120]

He also made a similar comment regarding the *mirabilia* of India:

Some write that in these Indies there are many types of human and animal monster[s], *but because few people believe these things, here I have made no note of them*, except for certain animals, such as the serpents which are said to have seven heads. Again, here there are ants so very large that – something I will not dare to say – they seem to be dogs. These could be a species of animals that are similar to ants.[121]

In Fra Mauro's work monsters are treated as a purely literary phenomenon. Although he recognized that monsters had been written about by *cosmografi* and *doctissimi homeni* (by which he meant, respectively, the cosmographers of the Classical age such as Pomponius Mela, Pliny, and Solinus and the humanists who had studied these cosmographers' works), nevertheless, the existence of these monsters had not been confirmed by the direct knowledge of Fra Mauro's contemporaries: "I have never found anyone who could give me information on what those men have written," he declared resolutely. This issue is raised again in the second legend: "Because few people believe in these things, I have made no reference to them here." This last passage makes it clear that Fra Mauro's decision not to include the teratological information reported by ancient and modern authors was determined by his sensitivity to the interests of his public, who had come to regard this topic with skepticism. This passage is worthy of note because it refers explicitly to the attitudes of the public for which the work was intended – a public whose tastes and concerns dramatically shaped the form of its contents, in terms of both the text and the iconography.

[120] "Perchè sono molti cosmographi e doctissimi | homeni i qual scriueno che in q(ue)sta affrica, ma|xime ne le mauritanie, esserui molti mo(n)struo|si homeni e a(n)i(m)ali, p(ar)me neccessario q(ui) notar el parer mio, no(n) | p(er)hò che io uogli (con)tradir a le autorità de ta(n)ti, ma p(er) dir la dilige(n)tia ho | habuta i(n) inq(ui)rir tute le nouità se à possudo i(n)uestiga(r) p(er) molti anni de | q(ue)sta Africa [...] no(n) troui | mai alguno me ne sapesse dar auiso de q(ue)lo io t(r)ouo sc(r)ipto da q(ue)li; | vunde no(n) ne sapia(n)do altro no(n) ne posso testificar, lasso | a çerchar a q(ue)li che sono curiosi de intender tal nouitade." *Mappamondo* (ed. 1956) Pl. XXIII, 23; the English translation derives from *Mappamondo* (ed. 2006) *1043.

[121] "Alguni scriueno che in queste | Indie sono molte diuersi|tà de mo(n)stri sì de homeni come de a(n)i(m) ali, | ma perchè a queste cosse pochi dano fede | q(ui) no(n) ne faço nota, saluo che pur à certo de | alguni a(n)i(m) ali, come sono serpe le qual se dice hauer | vij teste. Ancora de qui sono formige gra(n)dissime | e quasi che q(ui) dir no(n) ardisco pareno cani. Que|sto può ess(er) che'l sia specie de a(n)i(m)ali che sia simile | a le formige." *Mappamondo* (ed. 1956) Pl. XX, 68. The English translation above is mine. The bibliography for *mirabilia* in medieval and Renaissance literature is vast; on the influence of the so-called *Romanzo di Alessandro* upon literature of the Trecento and the Quattrocento, see the now classic studies by P. Meyer, *Alexandre le Grand dans la littérature française du Moyen Age*, 2 vols., Paris, 1886, and Olschki, *L'Asia di Marco Polo*, pp. 45-50, 304-350. For Mandeville see C. Deluz, J. Mandeville, *Le livre des merveilles du monde*, Paris: CNRS, 2000; and R. Tzanaki, *Mandeville's Medieval Audiences: a Study on the Reception of the Book of Sir John Mandeville (1371-1550)*, Aldershot: Ashgate, 2003, pp. 83-134. In the *mappa mundi*, the story of Alexander is recounted in nine legends: (XXXIII, 77); (XXVI, 62); (XXI, 17); (XXVII, 95); (VIII, 5); (XXVI, 69); (XXXIII, 77); (XXVIII, 114); and (XXXII, 86).

At first glance the viewer's expectations of finding in Fra Mauro's imagery concrete relationships with the real world are frustrated. There are, however, two exceptions to this general statement: the galleons and ships that sail over the Mediterranean Sea, the Sea of Darkness, and the Indian Ocean, as well as an extraordinarily realistic and perspectival representation of the Masjid Al Haram, the Islamic sanctuary housing the Ka'ba. The Masjid Al Haram is located in Mecca, Islam's holiest city and the destination of the annual Hajj pilgrimage; Fra Mauro must have seen a realistic pictorial representation of this Islamic landmark given the surprising correspondence it bears in its many details to the actual building. Unfortunately, any surviving documentary evidence cannot confirm this.

Upon closer examination of the *mappa mundi* it is possible to discern among its numerous drawings a few significant pictorial representations that, despite not having been observed directly by the cartographer, constitute painstakingly faithful visual transpositions of information he had culled from his sources, clearly surpassing the mere symbolic function of the majority of the map's imagery. In other words, Fra Mauro sought to synthesize and to depict faithfully what he found in literary and cartographic sources, as well as what he had learned from the eye-witnesses he interviewed. To express these findings he employed a pictorial realism that could be considered "conjectural"; in the absence of direct visual knowledge of the object he wished to depict he was obliged to imagine a visual form that could faithfully render the verbal knowledge at his disposal. This is especially evident when we compare his representation of Asia with the contemporary conceptions of the geographic region comprising *Cathay* (northern China), *Mangi* (southern China), India both above and below the Ganges, and the Indian Ocean. Some of the images that Fra Mauro and his assistants illustrated in this region of the map constitute veritable pictorial citations – sometimes in a literal fashion and other times in more creative ways – of the texts of the *Milione* and Book IV of *De varietate fortunae*.

When visually citing passages from Marco Polo's and Niccolò de' Conti's books Fra Mauro preferred to represent cities, bridges, boats and landscapes, while ignoring the inhabitants and fauna as well as the monsters and natural wonders. Only in one case is it possible to perceive a human figure: under the arch of a "noble building for idols" – that is, a temple inspired by a passage in the *Milione* – there appears a stylized and crudely drawn human figure that could also be interpreted as a statue.[122] Let us consider and analyze in detail five of the *mappa mundi*'s representations that at first sight would seem to be the fanciful inventions of its creator, but which turn out instead to be scrupulously faithful and realistic citations of specific passages from Marco Polo's and (in one case) Niccolò de' Conti's books.

1. The junks in the Indian Ocean

A prime example of Fra Mauro's faithful and realistic transposition of passages from Marco Polo's book can be found in the drawing of the four-masted junks in the Indian Ocean (*çonchi da India*). In the *mappa mundi* sixty-three ships of varying tonnages and conformations are depicted. Of these, the four-masted ships with a single rudder and a flat keel are an unmistakable reference to the Chinese junks described by Marco Polo (plate I.1). Their image is accom-

[122] "Hedificio no|bile di idoli" *Mappamondo* (ed. 1956) Pl. XXXII, 39; *Mappamondo* (ed. 2006) *2286.

panied by a legend that describes the particular characteristics of these ships and refers to the astronomical navigation used to direct these ships across the Indian Ocean:

> The boats – or junks – that navigate this sea have four masts, as well as two that can be raised and lowered; they have 40 to 60 cabins for the merchants, and they have a single rudder. These ships sail without a compass because they take on board an astrologer, who is set alone, high up in the ship, with an astrolabe in his hand, and he gives orders to the helmsman.[123]

A comparison of the drawing, Fra Mauro's legend, and the following passage from the *Milione* clearly indicates that the last of these served as the source:

> And first let us speak of the ships in which merchants go to and from amongst the Isles of India. These ships, you must know, are of fir timber. They have but one deck, though each of them contains some 50 or 60 cabins, wherein the merchants abide greatly at their ease, every man having one to himself. The ship hath but one rudder, but it hath four masts; and sometimes they have two additional masts, which they ship and unship at pleasure. [...]Each of their great ships requires at least 200 mariners. They are indeed of great size, for one ship shall carry 5000 or 6000 baskets of pepper.[124]

Fra Mauro, who surely never saw a Chinese junk, attempted to represent them realistically, following Marco Polo's description almost literally. The four masts, the single rudder, and the quadrangular form of the keel all serve as distinguishing characteristics of junks with respect to the bilge-keeled ships of the Western and Arabian marine traditions. Their proportions – oversized with regard to the rest of the map – enabled Fra Mauro to represent many of the details he culled from Marco Polo: for example, one clear indication of Fra Mauro's desire to remain as faithful as possible to the *Milione* is the fact that he went so far as to depict even the individual portholes of the merchants' cabins on these vessels.

Fra Mauro was not the first to give a visible form to the *Milione*'s description of the junks. Depicted in the Catalan Atlas, dating from around 1375, are curious vessels in the Indian Ocean with mat sails and multiple masts, perhaps inspired by the *Milione*'s account. The Catalan Atlas's draughtsman, however, made it difficult to recognize this distinctive vessel-type because he

[123] "Le naue ouer çonchi che nauegano questo mar por|tano quatro albori e, oltra de questi, do' che se può | meter e leuar et ha da 40 in 60 camerele p(er) i mar|chadanti e p(or)tano uno solo timo(n); le qual naue|ga sença bossolo, p(er)chè i portano uno astrologo el | qual sta i(n) alto e separato e co(n) l'astrolabio in | man dà ordene al nauegar." *Mappamondo* (ed. 1956) Pl. IX, 16; the English translation derives from *Mappamondo* (ed. 2006) *48.

[124] *The Book of Ser Marco Polo*, Vol. 2, Book III, Chapter I, "Of the Merchant Ships Of Manzi That Sail Upon The Indian Seas," pp. 249-250. See also Ramusio's version: "[P]oscia ch'abiamo contato di tante province terrene, com'avete udito, noi conteremo de le meravigliose cose che sono ne l'India. E coninceròvi a le navi, ove i mercatanti vanno e vegnono. Sapiate ch'elle sono d'un legno chiamato abeta e di zapino, ell'ànno una coverta, e 'n su questa coperta, ne le piúe, à ben 40 camere, ove in ciascuna può stare un mercatante agiatamente. E ànno uno timone e 4 àlbori, e molte volte vi giungono due àlbori che si levano e pognono: le tavole so' tutte chiavate doppie l'una sull'altra co buoni aguti. (...) Queste navi voglion bene 200 marinai, ma elle sono tali che portano bene 5.000 sporte di pepe, e di tali 6.000. E' vogano co remi; a ciascun remo si vuole 4 marinai, e ànno queste navi ta' barche, che porta l'una ben 1.000 sporte di pepe. (...) Or v'ò contato de le navi che vanno per l'India." Marco Polo, *Historia dei Tartari*, III.1, *Dell'India maggiore, minore e mezzana*, pp. 250-251.

coalesced its multiple sails into a single, large, square sail, probably in order to make its form resemble European ships with their single square sails, with which he was familiar.[125]

Fra Mauro retains such details, and therefore the *mappa mundi* can be considered the first realistic representation of Chinese junks in the West. Of particular noteworthiness is the indication of the junks' multiple masts. According to Laird Clowes, "European vessels were then (in the first half of the fifteenth century) essentially single-masted, and although some two-masted ships had existed and probably still existed in the Mediterranean, European vessels of three masts were then not known, nor have we any record of them until 1466, some fifteen years later than Fra Mauro's map."[126] The disproportionately large depiction of the junks in the Indian Ocean with their single rudders and, most importantly, their four clearly visible masts, demonstrated to the European public – or at least to a Venetian, Florentine or Portuguese public (in other words, those who had commissioned copies of the *mappa mundi*) – the radical diversity, the power, and the technical advancement of the Chinese, and with these indicators he simultaneously implied that maritime trade, even as far west as the Persian Gulf, was practically monopolized by the Chinese. It was the introduction of the three-masted ship to Europe, with its improved ability to contend with adverse winds, that ultimately rendered possible the great voyages of discovery in the late fifteenth century – including those of Columbus to the West Indies, Vasco da Gama to India, and the Caboto brothers to Newfoundland.

2. The palace in Cambaluc and the Khan's hunting tents

Another example of Fra Mauro's practically literal visual citation of Marco Polo's account can be found in the depiction of the city of Cambaluc (a derivation from the Turk "Khàùn bàlêq" meaning 'City of the Khan', present-day Beijing) straddling a river, and in the depiction of the Khan's hunting tents (plate I.2). These representations of Cambaluc and the three hunting tents take on greater significance in light of the following two passages from Marco Polo's text :

> The city of Cambaluc is placed on a huge river in Cathay (…) And I tell you that [this] huge river enters and leaves [from Cambaluc] and is so well controlled that not even a single fish can escape from it […][127] :

> And when he [the khan] has travelled till he reaches a place called Cachar Modun, there he finds his tents pitched, with the tents of his Sons, and his Barons, and those of his Ladies and theirs, so that there shall be full 10,000 tents in all, and all fine and rich ones. And I will tell you how his own quarters are disposed. The tent in which he holds his courts is large enough to give cover easily to a thousand souls. It is pitched with its door to the south, and the Barons and Knights remain in waiting in it, whilst the Lord abides in another close to it on the west side. When he wishes to speak with any one he causes the person to be summoned to that other tent. Immediately behind the great

[125] See G. S. Laird Clowes, "Ships of Early Explorers," *The Geographical Journal* 69, 3 (1927), pp. 216-230 (pp. 222-223).

[126] Laird Clowes, "Ships of Early Explorers," p. 223.

[127] This is my translation of "La città di Cambalú è posta sopra un gran fiume nella provincia del Cataio (…) E sí vi dico che un grande fiume v'entra e esce, e è sí ordinato che niuno pesce ne puote uscire." Polo, *Historia dei Tartari, ibid.*, II.7, *Della nuova città di Taidu*, pp. 164-165.

> tent there is a fine large chamber where the Lord sleeps; and there are also many other tents and chambers, but they are not in contact with the Great Tent as these are. The two audience-tents and the sleeping-chamber are constructed in this way. *Each of the audience-tents has three poles, which are of spice-wood, and are most artfully covered with lions' skins, striped with black and white and red, so that they do not suffer from any weather. All three apartments are also covered outside with similar skins of striped lions, a substance that lasts for ever. And inside they are all lined with ermine and sable, these two being the finest and most costly furs in existence.*[128]

These citations clearly demonstrate Fra Mauro's attempt to give a realistic visible form to Marco Polo's text: the details of the three richly decorated tents and the river that enters and leaves Cambaluc accurately reflect some of the specific information provided in Marco Polo's long descriptions of the capital of the Mongol empire and the hunting customs of the Khan. The descriptions of the Khan's wintertime residence in Cambaluc and his summertime hunting grounds are found in different chapters of Marco Polo's book (*Historia dei Tartari*, II.7-II.16); yet Fra Mauro's unrelenting process of synthesis transforms and then joins them in a single representation, so that the winter palace in Cambaluc and the sumptuous summertime hunting grounds become a single drawing.[129]

3. The bridge over the Polisanchin River and the roads of Mangi and Cathay

The region of Cathay, based entirely on Marco Polo's account, contains a remarkable example of visual citation: the bridge over the Polisanchin River (in contemporary nomenclature, the famous Lugou Bridge over the Yongding River) – one of the largest of the *mappa mundi*'s images – together with the road which passes over the bridge and proceeds southwards before forking towards the west and the Mangi region (plate I.2). Both the bridge and the road are literal representations of passages in the *Milione*:

> When you leave the City of Cambaluc and have ridden ten miles, you come to a very large river which is called Pulisanghin, and flows into the ocean, so that merchants with their merchandise ascend it from the sea. Over this River there is a very fine stone bridge, so fine indeed, that it has very few equals. The fashion of it is this: it is 300 paces in length, and it must have a good eight paces of width, for ten mounted men can ride across it abreast. It has 24 arches and as many water-mills, and 'tis all of very fine marble, well built and firmly founded. Along the top of the bridge there is on either side a parapet of marble slabs and columns, made in this way. At the beginning of the bridge there is a marble column, and under it a marble lion, so that the column stands upon

[128] "E quando ha uccellato per alquante ore, se ne viene ad un luogo chiamato Caczarmodin, dove sono le trabacche e i padiglioni de suoi figliuoli (...). Il padiglione veramente del signore, nel quale tiene la sua corte, è tanto grande e amplo che sotto vi stanno diecimila soldati (...) e v'è ancora un'altra tenda verso levante, a questa congiunta, dove è una gran sala dove stanzia il signore con alcuni suoi baroni (...) Sonvi molte altre tende e camere (...) e sale sono ordinate in questo modo, che ciascuna ha tre colonne di legno intagliate con grandissimo artificio e indorate. E detti padiglioni e tende di fuori sono coperte di pelli di leoni, e vergate di verghe bianche, nere e rosse, e cosí ben ordinate che né vento né pioggia li può nuocere; e dalla parte di dentro sono fodrate e coperte di pelli armelline e zibelline (...)," Polo, *Historia dei Tartari*, *ibid.*, II.16. *Del modo che va il gran Can a veder volare li suoi girifalchi*, pp. 178-181.

[129] In reference to the three tents, Fra Mauro wrote: "Qui l'imperador sta l'insta|de a sua diletevele uena|tion," *Mappamondo* (ed. 1956) Pl. XXXII, 89; *Mappamondo* (2006) *2343.

> the lion's loins, whilst on the top of the column there is a second marble lion, both being of great size and beautifully executed sculpture. At the distance of a pace from this column there is another precisely the same, also with its two lions, and the space between them is closed with slabs of grey marble to prevent people from falling over into the water. And thus the columns run from space to space along either side of the bridge, so that altogether it is a beautiful object.[130]

In addition to the drawing, a brief legend placed above the bridge summarizes Marco Polo's description: "Marvelous and famous bridge across the river Polisanchin. It has three hundred arches and six thousand images of lions bearing an equal number of columns with capitals."[131] This text legend and the image of the bridge in the *mappa mundi*, when compared with Marco Polo's description, demonstrate the existence of a relationship of "creative dependence" between the map and the *Milione*. Whereas Fra Mauro's legend exaggerates Marco Polo's description out of proportion – namely, the *Milione* speaks of twenty-four arches, but the *mappa mundi* legend speaks of three hundred arches – the drawing of the bridge, by contrast, is more faithful to the source. In the drawing of the bridge, twelve arches are shown, and though this is half the number given in the *Milione*, the drawing succeeds in conveying the exceptionality of the bridge's architecture; moreover, the viewer is able to perceive the details of the marble parapet and the ornamental statues. In addition, the drawing attentively reproduces the particular aspects of the column that Marco Polo described with great emphasis, and which Fra Mauro locates at the foot of the bridge.

Viewed from a broader perspective, the region depicted near the bridge over the *Polisanchin* River includes further elements that help demonstrate Fra Mauro's method of subjecting the *Milione* to a process of visual citation. When considering the representation as a whole, the proximity of the bridge to the city of *Cambaluc* is clearly apparent, and in fact Marco Polo had indicated this distance to be merely ten miles. Just a little to the south, beyond the cities of *Çianglu* and *Chaicianfu*, Fra Mauro drew a road which shortly thereafter divided into "the Road to Mangi" (*strada de Mango*)[132] and "the Road to Cathay" (*strada del Cataio*). Once again, this is an instance in which the visual representation conforms literally and chorographically to Marco Polo's account: "Thirty miles beyond the bridge there is a city called Jonju (Cho-chou). It contains many monasteries for idolaters. Beautiful gold and silver cloths and very fine sandals are made there." And further on in the text: "one mile out of Jonju, one finds two roads, one

[130] *The Book of Ser Marco Polo*, Book II, Chapter XXXV, "Here begins the description of the interior of Cathay, and first of the river Pulisanghin," pp. 3-4. See also Ramusio's version: "Come si parte dalla città di Cambalú e che s'ha camminato dieci miglia, si truova un fiume nominato Pulisangan (…) Sopra detto fiume è un ponte di pietra molto bello, e forse in tutt'il mondo non ve n'è un altro simile. La sua longhezza è trecento passa e la larghezza otto, di modo che per quello potriano commodamente cavalcare dieci uomini l'uno a lato all'altro. Ha ventiquattro archi e venticinque pile in acqua che li sostengono, ed è tutto di pietra serpentina, fatto con grand'artificio. Dall'una all'altra banda del ponte è un bel poggio di tavole di marmo e di colonne maestrevolmente ordinate (…) Ciascune colonne sono distanti l'una dall'altra per un passo e mezo, e a ciascuna è sopraposto un leone, con tavole di marmo incastratevi dall'una all'altra (…)." Polo, *Historia dei Tartari*, *ibid.*, II.27, *Del fiume Pulisangan e ponte sopra quello*, pp. 194-195.

[131] "Ponte mirabile e famoso co(n) tresento | archi e siemil a imagine de lioni | i qual reze tante collone co(n) i suo capi|telli a più suo adorname(n)to. | super el qual se trau(er)sa el fiu|me polisanchin." *Mappamondo* (ed. 1956) Pl. XXXII, 75; the English translation derives from *Mappamondo* (ed. 2006) *2315.

[132] *Mappamondo* (ed. 2006) *2287.

leading to the west, and the other to the south-east. The westerly road is that of Cathay, the other towards the great province of Man[g]i."[133] To all effects, the whole scene is a pictorial representation of Marco Polo's account. The trees along the road clearly correspond to the "nice houses, buildings, trees, and vineyards" described in the *Milione*. A little to the south, a brief inscription saying "noble building for idols"[134] – unmistakably inspired by the passage in the *Milione* that says, "here there are many temples for idols" – identifies an immense, four-story temple with arabesque decorations. Along with the bridge, the Chinese junks, and the fantastic sepulcher of the Khans near Mount Altai, it is among the largest and most eye-catching of the *mappa mundi*'s depictions.

4. The felt-covered carriages of the Tartars

Even Fra Mauro's drawings of nomads' "houses," felt-covered shelters that were mounted upon two- and four-wheeled carriages, both in Tartary and the northern region of *Rossia*, constitute visual citations of a passage in the *Milione* describing the customs of the Tartars (plate I.3). A comparison of these drawings with Marco Polo's text reveals beyond the shadow of a doubt the origin of these images :

> Their houses are circular, and are made of wands covered with felts. These are carried along with them whithersoever they go ; for the wands are so strongly bound together, and likewise so well combined, that the frame can be made very light. Whenever they erect these huts the door is always to the south. They also have waggons covered with black felt so efficaciously that no rain can get in. These are drawn by oxen and camels, and the women and children travel in them.[135]

Fra Mauro is not the only one to represent on a map the Tartars' felt-covered shelters that were mounted upon two- and four-wheeled carriages. The same carriages are engraved in the

[133] "Partendosi da questo ponte e andando per trenta miglia alla banda di ponente, trovando di continuo palagi, vigne e campi fertilissimi, si truova una città nominata Gonza (...). Partendosi da questa città e andando per un miglio si truovano due vie, una delle quali va verso ponente, l'altra verso scirocco : per la via di ponente si va per la provincia del Cataio, per la via di scirocco alla provincia di Mangi. E sappiate che dalla città di Gonza fino al regno di Tainfu si cavalca per la provincia del Cataio dieci giornate, sempre trovando molte belle città e castella, fornite di grand'arti e mercanzie, e trovando vigne e campi lavorati." Polo, *Historia dei Tartari*, II.27, *Delle condizioni della città di Gonza*, pp. 195-196.

[134] "Hedificio no|bile de idoli," *Mappamondo* (ed. 1956) Pl. XXXII, 31 ; *Mappamondo* (ed. 2006) *2286. The translation is mine.

[135] *The Book of Ser Marco Polo*, Book I, Chapter LII, "Concerning the customs of the tartars," pp. 251-252. See also Ramusio's version : "I Tartari non stanno mai fermi, ma conversano al tempo del verno ne' luoghi piani e caldi. (...) Hanno le case coperte di bacchette e feltroni e rotonde, cosí ordinatamente e con tale artificio fatte che le verghe si raccolgono in un fascio, e si ponno piegare e acconciar a modo d'una soma : quali case portano seco sopra carri di quattro ruote ovunque vadano, e sempre quando le drizzano pongono le porte verso mezzodí. Hanno oltre ciò carrette bellissime di due ruote solamente, coperte di feltro, e cosí bene che se piovesse tutt'il giorno non si potria bagnar cosa che fosse in quelle, qual menano con buoi e camelli. Sopra quelle conducono li loro figliuoli e mogli, e tutte le massarie e vettovaglie che li bisognano (...)." Polo, *Historia dei Tartari*, ibid., I.45, *Della vita de' Tartari (...) e delle lor case sopra carrette, ibid.*, pp. 136-137.

so-called "Borgia Mappa Mundi" (*c.* 1430-50).[136] Therefore, in the case of the Tatars' shelters, it is also possible to conjecture that, beside a direct reading of Marco Polo' text, Fra Mauro might have relied upon a well-known iconographic tradition.

5. The sepulchral shrine of the Khans and Andramania

A particularly curious case of visual citation – one that is also typologically distinct from the preceding examples – regards the mountain with a temple and the tombs of the Khans (plate I.4). In this instance the visual citation imaginatively expands upon a passage in Marco Polo's work. Whereas Marco Polo mentions only a mountain on which the Khans were buried :

> You should be told also that all the Grand Kaans, and all the descendants of Chinghis their first Lord, are carried to a mountain that is called Altay to be interred. Wheresoever the Sovereign may die, he is carried to his burial in that mountain with his predecessors ; no matter an the place of his death were 100 days' journey distant, thither must he be carried to his burial... Let me tell you a strange thing too. When they are carrying the body of any Emperor to be buried with the others, the convoy that goes with the body doth put to the sword all whom they fall in with on the road, saying : "Go and wait upon your Lord in the other world !"[137]

Fra Mauro instead imagined the sepulchers as being located in a magnificent temple – the largest of all the temples in the *mappa mundi* – situated on the peak of a mountain. Under the topographic rubric "Mountain called Alchai"[138] Fra Mauro follows with a "visual expatiation" on Marco Polo's account. He envisioned an imperial tomb, which was not described by Marco Polo, and he imagined it as being wrought in an improbable European Gothic style that dramatically conveys the magnificence and power of these emperors whom Marco Polo had characterized in other earlier passages of his book.

IV. – Reconsidering the "marvels of the East"

Analysis of the *mappa mundi*'s visual citations of the *Milione* is significant not only for understanding how Fra Mauro makes use of Marco Polo's work, but also for undertaking a critical re-evaluation of the interpretations formulated by art historians in the wake of Rudolf

[136] Vatican City, Biblioteca Apostolica Vaticana, Borgiano XVI, [Borgia World Map], engraved on two semicircular copper plates joined by nails : Austria, southern Germany or Bohemia, diameter *c.* 63 cm, South on the top, legends written in Latin, *c.* 1430. See Almagià, *Monumenta cartographica vaticana*, vol. 1, pp. 27-28, Plate 8 ; Destombes, *Mappemondes*, entry 52.10, pp. 212-214.

[137] *The Book of Ser Marco Polo*, Book I, Chapter LI,"Of those who did reign after Chinghis Kaan, and of the customs of the Tartars," p. 246. This is instead Ramusio's version : "E dovete sapere che tutti i gran Can e signori che descendono dalla progenie di Cingis Can si portano a sepelire ad un gran monte nominato Altay, e in qualunque luogo muoiono, se ben fossero cento giornate lontani da quel monte, bisogna che vi sian portati. E quando si portano i corpi di questi gran Cani, tutti quelli che conducono il corpo ammazzano tutti quelli che riscontrano pel cammino (...)." Polo, *Historia dei Tartari*, ibid., I.44, "(...) solennità che gli fanno quando li sepeliscono nel monte Altay," pp. 135-136.

[138] "Monte alchai," *Mappamondo* (ed. 1956) Pl. XXXVIII, 38.

Wittkower in regard to the pictorial representation of the East, especially the illustrations based on passages from the *Milione*. The miniatures that illustrate medieval manuscripts generally partake of long-established, codified iconographic traditions, and for this reason it is very difficult to detect instances in which artists have contributed something new to pre-existing codifications with their knowledge, aesthetic sensibility, or imagination. The conscious pursuit of innovation was not a characteristic of the medieval artisan's mentality, which was mostly focused on conforming to precedents rather than making new and original contributions. Only in the case of a few distinguished masters is it possible to identify the maker's independent creativity.[139] Taking this general context into account, even further challenges are posed by the miniatures that illustrate travel accounts such as the *Milione*, set in far-off places, because the illuminators of these texts did not have consistent iconographic traditions which to mine, and quite often they had no recourse to any iconographic tradition whatsoever – aside, of course, from the rich literary and pictorial tradition that was devoted to the monsters and natural wonders of the Orient. The question therefore deals with how miniaturists did approach the challenge of reproducing through a pictorial language those factual descriptive accounts of peoples, cities, fauna and flora that were unknown and exotic to them – things of which they knew only through written descriptions, and for which they had no iconographic precedent as a reference point.

According to Rudolf Wittkower, the two most celebrated illuminated manuscript of Marco Polo's work[140] – the so-called *Livre des merveilles*, Ms. fr. 2810 at the Bibliothèque nationale de France with eighty-four miniatures,[141] and Ms. Bodley 264, at the Bodleian Library in Oxford

[139] Only in the case of a few supremely distinguished masters is it possible to identify the maker's independent creativity. B. Brenk, "Originalità e innovazione nell'arte medievale," in E. Castelnuovo and G. Sergi (eds), *Arti e storia nel Medioevo*, vol. 1, *Tempi Spazi Istituzioni*, Turin : Einaudi, 2002, pp. 3-69.

[140] Nine illuminated manuscripts of the *Milione* have been recorded, but scholars have demonstrated that the majority of these present little interest with regard to the representative history of the Orient or to the study of relationships between text and image. This paucity of interest can be explained in a few cases by the scarce number of images contained in the manuscripts – such as Ms. fr. 5631 with only two miniatures at the Bibliothèque nationale de France, or Ms. 3511 with only a single miniature at the Bibliothèque de l'Arsenal in Paris. In other cases the reason is the late date of the manuscript ; for example, Ms. 5219 at the Bibliothèque de l'Arsenal, containing 197 miniatures, was executed in the first part of the sixteenth century. Finally, in yet other cases, the illuminators paid no attention to the particular details of the narrative, a situation witnessed by the nineteen miniatures of Ms. Royal 20 D.i at the British Library that bear no relation whatsoever to Marco Polo's text, ignoring it outright. For a codicological analysis of the nine manuscripts, see Benedetto, "Introduzione," pp. xxxiv-xlv. See also P. Ménard, "L'illustration du Devisement du monde de Marco Polo : étude d'iconographie comparée," in *Métamorphoses du récit de voyage*, Paris : Champion, 1986, p. 17-31.

[141] The Ms. 2810 at the Bibliothèque nationale de France is an anthology of eight voyages to the East entitled *Livre des merveilles*, compiled in Paris between 1410 and 1412 according to a commission by John the Fearless, Duke of Burgundy (Jean sans Peur, 1371-1419), who then gave the work in 1413 to his uncle, John of Valois, Duke of Berry (1340-1416). The miniatures were illuminated by the famous Parisian Boucicaut Master (the name Boucicaut derives from the identification of this artist's hand in a manuscript made for a patron named Boucicaut ; his real name remains a mystery to us). For a description of the manuscript, see F. Avril, *Das Buch der Wunder der Welt. Ms. fr. 2810 der Bibliothèque Nationale de France*, in *Das Buch der Wunder : aus "Le Livre des Merveilles du monde", Ms. fr. 2810 der Bibliothèque Nationale de France*, Übersetzung der Handschrift M-H. Tesnière. Kritischer Essay F. Avril. Bilderäuterungen M.-T. Gousset, Munich : Hirmen Verlag, 1999, pp. 197-216.

with thirty-eight miniatures[142] – owe their iconography primarily and simply to the Classical and Medieval literary and pictorial genre of *mirabilia* with their accounts of Oriental marvels, wonders and monsters, rather than to the specific details about the East that are found in the *Milione*.[143] In their illuminations – Wittkover argues – there emerges a stereotyped image of the "East" that bears no link to any concretely discernible geographic or cultural entity. Rather, this stereotype is rooted in representations that had been developed and utilized by European cultures in order to create, by implication, a contrasting sense of their own identity as the "West," through a process that reduced eastern cultures to conventional and generic clichés : that of the marvelous and the monstrous ; fabulous wealth ; as well as certain cruel and idolatrous practices. Even when Marco Polo offered especially detailed and realistic descriptions of the things he had observed – descriptions that very well could have served to inspire and guide an illuminator with no iconographic tradition from which to draw – instead it is the world of the marvelous and monstrous (represented according to the conventions of the Western canon) that predominates in these miniatures.[144]

[142] The Ms. Bodley 264 at the Bodleian Library in Oxford features Marco Polo's text (ff. 18r-217v) preceded by *The Romance of Alexander* and *Alexander and Dindimus* (*Alexander Fragment B*). The manuscript is available in its entirety in a high-resolution version on the internet : ‹http ://image.ox.ac.uk/show ?collection=bodleian&manuscript=msbodl264›. On f. 208r, the *scriptor* and the illuminator each placed his respective *subscriptio* indicating when the work was completed : "Chi define li romans du boin roi Alixandre | Et les vens du pauon les acomplissemens | Le restor du pauon et le pris qui fu perscript | le XVIIIe ior de decembre lan MCCCXXXVIII." Below, on the same page, is the *subscriptio* of the illuminator : "Che liure fu perfais de le enluminure | au XVIIIe jour dauryl per Jehan de | Grise. Lan de Grace MCCCXLIIII."

[143] R. Wittkover, "Marco Polo and the Pictorial Tradition of the Marvels of the East," in *Oriente poliano*, Studi e conferenze tenute all'ISMEO in occasione del vii centenario della nascita di Marco Polo (1254-1954), Rome : Ismeo, 1957, pp. 155-172 (p. 155) ; Id., "Marvels of the East. A Study in the History of Monsters," *Journal of the Warburg and Courtauld Institutes* 5 (1942), pp. 159-197, reprinted in M. Wittkower (ed.), *Allegory and the Migration of Symbols*, London : Thames and Hudson, 1977, pp. 75-92.

[144] A prime example of this approach – Wittkover argues correctly – regards the depiction of *leoncorni*, or unicorns, in the f. 85 of the Ms. fr. 2810 at the Bibliothèque Nationale de France. In various passages of the *Milione* (in particular in the chapter "Concerning the Island of Java the Less. The Kingdoms of Ferlec and Basma) Marco Polo describes an animal that was unknown in the West : the rhinoceros. Because it has a single horn, he called it *leoncorno*, a term that is a medieval variant of 'unicorn', referring to the famous mythological animal for which there existed a substantial literary and iconographic tradition in the form of a magical white horse with a single, long, twisting horn growing from its head that could only be tamed by a virginal maiden, with which the unicorn is almost always shown (cf. the late-fifteenth-century Flemish tapestries at the Musée de Cluny in Paris, known collectively as *The Maiden with the Unicorn* ; see A. Erlande-Brandenburg, *La Dame à la licorne*, Paris : M. Aveline, 1993). Aware of this literary and iconographic tradition of the unicorn and, at the same time lacking a visual representative model of the animal in question, Marco Polo first sought to assist his readership in imagining the true appearance of the animal he named after the unicorn by pointing out its resemblances to some of the animals known in Europe (such as elephants, water buffalos, and boars), and then by emphasizing that it was not to be identified with the unicorns of the tradition : "There are wild elephants in the country, and numerous unicorns, which are very nearly as big. They have hair like that of a buffalo, feet like those of an elephant, and a horn in the middle of the forehead, which is black and very thick. […]. The head resembles that of a wild boar, and they carry it ever bent towards the ground. They delight much to abide in mire and mud. 'Tis a passing ugly beast to look upon, and is not in the least like that which our stories tell of as being caught in the lap of a virgin ; in fact, 'tis altogether different from what we fancied" (cf. *The Book of Ser Marco Polo the Venetian*, Book 3, Ch. 9. See also Polo, *Historia dei Tartari*, III.12, *Del secondo regno di Basma*, pp. 260-261). Despite his precise and detailed description, that above all attempts to prevent the

In light of direct comparative examinations of Fra Mauro's visual citations of Marco Polo's book with the illuminations in the manuscripts in Paris and Oxford, it becomes clear that the widely held assumptions regarding this topic (particularly as expressed by Wittkower) require radical reconsideration: the pictorial strategies in the illustration of the *Milione* are far more complex and articulated than those indicated by the German scholar.[145] We have already seen that Fra Mauro rejected anything related to marvels and monsters; despite Wittkover's interpretation, even the analysis of the two most important illuminated manuscripts of the *Milione* reveals that the theme of the Orient as a place of marvels and monsters is only one of many different lenses through which the Medieval readers and illuminators understood Marco Polo's text. The miniaturists of the two manuscripts, as well as the patrons in concert with whom they determined their pictorial strategy, took a highly sophisticated approach not only to the text *per se*, but also to the descriptions contained therein of civilizations, kingdoms, cities, and their inhabitants. Comparison of the two manuscripts evidences substantial differences between them, due in part to the fact that they are separated by a considerable chronological gap. It is also fundamental to note that the manuscript at the Bibliothèque nationale de France contains more than twice as many miniatures as the Oxford manuscript. But, in addition to this quantitative comparison, their illustrations can also be differentiated in terms of execution and, above all else, of style. But despite these differences a comparative analysis specifically focused on the connection between the *Milione*'s text and the imagery in each of the two manuscripts ultimately reveals them to be largely alike in terms of their interpretive approach to Marco Polo's book and the formal language they deploy to illustrate its passages. The bond governing text and image in the two manuscripts under consideration entails three different fundamental strategies: a faithful rendition of the text, a supplementation of the text, and a deliberate departure from the text.[146] It should be underlined that, notwithstanding the absence of models, and despite sometimes the lack of precise details in the text, the miniaturists made relatively little recourse to the realm of the monstrous or marvelous. As in the case of Fra Mauro's visual citations, it is, instead, Marco Polo's text that predominantly shapes the images. In the case of these manuscripts and the *mappa mundi*, the illuminators (two of the most refined masters of the early decades of the fifteenth century, the Boucicaut Master and Jean de Grise) and Fra Mauro had ample opportunity to delve into the iconography of the Medieval imagination for their imagery; yet they chose to focus upon the goal of bringing the text to life, often surpassing the image's mere *titulus* and searching widely in Marco Polo's descriptions for those details which permitted them to animate the scenery. Whenever such details were not available they made use not of imaginary conventions, but of their own imagination.

reader from drawing a false conclusion, the illuminators depicted the unicorns as little white horses with a single, long, twisting horn growing from the head to accord with the tradition.

145 Ménard, "L'illustration du Devisement du monde de Marco Polo," pp. 17-31.

146 "Réflexions sur l'illustration du texte de Marco Polo dans le manuscrit français 2810 de la Bibliothèque nationale," *Mélanges in memoriam Takeshi Shimmura*, Tokyo, 1998, p. 81-93.

V. – "Fra Mauro was not a liar !"

This study affirms a divergence of method compared to what was recently written on the epistemic statute of the travel narratives, whether in terms of narrative structure, or in reference to models used to interpret faraway lands, or what was "new" for the Christian west.[147] In his popular study Stephen Greenblatt writes :

> The authors of the anecdotes with which this book concerns itself were liars – few of them steady liars, as it were, like Mandeville, but frequent and cunning liars none the less, whose position virtually required the strategic manipulation and distortion and outright suppression of the truth. But though they were liars, European voyagers to the New World were not systematic, so that we cannot have the hermeneutic satisfaction of stripping away their false representations to arrive at a secure sense of reality. Instead we find ourselves groping uneasily among the mass of textual traces, instances of brazen bad faith jostling homely (and often equally misleading) attempts to tell the truth.[148]

Using the same rhetorical register of Greenblatt this chapter could conclude with the epithet "Fra Mauro was not a liar." Not only was the Camaldolese not a 'liar' but, above all, the categories "truth" and "lie" used by Greenblatt are inadequate and *naïve* ; and moreover discreditable and coarse – whether from the heuristic or analytical point of views – when accompanied by the discursive tradition of travel narratives and of maps between the Low Middle Ages and Early Modernity.[149] The concrete study of the relationship between the eidetic sources of Fra Mauro and its cosmographic transcription demonstrates that Marco Polo (with the mediation of Rustichello da Pisa and of other anonymous mediators of the numerous drafts of the Venetian's travel account of his voyages), Niccolò de' Conti (with the mediation of Poggio Bracciolini), and Fra Mauro, their adroit reader, displayed an attitude that assumed forms and tonalities of attention to the real, and to something that was new, for the Christian West that escape the "conceptual" grid unraveled of 'truth' or 'lie' in Greenblatt. Other considerations in *Marvellous Possessions*, for example to assume the use and persistence of narrative models and rhetorical structures derived from the "satirical novel" as "proof" of this practice of systematic

[147] In my opinion, in that they embrace through comprehensive glances two thousand years of history of travel narratives, the following studies are insufficient : J. Elsner and J-P. Rubiés, *Voyages and visions : towards a Cultural History of Travel*, London : Reaktion Books, 1999 ; E. J. Leed, *La mente del viaggiatore. Dall'Odissea al turismo globale*, Bologna : Il Mulino, 1992. Further problematic are the studies that presume to be "theoretical," indicating moreover what is and is not a voyages or a travel narrative. See H. Schulz-Forberg (ed.), *Unravelling Civilisation : European Travel and Travel Writing*, Bruxelles : Peter Lang, 2005.

[148] S. Greenblatt, *Marvellous Possessions. The Wonder of the New World*, Oxford ; New York : Clarendon Press, 1991, p. 7. The idea of "Travel Liars" seems to originate from a study by P. G. Adams, *Travellers and Travel Liars, 1660-1800*, Berkeley : California University Press, 1962.

[149] A similar dispute occurred on the genuineness of Marco Polo's travel to China. See F. Wood, *Did Marco Polo go to China*, London : Secker & Warburg, 1995. For a clear discussion of the elements in favour and those against the authenticity of Marco's travel, see A. Barbieri, "Un Veneziano nel Catai : sull'autenticità del viaggio di Marco Polo," *Critica del testo* 3.3 (2000), pp. 993-1022 (reprinted in Id., *Dal viaggio al libro. Studi sul Milione*, Verona : Fiorini, 2004, pp. 9-43). Barbieri convincingly argues against the intrinsic weakness of Wood's arguments (among others). Also Wood's blaming of the genuineness of Marco Polo's travel seems to be heavily biased by a coarse dichotomy "truth-lie" that is uncapable to deal with the complexity of the medieval discursive tradition.

falsification, seem far flung. Aside from the evident errors in the identification of what are today defined as literary genres (in the *Milione* there was opportunely recognized above all the imprint of the chivalric stories and Arthurian novels, the '*roman cortese*', and not the 'satirical novel') the establishment of the *contaminatio*, used so awkwardly, it seems to me to simply 'discover' the obvious. Although I am not a scholar of the history of languages or of linguistic theory, I wonder nevertheless which alternatives to shared language and discursive traditions would there be for the formulation of any kind of narrative, including those related to travel? What possibilities would there be to describe what is new other than those that resort to 'shared languages and rhetoric (at least in part)'?[150] Much more effective continue to be the methodological premises outlined by Leonardo Olschki in his *Storia letteraria delle scoperte geografiche* (Literary history of geographical discoveries). As is evident from the title, Olschki expressed the "intention to apply methods of philological and, in a more narrow sense, literary criticism for the reconstruction and interpretation of travel contacts and descriptions of places and populations."[151] This study began with the consideration, only seemingly obvious, that what remained of the past, in terms of documents, assumes the form of discursive traditions; their analysis and interpretation, therefore, cannot free themselves of a philological comparison with, and among, these traditions if it wants to be closer to the historical truth of these narratives.

The study of the *mappa mundi* in its relation to fourteenth- and fifteenth-century travel accounts reflects, and is at the same time the consequence of, three large processes implicit in the cosmography of the time: the development of a network of people and knowledge from a great distance[152]; the foundation of a world economy in which Venice is only one of the principle centres[153]; and, lastly, the expansion and a mental opening to lands and seas once considered as inaccessible to man.[154] It is a question of a culture in which tradition, written knowledge, and experience all coexist and are integrated into it. On a Ptolemaic map, designed according to the syntax of nautical cartography, Marco Polo and Niccolò de' Conti are positioned next to

[150] R. Chartier, *Au bord de la falaise. L'histoire entre certitudes et inquiétudes*, Paris: Albin Michel, 1998 (in part. *L'histoire entre récit et connaissance*, pp. 67-106).

[151] "[...] proposito di applicare i metodi della critica filologica e, in più ristretto senso, letteraria alla ricostituzione e all'interpretazione delle relazioni di viaggi e delle descrizioni di terre e di popoli." L. Olschki, *Storia letteraria delle scoperte geografiche. Studi e ricerche*, Florence: Olschki, 1937, p. 1.

[152] On Venice as the world center of the receiving and transmission of information, see P. Sardella, *Nouvelles et spéculations à Venise au début du XVIe siècle*, Paris: Armand Colin, 1948, pp. 10-14; P. Burke, "Early Modern Venice as a Center of Information and Communication," in J. Martin and D. Romano, *Venice Reconsidered: the History and Civilization of an Italian City-State, 1297-1797*, Baltimore: Johns Hopkins University Press, 2000, pp. 389-419.

[153] F. Braudel, *Civilisation matérielle, économie et capitalisme XVe-XVIIIe siècle*, Paris: Armand Colin, 1979, t. II, *Les jeux de l'échange*, p. 129; I. Wallerstein, *The Modern World-System I: Capitalist Agriculture and the Origins of the European World-Economy in the Sixteenth Century*, New York: Academic Press, 1974; A. Molho and D. R. Curto, "Les réseaux marchands à l'époque moderne," *Annales* 58, 3, (2003), pp. 569-579. However, see also R. Morozzo della Rocca and A. Lombardo, *Documenti del commercio veneziano nei sec. XI-XIII*, 2 vols., Turin: Libreria italiana, 1940; R. S. Lopez, *Venezia e le grandi linee dell'espansione commerciale nel sec. XIII* in *La civiltà veneziana del secolo di Marco Polo*, Florence: Sansoni, 1955.

[154] See at least F. Fernández-Armesto, *Before Columbus: Exploration and Colonization from the Mediterranean to the Atlantic, 1229-1492*, Philadelphia: University of Pennsylvania Press, 1987; S. Pittaluga (ed.), *Relazioni di viaggio e conoscenza del mondo fra Medioevo e Umanesimo*, Genoa: 1993. G. Cavallo (ed.), *Cristoforo Colombo e l'apertura degli spazi*, 2 vols., Rome: Istituto Poligrafico e Zecca dello Stato, 1992.

each other, with references to Solinus, Pliny, and Pomponius Mela. This is a habit of mind that characterizes a cultural phase that does not dismiss either the miraculous or the marvelous. It welcomes ancient and modern authors, simultaneously taking great care to follow the experience of modern travelers by not only recording, but re-elaborating, it; by comparing it to the knowledge transmitted by ancient authors; and by ultimately adopting critical approaches with regards to it.[155] The dualisms "truth-lie," "novelty-tradition" do not help us in interpreting and understanding the historical truth of these texts.

[155] V. Branca, "Dal favoloso al realistico: esotismo fra pellegrini mercanti e Boccaccio," in R. Zorzi (ed.), *L'Europa delle scoperte*, Florence: Olschki, 1994, pp. 39-60.

CHAPTER VI

THE CULTURAL MATRICES AND NARRATIVE STRUCTURE OF FRA MAURO'S *MAPPA MUNDI*

> So I say that in my own days I have been careful to verify the texts by practical experience, investigating for many years and frequenting persons worthy of faith, who have seen with their own eyes what I faithfully report above.[1]
>
> Fra Mauro

I. – Language

1. The *mappa mundi* and the Venetian vernacular

One of the most significant characteristics of the *mappa mundi*'s narrative structure regards the language chosen by Fra Mauro : the Venetian vernacular. Indeed, it is essential to consider the language used in the preparation of the *mappa mundi* in order to understand the precise relationship between the author and his public. What difference is there between writing a cosmographic work in Latin versus the vulgate ? Does the author's attitude towards his readership alter as a result of the change in his audience ? Can we be certain that two different publics were implicated by this language choice ? Did writing in the vernacular necessarily result in an impoverishment of the content ? These are a few of the issues that will be tackled in the next paragraphs.

Loris Sturlese has drawn attention to an unspoken erroneous axiom in historiography which holds that in the process of transmission and transformation of information from Latin into a vernacular language also implies a vulgarization, that is, a debasement of the content. Based on this hypothesis, the texts written in medieval vernaculars are works of second rank, and without originality. According to the same line of thought, the decision to write in the vernacular implied the use of a language that was unacademic, unscientifically structured and non-international.[2] These assumptions call for a more subtle and precise clarification.

While it is undeniable that Latin was an essential component of the identity of Europe's cultural élite and its academic culture, it is still necessary to pay heed to the distinctions among

[1] "Perta(n)to dico che io nel te(m)|po mio ho solicitado uerificar la scriptura | cu(m) la experie(n)tia, i(n) uestiga(n)do p(er) molti a(n)ni | e pratica(n)do cu(m) p(er)sone degne de fede, | le qual hano ueduto ad ochio q(ue) lo che | qui suso fedelme(n)te demostro," *Mappamondo* (ed. 1956) Pl. XL, 49. The translation is mine. For a different version see *Mappamondo* (ed. 2006) *2834.

[2] N. Bray and L. Sturlese (eds), *Filosofia in volgare nel medioevo*. Atti del Convegno della Società Italiana per lo Studio del Pensiero Medievale (SISPM), Lecce, 27-29 September 2002, Turnhout : Brepols, 2003, pp. 1-14.

various vernaculars and to consider what their influence might have been in specific geographic areas and among the various social strata of European peoples.[3] The conviction that Latin was, in contrast to vernacular tongues, a universal scientific language both in its etymology and in its grammatical and syntactic structures, having "le privilège et l'exclusivité de l'expression savante," seems more and more to be a prejudiced assumption.[4] The formulas *competence in Latin = learned culture* and *use of the vernacular = impoverished culture* are now recognized to be insufficient – especially with regard to the late Middle Ages – to account for the enterprise undertaken by so many to "vernacularize science," long before Italian – or rather, the Florentine vernacular in its syntactical, grammatical, and lexical aspects – was recognized as an autonomous and fully valid language by the Venetian humanist and cardinal Pietro Bembo (1470-1547) in his *Prose nelle quali si ragiona della volgar lingua* (*editio princeps* : Venice, 1525).[5] In addition, it is necessary to examine the validity of the assumption that the vernacular pertained to a specific community of local readers that was more limited than the community of readers of Latin, the international language by definition.

2. The 'Question of Language' in Venice : the role of the Camaldolese Order

Fra Mauro's choice of the vernacular is highly meaningful in light of the history of Venetian culture, and above all in the context of the production of vernacular books that was characteristic of Murano's Camaldolese monasteries in the fifteenth and the early sixteenth centuries.[6] In the forty years spanning 1435 to 1475, the Camaldolese monasteries of San Mattia and San Michele in Murano redacted some of the most fundamental works to have ever emerged from the Venetian vernacular cultural production of the fifteenth century. In the same years that Fra Mauro was preparing his *mappa mundi*, a fellow friar, Mauro Lapi († 1482), began to edit a *Laudario* that preserves one of the most complete collections of praises in the vulgate to have ever been compiled in the history of Italian literature ; this friar's work was completed between 1475 and 1477.[7] The first print editions in the vulgate of the Bible and Jacopo da Varagine's

[3] F. Waquet, *Latin, or, The Empire of the Sign : from the Sixteenth to the Twentieth Century*, trans. J. Howe, London and New York : Verso, 2001, pp. 207-229 (original ed., *Le latin ou l'empire d'un signe, XVIe-XXe siècle*, Paris : Albin Michel, 1999).

[4] S. Lusignan, *Parler vulgairement : les intellectuels et la langue française aux XIIIe et XIVe siècles*, Paris : Vrin, 1987, p. 186 ; G. Tanturli, "Volgarizzamenti e ricostruzione dell'antico. I casi della terza e quarta Deca di Livio e di Valerio Massimo, la parte del Boccaccio (a proposito di un'attribuzione)," *Studi medievali* 27 (1986), pp. 811-888 ; and Id., "La cultura fiorentina volgare del Quattrocento davanti ai nuovi testi greci," *Medioevo e Rinascimento* 2 (1988), pp. 217-243.

[5] P. Bembo, *Prose della volgar lingua : l'editio princeps del 1525 riscontrata con l'autografo Vaticano latino 3210*, C. Vela (ed.), Bologna : CLUEB, 2001.

[6] M. Cortelazzo and I. Paccagnella, "Il Veneto," in F. Bruni (ed.), *L'italiano nelle regioni*, Turin : UTET, 1992, pp. 220-280, esp. 235-245. An excellent overview of the development of the bookmaking art, manuscripts, and the printing press in Renaissance Venice is M. Zorzi, "Dal manoscritto al libro," in A. Tenenti and U. Tucci (eds), *Storia di Venezia. Dalle origini alla caduta della Serenissima*, vol. IV, *Il Rinascimento. Politica e cultura*, Rome : Istituto della Enciclopedia Italiana, 1996, pp. 817-958.

[7] The autograph codex of the work is Venice, Biblioteca Nazionale Marciana, Ms. It. IX 182 (= 6284). For a description of the manuscript and a relevant bibliography, see E. Barbieri, *Il libro nella storia : tre percorsi*, Milan : CUSL, 1999, p. 83.

Golden Legend (composed *c.* 1260) were also carried out by a Camaldolese monk from the monastery of San Mattia di Murano, Nicolò Malerbi (1422-1481), who died while staying at the monastery of San Michele.[8] The *Laudario* of Mauro Lapi, the Vulgate Bible and Nicolò Malerbi's *Lives of the Saints* demonstrate the importance of the contribution made by the Venetian Camaldolese monastic community to cultural production in the vernacular.

Edoardo Barbieri, the scholar who has carried out the deepest investigation of this aspect of Venetian culture, has noted the connection between, on the one hand, the publication of the Bible and Malerbi's *Lives of the Saints*, and, on the other hand, the trend of vernacularization initiated in the 1430s by Mauro Lapi.[9] Fra Mauro's work ought to be contextualized and understood within this cultural *milieu* that chose the vulgate as the preferred language through which to disseminate knowledge. Considered in light of contemporary production in the vulgate among the Camaldolese monasteries in Venice, the *mappa mundi* can be recognized as an authentic cultural enterprise comparable to the vernacularizations of the Bible, the *Lives of the Saints* and the *Laudario*. Rather than a public "unlettered in Latin," Fra Mauro and the other Camaldolese translators seem to address a public composed of "readers in the vulgate." These are not interchangeable terms. In a bi-lingual system such as was the Medieval one, the two categories often coincide, but not always and not of necessity.

An initial analysis of Mauro Lapi's translation and anthologizing of vernacular texts serves to clarify some of the historical and cultural context of the *mappa mundi* that have been overlooked thus far. The principle sources for the history of the Camaldolese Order all agree in their description of Lapi's role at the Murano monasteries of San Michele and San Mattia as *scriptore*, which is to say he was considered a copyist and vernacularist translator rather than as an author. Mittarelli's *Annales Camaldulenses* and his *Bibliotheca codicum manuscriptorum Monasterii S.ti Michaelis Venetiarum* are two works which instead identify Lapi as an author; they also claim he copied over a thousand (*sic*) manuscripts.[10] However, putting aside this eighteenth-century Camaldolese mythopoetics, the recent study by Edoardo Barbieri on the Camaldolese book culture of the fifteenth century has shown that Lapi was one of the most prolific copyists and authors in that Order.[11] An analysis of Lapi and his activity in the vernacularization and

[8] [Biblia uulgatizata], Venice: [Vindelinus de Spira], In Kalende de Augusto 1471 (IGI 1697). *Incipit*: "Epistola De don Nicolo di Malherbi ueneto al Reuerendissimo professore dela sacra Theologia maestro Laurētio del ordine de sancto Francesco nella Biblia uulgatizata..." (f. 1v:); *incipit* from the *Biblia*. "El Principio Dio creo Il Cielo et La terra...." f. 13r. The *Vite de' Sancti*, the vernacularization of the *Legenda aurea* by Jacopo da Varazze, was published by Nicolas Jenson in 1475: *Le lege(n)de de tutti li sancti & le sancte dalla romana sedia acceptati & honorati* [Venice]: Nicolo ienson, [1475, after 1 July] (IGI 5037).

[9] Barbieri, *Il libro nella storia*, pp. 109-110. For bibliography and an account of Nicolò Malerbi's cultural background, see E. Barbieri, *Le bibbie italiane del Quattrocento e del Cinquecento: storia e bibliografia ragionata delle edizioni in lingua italiana dal 1471 al 1600*, 2 vols., Milan: Bibliografica, 1992, vol. I, pp. 15-35, 37-70; and C. Dionisotti, "Tradizione classica e volgarizzamenti," in Id., *Geografia e storia della letteratura italiana*, Turin: Einaudi, 1983, pp. 124-178.

[10] A. Fortunio, *Historiarum Camaldulensium pars posterior...*, Venice: ex typographia Guerraea apud Dominicum, et Io. Baptistam Guerraeos fratres, 1579 mense Septembris, pp. 174-175; see also the reprint in *Annales Camaldulenses*, vol. VII, pp. 299-302; and G. B. Mittarelli, *Bibliotheca codicum manuscriptorum S. Michaelis prope Murianum una cum Appendice librorum impressorum saeculi XV, opus posthumum Johannis-Benedicti Mittarelli Veneti, abbatis ex-generalis benedictino-Camaldulensis*, Venice: ex typographia Fentiana, 1779, col. 650-654.

[11] Barbieri, *Il libro nella storia*, pp. 44-48.

copying of texts offers an important means of understanding the motivations and the *modus operandi* of the friars engaged in literary activities with Venice's fifteenth-century Camaldolese monasteries. In this regard, it is rather interesting to consider what Lapi affirmed in his prologue (*Prologo del volgarizzatore*) preceding the translation of the *Stimulus amoris* (attributed by Lapi to Saint Bonaventure, although actually by Giacomo da Milano), a work he instigated in 1432 at the Monastery of San Michele in Murano, and concluded between 1451 and 1452 at the Monastery of San Mattia. After invoking the help of the Madonna, Lapi justifies the necessity of a vulgate edition with this explanation :

> And for these reasons I was and am inspired to work : so as not to be idle and to have material in which to exercise my scant genius and virtues, and to bring out of hiding this modicum of talent, in the hope that I shall not be judged according to that Evangelical maxim. And moreover so that this precious treasure should not be hidden from the illiterate, for I have never seen nor heard of any rendering of this work into the Italian vulgate.[12]

It is a statement that echoes what had been written by Fra Mauro to justify his choice of the vernacular as the language of his *mappa mundi*. The efforts of the Venetian Camaldolese Order towards the promotion of the vulgate would continue on through the successive century. In 1515, two Venetian Camaldolese friars, Paolo Giustiniani and Pietro (formerly Vincenzo) Quirini, in the context of the debate on the religious orders that gripped the early sixteenth-century Church, wrote their *Libellus ad Leonem X* in which they brought their concerns to the pope's attention and called for the reform of the Church.[13] Among their various proposals, the most significant – especially in light of what took place in the immediate future with Luther's reform – was the recommendation to adopt and make official the vulgate editions of the Holy Scriptures.[14] Following the model of fellow friar Nicolò Malerbi's *Biblia vulgarizata*, the proposal to sanction usage of the vulgate Bible official was sparked by a recognition of the widespread ignorance of Latin, on the part of both the clerics and the congregation. The two Camaldolese friars underlined that such an undertaking would only amount to following the example of the Church Fathers (*antiquorum Patrum mores*), who in fact had translated the Scriptures from Hebrew and Greek to Latin, in response to the fact that the majority of contemporary Christians did not know these languages. Giustiniani and Quirini therefore asserted that, with respect to the original languages of the Scriptures, the vulgate was no more unsuitable

[12] "E a questi sono e fui allora in spirito di fare, per non mi stare otioso e per avere materia d'essercitare il mio poco ingegno e virtute, e questo poco del talento non ascondere, acciò ch'io non sia giudicato da quella sententia evangelica. E ancora acciò ch'esso pretioso tesoro non sia occulto agl'illiterati, perciò ch'io non vidi mai né udii questa operetta essere volgarizata in taliano." The vernacularization of the *Stimulus amoris* is known through the codex Florence, Biblioteca Nazionale Centrale, Conv. Soppr., C.2.282, fols. 118r-209v ; the citation is from f. 118r. In 1521 a printed edition was prepared : *Libro intitulato Stimulo de amore : composto per il seraphico doctore sancto Bonauentura*. In Venetia per Ioanne Antonio & Fratelli da Sabbio, 1521 dil mese de septembrio. Barbieri, *Il libro nella storia*, p. 52.

[13] The *Libellus ad Leonem X* by Paolo Giustiniani and Pietro Quirini was published by Mittarelli and Costadoni in *Annales Camaldulenses*, vol. IX, col. 612-719.

[14] This regards an aspect that was the subject of a complex debate within the Roman Catholic Church, sparked by the imposition of Latin as the Church's single official language Chiesa by the Index of Paolo IV (1559) and the Tridentine Index (1564). G. Fragnito, *La Bibbia al rogo : la censura ecclesiastica e i volgarizzamenti della Scrittura*, 1471-1605, Bologna : Il Mulino, 1997, pp. 23-73.

(*peregrinus*) than Latin. At the same time, they indicated the need for studying the sacred texts directly in their original language, whether Hebraic, Greek or Aramaic.[15]

Pressures stemming from the internal affairs of the Church do not entirely explain the choices and cultural enterprises of Lapi, Malerbi, Quirini and Giustinaiani – or of Fra Mauro, for that matter. Additional factors included the fact that Latin's (arguably) sacred and providential status had been undermined, and its monopoly in scholarly ambits had eroded. Of great relevance to these developments are tendencies and attitudes strongly present in the Venetian monastic culture around 1440-1480 (even while they may have originated elsewhere) : in this particular context, the use of the vernacular – the justification for which could be found in the Holy Scriptures – presented an inevitable solution to the widespread, and long-standing quest to communicate to the masses in non-professional literary contexts. Even where the assertion of Latin's indispensability and universality held a stronger claim, such as early fifteenth-century Florence, the link with vernacular culture was more important and influential than what the protagonists of that same cultural movement would have believed. In light of recent research, the program of a cultural movement that sought to reinstall the exclusive use of Latin and Greek, such as Coluccio Salutati, Leonardo Bruni, and Niccolò Niccoli had proposed, fully exhibits its strongly ideological and political character, as opposed to growing naturally out of an historical praxis.[16]

3. A conscious choice

Employing arguments similar to Lapi's, Fra Mauro explained his reasons for the choice of the Venetian vulgate : "In this work I have of necessity decided to use the modern names of common speech because, to tell the truth, if I had done otherwise only a few learned men would have understood me. However, even they cannot always reconcile what the *auctores* say and what is current practice."[17] He justified his decision by making recourse to the now-classic *topos* of the diatribe between Latin and the Vulgate, following upon the ardent defense of the vulgate as a language for the diffusion of knowledge, as formulated in Dante's *Convivio*, beginning in chapter five of the first treatise. Among the reasons Dante proposed – reasons which, in the future, would have a profound cultural impact – the most fundamental was the affirmation that a summary in Latin would fail in its objective to facilitate understanding among those very

[15] The passage from the *Libellus ad Leonem Decimum* discussing the official use of vulgate translations of the Holy Scriptures is located in the *Annales Camaldulenses*, vol. IX, col. 681-685. See also A. Stussi, "Lingua, dialetto e letteratura", in *Storia d'Italia*, vol. I, *I caratteri originali*, Turin : UTET, 1972, pp. 677-728 (p. 699) ; and S. Tramontin, "Un programma di riforma della Chiesa per il Concilio Laterano V : il Libellus ad Leonem X dei veneziani Paolo Giustiniani e Pietro Quirini," in *Venezia e i Concili*, Venice : Fondazione Giorgio Cini, 1962, pp. 67-95.

[16] G. Tanturli, "Umanesimo civile, Umanesimo volgare : i sonetti di Coluccio Salutati," in M. Bendinelli Predelli (ed.), *Firenze alla vigilia del Rinascimento. Antonio Pucci e i suoi contemporanei*, Fiesole : Edizioni Cadmo, 2006, pp. 333-378 ; Id., "Coluccio Salutati e i letterati del suo tempo," in T. De Robertis, G. Tanturli, and S. Zamponi (eds), *Coluccio Salutati e l'invenzione dell'Umanesimo*, Florence : Mandragora, 2008, pp. 41-47.

[17] "In questa opera p(er) necessità ho co(n)uenuto usar no|mi moderni e uulgari p(er)ché al uero se io hauesse fa|to altrame(n)te pochi me haueria inteso saluo che q(u)alche li|terato, auegna che ancora lor no(n) possa acordar hi | autori cu(m) q(ue)l che hora se pratica," *Mappamondo* (ed. 1956) Pl. XXIX, 89. The English translation derives from *Mappamondo* (ed. 2006) *2202.

readers who, not knowing Latin, were deprived of the "seeds (*frumento*, literally, wheat) of wisdom" : "Returning to the original problem, it can thus be asserted that Latin would have offered its benefits to very few, while the vulgate was truly useful to many."[18] Those who knew Latin, asserted Alighieri, certainly had less need of a cultural mediator that did the much vaster public he wished to address. A commentary in Latin would have been inconvenient (*disconvenevole*), and thus useless : it would not have served its purpose, and, sooner or later, it would be translated into the vernacular by others. He added to this main reason the argument regarding his natural love for his own native language, which Dante held to be in no way inferior to those of transalpine lands, which had already become languages with cultural standing, and he added to this argument the awareness of the perfectibility of the vulgate, potentially equivalent to Latin in this aspect (Trattato I, cap. V).[19]

More than one-hundred-and-fifty years after Dante was working on the *Convivio*, Fra Mauro, in his *mappa mundi*, invoked similar arguments and attitudes. The Camaldolese friar clearly discerned that one of the distinguishing characteristics of intellectual life of his age was the emergence of a class of men with cultural interests despite having being educated outside of the Scholastic tradition ; men who desired that their own language serve as an instrument for learning and as vehicle for the highest levels of discourse. For this very reason, Fra Mauro was so firmly resolved about the necessity of using the vernacular, the language known to and used by the most people. The *mappa mundi* thus is a work which seeks to address a public of men beyond the class of clerics and magistrates, as well as learned humanists, the same public to which Dante, a century-and-a-half earlier, had directed his vernacular encyclopedia ; men who, in Dante's words, wish to satiate their "human hunger to understand."[20] Nevertheless, the international dimension of Venetian vulgate and the commission for a copy of the *mappa mundi* on the part of the Portuguese Court demonstrate that Fra Mauro's work had an impact far beyond the local sphere.

4. The Venetian vernacular : an international language ?

Since in the thirteenth century, the Venetian vulgate had been a crossroads of European culture, linking the Latin West and the Byzantine and Slavic East, and serving as a meeting place and watershed for multiple linguistic and cultural currents, whose circulation ranged over a vast territory.[21] The spread of the vulgate followed upon the political and economic expansion of Venice, in step with the affirmation of Venice as a marine power and with the building of its colonial empire.[22] The Venetian vernacular became one of the languages of international com-

[18] Dante, *Convivio*, I, ix.

[19] Dante opened a fierce polemic against "li malvagi uomini d'Italia, che commendano lo volgare altrui e lo loro proprio dispregiano" ; he condemned the "abominevoli cagioni" that induced the various "sette" of his enemies to regard as "vile" such a "preziosa" language whose only fault was to have echoed even "ne la bocca meretrice di questi adulteri". Dante, *Convivio*, I, v-xiii.

[20] Dante, *Convivio*, Trattato I, I.

[21] G. Folena, *Lingue e culture nel Veneto medievale*, Padua : Editoriale Programma, 1990, p. 299.

[22] There is a vast bibliography on this aspect of Venetian history. I have relied above all on the studies published in *Storia di Venezia. Dalle origini alla caduta della Serenissima*, II vols., Rome : Istituto della Enciclopedia

merce, from the shores of the Adriatic to the Eastern conquests of the Crusaders, then spreading further in the wake of the Venetian galleys and warehouses, in easterly as well as westerly directions.[23] This is documented, beginning in the last years of the thirteenth century, in extracts from ship logs, treatises on commerce translated from Arabic, merchants' handbooks (*tariffe* in the Venetian vernacular, and *pratiche di mercatura* in the Florentine), accounting books and *ricordanze*, as well as glossaries – such as those translating Venetian to German or Venetian to Arabic – for use by merchants or pilgrims who, in great numbers, embarked on voyages to the Holy Land from Venice.[24] These documents, which are conserved in the many archives that accompanied the diffusion of Venetian warehouses throughout the Mediterranean, yet also found far outside this region both to the east and to the west, confirm the role attained by the Venetian vernacular as an international language, even beyond commercial contexts.[25] The *mappa mundi* commissioned from Fra Mauro by the Portuguese Court of Afonso V and dispatched to Lisbon in April of 1459, undoubtedly written in Venetian, is a clear example of this fact. A careful look at the preparatory processes that led up to the work as reported in the registry of the monastery of San Michele do in fact speak of copyists (*scriptori*) who transferred to the new map the cartouches of an already existing *mappa mundi* which, as already explained in Chapter I, had been lost, along with Fra Mauro's cosmographic notebooks. This implies that even among other European courts, the Venetian vernacular was recognized as a lingua franca – one of several then current in the Mediterranean world – due to the leading role that Venetians played in contemporary international commerce.[26]

The *mappa mundi*, written in Venetian, thus allowed Fra Mauro to reach out to a rather diversified public: from the cult public of the *literati* and the courts that was educated in the universities, to the public represented by the patrician-merchants, less learned yet well informed and making first-hand use of knowledge that held particular currency ("che hora se pratica" as Fra Mauro put it), and for this reason they sought the kind of knowledge provided by Fra Mauro. A public which had difficulties similar to those experienced by Leonardo da Vinci, who, just a few years after the making of the *mappa mundi*, wrote that he found himself confronting the Classic authorities, when, with a slightly devious air, he admitted to being capable of read-

italiana, 1992-1998. See, in vol. II, G. Cracco and G. Ortalli (eds), *L'età del comune*, 1995: D. Jacobi, "La Venezia d'oltremare nel secondo Duecento," pp. 263-299; in vol. III, G. Arnaldi, G. Cracco, and A. Tenenti (eds), *La formazione dello stato patrizio*, 1997: B. Krekic, "Venezia e l'Adriatico," pp. 51-81; S. Borsari, "I Veneziani delle colonie," pp. 127-158; J.-C. Hocquet, "I meccanismi dei traffici," pp. 529-616 and A. Stussi, "La lingua," pp. 783-801; in vol. IV, A. Tenenti and U. Tucci (eds), *Il Rinascimento. Politica e cultura*, 1996: G. Gullino, "Le frontiere navali," pp. 13-111 and B. Doumerc, "Il dominio del mare;" pp. 113-180.

[23] Cortelazzo and Paccagnella, "Il Veneto," pp. 220-280, esp. 230, 235 and 242-245. Useful for following the development of routes during Venice's commerical expansion is the chart prepared by Alberto Tenenti and Corrado Vivanti for the *Annales*: A. Tenenti and C. Vivanti, "Le film d'un grand système de navigation: Les galères vénitiennes, XIVe-XVIe siècles," *Annales: histoire, sciences sociales* 16, 1 (1961), pp. 83-86.

[24] On the merchants' handbooks, see U. Tucci, "Tariffe veneziane e libri toscani di mercatura," *Studi veneziani* 10 (1968), pp. 65-108; and Id., "Manuali di mercatura e pratica degli affari nel Medioevo," in *Fatti e idee di storia economica nei secoli XII-XX. Studi dedicati a Franco Borlandi*, Bologna: Il Mulino 1977, pp. 215-231. See also *Storia della cultura veneta*, Vicenza: Neri Pozza, 1976, vols. 3.1 and 3.2.

[25] G. Folena, *Lingue e culture nel Veneto medievale*, Padua: Editoriale Programma, 1990, p. 237.

[26] A precise analysis of the preparation of the map for the Portoguese court is found below in Part II, Chapter I.

ing and thus of citing ("allegare" as Leonardo called it) the Classic authors, yet he did "not know how to express well that which he wanted to discuss;" in other words, Leonardo could understand what the Classical authors wrote, but he did not have the language skills to conceptualize that information in Latin when composing his own works.[27] Nevertheless, in choosing the vernacular, Fra Mauro was also addressing a public whose culture can be identified with what Carlo Maccagni has defined as a culture of "technicians," disseminated and nurtured in the craftsmen's workshops, in warehouses, and on board the galleys, almost always without recourse to Latin, rife with references to the abacus, based in practical experience and condensed into aphorisms and axioms standing apart from any kind of organic or structured field of intellectual inquiry.[28]

II. – Fra Mauro's knowledge base

1. The recollection, reading, and obscuring of sources

Addressing its readership directly, Fra Mauro's cosmographic treatise is structured in terms of a lively dialogue with the some of the great thinkers of the millennial history of natural philosophy and cosmography. Throughout this work, the Camaldolese friar continually cited, compared, critiqued, and censured authoritative writers. As shown above in Chapter II, Fra Mauro expertly handled the principles of physics derived from the Aristotelian corpus, in particular Albert the Great and John Buridan. In addition, although somewhat overlooked, Fra Mauro ought to be counted among the earliest to have taken an analytical and critical approach to Ptolemy's *Geography*. Furthermore, by means of the works of Marco Polo, Niccolò de' Conti, and the testimony of who "had seen with their own eyes," he succeeded in gathering together the most important geographic innovations and discoveries. In sum, Fra Mauro's cultural formation was in no way deficient or obsolete with regard to his contemporaries; he was receptive to the intellectual developments of the last two centuries – especially with respect to bibliographic and geographic discoveries – from which he mined all that might prove of benefit to his task of organizing and synthesizing cosmographic information.

Mentioned explicitly in the *mappa mundi* are about forty works. This "quantitative" index, based on the direct citation of the great authors, tells us rather little about what he had actually read or his direct knowledge of the texts and the authors, and for this reason this information is to be applied with great caution. While an uncritical examination of the *mappa mundi* impresses the reader with the vastness of Fra Mauro's learning and the numerous sources of information to which he made recourse, a more rigorous analysis suggests the number of the

[27] Leonardo da Vinci, *Codice Atlantico*, ff. 323 and 327v (formerly 117 r.b. and 119 v.a.). See also A. Marinoni, *Introduzione* in Leonardo da Vinci, *Codice Trivulziano*, Milan: Arcadia-Electa, 1980, p. xiv.

[28] C. Maccagni, "Leggere, scrivere e disegnare la 'scienza volgare' nel Rinascimento," *Annali della Scuola normale superiore di Pisa. Classe di lettere e filosofia* 23 (1993), pp. 631-676; Id., "Sapere e cultura dei tecnici nel Rinascimento," in *Piero della Francesca tra arte e scienza*, M. Dalai Emiliani and V. Curzi (eds), Venice: Marsilio, 1996. Of further interest is C. Bec, *Les marchands écrivains. Affaires et humanisme à Florence, 1375-1434*, Paris: Mouton, 1967.

works he cites is somewhat inflated. In the next paragraphs, an attempt will be made to determine what sources of information were directly consulted by Fra Mauro, what types of manuscripts he used, and how he might have procured them, first through a consideration of the traditional authorities and texts mentioned explicitly in the *mappa mundi*, followed by a discussion of texts that were undoubtedly consulted yet not explicitly mentioned, and concluding with a comparison to the *Bibliotheca codicum manuscriptorum Monasterii S.ti Michaelis Venetiarum* drawn up by the Camaldolese friar Giovanni Battista Mittarelli.

The dispersion of the contents of the library of San Michele in Murano rather complicates the endeavor to identify Fra Mauro's precise textual resources. Moreover, it has emerged from the study of the long cartouches posted in the four angles of the *mappa mundi* that Fra Mauro was a mouthpiece for a kind of circulated knowledge that cannot be easily associated with any one specific authoritative source due to the fact that it had become integrated into the body of notions shared by a vast and heterogeneous public of merchants, notaries, doctors, *litterati*, and poets. Thus, his obvious direct familiarity with a wide range of authoritative authors (*auctores*) is played off against passages that recapitulate the commonplaces of cosmographic discourse in an age marked by the diffuse omnipresence of the same great authors' ideas, invoked through *indirect* references and transformed by a richly expressive language that deployed metaphors and imagery not found in the original sources. Here below is a list in alphabetical order of the authors and works cited explicitly by Fra Mauro. In those cases where he noted only the author without also citing a title, or, inversely, when he cited a title without naming the author, these missing authors and titles have been supplied here between parentheses whenever it was possible to identify the lacking element with reasonable probability (see the Appendix to this chapter for the transcription of the legends in which these authors and works are quoted):

Albertus Magnus (1200-1280), *De causis et proprietatibus elementorum*.(one explicit quotation) ;—, *De Natura loci* (two explicit quotations) ; Arrianus, Lucius Flavius (*c.* 86-after 146), *Anabasis Alexandri :* Book VIII *Indica* (two explicit quotations) ; Aristoteles, *De Generatione et corruptione* (one explicit quotation) ;—, *Meteorologica* (one explicit quotation) ; Aristoteles (Pseudo), *De proprietatibus elementorum*. (one explicit quotation) ; Averroes (1126-1198), *Commentarium in Aristotelis De caelo et mundo* (two explicit quotations) ; Augustinus (354-430), *De civitate Dei* (three explicit quotations) ;—, *Quaestiones euangeliorum* or *Quaestiones in Heptateuchum* (one explicit quotation) ;—, *De passione Domini* (one explicit quotation) ;—, *De Genesi ad litteram* (one explicit quotation) ; Avicenna (Abu 'Ali al-Husayn ibn 'Abd Allah ibn Sina, 980-1037), *Canon medicinae* (one explicit quotation) ; Basilius Caesariensis (330-379), [*Hexaemeron*] (one explicit quotation) ; Beda Venerabilis (673 ? -735), [*Super Genesim*] ? (one explicit quotation) ; Boethius, Anicius Manlius Severinus (*c.* 480-524), *De consolatione philosophiae* (one explicit quotation) ; Dionysius Periegetes (sec. II), *De situ habitabilis orbis* (one explicit quotation) ; Euclides (*c.* 300 bc), *Elementa* (one explicit quotation) ; Fazio degli Uberti (1305-*post* 1367), *Il Dittamondo* (one explicit quotation) ; Hieronymus (*c.* 347-420), *Epistola ad Paulinum presbyterum* (one explicit quotation) ; Hrabanus Maurus (780-856), *De universo* (one explicit quotation) ; Iohannes Chrysostomus (*c.* 347-407), [*Homiliae super Genesim*] (one explicit quotation) ; Iohannes Damascenus (*c.* 675-749), *De fide orthodoxa* (one explicit quotation) ; Iohannes de Sacrobosco (John of Holywood, † 1256), *De sphaera* (one explicit quotation) ; Nicolaus de Lyra (1270-1339) *Postilla super totam Bibliam* (one explicit quotation) ; Pseudo Messalla, *De progenie Augusti Caesaris* (one explicit quotation) ; Petrus Lombardus (*c.* 1095-*c.* 1160), *Sententiae in IV libris distinctae* (one explicit quotation) ; Pomponius Mela (III sec. ce), *Cosmographia* (two explicit quotations) ; Plinius Secundus, Gaius

(23?-79) *Historia naturalis* (two explicit quotations); PTOLEMAEUS, CLAUDIUS (II sec. CE), "Libro de la disposition de la spera" (*Almagesta*? One explicit quotation):—, *Geography* (five explicit quotations):—, *Quadripartitum* or *Tetrabiblos* (one explicit quotation); SERVASANTO DA FAENZA OFM (SERVASACNTUS DE FAVENTIA, *c.* 1220/1230-*c.* 1300), *Liber naturalium rerum* sive *Summa de exemplis contra curiosos* (one explicit quotation); SOLINUS, CAIUS JULIUS (*c.* 250), *Collectanea rerum mirabilium* sive *Polyhistor* (four explicit quotations); STATIUS, PUBLIUS PAPINIUS (*c.* 45-96), *Thebaidos* (one explicit quotation); STRABO (*c.* 63 BC-*c.* 24 CE), *Geography* (one explicit quotation).

2. The Church Fathers. Aristotelian philosophers. Ancient and modern geographers. Travelers.

One of the most immediately striking aspects of this list is the impressively wide array of authors (twenty eight) and works which, in their totality, represented a sizeable portion of fifteenth-century learning, especially in the realm of natural philosophy. This extensive intellectual dominion is truly remarkable, and is unusual even in comparison with works similar in nature to Fra Mauro's. With this first approach to a survey of the extent of Fra Mauro's learning, it is necessary to emphasize that the entries in this list are determined solely on the basis of his explicit references and invocations, and, as we shall see, they cannot be presumed to correspond with the texts that he actually consulted and read. The whole question becomes much more complicated when one tries to move beyond this list towards a reconstruction of Fra Mauro's scholarly preparation to establish which authors and texts he read *directly*. Moreover, it will be demonstrated that Fra Mauro did not consult many of these texts directly; to the contrary, it is primarily the texts Fra Mauro didn't mention that he relied upon most heavily and that can be most easily recognized in the cartouches of the *mappa mundi*.

Heavily quoted and used throughout the work, yet never declared as sources, are Marco Polo's *Milione*, Book IV of Francesco Poggio Bracciolini's *De varietate fortunae* recounting the Indian travels of Niccolò de' Conti (and also, according to Ivar Hallberg's research, the narrative of Odorico da Pordenone[29]), and Thomas Aquinas' *Summa theologica* (in this case the work is cited, although not in relation to the long cartouche, "On the number of the heavens," that translates word for word into Venetian a part of article 68 of *Pars prima* of the *Summa theologica*). This curious circumstance could perhaps be explained by the fact that authors such as Thomas Aquinas, Marco Polo, and Niccolò de' Conti would have been immediately recognizable to readers of the *mappa mundi*'s legends, and therefore citation may have been regarded as redundant. In the case of the two Venetian travelers and Odorico da Pordenone, another explanation can be deduced: perhaps they were not identified and included among the *auctores* – not even the moderns – because they were instead perceived as 'witnesses', and, falling under this category, were cloaked in the anonymity of a remaining classification made up of those "who had seen with their own eyes."

Now that several crucial clarifications have been made, let us reconsider the authors and works cited by Fra Mauro. These can be further divided into four sharply defined groups. The

[29] I. Hallberg, *L'extrême orient dans la literature et la cartographie de l'Occident des XIIIe, XIVe, XVe siècles*, Goteborg: W. Zachrissons Boktryckeri, 1906. Hallberg identified possible derivations, including the accounts of the embassy in Tamerlane by Rui Gonzales de Clavijo of Castille. On this topic, see Chapter V.

first group, defined as 'theologians and exegetes', contains the Church Fathers and their sources: Augustine, Boethius, Jerome, John of Damascus, Basil the Great, John Chrysostom, Bede, and Rabanus Maurus. A second group is comprised of commentators on Aristotle, both Arab and Christian: Avicenna, Averroes, Albert the Great, Thomas Aquinas, Johannes de Sacrobosco, Johannes Campanus of Novara, and John Buridan (these last two were not explicitly cited but are nonetheless recognizable, as shown in Chapter II). A third group is composed of the geographers from the ancient world who laid the foundations of cosmography and founded the history of Western cosmography: Dionysius Periegetes, Arrian, Pliny, Solinus, Pomponius Mela, Claudius Ptolemy. Finally, there is a fourth group whose members are almost never directly invoked, and yet they can be recognized and identified: the traveling merchants and the modern geographers, in particular Marco Polo, Niccolò de' Conti, Fazio degli Uberti, and numerous anonymous travelers whom Fra Mauro occasionally refers to as "those who were really born here" (*queli proprij che sono nasudi qui*)[30] or "those who have seen with their own eyes" (*questi i qual hano uisto ad ochio*).[31] It is this last group to which the Camaldolese friar makes recourse when describing zones of the inhabited world nevertheless unknown to the ancient geographers, such as southern Africa, the Far East and the Indian Ocean basin.

Surprisingly, Fra Mauro even cites Strabo. Even though he makes only a passing reference to this author, it represents an important new development, since Strabo's *Geography* was known only as a title to the Latin West until Georgius Gemistus (later Plethon, Greek: Γεώργιος Πλήθων Γεμιστός, *c.* 1355-1452/1454) brought its contents to the attention of the Latin West during the Council of Florence, between 1439 and 1441. The work was then translated into Latin by Guarino Veronese and Gregorio Tifernate just a few years later, at the end of the 1450s.[32] Fra Mauro is thus one of the first to cite it (as a source for the description of the customs of several unstudied peoples of India).[33] Among the sources cited by Fra Mauro is Euclid, who does not easily fall into one of the four above-named categories. He is invoked twice in the sections which deal with the relationships among the dimensions of the elements.[34] What, then, is one to make of this selection of authors, many of whom were not actually read, and, by contrast, the omission of

[30] *Mappamondo* (ed. 1956) Pl. X, 30. Fra Mauro is referring to the Ethiopian monks who advised him on Africa and the sources of the Nile.

[31] *Mappamondo* (ed. 1956) Pl. XI, 2. Fra Mauro is referring to the Portuguese navigators who sailed off the African coast around the mid-fifteenth century.

[32] On this subject, refer to the discussion above in Chapter I.

[33] "[...] Q(ue)sto dico p(er)chè | q(ue)sti p(o)p(u)li, çoè bragmani, massa|geti, polibotri, pignei, pandi, | astiacani e de altri che de qui se | scriue esser, a nui no(n) sono noti mol|to, p(er)hò de li lor costumi et habiti | discrepatissimi chi ne uol hauer notitia | leça Ariano e Strabo, i qual de questa | india scriueno copiosamente" *Mappamondo* (ed. 1956) Pl. XXVI, 18: "[...] I say this because these peoples – that is, the Bragmani, the Massageti, the Polibotri, the Pigmei, the Pandi, and Astiacani and the others who are written about here – are not very well known to us. Those who want information on their very different customs and practices should read Arrianus and Strabo, who write extensively about this India." *Mappamondo* (ed. 2006) *1380.

[34] "Ma la raxon geometricha a q(ue)sto parlar co(n)tradice co(n)ciosiachè | quella proportio(n) è de uno circulo a l'altro, che è dal quadrato del diametro de uno de quelli circ(u)li | al quadrato del diametro de l'altro circulo, come chiaramente è provado nel duodecimo libro de Euclides | a la propositio(n) secu(n)da." in "De hi elementi, cioè quanto uno elemento | excieda l'altro in quantità. Rubrica." *Mappamondo* right lower corner (right). "But this argument is contradicted by geometrical calculation which defines the proportion between one sphere and another, saying that it is the square of the diameter of one sphere that is in ratio with the square of the diameter of the other

the citations of works which Fra Mauro indeed knew first-hand? Which books were found in Fra Mauro's library? Which authors did he read and which works did he draw inspiration? The reconstruction of the account of a research hypothesis that turned out to be erroneous, alongside with a study of Mittarelli's *Bibliotheca codicum*, permit us a better understanding of these aspects of the *mappa mundi* and offer some answers to these questions.

3. False genealogies: a methodological issue

Analyzed in Chapter II, the cartouche entitled "On the number of the heavens according to the authority of the sacred theologians" in the upper left corner of the *mappa mundi* takes up one of the *quaestiones* of the scholastic tradition: "On the number of the spheres, whether there are eight, nine, or more, or less." Several months of work were dedicated to the identification of its primary sources. The cartouche seemed to be organized into three thematic divisions, each with allusions to specific sources. In the first of these, Fra Mauro openly quoted and explained the cosmological doctrine of John Chrysostom (*Sermons on Genesis*), according to which there exists only one heavenly sphere. In the second part of the legend, the friar countered that doctrine with the theses of the existence of ten heavenly spheres of Basil the Great and John of Damascus. Ultimately, while celebrating the mystic light described in Basil the Great's *Homilies on the Hexameron*, and making a reference to Rabanus Maurus' *De universe*, Fra Mauro concluded the *quaestio* affirming that, despite the diversity in form, there exists a general unity of the heavens.[35]

Having determined the sources, I began to scrutinize Mittarelli's *Bibliotheca codicum manuscriptorum Monasterii S.ti Michaelis Venetiarum* in search of the codices that could have conceivably provided access to the relevant works referred to. The results were encouraging. Preserved in the library of San Michele are various codices containing the above-mentioned works by the Fathers of the Eastern Churches.[36] With caution, I proceeded on the assumption that Fra Mauro could have consulted these codices. Such a scenario was in perfect conformity with what we know about the cultural activity of the Camaldolese Order in the mid-fifteenth century. Ambrogio Traversari, the General of the Order in Fra Mauro's time, is best known for translating the *Lives of Philosophers* by Diogenes Laërtius; in addition he translated numerous

sphere. This is clearly demonstrated in the second proposition of Book Twelve of Euclid [The Elements]." The English translation is from *Mappamondo* (ed. 2006), p. 731.

[35] Rabanus Maurus, *De universo*, in *Patrologia Latina*, vol. 111, col. 9-614 (col. 258-279).

[36] Basil the Great [*Homelies on the Hexameron*], referred to by Fra Mauro as "Basilio": Mittarelli, *Bibliotheca codicum manuscriptorum* "Basilius, *Homiliae Graecae in Genesim*, cod membr., in fol., sec. XIV" (col. 111) – current location: Biblioteca Nazionale Marciana, Ms. gr. II, 183 (= 1000). John Chrysostom [*Sermoni sul Genesi*], referred to by Fra Mauro as "Crisosthomo": Mittarelli, *Bibliotheca codicum manuscriptorum*, n. 95 "Johannes Chrisostomus San Archiepiscopus Costantinopolitanum Opera" (cols 532-536). John of Damascus (referred to by Fra Mauro as "Damasceno"): Mittarelli, *Bibliotheca codicum manuscriptorum* nn. 46 and 784 "Damasus Sanctissimus, Epistola ad beatum Hieronymum Presbyterum super tribus quaestionibus libri geneseos" (col. 499) – current location: Biblioteca Nazionale Marciana, Ms. lat. II, 71 (= 2264). I am grateful to Lucia Merolla for having kindly indicated to me the present locations of the above-named codices.

other works by the Fathers of the Eastern Churches, especially those of Basil the Great.[37] The conclusion that followed was that Fra Mauro's citations of Basil the Great and the Greek Fathers pertained to a larger, cultural endeavor oriented towards humanistic studies of the works of the Greek Fathers and the translation of their works into humanist Latin, and the fundamental influences of Basil's theological doctrine upon Latin doctrinal exegesis as encountered in the work of Traversari, who ardently dedicated his fundamental writings to Basil.[38] As a result, the *mappa mundi* appeared to echo the rediscovery of the Greek Fathers of the Church, a cultural event in which the contribution of the Camaldolese humanists was decisive. *Tout se tien...* Fra Mauro's citations, the contemporary production of Latin translations of the works of the Fathers of the Eastern Churches, and the existence in the Library of San Michele of their codices seemed to offer sufficient evidence of a clear line of cultural transmission. On all counts, the preceding hypothesis turned out to be false. As demonstrated in Chapter II, Fra Mauro did not, after all, make use of the homilies of Basil the Great, John of Damascus, John Chrysostom, or Rabanus Maurus' *De universo*. Instead, he had translated literally *Articulus IV* ("Utrum sit unum caelum tantum") of *Quaestio LXVIII* ("De opere secundae diei in quatuor articulos divisa") in *Pars prima* of Thomas Aquinas' *Summa theologica* (see the documents I and Ia in the Documentary Appendix of Chapter II).[39] Our "exegesis" thus had centered upon the sources of Thomas Aquinas, rather than those of Fra Mauro.

What is the significance of all this? Above all, it furnishes a useful means of understanding the *mappa mundi* and Fra Mauro's working method, mainly in reference to the problem of sources. The legend "Del numero de hi cieli" gives clear proof that the authors who are cited are not the ones Fra Mauro actually consulted. Identifying these authors as sources of *mappa mundi* would thus be a misstep, a oversight, a distortion of the truth.[40] The source that Fra Mauro actually consulted was never revealed in the legend, and yet only by means of that hidden, unnamed work were explicit citations made of authoritative sources. This method of citation and invocation of authoritative sources could be termed "intertextuality." This method was used frequently throughout the *mappa mundi*. Just as Thomas Aquinas is the true source for the citations of those authors who have been designated here in the grouping "theologians and exegetes," we can presume that Albert the Great's *De natura loci* and his commentary on *De causis proprietatum elementorum*, together with a group of manuscripts with excerpts on cos-

[37] M. Cortesi and C. Leonardi (eds), *Tradizioni patristiche nell'Umanesimo*, Florence: Sismel, 2000, esp. the following essays: P. Viti, "Bruni e Traversari lettori di san Basilio," pp. 23-41 and C. Caby, "I Padri nell'osservanza camaldolese: uso, riuso, abuso," pp. 175-191.

[38] M. Cortesi, "Umanisti alla ricerca di Padri greci," in S. Gentile (ed.), *Umanesimo e Padri della Chiesa. Manoscritti e incunaboli di testi patristici da Francesco Petrarca al primo Cinquecento*, Milan, [s.e.], 1997, pp. 63-75; M. Cortesi, "Gli umanisti lettori di s. Basilio tra proposte pedagogiche, motivi ascetici e dottrina teologica," in *Basilio tra Oriente e Occidente*. Convegno internazionale 'Basilio il Grande e il monachesimo orientale', Cappadocia, 5-7 ottobre 1999, Magnano: Qiqajon, 2001, pp. 253-278.

[39] *S. Thomae Aquinatis Summa theologiae*, Pars I, q. 68, a. IV, Turin: Marietti, 1952, p. 334.

[40] As an example, Piero Falchetta is still convinced that Fra Mauro read and quoted Rabanus Maurus. See "Mappamondo di Fra Mauro," in G. Barbieri and G. Fiaccadori (eds), *Nigra sum sed Formosa. Sacro e bellezza dell'Etiopia cristiana*, Treviso: Terra-Ferma, 2009, p. 50: "le fonti letterarie [del mappamondo di Fra Mauro] comprendono autori classici, post-classici e contemporanei: tra essi, Solino, Arriano, Papinio Stazio, Girolamo, Rabano Mauro (...) e molti altri."

mographic topics that were held in the Library of San Michele – which we will analyze shortly – are most probably the texts that enabled Fra Mauro to make reference to various works in the Aristotelian *corpus* and commentaries. Much in the same way, the Fathers of the Eastern Churches and Rabanus Maurus, as well as Aristotle, Avicenna and Averroes, all mentioned explicitly in *mappa mundi*, are acknowledged by Fra Mauro even though he did not actually consult them.

4. The *Bibliotheca codicum manuscriptorum S.ti Michaelis* and the works cited by Fra Mauro

A promising means of determining which works Fra Mauro would have had at his disposition at the time he was composing the *mappa mundi* is an examination of the *Bibliotheca codicum manuscriptorum S. Michaelis prope Murianum*, the eighteenth-century catalogue of the library of San Michele compiled by Mittarelli. Comparing the works and authors likely to have been cited first-hand in the *mappa mundi* – and not merely acknowledged – with Mittarelli's catalogue of the codices of San Michele reveals that the library of the Camaldolese contained about ten of the works cited by Fra Mauro. Of course, it must be not be forgotten that we are dealing with a catalogue compiled in the eighteenth century and therefore it is probable that a portion of the manuscripts contained by the library at that point had been acquired at some point after the mid-fifteenth century. By the same token, it cannot be entirely excluded that some manuscripts in the library's possession in the fifteenth century might have thence been sold or lost, and thus were missing by the time Mittarelli undertook his catalogue.

Although mindful of the uncertainty of this enterprise and maintaining sound caution, we can nonetheless be assured that apparent concordances between the texts cited by Fra Mauro and the titles in Mittarelli's catalogue do hold real significance. It is almost unthinkable that an undertaking of the nature of the *mappa mundi* would have not left behind any 'trace' of its production – in other words, it is not likely that every single one of the authoritative texts (*auctoritates*) invoked by Fra Mauro should have gone missing from the library. A study of these correspondences indicates that around ten of the works among those cited in the *mappa mundi* were potentially owned by the Library of San Michele. Below is the list of the works cited by Fra Mauro and the corresponding codices in Mittarelli's catalogue (indicated with San Michele's original shelf number). In parentheses, the present-day cataloguing information is given whenever available.

Aristoteles (Pseudo), *De proprietatibus elementorum (Commento al Libro de le proprietà e n(atura) di eleme(n)ti).*

Mittarelli col. 15-16, San Michele 136 : "Albertus Magnus appellatus Philosophus locupletissimus, *Philosophia naturalis*, cod. chart., in fol., sec. XIV." (?)

Aristoteles, [*De caelo*].

Mittarelli col. 248, San Michele 312 : "Aristotele, *Tractatus de coelo.*"[41]

[41] Mittarelli also lists a Greek manuscript of both the *De caelo et mundo* and *De generatione et corruptione*, currently held at the Bodleian Library in Oxford. See *Mittarelli* col. 70-71, San Michele 70. Oxford, Bodleian Library, Bywater 4. This manuscript belonged to the Venetian humanist Francesco Barbaro.

Aristoteles, *Meteorologica.*

Mittarelli col. 73, San Michele 104: "Excerpta ex libris *Physicorum* et *Metheorum.*"

Beda Venerabilis (673?-735), [*Super Genesim*] (?)

Mittarelli col. 116, San Michele 707: "Beda Venerabilis Presbyter, *Homiliae*, cod membr., in 4°, ante sec. XV";

Mittarelli col. 117, San Michele 329: "Beda Venerabilis Presbyter, *Chronicon de sex mundi aetatibus*, cod membr., in fol., sec. XV."

Boethius, Anicius Manlius Severinus (*c.* 480-524), *De consolatione philosophiae.*

Mittarelli col. 145, San Michele 90, 113, 273: "Boethius Anitius Manlius Torquatus Severinus, *De consolatione Philosophiae libri V*, cod membr., in fol., sec. XIII; cod membr., in fol., sec. XIV; cod membr., in fol., sec. XIII."

Petrus Lombardus (*c.* 1095-*c.* 1160), *Sententiae in IV libris distinctae.*

Mittarelli col. 690, San Michele 104: "Lombardus Petrus Magister Sententiarum, ext. aliquae ex eius sententiis."

Iohannes de Sacrobosco († 1256), *De sphaera.*

Mittarelli col. 195, San Michele 297: "*Tractatus de Sphaera*, cod. membr., in 8°";

Mittarelli col. 1030, San Michele 919: "Sacrobosco overo Sacrobusco Giovanni, *Dichiarazione della sfera*, cod. in 4°";

Mittarelli col. 1030, San Michele 972: "Magistro Johanne de Sacrobusco Anglico, *Tractatus Latinus de Sphaera*, cod. membr., sec. XV." Current location: Osimo, Biblioteca del Collegio Campana, ms. 18.L.9.[42]

Mittarelli col. 1069-1070, San Michele 566: "*Tractatus de Sphaera*, ext. cum operibus Johannis Olivi".

Solinus, Caius Julius (*c.* 250), *Collectanea rerum memorabilium* sive *Polyhistor.*

Mittarelli col. 1143, San Michele 51: "*Tractatus de hominum & locorum & fluviorum nomine & natura*, cod. chart., in fol., sec. XV." Current location: Venice, Biblioteca Nazionale Marciana, lat. XIV, 128(= 4333).[43]

Ptolemaeus, Claudius (*fl.* 127-148), *Libro de la dispositio(n) de la spera.*

Mittarelli col. 77, S. Michele 153: "*Astrolabii tractatus ex libri Ptolomaei & aliorum Astrologorum collectus*, cod membr., in fol., sec. XIV, plura praeterea de Astronomica arte leguntur, necnon & Canones super Tabulas Toletanas". (?)

[42] This is Merolla's description: "S. Michele 972 = *OBIC, ms. 18.L.9. Membr.; fascicoli legati; 1401-1500 stimata; cc. 38; mm 172 × 124. [...] N. 2. pp. 49-76 Ioannes de Sacrobosco, Tractatus de Spera editus a magistro Iohanne de Sacrobusco Anglico." L. Merolla, *La biblioteca di San Michele di Murano all'epoca dell'abate Giovanni Benedetto Mittarelli. I codici ritrovati*, Manziana: Vecchiarelli editore, 2009, p. 291.

[43] This is Merolla's description: "San Michele 51 = Venice, Biblioteca Nazionale Marciana, Ms. lat. XIV, 128 (= 4333). "Cart., guardie cart.; fascicoli legati: 1401-1500 stimata: ff. 253 + I: mm 286 × 208. (...) N. 12: ff. 160r-194r Caius Iulius Solinus, Collectanea rerum memorabilium." Merolla, *La biblioteca di San Michele di Murano*, pp. 76-78.

Ptolemaeus, Claudius, *Quadripartitum.*

Mittarelli col. 974, S. Michele 645: "Ptolemaeus Claudius, *Itinerarium stellarum significationes per Nicolaum Leonicenum e graeco translate*, cod. in 8°, sec. XV." Current location: Camaldoli, Biblioteca del monastero di Camaldoli, S. Michele di Murano 645.

Among these codices, some seem especially key for understanding Fra Mauro's intellectual background and the books he most likely read. I am referring in particular to San Michele's codices 104, 153, 312, 297, 566, 919, 972. They contain summaries and excerpts including the *Astrolabii tractatus* compiled from Ptolemy and other "astrologers" (who are never further identified), the *De caelo*, and the *De generatione* with *excerpta ex libris Physicorum et Metheorum*, together with a collection of the Sentences of Peter Lombard and, the last four, Sacrobosco's *De Sphaera*.[44] The information available to us does not allow to determine with absolute certainty the degree to which Fra Mauro knew these texts and how much, instead they reached him through secondary sources that laid out the concepts and indicated the most renowned and significant *loci*, as in the collections and encyclopedic anthologies of the age. In light of these comparisons, it would appear that one fact emerges with clarity: besides the authoritative sources, Fra Mauro had at his disposition anthologies dealing with cosmography and natural science. Lucia Merolla's research on the library of San Michele have enabled us to identify about 600 of the 1200 manuscripts recorded by Mittarelli. Of the codices that could possibly correspond to those indicated by Fra Mauro in the *mappa mundi*, unfortunately almost all those treating cosmography (in particular codices 104 and 312, containing Peter Lombard's Sentences, the *De caelo*, the *De generatione*, as well as *excerpta* from *Physica*, and manuscripts 297, 566, and 919, with Sacrobosco's *De sphaera*) in both Latin and the vulgate have been lost. This suggests that here lie the limits beyond which, for the moment, research cannot progress.[45] Examination of the *Bibliotheca codicum manuscriptorum Monasterii S.ti Michaelis* shows that a few of the works that were fundamental for the creation of the *mappa mundi*, such as the *Milione*, Book IV of *De varietate fortunae*, Pomponius Mela's *De chorographia*, Ptolemy's *Geography* and *Almagest*, and Fazio degli Uberti's *Dittamondo* – all texts that Fra Mauro *undoubtedly* read and used extensively – were not in the possession of San Michele.[46] This is a circumstance that allows us to bring into to sharper focus another, equally important, aspect of Fra Mauro's *modus operandi* and the monastic culture of the era.

[44] Catalogo Mittarelli n. 104: "Lombardus Petrus Magister Sententiarum, ext. aliquae ex eius sententiis" (col. 690); cod. 153: "*Astrolabii tractatus ex libri Ptolomaei & aliorum Astrologorum collectus*, cod membr., in fol., sec. XIV" (col. 77); cod. 312: "*Syderibus & de Coelo & de Anima tractatus*, in 4°" (col. 78).

[45] I would like to thank Lucia Merolla for sharing with me the results of her research on the dispersion of the Biblioteca di San Michele di Murano even before their imminent publication by the Istituto Centrale Catalogo Unico (ICCU) at the Biblioteca Nazionale di Roma.

[46] The objection could be made that, while these works were not among those in the possession of the monastery at the end of the eighteenth century, they could have been sold at a previous time. Yet this is drawn into question by the fact that neither the *Bibliotheca codicum manuscriptorum* nor the *Annales Camaldulenses* by Anselmo Costadoni and Giambattista Mittarelli indicate the sale of codices, whereas the purchases of them are noted here. Cf. G. Mittarelli and A. Costadoni, *Annales Camaldulenses Ordinis S. Benedicti*, 11 vols., Venetiis, 1755-1773.

5. Fra Mauro's library: books owned and books borrowed

The information at our disposition indicates that Fra Mauro's 'personal' library of cosmographic works probably included about ten authoritative sources and a few historical and geographic epitomes. But there is one peculiar aspect to this situation, because in addition to the library of San Michele, Fra Mauro had access to other noteworthy library collections in Venice, the seat of important monastic and ecclesiastical institutions; thus, we must take into consideration the probability that he consulted other libraries, either monastic or lay, in Venice. Moreover, the loan of manuscripts was anything but exceptional. In the same years in which Fra Mauro worked on the *mappa mundi*, a confrere, the above-mentioned Mauro Lapi, had borrowed twelve manuscripts – first from San Michele and then from San Mattia in Murano – for the compilation of his important vernacular *Laudario*.[47] It is worthwhile to consider the compositional techniques Lapi used in the creation of his *Laudario*, because it is very likely that these are similar to those Fra Mauro used with the *mappa mundi*.

With the *Laudario*, Mauro Lapi endeavored to produce a vast collection of *laude* in the vernacular, ordered alphabetically. For such an undertaking, finding the sources was a central problem. The miscellaneous character of the collection showcased the ability of its Camaldolese author to get hold of a large number of manuscripts and, from these, to select, order and copy *laude*. The result was the *Laudario*, of which the autograph version can be found today in the Biblioteca Marciana, and indeed it is an impressive repertory of the laudatory compositions that circulated in Venice in the later fifteenth century. The difficulties and the characteristics of the undertaking completed by Fra Mauro in the preparation of his *mappae mundi* were – despite the diverse content – rather similar to those faced by Lapi during the making of the *Laudario*. The goal of constructing an encyclopedic gathering of *laude* confronted Mauro Lapi with the problem of tracking down so many source materials, which then had to be compared, sifted through, and finally recopied. Like Fra Mauro, Lapi also endeavored to present a synthetic overview of the codices he utilized. And, like Fra Mauro, he too drew up a list of the sources that were used in the preparation of the *mappa mundi*. In this regard, a few of Lapi's handwritten notes describe how he managed to assemble a great quantity of manuscript sources necessary for the creation of his encyclopedic achievement:

> I have tried as much as I could to makes sure that nothing appears in this book more than once, and yet I have made some occasional errors. I obtained a copy of Beato Iacopone [da Todi] from the Jesuates. And [I obtained] another book also from them, one of those made by their own friars. And I borrowed two other times [a manuscript of] the Beato Iacopone belonging to the nuns of Santa Chiara because theirs was lengthier. I also had two books of laude of Santa Maria dell'Orto two times; and one from San Salvadore; and from Father Giovanni and from master Francesco d'Amore. From all these [collections of laude] I took whatever I liked. Here at home I selected from our books and from those of Friar Bernardo, and from those in print, desiring that everything be contained in this book. But because I don't have a great number, and yet wanting to put other pious things into this book, I think that it would be too many: and these laude amount almost to infinity. [...] It is true that all this things would be better placed at the beginning of the book, but I had copies again by means of a book that I borrowed from our monastery in Chioggia, and all of these

[47] Venice, Biblioteca Nazionale Marciana, Ms. It. IX 182 (= 6284).

are in Latin, but the calligraphy is hard to read and not really [orthographically] correct : I removed many things that I did not approve of. And in vernacular [I removed] one thing alone, that is, the vision of Saint Elizabeth.[48]

It has already been pointed out that Lapi availed himself of at least twelve books, of which eleven were manuscripts and one was an incunabulum ; two of the manuscripts belonged to San Mattia, while the other nine were borrowed : two from private individuals (one priest and one layman) and seven from religious communities that for the most part were Venetian. This indicates a robust circulation of manuscripts which, however, was in no way out of the ordinary. For example, in the accounting book of San Michele for the years 1453-1464, documentation can be found for very similar situations ; in particular, Maffeo Gherardo himself, the abbot of San Michele, took note of book exchanges with the facing monastery of San Mattia.[49] The loan of manuscripts to San Mattia indicates the presence of a network of the exchange of manuscripts, from which Fra Mauro seems most probably to have benefited during the preparation of the *mappa mundi*. There is not enough information to confirm whether these were the circumstance under which Fra Mauro obtained some of the most important manuscripts he used, and which he is certain to have consulted directly : Ptolemy's *Geography*, Marco Polo's *Milione*, and *Book IV* of *De varietate fortunae*.

III. – The cultural matrices

1. Encyclopedism : proximity and antithesis

When he undertook to compile the *mappa mundi*, the task Fra Mauro set for himself was encyclopedic in nature. His work is a treatise on cosmography and universal geography that describes the populations of the entire *oikumene*, their traditions, customs, religions, and above all, their commercial activities, set in the context of the great trade networks that connected the Indian Ocean and Africa to the Mediterranean via Persia, the Middle East, and Armenia. The *mappa mundi*, which entails a knowledge of sublunary phenomena, as well as some references

[48] Venezia, Biblioteca Nazionale Marciana, Ms It. IX, 182 (= 6284), Mauro Lapi, *Laudario*, 1475-1477, f. 218r : "Ma ho guardato quanto l'ho potuto nulla sia in questo libro più ch'una volta, pure i'ho alcuna volta errato. Ho avuto la copia di beato Iacopone dagli gesuati. E un altro libro pur dalla loro, di quelle ch'hanno fatte e'loro fratelli. E ho avuto per due volte ancora quello d'esso beato Iacopone perch'era più copioso dalle monache di Santa Chiara. Honne ancora avuti due altri libri pur di laide da Santa Maria dell'Orto per due volte ; e uno da San Salvadore ; e un da pre' Giovanni e un da maestro Francesco d'Amore. Da tutti ho tolto quello che me piaciuto. Qui in casa ho tolto de' nostri libri e di quel di fra Bemardo, e di quello della stampa, volendo che ogni cosa sia in questo libro. Ma perch'io n'ho assai gran quantità, e volendo mettere altre cose pie in questo libro, veggo sarebbe troppo alto : e esse laide sono uno numero quasi infinito. [...] Vero è che queste cose tutte sarebbono state meglio tutte in capo del libro, ma i'ho avute queste copie ancora da un libro ch'io portai in presta dal nostro luogo di Chioggia, e tutte quelle sono in gramatica, ma una lettera non troppo buona ne ancora ben corretto : ho tolte più cose ch'io non credetti, e ancora in volgare una cosa sola, cioè quelle visione di sancta Helisabetta." The English translation is mine. Also see Barbieri, *Il libro nella storia*, p. 90.

[49] Rome, Archivio di Stato, *San Gregorio al Celio* (inv. 25/II, n. 9), n. 63, Maffeo Gherardo, *Entrata e Uscita, San Michele di Murano*, (1453-1460), f. 43.

to earthly paradise and the celestial cosmos, was – despite the fact that it was carried out in a confined conventual setting – undoubtedly designed to function as a *summa* of fifteenth-century cosmographic knowledge for the benefit of the Venetian Republic and then the Portuguese Court. Both of these economic powerhouses ardently desired to keep informed of cosmographic innovations – namely those deriving from literary and geographic discoveries in the first half of the fifteenth century – which were taking shape in Venice and Lisbon themselves, as well as Florence, Genoa, and the papal court (including its conciliar outposts, Constance and Florence).[50]

Fra Mauro's *mappa mundi*, in its essential characteristics, could be considered a *forma encyclopediae* particularly in light of its grounding in the concept of closed and complete epistemological order : each detail, whether written or drawn, was seemingly imagined within the context of a totality that explains and contains it. The Renaissance Latin term *encyclopaedia* derives from Plutarch's Greek expression *enkuklos paideia*, literally "the circle of learning," and Fra Mauro's work, even in its material aspect, evokes this characteristic. With its circular shape and its striking frames, it appears to contain, circumscribe and bring together knowledge that has been passed down and circulated in a literally encyclopedic fashion. An analysis of the cartouches and the hundreds of drawings that comprise this monumental *liber figurarum* does show that the map represents a large share of contemporary cosmographic and cartographic knowledge, nevertheless, this analysis also indicates that the work Fra Mauro constructed – and this ought to be strongly underlined – is not the equivalent of a Medieval encyclopedia.

Although it shares certain traits in common with the medieval *mappae mundi* and *summae*, Fra Mauro's work does not correspond to the specific sense that the term *encyclopedia* held between the Middle Ages and early modernity. In comparison to the encyclopedists and the creators of the two great fourteenth-century *mappae mundi* of Ebstorf and Hereford, Fra Mauro was not directed by the principle of *compilatio*, nor by a drive to harmonize conflicting information. Rather, he brought to light the contradictions and discrepancies among various ancient authors, as well those dividing the ancient authors from modern authors and from the knowledge that came by word of mouth regarding new navigations and discoveries. To borrow Fra Mauro's apt phrase, he subjected to scrutiny the whole conglomeration of "writings" and "experience," as well as the discrepancies between them testifying to the impossibility of achieving an encyclopedic *compilatio*. Put in the position of critically reconstructing a large, heterogeneous, and often contradictory mass of information, Fra Mauro became an author in the first person, critically intervening during the composition of his own work. To understand this issue better, it is necessary to clarify what the making of encyclopedias and medieval encyclopedic *mappae mundi* signified, and to explain the mentality that inspired them.

[50] On the relationship between literary and geographic discoveries, see A. Cattaneo, "La cosmografia del Rinascimento : una storia polifonica," in G. Ernst and A. Clericuzio, *Le scienze nel Rinascimento italiano*, vol. 4, G. L. Fontana and L. Molà (eds), *Il Rinascimento italiano e l'Europa*, 12 vols., Verona : Fondazione Cassamarca, Angelo Colla Editore, 2008, pp. 551-569. On the Council of Costance, D. Marcotte (ed.), *Humanisme et culture géographique a l'époque du Concile de Constance : autour de Guillaume Fillastre*, Turnhout : Brepols, 2002.

2. Encyclopedism, *imagines mundi*, and *mappae mundi*

Encyclopedism, *imagines mundi*, and *mappae mundi* have been at the center of an historiographic debate. Starting with the now-classic study by Anna-Dorothee von den Brinken, published in 1968, in which the relationship between *mappae mundi*, universal chronicles, and medieval encyclopedias was revealed,[51] David Woodward, Peter Barber, and Christian Jacob have sought to interpret medieval *mappae mundi* as chronicles and visual encyclopedias.[52] At the root of this was the observation that the vastness and diversity of contents transmitted by *mappae mundi* could be compared to those of encyclopedias. Recently, Margriet Hoogvliet, in her study of the Ebstorf *mappa mundi*,[53] rightly introduced into debate the idea that the relationship between *mappae mundi* and medieval encyclopedias should not be measured on the basis of a weak (at best) analogy between the masses of information they encompass, but by comparing their respective compositional methods.[54] This important theoretical proposition

[51] A-D von den Brincken, "Mappa Mundi und Chronographia : Studien zur imago mundi des abendländlischen Mittelalters," *Deutsches Archiv für Erforschung des Mittelalters* 24 (1968), pp. 118-186 esp. 122-124. This study has attracted recent interest : A-D von den Brincken, *Finis Terrae. Die Enden der Erde und der vierte Kontinent auf mittelalterlichen Weltkarten* (Schriften der MGH 36), Stuttgart, 1992 and "Mappe del Medio Evo : mappe del cielo e della terra," in *Cieli e terre nei secoli XI-XII. Orizzonti, percezioni, rapporti*. Atti della tredicesima Settimana internazionale di studio, Mendola, 22-26 August 1995, Milan : Vita e pensiero, 1998, pp. 51-73.

[52] Woodward, "Medieval *Mappae mundi*", pp. 288-290 ; P. Barber, "Visual Encyclopedias. The Hereford Map and Other Mappae Mundi," *The Map Collector* 48 (1989), pp. 2-8. C. Jacob, *L'empire des cartes. Approche théorique de la cartographie à travers l'histoire*, Paris : Albin Michel, 1992, pp. 327-328, 336-337. There are interpretations that have penetrated even popular publications. See for example Peter Whitfield's writings on the *mappa mundi* of 1457 in the Biblioteca Nazionale Centrale di Firenze (Portolano 1) : "The map retains the character of a visual encyclopedia, derived from the *mappae mundi* showing real or mythical all over the world, although the religious theme is now completely replaced by secular history and legend : the emperor of China, warring pygmies, mermaids, Prester John, and so on." P. Whitfield, *The Image of the World : Twenty Centuries of World Maps*, San Francisco : Pomegranate Artbooks, 1994, p. 40.

[53] The *Ebstorf Mappa mundi* measures 358 × 356 cm and is the largest and most lavishly decorated specimen from the thirteenth century. Its name stems from the Benedictine convent of Ebstorf (near Illzen on the Lüneburg Heath, in Northern Germany) where it was preserved until its public discovery in 1830. Fifteen years later it was moved to the Historisches Verein für Niedersachsen [Museum of the Historical Society of Lower Saxony] in Hannover, where it remained until 1888. It was then brought to Berlin for restoration. The map was separated into thirty vellum sheets (corresponding to the original pieces it was made of) and photographed. Brought back to Hannover, the *Ebstorf Mappa Mundi* went destroyed during World War II under the bombing of 1943. In the course of the nineteenth century two facsimile editions went to press, based on the pictures taken in Berlin : Ernst Sommerbrodt, *Die Ebstorfer Weltkarte, nebst Atlas von 25 Lichtdrucktafeln*, Hannover : 1891 ; a second facsimile was printed in colours five years later : Konrad Miller, *Mappae mundi : Die ältesten Weltkarten*. Vol. V : *Die Ebstorfkarte. Mit dem Facsimile der Karte in den Farben des Originals*, Stuttgart : 1896. For the last major publication on the *Ebstorf mappa mundi*, in two volumes, including a facsimile and a commentary, see H. Kugler, *Die Ebstorfer Weltkarte : Kommentierte Neuausgabe in zwei Bänden*. Band 1 : *Atlas*. Band 2 : *Untersuchungen und Kommentar*, Berlin : Akademie Verlag, 2007 ; P. Gautier Dalché, "A propos de la mappemonde d'Ebstorf," *Médiévales* 55 (2008), pp. 163-170.

[54] "I will argue that the mass of information displayed on medieval maps like the Ebstorf map was not just a meaningless collection of interesting and amusing encyclopedic knowledge. On the contrary, medieval maps of the world were the result of a deliberate process of compilation. The compilatory nature of the maps is analogous to the practice in medieval encyclopedias, although a map is not only a textual but also a visual compilation." M. Hoogvliet, "*Mappae Mundi* and Medieval Encyclopedias : Image versus Text," in P. Binkley (ed.), *Pre-Modern Encyclopaedic Texts*, Proceedings of the Second COMERS Congress, Groningen, 1-4 July 1996, Leiden and New York and Köln :

obviates the problem that was evident in the works of the other authors cited above, namely, that of transforming the analogy between encyclopedias and *mappae mundi* into a generalizing and homogenizing approach, one which unfortunately does not help to analyze medieval maps at a deeper level. This issue will now be examined more profoundly, affording an opportunity to bring to the forefront some crucial aspects of late medieval culture regarding the formation, transmission and institutionalization of knowledge.

Although there are no medieval works that explicitly use the term "encyclopedia" in their titles, the historiographic study of this term has identified a vast body of compilatory works that, despite variations in format, content, and the weight they give to the diverse fields of learning (not to speak of their diffusion and transmission), constitute together a literary genre that is distinguished by the way in which knowledge is accumulated and diffused. Particularly in the twelfth and thirteenth centuries, this literary genre was identified by means of titles like *imago*, *speculum*, *summa*, and *trésor*, or by means of allocutions referring to the encyclopedic content, such as *de proprietatibus rerum* o *de natura rerum*. Alexander Neckam's *De naturis rerum*,[55] Vincent de Beauvais' *Speculum maius*,[56] Bartholomaeus Anglicus's *De proprietatibus rerum* (*c.* 1240),[57] Thomas de Cantimpré's *De natura rerum* (*c.* 1237-1240),[58] Brunetto Latini's *Trésor*,[59] Domenico Bandini's *Fons memorabilium universi* (*c.* 1384)[60] and Domenico Silvestri's *De insulis et earum proprietatibus* (*c.* 1400),[61] all written in the period between the last quarter of the twelfth century and the beginning of the fifteenth century, have been shown to be uncontestable examples of medieval encyclopedism. Despite the fact that they were composed in diverse eras, under diverse circumstances, and in fulfillment of different aims, these works are linked by important affinities. Above all, the *summae*, *trésors*, and *specula* depict reality as a

Brill, 1997, pp. 63-74 (for the quotation: p. 67). The argument was taken up again by the author, though less effectively, in "Des chroniques et des encyclopédies visualisées?," in *Pictura et scriptura: textes, images et herméneutique des mappae mundi (XIIIe-XVe siècles)*, Turnhout: Brepols, 2007, pp. 154-168 (esp. 163-168).

55 *Alexandri Neckam De naturis rerum libri duo with the poem of the same author, De laudibus divinae sapientiae*, T. Wright (ed.), London: Longman & Co., 1863.

56 *Speculum maius* by Dominican friar Vincent of Beauvais (Vincentius Bellovacensis or Vincentius Burgundus, *c.* 1190-1264) and comprising three parts (*Speculum naturale*, in 32 books; *Speculum doctrinale* in 17 books; and *Speculum historiale*, in 31 books tracing the story of humanity since the creation of Adam; for a total of eighty books) was perhaps the most ambitious encyclopedic example of the late medieval and Scholastic period. The bibliography on the topic is vast. See the recent work of M. Paulmier-Foucart, *Vincent de Beauvais et le Grand miroir du monde*, Turnhout: Brepols, 2004.

57 Bartholomaeus Anglicus, *De rerum proprietatibus*, Frankfurt: 1606. John De Trevisa, *On the Properties of Things: John Trevisa's Translation of Bartholomaeus Anglicus, de Proprietatibus Rerum: A Critical Text*, Oxford: Oxford University Press, 1988.

58 Thomas of Cantimpré, *Liber de natura rerum*, H. Boese (ed.), Berlin: de Gruyter, 1973.

59 Brunetto Latini, *Li Livres dou Trésor*, F. O. Carmody (ed.), Berkeley: University of California Publications in Modern Philosophy 22, 1948.

60 On Domenico Bandini and the *Fons memorabilium universi*, in five parts divinded into 34 books, see T. Hankey, "The Successive Revisions and Surviving Codices of the 'Fons memorabilium universi' of Domenico di Bandino," *Rinascimento* 8 (1957), pp. 1-49; Id., "The Library of Domenico di Bandino," *Rinascimento* 8 (1957), pp. 177-207; Id., "Domenico di Bandino of Arezzo," *Italian Studies* 12 (1957), pp. 100-128.

61 Domenico Silvestri, *De insulis et earum proprietatibus*, C. Pecoraro (ed.), *Atti dell'Accademia di Scienze Lettere e Arti di Palermo*, s. IV vol. XI, 1954, pp. 5-270 e M. Milanesi, "Il De insulis et earum proprietatibus di Domenico Silvestri (1385-1406)," *Geographia Antiqua* 2 (1993), pp. 133-146.

finite totality, and thus as an entity subject to complete description. Moreover, they share the rationale of the systemic compilation of the totality of knowledge and the goal of transmitting (*tradere*) a great a mass of information that has been organized according to a logical and intelligible framework, with numerous references to authoritative sources, to a public more interested in practical applications than in speculation.[62] More importantly, they were all based on praxis of *tradere*. In the world of Scholasticism, the concept and praxis of *tradere* imply not only the transmission, but also the collection and verification of knowledge in a public forum – that is the ambit of the *scholae* – by using the method of *quaestiones*.[63] The *magister* and the encyclopedist (two roles which often overlapped) were not expected to elaborate new doctrines, but rather to organize and transmit the legacy of truth that had been accumulated through *traditio* ("tradition," meaning in the literal and etymological sense, "the achieved act of transmitting knowledge"). In this cultural framework, the truth emerges not from research or individual innovation, but from the assimilation of authoritative texts : the *auctoritates*, those sacred and profane texts that are institutionally recognized as bearing a "presumption to the truth." The special task of the encyclopedist and the Scholastic *magister* is to organize wisdom and knowledge according to a principle of "concordance", individuating among diverse and divergent traditions, and among a multiplicity of authors and sources, the one and only truth, which is unfurled for the sake of its perpetuation. Comprehension of the world, including the understanding of the *orbis terrarum*, is not conceived as a creative process, but rather as the result of recognizing transcendental truths and information that has already been accumulated. The aim of handbooks, *summae*, and encyclopedias thus consists in exposing, organizing, clarifying, and transmitting the legacy of truth that was created over time and according to the course of tradition.[64]

The majority of geographic medieval works (often entitled *imago mundi* or *image du monde*), whether in Latin or the vernacular, and including the *mappae mundi* of the period, were carried out with the prevailing purpose of transmitting knowledge that had been consolidated from a literary tradition. Gautier Dalché has shown that the distinctive characteristics of these representations stem from the adaption of the legacy of Roman geography towards the goal of legitimizing the written and visual description of the Christian *oikumene*. There are three basic functions and semantic levels that can be recognized in these medieval *imagines mundi* : the (monastic) *contemplatio* of God's creation, the spatial representation of the history of salvation, and a reminder of the religious foundation of ecumenical geography and cartography.[65] The encyclopedic *compilatio* guaranteed the accuracy of the representations of the Christian

[62] B. Ribémont, *De natura rerum : études sur les encyclopédies médiévales*, Orleans : Paradigme, 1995 ; Id., *Les origines des encyclopédies médiévales, D'Isidore de Séville aux Carolingiens*, Paris : Champion, 2001 ; also the essays (and their bibliographies) by M. Picone, J. Le Goff, M. Ciccuto, and C. Vasoli in M. Picone (ed.), *L'enciclopedismo medievale. Atti del convegno di San Gimignano del 1992*, Ravenna : Longo, 1994, pp. 23-39, 79-115, 363-380 respectively.

[63] One particularly useful work on Scholasticism with a good bibliography is R. Schönberger, *La Scolastica medievale : cenni per una definizione*, Milan : Vita e pensiero, 1997.

[64] E. R. Curtius, "Il libro come simbolo," in *Letteratura europea e Medio Evo latino*, Florence : La Nuova Italia, 1992, p. 361.

[65] P. Gautier Dalché, "L'héritage antique de la cartographie médiévale : les problèmes et les acquis," in R. J.-A. Talbert and R. W. Unger (eds), *Cartography in Antiquity and the Middle Ages*, Leiden : Brill, 2008, pp. 29-65.

oikumene and legitimized the three primary cultural functions that were associated with universal cartography.

3. Fra Mauro : a composite authorial image

Monastic *contemplatio*, the history of salvation, political and religious aspects continued to be key functions even with regard to Fra Mauro's representation. The *mappa mundi* was displayed in a chapel on the left side of the high altar of the monastic church of San Michele in Murano, the miniature of earthly Paradise, attributed to Leonardo Bellini, recalls in an eloquent manner the history of salvation, while the ecumenical dimensions of the representation evoke imperial iconography. Yet Fra Mauro and his work combine these diverse semantic levels pertaining to the oldest maps with a new function : the map also becomes, above all, a project that allows the cartographer first to create, and then to express, his personal vision of the world – one which is idiosyncratic by definition, selective, and self-conscious of its own limits. The main issues of Fra Mauro's universal cartography – as seen above in the chapters on cosmography and the travel writings of Marco Polo and Niccolò de' Conti – regard the navigability of the seas, the circumnavigation of Africa, and the stringing together into a single immense route all the principle ports and cities of Cathay, India and Persia, with those of the Middle East, Europe and the Mediterranean, by means of five great commercial routes. To carry out this project, Fra Mauro took inspiration – a process involving selection, citation, criticism and disputation – from ancient literary sources (including Pliny, Ptolemy, Pomponius Mela) ; he translated and paraphrased scholastic authors like Albert the Great and Thomas Aquinas ; he cited travel accounts and other maps, like Bianco's 1448 marine chart ; and, finally, he heard the testimonies (true or invented) from travelers and ambassadors who came to Venice.

In the historiography the strict interpretations of Fra Mauro's work tend towards a banalizing insistence on its presumed "medieval archaism" as opposed to "experimental modernity." The latter are rather useless categories and judgments that do not add to our understanding of his map ; to the contrary, they lock one into conceptual preconceptions. The analysis of the *mappa mundi* shows that Fra Mauro took inspiration on the one hand from medieval methods of reading and citing *auctores* (in other words, the *compilatio*), and on the other hand from new, more critical, humanistic attitudes towards sources. This produces an authorial image that is composite, not simple, and it is the outcome of the interaction of diverse modes of reading, selecting sources, and directing the writing towards the goal of rendering knowledge of the *orbis terrarum* applicable and useful to his contemporary readership and spectatorship.[66]

In close connection with the encyclopedias and the medieval *mappae mundi*, the narrative structure of the Fra Mauro's *mappa mundi* is undergirded by a particular desire to explain, define and order the world. The text in the majority of the legends intones the pedagogical

[66] The literature on this concept is vast. See *Lexicon des Mittelalters* I, col. 1189-1190, *s.v.* "Auctor" and "Auctoritas." Also, the 31 essays (in seven sections) collected in M. Zimmermann (ed.), *Auctor et auctoritas : invention et conformisme dans l'écriture médiévale*, Paris : École des Chartes, 2001 (in particular, P. Gaurier Dalché, "Sur l' "originalité" de la "géographie" médiévale," pp. 130-143).

synthesis of the medieval encyclopedic compilations, especially the vernacular brand.[67] Without a doubt, among the numerous textual entries, whether placed in the circular *mappa mundi* or in the four external angles, Fra Mauro filtered and rearranged concepts, ideas and theories, putting them forth in the form of a compendium or an epitome, but also in terms of the simplest didactic display. Following the flow of the discourse one has the clear impression – often with philological confirmation as seen above in Chapters II and V – that Fra Mauro freely moved and selected among heterogeneous materials (natural philosophy, notices drawn from travel accounts, from the personal or imaginary reports of navigators, merchants, and missionaries) which he then ordered and rearranged in the graphic space of the *mappa mundi* following the model of the *compilatio*.

It is nevertheless crucial to highlight that beside these legends, there are nearly thirty in which the narrative modality assumes traits and characteristics quite different to those of the encyclopedic works. Unlike the Scholastics and the compilers of medieval encyclopedias, Fra Mauro was not searching for concordances among authors, sources, and concepts. Instead, he seems to have been well aware of the disparities among the ancient authors, as well as the irreconciled differences between the ancients and the moderns. Moreover, he understood well that, occasionally, the work he was designing and writing stood in contrast to what had been written by contemporary "learned cosmographers" (*cosmographi docti*) – by which expression Fra Mauro probably meant those who, in his own time, focused specifically on the passive reproduction of notions deriving from Ptolemy's *Geography* or other classical geographic works) – whom he regards with a certain ironic distance:

> Regarding the divisions of the world – that is, of Asia from Africa and of Europe from Asia – cosmographers and historiographers give various opinions. Of these one could discuss at length, but because it is boring to dwell on this controversy, I will make a brief note with their opinions and leave the prudent to decide which one they should hold as best. Some follow the Ancients [...] Some claim that Ptolemy says that the chain of the Arabian Mountains [...] constitutes the division between Asia and Africa. Others – that is, the modern writers – observe that these divisions of Africa (either by the Nile or by those mountains) make Africa too small. They say that it is the Red Sea or Arabian Gulf that divides Africa from Asia [...] However, I advise those who are looking at this work not to worry themselves too much about discussing this division, given that it is not very important. Let them opt for that which seems to them most reasonable and probable, both to the eye and to the intellect. None the less, I remind them that it is a praiseworthy thing to follow the authority of the most veracious.[68]

[67] R. Casapullo and M. R. Policardo, "Tecniche della divulgazione scientifica nel volgarizzamento mantovano del "De proprietatibus rerum" di Bartolomeo Anglico," *Lingua e Stile* 2 (2003), pp. 139-176.

[68] "Circa la diuision de la terra, çoè de la affrica da l'a|sia e similiter de la europa da l'asia, ne trouo apresso | cosmographi et istoriographi diuerse opinio(n) de le | qual se poria parlar diffusamente, ma per esser materia | tediosa a demorar i(n) q(ue)sta controuersia farò qui un poco de no|ta de le opinio(n) de questi e q(ue)lo se de' tignir lasserò eleçer ai pru|denti. Alguni ch(e) siegue li antichi, di quali son Messala orator | che sc(r)iue la p(ro)genie de Otauia(n) Augusto e Po(m)ponio Mela e q(ue)li che'l | siegue, uuol che'l nilo diuida l'asia da l'affrica, et thanai la euro|pa. Alguni dice che Tolomeo uuol che q(ue)la costa de mo(n)ti de arabia, | che sono da ladi de nubia e tirano p(er) abassia e oltra q(ue)la ethyopia | austral, faça la diuisio(n) de l'affrica. Alguni çoè i a(uto)ri moderni ueda(n)do che q(ue)sta diuisio(n) de l'affrica o p(er) el fiume nilo o p(er) queli mo(n)ti fa l'affrica tro|po picola, fa alt(r)a diuisio(n) e dice che'l mar rosso ouer sino arabico diuide | questa affrica. Ite(m) ueda(n)do che'l fiume edil el q(u)al intra nel mar chaspio e uien | de più al dreto de uerso tramo(n)tana cha'l fiume thanai, dicono che q(ue)sto | fiume diuide

At the same time – and in contrast to the works of the encyclopedists – Fra Mauro remains conscious of the unsystematic way in which collected knowledge is then transmitted, and of the impossibility of distributing it perfectly amidst the graphic space of the *mappa mundi*, as demonstrated by the following legend:

> Note that these mountains that run across this part of Asia can be said to be the source of the noble rivers of Asia: the Quian, the Sumas, the Burdus, the Ganges, the Mandus or Timandus, the Indus, the Reus, the Theus and the Cambises; and in Armenia, of the Tygris and the Euphrates; and of many others that cannot be given special note because they are infinite in number.[69]

Thus, although close to compilatory practices, Fra Mauro's map also reveals an approach to the reading, selection and reworking of sources that is similar to the writing style that emerged with humanism, and which clearly influenced Fra Mauro's work. With the systematic use of the personal pronoun "I" (*io*) and participation in first person in the ecdoctic definition of his own work, Fra Mauro asserts himself boldly in the *mappa mundi* as an author, well aware of the impossibility of homogenizing all the various information at his disposition, and of describing and ordering his representation of the *imago mundi* in a complete and encyclopedic fashion. His work makes it apparent, that around the mid-fifteenth century, cosmography could no longer be the result of the *compilatio*, but depended instead on individual cultural projects carried out by means of an explicit, personal, and thus idiosyncratic, selection of subject matter.

4. Humanism and cosmography around Fra Mauro

Among the many studies that Vittore Branca, Luciano Gargan, Manlio Pastore Stocchi and Margaret King – among many others scholars – have dedicated to Venetian fifteenth-century Humanism and Humanists, there is no mention, whether direct or indirect, of Fra Mauro's work.[70] Fra Mauro does not even figure in their investigations of the system of the schools and

meglio la europa da l'asia, e q(ue)sta ultima opinio(n) par | che sia ap(er)ta et più manifesta et habi me(n) bisogno de linea imagina|ria, chome par che uoiano q(ue)li che fano le p(r)ime diuisio(n). Vnde | co(n)forto q(ue)li che uedeno questa opera che no(n) uogli tropo occu|parse i(n) desputar questa diuision no(n) essendo molto necessa|ria ma tegna quelo li par più rasoneuole e approbabile | e quanto a l'ochio e quanto a l'intel|leto, no(n) de men io ricordo esser lau|deuele acostarse a la autorità | de li più auten[ti]ci." *Mappamondo* (ed. 1956) Pl. XXXIV, 79. The English translation derives from *Mappamondo* (ed. 2006) *2489.

[69] "Nota che questo mo(n)te che trau(er)|sa per questa asia se può dir che sia la fonte de | i nobel fiumi de asia, çoè quia(n), Sumas, Burdus, Gange, | Mandus ouer timandus, Indus, Reus, theus, Cambi|ses, et i(n) armenia tygris et euphrates, e de molti al|tri di qual no(n) è possibi|le farne special nota | p(er)chè sono i(n)finiti." *Mappamondo* (ed. 1956) Pl. XXVI, 54. The English translation derives from *Mappamondo* (ed. 2006) *1422.

[70] V. Branca, *La sapienza civile. Studi sull'Umanesimo a Venezia*, Florence: Olschki, 1998; C. Griffante, "L'umanesimo," in *Storia di Venezia*, vol. IV, *Il Rinascimento. Politica e cultura*, pp. 723-755. For the contributors to Venetian and Venetan Humanism, I relied on L. Gargan, *Lo studio teologico e la biblioteca dei domenicani a Padova nel Tre e Quattrocento*, Padua: Antenore, 1971; Id., *Cultura e arte nel Veneto al tempo del Petrarca*, Padua: Antenore, 1978; M. Pastore Stocchi, "Scuola e cultura umanistica fra due secoli," in G. Arnaldi and M. Pastore Stocchi (eds), *Storia della cultura veneta. Dal primo Quattrocento al Concilio di Trento*, Vicenza: Neri Pozza, 1980, vol. 3.1, pp. 93-122; M. L. King, *Venetian Humanism in an Age of Patrician Dominance*, Princeton: Princeton University Press, 1986; here I have used the Italian translation, *Umanesimo e patriziato a Venezia nel Quattrocento*, 2 vols., Rome: Il Veltro, 1989.

the culture that flourished there.[71] These scholars have conducted impressive archival and bibliographic researches on the so-called "first generation" of humanists, patrons and recipients of the literary, oratorical and artistic cultural production between 1400 and 1460, in other words, Fra Mauro's generation. Though they have investigated figures such as Francesco Barbaro, Leonardo Giustinian, Giovanni Caldiera, Pietro Tommasi, and Jacopo Antonio Marcello – just to name those who were interested, among other things, in cosmography and geography, and particularly in Ptolemy and Strabo[72] – they have ignored every contact, reference, and participation in this culture and the related social networks that is represented by Fra Mauro and his works.

From a much broader perspective, it emerges clearly from all the work that has been done in the field of Venetian Humanism, especially in regard to the first half of the fifteenth century, that being a Venetian Humanist meant above all being an exponent of the patriciate, that is, a member of one of the two-thousand families that actively participated in the life of the Republic. As a consequence, here – in contrast with the situation in Florence, Mantua, Ferrara, and the Pontifical Curia – there was no phenomenon of humanistic literary patronage. With the exception of the instructors at the *scuola* of Rialto and the *scuola* of S. Marco – and particularly with respect to the professor of Aristotelian logic, Paolo della Pergola, and the humanist Guarino Guarini – there were almost no professional men of letters until at least the period of 1460 to 1470. Venice's patricians were thus also its leading humanists, a point demonstrated by the biographies of the students that attended the small, private humanistic school that Guarino Guarini (1374-1460) established in Venice between 1414 and 1419 – crucial years both for the formation of humanistic interests in Venice, and for the emergence of Venice's ruling class. Vittorino da Feltre (*c.* 1373-1446) – a patrician yet of humble origins – was the only one of Guarino's Venetian students to pursue humanistic studies by founding a school that conjoined humanistic ideals and Christian spirituality in Mantua under Gonzaga patronage. The rest of his students, such as Francesco Barbaro, Leonardo Giustinian, and even the future doge Francesco Foscari, were economically independent and authoritative patricians, who subordinated their humanistic formation to the needs of the State.[73]

The intermarriage of the patriciate and Humanism would strongly influence the inherent and the visible aspects of Venetian culture, including both its forms and its ideals. Margaret King has pointed out how in Venice, the individual held less value than the *societas* (meaning a unitary group upon which are focused all the energies of the ruling class, the latter being both

[71] Even though it was not home to a university, Venice boasted a fair number of private *scuole*. In the years corresponding to Fra Mauro's life, in a still-Gothic Venice, the most important ones were the *Scuola* of Rialto, which was founded in 1408 thanks to the testamentary gift left by the rich Venetian merchant Tomà Talenti, and headed from 1421 to 1454 by the Aristotelian mathematician Paolo della Pergola, student of Paolo Veneto. B. Nardi, "Letteratura e cultura veneziana del Quattrocento," in *La civiltà veneziana del Quattrocento*, Venice : Sansoni, 1957, pp. 99-145, esp. 103-120 ; F. Lepori, "La scuola di Rialto," in *Storia della cultura veneta*, vol. 3.2, pp. 539-605.

[72] See the biographic sketches and related bibliography in King, *Venetian Humanism in an Age of Patrician Dominance*. For Francesco Barbaro, refer to the recent and excellent biography by D. Romano, *The Likeness of Venice : A Life of Doge Francesco Foscari, 1373-1457*, New Haven and London : Yale University Press, 2007.

[73] R. Sabbadini, *Guariniana. 1, Vita di Guarino veronese, 2, La scuola e gli studi di Guarino veronese*, M. Sancipriano (ed.), Turin : Bottega d'Erasmo, 1964 (reprint of the Genoa edition : Tipografia del R. Istituto sordo-muti, 1891).

a cultural and a political protagonist). It follows from this that, in a most striking fashion, the cultural production assumed the characteristics of a group culture, as an expression of a ruling èlite that monopolized political, economic, religious, and cultural life. The ideals of the *societas* and of *unanimitas* triumphed in Venice's humanistic culture. While the close relationship between culture and knowledge and private, individual experience was a crucial characteristic of civic Humanism in Florence and elsewhere, this relationship took on a different shape in Venice due to the fact that culture and knowledge pertained above all to communal life, and they manifested themselves in social, shared experiences.

This is not the place to undertake a deep discussion or even to present a further overview of the studies on Venetian Humanism and, more specifically, early fifteenth-century Venetian culture, nevertheless the very existence of a document as impressive as Fra Mauro's *mappa mundi* does invite us to reflect on such issues. Though without denying their merit and import, it can be stated that the majority of existing studies on Venetian Humanism seem to center upon excessively rigid and "official" readings and interpretations that adhere too closely to the ideology inherent in the way the Venetian Humanists conscientiously portrayed themselves (as a cohesive and devout social group) and their culture (rigorously Latin and deeply infused with the study of rhetoric and morals).

In contrast to this "official" face of Venetian humanism, there are at least two other principle portals of analysis and interpretation that permit the evaluation of the humanistic aspects of Fra Mauro's world map, and which bear witness to the circulation of humanistic ideas and attitudes beyond the social circle of the patriciate: the first regards the selection of, and the relationship with, his sources; the second regards the way in which Fra Mauro emerged as an author of the cartouches in his *mappa mundi*. To better understand these two perspectives, it is worthwhile to recall some salient points that can be gleaned from the studies of Eugenio Garin, Giuseppe Billanovich, and, among more recent Humanistic scholars, Riccardo Fubini. These three authors all noted that in the convergence between the attempt to systematically recover the knowledge of Greco-Roman antiquity and new methods of study e translation of classical works (above all those which the West had lost direct contact with), one finds the essence of Humanistic culture, a concept that embraces not only the literary and rhetorical phenomenon, but also critical consciousness and the history of knowledge.

Beginning with Francesco Petrarch's rediscovery of *De chorographia* by the Roman geographer Pomponius Mela (first century BC) in the early years of the fourteenth century,[74] the desire to recuperate in the most integral form possible all the geographic learning possessed by the ancients, above all by means of the study and translation of the works of the Greek authors of whom the West had lost track, remained a constant of humanistic and philological activity.

[74] The manuscript with Pomponius Mela's *Chorography*, which belonged once to Petrarch, has been lost; surviving versions include the ninth-century manuscript from which it had been copied, namely the Ms. Vaticano lat. 4929, itself an apograph of a sixth-century manuscritp originally from Ravenna. From Petrarch's copy were derived the majority of transcriptions of Mela's work that circulated in the fifteenth century, such as Pl. XXX 21 at the Biblioteca Medicea Laurenziana in Florence, that belonged to Coluccio Salutati, who managed to obtain a copy of Petrarch's manuscript around 1379-1381. See G. Billanovich, "Dall'antica Ravenna alle biblioteche umanistiche," *Aevum* 30 (1956), pp. 319-353, revised version in *Annuario dell'Università Cattolica del Sacro Cuore*, 1955-1957, Milan, 1958, pp. 73-107, then reprinted in *Italia medioevale e umanistica* 36 (1993), pp. 107-174; see also S. Gentile, *Firenze e la scoperta dell'America. Umanesimo e geografia nel '400 fiorentino*, Florence: Olschki, 1992, pp. 45-46.

The history of the Latin translation of Ptolemy's *Geography* in the early fifteenth century, Strabo's *Geography* in the middle of the century, Poggio Bracciolini's composition of Book IV of *De varietate fortunae* based on the so-called "Voyage in India" by Niccolò de Conti, the *De Asia* and the *De Europa* by Pius II, and the editions and translations of Pliny's *Natural History* can be understood in this context and, at the same time, they offer eloquent proof of Humanists' interest in cosmography, above all in Florence and at the Papal Court.[75]

Fra Mauro, his work, as well as the context in which he lived, ought to be considered atypical and even eccentric with regard to the "mainstream" of humanism. Fra Mauro was not a philologist. He knew Latin but not Greek; perhaps he could transliterate it, thus managing to read the toponyms of a Ptolemaic map or the index of a work, but nothing more. His erudition and his interests do not compare to those that motivated the research and study of Greek and Latin among his Florentine (whether by birth or adoption) contemporaries (Leonardo Bruni, Niccolò Niccoli, Ambrogio Traversari, Carlo Marsuppini, Poggio Bracciolini, and Enea Silvio Piccolomini), his Venetian contemporaries (Francesco Barbaro, Pietro Tommasi, and Ermolao Barbaro), and his Veronese contemporaries (Guarino Guarini and Domizio Calderini). There is no trace in Fra Mauro of the passion for philology that inspired these Humanists to learn Greek and to undertake new Latin translations that adhered as closely as possible to the spirit of the ancient works confronting them. Unlike contemporaries such as Cristoforo Buondelmonti (*c.* 1380-*c.* 1430)[76] or Ciriaco d'Ancona (1391-1452),[77] whose antiquarian and

[75] On this issue, see the bibliography provided in Chapter IV and the essays of D. Woodward and P. Gautier Dalché in *History of Cartography*, vol. III, *Cartography in the European Renaissance*, Chicago: Univerisity of Chicago Press, 2008, pp. 3-24 and pp. 285-364, respectively. On the history of the translation of Strabo's *Geographia*, see A. Diller and P. O. Kristeller, "Strabo," in F. E. Cranz, P. O. Kristeller and V. Brown (eds), *Catalogus translationum et commentariorum: Medieval and Latin Translations and Commentaries: Annotated Lists and Guides*, Washington, D.C.: Catholic University of America Press, 1960-1986, vol. 2, 1971, pp. 225-233.

[76] Cristoforo Buondelmonti was born in 1386 to a powerful Florentine family with strong ties to the Levant. Most certainly he had a humanistic education in the *scuola* of Guarino da Verona and was close to the humanistic circle of Niccolò Niccoli, with whom he shared an interest in classical geography. Having left Florence, in 1414 he arrived in Rhodes to study the ancient Greek civilization; over the next 16 years he visited all the islands, cities, and monasteries of the Aegean Sea, Crete, Cyprus, Hellespont, and Costantinople. The fruits of these long travels were two historico-geographic works, *Descriptio insulae Cretae*, sent in 1417 to Florence to Niccolò Niccoli, and *Liber insularum Archipelagi*, dedicated to Cardinal Giordano Orsini in 1420. The *Liber insularum* underwent four manuscript editions (the first one has been lost). The definitive one, achieved in Constantinople and dating from *circa* 1430, was included and published in the late fifteenth and early sixteenth century in works such as the *Insularium illustratum* by Henricus Martellus Germanus (*c.* 1490), and printed by Bartolomeo de li Sonetti and Giovanni Bordone. The work was an immediate success, witnessed by the numerous manuscripts held in both European and American libraries. See G. R. L. von Sinner, *Chistophori Bondelmonti Florentini Librum Insularum Archipelagi a codicibus Parisinis regiis nunc primum totum adidit*, Leipzig, 1824; R. Weiss, "Cristoforo Buondelmonti," in *Dizionario biografico degli italiani*, vol. XV, pp. 198-200; Gentile, *Firenze e la scoperta dell'America*, pp. 107-110, entries 55 and 56; Christophorus Buondelmontius, *Liber insularum archipelagi: Transkription des Exemplars Universitäts und Landesbibliothek Düsseldorf Ms. G 13, Übersetzung und Kommentar*, K. Bayer (ed), 2 vols., Wiesbaden: L. Reichert, 2005 and 2007.

[77] Ciriaco d'Ancona traveled throughout the Mediterranean studying Roman, Greek and Egyptian ruins. During his numerous trips in Italy, Dalmatia, the Aegean Islands, Egypt, and Constantinople, he composed detailed descriptions of monuments and the ruins he encountered, accompanied by his own drawings. He was the first modern European to describe the Athenian Acropolis and the Parthenon, and the first to inform Europeans about the Egyptian hieroglyphs and pyramids. He learned Latin and Greek autodidactically and was a tireless collector of

archaeological quests in Italy, Dalmatia, Constantinople, the Aegean Islands, and Egypt were one of the channels by which the Latin West spatially reappropriated the Classical world, Fra Mauro did not nourish an antiquarian's interest in the ancient world. He was fascinated by the conquests of Alexander the Great, having been struck and perhaps even terrified by the more recent conquests of Tamerlane – which he described in terms of the atrocities and immane horrors – yet these were very common themes, and moreover they were dealt with in an unsystematic way, in brief passages and without any originality.

During Taversari's priorate (1431-1439), the Florentine Camaldolese monastery of Santa Maria degli Angeli became the primary gathering place for the city's Humanists ("an informal academy," as Cécile Caby defined it) visited daily by Niccolò Niccoli, Paolo dal Pozzo Toscanelli, Filippo Pieruzzi, and by Cosimo and Lorenzo de Medici, who all gathered around *Ambrosius monachus*.[78] By comparison, the Camaldolese monastery of San Michele in Murano was neither a base nor a meeting ground for humanistic studies until at least the second half of the fifteenth century. Traversari visited San Michele di Murano in 1433 and stayed as a guest there for forty days (as highlighted in Chapter I, Fra Mauro was staying at the monastery during this time and it is probable he also met the general). In his accounts of the visit, although he praised the governance of old abbot Paolo Veniér (1367-1448)[79] and the caring atmosphere of the entire community of the monastery, he did not devote a single word to the library, nor to the cultural activities that took place there – something that, in other circumstances, he never missed an opportunity to remark on. Instead, he exaggerated the dimensions of the structural repairs that were being undertaken at the monastery, saying it had been "rebuilt from its foundations."[80]

Greek codices; in particular he found and acquired manuscripts of the *Iliad*, the *Odyssey*, and Strabo's *Geography*. Except for the *Itinerarium* (a long letter to Pope Eugene IV from 1441-1442 in which he describes his own interests and travels to preserve a record of the past amidst the destructive forces of time and human history) which is conserved in its entirety, the *Commentarii*, his principle achievement, originally written in six volumes in which he presented together all he had seen, recorded, described and drawn during his travels, was lost in the fire of 1514 that destroyed the library of Alessandro and Costanza Sforza in Pesaro in which it was kept. See G. Paci and S. Sconocchia (eds), *Ciriaco d'Ancona e la cultura antiquaria dell'Umanesimo*, Reggio Emilia: Diabasis, 1998; G. A. Possedoni (ed.), *Ciriaco d'Ancona e il suo tempo*. Ancona: Canonici, 2002. Ciriaco di Ancona (Kiriaki Anconitanis), *Itinerarium*, ed. L. Mehus (Editionem recensuit, animadversionibus, ac praefatione illustravit, nonnullisque ejusdem Kyriaci epistolis partim editis, partim ineditis locupletavit Laurentius Mehus), Florence: 1742 (reprint, Bologna: Forni, 1968); Ciriaco di Ancona, *Later Travels*, E. W. Bodnar and C. Foss (eds), Cambridge, MA and London: I Tatti Renaissance Library and Harvard University Press, 2003.

[78] C. Caby, "Culte monastique et fortune humaniste: Ambrogio Traversari, *vir illuster* de l'ordre camaldule," in *Mélanges de l'École française de Rome. Moyen Âge* 108 [1], Rome: École française de Rome, 1996, pp. 321-354; and Ead., "Entre observance et humanisme. Définition et pratiques d'une orthodoxie culturelle dans l'Ordre Camaldule," in *Orthodoxie, christianisme, histoire*, Collection de l'École française de Rome 270, Rome: École française de Rome, 2000, pp. 3-22.

[79] On Paolo Venier (abbot of San Michele between 1392 until 1448) see Meneghin, *San Michele in Isola*, pp. 21-35.

[80] "Continuo monasterium ipsum nostrum [San Michele di Murano], ubi hospitabamur, visitare aggressi, paucis diebus absolvimus maxime delectati religione summa et sanctitate vitae quam ibi comperimus, mira patris peritia et solertia singulari. Indicem quoque confecimus rerum omnium, admirati nimium ex quibus ruinis in quantam amplitudinem pater ipsum monasterium auxerat, cum, praeter fratrum optimorum numerum, praeter extructionem parietum (nam ipsum ferme a fundamentis erexerat) tanta e tam varia suppellex ubique enitebat ut stupori nobis esset. Quaedam tamen illic constituere ad incrementa decoris volentes pater grate admodum admisit." S. Iaria,

At the time of Traversari's visit, Fra Mauro had just initiated his scholarly activity and probably had not yet completed any work comparable to his world map.[81] The library of the monastery, which fifty years later would become one of Venice's most important libraries, was then of modest dimensions and possessed a Scholastic collection that was completely lacking in classical texts.[82] Despite this, the *Hodoeporicon* by Traversari throws light on an interesting fact regarding the monastery : it received visits several times from "many nobles and dear friends" (*molti nobili e di amici carissimi*), including Francesco Barbaro, Lorenzo, Leonardo, Marco Giustiniani, the physician and Humanist Pietro Tommasi, and Ciriaco d'Ancona, and these visitors engaged in discussions with Traversari about, among other things, Greek codices and their translation into Latin.[83] Francesco Barbaro (1395-1454), perhaps the most illustrious representative of the so-called "first generation" of Venetian Humanists, made a testamentary gift of his own library to the monastery of San Michele, a gift which included a chartless Greek codex of Ptolemy's *Geography*, now located in the Newberry Library of Chicago as codex Ayer 41.[84]

The monastery was not a gathering place for Humanists (at least not before the Barbaro library passed to it), it did not contain any relevant manuscripts, and however it was nonetheless commonly visited by at least some of the Venetian Humanists of the period. It was thus in contact with this Humanist culture, perhaps even more so than the bibliographic records and histories of book circulation have indicated. At the same time, the Humanistic movement is more permeable in terms of communication networks, circulation and exchange than the epistolary especially would have us believe.

A critical analysis of Ptolemy's *Geography*, the citation of Pomponius Mela's *De chorographia*, and the precocious reference to Strabo (a work which, in the very years in which Fra Mauro compiled his *mappa mundi*, Guarino Veronese and Gregorio Tifernate had only just begum to translate at the commission of Pope Nicholas V) all demonstrate Fra Mauro's contact with the principle literary discoveries of Humanist circles. This points to an often undervalued aspect of the history of Humanism, and in particular of the history of the reception, translation and circulation of Greek and Latin texts rediscovered since the fourteenth century : here we are confronted by a circulation and diffusion of such works across a wide range, or at the very least, a rather heterogeneous range. In any case, it proves that circulation was not confined to the "official" Humanist sphere. Yet this aspect is less revealing with respect to other facts that have emerged as a result of this analysis of Fra Mauro's world map. Whenever we probe the origins

Ambrogio Traversari, Hodoeporicon : edizione critica e commento, Ph.D. dissertation, Florence : Università degli Studi di Firenze, 2004, pp. 62-63.

[81] In 1437 Fra Mauro designed a map of the land belonging to the Camaldolese monastery of San Michele di Leme, in Istria. See at this regard Chapter I.

[82] Caby, *De l'érémitisme rural au monachisme urbain*, pp. 625-631.

[83] See in this regard the letter sent by San Michele to Niccolò Niccoli on 30 May 1433 : "Ibi quanto studio acceperint nos amicissimi nostri, Barbarus, Leonardus et nobiles plurimi adolescentes, evolvere non opis est nostrae… Offendi Cyriacum Anconitanum hominem antiquitatis studiosum… Libros Barbari nostri atque Iustiniani necdum vidi… Venetiis, ex nostro monasterio Sancti Michaelis de Muriano, III Kalendas Iunias," in A. Favi, *Ambrogio Traversari. Lettere a Niccolò Niccoli. Edizione critica e commento*, Ph.D. dissertation, Florence : Università degli Studi di Firenze, 2003-2004 : "Letter 46" (May 30, 1433), pp. 86-87 (pp. 192-194 for the commentary).

[84] A. Diller, "The Library of Francesco and Ermolao Barbaro," *Italia Medioevale e Umanistica* 6 (1963), pp. 253-262.

of Humanism or its nature as a distinctively modern cultural movement, it is unlikely that the mere study of reception, recovery, and use of the ancient literary tradition should furnish us with exhaustive answers. Rather, it is in the study of the ways in which people began to read, cite, and, in more general terms, relate to the *auctoritates*, that we can grasp the essential differences between Scholastic culture and this new phenomenon.

In a recent article that defined Humanism in contrast to Scholasticism, Riccardo Fubini identified in the manners of recounting Florence's history by Leonardo Bruni (1374-1444) a revealing instance in the passage from one culture to the other. Bruni's *Historiae Fiorentini populi* begins *ex abrupto*, while Scholastic authors would have instead listed all the testimonials on the foundation of the city. For Bruni relaying in the authoritative sources was sufficient and suppressed other spurious sources, thus allowing proceeding straight to the enunciation of a fact, even if this fact was contradicted or even denied by the *tradition*.[85] Fubini's analysis is useful for understanding the spirit in which Fra Mauro approached his work. It is in Fra Mauro's very approach to his sources (often without citing them, in other places synthesizing them, and quite often contradicting them) and in his writing that we are best able to grasp the humanistic quality of his approach to knowledge.

The *mappa mundi* situates itself outside of the Scholastic compositional method – in other words, that institutional system for the formation and transmission of knowledge that scrutinizes the reliability of individual claims, weighing in on their greater or lesser probability, just as it distances itself from medieval encyclopedism – by which we mean that institutional epistemology that orders and assimilates knowledge by seeking to establish agreement among heterogeneous sources and ideas. The study of the *mappa mundi* demonstrates that Fra Mauro read the *auctores* and composed his work taking inspiration from various principles that were characteristic of Humanist attitudes – the most important of which was critical exemplarity. Moreover, the narrative structure reveals an author who writes and thinks in a critical way and who, most importantly, emerges in first person through the reiteration of the personal pronoun "io," "I," in the composition of the work, a unique phenomenon in the cosmography of his time. Yet another, highly relevant instance of his Humanist leaning can be detected primarily in the way Fra Mauro positions himself with regard to his sources : for Fra Mauro, the *auctori* are not texts in the Scholastic sense of the term *auctoritates* (and we might recall here that such texts, once accepted by tradition, were invested with a "presumption of truthfulness") but rather historically determined individuals who lived and who composed their works in historically determined contexts and who had specific agendas ; it is with these individuals that Fra Mauro, as a person and as a man of his own times, confronts them, disputes with them, and dialogues with them, in light of *his own* knowledge. Considered together, all of these characteristics results in the emergence of the modern figure of the cartographer as a sovereign and rational judge of the world who selects and arranges an enormous and variegated mass of information which is diverse in nature, origin and value.

[85] R. Fubini, "Umanesimo e Scolastica. Saggio per una definizione," in *Medioevo e Rinascimento. Annuario del Dipartimento di studi sul Medioevo e il Rinascimento dell'Università di Firenze*, XVII/n.s. XV, 2004, pp. 165-174 (English translation,"Humanism and Scholasticism. Toward an Historical Definition," in A. Mazzocco (ed.), *Interpretations of Renaissance Humanism*, Leiden : Brill, 2006, pp. 127-136.

5. Critical exemplarism

In the *mappa mundi* the graphic and written text are the product of a personal, idiosyncratic and altogether extraordinary selection applied to heterogeneous languages and fields of knowledge. The way that the Camaldolese friar transcribes, modifies, and critiques in first person the authors whom he cites can be identified with the exemplaristic praxis of the Humanists. Humanistic critical exemplarism constituted an *episteme*, a 'spirit of the times' and, contemporaneously, it was also a cultural practices, a way of seeing the world and interacting with it.[86] In contrast to a more antiquated approach which regarded the ancient sources as a normative *corpus*, Fra Mauro saw his sources as offering specific solutions and answers, which in light of specific problems, he compared, verified, and selected from. Fra Mauro broke down into segments and fragments a rather vast number of discourses and texts, which, when reassembled, are used to construct a decidedly modern and original cosmographic synthesis. By means of a compositional method that could be termed "citational," he re-elaborated the informational and expressive potentialities of diverse languages and authors in order to express his *own* cosmographic concepts – in which case the innovative aspect nevertheless was rooted in a solid base of doctrinal legitimation.

The *mappa mundi* is based upon a technique for the citation and transmission of knowledge that reconfigures a rich catalogue of citations and ideas : the Church Fathers (via Thomas Aquinas) ; Aristotelian natural philosophy (mostly via Albert the Great) ; the classical geographers (Solinus, Pomponius Mela, Pliny, and, above all, Ptolemy ; the accounts of recent travelers like Piero Quirini, as well as the two most important sources for the representation of Asia, namely Niccolò de' Conti and Marco Polo ; and the personal testimonies of travelers (such as the Coptic monks whom he questions regarding Ethiopia, and perhaps Niccolò de' Conti himself and Ciriaco d'Ancona) ; merchants handbooks ; and the marine charts available to him (those by Andrea Bianco's as well as Portuguese ones, according to the Camaldolese friar). For Fra Mauro these all are like a building that can be disassembled and adapted to new priorities and expressive exigencies. Fra Mauro, in the role of author, determined which debates, ideas, and arguments were interesting and worth pursuing, and which ones are not. He defended or challenged the value of his sources. He indicated which opinions ought to be accepted and which should be discounted. He declared which concepts he was able to verify, which ones he could not, and which ones were not subject to verification. This approach to source material, although an essential practice within the humanistic culture, was transformed by Fra Mauro into a practical and systematic compositional method, above all in regard to the criticism and comparison of his sources.

[86] On citation and critical exemplarism, M. Fumaroli, *L'età dell'eloquenza : retorica e 'res literaria' dal Rinascimento alle soglie dell'epoca classica*, Milan : Adelphi, 2002, pp. 95-97 (original French edition, Geneva : Droz, 1980) ; on the extension of the concept towards non-literary practices such as architecture, see M. Carpo, *Metodo ed ordini nella teoria architettonica dei primi moderni : Alberti, Raffaello, Serlio e Camillo*, Geneva : Droz, 1993.

6. Writing (*scriptura*), oral knowledge and experience (*experientia*)

Written culture is but one of the means by which Fra Mauro's cosmography is articulated. The text and images of the *mappa mundi* are the result of a synthesis of many fragments of books, but at the same time, the Camaldolese friar indicates that his legends contain knowledge and information that is not otherwise preserved in written form. Fra Mauro says, and he wants his readers to know, that he has also collected, selected and transcribed testimonies of those "who have seen with their own eyes" (*chi ha visto a ochio*), of "those who sail" (*queli che navegano*), and of "native people" (*queli che sono nasudi qui*), who are defined as "people worthy of belief" (*persone degne di fede*).

How and why did Fra Mauro use – or, in other cases, would have us believe that he used – oral testimonies ? How does he make simultaneous use of the texts of classical and medieval authorities, often in Latin, and new information that had eluded the systematization and the codification of writing ? How are we to interpret this particular process of exegesis ? We will tackle these questions regarding Fra Mauro's work by analyzing the way certain place names in Asia that were know in Europe in their Latin morphology after the publication of Book IV of *De varietate fortunae*, were instead transcribed by Fra Mauro following a different and to certain extent more correct spelling ; later we will consider some legends where explicit reference is made to use of orally transmitted knowledge, knowledge considered authoritative and reliable to the same degree as that acquired from books. These two ideas will bring into focus the complementary relationship between writing and experience, not only in the work of Fra Mauro, but also in the particular cultural context in which he lived.

As already mentioned in Chapter V, in addition to his reading of Poggio Bracciolini's *De varietate fortunae* from which he derived many of the legends and place names of southern and eastern Asia, some toponyms, which are plainly traceable to Book IV of the *De varietate fortunae*, are spelled in such a way as to show that Fra Mauro was not exclusively dependent on the linguistic transformations introduced by Poggio Bracciolini in his attempt to Latinize the names of the cities of the Indian Ocean and of southern and eastern Asia. The studies of the French orientalist Paul Pelliot and of Outi Merisalo indicate that the morphology of some of the Asian place names handed down by Fra Mauro has significant similarities with the morphology in which these toponyms were known in the languages of southern Asia of the fifteenth century, prior to their Latinization by Bracciolini.[87] For example, when the *De varietate* mentions *Pauconia* (or *Panconia*), Fra Mauro uses the form *Paigu*,[88] most plausibly identified with the

[87] According to Paul Pelliot, Fra Mauro must have had direct contact with Conti and suggested that the "a person worthy of trust (*persona digna de fede*) mentioned in the following legend was Niccolò de' Conti, but this is merely a guess : "Anchora io ho parlato cum persona digna de fede, che afferma hauer scorso cum una naue de india per rabia de fortuna de trauersà per zorni 40 fuora del mar d'india ...." *Mappamondo* (ed. 1956) Pl. XI, 2 ; "What is more, I have spoken with a person worthy of trust, who says that he sailed in an Indian ship caught in the fury of a tempest for 40 days out in the Sea of India, beyond the Cape of Soffala and the Green Islands towards west-southwest ; and according to the astrologers who act as their guides, they had advanced almost 2.000 miles...." *Mappamondo* (ed. 2006) *149. See P. Pelliot, *Notes on Marco Polo*, 2 vols. and *Index*, Paris : Imprimerie Nationale, Librairie A. Maisonneuve, 1959-1963, 1973 : vol. 1, pp. 176-182 and 275-278, esp. p. 181.

[88] *Mappamondo* (ed. 1956) Pl. XIV, 69 : *Mappamondo* (ed. 2006) *226.

city of Pegu, on the Pegu River east of Rangoon, where *Paigū* retains the Malese morphology.[89] *Scierno*,[90] present day Guar, for which Poggio Bracciolini in the *De varietate* uses the form *Cernove*, presents another case of transliteration that differs from the Latin of Bracciolini and is closer to the presumed original morphology *Shahr-i-Naw* ("New City"), as it was known in the fourteenth century.[91] Thus Fra Mauro would seem to have transcribed some toponyms from direct and personal conversation with Niccolò de' Conti, who he believed to be more reliable, in certain circumstances, than the written text in Latin of the *De varietate fortunae*.

Fra Mauro distinguishes, even if not in a systematic way, between things seen and heard by eye-witnesses and those derived from written sources : the voices that Fra Mauro heard complement the written sources. This is one of the traits that distinguishes Fra Mauro's map from more learned cosmographic works such as Pius II's *De Asia* : the *mappa mundi* contains, in addition to a dense systematization of written sources (whether *auctores veteres* or *novi*) a complementary set of knowledge that, in the case of Niccolò de' Conti and of other merchants, was the result of direct experience. In order to better understand how, in the mid-fifteenth century, assertions of ancient authors were evaluated together with new information that eluded the systematicization of writing, we will consider three legends on the *mappa mundi*. In the first legend Fra Mauro explains how, in comparing the texts of ancient authors with the travel experiences of his contemporaries – that is, in comparing writing and experience (*scriptura* and *experientia*) in Fra Mauro's effective expression – he had come to the decision not to follow Ptolemaic cartography, either in shape or in measurements :

> But since I am aware of that he [Ptolemy] had not been able to verify his cosmography thoroughly, because it is a long and difficult thing, and life is short and experience fallible, he is actually admitting that with the passage of time the work could be improved – that is, one could have more certain information than that available to him. So I say that in my own day I have been careful to compare the texts with practical experience, investigating for many years and frequenting persons worthy of faith, who have seen with their own eyes what I faithfully report above.[92]

Equally significant is the legend in which Fra Mauro justifies himself for having placed the city of Babylon on the river Tigris, thus moving away from the tradition of ancient authors who had placed it on the Euphrates, and following instead the reports from travelers of his era :

> If to some it seems that I have not located Babylon well, because I have shown it on the Tigris and not the Euphrates, as the Authors write, may it please them to look first at the drawing and then

[89] O. Merisalo, "Il Commento," in Poggio Bracciolini, *De varietate fortunae*, O. Merisalo (ed.), Helsinki : Suomalainen Tiedeakatemia, 1993, p. 235, *s.v.* "Pancouiam."

[90] *Mappamondo* (ed. 1956) Pl. XX, 51, 56, 61, 63. *Mappamondo* (ed. 2006) *683, *684, *685, *692.

[91] Pelliot, *Notes on Marco Polo*, vol. 1, p. 182. Merisalo, "Il Commento," pp. 232-233.

[92] This is my translation of this legend : "[...] Però | [io] intendando lui [Tolomeo] no(n) haver possudo in tuto verificar la sua cosmogra|phia, sì p(er) la cossa longa e difficile, e p(er) la vita brieve, e l'experime(n)to fal|lace, resta che'l co(n)ciede che cum longença de te(m)po tal opera se possi meglio | descriver, over haverne più certa noticia de | quel habuto lui. Per ta(n)to dico che io nel te(m)|po mio ho solicitado verificar la scriptura | cu(m) la experie(n)tia, i(n)vestigando p(er) molti an(n)i | e pratica(n)do cu(m) persone degne de fede | le qual hano veduto ad ochio q(ue)lo che | qui suso fedelme(n)te demostro." This fragment is part of a legend that can be found in *Mappamondo* (ed. 1956) Pl. XL, 49 and for the largest part in the plate XLVI. A different translation and interpretation are provided in *Mappamondo* (ed. 2006) *2834.

ask those who have seen with their own eyes; they will thus understand that I do not stray from the truth.[93]

Finally, in the third legend Fra Mauro justifies his innovations in the design of the southern regions of Africa, which were unknown to the ancients and declared *terra incognita* by Ptolemy. For the preparation of this part of the map Fra Mauro says that he used the narrative and maps of some religious men from Ethiopia who had come to visit San Michele di Murano:

> Because to some it will appear as a novelty that I do speak of these southern parts, which were almost unknown to the Ancients, I will reply that I received this entire drawing, from Sayto upwards, from those who were born there. These people are clerics who drew for me, with their own hands, these provinces and cities and rivers and mountains with all their names; I have not been able to put all these things in due order for lack of space.[94]

Regardless of their specific contents, these three legends show that Fra Mauro compared written information with complementary and partly alternate information which claimed to be the result of direct experience. Through this comparison Fra Mauro evaluated and confirmed or denied what was written by authors, and vice versa. The issue of the close and interdependent relationship between knowledge and experience, and of the interaction between written knowledge and oral knowledge, is very clear in the writing of Fra Mauro. Half a century before the officers of the *Casa de Contratación de Indias* (also known as *La Casa y Audiencia de Indias*)[95] in Seville and then Cadiz, and l'*Armazém da Guiné e da India* (the Arsenal) in Lisbon,[96] began collecting in a systematic way the narratives of navigation and astronomical data in order to

[93] "Se'l parerà ad algun che io | no(n) habi ben posto Babilo|nia per hauerla descrita | sopra Tygris e non Eufra|tes, come scriueno li autori, piaquali prima co(n)|siderar q(ue)sto desegno e da poi do|mandi queli hano ve|duto ad ochio, e i(n)tenderano | che io no(n) me parto da la | verità." *Mappamondo* (ed. 1956) Pl. XXII, 93. The translation derives from *Mappamondo* (ed. 2006) *957.

[94] This is my translation of this legend: "Perché ad alguni par da nuovo che io parli | de questa parte meridional, la qual quasi està | incognita a li antichi, perhò io respondo che | tuto questo desegno da Sayto in suso io l'ò ha|buto da queli proprij che sono nasudi qui | che sono stà religiosi i qual cu(m) le suo man | me hano desegnato tute q(ue)ste p(ro)vi(n)cie | e citade e fiumi e mo(n)ti cum li | suo nomi, le qual tu|te cosse no(n) le ho possu|do meter cu(m) el debi|to ordine per | no(n) esservi logo." *Mappamondo* (ed. 1956) Pl. X, 30. For a slightly different translation see *Mappamondo* (ed. 2006) *98.

[95] E. Vila Vilar, A. Acosta Rodríguez, and A. L. González Rodríguez (eds), *La Casa de Contratación y navegación entre España y las Indias*, Seville: Universidad de Sevilla – Consejo Superior de Investigaciones Científicas (CSIC), 2004, in particular the following essays: R. Cerezo Martínez, "Los padrones reales del primer cuarto del siglo XVI," pp. 605-637; A. E. Buttinger, J. D. Buttinger, and G. M. Rivera, "Mateo Jorge, a Pilot of the Casa de la Contratación in Sixteenth Century Seville: A Study in the Transmission of Science and Technology as Expressed in the Graphics of a Rutter of Practical Navigation," pp. 639-665; and L. Martín-Merás, "Las enseñanzas náuticas en la Casa de la Contratación de Sevilla," pp. 667-693.

[96] On the translation of the technical terms connected to the Portuguese expansion, see P. E. H. Hair, "Portuguese Documents on Africa and Some Problems of Translation," *History in Africa* 27 (2000), pp. 91-97. Regarding the Armazém: "An Armazém is a structure (both material and social) that combines the concepts of a warehouse and an army base. Perhaps the best translation would be "arsenal," but only in the now-obsolete broader meaning of earlier centuries (e.g., the Arsenal of Venice)." See also A. Cattaneo, "From the Mediterranean to the Atlantic. Venice, Florence and Lisbon. Commercial Routes and Networks of Knowledge, 1300-1550," in J. A. Levenson (ed.), *Encompassing the Globe: Portugal and the World in the 16th & 17th Centuries. Essays* [Vol. 3], Washington, D.C.: Arthur M. Sackler Gallery, 2007, pp. 13-21.

update the *padrones reales* or *padrões reais* (master-charts) in the context of the Iberian Empires, Fra Mauro claims to have relied upon and transcribed in both literary and cartographic form the testimony of several travelers who passed through Venice. Merchants, pilgrims traveling to and from the Holy Land, missionaries, and ambassadors who came to Venice and from there embarked, became his interlocutors.

The legends of the *mappa mundi* make it possible, in accordance with the explicit intention of Fra Mauro, for us to hear the thoughts of social groups who were often "voiceless," who did not leave, except in a few rare cases, written testimonies, but who nevertheless played a crucial role in the formation of cosmographic and geographical knowledge in the medieval and early modern ages. The information that Fra Mauro acquired from his anonymous interlocutors is highlighted as trusted and valuable knowledge. Fra Mauro was aware that in order to update the representation of the *imago mundi* the cartographers of his time should also rely upon oral knowledge that most of the time was not systematic and quite often imprecise : the task of the cartographer was thus to select the information, transcribe it, and compare it with what was handed down in books and other maps.

The oral testimony of travelers allows Fra Mauro to found on first-hand experience his criticism of part of the ancient geographical tradition and also – as the third of the legends examined above well illustrates – to fill in certain gaps. But it would be a vulgar mistake to see in this a sign that Fra Mauro was breaking with the ancient or medieval tradition in the name of the superiority of "modern experience." Fra Mauro criticizes the cosmographic knowledge of the ancients and the moderns, not only through the use of modern experience, but also based on his critical reading of works such as the *Cosmographia* of Pomponius Mela, the *Geography* of Ptolemy, the *Liber de causis* and *De natura loci* of Albert the Great, and the *Sphera* of Sacrobosco, from which clearly discordant or contradictory data emerge.

Fra Mauro, like other fifteenth cosmographers, was aware that the only systematic representations of the *oikumene* that were available were those received through the writings of ancient authors who, however, proposed representations and conceptions that differed in some points and need to be updated. For example, since the early decades of the fifteenth century Guillaume Fillastre and Pierre d'Ailly had discovered that the text of *Geography*, available in Latin from 1409 and known to them during the Council of Constance (1414-1418), differed on matters of primary importance – such as the size of the *oikumene*, the shape and extent of the seas – from other ancient texts such as the *Almagest* and *Tetrabiblos* of Ptolemy himself, the *Cosmographia* of Mela, and the *Polyhistor* of Solinus.[97]

In the *mappa mundi*, contrary to what has recently been suggested, there is no celebration of epistemological breaks with the past or "revolutions," as there are no epistemological revolutions in early modern cosmography. Recourse to experience in the work of Fra Mauro, as well

[97] P. Gautier Dalché, "L'œuvre géographique du Cardinal Fillastre († 1428). Représentation du monde et perception de la carte à l'aube des découvertes," in D. Marcotte (ed.), *Humanisme et culture géographique à l'époque du Concile de Constance : autour de Guillaume Fillastre*, Actes du Colloque de l'Université de Reims, 18-19 novembre 1999, Turnhout : Brepols, 2002, pp. 293-355 ; Id., "The Reception of Ptolemy's *Geography* (End of the Fourteenth to Beginning of the Sixteenth Century)," in D. Woodward (ed.), *The History of Cartography*, vol. 3, *Cartography in the European Renaissance*, Chicago : University of Chicago Press, 2007, pp. 285-364, esp. pp. 299-306 ; and Id., *La Géographie de Ptolémée en Occident (IVe-XVIe siècle)*, Turnhout : Brepols, 2009, pp. 168-180.

as in the culture of the time, has absolutely nothing to do with "experimentalism." The world of experience mentioned by Fra Mauro is nothing "experimental."[98] Rather, it is the world of travelers, merchants, artists and artisans, architects, nautical cartographers, and also of sailors and officers of the fleet – all of these have in common, apart from the specific scope of their actions, a knowledge which is the direct result of experience, which increases when new answers to practical problems emerge. By taking into consideration the geographic and cartographic discoveries made by those who, in traveling, had had direct experience of the places described by ancient authors, or who had even gone beyond, in some cases, the boundaries those authors had set to the *oikumene*, Fra Mauro was using precisely the *modus operandi* of the cartographers of marine charts. At least from the twelfth century this had become an important practice in western culture. Fra Mauro was not doing anything new, then, and certainly nothing "experimental." At the same time it is important to remark that also the written works of authors such as Pliny, Pomponius Mela, Ptolemy, were believed to be based upon and convey experience.

His innovation was of a different nature. Fra Mauro not only knew how to listen, consider, and evaluate the testimony of travelers, and not only knew how to transform that information into a cartographic design – as many other cartographers, at least from the thirteenth century, also knew how to do – but he also knew how to compare that information with the geography of cultured authors. The Camaldolese friar could use and insert the new information (oral knowledge) into a context of bookish learning, with respect to which the new information was not only not contradictory, but rather complementary. Further, he realized the need, in a document written in the vernacular, to include the experience of moderns among the most important sources to draw the map of the world, sometimes beside the classical authorities, some other times as their replacement. Part of the information needed to renew the *imago mundi*, Fra Mauro insists, was produced by the direct experience of travelers, not only of western Christians, but also of Arabs and "Indians."

The task of the cartographer was precisely to collect their stories and, after comparing them with what was handed down by the authorities, put them down in both written and graphic forms in the world map. In so doing, Fra Mauro was not being "experimental," but was rather an important part of the learned cultural heritage of his time, and specifically of Aristotelian empiricism. From the twelfth century, the appeal to experience is an important feature of Aristotelianism, especially in the Veneto.[99] In cities such as Venice where there was – as Marica Milanesi has clearly pointed out – a notable production of cartographic and geographic knowledge related to the experience of voyages (made by merchants, travelers, and then realaborated by makers of nautical charts) even cultural milieux that were generally very conservative such as universities and monasteries were willing to consider and often to receive the latest news that emerged from the world of artisans, travelers, and merchants.[100] This is the context that explains how a monk and cultured cosmographer like Fra Mauro, in order to update the *imago mundi*, theorized the need to integrate writing and experience, that is, to take into account both sciences, that of the Latin authors which was more systematic and structured, and also the results, or

[98] For a contrary view, see Falchetta, *Fra Mauro's World Map*, p. 123.

[99] A. Poppi (ed.), *Scienza e filosofia all'Università di Padova nel Quattrocento*, Padua: Lint, 1983.

[100] M. Milanesi, "G. B. Ramusio e i piloti portoghesi," in L. Cusati (ed.), *Atti del Congresso Internazionale 'Portogallo e i mari: un incontro tra culture'*, Naples: Guida, 1998, pp. 231-248 (p. 237).

rather the written or more frequently oral testimonies, of the vulgar science. It was in cultural contexts characterized by familiarity with the procedures of both scientific fields and by the ability to compare and integrate their results, that is, in Venice, Barcelona, Genoa, Majorca, Ancona, and a few decades later, also in Portugal, that the cartography of the fifteenth and sixteenth century took shape.[101]

IV. – The Narrative structure of Fra Mauro's writing

In addition to around 2,500 toponyms (kingdoms, provinces, cities, seas, gulf, rivers, lakes, mountains, etc.), there are roughly 300 legends in the *mappa mundi*'s graphic space that can be subdivided in three principle groups : a first group of didactic legends written in third person, comprising 260 legends along with seven additional long cosmological notes posted on the exterior of the circular *mappa mundi* ; a second group composed of twenty-eight *notabilia* ; and finally, thirty-four legends – which in part coincide with the preceding group – in which Fra Mauro, using explicitly the pronoun '*io*' ('I'), speaks directly to the reader in first person around eighty times. Fra Mauro's *mappa mundi* is the oldest surviving cartographic document in which the latter narrative modality can be observed.[102]

1. Didactic legends in third person

The legends in third person are introduced by the subject they regard or, otherwise, they begin with the phrase, "In this (island, mountain, kingdom, province, etc)".[103] Other, less frequent forms of introduction include, "Here are born..." (nine legends), "Here are..." (five legends), "Here die..." (two legends), "Here they gather" (two legends), "Here are found..." (two legends), "Here they pay..." (one legend), "Here they make..." (one legend), "Here begins..." (one legend), and also "They say..." (two legends).[104] Fra Mauro utilizes didactic

[101] On this subject see also R. Pujades, *Les cartes portolanes. La representació medieval d'una mar solcada*, Barcelona : Institut Cartogràfic de Catalunya, 2007, pp. 30-58 (English translation, pp. 414-420) and pp. 290-334 (English translation, pp. 506-516).

[102] P. Gautier Dalché, "Weltdarstellung und Selbsterfahrung : Der Kartograph Fra Mauro," in H.-D. Heimann and P. Monnet (eds), *Kommunikation mit dem Ich*, Bochum : Dieter Winkler, 2004, pp. 39-51.

[103] See for example : "Mahal isola habita[da] per christiani, ne la qual è uno | archiuescouado, et in la dita se troua gran qua(n)tità de ambra e i chorsari de questi mari ha|no suo reduto qui, e qui spaçano e in deposi|to meteno le sue robarie. El luogo doue | habita el ueschouo se chiama schoria(n)," *Mappamondo* (ed. 1956) Pl. III, 6 ; "Mahal, an island inhabited by Christians, where there is an archbishopric. On said island there is a lot of amber to be found, and the pirates of these seas have their stronghold here, and here they sell and store their loot. The place where the bishop lives is called Schorian," *Mappamondo* (ed. 2006) *4. "Queste do' isole sono habitade p(er) christiani. In una | de le qual çoè in nebila habita le done e in l'altra dita | mangla habita li lor homeni, i qual solamente tre | mesi de l'ano stano con le done," *Mappamondo* (ed. 1956) Pl. III, 1 ; "These two islands are inhabited by Christians. In one of them, called Nebila, live the women ; in the other, called Mangla, leave the men, who pass only three months a year with the women," *Mappamondo* (ed. 2006) *6.

[104] For instance : "Qui nasce | peuere," *Mappamondo* (ed. 1956) Pl. XV, 36, "Here pepper grows," *Mappamondo* (ed. 2006) *291 ; "Qui sono | molte simie," *Mappamondo* (ed. 1956) Pl. XV, 11, "Here there are many monkeys," *Mappamondo* (ed. 2006) *301 ; "Qui muor questa | aqua," *Mappamondo* (ed. 1956) Pl. IX, 21, "Here this watercourse

descriptions in third person, when, either explicitly or implicitly, he cites or synthesizes written sources. In most cases, these legends – already analyzed in respect to their specific contents in earlier chapters – are derived from the accounts of Marco Polo and Niccolò de' Conti, or from Pliny, Solinus and Pomponius Mela, Ptolemy. However, the long discussions of natural philosophy placed in the four angles of the *mappa mundi* are reworking of, primarily, passages from Albert the Great's *De natura loci* and his *De causis proprietatum elementorum*.

It is important to note that there is no compositional difference between the 262 didactic legends in third person placed on the interior of the circular *mappa mundi* dealing with geography, geographic etymology, local customs, commerce, historical events (especially Alexander the Great and Tamerlane) and the long, complex legends on cosmological and cosmographic matters in the exterior angles of the *mappa mundi*.[105] Fra Mauro's compositional method did not change, no matter whether he was citing, summarizing or reworking passages from Book IV of *De varietate fortunae*, or the *Milione* (as is noted in Chapter V, he probably had seen a Latin version from the textual recension Z and a version in the Venetan vernacular), or, by contrast, Albert the Great's commentary on *De causis* and *De natura loci*.

The legends of the *mappa mundi* take a stance with regard to their original sources that is marked by an untrusting usage, analogous to other vernacularizations of the time.[106] As already shown in the chapter dedicated to Fra Mauro's cosmographic learning, there was a reawakened interest in natural philosophy even outside the university on the part of the *scholae* of the religious orders, the doctoral colleges, and in cultural environments where there thrived a passive knowledge of Latin while privileging the use of the vulgate. Similar to the strategies of vernacularization that can be observed in the *De proprietatibus rerum* and the *Metaura*, in Fra Mauro's vernacularization and adaptation of a few passages from Albert the Great's *De causis* and *De natura loci*, and from the *Milione* and *De varietate fortunae*, the hierarchical structure and the order of arguments – whether in Latin or the vernacular – is broken apart, and parts that are not absolutely necessary to the topic at hand are eliminated. Fra Mauro did not limit himself to omitting secondary information; he additionally he subjected all the selected texts to a process of general simplification that privileged information that he believed, in an entirely idiosyncratic fashion, to be more important or better suited to his overall narrative

ends." *Mappamondo* (ed. 2006) *47; "Qui se fano lavo|ri de porcel|lane." *Mappamondo* (ed. 1956) Pl. XXV, 36, "Here porcelain work is made." *Mappamondo* (ed. 2006) *1344; "Qui co|menza | el mar scuro." *Mappamondo* (ed. 1956) Pl. III, 11, "Here begins the dark sea." *Mappamondo* (ed. 2006) *2; "Qui se pa|ga el da|tio." *Mappamondo* (ed. 1956) Pl. XVI, 41, "Here tolls and levies are paid." *Mappamondo* (ed. 2006) *411; "Qui se recoie | oro." *Mappamondo* (ed. 1956) Pl. XVIII, 29, "Here gold is found." *Mappamondo* (ed. 2006) *575; "Qui se troua | oro de paiu|ola." *Mappamondo* (ed. 1956) Pl. XXVI, 91, "Here there is gold dust." *Mappamondo* (ed. 2006) *1426; "Se dice che presto Jane ha più de 120 regni soto el suo dominio...." *Mappamondo* (ed. 1956) Pl. X, 21, "It is said that Prester John has more than 120 kingdoms under his dominion...." *Mappamondo* (ed. 2006) *77.

105 An opposing opinion was expressed by Piero Falchetta, who detected a strong difference between the legends within the circular *mappa mundi* which he termed "experimental" and saw as being inspired by a critical rationality, and the cosmological ones on the exterior, which he termed "traditional" and "doctrinal," and which he considere to have therefore little significance to intellectural history. I do not agree with this interpretation. See the last part of Chapter II.

106 R. Gualdo (ed.), *Le parole della scienza: scritture tecniche e scientifiche in volgare, secoli XIII.-XV*, Galatina: Congedo, 2001; R. Librandi, *La Metaura d'Aristotile: volgarizzamento fiorentino anonimo del XIV secolo*, critical edition, 2 vols., Naples: Liguori, 1995.

structure. Sometimes, in fact, there are instances of redundancy, with repetitions of some key concepts such as references to the opulence and immensity of the cities of Asia (as opposed to, for example, Aristotelian concepts about the sublunar sphere). The legends written in third person by Fra Mauro clearly show the way in which, during the late medieval period, the relationship between a Middle Latin source and a vernacularization was structured in regard to the selection, the ordering of the information, and the vocabulary, with the intention of attaining an equilibrium between the need to preserve the rigor demanded by the contents, and the tendency to adapt texts to the culture of the non-specialist readership.[107]

In their respective studies of the *Metaura* (that is, the fourteenth-century Tuscan vernacularization of Aristotle's *Meteorology*) and the translation of Bartholomeus Anglicus's encyclopedia, *De proprietatibus rerum*, into the vernacular of Mantua by the notary Vivaldo Belcalzer between 1299 and 1309, Rita Librandi and Rosa Casapullo have demonstrated that the Latin text was treated only as an intermediary stage – a pre-text, so to speak – for works that demonstrate the faint links with the originals that they were, in turn, translated from or inspired by.[108] In a similar way, Fra Mauro produced something quite new in comparison to the texts from which he took his departure : the *mappa mundi*'s cartouches in third person give shape to a work that has an independent identity that, despite its dependence upon Latin commentaries on Aristotle, the works of Marco Polo, Poggio Bracciolini, Pliny, Pomponius Mela, Ptolemy, as well as the *florilegia*, remains autonomous with respect to its sources.

Fra Mauro was not seeking to write a treatise on natural philosophy along the lines of those written by contemporary Venetian scholars of Aristotle like Paolo Veneto, Gaetano of Thiene, or Giovanni Fontana.[109] His goal was to extrapolate from the Aristotelian commentaries available to him that which served his need to contextualize the description of the earth and sea within an understanding of the structure of the sublunary and celestial worlds. What results is a cosmographic knowledge that is distinct both from that found in the Aristotelian commentaries and that was produced and commented upon in the *studia*, for the simple reason that this was not created for the climate of rigorous study in the *scholae*, but rather for the curious readership made up of individuals who participated in the highest emanation of culture while avoiding the linguistic obstacles presented by Latin and the most complex and theoretical aspects of Aristotelian natural philosophy. Therefore, the cosmographic cartouches of the *mappa mundi* transmit only a minimal part of the Aristotelian text, leaving significantly more space

[107] For a general overview, R. Librandi, "Il lettore di testi scientifici in volgare" in *Lo spazio letterario del medioevo*, vol. 3 : *La ricezione del testo*, P. Boitani, M. Mancini and A. Varvaro (eds), Rome : Salerno, 2003, pp. 125-154. Ead. "Auctoritas e testualità nella descrizione dei fenomeni fisici," in *Le parole della scienza. Scritture tecniche e scientifiche in volgare*, pp. 99-126.

[108] Librandi, "Auctoritas e testualità nella descrizione dei fenomeni fisici," pp. 103-110. R. Casapullo and M. R. Policardo, "Tecniche della divulgazione scientifica nel volgarizzamento mantovano del 'De proprietatibus rerum' di Bartolomeo Anglico," *Lingua e stile* 38 (2003), pp. 139-176. See also R. Casapullo, "Segmentazione del testo e modalità d'uso delle enciclopedie tra latino e volgare," in *Le parole della scienza. Scritture tecniche e scientifiche in volgare*, pp. 153-185.

[109] On Paolo Veneto, Gaetano di Thiene and Giovanni Fontana in the context of Paduan Aristotelianism, see F. Lepori, "La scuola di Rialto dalla fondazione alla metà del Cinquecento" and F. Dupuigrenet Desroussiles, "L'Università di Padova dal 1405 al Concilio di Trento," in *Storia della cultura veneta*, vol. 3.2, pp. 539-605 and pp. 607-647 ; L. Olivieri (ed.), *Aristotelismo veneto e scienza moderna*. Atti del XXV anno accademico del Centro per la storia della tradizione aristotelica nel Veneto, Padua : Antenore, 1983, pp. 205-220. On Fontana see Chapter IV.2.

for the commentators, in particular Albert the Great. Moreover, the Aristotelian text occupies only the smallest percentage of space among the comments and vernacularizations ; it is deliberately marginalized with regard to the expository writings of commentators and translators, whether these were from the Muslim tradition or the Christian one. In fact, the codices that transmitted Albert the Great's and Thomas Aquinas' commentaries on Aristotle's works, by then considered to be autonomous works, sometimes are cited for their commentaries alone, without inclusion of the Aristotelian text they explicate.[110]

Closely relevant to this is the point made by Giovanna Murano in her recent study of the *modus operandi* of medieval copyists : in the workshops specialized in the production of manuscripts for use in the universities, there were both the *exemplaria* of Aristotle's works as well as *exemplaria* of commentaries on those same Aristotelian works by authors such as Albert the Great, Thomas Aquinas, Adam de Buckfield, etc. The transcriber of the codex could choose from either of these categories : in the case of the vernacularization of the *Metaura*, Rita Librandi has demonstrated that the translator chose to base part of his translation on a work by Aristotle and part on the citations of Aristotle found in a commentary, thus creating a "hybrid".[111]

Utilizing compositional techniques identical to those employed in the long, external legends, Fra Mauro did not include lengthy, direct citations from the *Milione* or Book IV of *De varietate fortunae* within the circular *mappa mundi.* Marco Polo's and Poggio Bracciolini's texts were instead summarized, adapted, occasionally condensed into just a few words, or even illustrated with ample creative license in order to create a graphic image of the entirety of Asia emphasizing its great cities and its wealth. The limited space for graphic description obliged the cartographer to restrict his imagery to a small selection of themes, just as he was elsewhere constrained to synthesize his verbal descriptions : the exposition is abbreviated as well as focused (or so it was intended). Because of this contingent factor, and in close connection with the medium he was employing, Fra Mauro moves directly to the conclusions of the Aristotelian commentators, and likewise he jumps directly to the conclusions – as he defines them, at least – of the other sources he availed himself of. "Tradition," "doctrine," and "experimentalism" have no bearing upon this compositional and narrative technique.

2. *Notabilia*

A second group, composed of twenty-eight legends, is introduced by the words "Take note that," followed either by descriptions of places or events that Fra Mauro condenses from his sources, or by observations that throw light upon particular aspects and specific interpretations of the narrative, in which he declares explicitly whether or not he is in agreement with other competing interpretations. These are undoubtedly a form of *notabilia*, a literary genre whose name derives from the observations located in the margins of Scholastic books on law or philosophy above all. These annotations began with the stock formula, "*Nota quod*...," which Fra Mauro translates literally as "*Nota che*..." (Take note that...). The function of the *notabilia*

[110] A. J. Minnis, *Medieval Theory of Authorship : Scholastic Literary Attitudes in the Later Middle Ages*, London : Scolar Press, 1984, p. 3.

[111] G. Murano, *Opere diffuse per exemplar e pecia*, Turnhout : Brepols, 2005, pp. 22-38.

was to draw the reader's attention to important themes and points that were either in the textual source itself, or deduced from its contents. Based on the practice of writing these annotations along the margins there arose from these annotations an independent writing form.[112] The presence of these *notabilia* indicates that, as he was drafting the map, Fra Mauro reviewed not just the sources, but also the arguments they contained. These twenty-eight legends are a clear proof of the existence of cosmographic syllogisms prepared by Fra Mauro, the so-called *scritture* mentioned by the Abbot of San Michele in Murano, Gherardo Maffeo, in his account book *Entrata e uscita di San Michele* for the year 1464 ; these *scritture* have probably been destroyed, or perhaps they have still not been found. The map thus makes public those personal annotations Fra Mauro made in the course of his reading, allowing his readership to partake in his learning and his thoughts.

These *notabilia* can be divided into two groups : those in which Fra Mauro makes evident some geographic precepts or facts, And those in which he expresses his own opinion on concepts and controversies, and above all those observations that he found to be lacking in some way, contradictory, or (to his mind) erroneous. In the first category, these *notabilia* are often taken from a source whose identity is usually hidden by Fra Mauro and certainly not easy to deduce. Take, for example, this legend in which the origin of the name of England is discussed :

> Note that in ancient times Anglia was inhabited by giants, but some Trojans who had survived the slaughter of Troy came to this island, fought its inhabitants and defeated them ; after their prince, Brutus, it was named Britannia. But later the Saxons and the Germans conquered it, and after one of their queens, Angela, called it Anglia. And these peoples were converted to the Faith by means of St. Gregory the pope, who sent them a bishop called Augustine.[113]

Its direct source – although not named here – is probably the *Dittamondo* by Fazio degli Uberti (*Dittamondo*, XXIII, vv. 1-6),[114] invoked explicitly once, along with Pliny, in a legend that refers to the circumnavigation of Africa in the cartouches of the *mappa mundi*, and a direct source for at least three others including the present one cited here.[115]

3. Legends in first person : the cosmographer as an authoritative narrator

The third and final group consists of thirty-three legends – in part overlapping with the preceding group – that contain Fra Mauro's comments and personal observations, expressed in first person, introduced by the pronoun '*io*' ('I'), often used in an adversative sense with respect

[112] P. Erdó, *Storia della scienza del diritto canonico : una introduzione*, Rome : Editrice Pontificia Università Gregoriana, 1999, pp. 70-71.

[113] "Nota che la insula anglia fo anticham(en)te habitada p(er) çiga(n)ti, | ma alguni troiani che rimase da può l'excidio de troia ue|ne a q(ue)sta i(n)sula (et) haue (con)flicto cu(m) hi habita(n)ti e sup(er)oli e | dal suo p(r) incipe Bruto la nominò britania, ma da poi i | saxonij e germani la (con)q(ui)stò e da una sua regina dita A(n)gela la nominò a(n)glia, e q(ue)sti populi fono a la fede co(n)uertidi p(er) meço | de sa(n) Gregorio papa, el qual lì ma(n) dò uno uescouo dito Augustino," *Mappamondo* (ed. 1956) Pl. XXXVI, 16 ; the English translation derives from *Mappamondo* (ed. 2006) *2706.

[114] Fazio degli Uberti, *Il Dittamondo*, XXIII, vv. 1-6 and XXIV. See Falchetta's comment to the legend *2706, *Mappamondo* (ed. 2006), p. 672.

[115] See Falchetta's comment to the legends *1886 *2845, and *2907 in *Mappamondo* (ed. 2006), p. 530, 702, 714.

to the other authors that the Camaldolese friar was criticizing, correcting, and contradicting. These legends animated an intense dialogue that Fra Mauro took up not only with the authors, but also with his own readership.[116] The cartouches in which the first person is employed are located all throughout the *mappa mundi*, with a few areas of special concentration. The vast marine spaces that surround the *oikumene*, North Africa, central and northern Asia contain the greatest number, while at least three fundamental cartouches of this kind can be found in Europe, in a very central position with respect to the entire map, highly visible and legible to spectators. The greater availability of graphic space in these sections, along with the awareness that knowledge of these territories was contradictory and uncertain – thus pointedly requiring the intervention of the cosmographer – explain the dislocation of the legends written in first person in these areas. In the concentrated dialogues with his readership, Fra Mauro completely reveals himself, referring on the one hand to his diligence, and on the other hand to his own limits, as in the following legend regarding the design of the Mediterranean coastlines : "Though I have been most diligent in trying to put all the coastlines of this sea in accordance with the most accurate map that I possess, those who are experts should not take it amiss if I am not always consistent. Because it is not possible to put everything accurately."[117] Then there are the instances where he corrects and brings into sharper focus the information transcribed in other cartouches : "Although in the nearby note I say that the spices travel as far as the Black Sea, today the roads are in such a poor state that they no longer reach that far."[118] Above all, it can be determined that Fra Mauro makes recourse to the pronoun '*io*' ('I') when he wants to distinguish and emphasize his own opinion and his own free thought with respect to a multiplicity of *auctoritates* that are cited from time to time in the *mappa mundi*. The '*io*' of the cosmographer, raising him up to the level of the *auctores* with whom he engages either to underline or contradict their opinions in a debate that the Camaldolese friar seems to wish to extend even to include the cosmographers of his own time, though cloaked in anonymity.[119] In this specific context, it will suffice to give a few paradigmatic examples in which Fra Mauro offers his own thoughts in first person by means of the emphatic use of the pronoun 'io' :

[116] Refer to this chapter's Documentary Appendix for the transcription of the legends in which Fra Mauro utilizes the first person.

[117] "Benchè io habi s(er)uato ogni diligencia i(n) meter le staree | de q(ue)sto mar seco(n)do la più iusta carta ho possudo, no(n)dime(n) | queli che sono exp(er)ti no(n) façа gra(n) caso se io me discordo | i(n) qual cossa, perochè no(n) è possibile meter tuto a ponto." *Mappamondo* (ed. 1956) Pl. XXIII, 153 ; the English translation derives from *Mappamondo* (ed. 2006) *1888.

[118] "Benchè i(n) q(ue)sta | proxima rubri|ca io habia dito | che le spetie uano | fina al mar de po(n)to, | ahora sono le stra|de sì rote che le | no(n) uano ta(n)to aua(n)|ti." *Mappamondo* (ed. 1956) Pl. XXI, 34 ; the English translation derives from *Mappamondo* (ed. 2006) *749.

[119] For example : "Nota che i cosmographi fano | diuisio(n) de le mauritanie e chiamano una cesariensis e l'altra stifensis e la | terça tingintanea, e q(ue)sti nomi sono deriuadi da | cesaria, stifi e tingi, i qual erano casteli cussì nomi|nati, ma ahora q(ue)sti nomi a nui no(n) sono noti molto, | perhò no(n) ne façо q(ue)sta distintion." *Mappamondo* (ed. 1956) Pl. XXIV, 32. "Note that the cosmographers divide the Mauritanias, calling one Cesariense, one Stifensis and the third Tingintanea. These names derived from those of three fortified towns : Cesaria, Stifi and Tingi. However, to us, they are not very well known, so I have not made this distinction." *Mappamondo* (ed. 2006) *1219. These cosmographers are identified as being all *auctores veteres* except for one, who is proposed as being a cosmographer contemporary to Fra Mauro, in Gautier Dalché, "Weltdarstellung und Selbsterfahrung," pp. 47-48.

> Some authors write that the Sea of India is enclosed like a pond and does not communicate with the ocean. However, Solinus claims that it is itself part of the ocean and that it is navigable in the southern and south-western parts. And I myself say that some ships have sailed it along that route.[120]

Another example is this: "Some write that in these Indies there are many types of human and animal monsters, but because few people believe these things, here I have made no note of them [...]"[121] The use of the personal pronoun 'io' in a deliberate and emphatic way is a conscious means of reclaiming for the role of the cosmographer an authoritative agency in the process of selection. The pronoun 'io' recognizes and reinforces the authenticity of his declarations and, at the same time, affirms his ability to criticize and dispute the *auctoritates* – those sacred and profane texts that were institutionally recognized as having a presumption of truth.[122] The modern author is *arbiter* and guarantor of the knowledge that permits him to decide which among his sources are worthy of citation.

In this context, even the position of the reader undergoes a notable transformation. The reader becomes a witness, one to whom the cosmographer makes direct recourse to obtain approval of the choices he has made. The reader is also called upon to compare various versions of the traditional *lectiones*. In other words, in the thirty-three legends written in first person, Fra Mauro critically reviews the sources and organizes a close comparison among them for the reader himself to weigh in on the matter as a third-party witness or judge. With these legends written in first person, Fra Mauro reveals the process of selection he applied to his sources. In this sense, the *mappa mundi* is among the first texts in which the readers are made aware of the critical choices made by the author during the composition of the work. On the one hand, therefore, Fra Mauro confronts the problem of justifying his choice of which sources to utilize, while on the other hand, he wishes to make manifest to his readership the process by which he had arrived at his choices. All this is enacted by Fra Mauro in various cartouches, almost as a kind of theater, as suggested in the following legend:

> If to some it seems that I have not located Babylon well, because I have shown it on the Tigris and not the Euphrates, as the Authors write, may it please them to look first at the drawing and then ask those who have seen with their own eyes; they will thus understand that I do not stray from the truth.[123]

120 "Alguni autori sc(r)iue del mar d'i(n)dia che'l sia serado come un stagnon | e che'l mar occea(n) no(n) li entri, ma Solin uol che'l sia occea(n) e ch(e) q(ue)la p(ar)te austral | e del garbi(n) sia nauigabile, et io affermo che algune naue açira e uol|ta quel cami(n)," *Mappamondo* (ed. 1956) Pl. X, 26. The English translation derives from *Mappamondo* (ed. 2006) *53.

121 "Alguni scriueno che in queste | Indie sono molte diuersi|tà de mo(n)stri [...] qui [io] no(n) ne faço nota [...]," *Mappamondo* (ed. 1956) Pl. XX, 68. The English translation derives from *Mappamondo* (ed. 2006) *707.

122 Discussions of, and thorough bibliographies for, the concepts of *auctor* and *auctoritates* are found in M. Teeuwen, *The Vocabulary of Intellectual Life in the Middle Ages*, Turnhout: Brepols, 2003, pp. 222-223.

123 "Se'l parerà ad algun che io | no(n) habi ben posto babilo|nia per hauerla descrita | sopra tygris e non eufra|tes, come scriueno li au|tori, piaquali prima co(n)|siderar q(ue)sto desegno e da poi do|ma(n)di queli hano ue|duto ad ochio e i(n)tenderano | che io no(n) me parto da la uerità," *Mappamondo* (ed. 1956) Pl. XXII, 93. The English translations derives from *Mappamondo* (ed. 2006) *957.

Here, then, are the actors in this scene: the primary role is that of the cosmographer, that is, Fra Mauro, authoritatively introduced by the pronoun "*io*"; a second, collective role is played by the *auctoritates*, "the Authors"; the third, also a collective role, is played by the merchants and travelers, "those who have seen with their own eyes." The final actor to enter the stage are the readers, indicated as "some", representing a specific category of well-informed individuals, whether they be humanists, theologians, or learned cosmographers. The scene unfurls as follows: if the informed readers would wish to dispute the cosmographer, that is, Fra Mauro, on account of his choices and the cosmographic design made in contradiction to the claims of "the Authors", these (humanists, theologians, or learned cosmographers) are invited to ask "those who have seen with their own eyes" who is right. It is nevertheless clear that the "opinion" of these witnesses does not occupy a literary space with any real existence beyond Fra Mauro's authoritative *reportatio*, and those who are called to verify the contrasting opinions of the cosmographer and the *auctoritates* have no choice but to accept the decision made by the former.

In this regard the *mappa mundi* is clearly a humanistic work, given the critical selection from among the results of an erudite program of research, combined with a growing interest in the direct experiences of the author or his oral sources deemed "trustworthy" (*degne di fede*). Marica Milanesi identified in the method of Humanistic textual critique outlined by Petrarch "the foundation of this new type of exegesis of the geographic sources"; a method that "compares various texts dealing with the same topic, not for the sake of amalgamating all available information, but rather in order to select out the most reliable information while rejecting the rest as false. A method that places in center stage the problem of the reliability of sources, beginning with those that – given their age and the prestige of their authors – do not supply guarantees of their proven value".[124]

With this very same frame of mind, Poggio Bracciolini questioned the Ethiopian monks about the sources of the Nile, and then Niccolò de' Conti during the Council of Florence; it is also the frame of mind exhibited by Domenico Silvestri in his selection of preferred information from among all the cited versions of *De insulis et earum proprietatibus*. Without getting too deeply involved in fundamental questions about the relationship between tradition and experience that is so closely linked to the relationship between Scholasticism and Humanism already analyzed above, it is necessary to point out that not all Humanistic works took inspiration from these principles of research and composition. Highly revealing is the case that Biondo Flavio recalls in Book IV of *Decades* regarding the three cardinals who, although surely not unaware of Humanistic culture, did not believe the accounts of the Ethiopian monks who were questioned regarding the geography of Africa on the occasion of the Council of Florence in 1439 (inasmuch as their claims contradicted what was stated in Ptolemy's *Geography*).[125] Moreover, already beginning in the fourteenth century, Petrarch, Boccaccio, and even Fazio degli Uberti in *Dittamondo* – foreshadowing certain aspects of Humanism – clamored for the need to compare the often inconsistent opinions and wisdom of the *auctores*, and to judge these in light of new discoveries.

A reading of the thirty-three legends in question suggests a second fundamental significance of Fra Mauro's use of the pronoun '*io*'. It can be observed that Fra Mauro makes use of

[124] M. Milanesi, "Il 'De insulis et earum proprietatibus' di Domenico Silvestri (1385-1406)," *Geographia Antiqua* 2 (1993), pp. 133-146 (p. 135).

[125] Gentile, *Firenze e la scoperta dell'America*, pp. 168-170, entry 81.

it when he is indicating information of absolute novelty in the West, such as in the description of the Ethiopian regions in which he affirms that "I have head from none other than those who were born here" (*ho habuto da queli proprij che sono nasudi qui*), or when he criticizes Ptolemy's *Geography* on the base of novelties whose origin he ascribes to Portuguese and Arab navigation in the Sea of the Shadows. To clarify this aspect of the *mappa mundi*, let us look closely at the cartouche that addresses the possibility of circumnavigating Africa.

> Many opinions and many texts claim that in the southern regions the water does not surround this whole inhabited and temperate area. But I have heard many opinions to the contrary, above all from those who were sent by his Majesty, King of Portugal, in caravels so that they might explore and see with their own eyes. These men say that they sailed around the coast from the south-west some 2,000 miles beyond the Strait of Gibraltar. And following that route, they then decided to sail south-southeast until they came to Tunis and almost as far as Alexandria ; and at each place they found good shores, with deep water and good navigation conditions without any hindrances. These men have drawn new navigation maps and have given names to rivers, gulfs, capes and ports, of which I have had a copy. So, if one wants to disagree with these men, who have seen with their own eyes, then there is all the more reason to dissent from and not believe those who have left writings on things they did not see with their own eyes but only believed to be the case. What is more, I have spoken with a person worthy of trust, who says that he sailed in an Indian ship caught in the fury of a tempest for 40 days out in the Sea of India, beyond the Cape of Soffala and the Green Islands towards west-southwest ; and according to the astrologers who act as their guides, they had advanced almost 2,000 miles. Thus one can believe and confirm what is said by both these and those, and that they had therefore sailed 4,000 miles. In the third book of his *Cosmography*, Pomponius Mela says that a certain Eudossus, fleeing from Lathmin, king of Alexandria, sailed out of the Arabian Gulf and navigated though those southern regions until he arrived at Gades, which is the Strait of Gibraltar. One can therefore claim without any doubt that this southern and south-western part is navigable, and that the Sea of India is an ocean and not an inland sea. This is what is said by all those who sail this sea and live in those islands.[126]

[126] "Molte opinio(n) e leture se troua che i(n) le parte meridional l'aq(u)a | non circunda questo n(ost)ro habitabile e temperado çona, ma | aldando molte testimonia(n)çe i(n) contrario e maxime q(ue)li i qual | la maiestà del Re de portogallo à mandato cu(m) le suo carauele | a çerchar e ueder ad ochio, i qual dice hauer circuito le spiaçe de | garbi(n) più de 2000 mia oltra el streto de çibelter i(n) tanto che a uoler | seguir q(ue)l cami(n) hano (con) uenuto dar la p(ro)da quarta d'ostro i(n)uer sirocho | e p(er) suo çudisio hano passato l'i(n)dromo de tunisto e q(u) asi son ço(n)ti a q(ue)l d'alexa(n)dria, p(er) tuto troua(n)do bone spiaçe cu(m) puoco fondo e nauegar assai bo(n) e se(m)pre | se(n)ça fortuna. E i diti hano fato nuoue carte de quel nauegar e hano posto | nomi nuoui a fiumere, colfi, caui, porti, di q(u)al ne ho habuto copia. Unde se'l se uorà | (con)tradir a q(ue)sti i qual hano uisto ad ochio, maçorme(n)te se porà no(n) asse(n)tir nè creder a q(ue)li | che hano lassato i(n) scriptis q(ue)lo hi no(n) uete mai ad ochio, ma cusì hano opinado esser. | Anchora io ho p(ar)lato cu(m) p(er)sona dig(n)a de fede, che afferma hauer scorso cu(m) una naue de i(n)dia | p(er) rabia de fortuna de trauersa' p(er) zorni 40 fuora del mar d'india oltra el cauo de soffala | e de le i(n)sule uerde e qui pur al garbi(n) e al pone(n)te e p(er) lo arbitrar de i suo astrologi i q(u) al son | lor guida i scorse circa 2000 mia. Unde certamente el se può affermar e cre|der cussì a questi come a queli i qual uien hauer scorso mia 4000. Dice | ancora Po(m)ponio Mela nel terço libro de la sua Cosmographia | che uno hauea nome Eudoxo, el qual sca(m)pando Lathmin, Re de alexandria, usì del colfo arabico e nauegò quela | parte austral e uene fin a gades ch'è al streto de çi|belterra. | Adoncha sença alguna dubitatio(n) se può affermar | che questa p(ar)te austral e de garbin sia nauigabile e ch(e) | quel mar indiano sia occeano e no(n) stagnon, e cusì | affermano tuti queli che nauegano quel mar | e che habitano quele i(n)sule." *Mappamondo* (ed. 1956) Pl. XI, 2. The English translation derives from *Mappamondo* (ed. 2006) *149.

In this dense and complex legend, Fra Mauro again invokes a kind of theater in which opinion, reading, and eye-witness testimony alternate in a tense dialogue. "These men, who have seen with their own eyes" and "those who have left writings on things they did not see with their own" are the main protagonists; the informed reader – a privileged spectator of the struggle, is called upon to resolve the diatribe between them. When there were contradictions and discrepancies between *auctoritates* and witnesses who "have seen with their own eyes," Fra Mauro, although not openly dissenting from the *autori* or disputing their wisdom, did however suggest that it was necessary to side with the witnesses. While this is not a sign of a rupture, it does clearly attest to Fra Mauro's acknowledgement that the wisdom of the ancients, while remaining important and essential, was historically contingent. Also, insofar as it had been traditionally held as reliable and valuable, it ought to be – or at least it could be – subject to verification. What followed was a complex, non-linear, and (in effect) dialectical development that determined the *mappa mundi*'s structural and cultural framework and led to a reworking of the cosmographic culture of that age. Even as he adopted from the *auctoritates* of his own time those doctrinal blocks that are found, for example, in Pierre d'Ailly's *Ymago mundi*, Fra Mauro stood apart from his contemporaries, driven by a personal sense of the need to choose among diverse languages and viewpoints.[127] It is important to recall in this context that the legends written in the form of a dialogue involving the cosmographer's voice in first person are placed near the far more numerous didactic legends that derive from and (either explicitly or implicitly) re-elaborate their written sources.

Patrick Gautier Dalché has justifiably noted that behind the apparently casual use of the cosmographer's "*io*" that authoritatively gives shape to his own work, there is more than just an attempt to demonstrate and prove his competence: there is as well a deliberate (and sometimes rather ironic) positioning with respect to his potential critics, whom he anticipates and refutes *ab origine*, even before they can formulate their objections. This surprising cartouche is found on the Nile:

> I think that many will be amazed that here I put the source of the Nile. But certainly if they approach the question rationally and undertake the same investigations that I have – and with the diligence that I cannot here describe – they will see that here I am undertaking to demonstrate this thanks to the very clear evidence I have had.[128]

Criticism on this point is obviated: "if they approach the question rationally" they will agree with him, writes Fra Mauro, who nonetheless, with great ease, hides his own sources behind the laconic and enigmatic reference to "the very clear evidence I have had". The reader finds no trace of those sources; the cosmographer's "*io*" affirms his own authority and his independent judgment, and these are enough. It should be underlined that this constitutes a novelty in the history of cosmographic representation. A legend regarding "miraculous things and other

[127] E. Buron, *Ymago mundi: texte latin et traduction française des quatre traités cosmographiques de d'Ailly et de notes marginales de Cristophe Colomb. Etude sur les sources de l'auteur*, 3 vols., Paris: Maisonneuve, 1930.

[128] "Credo che q(ui) molti se | meraueierà p(er)ché io | meto el nascime(n)to del | nilo, ma certo si se mo|uerano cu(m) raxio(n) et | uorano i(n)uestigar | q(uan)to ho fato mi e cu(m) | q(ue)la dilige(n)tia che q(ui) dir | no(n) posso, i uederano | che io me mouo a | demostrar q(ue)sto p(er) evide(n)|tissime chiareçe | ho habuto." *Mappamondo* (ed. 1956) Pl. XVI, 58. The English translation derives from *Mappamondo* (ed. 2006) *389.

monstrous things" reveals another aspect of Fra Mauro's subtly ironic way of treating the *auctores*:

> In this island of Hibernia, which is most extraordinarily fertile, it is said that there is a water in which, if you immerse wood, after a while that part of the wood which is in the earth becomes iron, whilst that in the water becomes stone, and that above the water remains wood. And if one believes this thing, one can also believe in the lake of Andaman. Those who wish to have plentiful information on these wonderful and these monstrous things should read Julius Solinus's Polyhistor, Pomponius Mela, St. Augustine, Albertus Magnus and St. Thomas Aquinas in his book against the curious. Similarly, they can read Aristotle's Meteorology and Pliny on the wonders of the world, and they will see thousands of things of which I have not mentioned one.[129]

Solinus, Pomponius Mela, Augustine, Albert the Great, Thomas Aquinas, Aristotle, and Pliny are all called into question, with – at the very minimum – the skepticism of one who cannot completely extricate himself from a still-authoritative tradition yet, at the same time, who seeks to distance himself from that tradition with the tacit awareness of his capacity to do just this.

4. Narrative convergences and connections

In Chapter V, which analyzed the links between the *mappa mundi* and travel accounts, it was pointed out that the presence of the pronoun *'io'* is found, significantly, in the narrative structure of the *Milione*, one of the map's most important sources. In this regard, Cesare Segre, in a study of Marco Polo's account, laid out a brief yet incisive history of the use of the pronoun *'io'* in late-medieval texts, one which is quite useful for understanding some aspects of Fra Mauro's first-person narration.[130] Segre demonstrated that, while the majority of medieval travel accounts are structured in the form of extremely neutral descriptions in third person, almost as if the authors wished to hallow with a certain objectivity the things they had seen, the people they had encountered, and the experiences they had undergone, by contrast, in the *Milione*, Marco Polo furnishes first-person descriptions of some episodes in which he was a protagonist or a witness during the course of his long voyage in Asia.

In his study of a peculiar, and often overlooked category of travel literature – namely the accounts of visions and travels in the afterlife – Cesare Segre revealed that travels and visions reported by others were narrated in the third person (for example, the travels of St. Brendan), whereas those reported by someone who had experienced them were usually told in the first

[129] "In questa insula de hibernia, la qual oltra modo è | fertillissima, se dice esser una aq(u)a ne la qual chi li me|te leg(n)o q(ue)la p(ar)te che è ne la terra cu(m) te(m)po deue(n)ta ferro e | q(ue)la è circu(m)dada da l'aq(u)a deue(n)ta piera e q(ue)la è sopra aq(u)a rima(n) legno, e se q(ue)sto se crede se pò ancor creder de lago de | andama(n), e queli che de q(ue)ste cosse mirabile e de al|tre mo(n)struose desiderano esser copiosi, leçano | Julio Soli(n) polistoro, Po(m)ponio Mela, s(an)c(t)o Agustino, Alber|to Magno, s(an)c(t)o Thomaso de Aqui(n) nel libro co(n)tra i cu|riosi, ite(m) le Metaure d'Aristotile e Plinio de le me|raueie del mo(n)do e uederano che de mile cosse | no(n) ne ho dite una." *Mappamondo* (ed. 1956) Pl. XXX, 120*. The English translation derives from *Mappamondo* (ed. 2006) *2212.

[130] C. Segre, "Marco Polo: filologia e industria culturale," in *Il menante filologo. Bullettino dell'Istituto di filologia moderna*. Parma: Università degli Studi di Parma – Edizioni Zara, 1983, pp. 9-20.

person. "Thus," concluded Segre, "the first persona serves to confirm the authenticity of what is said, and means that it has not been recounted because it was merely heard (*per udito*), but because it had been experienced." In other words, the author who makes use of the pronoun '*io*' enhances the veridicity of his account. This is the same narrative device used in the *Milione*: the recourse to first-person utterances is made when dealing with novelties and things that were unknown in the West. Rustichello introduced the pronoun '*io*', referring to Marco Polo, when he described the fact that asbestos was fireproof, something unknown in the West.[131] The same narrative form is found in the description of paper money, with a prominent use of the pronoun '*io*' serving to confirm what was strange or unfamiliar in the account.[132]

In the legends that describe new facts that were not handed down by classical literary sources, or which contradicted that which had been reported in other sources with a reputation for reliability, Fra Mauro made recourse to a narrative structure similar to the one found in the *Milione*, one which had already been employed in the accounts of visions and voyages in the afterlife. In passages of the *mappa mundi* that introduced to the West facts that, up until then, were unheard of in the West, the voice of the cosmographer-narrator dominates over the others. The cartographer's '*io*' becomes indispensable: the novelties of distant worlds, whether drawn from literature or from the accounts of those who "have seen with their own eyes," needed to be vouchsafed. This knowledge, whether entirely unknown in his culture, or otherwise found only in a few sources and in opposition to mainstream ideas, required a special guarantee to certify its accuracy. There was only one kind of guarantee possible: the author and his capacity to command authority, underlines by the epithet by which Fra Mauro has come to be known in history: *incomparabilis*.

[131] "In the same mountain [in the province of Ghinghintalas] there is a vein from which salamander is obtained. For you must know that salamander is not an animal, as people say, but is such as I shall tell you. It is a truth that you all know, that by nature no animal, no living creature, can live in the fire, as every animal is composed of all four elements. As people had no certain information concerning the salamander, they though, and still do think, that it is an animal, but this is not the truth, which is as I will now tell you. You must know, then, that I had a comrade, Zurficar by name, who was a very wise Turk, and who lived three years in that province, by order of the Great Kaan, to extract the salamander, the steel, the ondanique, and so forth. For it is the Great Kaan's custom to send a Lord to rule over that province for three years at a time, and to oversee the mining of the salamander. My friend explained the thing to me, and I myself witnessed it." *The Travels of Marco Polo*, trans. A. Ricci (from the text of L. F. Benedetto), London: Routledge, 1931, pp. 73-74. For a fourteenth-century Tuscan edition: Marco Polo, *Milione; versione toscana del Trecento*, V. Bertolucci Pizzorusso (ed.), descriptive index by G. R. Cardona, Milan: Adelphi, 1994: ch. 49, "Chinchitalas," pp. 80-81.

[132] "When this paper money has been prepared in the way I have told you, he has all payments made with it [...] Yet I assure you that all his subjects, of all countries and peoples, readily accept that paper in payment, for wherever they go they can pay for anything with it [...]. Further, many times a year merchants arrive in Cambaluc in groups, bringing pearls, precious stones, gold, silver, and other things [...] and they offer all these things to the Great Kaan. The latter sends for twelve wise men [...] and he bids them examine what the merchants bring, and have it paid what they consider the proper price. [...] then these twelve wise men look at the things and, appraising them according to their consciences, straightaway have the value paid, to the advantage of the merchants, in the paper money I have told you. [...] And I will also tell you that several times a year a proclamation is issued in the cities, to the effect that whoever possesses precious stones, pearls, [...]is to bring them to the Great Lord's mint [...] and all do so [...]. And they are all paid in paper." *The Travels of Marco Polo*, pp. 148-149; *Milione; versione toscana del Trecento*: ch. 95, "La moneta del Gran Chan," p. 151.

II
PATRONAGE. MARKETPLACES. THE HISTORY OF CARTOGRAPHY

CHAPTER I

AT THE ORIGIN OF THE MAP MARKET IN FIFTEENTH-CENTURY VENICE, FLORENCE AND BARCELONA

I. – The documentation on fifteenth-century cartographic production

1. Research agenda and selection of the documents

How much did a *mappa mundi* like Fra Mauro's cost in Venice ? What was it worth compared to other forms of cartographic representation such as a Florentine Ptolemaic manuscript, or a nautical chart made in Barcelona ? What factors determined the price of the maps ? What was the value of a map compared to luxury items *par excellence* in the early modern period such as brocades, silk, or ceremonial gowns ? And how many days of manual labor in the mid-fifteenth century would correspond to the cost of a *mappa mundi* ? We will address these questions by drawing on Venetian, Florentine and Catalan documents of the fifteenth century, in particular mercantile correspondence, account books, judiciary proceedings and *ricordanze*, a Florentine genre of "private memoirs treating laconic personal and family memoirs through account books," as Lauro Martines describes them.[1]

In order to be able to compare the prices paid for maps in Venice, Florence, and Barcelona I will also focus on the actual exchange rates (the "price of the money" as Giulio Mandich called them) among all the relevant currencies as recorded in merchants' accounting books and handbooks (*pratiche di mercatura*).

Although Florence and Venice were two very different worlds they nevertheless shared similarities in terms of their political power and economic and social life when compared to many other Renaissance cities. At least as early as 1358 the values of the ducat and the florin were equal, differing only in terms of milligrams of gold.[2] This fact naturally facilitates comparisons and is the most reliable indicator of the homogeneity at the base of the two societies and their economies. Moreover, a key issue for this study is how the perception of luxury and the value of *objets-d'art* (not necessarily reflecting good taste and the culture that shaped it) were indeed comparable.[3] Barcelona, the capital of a vast commercial network that included

[1] This is the definition of *ricordanze* suggested by Lauro Martines, "Review of 'Una Famiglia e le sue Ricordanze : I Castellani di Firenze nel Tre-Quattrocento'," *Speculum* 71, 3 (1996), pp. 707-708.

[2] "With both Venice and Florence issuing gold coin, the rate of exchange between them was not affected by changes in the bimetallic ratio. In the transmission of funds from one to the other by means of bills of exchange, rates based on those gold coins were used, and they varied slightly, according to market conditions, from counting 1 ducat as equal to 1 florin". See F. Lane and R. C. Mueller, *Money and Banking in Medieval and Renaissance Venice*, vol. 1, *Coins and Moneys of Account*, Baltimore and London : Johns Hopkins University Press, 1985, pp. 174-176 and for the citation, p. 308.

[3] For the fundamental role of the trade between Venice and Florence as well as the functioning of their markets of exchange, R. C. Mueller, "'Chome l'ucciello di passagio' : la demande saisonnière des espèces et le marché des

the Balearic Islands, the kingdom of Aragon, Valencia, Perpignan, and Montpellier, had close commercial ties to both Venice and Florence. It was there that predominantly Florentine merchants bought leather as well as wool from Majorca, renowned as the best in the Mediterranean. Until the end of the fifteenth century Florentine merchants and banks controlled the insurance market as well as other financial services, which allowed them to defeat their competitors and enjoy a monopoly in trade of wool and hides.[4]

The fifteenth-century documents that refer to the production of work of art are well diffused and relatively available. Contracts and *ricordanze* for the production of frescoes, altar pieces, paintings, illuminated manuscripts and statues are in many European archives still conserved in the accounting books of the workshops, tax declarations (*portate al catasto*), judicial proceedings, notarial acts and the documents of matriculation of painters into the Guild of Physicians and Chemists, and those of sculptors into the Guild of Masters of Stone. These documents have enabled art historians to reconstruct with remarkable precision the ways in which the workshops functioned, the artistic techniques, the professionals involved, the methods of apprenticeship, and the cost of the works produced.[5] But with regard to the cartographic works that, unfortunately, underwent a greater dispersal compared to other historical artifacts, the contracts that illustrate the costs and phases of realization for the different types of maps, above all manuscript maps (*mappae mundi*, nautical charts, illustrated manuscripts of Ptolemy's *Geography*, chorographies, city maps), are rather rare.[6] These documents are among the few that shed light onto the often obscure world of the cartographic workshops, the techniques of

changes à Venise au Moyen Age," in J. Day (ed.), *Etudes d'histoire monétaire, XIIe-XIXe siècles*, Lille : Lavoisier, 1984, pp. 195-219 ; R. A. Goldthwaite, *Studi sulla moneta fiorentina : secoli XIII-XVI*, Florence : Olschki, 1994. See S. Bertelli, N. Rubinstein and C. H. Smyth *Florence and Venice, Comparisons and Relations. Acts of two Conferences at Villa i Tatti in 1976-1977*, 2 vols., Florence : La Nuova Italia, 1979.

[4] C. Carrère, *Barcelone, centre économique à l'époque des difficultés, 1380-1462*, 2 vols., Paris and La Haye : Mouton, 1967 ; B. Dini, "La vita e l'attività professionale di Ambrogio di messer Lorenzo de' Rocchi," in Ambrogio di Lorenzo de' Rocchi, *Una pratica di mercatura in formazione : 1394-1395*, Florence : Le Monnier, 1980, pp. 5-33 ; and S. Tognetti, *Il banco Cambini : affari e mercati di una compagnia mercantile-bancaria nella Firenze del XV secolo*, Florence : Olschki, 1999, pp. 157-201.

[5] The relevant bibliography is very extensive. Some useful references include P. Paoletti, *Raccolta di documenti inediti per servire alla storia della pittura veneziana nei secoli xv e xvi*, Padua : Tip. Prosdocimi, 1894 ; G. Milanesi, *Nuovi documenti per la storia dell'arte toscana dal XII al XV secolo per servire d'aggiunta all'edizione del Vasari edita da Sansoni nel 1885 raccolti e annotati da G. Milanesi*, Florence : G. Dotti, 1901 ; P. Bacci, (ed.), *Documenti toscani per la storia dell'arte*, Florence : Gonnelli, 2 vols., 1910-1912 ; R. G. Mather, "Documents Mostly New Relating to Florentine Painters and Sculptors of the Fifteenth Century," *The Art Bulletin* 30, 1 (1948), pp. 20-65 ; M. Levi d'Ancona, *Miniatura e miniatori a Firenze dal XIV al XVI secolo : documenti per la storia della miniatura*, with a preface by M. Salmi, Florence : Olschki, 1962 ; A. Garzelli, *Miniatura fiorentina del Rinascimento, 1440-1525 : un primo censimento*, Florence : Giunta regionale toscana ; Scandicci : La nuova Italia, 2 vols., 1985.

[6] D. Woodward, "The 'Two Cultures' of Map History : Scientific and Humanistic Traditions : a Plea for Reintegration," in D. Woodward, C. Delano-Smith and C. D. K. Yee (eds), *Plantejaments i objectius d'una història universal de la cartografia – Approaches and challenges in a worldwide history of cartography* : 11e curs organitzat per l'Institut Cartogràfic de Catalunya i el Departament de Geografia de la Universitat Autònoma de Barcelona (Barcelona, 21-25 de febrer de 2000), Barcelona : Institut Cartogràfic de Catalunya, 2001, pp. 49-67. The largest collection of data regarding the cost of marine chars between 1294 and 1485 is published by Ramon Pujades in *Les cartes portolanes. La representació medieval d'una mar solcada*, Barcelona : Istitut Cartogràfic de Catalunya, 2007, pp. 279-280. Pujades records 39 catalan documents that refer the price of marine charts. For the transcription of the documents see pp. 84-106 in his book.

map production, and the day-to-day lives of the cartographers. In the fifteenth century, although there was significant specialization in the field with professionals called *maestri* (masters) and *dipintori di carthe da navichare* (painters of nautical charts), at the same time work was done by artisans who were not specialists in the production of maps. Copyists of manuscripts, painters, illustrators and – with the advent of the printing press in the last years of the century – also engravers and goldsmiths, were placed alongside "masters" and "painters of marine charts," with whom they shared instruments, techniques and commissions, often without understanding the geometric properties of the drawings that they reproduced.[7]

The literature that focuses on the commission and production of medieval maps and on their social-economic value is very scarce, as has been noted: the few existing contributions primarily (if not exclusively) discuss printed maps, neglecting the production of manuscripts.[8] The study of the economic and material aspects of the production of fifteenth-century manuscript cartography has mostly been neglected. With regard to Fra Mauro's *mappa mundi*, the Account Book of San Michele di Murano (*Entrata e uscita*) for the years 1453-1460,[9] fundamental for the dating of the map, is also essential for the analysis of its production as well as its economic and social value. This item, as we have seen, contains information that pertains to the *mappa mundi* commissioned from Fra Mauro by the Portuguese King Afonso V in 1457. Analysis of Maffeo Gherardo's annotations suggested that this *mappa mundi* was a copy of a pre-existing map that can be possibly be recognized as the one exhibited today in the Biblioteca Nazionale Marciana (or to another one very similar to this). To paint the copy of the *mappa mundi* for the Portuguese court took twenty days of work by three copyists (ignoring the work that Fra Mauro had invested in creating it). From the analysis developed in Chapter I, the costs, methods and preparation time reported by Maffeo Gherardo, although they do not deal directly with the *mappa mundi* held today at the Biblioteca Marciana, refer to a copy that must have been very similar to that map in size as well as in its graphical and cartographic structure.

This study will focus on the production costs of a copy Fra Mauro's *mappa mundi* in comparison with the costs of other cartographic and pictorial works of the period. Among the cartographic works we will consider a Florentine manuscript of Ptolemy's *Geography* painted by Piero del Massaio and another painter or copyist, Giovanni del Maestro Antonio, between 1460 and 1461; a *mappa mundi* made by the cartographer and Venetian humanist Antonio Leonardi and given to Pope Pius II in 1462; and four *mappae mundi* commissioned by Baldassarre

[7] See the study by Marica Milanesi on the illustration of four manuscripts of the commentary on the *Dittamondo* by Guglielmo Capello: in particular, regarding the difficulty of the non-Florentine illuminators in copying the Ptolomaic planispheres, M. Milanesi, "Il commento al 'Dittamondo' di Guglielmo Capello (1435-1437)," in M. Bertozzi (ed.), *Alla corte degli Estensi. Filosofia, arte e cultura a Ferrara nei secoli XV e XVI*, Ferrara: Università degli Studi, 1994, pp. 365-388.

[8] See for example David Woodward's study on Francesco Rosselli's printed cartographic works, *Maps as Prints in the Italian Renaissance. Makers, Distributors & Consumer*, The Panizzi Lectures, London: British Library, 1996. See also R. Shirley, *The Mapping of the World. Early Printed Maps 1472-1700*, London: Holland Press, 4th edition, 2003; D. Woodward, "The Italian Map Trade, 1480-1650," in *The History of Cartography*, Vol. 3: *Cartography in the European Renaissance*, Chicago and London: University of Chicago Press, 2007, pp. 773-803; and C. Hofmann, "Publishing and the Map Trade in France, 1470-1670," *ibid.*, pp. 1569-1588.

[9] Rome, Archivio di Stato, San Gregorio al Celio, inv. 25/II, n. 9, n. 63, Gherardo Maffeo, *Entrata e Uscita. San Michele di Murano* (1453-1460), ff. II, 41r, 42r, 44r, 123v-125r, 128v, 129r, 137v, 139r, 144r, 169r-171v, 172v, 173v, 174r, 175r, 180v, 181r. See the transcription in the Documentary Appendix of Chapter I.

di Simone degli Ubriachi, a Florentine merchant active in Venice, produced in Barcelona between 1399 and 1400 by Jacme Ribes and Francesco Becaria. The pictorial works to be analyzed are the fresco portraying *The Last Supper* in the refectory of the convent of Sant'Onofrio, known as "di Fuligno," produced in 1462 by the Florentine painter Neri di Bicci; and the decoration in the British Library of the *Promissione dogale di Cristoforo Moro* (Official Oath of Cristoforo Moro), painted in 1463 by the Venetian illuminator, Leonardo Bellini, author of the earthly paradise of the *mappa mundi*. The choice of these artifacts was based on two criteria: that, apart from the Barcelona maps, they were strictly contemporary with Fra Mauro's *mappa mundi*, and that documents existed regarding the cost of their production.

II. – Venice, Florence and Barcelona

1. Venice: the cost of a copy of Fra Mauro's *mappa mundi*

The study of the account book of San Michele conserved in the Archivio di Stato in Rome – a register compiled with remarkable precision – is the fundamental point of departure for calculating the economic value and understanding the methods of preparation of a *mappa mundi* similar to that of Fra Mauro. Organized chronologically, it periodically offers balance sheet summaries in which Maffeo Gherardo, treasurer and then abbot of the monastery, grouped together the incomes and expenditures related to specific items. The first issue that we need to consider is the reasons for which Fra Mauro employed external copyists and painters to make a copy of his work for the Portuguese court of Afonso V. Money paid to copyists for parchment and colors appears regularly in the account books of the Camaldolese monasteries and libraries.[10] This detail raises questions regarding the activity of the *scriptorium* of San Michele. The documents that we possess, which have been studied by Caby and Merolla, do not clearly indicate what its activity might have been, particularly in the fifteenth century. Nevertheless, the available evidence leaves room for the hypothesis that at the time of Fra Mauro the *scriptorium* saw limited participation by the monks, and depended on external copyists and illuminators for illumination and calligraphic decoration. At least until 1460-1470, before Pietro Dolfin and Bernardino Gadolo took over the abbey, among the Camaldolese Benedictines of San Michele of Murano the decoration of the manuscripts, psalm books, books of hours, and also often the copying of manuscripts, was generally entrusted to scribes and illuminators hired from outside the monastery. These scribes and illuminators could be secular, as in the case of Leonardo Bellini who painted the earthly paradise of the *mappa mundi*, or *confrères* as in the case of Lorenzo Monaco of the Camaldolese monastery of Santa Maria degli Angeli in Florence.[11] These artisans could work within the *scriptorium* of San Michele; however, as opposed to the situation in

[10] See C. Caby, *De l'érémitisme rural au monachisme urbain. Les Camaldules en Italie à la fin du Moyen Âge*, Rome: École Française de Rome, 1999, p. 629; and L. Merolla, "Un corale di San Michele di Murano," *Nuovi Annali della Scuola speciale per archivisti e bibliotecari* 11 (1997), pp. 111-127, esp. 120-122.

[11] For the commissions to the Camaldolese Monastery of Santa Maria degli Angeli in Florence of illuminated manuscripts and the hiring of external copyists of breviaries, a practice that is also recorded by Ambrogio Traversari in his letters, see the documentation published by Merolla, "Un corale di San Michele di Murano," pp. 111-127.

Cistercian monasteries, they did not belong to the congregation of San Michele.[12] Documentary evidence demonstrates that around the middle of the fifteenth century, while external copyists and painters worked on a copy of Fra Mauro's *mappa mundi*, further special allocations were made to San Michele to pay them to make breviaries, and also for the acquisition and loan of books.[13] This could explain why Fra Mauro relied on of the work of external copyists for the production of his *mappa mundi* for the Portuguese court.

As is reported in the register, the total sum of money paid in advance by the Portuguese ambassador João Fernandes da Silveira at the monastery of San Michele of Murano between February 8th, 1457 and April 24th, 1459 for the copy of the *mappa mundi* was 28 Venetian gold ducats.[14] The money left in deposit, intended to cover what the monastery would sustain in preparing the copy, was most likely based on the estimated costs of parchment, colors, and gold, as well as the labor necessary to complete the copy of the *mappa mundi*. The quantity of ducats and *soldi di piccioli* (silver sous) reported in the register for the realization of the *mappa mundi* amounts, in fact, to 18 ducats and 49 *soldi di piccioli*. These include 2.2 ducats for 20 days' salary of a *scriptore* or copyist, that is, a professional writer, who was paid 14 *soldi di piccioli* a day ; 2 ducats paid to the Venetian galley officer (*comitus*) and cartographer Andrea Bianco for his work (*premio del lavorier lui face al dicto mapamundi*) ; 13 ducats for general expenditure made by Fra Mauro that probably includes the costs of the parchment and colors ; and 1 ducat for *azzurro*, an expensive mixture made from powdered lapislazzuli.[15] Fra Mauro, being a monk, unlike the external painters and copyists from Venice, did not receive a personal salary, and his compensation most likely ended up in the monastery's register.

As it was shown in Chapter I, a receipt (*carta de quitação*) dated February 3, 1462 and drawn up by the chancellery of King Afonso V of Portugal in favor of João Fernandes da Silveira, Portuguese ambassador to the Pope and the Italian courts, registers the cost of 30¾ ducats for "the painters who painted the world map in Venice" (*aos pintores que pyntaram o mapa mundo*

[12] Kenneth W. Humphreys refers to them as "visiting scribes" ; see K. W. Humphreys, *The Book Provisions of the Mediaeval Friars, 1215-1400*, Amsterdam : Erasmus Booksellers, 1964 and Id., *The Library of the Franciscans of the Convent of St. Anthony, Padua, at the Beginning of the Fifteenth Century*, Amsterdam : Erasmus Booksellers, 1966. Regarding the book manufacturing in the Venetian Camaldolese monasteries, in the same years in which Fra Mauro worked on the *mappa mundi*, see Chapter VI.

[13] Rome, Archivio di Stato, *San Gregorio al Celio* (inv 25/II, n. 9), n. 64 ; *Entrata e uscita. San Michele di Murano* (1469-1478/1541-1573), f. 169v : "1457 adì 10 septembre. Ser Leonardo alinchonto die dar a di sopradecto per chasa che frar Michiel | auea contadi per dar a uno scriptor per comprar carta per il suo brevia|rio ducati 2 val a S. 180."

[14] *Entrata e uscita. San Michele di Murano* (1453-1460), f. 42v : "1457 adì primo zener. Adì 8 dictto perche io avi contadi da don Benedetto Miani per nome del Signor de Portogal in suma ducati 28 a val. 129." ("1457, January 1st. On December 8th I received the sum of 28 ducats valued 129 *soldi* from don Benedetto Miani on behalf of the King of Portugal"). On page 129r another note that refers to February 8th, 1457 better explains what happened : "La maiestad del Re de Portogal diè auer adì sopradictto perchè don Benedetto Miani me dè contadi in horo ducadi 28 i qual sono lasadi qui in nome del detto Segnor in depoxitto per suplir a certe spexe lequal erano de bixogno far per comprar l'opera del suo *mapamundi* el qual lauora frate Mauro val in chasa." ("With reference to the above mentioned day, I register in favor of His Majesty the King of Portugal 28 gold ducats given to me by Don Benedetto Miani, that were left here in deposit by this Majesty to supplement several costs that were needed in order to buy the work of his *mappa mundi* on which Fra Mauro worked").

[15] *Entrata e uscita. San Michele di Murano* (1453-1460), f. 124v.

em Veneza).[16] This document of Afonso V's chancellery, even though it confirms the commission of, and final payment for, the *mappa mundi*, in comparison with the *Entrata e uscita* of San Michele, reports an increase of 2¾ ducats. The difference is explained by f. 172v of the *Entrata e uscita*, in which Maffeo Gherardo notes that on April 24, 1459, Stefano Trevisan came to pick up the *mappa mundi* and gave 2 ducats in charity to San Michele on behalf of the King of Portugal, Afonso V (*de per l'emosena per nome do Segnor Re de Portugal in suma ducati 2 val a soldi 125*).[17] The Portuguese chancellery also records the donation of two ducats to the monastery. Thus the Portuguese records agree with Maffeo Gherardo's register of 28 ducats as the cost of the *mappa mundi*.

Gherardo's register permits us to assess the value of the *mappa mundi* in both monetary and social terms. As one ducat was declared by Maffeo Gherardo to be equivalent to 124 *soldi di piccioli* the price of 28 ducats, paid by the Portuguese crown for the *mappa mundi*, corresponded to 3.472 *soldi di piccioli*. The *Entrata e uscita* of San Michele thus supports Frederic Lane's and Reinhold Mueller's hypothesis that in the year 1454 the rate of exchange between the ducat and sous was 1 to 124, as this is the rate that Maffeo Gherardo recorded in the payments for the *mappa mundi* between 1457 and 1459.[18] Maffeo Gherardo says that the salary earned by a professional writer, on the same level of those who worked on the *mappa mundi*, was 14 *soldi* a day, so the 3.472 *soldi* or 28 ducats paid by the Portuguese court for the *mappa mundi* corresponded to 248 days, or about a year of work by a copyist (3.472 soldi/14 soldi per day = 248 days). The sum paid by the monastery of San Michele for the preparation of the *mappa mundi* amounted instead to 18 ducats and 33 *soldi*, equivalent to 2.315 *soldi di piccioli*, corresponding to the salary of a professional writer for 165 days of work (2.315 soldi/14 soldi per day = 165.3 days). During the mid-fifteenth century a *mappa mundi* was therefore a relatively costly object, perhaps even a luxury item. But how costly was it compared to other types of maps ; for example, to those in a manuscript of Ptolemy's *Geography* ? Before making this comparison it is worth considering the case of a *mappa mundi* designed by the Venetian cartographer and humanist Antonio Leonardi and donated to Pope Pius II in 1462. The importance of both the author and the recipient of the work, together with the existence of documents that describe the exchange, provide valuable information regarding the prices and social value of maps around the mid-fifteenth century.

2. Pope Pius II's donation of twenty-five gold ducats for Don Antonio Leonardi's *mappa mundi*

From the few and isolated accounts regarding the Venetian priest and cosmographer Antonio Leonardi, Rossella Bianchi has assembled a complete picture with a detailed discussion

[16] Lisbon, Archivo do Torre do Tombo, *Chanceleria de D. Afonso V*, liv. 1, f. 2. Published in *Monumenta Henricina*, XIV (1460-1469), edited by Comissão Executiva dos Comemorações do V Centenário da Morte do Infante D. Henrique, Coimbra, 1973, n. 73, pp. 199-204, esp. 202.

[17] *Entrata e uscita. San Michele di Murano (1453-1460)*, f. 172v.

[18] "The rate of 124 soldi per ducat (6 lire 4 soldi) was firmly established by the market in 1455-1456 after a sharp rise beginning in 1452, a development made very clear by the private account books of the Giustinian family and of Marino Sanuto's father and uncle". See Lane and Mueller, *Money and Banking in Medieval and Renaissance Venice*, p. 567.

that has made it possible to establish Leonardi's activities from 1462 to 1496.[19] We will now discuss a map, considered one of the earliest world map designed by Leonardi, which was contemporary with that of Fra Mauro's, donated to Pius II.

The wide-ranging and well-known geographical interests of Enea Silvio Piccolomini, later Pope Pius II, which are especially evident in his *De Europa*, *De Asia*,[20] and the *Germania*, as well as his refined taste for travel description, had notable influence on don Antonio Leonardi, the Venetian cartographer and humanist who was active from mid-fifteenth century to 1496 and no doubt inspired his ambition to reach the papal court in Rome in 1462, with the goal of entering into the intellectual circle and protection of the Pope and the Piccolomini family. On the occasion of his visit Leonardi donated a large world map to Pius II, which is now thought to be the Venetian cartographer's first important work. An accounting document of the *Camera apostolica* of January 19, 1462 registers the payment of 25 ducats by order of the Pontiff to "misser Antonio Nardes vinitiano" for the gift of a *mappa mundi*.[21] Upon Pius II's death the *mappa mundi* passed to his grandson, Francesco Piccolomini Todeschini (1439-1503), Archbishop of Siena and Cardinal of S. Eustachio, later Pope Pius III, and was placed in the Palazzo Piccolomini in Siena. Although the dimensions of Leonardi's *mappa mundi* designed for Pius II are unknown, the testament of Francesco Piccolomini of 1493 (copied in 1503) – that bequeaths the map to the sacristy of the Cathedral of Siena – describes it as a circular map of the world, painted on linen canvas with Pius's coat of arms : "relinquo etiam eidem Sacristiae Cosmographiam Tolomei, quam mappam mundi appellant, linea tela depicta‹m› a clarissimo cosmographo Antonio Leonardi, presbytero veneto, cum insigni‹i›s Pii II, in forma rotunda".[22] The fact that the *mappa*

[19] R. Bianchi, "Notizie del cartografo veneziano Antonio Leonardi. Con una appendice su Daniele Emigli (o Emilei) e la sua laurea padovana," in V. Fera and G. Ferraù (eds), *Filologia umanistica. Per Gianvito Resta*, Padua : Antenore, 1997, pp. 165-211. In her documented study Bianchi was able to confirm that Leonardi was the author of at least six world maps (for Pope Pius II, from whom he received a payment January 19, 1462 ; for Borso d'Este in 1463 ; for Cardinal Francesco Piccolomini, nephew of Pope Pius II, at the beginning of 1465 ; a fourth, finished in February 1466, likely a gift to Cardinal Francesco Piccolomini ; and a *mappa mundi* for the Palazzo Ducale in Venice finished in November 1476. He is also the author of at least two monumental maps of Italy from 1479 and 1485 ; and he gave a *mappa mundi* to Agostino Patrizi, who thanked him October 22, 1476. The entire cartographic *corpus* of don Antonio Leonardi has apparently been lost.

[20] In 1509 the humanist Geofroy de Tory de Bourges published the *De Europa* and *De Asia* as a single work under the title of *Cosmographia*, and subdivided it into numbered chapters (100 for the *De Asia* and 65 for the *De Europa*). He also modified the last phrases of the *De Asia*, introducing a passage that announces the beginning of the *De Europa* : "…et gentes percurrere decrevimus. Nunc de Europa dicemus." See *Cosmographia Pii Papae. in Asiae et Europae eleganti descriptione Asia Historias rerum ubique gestarum cum locorum descriptione complectitur. Europa temporum authoris varias continet historias. Impressa per Henricum Stephanum impressorem diligentissimum Parisiis e regione schole decretorum sumptibus eius Henr. et Ioh. Hogonti VI Id. Octobris 1509*. On the *Cosmographia* see N. Casella, "Pio II tra geografia e storia : la 'Cosmographia'," *Archivio della Società Romana di storia patria*, s. III, 26 (1972), pp. 35-112.

[21] This information, which I obtained from Rossella Bianchi, comes from E. Muntz and P. Fabre, *La Bibliothèque du Vatican au XVe siècle d'après des documents inédits*, Paris : E. Thorin, 1887, p. 126 (Bibliothèque des Écoles françaises d'Athènes et de Rome, 48) ; and E. Casanova, "Un anno della vita privata di Pio II," *Bullettino senese di storia patria* 2 (1931), p. 24.

[22] Bianchi, "Notizie del cartografo veneziano Antonio Leonardi," p. 172. See also G. Uzielli, "La vita e i tempi di Paolo dal Pozzo Toscanelli," in *Raccolta di documenti e studi pubblicati dalla r. commissione colombiana pel quarto centenario della scoperta dell'America*, Rome : Ministero della Pubblica Istruzione, 1894, pp. 191 and 304 ;

mundi was hanging in the Piccolomini Palace in Siena and that it then went to the Sacristy of the Duomo of Siena suggests that its dimensions were not very different from those of Fra Mauro's map, which from August 1460 until 1655 was hanging in the Sacristy of the Monastery of San Michele of Murano.[23] For the purposes of this study it is important to note the close correspondence between the 28 ducats paid by the Portuguese crown for the copy of Fra Mauro's *mappa mundi* and the implicit evaluation that Pope Pius II or, more likely, the secretaries of the curia, made of the *mappa mundi* donated to the Pontiff from Leonardi : 25 gold ducats.[24]

3. Florence : the cost of Ptolemy's *Geography* in 1461 and a reference to 1457

In the archives of the Cambini company, a Florentine business that had strong commercial interests in Portugal and maintained close contacts with the Portuguese court, there remain the receipts of payments made between 1460 and 1461 to accompany the maps for a manuscript of Ptolemy's *Geography* : probably the classic 27 maps of the planisphere and 26 regional maps (12 of Europe, 4 of Africa, and 10 of Asia).[25] The buyer of the manuscript was Álvaro Alfonso the Bishop of Algarve, who spent many years in Florence in the service of Iacopo Insulani, Cardinal of Sant'Eustachio, before returning to Portugal. The *dipintore* of the manuscript was Piero del Massaio, who is cited in the receipts together with most likely a copyist, one Giovanni del Maestro Antonio.[26] Sebastiano Gentile has already called attention to these payments "because they allow us to capture Piero del Massaio while working on the 'painting' (*dipintura*) of

G. Zippel, "Cosmografi al servizio dei papi nel Quattrocento," *Bollettino della Società geografica italiana* 4, 2 (1910), p. 844 ; R. Gallo, "Le mappe geografiche del Palazzo ducale di Venezia," *Archivio Veneto* 5, 32-33 (1943), pp. 47-113, esp. 50 ; R. Almagià, "Uno sconosciuto geografo umanista : Sebastiano Compagni," in *Miscellanea in onore di Giovanni Mercati*, IV, Vatican City : 1946, pp. 442-473, reprinted in his *Scritti geografici (1905-1957)*, Rome : Cremonese, 1961, pp. 527-552, esp. 529 ; and Casella, "Pio II tra geografia e storia," p. 79, note 127.

[23] This is demonstrated in a note written December 20, 1655 and signed by an abbot of San Michele, Francesco Gherardo Erizzo, which is placed half-way up on the right side of Fra Mauro's *mappa mundi*, stating that it was exhibited from 1460 in the sacristy of the church of San Michele, and then moved into the library of the monastery in 1655 : "Haec Tabula Geographica cum per centum et nonaginta quinque annorum curricula partim in Ecclesia partim vero in Aula, quae suo nomine dicta erat et dicebatur *Mappamondo*, fuisset appensa, tandem iussu Rev.mi P. D. Francisci Gherardi dicti Errici, Abbatis huius Monasterii, in hac Bibliotheca ab ipso instaurata, ditata et exornata translata et collocata fuit Anno 1655 Die 20 Decembris."

[24] The arguments of Giuseppe Zippel on this subject are misleading. Zippel posed that the stipend of 25 ducats was proof of the parsimony of the Piccolomini family towards artists and, in general, towards those who were under its protection. See Zippel, "Cosmografi al servizio dei papi nel Quattrocento."

[25] Florence, Archivio dello Spedale degli Innocenti, Fondo Estranei 248, f. 162r. On the Cambini see S. Tognetti, *Il banco Cambini : affari e mercati di una compagnia mercantile-bancaria nella Firenze del XV secolo*, Florence : Olschki, 1999 ; on the presence of Italian merchants in Lisbon during the fifteenth century see L. D'Arienzo, "La presenza italiana in Portogallo e nella Spagna meridionale all'epoca di Cristoforo Colombo," in G. Cavallo (ed.), *Cristoforo Colombo e l'apertura degli spazi*, Rome : Istituto Poligrafico e Zecca dello Stato, 1992, pp. 535-565 ; Ead., "La lettera Toscanelli – Martins e i mercanti fiorentini : la cultura toscana nel Portogallo delle scoperte," in D. Marrara (ed.), *Toscana e Portogallo. Miscellanea storica nel 650° anniversario dello Studio generale di Pisa*, Pisa : ETS, pp. 11-55.

[26] With regard to Piero del Massaio and his production of Ptolemaic maps, see P. Gautier Dalché, *La Géographie de Ptolémée en Occident (IVe-XVIe siècle)*, Turnhout : Brepols, 2009, pp. 224-225 ("Piero del Massaio, ou la production en série").

Ptolemaic maps in 1460-1461," more or less the same years in which Nicolaus Germanus was preparing his first manuscripts of Ptolemy's *Geography*.[27] These documents are particularly interesting for this research as they relate to a Florentine manuscript of the *Geography* contemporary with Fra Mauro's *mappa mundi*, and thus they lend themselves to a comparative study with the data in the *Entrata e uscita* of San Michele. The payments registered total 63 florins and 33 *soldi* and refer specifically to the "part of a book with the maps of Ptolemy" (*per parte d'uno libro delle tavole di Tolomeo*).[28] While Piero del Massaio's job was clearly to design and paint the Ptolemaic maps one could reasonably hypothesize that Giovanni del Maestro was a scribe, with the job of transcribing the toponyms and brief cartographic descriptions from Book VIII of the *Geography* that, beginning with Nicolaus Germanus and Piero del Massaio's manuscripts of the years 1456-1460, were separated from the other seven books and copied on the *versi* of the folios on which the corresponding maps were painted.[29] It is therefore possible to conclude that the payments to Piero del Massaio and Giovanni del Maestro Antonio that appear in the Cambini registers were just for the preparation of the maps with the text of Book VIII, and that they did not include funds for copying the text of the first seven books of the *Geography*.

4. The comparative cost of the *mappa mundi* and of Ptolemy's *Geography*

In Florence there were two types of florins : minted money called *fiorini a oro*, and unminted money called *fiorini di conto*. In order to compare the prices of the copy of Fra Mauro's *mappa mundi* and the manuscript maps for Ptolemy's *Geography*, it is therefore necessary to determine to which type of florins the documents were referring, and their price ; moreover, it is necessary to establish what was their exchange rate with the Venetian ducat. The unminted *fiorini di conto* was a type of money that was used exclusively for accounting (*moneta di conto*), and existed side-by-side with *monete sonanti* (ready cash).[30] The exchange rate between accounting money

[27] Gentile, *Firenze e la scoperta dell'America*, entries 100, p. 200-207 and F. Melis, *Documenti per la storia economica dei secoli XIII-XVI*, Florence : Olschki 1972, p. 125 ; and Id., "Di alcune figure di operatori economici fiorentini attivi nel Portogallo nel XV secolo," in L. Frangioni (ed.), *I mercanti italiani nell'Europa medievale e rinascimentale*, Florence : Le Monnier, 1990, pp. 1-18. For a critical analysis of Nicolaus Germanus's reception of the *Geography*, see Gautier Dalché, *La Géographie de Ptolémée en Occident*, pp. 219-224 ("Nicolaus Germanus, ou la fausse exactitude").

[28] Florence, Archivio dello Spedale degli Innocenti, Fondo Estranei 248, f. 162r. The first payment of 23 florins and 17 *soldi* on December 2, 1460 is in favor of "ser Giovanni del Maestro Antonio and Piero del Massaio [...] for part of a book with maps of Ptolemy ;" the second of 14 florins and 8 *soldi* on February 11, 1460 only to Giovanni, "for one part of the book of Ptolemy ;" the third of 18 florins on March 7, 1460 "to Piero del Massaio *dipintore* [...] on the part of a painting ‹of the maps› of Ptolemy ;" and a last payment of 8 florins and 8 *soldi* from July 16, 1461 ("per Piero del Massaio *dipintore*, portò chonti sono per resto di dipintura d'uno Tolomeo a dipinto a detto veschovo" ; see Florence, Archivio dello Spedale degli Innocenti, Fondo Estranei 250, f. 42). Regarding these payments, the sum arrived at by Sebastiano Gentile, 53 florins and 33 *soldi*, needs to be corrected : this was probably a *lapsus calami*. See *Firenze e la scoperta dell'America*, entry 101, pp. 202-207.

[29] Gentile, *Firenze e la scoperta dell'America*, entries 100-103, pp. 200-215 and the bibliography cited there.

[30] Richard Goldthwaite's studies show that the capacity to distinguish between *monete sonanti* and *monete di conto* was not something exclusive to bankers and the wealthy but, on the contrary, was a widely diffused practice. From the late 1300s to the early 1500s, different *monete di conto* were used in continuous exchange with the gold and

and ready cash depended on changes in the relative prices of gold and silver in the international precious metals market, and on the state of public finances, changing almost daily, in a way similar to today's stock market fluctuations.[31] It is therefore important, when studying historical Florentine accounting documents, to determine which types of florin are registered. The Cambini company's registers deal with the *fiorini di conto*.[32] The studies of Federigo Melis, Richard Goldthwaite, Peter Spufford, Giulio Mandich and Anthony Molho (based on direct analysis of accounting sources), confirm that, according to the relative prices of *fiorini sonanti* and *fiorini di conto* for the years 1460-1463, the 63 florins and 33 *soldi di conto* – which were valued at some 88 *soldi* per florin – paid by the Bishop of Algarve to Piero del Massaio and Giovanni del Maestro Antonio, were equal to 51 florins and 71 *soldi* of *moneta sonante*, at the price of 108 *soldi di piccioli* per florin.[33] It is this last value that we will compare with the 28 ducats of gold paid for the copy of Fra Mauro's *mappa mundi*.

At the end of the fifteenth century, the exchange rate between the Venetian ducat and the Florentine *florino a oro* was one of complete equality (the gold ducat weighed 3.559 grams, the gold florin weighed 3.537; both were minted of 24 carat gold).[34] Given the equality of the Venetian ducat and the Florentine *fiorino a oro*, around the year 1460 the cost of the twenty-seven maps for a manuscript of Ptolemy's *Geography* was a little more than double that of a *mappa mundi* similar to Fra Mauro's. What caused this difference in the costs of these two forms

silver *monete sonanti* in order to keep everyday balances, to finance loans to underqualified laborers, and for small monetary transactions. See R. A. Goldthwaite, *La costruzione della Firenze rinascimentale : una storia economica e sociale*, Bologna : Il Mulino, 1984, pp. 424-458 (p. 433-439).

[31] From the time of its first minting in the year 1252 the gold florin experienced a progressive rise in value against the *lire di piccioli* (silver money). From the original 1252 value of 1 gold florin = 1 *lira* = 20 *soldi* (silver sous) = 240 *denari* (deniers), it arrived at 167 *soldi* when it was last minted in 1542. For this is reason, Giulio Mandich speaks of the "price" of the florins, in place of the more generic "value." See G. Mandich, "Per una ricostruzione delle operazioni mercantili e bancarie della compagnia dei Covoni," in A. Sapori (ed.), *Libro giallo della compagnia dei Covoni*, Milan: Ist. Editoriale Cisalpino, 1970, pp. clxxviii-cxciii. The *moneta di conto* (whose value was fixed at various times by the officers of the mint) was introduced in an attempt to assure a certain stability and rationality to monetary transactions, and particularly in order to protect credit. The difference between the price fixed for the *fiorino di conto* by the officers and the actual acquisition price on the market of the *fiorino a oro* was defined as an exchange premium (*aggio*). See Goldthwaite, *Studi sulla moneta fiorentina : secoli XIII-XVI*, pp. 29-33.

[32] Tognetti, *Il banco Cambini*, pp. 9-19.

[33] Goldthwaite, *Studi sulla moneta fiorentina : secoli XIII-XVI*, pp. 94-95, 176, and 180-182 ; Melis, *Documenti per la storia economica dei secoli XIII-XVI*, p. 268 ; P. Spufford, *Handbook of Medieval Exchange* ; with the assistance of W. Wilkinson and S. Tolley, London : Royal Historical Society ; Woodbridge, Suffolk ; Wolfeboro, N.H., distributed by Boydell & Brewer, 1986, pp. 23-27 ; A. Molho, *Florentine Public Finances in the Early Renaissance, 1400-1433*, Cambridge MA, Harvard University Press, 1971, pp. 209-213.

[34] Lane and Mueller, *Money and Banking in Medieval and Renaissance Venice*, vol. 1, *Coins and Moneys of Account*, pp. 174-176. To emphasize this equality, and with a clear goal of commercial expansion, the Florentine florin was given a new form in 1422 and called the *fiorino largo*. It had the same weight and alloy as the original, but its diameter was increased in order to make it resemble the Venetian ducat. Following the Florentine conquest of Pisa in 1406 and the acquisition of the ports of Leghorn and Porto Pisano from the Genoese in 1421, this change was part of a general strategy through which Florence aimed to increase both its commercial and maritime power in order to better compete with Venice in the Near East. See P. Grierson, "The Weight of the Gold Florin in the Fifteenth Century," *Quaderni ticinesi di numismatica e antichità classiche* 10 (1981), pp. 421-431. The literature on the Florentine expansion in the fifteenth century is vast ; a useful overview can be found in B. Dini, *Saggi su una economia-mondo : Firenze e l'Italia fra Mediterraneo ed Europa, secc. XIII-XVI*, Ospedaletto, Pisa : Pacini, 1995.

of cartographic representation ? It might be tempting to suggest that this difference resulted from the perceived "higher scientific value" of Ptolemaic cartography, which better satisfied the desire of scholars and patrons to see the world represented using a system of scale, that maintained the correct proportions, and whose linear style created the illusion of precision.[35] But I would like to suggest a more objective approach to the problem, which in fact will generate a very different conclusion : that is, an analysis of the material aspects of the maps, particularly of the dimensions of their painted surfaces.

Fra Mauro's *mappa mundi* measures 223 × 223 cm. Subtracting 23 cm from these dimensions to remove the area occupied by the large square wooden frame, we find that the surface painted and written is a little less than four square meters. The manuscript of Ptolemy's *Geography* that we discussed above has been lost, and thus we do not know the dimensions of the maps painted by Piero del Massaio. But by studying the material aspects of other manuscripts painted and signed by Piero del Massaio or unsigned by, but attributed to, him in the following five libraries, we can nevertheless draw interesting conclusions. These, with their descriptions, are the manuscripts we will consider :

Rome, Biblioteca Apostolica Vaticana, Vaticanus Latinus 5699, manuscript on vellum, 599 × 444 mm, 27 Ptolemaic maps, 7 *tabulae modernae* : Spain, Gallia, Italy, Etruria, Peloponnesus, Candia and Egypt, with perspective views of the following cities : Milan, Venice, Florence, Rome, Constantinople, Damascus, Jerusalem, Cairo, and Alexandria. Dated 1469 (fol. 64r) and signed : "e numero scriptorum minimus Hugo Nicolai de Comminellis, natione Francus, me una cum Tabulis sequentibus ad instantiam Petri del Massaio qui me picturis decoravit" (Fischer L27).[36]

Rome, Biblioteca Apostolica Vaticana, Urbinas latinus 277, manuscript on vellum, 593 × 435 mm, 27 Ptolemaic maps, dated 1472 (fol. 69r) and signed "Claudii Ptolemaei Cosmographiae testus explicit per me Ugonem Comminelli de Meceriis supra Mos. in Francia" (Fischer L28).

Paris, Bibliothèque nationale, Latinus 4802, manuscript on vellum, 595 × 440 mm, 27 Ptolemaic maps, 8 *tabulae modernae* and perspective views (the 7 maps of Vatican 5699, with the addition of Adrianopoli ; undated ; signed : "Scripsit Ugo Comminelli ex Francia natus. Composuitque Petrus Massarius Florentinus" (Fischer L29).

San Marino, Huntington Library, Wilton-Huntingtonianus H.M. 1092, manuscript on vellum, 457 × 303 mm, 27 Ptolemaic maps, text by Berlinghieri from 1482, undated and unsigned, but attributed to Piero del Massaio (Fischer L30).

[35] D. Woodward, "Il ritratto della Terra," in F. Camerota (ed.), *Nel segno di Masaccio. L'invenzione della prospettiva*, exhibit catalog, Firenze, Uffizi, maggio 2001-febbraio 2002, Florence : Giunti, 2001, pp. 259-261.

[36] The notes "Fischer" followed by the letter "L" and by a number refer to the descriptions of the manuscripts of Ptolemy's *Geography* with maps, listed and described in *Claudii Ptolemaei Geographiae codex Urbinas graecus 82*, phototypice depictus consilio et opera curatorum Bibliothecae Vaticanae, Vatican City, Biblioteca Apostolica Vaticana, 1932 : Tomus prodromus : J. Fischer, S.J., *De Cl. Ptolemaei vita, operibus, Geographia praesertim eiusque fatis*, Leyden-Leipzig, 1932, First part, pp. 213-219. For an analysis of Piero del Massaio's *studio* between 1469 and 1479, and of his relationship with the French copyist Hugues Commineau de Mézières, and possibly with Francesco Rosselli, see L. Duval-Arnould, "Les manuscrits de la Géographie de Ptolémée issus de l'atelier de Piero del Massaio (Florence, 1469-vers 1478)," in D. Marcotte (ed.), *Humanisme et culture géographique à l'époque du Concile de Constance : autour de Guillaume Fillastre ; actes du Colloque de l'Université de Reims 18-19 novembre 1999*, Turnhout : Brepols, 2002, pp. 227-243.

Florence, Biblioteca Medicea Laurenziana, Pluteo XXX.2, manuscript on vellum, 558 × 416 mm, 27 Ptolemaic maps; text by Iacopo Angeli, *c.* 1460 (Fischer L37).

With the exception of the slightly smaller Wilton-Huntingtonianus H.M. 1092, the other manuscripts painted by Piero del Massaio have very consistent dimensions. Pluteo XXX.2 of the Biblioteca Medicea Laurenziana merits particular attention among the manuscripts listed above, as Sebastiano Gentile has recently and correctly attributed it to Piero del Massaio on the basis of its striking similarities to Urbinas latinus 277 of the Biblioteca Apostolica Vaticana. Measurement of the illustrated part of each map eloquently demonstrates the derivation of Pluteo XXX.2 from Urbinas Lat. 277, or the derivation of both from a common source.[37] This correspondence, which was not mentioned by Joseph Fischer, pertains to all of the maps in these two manuscripts of the *Geography*: the differences in their dimension are around a few millimeters.[38]

The existence of a series of similar manuscripts by the Florentine *maestro* allows us to hypothesize that the dimensions and other physical aspects of the manuscript painted for the Bishop of Algarve between the years 1460-1461 were in all likelihood very similar to those of Pluteo XXX.2 and Urbinas latinus 277.[39] The average dimensions of the folios painted by Massaio in these two works (and also, with minimal variations, in the other manuscripts of the *Geography* by his hand) are 56 × 82 cm. When we take into account the blank margins of the parchment in Pluteo XXX.2 and Urbinas latinus 277 we find that the dimensions of the folio surfaces actually painted are about 46 × 60 cm. To judge from other manuscripts painted by Piero da Massaio it is reasonable to imagine that the total painted surface of the maps in his manuscripts of the *Geography*, around the year 1460, were about 7.5 square meters.[40]

The hypothesis that the manuscript included several modern maps, perhaps seven or at the most eight, as in the case of two of the later manuscripts by Piero del Massaio listed above, appears improbable in a manuscript of 1460. But if, in addition to the twenty-seven traditional maps, it did include seven *tabulae modernae* the painted surface would be slightly larger, around 8.1 square meters, or 8.4 square meters if it included eight *tabulae modernae*[41]; but this difference in area would not substantially alter our conclusions. Comparing the area of the *mappa mundi* (4 square meters) with the combined painted area of the twenty-seven or thirty-four maps of the *Geography* in the manuscript of Massaio, we note that the latter (with or without

[37] Joseph Fischer places this manuscript among those that cannot be assigned to a specific author. See *Firenze e la scoperta dell'America*, pp. 202-207, esp. 205. I have been able to confirm the correspondence between the maps of Pluteo XXX.2 and those of Vaticanus latinus 277 in the Biblioteca Medicea Laurenziana, which has a facsimile of Vaticanus latinus 277: *La cosmografia di Claudio Tolomeo: codice Urb. Lat. 277 1472-73*, facsimile edition, Milan: Jaca Book, 1982 (Codices e Vaticanis selecti quam simillime expressi iussu Pii XI P.M. consilio et opera curatorum Bibliothecae Apost. Vaticanae; LIII).

[38] See Table 1 in the Appendix to this chapter for the dimensions of each map.

[39] P. Gautier Dalché, *La Géographie de Ptolémée en Occident (IVe-XVIe siècle)*, Turnhout: Brepols, 2009, pp. 224-225 ("Piero del Massaio, ou la production en série"). See also G. Aujac, "Le peintre florentin Piero del Massaio et la 'Cosmographia' de Ptolémée," *Geographia antiqua* 3-4 (1994-1995), pp. 187-204 and figs 1-10.

[40] This is the formula: [(46 cm × 60 cm × 27) / 10,000 cm = 7.45 square meters]. For these calculations see Tables 2 and 3 in the Appendix to this chapter.

[41] These are the two formulas: [(40 × 60 cm) × (27 + 7) / 10,000 = 8.1 square meters]; [(40 × 60 cm) × (27 + 8) / 10,000 = or 8.4 square meters].

tabulae modernae) is roughly double that of the *mappa mundi*, and this proportion is very similar to that between the costs of 28 ducats for the *mappa mundi* and 51 florins and 71 *soldi* (of *moneta sonante*) for Massaio's manuscript. This coincidence is remarkable and allows us to imagine that the costs of the two works were based on the painted area, and less on differences in content or cartographic syntax employed.

Comparing the elements of Tables 2 and 3 in the Appendix of this study, which summarize the calculations of the costs and dimensions studied here, we can deduce that the average cost per square meter of the *mappa mundi* and of the Florentine manuscript of the *Geography* of 1460 was around 7 florins or gold ducats: 7 ducats net for each square meter of the *mappa mundi* and 6 florins and 8/10 for each square meter of the maps of the *Geography*. The difference in total cost does not reflect a difference in scientific value or prestige between the two forms of cartographic representation, but rather the cost of the primary materials and the greater amount of labor. With a total surface to be painted measuring almost eight square meters the maps of the luxury manuscripts of the *Geography*, like those prepared in the studio of Piero del Massaio, required almost double the quantity of parchment and colors, such as the very expensive powdered *azzurro*, *porpora* (purple), and gold, as was necessary for the dimensions of Fra Mauro's *mappa mundi*.

An addition to the Florentine documents examined here is a letter from Naples by Giannozzo Manetti to his close friend Vespasiano da Bisticci in Florence, recently edited by Luca Boschetto, which refers to a manuscript of Ptolemy's *Geography* produced in Vespasiano's workshop "valued at 50 ducats" and sold in Naples between July 1457 and 1459.[42] Albinia de la Mare, manuscript librarian at the Bodleian Library at Oxford, proposed the identification of the manuscript Madrid, Biblioteca Nacional, Res. 255 as possibly the specific copy of Ptolemy's *Geography* in Jacopo Angeli's translation, "cholla pictura bellissima," with 27 maps (the planisphere and 26 regional maps of Europe, Asia and Africa), as being the object referred to in Manetti's letter to Vespasiano.[43] The size of the manuscript is 56 × 41 cm and therefore most of the 27 maps measure nearly 56 × 82 cm. The 50 ducats is a figure that includes the complete item (manuscript text and maps, illuminations and binding) and does not allow for the specific calculation of the cartographic component; still, the analysis of the manuscript allows to add a few points to our discourse. Unlike the majority of coeval luxurious manuscripts of the *Geography* crafted in Florence using precious colors, the Madrid manuscript is of a poor quality (as an example, the sea is brown, maps lack gold for highlighting of cities and borders, and place names are in sub-standard calligraphy). Apart from superior quality colors, applied also

[42] L. Boschetto, "Una nuova lettera di Giannozzo Manetti a Vespasiano da Bisticci. Con alcune considerazioni sul commercio librario tra Firenze e Napoli nei decenni centrali del Quattrocento," *Medioevo e Rinascimento* 27 (n.s. 15) (2004), pp. 175-205 (the letter is edited at p. 191). Manetti went to Naples in July 1457 in order to resolve a disagreement between Vespasiano and his agent in Naples, Piero di Ser Francesco da San Gimignano, who not only ignored his patron's instructions not to sell any book on credit but also omitted to send to Florence the 200 florins expected from the sale of the batch of books that Vespasiano sent before July 1457. Manetti managed to remedy the situation, and sent Vespasiano a very detailed letter that enables us to identify another contemporary item for comparison of costs.

[43] A. de la Mare, "Notes and Documents. Observations on Two Italian Manuscripts from Madrid Recently Exhibited in the Bodleian," *Bodleian Library Review*, pp. 242-247 (p. 243). Madrid, Biblioteca Nacional, Res. 255, manuscript on vellum, ff. 110, including 27 maps, 560 × 410 mm. *c.* 1455-1457.

to the non-geographic illuminations, a complete luxurious manuscript *Geography* would also exemplify better quality vellum, ink and calligraphy, and binding (e.g. blind stamped or gilt etc.). In short this item is a good example of mundane or more ordinary quality, and consequently gives us a better appreciation of the production and reception of Ptolemy's *Geography* in the fifteenth century, of which the high quality of the exemplars presented to patrons such as Lorenzo di Pierfrancesco de' Medici,[44] Lorenzo de Medici,[45] Camillo Maria Vitelli,[46] was not truly representative of an average production, such as the case of the Madrid exemplar. The poorer cartographic depiction on some eight square meters of parchment in the Madrid copy of the *Geography* probably explains why the overall cost of the Madrid manuscript (50 ducats) was relatively low. In the light of these consideration, it is therefore possible to conclude that also Manetti's letter confirms the hypothesis that among worldly costly goods, basic copies of Ptolemy's *Geography* were relatively less expensive items.

5. A contract for the production of *mappae mundi* in Barcelona around 1400

Useful comparative information regarding the material aspects of cartographic production, the dimensions, the number and quality of decorations added, also emerge in other documents. These relate specifically to contracts for the production four *mappae mundi*, and in the records of the subsequent legal dispute over problems of payment between a merchant, a "painter and decorator" of nautical charts (*dipintore di charte da navichare*), and a cartographer or "master of nautical charts" (*maestro di charte da navichare*), in Barcelona at the end of the fourteenth century. While in Barcelona in June of 1399 Baldassarre di Simone degli Ubriachi (*c.* 1360-1408) commissioned from Jacme Ribes of Majorca, called *maestro di charte da navichare*, four world maps and from Francesco Becaria of Genoa, called *dipintore di charte da navichare* their decoration with flags, castles, ships and fish.[47] Baldassare made jewels and objects of ivory and bone and sold them all over Europe. After being banished from Florence he moved to Venice where he opened a very notable, and possibly the most important, European workshop of ivory and bone.[48] The maps that Baldassare commissioned in Barcelona were intended for the King

[44] Florence, Biblioteca Medicea Laurenziana, Pl. XXX.2, owned by Lorenzo di Pier Francesco de' Medici ("Liber Laurentii Petri Francisci de Medicis n. 1") and later by Cosimo I ("Dell'Illustrissimo et excellentissimo S. Duca di Firenze Cosmo de' Medici").

[45] Milan, Biblioteca Nazionale Braidense, AC XIV 44, owned by Lorendo il Magnifico ("Liber Cosmografie quem composuit Franciscus de Berlinghieris et donavit Laurentio de Medicis ornatus minio et aliis").

[46] Florence, Biblioteca Nazionale Centrale di Firenze, Magl. XIII, 16, owned by Camillo Maria Vitelli.

[47] See R. A. Skelton, "A Contract for World Maps in Barcelona, 1399-1400, " *Imago Mundi* 22 (1968), pp. 107-113 ; I. Origo, *The merchant of Prato : Francesco di Marco Datini*, London : J. Cape, 1957 (Italian translation I. Origo, *Il mercante di Prato : Francesco di Marco Datini*, preface by L. Einaudi ; transl. by N. Ruffini, Milan : BUR, 1988) ; and R. J. Pujades y Bataller. *Les cartes portolanes. La representació medieval d'una mar solcada*, Barcelona : Istitut Cartogràfic de Catalunya, 2007, pp. 30-59. See also M. Donattini, *Spazio e modernità. Libri, carte, isolari nell'età delle scoperte*, Bologna : CLUEB, 2000, p. 125.

[48] J. von Schlosser, "Die Werkstatt der Embriachi in Venedig," *Jahrbuch der kunsthistorischen Sammlungen* [...] *in Wien* 20 (1899), pp. 220-282 ; R. C. Mueller, "Mercanti e imprenditori fiorentini a Venezia nel tardo medioevo," *Società e storia* 55 (1992), p. 49 ; on Baldassarre di Simone degli Ubriachi, see R. C. Trexler, "The Magi enter Florence : The Ubriachi of Florence and Venice," *Studies in Medieval and Renaissance History* 1 (1978), pp. 127-218.

of Aragon and Navarre as well as for the King of England, to be given with the goal of obtaining the right to enter their territory with jewels and pearls without paying taxes. The contracts and receipts from the beginning of June 1399 and the end of May 1400 are held at the Datini Archive in Prato and the State Archive in Barcelona. They are registered under the name of Simone d'Andrea Bellandi, the agent of Ubriachi's warehouse in Barcelona, who was responsible for the commerce between Majorca and the Catalan capital for the Company of Francesco Datini in Prato.[49] These documents allow us to analyze the production cost, the economic value, and also the preparation methods of the *mappae mundi* in question. The analysis that follows is based on the studies of these documents published by Giovanni Livi (1910), Marina Mitjià (1962) and Raleigh Ashlin Skelton (1968).[50]

Jacme Ribes (or Riba) is the christianized name that the important Jewish cartographer from Majorca, Jafuda Cresques (*c.* 1360-*post* 1420), received following his conversion. Jafuda was the son of Abraham Cresques (d. 1387) and, together with his father made the celebrated Catalan Atlas, most likely created in 1375 for the King of France, Charles V.[51] Following the persecution of the Jews in Aragon in 1391 Jafuda left Majorca for Barcelona, where he continued his career as *maestro di charte da navichare* under the protection of the King of Aragon. He was one of the most prestigious fifteenth-century cartographers from the Majorcan school and also worked for Henry 'the Navigator' and the French Court.[52] Francesco Becaria, on the other hand, is a more enigmatic figure. A citizen of Genoa, who is mentioned in documents as a *dipintore di charte da navichare*, he was most likely an illustrator of manuscripts and maps who belonged to the Genoese Becaria family, and had his own workshop.[53] Only two maps signed by "Francesco Becaro" survive: one is a nautical chart made in Savona in 1403, and now in the Beinecke Rare Book and Manuscript Library at Yale University (Art Object 1980.158)[54]; and the other is a map in the British Library (MS. Egerton 73, map 10[55]), in the so-called "Cornaro

[49] On Simone d'Andrea Bellandi see G. Nigro, *Mercanti in Maiorca. Il carteggio datiniano dall'isola (1387-1396)*, vol. II, *Documenti* (Tome 1), Florence: Le Monnier, 2003.

[50] G. Livi, *Dall'Archivio di Francesco Datini mercante pratese*, Florence: F. Lumachi, 1910; M. Mitjià, "Albandò de les Illes Canàries per Joan I d'Aragò," *Anuario de Estudios Atlánticos* 8 (1962), pp. 325-353 (doc. 7 and doc. 9); R. A. Skelton, "A Contract for World Maps in Barcelona, 1399-1400," *Imago Mundi* 22 (1968), pp. 110-113.

[51] The Catalan Atlas of 1375 is one of the most important cartographic documents of the Late Middle Ages; it is in the manuscript collection of the Bibliothèque Nationale de France (MS Espagnol 30). There are a number of facsimile reproductions of the map, including G. Grosjean, *Mapamundi: der katalanische Weltatlas vom Jahre 1375*, herausgegeben und kommentiert von Georges Grosjean, Dietikon, Urs Graf Verlag, 1977; also see the CD-ROM, *Atlas Catalan*, edited by M. Pellettier, Bibliothèque Nationale de France, 1998.

[52] G. de Reparaz, Jr., *Mestre Jacome de Malhorca, cartógrafo do Infante*, Coimbra: Imprensa da Universidade, 1930, pp. 31-45; J. Rey Pastor and E. García Camarero, *La cartografía mallorquina*, Madrid: CSIC, 1960, pp. 59-60.

[53] Skelton, "A Contract for World Maps in Barcelona, 1399-1400," pp. 107-108.

[54] This nautical chart is mentioned by Skelton, "A Contract for World Maps in Barcelona, 1399-1400," p. 107; also see R. Almagià, "Intorno ad alcune carte nautiche italiane conservate negli Stati Uniti," *Atti della Accademia Nazionale dei Lincei, Classe di scienze morali, storiche e filologiche*, Series 8, 7 (1952), pp. 356-366, esp. 360-363.

[55] London, British Library, MS Egerton 73. [Manuscript nautical atlas on vellum], closed: 440 × 295 mm; wooden binding covered in leather, with *a secco* impressions and metallic studs on two plates. Folios 3-38 contain 35 maps that are copies of older fifteenth-century maps, mostly of Venetian origin, made between 1480 and 1492 by Francesco Cexano, Zuan Soligo, Alvixe Cexano, Gracioxo Benincaxa, Nicolo de Pasqualin and Francesco Becario. In addition to the maps, folios 39r-79r contain an extensive cosmographic treatise.

Atlas" which is a collection of Venetian maps from the first half of the fifteenth century assembled in the years 1480-1492.

The contracts and judicial and notarial acts of Baldassarre degli Ubriachi, in which he demands completion of the commissioned works as well as those of Francesco Becaria to solicit payment for the four *mappae mundi*, refer to two *mappae mundi* which are around 310 centimeters long and 155 centimeters tall ("sixteen palms in length and seven palms and ¾ in height, according to the measure of the *cana* in Barcelona") and to two others much larger, square in form, with the enormous dimensions of 368 centimeters by 368 centimeters (about 13.5 square meters, larger than the Ebstorf *mappa mundi*). Unfortunately none of these maps seems to have survived.[56] In a notarial act dated May 26th, 1400, by which he demanded complete payment for the work, Francesco Becaria left a precise description of the contents of the smaller maps mentioned in the contract. Like the *Entrata e uscita* of San Michele di Murano, Becaria's notarial act is an important document that allows a better understanding of the practices, methods, and professionals involved in the production of maps at this period. The smaller *mappae mundi* – for each of which 60 florins of Aragon were paid – contained 165 designs of cities and animals, 25 ships and galleys, 100 large and small fish, 340 painted banners near castles and cities, and 140 trees, totallling of 770 illustrations : triple those originally estimated. In the other maps a smaller number of images were inserted, but still more than in the model provided by Baldassare degli Ubriachi. According to the contracts signed on June 7th, 1399 by Francesco Becaria and on October 17th, 1399 by Simone d'Andrea Bellandi, Jacme Ribes and Francesco Becaria were to receive as total compensation 320 florins of Aragon, of which they already had a down payment of 192. Of the total of 320 florins, 120 were for the two smaller *mappae mundi*, and 200 for the two larger ones. Between June 19th, 1399 and April 21st, 1400, the Datini warehouse in Barcelona paid 73 florins to Ribes and 38 to Becaria on Ubriachi's behalf.[57]

By the spring of 1400 only two maps had been completed, instead of the four that had been agreed upon, and Baldassarre degli Ubriachi decided to pay a slightly lower price : 303 florins of Aragon instead of the 320 agreed upon in the contract. Becaria, however, perhaps because he was held responsible for the delay, decided to take Baldassare degli Ubriachi to court.[58] From the text of the complaint we learn that the *maestro* and the *dipintore* had very different roles. The contract clearly differentiates the jobs of Jafuda Cresques and Francesco Becaria : the first is a cartographer, while the second is a decorator, specialized in embellishing nautical charts. The contract clearly specifies that once Cresques finished the cartographic design, the *mappae mundi* were to be delivered to Becaria, whose workshop was separate from Cresques'. The *maestro* and the *dipintore* therefore worked independently. Another significant aspect is that the *maestro* and the *dipintore* were required to design the *mappa mundi* based on a model supplied by the same Baldassarre through Simone d'Andrea Bellandi in Barcelona. This meant

[56] It is important to point out that, although the documents use the term *mappae mundi*, we cannot rule out the possibility that they were marine charts. In the period under consideration here, the term *mappamondo* was used in an indistinct way to refer to works such as those of Fra Mauro, or to nautical charts ; and following the recovery of Ptolemy's *Geography*, to Ptolemaic planispheres as well.

[57] Arxiu Històric de Protocols Notarials de Barcelona, Not. Tomás de Bellmunt, Reg. 30 abril 1399 – 28 novembre 1399, 7 giugno 1399. Transcribed in Skelton, "A Contract for World Maps in Barcelona, 1399-1400," p. 110.

[58] Arxiu Històric de Protocols Notarials de Barcelona, Not. Guillelmo Donadeu, Reg. (...), 28 maggio 1400. Transcribed in Skelton, "A Contract for World Maps in Barcelona, 1399-1400," pp. 111-113.

reproducing, or at most updating, a map already in existence. Given that the dimensions of the new maps were greater than those of the model, the *dipintore* was called upon to assure the compositional consistency of the copied maps by increasing the number of decorations, whether fish, trees, vessels, or ships.

Despite the fact that the Florentine and Venetian commercial companies had rich and continuous exchanges with their Catalan branches, the instability and decrease in value of the Aragon, in relation to the Florentine, florin and the Venetian ducat make a costs comparison of maps in the three cities a laborious operation.

The *florí d'or d'Aragó*, which was tied to the florin-ducat monetary system, was introduced by Peter IV of Aragon (called *El Ceremonioso*, "The Ceremonious") in 1346. The Aragonese florin was originally coined at Perpignan of 24 carat gold. The standard of 24 carat gold was not maintained for long: the wars against the Majorcans, and the continuous clashes with the Pisans and Genoese for possession of Sardinia, resulted in a notable weakening of the Aragonese economy; so much so that in 1352 the purity of the gold florin was reduced to 22¾ carats, fixing its value in gold at 94.79% of the Florentine standard. In 1362 the purity of gold was diminished to 22 carats, fixing its value in gold at 91.67% of the Florentine standard. Finally, in 1365, the general assembly of the Catalan States authorized Pedro IV to reduce the florin's purity to 18 carats. By royal edict the Aragonese florins then in use were removed from circulation and the new money was coined; its value was equivalent to 75% of the Florentine and Venetian currencies.[59] This exchange rate between the Aragonese florin and the Florentine and Venetian gold money is confirmed by a description of Florentine trading practices written by Ambrogio di messer Lorenzo de' Rocchi, agent of the Datini company in Majorca and Valencia between 1394 and 1395.[60] On page 58, Ambrogio describes in detail the Aragon florin, indicating its exchange rate with the principal currencies circulating in the Mediterranean basin, among them the Florentine florin: his notes that read "Florentine florins by weight: *soldi* 20 in 19½" and "Florin of Aragon: *soldi* 14 *denari* 10" imply an exchange rate of 75%).[61]

Accordingly, the amount agreed upon in the contract of June 7, 1399, for the smaller *mappae mundi*, namely 60 *florí d'or d'Aragó*, was equivalent to around 45 ducats or florins each;

[59] See Earl J. Hamilton, *Money, Prices, and Wages in Valencia, Aragon, and Navarre*, 1351-1500, Philadelphia: Porcupine Press, 1975 (reprint of the edition published in Cambridge MA, Harvard University, 1936), pp. 12-23, and the documents there cited, which are in the Archivio della Curia in Barcelona; see also Spufford, *Handbook of Medieval Exchange*; pp. 148-152.

[60] Prato, Archivio Datini, n. 960, *Quaderni mutili, scartafacci, ricordanze, ecc.*, ins. IX, X, XI. The three notebooks were edited with commentary by B. Dini. Ambrogio di Lorenzo de' Rocchi, *Una pratica di mercatura in formazione: 1394-1395*, Florence: Le Monnier, 1980.

[61] "Mccclxxxxiiii. Chanbi. Per tutto il Regno di Valenza, d'Araghona, e Catalogna e una medesima moneta, cioè fiorini d'Araghona, che vale s. 11 di moneta barzalonese; e cosi si spenda per tutta la terraferma del Re, e non vale mai ne piu ne meno. [...] I fiorini d'Araghona, si chanbia a Maiolicha per s. 5 maiolichini. | A prendere danari a canbio in Valenza per Barzalona, si fa tanti per tanti lire o fiorini, sichondo s'acordano. Se si dice lire, s'intende darli i danari in Tavola; se si dice fiorini, s'intende contanti in oro. Vale melglo far dare lire per canbio, che fiorini; pero che fiorini talvolta s'anno a chonprare o trarli di Tavole contanti, e costano 1 in 2 per cento. | Per Monpulieri e Vignone, si canbia a franchi, e vale comimemente in '1 franco, da s. 15 in 16 barzalonesi. [...] | Per Genova s'intende a fiorino di Genova, e vale da s. 14½ in 15½ barzalonesi. | Per Pisa, quello medesimo, o d. 2 in 4 meno per fiorino. | Per Firenze, quello medesimo, o d. 6 in 10 per fiorino meno che quelli di Genova." Ambrogio di Lorenzo de' Rocchi, *Una pratica di mercatura in formazione: 1394-1395*, p. 188.

while the cost of the larger *mappae mundi*, namely 100 *florí d'or d'Aragó*, was equivalent to 75 ducats or florins. Calculating the areas of the two different sizes of maps (155 × 310 cm for the smaller and 368 × 368 cm for the larger), we find that Jafuda Cresques and Francesco Becaria worked 36.69 square meters of parchment, 4.8 for each of the smaller *mappae mundi* and 13.54 for each of the larger ones. The smaller *mappae mundi* were therefore valued, according to the agreement, at 12.5 florins of Aragon per square meter, equivalent to 9 and 4/10 Venetian ducats; and the larger ones at 7 4/10 florins of Aragon per square meter, corresponding to 5½ ducats per square meter. The higher per-square-meter cost of the smaller maps was due to the larger number of decorations that adorned them; one can clearly see the difference in the notarial act of June 7th, 1400. The average cost of every square meter of the maps prepared by Cresques and Becaria amounted to 6½ ducats (see Tables 2 and 3 in the Appendix). If we repeat the same calculations on the amount actually paid by Baldassarre degli Ubriachi to the *maestro* and *dipintore*, i.e. 303 florins of Aragon, with an exchange rate of 1 to ¾, we find that this is equivalent to 227¼ ducats, and that the actual average cost of each square meter of the charts by Cresques and Becaria was 6 1/5 ducats.[62]

6. Interpretation of the results: a significant convergence

The paucity of documentary evidence available for this analysis entails a certain caution in the interpretation of the results. The maps and documents chosen for comparison appear entirely heterogeneous: they were of different types, languages, dimensions, and costs. They are so different, in fact, that a glance at the elements listed in Table 2 seems to suggest that comparison would be impossible. Nevertheless, by analyzing the material aspects of the maps, particularly their areas, and the relationships of the relevant currencies, it was possible to determine the average cost per square meter of the painted surfaces of each map. The results of these calculations, which are summarized in Table 3, demonstrate a significant convergence of the costs of different cartographic representations, in my opinion not ascribable to a simple coincidence – particularly in view of the apparent heterogeneity of the maps surveyed. The average per-square-meter cost of the painted surfaces, expressed in ducats or gold florins, in 1460 at the value of 124 *soldi* per ducat and 108 *soldi* per florin, ranges from 6 1/5 ducats to 9 3/5 ducats. Taking the average of all of the maps considered, we observe that the prices per square meter converge around 6½ ducats. The difference between this average value and the average calculated for each map is generally between ½ ducat and 1½ ducat, a minimal variation that expresses rather clearly a convergence of prices. The similarity of the costs per square meter of Fra Mauro's Venetian *mappa mundi* and the Ptolemaic maps painted in Florence by Piero del Massaio between 1460 and 1461 is very noteworthy in my opinion – and until now ignored in the literature.

The costs and methods of making great cosmographic documents at the beginning of the second half of the fifteenth century seem to have been remarkably similar in the different centers of cartographic production of the Mediterranean basin, such as Venice (with Fra Mauro), Majorca and Barcelona (with Jafuda Cresques) and Genoa (with Francesco Becaria). The cru-

[62] See Tables 2 and 3 in the Appendix for the calculations.

cial variables in the process seem to have been the material aspects of the maps : first, the dimensions, then the materials utilized (the documents analyzed are all parchment), and then the decorations. The cartographic content and the graphic syntax do not appear to influence the cost of the works. None of the documents studied here connects the amount paid to the *dipintori* with the cartographic language employed. The decorative elements of the map, that is, the pictorial content, have great importance for the establishment of the price. Becaria, *dipintore di charte da navichare*, in order to claim the money owed to him, listed with great precision the number of cities, castles, banners, fish "large and small," ships and galleys, painted in his *mappae mundi*, emphasizing (apparently in order to justify the delay and to show his good will) that he had planned more ; he based his claim for payment neither on his own skill, nor on the intrinsic value of his work, nor on the fact that he was decorating a nautical chart, or a *mappa mundi*. The emphasis is evidently placed purely on the "quantitative" aspects of the map.

III. – Maps, *objets d'art*, labor and luxuries

1. *Mappae mundi*, frescoes and illuminated manuscripts

Having concluded the comparison among different types of fifteenth-century maps we will now briefly compare the prices of the cartographic representations examined with those of some pictorial and illuminated works of the mid-fifteenth century for which there are documents that indicate the compensation received by the *maestri* for their work. We will begin with a fresco representing *The Last Supper* (*Ultima cena*), now lost, carried out in 1462 by Florentine painter Neri di Bicci (1419-1491) in the refectory of the Convent of the Nuns of Sant'Onofrio di Fuligno in Florence. The choice to analyze this work of art is justified by several factors : the rich documentation and the precise description of the work signed by the painter himself ; the fact that the work is contemporary with both Fra Mauro's *mappa mundi* and Piero del Massaio's Ptolemaic maps ; and the fact that we know the precise relationship between the amount paid to Neri di Bicci and the formal and material characteristics of his fresco. Neri di Bicci was one of the most prolific painters in one of the most active Florentine workshops in the mid- and late fifteenth century. He recorded details of his life and artistic ventures in the *Ricordanze*, written between March 1453 and April 1475 (Library of the Galleria degli Uffizi, MS 2) supplying precise information on the his works, their buyers and destinations, the artists with whom he was in contact, and the types of jobs that he was capable of fulfilling.[63]

The nuns of the Convent of Sant'Onofrio in Florence commissioned the fresco from Neri di Bicci in 1462, and the *Ricordanze* document the contractual terms and include a brief description of the work. There is no trace of the fresco today, as it was covered by a fresco painted by Perugino (Pietro Vannucci, *c.* 1450-1524) during the last decade of the fifteenth century, probably because the earlier work was by then regarded archaic. Neri di Bicci thus records the beginning of the work on March 20, 1461 (1462 according to modern chronology) :

[63] Neri di Bicci, *Le ricordanze : 10 marzo 1453 - 24 aprile 1475*, B. Santi (ed.), Pisa : Marlin, 1976, pp. 180-181 and 186-187.

I remember that the above mentioned day, I, Neri di Bicci, painter, began to paint one wall in the refectory of the nuns of the monastery of Fuligno della Porta of Faenza in which I was contracted to paint the Last Supper and, above it, Christ in the Garden of Gethsemane with a frame around the height of 6½ *braccia* and width of 13½ *braccia*, and I will make all of the haloes in gold and the ornaments in *azzurro di Magnia* [powdered *azzurro* from Germany] where necessary and in general everything will be made with good quality colors at my expense as agreed for the value of 24 *fiorini di suggello* [sealed florins]. Therefore they have to pay me 24 florins valued at 86 soldi.[64]

The final payment of 10 *fiorini di sugello* was recorded on August 4 of the same year.[65] What is important to note in this context is the economic value of the fresco: 24 *fiorini di suggello* (sealed florins) in all, for the fresco with thirteen figures, "di bracc[i]a sei ½ in circha e di largheza di bracc[i]a 13½ in circha," with "chorone d'oro fine e gli ornamenti e d'azuro di Magnia... di tuto fornito di buoni cholori a mia ispese," so that the colors and other materials bought by Neri di Bicci are included in the compensation for the fresco.

To have a more precise idea of the amount paid by the nuns of the convent of Sant'Onofrio, and in order to be able to compare that with the amounts paid for the contemporary cartographic documents analyzed above, we must determine the value of the *fiorino di suggello* (sealed florin) compared to that of the *fiorino a oro* of the time (1462-1463). In spite of the fact that the weight and alloy of the Florentine florin remained unaltered, except in extremely rare cases, from its first coining in 1252 until the last issue in 1542, clipping and erosion reduced the weight of many circulating coins. In 1294 an office called the *Ufficio del maestro del saggio* was created that allowed the public to officially weigh the gold florins that circulated, with the goal of thus guaranteeing their value. The office, in addition to weighing the circulating coins on request, issued florins in small leather purses sealed with a certification of the value of the coins inside. Florentine accounting documents refer to this type of certified florins in purses by the terms *fiorino di suggello*, *di suggello nuovo*, *in suggello*, or *in borsa*.[66] In 1463, the price of the sealed florin was 91 *soldi*, while the price of the *moneta d'oro* was 110 *soldi*.[67] Thus the 24 *fiorini*

[64] "Richordo ch'el sopradetto dì io Neri di Bicci dipintore ò tolto a dipignere dalle monache del munistero di Fuligno dalla Porta a Faenza una facc(i)a di refettorio nella quale ò fare la cena degli Apostoli e di sopra meser Domenedio quando adora nell'orto, chon uno freg(i)o d'atorno la quale facc(i)a e spazio è di sopra tondo, d'alteza di bracc[i]a sei ½ in circha e di largheza di bracc[i]a 13½ in circha, mete[re] tute le chorone d'oro fine e gli ornamenti e d'azuro di Magnia dove achedessi e chosì di tuto fornito di buoni cholori a mia ispese ned ò'avere d'achordo f. 24 di sugello. Posto che deon dare f. 24 a libro s[egnat]o D a f. 86." Neri di Bicci, *Le ricordanze*, pp. 180-181, ricordanza 357. See also S. Padovani, *Il cenacolo del Perugino detto 'di Fuligno'*, Ministero per i beni culturali ambientali, Soprintendenza per i beni artistici e storici di Firenze, Florence: Tipografia Coppini, 1990, p. 19.

[65] "Richordo ch'el detto dì ebi per lle donne di Fulignio da G[i]ovanni Aldobrandini f. dieci di suggello e chossì chonfesai avere auti in questo modo...," Neri di Bicci, *Le ricordanze (10 marzo 1453 - 24 aprile 1475)*, pp. 180-186, ricordanza 370: "Mecholedì a dì 4 d'aghosto 1462. Danari ebi da G[i]ovanni Aldobrandini per lle donne di Fulignio. Richordo ch'el detto dì ebi per lle donne di Fulignio da G[i]ovanni Aldobrandini f. dieci di suggello e chossì chonfesai avere auti in questo modo...."

[66] For a detailed analysis of the process of issuing *fiorini di suggello* see M. Bernocchi, *Le monete della Repubblica fiorentina*, Florence: Olschki, 1974-1978, vol. 3; and G. Targioni Tozzetti, "Del fiorino di sigillo," G. A. Zanetti (ed.), *Nuova raccolta delle monete e zecche d'Italia*, vol. 1, Bologna: Istituto delle Scienze, 1775, pp. 249-274. See also Spufford, *Handbook of Medieval Exchange*, pp. 25-26.

[67] For a brief, but useful, analysis that supplies additional details on the Florentine money discussed here, see Goldthwaite, *Studi sulla moneta fiorentina: secoli XIII-XVI*, pp. 31-33, 83 (fig. 5), and 95.

di suggello paid to Neri di Bicci in 1462 were valued at 28 *fiorini a oro* and 14 *soldi*. Given the parity of the florin and the ducat a large fresco (about 10 × 4 meters), with the representation of thirteen human figures and a smaller insert with the *Oration in the Garden*, painted by a maestro of average caliber (or at least not among the most prestigious) could cost as much as a *mappa mundi* similar to Fra Mauro's. It was also worth approximately half as much as the twenty-seven maps of a Florentine manuscript of Ptolemy's *Geography* painted by Piero del Massaio.

2. Mid-fifteenth-century Venetian illuminations : Leonardo Bellini

Having analyzed the cost of a pictorial representation of large dimensions we will now consider a contract for a prestigious manuscript illumination drafted in Venice on December 7th, 1463 by the notary Ulisse Aleotti. The document declares that the *Ufficio delle razon vecchie* gave to Leonardo Bellini – the painter of the earthly paradise in Fra Mauro's *mappa mundi* – four ducats for an illumination of an official oath.[68] Lino Moretti – as it was already discussed in Chapter III – identified the *Promissione dogale di Cristoforo Moro* (the official oath of fidelity and service to the *Serenissima* of the patrician Cristoforo Moro, on the occasion of his election to the seat of *Doge* in 1462) as the work cited in the contract.[69] On the opening page of the *Promissione*, within the initial letter (*capolettera*), Leonardo Bellini depicted a miniaturized Venetian *pala d'altare* of the time in accordance with the style and methods of his uncle and *maestro*, Jacopo Bellini, with the Virgin and Baby Jesus on a throne and the Doge kneeling with the patron saints Marco and Bernardino at his side.[70]

Leonardo Bellini was trained in the *bottega* of his uncle, Jacopo Bellini. A contract, signed August 23rd, 1443, relates to the working conditions of Leonardo, who at this time had already been living with his uncle for twelve years. Jacopo emphasized Leonardo's talent for painting and agreed to take him into his studio as an apprentice for two years, paying him a total of 25 gold ducats.[71] For the purposes of this study there are two important conclusions that it is possible to draw : first, that a painter's apprentice in Bellini's studio, the most prestigious of Venice, could earn in two years the equivalent of the value of a *mappa mundi* like Fra Mauro's ; and second, that an elegant manuscript illumination, with the Virgin, Baby Jesus, two saints on the side, and two small angels at the bottom of the page, completed by one of the best illuminators in Venice in the 1460s for a very prestigious buyer, cost only a fifth of a *mappa mundi* of the dimensions of Fra Mauro's.

[68] P. Paoletti, *Raccolta di documenti inediti per servire alla storia della pittura veneziana nei secoli XV e XVI*. Fascicolo I. *I Bellini*, Padua : R. Stabilimento P. Prosperini, 1894, pp. 7-8.

[69] L. Moretti, "Di Leonardo Bellini, pittore e miniatore," *Paragone : Arte* 99 (1958), pp. 58-96 ; and E. Billanovich, "Per la storia della pittura nel Veneto," *Italia medioevale e umanistica* 16 (1973), pp. 362-367.

[70] London, British Library, Add. MS 15816, f. 5r ; miniature on parchment, 340 × 235 mm. ; For reproductions see G. Mariani Canova, "Per Leonardo Bellini," *Arte veneta* 22 (1968), p. 10 ; and G. Mariani Canova, "Leonardo Bellini," in *Storia della miniatura veneta*, Padua, 1968, figs. 10 and 11. L. Moretti, "Di Leonardo Bellini, pittore e miniatore," *Paragone : Arte* 99 (1958), pp. 58-96 ; *Mappamondo* (ed. 2006), Fig. 16, p. 169.

[71] At this regard, see Chapter III.

3. The great luxuries (clothes and slaves). The cost of manual labor

In the fifteenth century one of the most important luxury items was fine materials clothing for the great Renaissance court families. The history of clothes, as Fernand Braudel wrote in his classic study, is less anecdotal than it appears. "There are at least two fundamental issues at stake : the production and costs of primary materials and their manufacturing process, as well as the political and social use of clothes in ceremonies and their function in marking social hierarchies."[72] The history of clothing and the value of the clothes in the Renaissance is indeed very complex ; therefore we will focus our attention on the value of clothes of the highest quality. Money spent on the acquisition of silks, velvet, gold and silver brocades[73] represented a significant proportion of the expenses registered in the domestic account books of wealthy families. Inlays, gold brocades, and velvets were acquired at prices between 4 and 8 ducats (or florins), a sleeve and a gown could easily cost 40 or 50 ducats, and the price for a woman's ceremonial dress could rise above 100 ducats.[74] In the mid-fifteenth century the prices of satin, velvet, and brocade rose so high as to generate scandal and, as a result, numerous sumptuary laws were passed in an ineffectual attempt to limit the exorbitant spending on clothes. Given this pricing information, we can see that a *mappa mundi* like that of Fra Mauro's cost about as much as 5 *braccia* of a high quality material that would be used in clothing suitable for members of the higher classes. In the mid-fifteenth century, the dress for a woman of this class would necessitate 6 or 7 *braccia* of fabric, and so a *mappa mundi* like that of Fra Mauro in fact cost less than a single ceremonial gown, and much less than one of the most sumptuous dresses of the period.[75]

Among the numerous goods that arrived in Venice, which Francesco Filelfo described in 1461 as *opulentissima e populosissima* (very opulent and overcrowded), there were also slaves. Most of the slaves, along with other merchandise that the Venetians stored in their warehouses in the city of Tana at the northeastern corner of the Black Sea, reached Venice from Tartary but also from Trebizond and Constantinople. Slaves also came to Venice from Central Asia together with furs, salt fish, and silks. In addition to their practical functions, the slaves represented a conspicuous display of personal and familial wealth. Sold at the Rialto in the Campo San Giacomo, they were normally owned only by noble families and the most prominent merchants who bought them for domestic service and for heavy work such as the digging and maintenance

[72] F. Braudel, *Civiltà materiale, economia, capitalismo (secoli XV-XVIII) : Le strutture del quotidiano*, Turin : Einaudi, 1982, pp. 282-301, particularly p. 282.

[73] Brocades were rich silk fabrics with a satin background on which designs were woven of silk, silver and gold.

[74] The bibliography on women's clothing in Renaissance Florence and Venice is extensive. See the classic study by E. Polidori Calamandrei, *Le vesti delle donne fiorentine nel Quattrocento*, Florence : La Voce, 1924, reprinted Rome : Multigrafica, 1973 ; C. C. Frick, *Dressing Renaissance Florence : Families, Fortunes, & Fine Clothing*, Baltimore and London : Johns Hopkins University Press, 2002 ; and P. Allerston, "Wedding Finery in Sixteenth-Century Venice," in T. Dean and K. J. P. Lowe (eds), *Marriage in Italy, 1300-1650*, Cambridge : Cambridge University Press, 1998, pp. 25-40.

[75] On the value, costs of production, and rituals relating to the use of clothing in the Early Modern Period, see M. G. Muzzarelli, *Guardaroba medievale. Vesti e società dal XII al XVI secolo*, Bologna : Il Mulino, 1999 and the bibliography cited there. On the problem of cost, see M. Spallanzani, *Ceramiche orientali a Firenze nel Rinascimento*, Florence : Cassa di Risparmio di Firenze, 1978.

of the canals and water procurement. A slave could cost between 30 and 60 ducats, a considerable sum around the mid-fifteenth century when a Venetian commoner (*del popolo*) could live for an entire year on 20 ducats, and a highly esteemed naval architect at the Arsenale earned around 130 ducats.[76] A Pisan document from 1398 confirms these calculations : a certain Donna Iacopa, widow of Albizzo da Vico, sold to a certain "Ciurlo del fu Cristoforo da Pratovecchio abitante in Peccioli" her slave, Lena, 38 years old, of Tartar or Greek origins, for 25 gold florins.[77] Thus the copy of Fra Mauro's *mappa mundi*, which cost about 28 gold ducats, was valued substantially less than a slave at the same period.

After having briefly compared the costs of fifteenth-century maps with those of some luxury items, I will now complete the comparative picture by comparing those costs with the price of manual labor by minor artisans, laborers, wool and silk workers, construction workers, and others of modest economic level. In the context of this study we will focus exclusively on the average value of a working day for some categories of manual labor for which there is substantial data available. Thus the 28 gold ducats paid for Fra Mauro's *mappa mundi* was equal in value to about 390 working days of a non-specialized laborer (around the mid-fifteenth century generally paid around 8 *soldi* a day), or 220 working days of a carpenter, or 250 of a stonecutter (who at this same period were paid 14 and 12 *soldi* a day respectively). In the domestic sector 1 ducat was the monthly wage of a wet-nurse, while a cook had an average monthly salary of around 0.5 to 0.8 ducats, so the cost of the *mappa mundi* was equivalent to more than two years of labor by a wet-nurse, and three years of labor by a cook.

IV. – Some final considerations

Undoubtedly the cost comparisons made in this chapter are based on a limited, and not necessarily representative, set of data that includes several different types of objects while ignoring others that are no less interesting. While conscious of these limitations we can nonetheless draw some significant conclusions. The objects and their costs that we have studied, beginning with the Venetian *mappa mundi* by Fra Mauro and the Florentine Ptolemaic maps by Piero del Massaio, show that maps, though clearly prestigious products, were at the lower end of this category ; the relationships demonstrated between the prices of maps and cost of the manual work of artisans and laborers confirms this. Although maps were certainly destined for elite customers they nevertheless are not in the same cost range as the true luxury items so abundant in the courts and the Venetian and Florentine palaces of the fifteenth century, such as silverware and jewelry, carved wood, antique sculptures and, above all, precious clothes. There are documents that demonstrate that for celebrations, furnishings and clothes, noble families spent

[76] B. Doumerc, "Il dominio del mare," in *Storia di Venezia*, vol. IV : A. Tenenti and U. Tucci (eds), *Il Rinascimento. Politica e cultura*, Rome : Istituto della Enciclopedia Italiana, pp. 113-180 (p. 146).

[77] "23.V.1398. Vendidit… sclavam unam de progenie tartarorum seu de progenie grecorum nomine Lenam annorum 38 vel circha… nomine certi pretii flor. 25 auri" ; Firenze, Archivio di Stato, *Nuove Accessioni* 1996, fasc. I, f. 30v.

hundreds, if not thousands of ducats and florins.[78] A slave, for example, was worth on average about twice the amount of a *mappa mundi* similar to Fra Mauro's.

Nevertheless, as we have seen in the cases of the maps commissioned by Baldasarre degli Ubriachi for the Kings of England and of Aragon, of the copy of Fra Mauro's *mappa mundi* commissioned by the Portuguese King Afonso V, and finally of the manuscript of Ptolemy's *Geography* ordered by the Bishop of Algarve Álvaro Afonso, maps were held in high regard and enjoyed great prestige.[79] The passion of the powerful figures of the fifteenth century for cartography and their desire to possess maps was cultivated chiefly for erudition and for the pleasure of collecting. Poggio Bracciolini, in the *proemio* to one of his dialogues, the *De infelicitate principum*, describes how, when he went to see the humanist Niccolò Niccoli in his library, he found him in the company of Cosimo de' Medici and Carlo Marsuppini, all three kneeling over a manuscript of Ptolemy's *Geography*.[80] For political figures such as Cosimo the Elder, Pius II, Duke Borso d'Este, and Afonso V of Portugal, cosmography and the use of maps constituted not only a fundamental interest and a way to symbolize their roles as leaders; they were also expressions of a real and important need for them to know the world.[81] In my opinion, this suggests a consideration of the practical role of maps, particularly in the hands of princes.

Cosmography, from Cresques to Fra Mauro, and also in Gerard Mercator, is a complex science that studied both space and time, developing and producing a synthesis of at least three discourses: the cosmologic, cosmographic and chorographic. It narrates the sacred time of creation, the biblical story of humanity, and the history of peoples, their habits, trade and locations. Among the highest expressions of cosmographic knowledge, the *mappae mundi*, told the viewer about the world and described it by bringing together images and texts from heterogeneous sources. The case of Baldassarre degli Ubriachi, merchant of jewelry, ivory, and inlaid bone, is revealing: he does not pay homage to the Kings of Aragon and England with precious silver, ivories or pearls. Instead he chose to present them *mappae mundi*, no doubt similar in their dimensions, and probably also in their contents, to Fra Mauro's. Although relatively costly (45 and 75 ducats), they were worth less than the inlays and sculptures in ivory and bone that, prepared in Baldassarre's workshop on ca' Zane, were sold all over Europe. It is evident that

[78] M. Spallanzani, *Ceramiche orientali a Firenze nel Rinascimento*, Florence: Cassa di Risparmio di Firenze, 1978; M. Spallanzani (ed.), *Inventari medicei, 1417-1465: Giovanni di Bicci, Cosimo e Lorenzo di Giovanni, Piero di Cosimo*, Florence: Associazione Amici del Bargello, 1996; M. Spallanzani and G. Gaeta Bertelà (eds), *Libro d'inventario dei beni di Lorenzo il Magnifico*, Florence: Associazione Amici del Bargello, 1992.

[79] On this subject see also M. Belozerskaya, "Jan van Eyck's Lost *Mappa mundi*: A Token of Fifteenth-Century Power Politics," *Journal of Early Modern History* 4, 1 (2000), pp. 45-84.

[80] On this subject, which I only touch on briefly here, see S. Gentile, "L'ambiente umanistico fiorentino e lo studio della geografia nel secolo XV," in L. Formisano, G. Fossi, P. Galluzzi, S. Gentile and R. Pasta (eds), *Amerigo Vespucci. La vita e i viaggi*, Florence: Banca Toscana, 1991, pp. 9-63; for the episode narrated by Bracciolini, see Gentile, *Firenze e la scoperta dell'America*, entry 51, p. 101; and P. Bracciolini, *De infelicitate principum*, D. Canfora (ed.), Rome: Edizioni di Storia e Letteratura, 1998, pp. 7-8.

[81] On the Ferrarese court and its interest in cosmography, see Ch. Rosenberg, "Arte e politica alle corti di Leonello e Borso d'Este," in *Le Muse e il Principe. Arte di corte nel Rinascimento pagano*, 1991, 2 vols., Modena: Panini, vol. 1, pp. 39-52; and M. Milanesi, "Il commento al 'Dittamondo' di Guglielmo Capello (1435-1437)," in M. Bertozzi (ed.), *Alla corte degli Estensi. Filosofia, arte e cultura a Ferrara nei secoli XV e XVI*, Ferrara: Università degli Studi, 1994 pp. 365-388. On Pope Pius II's interest in cosmography see N. Casella, "Pio II tra geografia e storia: la 'Cosmographia'," *Archivio della Società Romana di storia patria* 3, 26 (1972), pp. 35-112.

the value of the gift resided not just in the expensive materials (powdered *azzurro*, gold, etc.) used but also in the cosmographic contents. Shortly after this period the discovery and conquest of the "New World", and the cosmographic division of the world that followed the Treaties of Tordesillas of 1494, accentuated the value and importance of cosmography, which became essential strategic knowledge protected by a state policy of secrecy.[82]

The convergence of the costs of maps revealed in the Catalan, Venetian, and Florentine documents analyzed here invites us to reconsider an important topic in the historiography of early modernity : the forms of the market for artistic goods and their relationship with patronage systems. The history of the commission, production, and sale of maps in the fifteenth century seems to me to offer a case study whose significance and analytical implications go beyond the domain of medieval and Renaissance cartography, inviting for comparison with what is known of the relationship between patrons and artists in the Renaissance and Baroque courts. This, however, is a vast subject whose historiography is daunting.[83] While conscious of the opportunities presented by this avenue of research, that is, by the comparison with the forms of commission, production and sale of other valuable goods such as paintings, statues and manuscripts, we recognize that this would entail detailed discussion of the dynamics and methods of interaction between the market and systems of patronage, which is well beyond the scope of this research. This remains, however, a promising and inviting perspective for future exploration.

From this research there emerges an important fact that is worth highlighting. The maps that we have considered were commissioned and made under a system of patronage, but this system functioned according to recognized and shared pan-European market rules. In other words : cartographic production was organized locally but production costs, mechanisms of price formation, and marketing strategies in different locations tended to converge. A *mappa mundi* commissioned in Venice rather than in Barcelona, or a manuscript of Ptolemy's *Geography* commissioned in Florence – although under different systems of patronage involving individual cartographers, scribes and illuminators who were probably mutually ignorant of each other – were evaluated on the basis of common economic criteria. This implies the existence of an internationally shared awareness of the mechanisms for the pricing of maps, both by the cartographers (producers) and the clients (the patrons and buyers). The *circa* 28 gold ducats paid by the Portuguese Ambassador João Fernandes da Silveira to the Venetian Fra Mauro for a copy of his *mappa mundi*, and Pope Pius II's donation of 25 gold ducats for a *mappa mundi* drawn by the Venetian cartographer Don Antonio Leonardi – rather than the 110 and 60 florins of

[82] A. Teixeira de Mota, "Reflexos do Tratado de Tordesilhas na cartografia nautica do seculo XVI," in *El Tratado de Tordesillas y su projección*, Serie Americanista del Seminario de Historia de América, Universidad de Valladolid, Valladolid, 1973, vol. 3, pp. 119-136.

[83] The literature is really vast. See for example : M. Baxandall, *Pittura ed esperienze sociali nell'Italia del Quattrocento*, Turin : Eiunaudi, 1978 ; F. Ames-Lewis (ed.), *Cosimo 'il Vecchio' de' Medici, 1389-1464*, Oxford : Clarendon Press, 1992 ; R. A. Goldthwaite, *Wealth and the Demand for Art in Italy, 1300-1600*, Baltimore and London : Johns Hopkins University Press, 1993 ; M. Hollingsworth, *Patronage in Renaissance Italy from 1400 to the Early Sixteenth Century*, John Murray, London, 1994 ; D. Thornson, *Renaissance Architecture : Critics, Patrons, Luxury*, Manchester : Manchester University Press, 1993 ; A. Esch and C. L. Frommel (eds), *Arte, committenza ed economia a Roma e nelle corti del Rinascimento (1420-1530)*, Turin : Einaudi, 1995 ; F. Haskell, *Mecenati e pittori : studio sui rapporti tra arte e società italiana nell'età barocca*, Florence : Sansoni, 1966 ; M. Fantoni and S. Matthews-Grieco (eds), *The Art Market in Italy, 15th-17th Centuries*, Ferrara : Istituto di Studi Rinascimentali, Modena : Panini, 2003.

Aragon paid by Baldassarre degli Ubriachi through his Florentine agent to Jacme Ribes, a Mallorcan cartographer, and Francesco Becario, a Genoese decorator, in Barcelona, to buy *mappae mundi* of different sizes – these prices are not "arbitrary." On the contrary, even when they constituted "gifts" and despite being denominated in currencies that continuously changed value, they indicate the existence of common criteria for the pricing of the fifteenth-century maps : criteria that were shared and tended to be similar in different regions, and which were especially based on the material costs of production. Inequality or heterogeneity in prices, therefore, is only apparent. Map size, the quantity of parchment, the number of decorations, color quality, and the presence or absence of gold, are the main variables in price formation. Their value was known and more or less the same all across Europe. The commission system was that of patronage ; the production, sale, and purchase of maps – as was common with other *objets-d'art* – were instead set in the context of an operative Europe-wide embryonic market system which, joining in a large network of the main centers of cartographic production of the Mediterranean Sea basin.

CHAPTER II

THE CONFLUENCE OF POLITICS AND SCHOLARSHIP THE *MAPPA MUNDI* WITHIN THE DAWNING OF THE HISTORY OF CARTOGRAPHY

I. – Beyond the history of geographical discoveries

In this last chapter we will turn our attention to the historic reaction to the *mappa mundi* from the mid-eighteenth century to the mid-nineteenth. In this relatively brief time period Fra Mauro's work lay at the centre of four distinct cultural scopes; their common link being an intense interest in the study and reproduction of ancient maps, analyzing the said documents in their own right, and not exclusively in the context of the history of geographical discoveries.

Between 1761 and 1762 the *Riformatori allo Studio di Padova* (the Reformers of the University of Padua) recognized the *mappa mundi* of Fra Mauro as one of the map to be used as a model to restore the maps (*teleri*) adorning the Sala dello Scudo inside Venice's Palazzo Ducale, by that time deteriorated to the point of illegibility. For the first time, an ancient map, considered the most faithful representation of the world and of the voyages described at the beginning of the fourteenth century by Marco Polo, would form the basis of constructing and celebrating a collective historical memory of the Venetian *patria.*[1]

Roughly half a century later, during the same period that witnessed the first monographic publications on Martim Behaim's *Erdapfel* (literally 'potato', that is, terrestrial globe, *c.* 1492)[2] and the so-called "Borgia *mappa mundi*" (*c.* 1430-1450),[3] Fra Mauro's map became one of the

[1] Venice, Archivio di Stato, Riformatori allo Studio di Padova, filza 29, ff. 578-582, *Memoria per V.V.E.E.*, published by R. Gallo, "Le mappe geografiche del Palazzo Ducale di Venezia," *Archivio Veneto* 32-33 (1943), pp. 47-113 (for the *Memoria*: pp. 93-100).

[2] Behaim's globe was studied by Murr, Christoph Gottlieb von in *Diplomatische Geschichte des portugesichen beruehmten Ritters Martin Behaims. Aus Originalurkunden*, Nuerenberg: Johann Eberhard Zeh, 1778. French translation: *Histoire diplomatique du chevalier portugais Martin Behaim de Nuremberg avec la description de son globe terrestre*, traduite de l'allemand par le citoyen H.-J. Jansen. 3rd edition, Strasbourg & Paris: Treuttel et Würtz, 1802. The work had already been published in the appendix of Antonio Pigafetta's *Premier voyage autour du monde par le Chevr. Pigafetta sur l'escadre de Magellan, pendant les années 1519, 20, 21 et 22, suivi de l'extrait du Traité de navigation du même auteur et d'une Notice sur le chevalier Martin Behaim, avec la description de son globe terrestre* [par M. de Murr]; [traduit de l'allemand par H. Jansen], Paris: H.-J. Jansen, an IX [1800-1801].

[3] The "Borgia Mappa Mundi", a disc of copper incised with a graver, was studied and reproduced for the first time in 1797, then again in 1808: *Apographon descriptionis Orbis terrae, figuris et narratiunculis distinctae, Manu Germanica opere nigelliari discolorio circa medium Saec. XV. Tabulae aeneae Musei Borgiani Velitris consignatae, Quod Camillus Joh. Paulli F. Borgia, Cruce Hieros. ornatus, ab intimo cubiculo Electoris Bavarici, Patrui Cardinalis exempla imitatus Summa fide, maximoque artificio expressum, recognitumque Eruditis spectandum proponit* A. C. MDCCXCVII. (1797); Heeren, Arnold Hermann, "Explicatio planiglobii orbis terrarum faciem exhibentis, ante medium seculum XV. summa arte confecti, Musei Borgiani Velitris," *Commentationes Societatis Regiae Scientarum Gottingensis* 16 (1808), pp. 250-284.

first cartographic works to be an object of monographic analyses. As we have already noted in earlier chapters, the *mappa mundi* was studied in great detail by Bartolomeo Alberto Cappellari (1765-1846) – first called Mauro, in honor of Fra Mauro as a Camaldolese monk and, from 1831 to 1846, Pope Gregory XVI – and by another Camaldolese monk, Giacinto Placido Zurla (1769-1834), in 1823 named Cardinal-Vicar of Rome. While Cappellari's research was never published[4] that of Zurla, an impressive *summa* of more than one hundred years of Camaldolese historiography on the *mappa mundi*, was printed in 1806.[5]

Between 1804 and 1805, the same period in which the map was being studied at the Monastery of San Michele of Murano, the English painter William Frazer, using the transcriptions of the map's legends prepared by Zurla for his own monograph, executed a same-sized manuscript copy of the *mappa mundi* on vellum. This copy was commissioned by William Vincent, Dean of Westminster jointly with the Earl Macartney, the 2nd Earl Spencer, and the 4th Earl of Buckinghamshire – all first-tier politicians with important duties in the British colonies ruled by George III (1760-1820) – in addition to the financial support of the East India Company. Vincent, Macartney, Spencer, and the East India Company's Board believed to have finally found the original map by Marco Polo, so closely did the contents of the *mappa mundi* reflect those of the alleged map brought back by the Venetian merchant. Before the world map was copied, however, the group of financiers discovered that the true author of the work was, in fact, Fra Mauro. They nevertheless decided to proceed with their plan to reproduce it, and deposited the resulting copy in the British Museum in 1807. This copy is now in the Manuscripts Reading Room of the British Library as Additional MS 11267 (plate IX).[6]

Finally, in April 1849 in Paris, where he had been in exile since 1820, Manuel Francisco de Barros e Sousa de Mesquita de Macedo Leitão e Carvalhosa (Lisbon, 1791 - Paris, 1855), known as the Second Viscount of Santarém (*Segundo Visconde de Santarém*) and already Minister of the Kingdom of Portugal and historian of international fame, wrote in the *Bulletin de la Société de géographie* a brief "Note sur la Mappemonde du cosmographe Fra-Mauro."[7] The Viscount's note announced, with evident satisfaction, the imminent publication of "a faithful copy of that large world map" ("une copie fidèle de cette grande mappemonde"). Santarém was proud to have succeeded in reproducing a document which, by virtue of the copy commissioned by the Portuguese king Afonso V in 1457, having unfortunately gone missing, he did not hesitate to declare of Portuguese origin (plate X). Through his friend John Holmes, an Assistant Keeper in

[4] Camaldoli (Arezzo), Biblioteca del Monastero, Fondo San Gregorio, n. 19, M. Cappellari, *Scritti del P. Mauro Cappellari "Gregorio XVI" sul Planisferio di Fra Mauro Camaldolese*, manuscript on paper, middle of the eighteenth century. For Cappellari's biography see *Gregorio XVI. Miscellanea commemorativa*, Rome : Università Gregoriana, 2 vols., 1948.

[5] P. Zurla, *Il mappamondo di Fra Mauro camaldolese descritto ed illustrato da D. Placido Zurla dello stess'ordine*, Venice, s.n., 1806. Zurla dedicated further writings to the *mappa mundi* in P. Zurla, *Di Marco Polo e degli altri viaggiatori veneziani*, 2 vols., Venice : G. Fuchs, 1819, vol. 1, pp. 230, 343-349. For Zurla's biography see : V. Meneghin, *San Michele in Isola di Venezia*, 2 vols., Venice : Stamp. di Venezia, 1962, vol. 1, pp. 231-243, 405-410 ; vol. 2, pp. 7-27.

[6] London, British Library, Additional MS 11267, [The Fra Mauro World Map copied by William Frazer], manuscript on vellum, 239 × 229 cm, 1804. The history is deduced in all its particulars from the correspondence between William Vincent and the 2nd Earl Spencer, held at the British Library in London. See note 33 below for the complete bibliographical details.

[7] M. F. Santarém de Barros e Sousa, "Note sur le mappemonde de Fra Mauro," *Bulletin de la Société de Géographie de Paris*, 1846, pp. 251-252.

the British Museum's Department of Manuscripts, the Viscount had procured the manuscript copy of the map deposited there by William Vincent, Lord Hobart, and Earl Spencer in 1807.[8]

In the following paragraphs we will investigate in detail these cultural and editorial undertakings in which the protagonists, though separated by space and time, and motivated by diverse interests, were nevertheless bound tightly together by other factors and relationships. The different ways in which Fra Mauro's world map was utilized, contextualized, and understood in four western European intellectual centers – Venice (1761-1762, 1806 and 1818), London (1804-1807), Paris, and Lisbon (1849-1855) – sheds significant light on the birth and steady growth of scientific interest in ancient maps in Europe over nearly a century. Understanding the response to Fra Mauro's work also allows us to observe, from a clearer perspective, the circulation of geographical knowledge within Europe at that time; a circulation which engaged – often in a highly competitive manner – some of the most eminent cultural institutions and personalities of the Italian peninsula, Britain, Portugal, and France. Studying the way the *mappa mundi* was received – the reactions it engendered and initiatives it inspired – through the observations of four separate groups of readers, spectators and scholars of the work, enables us to illuminate a broader image of the map's historical significance. Between the late eighteenth and early nineteenth centuries, in the wake of the *mappa mundi*, the history of ancient cartography was born in Europe – amidst erudition, construction of a collective memory, orientalism and colonialism – as a distinct discipline. In this vein, the way in which Fra Mauro's work was awarded a place in European history offers an exceptionally clear example of how a collective historical consciousness could be constructed through the study and the reproduction of maps.

II. – Venice, 1761-1763

Maps and the construction of a collective historical memory

The story of the maps in the Ducal Palace of Venice, and their vicessitudes over the centuries has been the subject of accurate and detailed analysis by Rodolfo Gallo and, recently, Marica Milanesi.[9] From their timely analysis, we know that in 1761 the *Riformatori allo Studio di Padova* – a magistracy of the Venetian Republic charged with the management and control

[8] *Correspondência do 2° Visconde de Santarém, colligida, coordenada e com anotações de Rocha Martins; publicada pelo 3° Visconde de Santarém.* 8 vols., Lisbon, Alfredo Lamas & Motta, 1918-1919: Letter from John Holmes to Santarém, London, 22 August 1841, vol. VI, 1919, p. 211; letter from Santarém to the Portuguese Minister of Foreign Affairs, 28 June, 1844, vol. VI, pp. 458-459; *idem*, Paris, 22 May 1844, vol. VI, p. 446-447; letter from Santarém to Mr. Wright, Paris, 16 August 1845, vol. VI, p. 512; letter from Santarém to the engraver Schwaertzlé, Paris, 30 November 1845, vol. VI, p. 556.

[9] Gallo, "Le mappe geografiche del Palazzo Ducale di Venezia," pp. 47-113; M. Milanesi, "Nelle stanze di palazzo. Venezia, Firenze, Roma: qualche confronto," Proceedings of the Symposium "Cristoforo Sorte e il suo tempo," Verona, 31 October 2008, in press. Ead. "Le regard de la postérité. L'âge des découvertes vu un siècle après." *Médiévales* 58 (2010), pp. 11-26 (Actes du Séminaire "Humanisme et découvertes géographiques," organized by N. Bouloux, A. Cattaneo and P. Gautier Dalché, Tours 2009). These two articles by Milanesi are the most updated studies on the complex history of the maps of the Sala dello Scudo, from their sixteenth-century origins until their eighteenth-century renovations. I am grateful to Marica Milanesi for sharing with me the results of her research.

of the entire cultural-didactic sector within the Venetian State[10] – sent a report to Doge Francesco Loredan (doge from 1752 to 1762) expressing their wishes for the restoration of the Sala dello Scudo, the antechamber of honor to the doge's apartments, in Venice's Palazzo Ducale. In particular, the report detailed the reconstruction of the four large canvases (*teleri*) displaying four painted mural maps, which illustrated – to quote the letter that accompanied the report – "the vast extent of ancient maritime and terrestrial Commerce of the happy subjects of this August Republic in all the known world, and particularly representing the remote countries, into which, more than any other Nation of Europe, [the Republic] came to penetrate."[11] The four maps, regarded by the *Riformatori* as the largest existing in the world, were believed to have been based on the "Memoirs of the famous Traveler and Noble Venetian, Marco Polo" (*Memorie del famoso viaggiatore patrizio veneziano Marco Polo*), and were considered the clearest evidence in existence of Venetian scientific and cartographic excellence – a direct consequence of the power and wealth of the *Serenissima*.

The maps, however, were in such a poor state of conservation that it was not actually possible to restore them. The only way to recover their contents, function, and beauty, then, was to create new, oil-painted copies, using the original *teleri* as models. The contents that were still legible would be copied to the new canvases using the *tecnica dello spolvero* : a method that consisted of pricking through tiny dots on the original canvas, which would then be covered with a dusting of soot, thus transferring the outline of the original drawings onto the new material. The man chosen to perform the reproduction was Francesco Grisellini (1717-1787), a man of eclectic interests and skills : naturalist, agronomist, traveler, founder and illustrator of scientific reviews, member of various scientific academies, draftsman, engraver, and cartographer.[12] For a total of 400 ducats, Grisellini was hired to carry out – as an experiment – the reconstruction of the map illustrating India, Tartary, and China, in which "all of the countries penetrated by Marco Polo are clearly delineated, exalting the name Veneto, and the power of this glorious Republic."[13] If the result of this first copy fulfilled expectations Grisellini would then replicate the other maps within two years under the direction and strict surveillance of Antonio Maria

[10] The *Riformatori allo Studio di Padova*, originally a temporary commission founded in 1517 to hire new professors at the University of Padua, became permanent in 1528 with gradually broadened responsibilities, eventually regulating the entire cultural sector both in Venice itself and within the Venetian State. See A. Da Mosto, *L'Archivio di Stato di Venezia. Indice generale, storico, descrittivo ed analitico*, Venice : Archivi dell'amministrazione centrale, vol. I, 1937, p. 217.

[11] "[...] la vasta estensione dell'antico Commercio marittimo, e terrestre dei felici suditi di questa Augusta Repubblica per tutta la terra nota, e che particolarmente anche rappresentassero i rimoti paesi, ne quali prima d'ogni altra Nazione d'Europa era loro succeduto di penetrare.") Venice, Archivio di Stato di Venezia, Riformatori allo Studio di Padova, filza 29, ff. 578-582, *Memoria per V.V.E.E.*, published by Gallo, "Le mappe geografiche del Palazzo Ducale di Venezia," p. 93. For the full reconstruction of this event, see Milanesi, "Nelle stanze di palazzo. Venezia, Firenze, Roma : qualche confronto" and Ead. "Le regard de la postérité. L'âge des découvertes vu un siècle après," pp. 13-19.

[12] See P. Preto, "Grisellini Francesco," *Dizionario biografico degli italiani* 59 (2002), pp. 691-696. G. B. de Toni, "Francesco Griselini [*sic*] viaggiatore e naturalista veneziano del secolo XVIII," *Archivio di storia della scienza* 1 (1919), pp. 1-21. For the maps drawn and engraved by Grisellini, see : Gallo, "Le mappe geografiche del Palazzo Ducale," pp. 91-92.

[13] "Ben delineati si vedono i paesi tutti ne quali penetrò Marco Polo prima d'ogni altro europeo facendovi conoscere il nome Veneto, e la possanza di questa gloriosa Repubblica," *ibidem*.

Zanetti, director of the Biblioteca Marciana. Zanetti's role would be to verify Grisellini's adherence to the techniques indicated by the *Riformatori* and approved by the Doge, and to judge the quality of the resulting reproductions.[14]

Indeed, the *Riformatori's* report specified the minutest detail of the procedure Grisellini was to follow. The original *teleri* would be removed from the walls, rolled up, and taken to a laboratory where Grisellini would reproduce those parts still legible onto paper board (*cartone*). As for those parts missing or no longer legible, the *Riformatori* stipulated:

> [we] will utilize observations made on the *Planisphere of S. Michele of Murano* [Fra Mauro's *mappa mundi*], which have already been used in the Memoirs of Marco Polo, which book shows in miniature all that which is shown in more detail on these very maps. From this planisphere, we will copy with the utmost imaginable diligence the places, and all the missing pieces of the great Map, which will then be transferred and geometrically enlarged with the guidance of the scale of distances on the prepared *cartone*.[15]

In particular Grisellini would have to adhere to the

> geometric structure and contents deduced from the well-known Planisphere of Saint Michel of Murano addressed in the aforementioned century by a Layman of that Monastery, likewise in the memoirs of Polo, and the directions regarding mechanical skill, which the named Grisellini will maintain in the restoration of these illustrious Geographic Works.[16]

As for the extent of our research, we will limit ourselves in noting that the *Riformatori* identified the *mappa mundi* by Fra Mauro as one of the models which, in virtue of its age and contents, allowed them to restore the ancient splendor of the maps in the Sala dello Scudo, adding to the prestige of Venice and of its Doge. In the eyes of the *Riformatori*, the work of Fra Mauro constituted a historic monument, narrating, with fitting eloquence, of Venice's past glory – particularly its commercial and scientific achievements. The maps of the Palazzo Ducale, then – which the *Riformatori* and Grisellini believed to date from the fourteenth century, but

[14] Antonio Maria Zanetti, in response to the request of the *Riformatori*, sent six reports tracking the work that Grisellini carried out from 1761 to 1763 (Venice, Archivio di Stato di Venezia; Riformatori Studio di Padova, filza 29, ff. 327-329, report on the first map copied by Grisellini, depicting Asia (Gallo, pp. 107-108, doc. VI); Venice, Archivio di Stato di Venezia; Riformatori Studio di Padova, filza 29, f. 176, 10 October, 1761: Zanetti's progress report on the second map copied by Grisellini, depicting the Mediterranean and Cyprus (Gallo, p. 108, doc. VII); Venice, Archivio di Stato di Venezia; Riformatori Studio di Padova, filza 29, ff. 178-179, Zanetti's final report on the second map (Mediterranean and Cyprus) copied by Grisellini (Gallo, pp. 109-110, doc. VIII); Venice, Archivio di Stato di Venezia; Riformatori Studio di Padova, filza 29, ff. 17-18, Zanetti's report on the third map copied by Grisellini, illustrating the coasts of the Mediterranean from Italy to Constantinople... (Gallo, pp. 110-111, doc. IX).

[15] "Si ricorrerà all'osservazione del Planisferio di S. Michiele di Murano già dirizzato anch'egli come si disse, sulle Memorie di Marco Polo, il quale in piccolo ha tutto ciò, che in grande fu espresso e rappresentato su di queste mappe stesse. Da questo planisferio si copieranno colla più immaginabile diligenza i siti, ed i pezzi tutti mancanti nella gran Mappa, i quali poi si riporteranno e ridurranno geometricamente in grande colla guida dellla scala delle distanze sul preparato Carton...," *Memoria per V.V.E.E.*, Gallo, "Le mappe geografiche del Palazzo Ducale," p. 98.

[16] "[...] modi Geometrici e di osservazione desunti dal notissimo Planisferio di San Michele di Murano dirizzato nel detto secolo da un Laico di quel Monistero, parimenti sulle memorie del Polo, e le direzioni d'industria meccanica, che dal nominato Griselini verrann serbate nel ravvivare queste insigni Opere Geografiche," *ibidem*.

which are actually products of the sixteenth century[17] – not only celebrated Venice's past, but also offered a model for the future.[18] These maps illustrated, according to the *Riformatori*, not only the courage and resourcefulness of Marco Polo and other Venetian merchants and explorers, but also the advanced scientific competence of the Venetians, eloquently expressed through the skill of drafting innovative maps of the world.

Despite the Magistrate's specific instructions, it became immediately clear to Grisellini that Fra Mauro's map was not adequate to fill in the pieces, worn or missing, in the *teleri* of the Sala dello Scudo : Antonio Zanetti's reports on the actions of Grisellini tell a different story from that expected by the *Riformatori*.[19] Though Fra Mauro's map was of little service in the restoration of the Sala dello Scudo's sixteenth-century *teleri* – the mention of the attempt to use the *mappa mundi* would seem, if anything, simply an effort by Zanetti to pay court to the *Riformatori* – it is very relevant for us to underline that the *Riformatori* already perceived it as a monument to Venetian history. In the process of the restoration of the maps in the Sala dello Scudo, within the political and academic circles of mid-eighteenth-century Venice, there emerged a new regard for such works as the map by Fra Mauro. The reports of the *Riformatori* set an important precedent of referring – even deferring – to ancient cartography that would influence the future perception of the function of maps in the political and civic spheres. Ancient maps – like architectural ruins – became monuments to history, and were potentially useful in constructing and celebrating a nation's collective historical memory. Therefore, the *mappa mundi* of Fra Mauro, together with the maps of the Sala dello Scudo, became instruments vital to the transmission of a Venice-centered national historical memory. The maps of the Palazzo Ducale of Venice, and their presumed models, together with the rest of the Palazzo's extraordinary iconography, contributed to constructing and maintaining the civic identity of the *Serenissima*.[20]

III. – Venice, 1806 : maps as a funeral epitaph to lost greatness

Beginning in the 1730s two Camaldolesi brothers and chroniclers, Giovanni Benedetto Mittarelli (1708-1777) and Anselmo Costadoni (1714-1785), began a systematic inventory of the

[17] Francesco Grisellini, *Succinta descrittione delle bellissime tele geografiche ora rinnovate ed accresciute nella sala del palazo ducale di san Marco detta dello Scudo ed esposte alla pubblica vista il dì 24 dicembre MDCCXII*, Venice : Modesto Fenzo, 1763, p. 1 (reprinted under the same title as "Pubblicazione per Nozze Zajotti-Antonini," Venice, 1880).

[18] In 1818 Placido Zurla had already begun to correctly attribute the maps to Jacopo Gastaldi. Marica Milanesi has recently connected the maps to Domenico Zorzi di Modone, in addition to Gastaldi. See M. Milanesi, "Sulle mura di palazzo. Firenze, Roma, Venezia : qualche confronto."

[19] For example, when Grisellini attempted to replace the missing contents in the northern part of Cyprus in the map of the Mediterranen Sea (the second that he restored), the scale difference between the *telero* and Fra Mauro's map permitted only a very limited use of the *mappa mundi*. Zanetti's second report confirms, in fact, that it was impossible for Grisellini to use Fra Mauro's map and that they relied instead on a map by Ortelius. Cf. Venice, Archivio di Stato di Venezia, filza 29, f. 176 [Letter from Antonio Zanetti to the *Riformatori dello Studio di Padova*], 18 October 1761, and ff. 178-179 [*idem*], undated but after 18 October and before 11 June 1762. Both published by Gallo, pp. 108-110, doc. VII and doc. VIII.

[20] W. Wolters, *Storia e politica nei dipinti di Palazzo Ducale. Aspetti dell'autocelebrazione della Repubblica di Venezia nel Cinquecento*, Venice : Arsenale, 1987.

historical Camaldolese documents in all the Order's monasteries. The result was the publication, between 1755 and 1773, of the *Annales Camaldulenses* in nine volumes, ordered chronologically and based on the Benedictine Annals covering seven and a half centuries of history from 907 to 1764.[21] Mittarelli, librarian and later abbot of San Michele, followed the *Annales* with the publication of the volume *Bibliotheca codicum manuscriptorum Monasterii Sancti Michaelis prope Murianum*, which listed and described the manuscript codices and the *incunabula* of the San Michele Library, dedicating a long and detailed passage to the *mappa mundi*, and assessing documentary sources on Fra Mauro and his map up to the eighteenth century.[22] Following in Mittarelli's footsteps, within the context of these monumental studies, the monks Anselmo Costadoni and Mauro Cappellari dedicated manuscript monographic studies to the *mappa mundi*, that culminated in the renowned monograph published by Zurla in 1806.

The first twenty-five sheets of the 1785 work by Costadoni entitled Memorie per servire ad una dissertazione da farsi sul planisfero antico che trovasi nella Biblioteca de' monaci di S. Michele di Murano (Memoirs for the purpose of a dissertation on the ancient planisphere found in the Library of the monks of San Michele of Murano), transcribe a good many of the legends of the mappa mundi.[23] On folio 1v there is also a copy of the list of expenses accrued by the Monastery of San Michele, as well as other respective payments made by the king of Portugal toward the reproduction of the *mappa mundi* between 1457 and 1459, taken from the *Libro di entrata e uscita di San Michele* (Account Book of San Michele), today in the State Archives in Rome.[24]

At the end of the eighteenth century the Camaldolese friar Mauro Cappellari – (1765-1846; became Pope Gregory XVI from February 2nd, 1831 to June 6th, 1846) – prepared a manuscript study of Fra Mauro's work while at San Michele with Placido Zurla.[25] Though unfinished, Cappellari's study is the first attempt of a comparative analysis of Fra Mauro's map with another of the fifteenth-century, the so-called "Borgia World Map,"[26] engraved on copper around 1430,

[21] G. B. Mittarelli, A. Costadoni, *Annales Camaldulenses ordinis Sancti Benedicti quibus plura interseruntur tum ceteras Italico-monasticas res, tum historiam ecclesiasticam remque diplomaticam illustrantia*, vols. I-IX, Venetiis, Giambattista Pasquali, 1755-1773. For the pages dedicated to Fra Mauro, see: *Annales Camaldulenses*, Tomus VII, p. 252 "ad annum 1457" [Col. 760-]. The maps not included in the *Annales* were collected in the *Miscellanea Camaldulensis*, in eleven volumes, the manuscript of which is conserved at the Library of the Monastery of the Camaldoli (Biblioteca del Monastero di Camaldoli). Camaldoli (Arezzo), Biblioteca del Monastero, Cod. 1339. See V. Meneghin, *San Michele in Isola*, vol. 1, pp. 200-293.

[22] G. B. Mittarelli, *Bibliotheca codicum manuscriptorum Monasterii S.ti Michaelis Venetiarum*, Venice: ex Typographia Fentiana, 1779, col. 756-760. For an analysis of Mittarelli's description and for the transcription of said analysis, see Chapter I.

[23] Camaldoli (Arezzo), Biblioteca del Monastero, Cod. 1680 (427), Anselmo Costadoni, *Memorie per servire ad una dissertazione da farsi sul planisfero antico che trovasi nella Biblioteca de' monaci di S. Michele di Murano*, manuscript, paper, 1785.

[24] Rome, Archivio di Stato, *San Gregorio al Celio*, inv. 25/II, n. 9, n. 63, Maffeo Gherardo, *Entrata e uscita. San Michele di Murano* (1453-1460). See Chapter I.

[25] Camaldoli (Arezzo), Biblioteca del Monastero, Fondo San Gregorio 19, (Bartolomeo Alberto) Mauro Cappellari, *Scritti del P. Mauro Cappellari "Gregorio XVI" sul Planisferio di Fra Mauro Camaldolese*, manuscript, paper, end of eighteenth century.

[26] Vatican City, Biblioteca Apostolica Vaticana, Borgiano XVI [Borgia World Map], engraving on a copper plate, diameter *c.* 63 cm, *c.* 1430. The map was probably engraved in Austria, southern Germany, or Bohemia. The world map has south at the top; the legends are in Latin. The first study on this map was published in 1797 (see note 4 for

and owned in the eighteenth century by Cardinal Stefano Borgia (1731-1804). A famous theologian, antiquarian, and historian, Borgia displayed the map in the *Museo Borgiano*, the museum he founded in Velletri, near Rome, to promote the study of his valuable collection of antiques, and in particular of Coptic manuscripts.[27] A note on the verso of the second page attests that Cappellari's unfinished comparative study of Fra Mauro's map was inherited by Cardinal Placido Zurla and was subsequently found on his nightstand the day he died. It is important to note that the handwritten works by Costadoni and Cappellari constitute the principal sources from which Placido Zurla later drew information for the preparation of his own monograph, printed in Venice in 1806, on Fra Mauro. It should be emphasized that Mittarelli, Costadoni, Cappellari, and Zurla could consult and benefit from a remarkable collection of manuscript documents composed by anonymous Camaldolese monks, in which were transcribed many legends and information about the *mappa mundi*.[28]

While Costadoni's research was rooted in the monastic scholarship of the *Annales*, and sought – in the tradition of the celebrated French Benedictine monk and scholar, Jean Mabillon (1632-1707) – to exhaustively complete the documentary collection of the Camaldolese Order, the studies carried out by Cappellari and Zurla displayed different objectives. Though inspired by Camaldolese scholarship, Cappellari and Zurla studied the *mappa mundi* with such intensity for two principle motives. Their first aim was to celebrate – and thus preserve – the lost grandeur of the Venetian State, only recently fallen from glory. During the period in which Costadoni, Cappellari, and Zurla prepared their studies on the *mappa mundi*, Napoleon's armies in 1796

the bibliographical details). This publication may have been at the origin of Cappellari's interest in the comparison between the Borgia Map and Fra Mauro's work. A facsimile edition of the Borgia Map was then published by Manuel Francisco de Barros e Sousa, Visconde de Santarém, *Atlas composé de mappemondes, de portulans et de cartes hydrographiques et historiques depuis le VIe jusqu'au XVIIe siècle*, Paris : Maulde et Renou, 1849, Plate 23.

27 The studies on the Borgia Map, from 1797 and 1808, were carried out contemporaneously to the editing of the catalogue of Borgia manuscripts, by the Danish archeologist Jørgen Zoëga (1755-1809), and published posthumously in 1810 : *Catalogus codicum Copticorum manuscriptorum qui in Museo Borgiano Velitris adservantur*… Rome : typis Sacrae Congregationis de Propaganda Fide, 1810. The catalogue actually deals with the entire body of manuscripts owned by Stephano Borgia, and preceded the 1816 separation of the biblical manuscripts (which went first to the library of the *Propaganda Fide* and then to the Biblioteca Apostolica Vaticana) and the rest of the manuscripts (which, in 1816, joined the Biblioteca Borbonica in Napoli, known today as the Biblioteca Nazionale Vittorio Emanuele III, when the antiquities of the Borgia Museum of Velletri were acquired by King of Naples Ferdinando I of Bourbon).

28 Camaldoli, Biblioteca del Monastero, Cod. 607, parte n. 7, *Miscellanea Camaldolese. Raccolta di documenti e notizie da servire per la compilazione degli Annales Camaldulenses*. Tomo I., n. 7., sec. XVIII : preparatory sketches for the *Annales Camadulenses* in which the legend from the mappa mundi is transcribed : "Come per la uirtù atractiua de la luna le aque cresce e descresce." Cod. 610, *Miscellanea Camaldolese*. Tomo IV, n. 4, *Ranunzio* (ndr. Ramusio) *circa il planisferio di Fra Mauro*, p. 57. *Incipit* : "Della Dichiarazione posta in fronte al secondo tomo Delle navigazioni di Ramusio alla pg. 17. Così parla lo stesso Ramusio…". Cod. 613, *Miscellanea Camaldolese*. Tomo VII, parte n. 17, *De Planisferio di Fra Mauro*, pp. 429-430 : an anonymous note, which summarizes the history of the *mappa mundi*. It contains references to the description of Ramusio references, the links with Marco Polo, the shipwreck of Piero Querini, and the commemorative coin minted in 1459. Cod. 627, parte I, n. 9, *Anedocta Camaldulensia*. Tomo XII. Parte I, *Osservazioni in lingua veneziana sul suo Planisferio*, ff. 81-82 : raccolgono il lavoro preparatorio e le note in italiano di Anselmo Costadoni sulla *mappamundi*, preludio alle più ricche *Memorie per servire ad una dissertazione da farsi sul Planisferio che trovasi nella Biblioteca de' monaci di S. Michele di Murano*. Cf. di seguito il Cod. 1680.

irrevocably ended the secular independence of the *Serenissima*. The laudatory study of Fra Mauro's world map (followed in 1818 by two volumes dedicated to well-known Venetian travelers, beginning with Marco Polo) was meant as a fundamental testimony to Venice's crucial role in world history as well as irrefutable proof of the heights of knowledge reached by the *Serenissima*. Zurla himself was equally eloquent on the matter, introducing his 1806 monograph with these words: "... above all in the revival of Geography, new importance has been given also to those maps destined to represent it, principally Venetian, which, because with all arts pertaining to navigation they [the Venetians] had supremacy, thus they attained even [such a status] with Geography, and in the related maps." In such a context, the *mappa mundi* was defined "a glorious monument to the supreme influence of the Venetians in the revival of Geography, no less than to analogous research and arts."[29]

To this was added an ulterior and even more important motive in the Camaldolese monks' scholarly undertaking: the crucial role held by the monastic orders in promoting science and culture, against the harsh anticlerical campaign of Illuminism. It is a responsibility which, already evidenced in the *Annales Camaldulenses* and the *Bibliotheca codicum manuscriptorum* by Mittarelli and Costadoni, became more pressing, even compulsory, with Cappellari and Zurla, whose research on the *mappa mundi* was unfolding just as the Napoleonic government prepared to suppress San Michele's Monastery; this occurred in 1810, causing the dispersal of the monastery's library and archives, among the most copious and valuable in Venice.

The *Annales camaldulenses*, the *Bibliotheca codicum manuscriptorum* by Mittarelli, the manuscript monographs by Cappellari and Costadoni, and the printed monograph by Zurla are, in fact, all part of the same cultural undertaking. By studying and celebrating the *mappa mundi*, these scholars were affirming the crucial role of the Camaldolese brothers of San Michele within Venetian culture and history, and, at a larger scale, the role held by the Camaldolese and other monastic orders in the formation of modern culture. To this purpose, in 1822 Zurla published an apologetic pamphlet, entitled "Advantages Brought by Catholic Religion to Geography and the Associated Sciences" (*Dei vantaggi dalla Cattolica Religione derivati alla geografia e scienze annesse*).[30] The pamphlet's very title explicitly communicated the ideological purpose implicit in the historical research of the Camaldolese brothers since the mid-eighteenth century. In this context, the *mappa mundi* studies, too, became instrumental in celebrating the role held by the Church in the establishment of modern culture and knowledge.

[29] "...sopra tutto nel risorgimento della Geografia nuovo pregio si aggiunse anche alle carte a rappresentarla destinate, tra Veneti principalmente, che siccome dell'arti tutte alla nautica attinenti allora ebbero il primato, così a diritto l'ottennero anche per la Geografia, e nelle tavole ad essa relative"; "un glorioso monumento della primaria influenza de' veneti al risorgimento della Geografia, non men che de' studi ed arti analoghe," Zurla, *Il mappamondo di Fra Mauro*, p. 6 and p. 14.

[30] P. Zurla, *Dei vantaggi della cattolica religione derivati alla geografia e scienze annesse Dissertazione letta nell'Accademia di religione cattolica nel giorno 23 di maggio 1822 dal padre d. Placido Zurla abate camaldolese*, Rome: Tipografia della Sagra Congregazione de Propaganda Fide, 1822.

IV. – London : orientalism and the search for the map by Marco Polo

During the same period in which Placido Zurla was preparing his monograph on the *mappa mundi* in Venice, William Vincent, Dean of Westminster (1739-1815) convinced George Macartney, Earl Macartney (1737-1806), George John Spencer, 2nd Earl Spencer (1758-1834), and Robert Hobart, 4th Earl of Buckinghamshire (1760-1816) in London in 1804 to finance – and at the same time to support a financial request, totaling 50 guineas, to the Board of Governors of the East India Company – a manuscript reproduction on parchment of Fra Mauro's *mappa mundi*.

William Vincent, renowned Hellenist, translator and editor of *The Voyage of Nearchus* (1797) the *Periplus of the Erythrean Sea* (1800-1805),[31] and author of *The commerce and navigation of the Ancients in the Indian Ocean* (1807), solicited subscriptions for what he termed to interested parties – in particular to the Board of Directors of the East India Company – as the "map of Marco Polo." Thus, on 22 February, 1804, "On a motion, [it was] resolved that this Court [of the East India Company] subscribe the sum of Fifty Guineas towards a fund now raising for the purpose of obtaining a Copy of the original Map of Marco Polo which has been discovered in Venice."[32]

Attributing to Marco Polo the map held in San Michele was most likely caused by Vincent's adherence to a widely-held interpretation of it, in turn based on a "over-interpretation" of the description of Fra Mauro's work written by Ramusio in 1553 and published in 1559 in the second volume of the *Navigationi et viaggi*. In Ramusio's opinion Fra Mauro's map – as previously and extensively discussed in Chapters I and V – must have been "diligently drawn and copied from a most beautiful and very old marine chart and a world map that had been brought from Cathay by the most honorable Messer Marco Polo and his father."[33] Ramusio, then, believed that the *mappa mundi* of San Michele represented a first attempt to faithfully reproduce Marco Polo's representations of the world, among which Ramusio included not only the travelogue, *Milione*, but also the two maps brought back from China. This interpretation was often repeated unquestioningly, and just as often misunderstood. For example, the Benedictine father Vitale Terrarossa (1623-1692), a philosophy lecturer at the University of Bologna, and later abbot of the Monastery of Santa Giustina of Padua, introduced a significant modification to Ramusio's "reconstruction"

[31] Nearchus (*c.* 360-300 BC) was one of the officers, a navarch, in the army of Alexander the Great. His celebrated return voyage from the Indus, via the Persian Gulf, to Susa, is preserved in Arrian's account of Alexander's expedition to India, the *Indica*. After many vicissitudes, Nearchus arrived in Carmania, meeting up with Alexander after the latter's crossing of the Gedrosian desert ; later he went as far as the Euphrates before turning back to rejoin Alexander at Susa, in early 324 BC.

[32] London, British Library, India Office Records, B/137, Court Book (13 April 1803 – 28 September 1803) Index – sub vocem "Publications" ; B/138, Court Book (5 October 1803 – 11 April 1804), Minutes of the Court, Board, 22 February 1804, p. 1214.

[33] "[ ] diligentemente tratto e copiato da una belissima [sic] e molto vecchia carta marina e da un mappamondo, che già furono portati dal Cataio per il magnifico messer Marco Polo e suo padre ; il quale, cosí come andava per le provincie d'ordine del gran Can, cosí aggiugneva e notava sopra le sue carte le città e luoghi che egli ritrovava. [come vi è sopra descritto]," G. B. Ramusio, "Dichiarazione d'alcuni luoghi nè libri di messer Marco Polo, con l'istoria del reubarbaro," in *Navigationi et viaggi*, Venice : Giunti, 1559, vol. II, f. 17r ; modern edition, *Navigazioni e viaggi*, edited by M. Milanesi, Turin : Einaudi, vol. III, 1980, pp. 69-71. For an in-depth discussion of Ramusio's description and of the relationship between the *mappa mundi* and Marco Polo, see Chapters I and V.

of events, sustaining that the *mappa mundi* conserved in San Michele was actually a faithful copy of the original map brought back by Marco Polo.[34]

There existed, then, an influential tradition of attributing the world map of San Michele to Marco Polo; and it is upon this tradition that William Vincent was able to convince his British financiers to commission a manuscript copy.[35] The misunderstanding of the map's origin, however, soon came to light. In March 1804 a report reached Lord Hobart, one of the underwriters of Vincent's project, informing in no uncertain terms that the world map destined to be copied was not actually a work by Marco Polo, but rather a map designed by the Camaldolese Fra Mauro in the mid-fifteenth century.[36] Despite this revelation – which caused not a little embarrassment – with the others' agreement, Vincent insisted that obtaining the map from San Michele remained worthwhile, even if it was not authored by the most celebrated Venetian traveler of all time.[37]

Costing a total of two hundred guineas (or £210), the copy prepared for the East India Company and its associated British financiers is a colored reproduction, scaled identically to the original. William Frazer, the painter responsible for the reproduction, spent nine months at San Michele of Murano to complete it, copying the map's legends one by one, and relying principally on the transcriptions compiled by the Camaldolese fathers Costadoni, Cappellari, and Zurla.[38] When completed, the copy was sent to Vienna and was to be forwarded, from there,

[34] V. Terrarossa, *Riflessioni geografiche circa le terre incognite distese in ossequio perpetuo della nobilta veneziana... dal... Vitale Terra Rossa...*, Padua: Giacomo Cadorin, 1686, p. 19.

[35] The story is detailed in the correspondence in the British Library between William Vincent and the 2nd Earl Spencer. We have also transcribed it in entirety in this chapter's Appendix. For the original documents see London, British Library, Add. MS 76092, Earl G. J. Spencer, *General Correspondence* (January-June 1804), letters of 24 February, 3 and 4 March from W. Vincent; the last letter is accompanied by the *Translation of a paper communicated by Mr Stuart to Lord Hobart relative to the Map now existing in the Church of St. Michael di Murano at Venice*, 8 sheets (in 2 folders), 230 × 325 mm, paper, one column 158 × 325 mm, with numerous marginal notes; Add. MS 76094, Earl G. J. Spencer, *General Correspondence* (January-June 1805), letter of 19 April from W. Vincent; Add. MS 76096, Earl G. J. Spencer, *General Correspondence* (January-June 1806), letter of 26 June from W. Vincent.

[36] London, British Library, Add. MS 76092, 2nd Earl Spencer, *General Correspondence* (January-June 1804), letter from William Vincent of 4 March 1804, attached to which is the *Translation of a paper communicated by Mr Stuart to Lord Hobart relative to the Map now existing in the Church of St. Michael di Murano at Venice*, 2 folders (8 sheets), 230 × 325 mm, paper, one column 158 × 325 mm, with numerous marginal notes. The report translates into English the long description of Fra Mauro's map prepared by Mittarelli and published posthumously in Venice in 1779 in the *Bibliotheca codicum manuscriptorum Monasterii S.ti Michaelis Venetiarum*. See the transcription in this chapter's Appendix.

[37] London, British Library, Add. MS 76092, letter from William Vincent of 4 March 1804: "I send your Lordship enclosed the paper I had from Lord Hobart, by which it appears that the Map of St. Michael at Murano, thus not Marco Polo's, has been perpetually mistaken for it. And as it is the Map copied for the King of Portugal on which the Portuguese discoveries depended, a copy of it may be worth bringing to England. The India Company has subscribed £50, and by accepting £20 as your Lordship's subscription, I have now £150. And should I obtain your opinion in favour of procuring the Copy, be the map what it may, I would solicit to other no other judgements but Lord Macartney's to determine me in pursuing the research always submitting the truth to the other subscribers before I take their money."

[38] According to what Vincent writes, William Frazer was a talented miniaturist, however unlucky: "In short, he [William Frazer] has completed the work highly to his own honour, and to the satisfaction of his employers: and it is but justice to add, that the specimens he has produced of his excellence as a miniature painter place him high in his profession, and in hopes of contributing to his success at home, I think myself bound by a debt of gratitude

to London. However, it remained in the then capital of the Austrian-Hungarian Empire for almost two years, unavoidably detained by the Napoleonic Wars. Urgently requested, it finally reached London in 1807 and, through a unanimous motion by its underwriters, was deposited in the British Museum. As already mentioned, today the copy is in the British Library's Manuscripts Room (Additional MS 11267).

What were the motives which led to commissioning an undertaking so demanding, both logistically (due largely to the instability within Europe during that period caused by the Napoleonic Wars) and financially ? What were the specific considerations which compelled the Dean of Westminster, the Earls Macartney, Spencer, and Buckinghamshire, and the East India Company's Board of Directors to interest themselves in a hand-rendered copy of the alleged "original Map of Marco Polo ?" And lastly, why, even after discovering that the attribution was erroneous, was the project still brought to fruition ? The cultural-ideological stage upon which the protagonists of this story moved would seem to have been that of Orientalism – according to the meaning attributed to the term after the publication of Edward Said's renowned book on the subject in 1978, though from a reversed perspective. A look at the biographies of the principal players behind the project, then – in particular those of William Vincent, Earl Macartney, and Hobart – is revealing.

William Vincent, Dean of Westminster, theologian and scholar of the Greek Classics, was particularly interested in ancient geography and the history of navigation in the Indian Ocean.[39] His scholarly reputation rests on two major works. *The Voyage of Nearchus from the Indus to the Euphrates* (1797) is a commentary on an expedition recorded by Arrian of Nicomedia in his *Indica*, which Vincent termed "the first event of general importance to mankind in the history of navigation." The voyage was conceived by Alexander the Great, about whom Vincent wrote with an admiration unusual for the time. His commentary drew on a wide range of sources and he was assisted by Samuel Horsley,[40] Dean of Westminster, who loaned Vincent two astronomical treatises, and by Alexander Darlymple,[41] hydrographer to the Admiralty, who prepared charts for him. The subject was pursued further in *The Periplus of the Erythraean Sea*, issued in two parts, in 1800 and 1805, respectively. He argued against the prevailing view that this too was the work of Arrian, attributing it instead to an unidentified Alexandrian merchant of the first or second century BC. In 1807, these erudite commentaries were brought together as a second edition (in two volumes) under the title *The Commerce and Navigation of the Ancients in the Indian Ocean*. English translations of the original Greek texts of the *Voyage of Nearchus* and *The Periplus* followed in 1809.[42]

to advertise the public, that his address is no. 5, Suffolk-street," W. Vincent, *The Commerce and Navigation of the Ancients in the Indian Ocean*, 2 vols., London : Printed for T. Cadell and W. Davies, 1807, vol. 2, p. 679. The studies conducted at the British Library have not yet aided me in painting a more precise picture of who Frazer was and what other works he executed in his lifetime.

[39] T. Trowles, "Vincent, William (1739-1815)," in *Oxford Dictionary of National Biography*, ed. H. C. G. Matthew and B. Harrison, Oxford : Oxford University Press, 2004 (on-line edition).

[40] For bibliography and works, see : Robert Hole, "Horsley, Samuel (1733-1806)," in the *Oxford Dictionary of National Biography*, 2004 (on-line edition).

[41] For bibliography and works, see : Andrew S. Cook, "Dalrymple, Alexander (1737-1808)," in the *Oxford Dictionary of National Biography*, 2004 (on-line edition).

[42] See the highly detailed biography of Vincent by T. Trowles, "Vincent, William (1739-1815)."

The conquests of Alexander, the prodigious navigation of the navarch Nearchus, and ancient maps, such as the *mappa mundi* by Fra Mauro, offered a solid, even prestigious, basis for both celebrating and justifying, from a cultural-historical perspective, the expansion of European (above all, British) dominion in Asia. The maps – in particular the imaginary one by Marco Polo and, in its absence, the real one by Fra Mauro – embodied the historic ties between Europe and Asia, most notably the "European" conquests in the Orient. In Vincent's work, literary and cartographic testimonies of such conquests were meant to outline the ideal basis for plans of European – and especially British – discovery and colonization of Asian civilizations. This excerpt from *The Commerce and Navigation of the Ancients in the Indian Ocean* is an eloquent example of such intentions :

> The map of Fra Mauro and the letter of Covilham both influenced the success of Diaz and Gama ; but who ever thought it derogatory to their honour, that they proceeded to their discovery upon previous information ? The union of theory and practice is the most solid foundation of all real achievements ; and though it is true that more honour will accrue to the performer of great exploits than to the theorist who matured the plan, it is injurious to detract from their respective merits. The victories of Philip and Alexander receive no diminution of their splendor, by referring the constitution of their armies to the tactics of Epaminondas.[43]

Therefore, if Fra Mauro and Pero Covilham had contributed to the success of Bartolomeu Dias and Vasco da Gama, and Epaminondas in the conquests of Alexander the Great, the role that Vincent expected for himself and the intellectuals in projects of George III's empire was equally clear. But let us also examine the biographies of the other figures who had important roles in the reproduction of the *mappa mundi*.

George Macartney (1737-1806) distinguished himself as a diplomat and governor of English colonies, beginning his career in 1767 as British ambassador to Russia. After a stint as chief secretary in Ireland, in 1781 Macartney was offered the position of Governor of Madras by the East India Company, and remained there until 1786. After returning to England, in 1788 he became a member of the Chamber of Lords. In 1792, the British governor (with the support of the East India Company) hoping to improve relations with China, chose Macartney as head of the embassy being sent to Peking, at the same time making him a viscount. Returning from China in 1794, he was nominated Earl of Ireland. Macartney ended his career as governor of Cape Town, from 1796 to 1798. In 1801, he declined membership to the Board of Control of the East India Company.[44]

The Irishman Robert Hobart (1760-1816), an active book collector, began his political career in 1784 as Member of Parliament in Dublin. In October 1793, he was nominated governor of Madras, arriving there in September of 1794 and remaining until 1798. Hobart then succeeded his father, acquiring the title of Fourth Earl of Buckinghamshire in 1804. In 1812 the Prime Minister, Lord Robert Jenkinson Liverpool, named him as President of the Board of Control for

[43] W. Vincent, *The Commerce and Navigation of the Ancients in the Indian Ocean*, vol. 2, "The Periplus of the Erythrean Sea," Part I, Book II.XXVI, "Discoveries of the Portuguese, Covilham, Marco Polo," pp. 214-234, in particular p. 228.

[44] R. Thorne, "Macartney, George, Earl Macartney (1737-1806)," in the *Oxford Dictionary of National Biography*, 2004 (on-line edition).

India – the governing commission in control of the operations of the East India Company – an employ he retained until his death in 1816.[45]

The East India Company, according to its charter, was not officially involved in acts of cultural patronage. Nevertheless, between the late eighteenth and early nineteenth century, with the creation of a museum directly financed by the Company, an intense search for cultural works, especially scientific and naturalistic pieces, began.[46] Therefore, if Fra Mauro and Pero Covilham had contributed to the success of Bartolomeu Dias and Vasco da Gama, and Epaminondas in the conquests of Alexander the Great, the role that Vincent expected for himself and the intellectuals in projects of George III's empire was equally clear. But let us also examine the biographies of the other figures who had important roles in the reproduction of the *mappa mundi*.[47] George Macartney (1737-1806) distinguished himself as a diplomat and governor of English colonies, beginning his career in 1767 as British ambassador to Russia. After a stint as chief secretary in Ireland, in 1781 Macartney was offered the position of Governor of Madras by the East India Company, and remained there until 1786. As for the Company's interest in maps, there was already the significant precedent set by Alexander Dalrymple (1737-1808).[48] Official hydrographer to the East India Company and (from 1795) to the British Admiralty, between 1774 and 1775 Dalrymple published privately an imposing six-part compendium entitled *A Collection of Plans of Ports in the East Indies*, the most important underwriter and acquirer of which happened to be the East India Company.[49] The intimate connection between Macartney and Hobart – both avid collectors, like George John Spencer, of books and antique documents[50] – and the East India Company, was certainly facilitative as long as the Board of Directors of the company voted in favor of financing a copy of the "original map of Marco Polo which has been discovered in Venice."

The interest of the distinguished politicians involved by Vincent in the project, as well as that of the East India Company, clearly reveals one of the key points of the classic analysis of orientalism by Edward Said : in particular, the connections which, in Europe, tied the production of knowledge, history, and theory concerning the "Orient" to the active expansion of colo-

[45] R. Thorne, "Hobart, Robert, fourth earl of Buckinghamshire (1760-1816)," in the *Oxford Dictionary of National Biography*, 2004 (on-line edition).

[46] See the descriptions by J. F. Watson, *On the Establishment in Connection with the India Museum and Library of an Indian Institute for Lecture, Enquiry, and Teaching, and on its Influence on the Promotion of Oriental Studies in England…*, London : 1875 ; R. G. C. Desmond, *The India Museum, 1801-1879*, London : HMSO, 1982. The bibliography of works on the East India Company is vast. Consult the indicative title by J. Keay, *The Honourable Company : A History of the English East India Company*, London : Harper Collins, 1991 ; K. N. Chaudhuri, *The Trading World of Asia and the English East India Company, 1600-1760*, Cambridge : Cambridge University Press, 1978.

[47] M. Archer, *British Drawings in the India Office Library*, 2 vols., London : Foreign and Commonwealth Office, 1969, vol. 1, pp. 15-21.

[48] For bibliography, see : A. S. Cook, "Dalrymple, Alexander (1737-1808),", in the *Oxford Dictionary of National Biography*, ed. H. C. G. Matthew and B. Harrison, Oxford : Oxford University Press, 2004 (on-line edition). See also A. S. Cook, *Alexander Dalrymple (1737-1808), hydrographer to the East India Company and to the admiralty, as publisher*, Ph.D. dissertation, St. Andrews : University of St. Andrews, 1992.

[49] A. S. Cook, "Alexander Dalrymple's 'A Collection of Plans of Ports in the East Indies' (1774-1775) : A Preliminary Examination," *Imago Mundi* 33 (1981), pp. 46-64.

[50] M. Lester, "Spencer, George John, second Earl Spencer (1758-1834)," in the *Oxford Dictionary of National Biography*, H. C. G. Matthew and B. Harrison, Oxford : Oxford University Press, 2004 (on-line edition).

nial power. In the name of European expansion, cultural study of the Orient – its history, languages, archeology, etc. – was transformed into a conscious and effective instrument of imperialism, part of a circular relationship with the keepers of political and economic power.[51] The antiquarian interest in the commission of a copy of Marco Polo's/Fra Mauro's map, as shown in the correspondence exchanged by Vincent and the other subscribers as well as by the accounting documents of the East India Company, is clearly motivated by increasing British intervention in the southern and eastern Asia : the map served to validate British intervention by implicitly suggesting that they were the heirs of Marco Polo and of Alexander the Great.

Vincent, an intellectual, saw an opportunity to make a name for himself at the height of the Empire's power by studying the ancient documents related to expeditions to Asia and within the Indian Ocean, thereby assuring himself at least a small role in the British Empire. Alexander the Great, Marco Polo, and a long-forgotten Camaldolese friar – *cosmographus incomparabilis* – all became instruments used by intellectuals to approach, flatter, and cajole the governors of the Empire. Thus, in this case, contrary to Said's classic analysis, Orientalism was born of the instrumentalization of those in power by intellectuals.

V. – Paris and Lisbon : maps and colonial conflict

Manuel Francisco de Barros e Sousa de Mesquita de Macedo Leitão e Carvalhosa (1791-1856), Second Viscount of Santarém, is a central figure in Portuguese and European nineteenth-century political and cultural history. Nominated Head of Wardrobe of the future king, Pedro IV, in 1807 Santarém accompanied the Court to Rio de Janeiro when it fled Napoleon's armada. Afterward, Santarém acted several times as Minister of the Kingdom and ambassador, and in 1821 he became Director (*Guarda-Mor*) of the *Torre do Tombo* Royal Archive, the archive of the Portuguese state, and partner and correspondent of numerous scientific academies within Europe. In 1809, Santarém began to systematically collect documents that traced the political and diplomatic history of Portugal. In 1831, following a diplomatic incident with France (he was involved in the "Questão Roussin" – the "Roussin Case" – in which the French war fleet cornered the Portuguese fleet at the mouth of the Tagus River), Santarém was obliged to resign from all official positions and decided to go into exile in France. However, in Paris, not only did his interest in Portuguese history not falter, but he conceived of a project with the aim of constructing a new Portuguese history, one which would emphasize Portuguese maritime expansion. The project proposed to methodically collect and publish all of the maps that aided in demonstrat-

[51] E. Said, *Orientalism*, New York : Pantheon Books, 1978. There are several reissues and translations. Though it remains epistemologically valid and creative enough to stimulate numerous international debates, the work of Said appears lacking from the historiographical perspective, with evident shortcomings as a result of not having been able to determine the chronological limits of the debate. The argument contains too little differentiation in addressing the thirteenth through the twentieth centuries, when the Orientalism of the thirteenth or fifteenth centuries is unquestionably different from that of the following centuries. Said's approach also contains other limitations. The historical analysis of *Orientalism* is, in large part, circumscribed to French, British, and German contexts, with few elaborations from Portuguese, Spanish, Dutch, or Italian sources. For a recent critical revision, at times even eccessively caustic, of Orientalism, see : R. Irwin, *For Lust of Knowing : The Orientalists and Their Enemies*, London : Penguin and Allen Lane, 2006.

ing the precedence of Lusitanian colonial discoveries and conquests over those of the discoveries of other European powers.[52] It is in this context and with these objectives that the Viscount of Santarém began to take an interest in ancient maps (he was, erroneously, even credited with coining the neologism "cartography" in 1839).[53]

Underlying the viscount's cultural initiative was the growing colonial conflict among Portugal, France, and England over the possession of Africa's western coast, near the so-called *Rios da Guiné* (the Guinea Rivers), or the Bissau, Bolama, Bijagós, and Casamansa rivers.[54] Even in exile, Santarém received the official and financial support of the Portuguese crown to publish – in lavish editions, in both French and English – all of the maps attesting to the precedence of Portuguese presence in Africa and Portugal's overseas colonies. Together, this collection of maps would have enabled the Portuguese to better reclaim their colonial rights, including the "legitimacy" of the slave trade, which other European powers had begun to seek an end to.

The aforementioned political and diplomatic connotations, then, lay at the base of Santarém's entire scientific enterprise: his studies on Amerigo Vespucci role in Portuguese discoveries (published in 1836 by the invitation of Martín Fernández de Navarrete, president of the Real Academia de la História de Madrid), those on Henry the Navigator – the mythical personage of whom he created (1840) – to the edition of the *Chronica do descobrimento e conquista da Guiné* (Chronicle of the geographic discovery and conquest of Guinea) and the *Memoria sobre a prioridade dos descobrimentos dos portugueses na Costa d'Africa Occidental* (Essays attesting to the precedence (*prioridade*) of the Portuguese discoveries made along the western coast of Africa) – both published in 1841 – and finally, the three editions of the *Atlas* (1841; 1842-1844; 1849-1855) and three volumes of the *Essai sur l'histoire de la cosmographie et de la cartographie pendant le Moyen Âge* (1849-1852).[55] Such was the importance attributed

[52] For a complete biographical and bibliographical study on the Viscount of Santarém, see the researsch by João Carlos Garcia, Daniel Estudante Protásio, and Angelo Cattaneo, as well as the bibliographic write-up and cataloguing of works by the viscount coordinated by Maria Joaquina Feijão, in J. C. Garcia and M. J. Feijão, *O 2º Visconde de Santarém e a História da Cartografia*, Lisbon: Biblioteca Nacional de Portugal, 2006.

[53] In his letter to Francisco Varnhagen, 8 Dicember 1839 (*Algumas cartas ineditas do Visconde de Santarém*, introdução e notas de Vicente Almeida d'Eça, Lisbon: Sociedade de Geographia de Lisboa, 1906, p. 30). But the inter-related terms 'cartographie', 'cartographes', and 'cartographiquement' appeared eight years before in *Traité élémentaire de géographie, suivi d'un aperçu de la géographie ancienne… par Malte-Brun; terminé… par ses collaborateurs MM. Larenaudière, Balbi et Huot… Tome second*, Paris: Aimé-André; Bruxelles: à la Librairie parisienne, 1831, pp. 669 and 675-676. See F. Herbert "Jacob Gråberg af Hemsö, the Royal Geographical Society, the Foreign Office, and Italian Portolan Charts for the British Museum," in G. Bäärnhielm, F. Sandgren, and A. Burius (eds), *Accurata descriptio: Papers in Cartography, Numismatics, Oriental Studies and Librarianship Presented to Ulla Ehrensvärd*, Stockholm: Kungliga Biblioteket, 2003, pp. 269-314.

[54] R. Pélissier, *História da Guiné: Portugueses e Africanos na Senegâmbia (1841-1936)*, Lisbon: Editorial Estampa, 1989, p. 47 and pp. 64-65.

[55] For the complete bibliography, see "Os Descobrimentos Portugueses e a Cartografia Antiga," in particular, docs 26-44; for a detailed study on the ideological structure of the work by the Second Viscount of Santarém see J. C. Garcia, "Um Castelo de Cartas Antigas: construir e comemorar o Império," in T. Pinto Coelho (ed.), *Os Descobrimentos no mundo de língua inglesa, 1880-1972*, Lisbon: Colibri, 2006, pp. 167-174 and Id., "Mapas e Atlas do Visconde de Santarém. A prioridade no descobrimento da África ocidental," in *O 2º Visconde de Santarém e a História da Cartografia*, pp. 7-16.

to Santarém's work by the Portuguese government, that in 1841, despite being in exile, he was nominated Director of the Torre do Tombo Royal Archives, a position he held until his death.

As for the study and publication of ancient maps, in 1841 the Viscount of Santarém prepared the first edition of a collection of facsimiles, entitled *Atlas composé de cartes des XIVe, XVe, XVIe et XVIIe siècles, pour la plupart inédites, et devant servir de preuves à l'ouvrage sur la priorité de la découverte de la côte occidentale d'Afrique au-delà du Cap Bojador par les portugais*. On one hand, the work was intended to be a study of Portuguese discoveries and expansion. At the same time, it was also meant to reproduce those maps – particularly those never before printed – which gave testimony to Portugal's scientific and cultural legacy, beginning with Henry the Navigator. The first two editions of the *Atlas* were accompanied by the *Essai sur l'histoire de la cosmographie et de la cartographie pendant le Moyen Âge* – a three-volume study meant to explain the maps – both printed in Paris on behalf of the Portuguese government between 1849 and 1852.[56] Beginning with the third edition, the *Atlas* was published as an autonomous work, using ancient maps as a means of explaining the history of cosmography and medieval and early modern geography, and of illustrating the progressive discovery and conquest of the Earth, particularly highlighting Portuguese and Spanish maritime expansion.

As for the *mappa mundi* of Fra Mauro, it was Santarém to explain with clarity the motivation behind this project: "While I was preparing my *Recherches sur la priorité des découvertes des Portugais*, I immediately understood that Fra Mauro's world map was one of the most indisputable and precious documents [attesting to the history of Portuguese expansion] because we find there a written account of Portuguese discoveries."[57] Thanks to the legends in the *mappa mundi* testifying to Portuguese fifteenth-century navigations, the Viscount of Santarém had found confirmation of his theory of Portuguese precedence (*prioridade*) in the European expansion – a theory which characterized, even drove, Santarém's entire scientific collection.

The *mappa mundi*, then – already transformed into a historic monument to Venice and to the religious orders, as well as a monument which celebrated, and even foretold, British expansion in Asia – in Santarém's hands, became a monument to the history of Portuguese discoveries and conquests, and proof of their precedence among all European nations. Santarém conceived the publishing of the *mappa mundi* as a continuation of the vast collection of manuscripts, *Symmicta Lusitana*, which, according to its patron, João V de Bragança (John V of Bragança, King of Portugal from 1706 to 1750), was intended to encompass the entire corpus of documents concerning relations between Italian city-states and Portugal, dating back to the Middle Ages: "When John V, at the beginning of the last century, ordered to copy all of the documents related to Portugal that were held in Italian libraries, in a precious collection known as *Symmicta Lusitana*, which is composed of 200 enormous in-folio volumes, this great king

[56] *Essai sur l'histoire de la cosmographie et de la cartographie pendant le moyen-âge et sur les progrès de la géographie après les grandes découvertes du XVe siècle, pour servir d'introduction et d'explication à l'Atlas composé de mappemondes et de portulans, et d'autres monuments géographiques, depuis le VIe siècle de notre ère jusqu'au XVIIe*, Paris: Imprimerie Maulde et Renou, 1849, 1850 and 1852.

[57] "Lorsque j'ai rédigé mes *Recherches sur la priorité des découvertes des Portugais*, j'ai tout d'abord compris que, parmi les pièces justificatives, une de plus positive et de plus précieuse était la mappemonde de Fra-Mauro où l'on trouve des légendes relatives aux découvertes des Portugais." Viscount of Santarém, "Note sur le mappemonde," p. 251.

also had the idea to make a copy of Fra Mauro's map. However, this project was left incomplete."[58] Santarém did not conceal his pride in succeeding to follow through with the plan that Manuel Pereira de Sampaio, minister and ambassador to Rome under João V (1689-1750), had failed to realize.

The Viscount was informed of the existence of a copy of Fra Mauro's work in the British Museum in 1841 by John Holmes, an Assistant Keeper in the Museum's Department of Manuscripts.[59] In May 1844 Santarém wrote to the Portuguese Minister of Foreign Affairs from Paris, informing him both of the necessity of reproducing Fra Mauro's map and of the inevitable difficulties this would entail. He explained to the minister that they could, instead, attempt to create a printed reproduction from the manuscript copy deposited by William Vincent in the British Museum, for a considerable, but not unaffordable, fee. They would be attempting to undertake what, at that time, no one else had ever tried. Their success, argued Santarém, would be a great benefit to science and, of course, "in [illustrating] the history of Portuguese discoveries, of which [the *mappa mundi*] was the most authentic testimony."[60] Santarém's familiarity with Richard Major, a staff member of the British Museum made obtaining a copy of the manuscript reproduction held in London a simpler and more efficient process than gaining access to the original, by that time held in the custody of the Palazzo Ducale of Venice.[61] Moreover, the legends of the Museum's copy were all accurately and legibly transcribed in modern, if vaguely gothic, calligraphy – a notable advantage which greatly simplified the work

[58] "Lorsque Jean V, au commencement du dernier siècle a fait copier tous les documents concernant le Portugal qui se trouvent dans les bibliothèques de l'Italie, compilation précieuse qui, sous le titre de *Symmicta Lusitana*, se composent de 200 énormes volumes in-folio, ce grand roi eut l'idée de faire exécuter une copie de la mappamonde de Fra-Mauro. Ce dernier projet n'eut pas d'effet," Viscount of Santarém, "Note sur le mappemonde," p. 251. The collection of documents entitled *Symmicta Lusitana* is held in the Biblioteca da Ajuda of Lisbon. The collection principally consists of documents originally from the Biblioteca Apostolica Vaticana. The Portuguese king did not try to hide his deep admiration for papal Rome, considered an infinite source of images and art forms with which to confer splendor and greatness on its capital. See R. Ortigão, "A Symmicta Lusitanica. Inédito de Alexandre Herculano. Relatório apresentado a D. Manuel II," *Archivo historico portuguez*, vol. I, n.° II, Lisboa, Novembro de 1903, pp. 369-370 ; G. Borghini and S. Vasco Rocca (eds), *Giovanni V di Portogallo (1707-1750) e la cultura romana del suo tempo*, Rome : Argos, 1995.

[59] Viscount of Santarém, *Correspondência do 2° Visconde de Santarém*, Letter from John Holmes to Santarém of 22 December 1841 : "[...] of maps of Africa before the year 1500 we have first a facsimile copy of the Mappa Monde of Fra Mauro which I do not mention further as you no doubt know it well, from the description of Mittarelli, Zurla, and Vincent," vol. VI, p. 211.

[60] "Bem desejava eu poder publicar o mais interessante de todos os monumentos geographicos o Mappa Mundi de Fra Mauro de 1460 (...) O Governo inglez mandou tirar um fac-simile em 1804 no Ministerio de Lord Hobar[t], então Ministro dos Negócios Estrangeiros. Algumas porções reduzidas deste Mappa Mundi foram já publicadas pelo sabio cardeal Zurla e pelo Dr. Vincent, mas o monumento inteiro e em fac-simile nunca se publicou. Esta publicação interessaria pois em summo gráo a sciencia em geral e a historia das nossas descobertas pois é um dos titulos mais authenticos d'ellas. E todavia é este Mappamundi de grande dimensão e no caso de se gravar só se poderia fazer em 6 ou 8 folhas. A copia custou ao Governo Inglez 100 libras, como se vê na conta que sobre este negocio deo o Dr. Vincent, e a gravura, tiragem, colorido, etc. custaria mais de 150 libras." Viscount of Santarém, *Correspondência do 2° Visconde de Santarém*, vol. VI, 1919, pp. 446-447.

[61] In Santarém's correspondence there are also regular letters between him and Richard Major, who had collaborated with Holmes on the *Atlas* and on the reproduction of manuscript maps in London. See J. C. Garcia, "Mapas e Atlas do Visconde de Santarém," pp. 14-15. On the presence of Fra Mauro's map in the Palazzo Ducale, see chapter I note 6 and *Mappamondo* (ed. 1956), pp. 11-16.

of engraving, significantly lowering the costs of the project, as well as the time used for preparation and printing.

There is no doubt of a sense of anxiety present in the missive, the desire for a rapid decision that would ensure that Portugal was the first nation to publish the *mappa mundi*.[62] Santarém was fully aware that Edmé-François Jomard (1777-1862), who was, at the time, the editor of *Les Monuments de la Géographie* – a work financed by the French government, similar in both objectives and contents to the *Atlas* – was also working to reproduce and print Fra Mauro's map.[63] The rivalry between the two editors effectively overshadowed the emerging conflict between European nations over the colonial dominion of Africa's western coast. Indeed, this was one of the reasons – if not the principal one – why the Portuguese, French, and English Ministers of Foreign Affairs financed works such as the *Atlas* and *Les Monuments de la Géographie*.

Finally, in April 1849 in the *Bulletin de la société de géographie*, the Viscount of Santarém wrote a brief "Note sur la Mappemonde du cosmographe Fra-Mauro," announcing with evident satisfaction the imminent publication of "a faithful copy of that great world map ..." ("une copie fidèle de cette grande mappemonde ; copie que je viens d'obtenir, et que j'ai fait graver en ce moment"). Santarém also took care to remind the readers that, "in any case, while waiting until the preparation of the printed copy is completed, I feel happy not only to announce this news to you, my dearest compatriots, but also to tell you that I have in my hands a manuscript copy of Fra Mauro's map, which is at the disposal of all those who wish to consult it."[64] Thus, a copy of the world map, perhaps the very manuscript reproduction derived from that in the British Museum, once received by Santarém, was shown to King Pedro V (1837-1861) during his visit to the Royal Library of France in 1855.[65]

[62] "Não me atrevo, pois, a propôr esta despeza, mas verei com grandissimo pezar e talvez em poucos annos um monumento tão portuguez, ser publicado aqui por um homem que não póde tragar Portugal fosse a 1ª nação que publicasse e levantasse um tão grande monumento à História da Geographia e das Sciencias." Viscount of Santarém, *Correspondência do 2° Visconde de Santarém*, vol. VI, 1919, p. 447.

[63] Edme-François Jomard, *Les Monuments de la Géographie, ou Recueil d'anciens cartes Européennes et Orientales accompagnés de Sphéres Terrestres et Célestes, de Mappemondes et Tables Cosmographiques, d'Astrolabes et autres instruments d'observation, depuis les temps les plus reculés jusqu'à l'époque d'Ortelius et de Gérard Mercator, publiés en facsimile de la grandeur des originaux...*, Paris : 1854. Jomard was president of the *Société de Géographie* and founder of the Département des Cartes et Plans (Map Room) of the Bibliothèque Royale, known today as the Bibliothèque nationale de France, of which he was director and administrator from 1838. The publication of the *Monuments de la Géographie* followed the monumental editions on Egyptian antiquities, which came out in the three decades following the inclusion of the pieces in the Napoleonic *Expédition d'Égypte* of 1798 (*Description de l'Égypte ou Recueil des observations et recherches qui ont été faites en Égypte pendant l'expédition française*, 10 vol. Paris : Imprimerie Impériale, 1809-1828). See Y. Laissus, *Jomard – le dernier Égyptien*, Paris : Fayard, 2004.

[64] "Toutesfois, en attendant qu'elle soit entièrement terminée, je me crois heureux de pouvoir non seulement annoncer cette nouvelle à mes honorables confrères, mais aussi de leurs dire que ce même monument est entre mes mains, à la disposition de tous ceux qui voudront le consulter." M. F. Santarém de Barros e Sousa, "Note sur le mappemonde de Fra Mauro." *Bulletin de la Société de Géographie de Paris*, 1849, pp. 251-252. See the transcription in this chapter's Appendix.

[65] R. Andresen Leitão, *Diário de viagem de França del-rei Dom Pedro V (1855)*, Paris : Fundação Calouste Gulbenkian, 1970, p. 189 : "8 de Junho de 1855. A parte geográfica [da Biblioteca Real] é muitíssimo interessante ; contém cartas do XIVº século : o famoso mapa de Fra Mauro ; cartas chinesas interessantíssimas ; e achase colocada debaixo da direcção de um dos raros restos da expedição científica que acompanhou Bonaparte ao Egipto."

In 1854, the Parisian lithographic printers Lemercier and Kaeppelin, the engraver J. Schwaerzlé ("Schwaerzlé Sculp") and the lithographer Jean Feuquières ("J. Feuquières Lith.") ran off, on six large sheets, a reproduction of Frazer's manuscript copy of the *mappa mundi* that he had executed at San Michele from 1804 to 1805 (plate X). This six-sheet lithographic map was included in the third edition of the gigantic *Atlas* (1849-1855), a composite of reproductions of ancient maps selected by Santarém, and inaugurating its "Troisième partie" entitled "Série de mappemondes à partir de celle du célèbre Cosmographe Fra-Mauro, de 1459 jusqu'au XVIIe siècle…".[66] Also in 1854, a reprinting of the *mappa mundi* was prepared separately, so that the six sheets could be assembled and mounted as one in order to reproduce the map of Fra Mauro in its original dimensions.[67]

VI. – The cultural costruction of a documentary *corpus*

The scientific and editorial activities of the Viscount of Santarém brought about a profound revival of interest in the history of geography : according to the viscount, the study of medieval and early modern geography should be based principally upon ancient maps. If van Murr, Heeren, Costadoni, Cappellari and Zurla had compiled the first monographic studies on ancient maps – with aims that might be termed apologetic – the Viscount of Santarém was one of the first promoters of the systematic study of ancient cartography as a scientific discipline. Such a discipline had its roots in the cultural construction of a homogeneous and coherent documentary *corpus*, whose separate components must be tracked down, defined, and catalogued in each library and archive, public and private, in Europe. Thus the conceptual and methodological seeds of the history of cartography, as a distinct discipline and field of historic-scientific research were sown.

The 1761 project of the *Riformatori allo Studio di Padova* to simultaneously construct and celebrate a collective memory of the Venetian State through the maps encompassing the world of Venetian travelers such as Marco Polo, the Zeno brothers, Niccolò de' Conti, Alvise Cadamosto, and the Cabotos would be taken up again roughly forty years later by the Camaldolese monk Placido Zurla ; even after the Venetian Republic had ceased to exist and the monastic order inside of which Fra Mauro's *mappa mundi* had taken form was about to be suppressed. Thus, the construction of a collective memory of the history of Venice, by celebrating its ancient merchant-travelers and cartographers, became a sort of funeral epitaph to a disappearing world – a world which historians and politicians were determined to remember, even if they could not save it.

[66] The numbering of the plates in the *Atlas* varies considerably according to the exemplar seen. See the "Préfazio" in the *Atlas du Vicomte de Santarém*, org. et préf. par Martim de Albuquerque, Lisbon : Administração do Porto de Lisboa, 1989, p. 8 and J. de Freitas, *O 2° Visconde de Santarém e os seus atlas geographicos*, Lisbon : Officina Typographica, 1909.

[67] *Mappemonde dressée en 1459, par Fra Mauro, cosmographe vénetien, par ordre d'Alphonse V roi de Portugal : publiée pour la première fois de la grandeur de l'original avec toutes les légendes par le Vicomte de Santarém*, Lith. J. Feuquières, Paris : Kaeppelin, Lemercier, 1854. Map on six sheets, each 81 × 105.5 cm ; when joined, 199.50 × 201.00 cm.

The work of transcribing and studying Fra Mauro's map, as realized by Zurla and Cappellari, was used, in turn, by William Frazer to prepare the previously-mentioned copy sponsored by the East India Company, at the request of a group of British scholars and politicians. The reproduction of the *mappa mundi*, believed to be the closest illustration of the contents of the lost map allegedly brought to Venice by Marco Polo (the very one for which the *mappa mundi* was initially mistaken), celebrated, in the minds of those who sponsored its duplication, the dawning of the British Empire by tracing a direct history from the conquests of Alexander the Great to those of Marco Polo, and finishing with the endeavors of the East India Company. In the course of a few years, and with a complete reversal of prospects, the world map of Fra Mauro was transformed from epitaph of a fading world to the backdrop on which to project British colonial conquests.

About 40 years later, around 1850, the Second Viscount of Santarém, backed by the Portuguese government, managed to obtain a reproduction of the English copy of Fra Mauro's world map. After procuring the work over Jomard – the latter supported by the French government – the satisfied viscount realized a printed edition. With this foray into publishing on Santarém's part, the *mappa mundi* became the most important document among those considered to celebrate Portuguese geographical discoveries (*descobrimentos*) and colonialism. In fact, the viscount's mind, ancient cartography provided the most reliable and relevant documents with which to defend the colonial rights of Europe, therefore making it essential to study the maps, reproduce them, and diffuse the "scientific" results internationally. The direct result of said conviction are the several editions, of pantagruelic proportions, printed at the expense of various governments (Portuguese, French, Italian…), thanks to which fourteenth-, fifteenth-, and sixteenth-centuries manuscript cartography became known outside of the various maps' places of origin and conservation.

It is through this deliberate dissemination, then, in addition to the documentary research which emerged in its aftermath, that the history of cartography became recognized within Europe as a discipline in its own right. Quite remarkably, Fra Mauro's *mappa mundi* played a crucial role within this composite cultural enterprise.

EPILOGUE

"It is a beautiful day to land at Fra Mauro"

The story of the cultural reception of Fra Mauro's *mappa mundi* finds an interesting moment in the middle of the nineteenth century, when Johann Heinrich von Mädler produced the first complete map of the visible side of the Moon made on the basis of telescope observations and edited the *Mappa selenographica*, published in four volumes in 1834-183.[1] Von Mädler (1794-1874), one of the eminent astronomers of the nineteenth century, in 1824 met and became a close friend of Wilhelm Beer (1797-1850), a wealthy banker who in 1829 set up for Mädler a private observatory with the best technology available at the time, a 95mm refractor telescope whose lens had been made by Joseph von Fraunhofer (1787-1826), one of the best lens makers of the period. In 1837 a description of the Moon was also published by Mädler and Beer with the title *Der Mond*.[2] The *Mappa selenographica* and *Der Mond* were the best descriptions of the Moon for many decades, not superseded until the map of Johann Friedrich Julius Schmidt in 1870. On the surface of the Moon, east of the area dedicated to ancient astronomers and identified as the "Ptolemy-Alphonsus-Arzachel area," J. H. von Mädler named a vast crater "Fra Mauro" in honor of the Venetian Camaldolese monk (coordinates : 6.0°S 17.0°W ; diameter 95 km).[3] It was very probably the publication of the book by Zurla in 1806 that inspired Mädler to name the crater after Fra Mauro.

About a hundred years later, on January 31, 1971 – a year and a half after the historic moon landing of Apollo 11 on July 21, 1969, piloted by Neil Armstrong – NASA launched the Apollo 14 mission. Led by Commander Alan Shepard (1923-1998), it had as its goal a lunar landing in the Fra Mauro Crater. Shepard was one of the first seven astronauts selected by NASA in April 1959 as part of Project Mercury, and on May 5, 1961, he became the first American to fly in space aboard the *Freedom 7* spacecraft. Ten years later, Shepard made his second space flight as commander of Apollo 14, from 31 January to 9 February 1971, the third mission which had as its goal the landing on the Moon, thus being the first attempt following the failure of Apollo 13. The other crew members were the pilot Stuart Roosa (1933-1994) and the lunar landing module pilot Edgar Mitchell. The mission was not at all simple. During the flight towards the Moon, several failures occurred, including that of the computer that operated the flight system and controlled automatic landing operations. With the help of NASA engineers in Houston, Shepard was able to repair and solve each technical problem that arose.

[1] W. Beer and J. H. von Mädler, *Mappa Selenographica totam Lunae hemisphaeram visibilem complectens observationibus propriis secundum projectionem orthographicam quatuor sectionibus constructa…*, Berlin, 1834. Apart from mapping the complete visible surface of the Moon in great detail, Beer and Mädler drew the firm conclusion that the features on the Moon do not change, and that the Moon has no atmosphere or water.

[2] W. Beer and J. H. von Mädler, *Der Mond nach seinen kosmischen und individuellen Verhaltnissen oder Allgemeine vergleichende Selenographie mit besondrer Beziehung auf die von den Verfassern herausgegebene Mappa selenographica*, Berlin : Schropp & Co., 1837.

[3] Beer and von Mädler, *Der Mond*, p. 38.

The Fra Mauro Crater was chosen because, from photographic surveys carried out by NASA before the mission, it seemed an ideal place for landing and to perform exploratory missions on the lunar surface: it was flat and wide. But when Shepard and Mitchell began the landing – Roosa remained in the spacecraft orbiting around the Moon – and could see through the windows the landing site to which the onboard computer was guiding them, they realized that the crater was by no means flat; on the contrary, it was steep and rugged. They decided to turn off the autopilot and make a manual landing, which would ensure greater accuracy and freedom to manoeuvre. Exactly 108 hours and 25 seconds into the mission, on February 3, 1971, Shepard, who was in radio contact with Houston, was finally able to announce:

"All righty; it's a beautiful day to land at Fra Mauro. (Long pause)."[4]

[4] For all of the details of the mission and the official transcript of the radio communications see "Landing at Fra Mauro." Corrected transcript and commentary by Eric M. Jones, 1995 (last revised 12 February 2009): ‹http://www.hq.nasa.gov/office/pao/History/alsj/a14/a14.html›.

DOCUMENTARY APPENDICES

I

CHAPTER I — DOCUMENTARY APPENDIX

A critical survey of sources related to Fra Mauro and his *mappa mundi* to *c.* 1600

Context and transcription notes

The documentary appendix unites in chronological order manuscripts and printed testimonies that regard, or were credited to regard, Fra Mauro and his *mappa mundi* up to *circa* 1600. A significant amount of the documents listed below have already been noted by Gasparrini Leporace (1956) and by Antonio Ratti (1987). However, the list below adds texts not previously taken into consideration in a systematic manner. Some, such as document 8, were previously considered lost but have been recently rediscovered; others have been ignored, in particular certain texts relating to Portugal, as well as to the literary reception of Fra Mauro's map. The most relevant documents are introduced by brief contextual descriptions. In some cases, partial or complete transcriptions are provided. In particular, from the accounting register of San Michele di Murano (*Entrata e Uscita. San Michele di Murano 1453-1460*, known before only through Zurla 1806 partial edition), from a letter of the general of the Camaldolese Pietro Delfino (May 26, 1494), from a manuscript collection of travel accounts compiled by the erudite Venetian, Alessandro Zorzi (*c.* 1519-1524), from Giovanni Battista Ramusio's *Navigazioni e viaggi* (1559), from the *Tratado dos diuersos & desuayrados caminhos* of the Portuguese chronicler António Galvão (1563), from the *Historiarum Camaldulensium pars posterior* of the Camaldolese historian, Agostino Fortunio (1579).

Doc. 1. July 9, 1409. Venice, Archivio di Stato, *S. Michele di Murano*, B. 3, loose parchment.

In this proxy of the Chapter to the abbot Paolo Venier and Nicolò da Modena, drafted by the notary Prosdocimo Domenico da Padova, "Frater Maurus de Veneciis conversus" is recorded as being part of the chapter of the Monastery of San Michele of Murano as a lay brother (*converso*).[1]

Doc. 2. April 15, 1434. Venice, Archivio di Stato, *S. Michele di Murano*, B. 1, fasc. I, f. 20r.

"Fra Maurus de Vinetiis" is listed among those who participated in the chapter at San Michele of Murano. The presence of Francesco da Cherso is also recorded, a figure who will appear numerous times in the *Entrata e Uscita. San Michele di Murano* (1453-1460) (see doc. 8) as a courier for the payment of the workers on the *mappa mundi* commissioned by the Portuguese court in 1457.

Doc. 3. 1437. Map of the territory of the Camaldolese monastery of Leme, in Istria.

This map was drawn by Fra Mauro in the context of a dispute between the Camaldolese monastery of San Michele of Leme and a local feudatory regarding the bounds between the land of the feud and that belonging to the monastery. The map drawn on vellum was still in existence around mid-eighteenth

[1] *Mostra dei navigatori veneti del Quattrocento e del Cinquecento*. Catalogue curated by the Biblioteca Nazionale Marciana and the Archivio di Stato of Venezia, May-June 1957, Venice, p. 79.

century, but was in such poor condition that in 1762 Mittarelli and Costadoni decided to engrave and print a copy for volume VII of the *Annales Camaldulenses* with the title of *Tabula hanc topographie | comitatus, divi Michaelis Lemni | in Histria camaldolensi abbatiae divi. Mathiae prope. Murianvm, Venetiarum Adiecti | a Mauro monacho et cosmographo | inlustri | medio. Recurr. seculo XV (elaboratam) ne. ulterius. temporis. iniuria. vitiaretur aere. incidit curavit. Maurus.*[2] The original copy seems today to be lost.

Doc. 4. August 2, 1444. Venice, Archivio di Stato, *Savi ed Esecutori alle Acque*, Capitolare I, B. 342, f. 47v.

Fra Mauro appears among the experts consulted by the magistrate of the lagoon (*magistrato alle acque*) in merit of the diversion of the river Brenta. Fra Mauro declared his support for the diversion through the construction of an artificial canal that would have moved the estuary of the Brenta to Lizza Fusina, on the Adriatic coast, west of Venice.[3]

Doc. 5. July 18, 1445. Venice, Archivio di Stato, *Commemoriali*, Reg. 13, ff. 171r-174r.[4]

Fra Mauro is again cited as being among the witnesses, in the *Monastero dei Crociferi* in Venice (Monastery of the Cross-bearers), on the occasion of the drafting of the *Instrumentum Concordiae* with the Patriarch of Aquileia.

Doc. 6. 1448-1449. Register 451 of the Monastery of San Michele di Murano, *Spese di D. Niccolò Monaco economo del Monastero dal 1448 fino al luglio 1449* (Expenses of Niccolò Monaco treasurer of the Monastery between 1448 until the end of July 1449).[5]

According to Placido Zurla's testimony, many of the expenditures were annotated on this register and attributed to Fra Mauro for the acquisition of colors, parchment and gold for the preparation of maps. The register has been lost, most likely during the forced moving of the archival documents of San Michele from Venice to Rome in 1811.

Doc. 7. May 26, 1449. Camaldoli, Biblioteca del Monastero di Camaldoli, Cod. 393, *Indice delle pergamene esistenti nell'archivio di San Michele di Murano* (Index of the parchments extant in the archive of San Michele of Murano), f. 5v.

This list of the manuscript documents on vellum that were held at San Michele includes a reference to the "Pergamena CXXIV (original numeration : Register 13 n.4.-.C)" entitled "Possesso dato da Martino

[2] G. B. Mittarelli and A. Costadoni, *Annales camaldulenses ordinis Sancti Benedicti*..., Venice : Giambattista Pasquali, vol. 7, 1762, pp. 252-256. See G. Marinelli, *Saggio di cartografia della regione veneta*, Venice : R. Deputazione veneta di storia patria, 1881, p. 31 and R. Almagià, "Presentazione," in *Mappamondo* (ed. 1956), p. 5, 10.

[3] B. Zendrini, *Memorie storiche delle acque di Venezia*, 1726, second part, pp. 172-180 (for the period 1444-1452), pp. 148-186 (for the period 1424-1468) ; *Mostra dei navigatori veneti del Quattrocento e del Cinquecento*, p. 79.

[4] *I Libri Commemoriali della Repubblica Veneta. Regesti*, R. Predelli (ed.), vol. IV, Venice, 1896, pp. 292-293 ; *Mostra dei navigatori veneti*... cit., p. 79.

[5] P. Zurla, *Il mappamondo di Fra Mauro camaldolese descritto ed illustrato da D. Placido Zurla dello stess'ordine*, Venezia, 1806, p. 79.

Arcivescovo di Corfù del Monastero di San Michele a Don Maffeo Gherardo Abate." This document refers that Fra Mauro in May 1449 was still registered among the lay brothers of San Michele. The original document seems to be lost.

Doc. 8. 1453-1464. Rome, Archivio di Stato, *San Gregorio al Celio* (inv. 25/II, n. 9), n. 63, Maffeo Gherardo, *Entrata e Uscita. San Michele di Murano* (1453-1460), ff. II, 41r, 42r, 44r, 123v-125r, 128v, 129r, 137v, 139r, 144r, 169r-171v, 172v, 173v, 174r, 175r, 180v, 181r.

This autographed register of don Maffeo Gherardo, Abbot at San Michele from 1448 to 1466, was considered lost and recently rediscovered in the Archvio di Stato in Rome by Cecile Caby.[6] It gives us pertinent information on the expenditures for the making of a *mappa mundi* commissioned from Fra Mauro by the king of Portugal Afonso V. This map was prepared with the help of the cartographer and official of the galleys, the Venetian Andrea Bianco, and a few copyists between 1457 and 1459. Following are the transcriptions of the passages that refer to this event derived directly from the manuscript.[7]

F. II : 1459 adì 20 octubre. Memoria fazo chomo le copie de mapa|mondi e desegni e scripture de Frar Mauro ho depositado al monastier de frari da Zuane de la Zudecha in man del prior del detto monastier zoè don An|drea, le qual scripture e desegni tuti sono posti in una chassa ouer bancho e con uno luchetto, la chiave del qual ho qui apresso de mi. Ho avuto tutto in drietto e però adì 25 octubrio 1464.

F. 42r : L'officio conscripto dio auer adì sopradetto che ser Alvise Sandeli me mando contadi in monede ducati 10 ‹ › paga de marz 1446 in monede d. 186 per el pro de Ossoduro paga de setembre 1445 | portali Frate Mauro converso.

F. 42v : mcccclvii adì primo zener. Adì 8 dictto perche io avi contadi da don Benedetto Miani per nome del Signor de Portogal in suma ducati 28 a val. 129.

F. 124r : mcccclvii adì iii zener. Adì 8 febrier contadi pro Lio Roso per nome del magnifico Signor de Portogal per pagar pentori per lavurar el suo mapa|mundi e per altre spese in suma ducati 11 val soldi 129.

F. 124v : mcccclvii adì iii zener. Adì 8 dictto dati a Lio Roso per nome dela maiestad del Re de Portogal in suma ducati 11 val soldi 124.

F. 125v : mcccclvii adì viii febrier. La maiestad del Segnor Re de Portogal die dar adì sopra|dictto perche io ho dato a messer Lio Roso contad in horo per suo nome | per pagar pentori per lavorar al suo mapamundi e per altre spe|xe in suma ducati xi val a 124.

Adì 21 hoctubre 1457 che io ho dado contadi a frar Mauro per | pagar uno scriptor a laurado ouer scripto su il mapa|mundi zorni 17 a raxon da soldi 12 al zorno monta L. 17 S. 4 | val a S. 124 per ducato.

Adì 7 hoctubre 1458 contadi a don Francesco da Cherso per pagar quondam | scriptor el qual scripse al dictto mapamundi | zorni 4 a soldi 14 al giorno monta soldi 56 val 124 soldi per ducato.

Adì 9 dictto contadi al dictto per pagar uno maistro scripse ala | dictta opera zorno uno soldi 14 per val.

Adì 8 zener 1459 contadi al dictto per pagar maistro uno | scripse ut supra per zornada una soldi 15 per val.

[6] Caby, *De l'érémitisme rural au monachisme urban*, p. 50.

[7] Cf. Zurla, *Il mappamondo di Fra Mauro*, pp. 79-80.

Adì 5 febrier dadi al dicto maistro per certa scritpura facta ala dicta | opera in suma soldi 8.

Adì 17 dictto contadi a don Francesco da Cherso per fra comprar | azuro per la detta opera val.

Adì 10 marzo 1459 per ducati 2 tolti in nui per messe per nome | de ser Andrea Biancho che lui douea hauer per suo premio del lauorare lui | fece al detto mapamundi val in chassa.

Adì dictto per saldo de questa chomo appar per reporto factto.

F. 129r : mccccclvii adì viii febrier. La maiestad del Re de Portogal diè auer adì sopradictto perchè don Benedetto Miani me dè contadi in horo ducadi 28 i qual sono lasadi qui in nome del detto Segnor in depoxitto per suplir a certe spexe lequal erano de bixogno far per comprar l'opera del suo mapamundi el qual lauora frate Mauro val in chasa.

F. 139r : mccccclvii adì xvii agosto. Adì 21 hoctubre 1457 che io ho dado contadi a frar Mauro per | pagar uno scriptor a laurado ouer scripto su il mapa|mundi zorni 17 a raxon da soldi 12 al zorno monta L. 17 S. 4 | val a S. 124 per ducato.[8]

F. 169v : mccccclvii adì primo luglio. La maiestad del segnor Re de Portogal die dar adì sopradictto per | chassa che io ho dato a don Francesco da Cherso che frar Mauro | mandò a domandar per certe spexe lui disse haver factte per el | mapamundi ducati 2 val in chassa

E adì 24 april 1459 que dom Nicholò nostro mea dictto che essendo | io a capitolo a Camaldoli è stato saldà questa raxon a messer Stefano Trevisan | per nome del dictto Segnor grando per el dectto Stefano li fo | mandatto el suo mapamundi. Meto per saldo in suma.

mccccclvii adì x – 20 septembre. Ser Leonardo alinchonto die dar a di sopradecto per chasa che frar Michiel | auea contadi per dar a uno scriptor per comprar carta per il suo brevia|rio ducati 2 val a S. 180.

F. 172v : mccccclviiii adì xxiiii april. Adì 26 dictto che ser Stefano Trevisan de per lemosena per nome | do Segnor Re de Portugal in suma ducati 2 val a soldi 125.

E adì [4 marzo] per denari hauti da frar Mauro nostro in suma ducati 8 s. 12

F. 174r : mccccclviiii adì xxiiii april. Adì [4 marzo] per denari hauti da frar Mauro nostro in suma ducati 8.

F. 175r : mccccclviiii adì 26 april. Adì datto che ser Stefano Trevisan per la lemonsina per nome | del segnor Re de Portugal chomo in chassa appar.

F. 175v : mccccclviiii adì 12 hoctubre. Don Mauro monacho de Sancto Mathia dee dar adì sopraddetto che io lio | prestado contadi per mandar a fiorenza ducadi 2 doro val in chassa.

Doc. 9. February 3, 1462. Lisbon, Archivo da Torre do Tombo, *Chanceleria de D. Afonso V*, liv. 1, f. 2. *Carta de quitação* (quittance) by Afonso V to the ambassador João Fernandes da Silveira.

This document reports the quittance released by the Portuguese court for the expenditures sustained by João Fernandes da Silveira (1431-1484), Portuguese ambassador to Pope Pius II and the Italian courts during his stay in Italy between 1456 and 1460. The ambassador visited Venice, Milan and Florence also to negotiate the Portuguese intervention in the crusade proclaimed by Pope Nicholas V for the recapture of Constantinople. Among the many registered payments, there is also a receipt for the amount of 30 and ¾ ducats for the payment of Fra Mauro's map : "E deu e despendeu trimta ducados tres quartos aos pintores que pyntaram o mapa mundo em Veneza ; e esto por sse nom perder o que ja em elle era fecto"

[8] The same record appears on f. 125v.

("I gave as a payment 30 ducats and ¾ to the painters who depicted the map of the world in Venice; and I did this in order not to lose what has already been done in the map"). This document of the Alfonso V's chancellery confirms both the commission and the definitive payment of the *mappa mundi* by the Portuguese court.

Doc. 10. May 13, 1483. Felix Fabri, Evagatorium Fratris Felicis in Terrae sanctae, Arabiae et Egypti peregrinationem.

In a note dated May 13, 1483, the Dominican Felix Fabri, in the *Evagatorium Fratris Felicis in Terrae sanctae, Arabiae et Egypti peregrinationem*, the diary of his pilgrimage to the Holy Land, conveys the following information: "Est enim inter Venetias et Murianum insula, in qua est ecclesia nova et pulchra S. Cristophori cum monasterio Ordinis albi. In illo monasterio est depicta una mappa mundi valde pulchra"[9]; (Then there is between Venice and the island of Murano an island in which there is the new and beautiful church of Saint Christopher close to a Camaldolese monastery. In that monastery a very beautiful map of the world is depicted). This is one of the first extant document attesting that the *mappa mundi* was well known outside of the Camaldolese Order and considered one of the "miracles" of Venice, as Giovanni Battista Ramusio would write fifty years later.

Doc. 11. May 26, 1494. Venice, Biblioteca Nazionale Marciana, Ms. Lat. XI, 92 (= 3828), pp. 502-503. Delfinus Petrus Venetus Patritius Abbas S. Michaelis de Muriano inde Prior & Abbas generalis Camaldulensis, *Epistula ad Bernardinum Gadulum* in *Epistularum libri tres manu ipsius scripti*.[10]

This letter was sent to Bernardino Gadolo, monk at Camoldoli, by the general of the Camaldolese Order, Pietro Delfino. The letter refers that in Delfino's cell in the Monastery of San Michele an autographed book could be found in which Delfino drafted the translation into Latin of the legends of Fra Mauro's map. Delfino had prepared this Latin translation of the map's legends to be used in the preparation of a copy of Fra Mauro's map commissioned by the Medici family around 1480, at the time of Lorenzo il Magnifico. In order to do this, Florentine painters were sent to San Michele. The copy for the Medici was still extant in the Medici Palace in via Larga in Florence in 1494, Delfino confirms in his letter. Both the manuscript book with the translation and the map have been lost. Following is the transcription of relevant the passages:

Bernardino Priori. Redierunt nostri Venetiis, cum sarcinulis, incolumes, Deo gratias. Miratus sum, tot, tantaque volumina, prae oculis exposita, tam parvo coempta fuisse pretio: ut tantidem profecto, aut pluris etiam, vel unum ex iis debuerit constitisse. Magna nostrorum temporum foelicitas, ut suppellex tam pretiosa tam vili veneat. Benedictus Deus, a quo omne datum optimum, et omne donum perfectum. Putabam esse inter alias cartulas nostras, quae olim in pictura orbis annotata, latina feceram. Si perquisieris illa diligentius, forte in cellula tua omnia invenies, conscripta manu mea: alias non facile speres me ex domo medice, ut petisti, exemplar habiturum. Redacta in codicem, haud dubie commodarentur.

[9] F. Fabri, *Les errances de Frère Felix, pèlerin en Terre sainte, en Arabie et en Egypte (1480-83)*. Texte latin, traduction, introduction et notes sous la direction de J. Meyers et N. Chareyron. Montpellier: Publications du Cercam, 2000, p. 136. *Mappamondo* (ed. 2006), p. 24.

[10] G. B. Mittarelli, *Bibliotheca codicum manuscriptorum*, col. 360; See also P. Zurla, *Il mappamondo di Fra Mauro*, pp. 150-151. Note that Placido Zurla, as opposed to Mittarelli, erroneously cites manuscript 729, f. 129v.

mihi transcribenda. Quod vero petere habeam, ut quis meo nomine in domo illa permittatur, annotationes ipsas ex eo orbe quem quondam Florentini pictores ex archetypo nostro effinxerunt, describere, non ego sane id auderem ; hoc presertim tempore quo non penitus vacat intestinis seditionibus civitas, relegatis extra urbem duobus illis tantae autoritatis fratribus gentis Medicis, primariis utique civibus et praepotentibus, et universo populo quam carissimis. Excusationem meam admittes tu cum caeteris illis, quorum nomine hoc a me postulaveras. Videte inquit apostolus quomodo caute ambuletis, quonium dies mali sunt. (...) Vale ex Fonte bono [Camaldoli] die XXVI maji 1494.[11]

Doc. 12. November 1494. *Itinerarium siue peregrinatio excellentissimi viri artium ac vtriusque medicine doctoris Hieronimi Monetarii de Feltkirchen ciuis Nurembergensis.*[12]

"Eodem die ascendimus Castellum et vidimus duos fortissimos leones adeo pulcros sicut nunquam vidi. Similiter cosmographiam in maxima et bene descripta tabula deaurata cuius dyameter erat 14 palmorum."

Doc. 13. Circa 1519-1524. Florence, Biblioteca Nazionale, B.R. 236 (Magl. XIII, 84), ff. 38r-43v (previous numeration ff. 25r-28bis v). Alessandro Zorzi, *Aviso di Fra Nicola inscripto fra di S. Michiel di Muran* (Notice by Fra Nicola of San Michele of Murano).[13]

Zorzi transcribes four legends and toponyms displayed in the African part of the map for one of the four itineraries to Ethiopia described in his geographic *zibaldone*. As the transcription clearly shows, the information attributed by Zorzi to a certain "Fra Nicola of San Michele of Murano" was instead literally derived (if not copied) from some legends of the *mappa mundi*. This is the first documented case in which Fra Mauro's map was used as a source in order to describe far away lands. Following is the transcription of folio 43v.

Zorzi : Se dice che il presta Janni ha piu di 120 regni dei quali piu di 60 sono di diferenti lengue ; et 72 di questi regni sono potenti ; li altri non sono da far conto. El Regno suo di Saba over Sadai è una gran cita.

Source : Se dice che presto Jane ha più de 120 regni soto el suo dominio, di qual più de 60 sono de differente lengue. E de tuto questo numero, zoè 120, se dice che 72 sono potenti signori, el resto non è da far conto. *Mappamondo* (ed. 1956) Pl. X, 21 ; *Mappamondo* (ed. 2006) *0077.

Zorzi : La provincia di Benicilep e habitata di gente fortiss(im]a et sono gran popoli ; lianno il volto come chagneschi et questi tali non funo mai subiugati da Romani.

[11] For the English translation, see Chapter I, paragraph IV.

[12] The work survives in just one manuscript, which belonged to the famous author and editor of the *Nuremberg Chronicle*, Hartmann Schedel (1440-1514), and is now in Munich, Bayerische Staatsbibliothek, Clm 431, ff. 96-274v. The manuscript has still not been published in its entirety in Latin, although there are two Spanish translations : Hieronymus Münzer, *Viaje por España y Portugal (1494-1495)*, introd. M. Gómez Moreno, trasl. J. López Toro, Madrid : Almenara, 1951 and Id. *Viaje por España y Portugal (1494-1495)*, Madrid : Polifemo, 1991 and two partial translations into Portuguese. I did not manage to consult the manuscript in Munich.

[13] Modern edition, transcription and translation in English in O. G. S. Crawford (ed.), *Ethiopian itinineraries circa 1400-1524 Including Those Collected by Alessandro Zorzi at Venice in the Years 1519-24*, Cambridge : Cambridge University Press, 1958, pp. 108-125.

Source : Benichileb. Questo nome è interpretado fiol de can. p(er)ché questi populi hano j uolti cagnesci. *Mappamondo* (ed. 1956) Pl. XI, 11 ; *Mappamondo* (ed. 2006) *155.

Zorzi : Dal 1420 una nave scorse de India alla via delle insole deli homeni et donne sopra dil capo de Diab alla via dil ponente per 40 giornate da circa a 2000 meia et ritornorino in 70 giornate fino a dito capo, dove alla ripa troveron uno grandissimo ucello con gran ovo dicon cosa grande che par fabule a dirle, le qual non è verisimile che ( ) al presente Portogalese ne havian viste.

Source : Circa hi ani del Signor 1420 una naue ouer çoncho de india discorse per una trauersa per el mar de india a la uia de le isole de hi homeni e de le done de fuora dal cauo de diab e tra le isole uerde e le oscuritade a la uia de ponente e de garbin per 40 çornade, non trouando mai altro che aiere e aqua, e per suo arbitrio iscorse 2000 mia e declinata la fortuna i fece suo retorno in çorni 70 fina al sopradito cauo de diab. E acostandose la naue a le riue per suo bisogno, i marinari uedeno uno ouo de uno oselo nominato chrocho, el qual ouo era de la grandeça de una bota d'anfora, e la grandeça de l'oselo era tanta che da uno piço de l'ala a l'altro se dice esser 60 passa, e con gran facillità lieua uno elefante e ogni altro grando animal e fa gran dano a li habitanti del paexe et è uelocissimo nel suo uolar. *Mappamondo* (ed. 1956) Pl. IV, 6 ; *Mappamondo* (ed. 2006) *19.

Zorzi : Nota che li Abbasini dicono haver piu territorio di sopra il nasimento dil Nilo che di soto, cioé in verso nui, et dicono haver magior fiumane che il Niro [sic] il quale fra noi e tanto nominato e(ss) er grande et molti fiumi come aparl [sic] li intrano che 'l fa e[ss]er grande : onde dice che al tempo di so I[n]verno [sic] che è il Magio et Zugno per la gran piogie che sono de quelli fiumi cresono molto et agumenteno il ditto Nilo si che crese maraviosamente.

Source : Nota che abassini dicono hauer più teritorio de sopra el nascimento del nilo che de soto, çoè inuer nui. E dicono hauer maçor fiumi del nilo, el qual fra nui è tanto nominato de esser grando. Ma molti fiumi come apar li entra che'l fa esser grando. Unde i dice che al tempo del suo inuerno ch'è el maço e çugno, per le gran pioçe che sono queli fiumi chrescono molto e dano augmento al nilo, per modo che l'abunda e cresce tanto che'l inunda l'egypto al tempo come è noto." *Mappamondo* (ed. 1956) Pl. XVII, 1 ; *Mappamondo* (ed. 2006) *155.

Doc. 14. Giovanni Battista Ramusio, *Dichiarazione d'alcuni luoghi ne' libri di messer Marco Polo, con l'istoria del reubarbaro in Di messer Giovanbattista Ramusio prefazione sopra il principio del libro del magnifico messer Marco Polo. All'eccellente messer Ieronimo Fracastoro, in Navigationi et viaggi*, in Venetia, appresso i Giunti, 1559, vol. II, f. 17r.[14]

This is the first description of Fra Mauro's *mappa mundi* defined by Ramusio as "one of the miracles" of Venice. According to Ramusio's interpretation, among Fra Mauro's sources there were a *mappamondo* and a marine chart brought to Venice by Marco Polo one hundred and fifty years earlier. Following is the transcription of Ramusio's description.

Resta ch'io dica ancora in generale alquante cose sopra questo libro, ch'io già essendo giovane udi' piú volte dire dal molto dotto e reverendo don Paolo Orlandino di Firenze, eccellente cosmografo e molto mio amico, che era priore del monasterio di Santo Michele di Murano a canto Venezia, dell'ordine de

[14] Modern edition Giovanni Battista Ramusio, *Dichiarazione d'alcuni luoghi ne' libri di messer Marco Polo, con l'istoria del reubarbaro*, in Giovanni Battista Ramusio, *Navigazioni e viaggi*, M. Milanesi (ed.), Turin : Einaudi, vol. III, 1980, pp. 69-71.

Camaldoli, che mi narrava averle intese da altri frati vecchi pur del suo monasterio. E questo è come quel bel mappamondo antico miniato in carta pecora, e che oggidí ancor in un grande armaro si vede a canto il lor coro in chiesa, la prima volta fu per uno loro converso del monasterio, quale si dilettava della cognizione di cosmografia, diligentemente tratto e copiato da una bellissima e molto vecchia carta marina e da un mappamondo, che già furono portati dal Cataio per il magnifico messer Marco Polo e suo padre; il quale, cosí come andava per le provincie d'ordine del gran Can, cosí aggiugneva e notava sopra le sue carte le città e luoghi che egli ritrovava, come vi è sopra descritto. Ma per ignoranzia d'un altro che dopo lui lo dipinse e forní, aggiugnendovi la descrizione d'uomini e animali di piú sorti e altre sciocchezze, vi furono aggiunte tante cose piú moderne e alquanto ridiculose, che appresso gli uomini di giudicio quasi per molti anni perse tutta la sua auttorità. Ma poi che non molti anni sono per le persone giudiciose s'è incominciato a leggere e considerare alquanto piú diligentemente questo presente libro di messer Marco Polo che fin ora non si avea fatto, e confrontare quello ch'egli scrive con la pittura di lui, immediate si è venuto a conoscere che 'l detto mappamondo fu senza alcuno dubbio cavato da quello di messer Marco Polo, e incominciato secondo quello con molto giuste misure e bellissimo ordine: onde fin al presente giorno è dapoi continuamente stato in tanta venerazione e precio appresso tutta questa città, e coloro massime che si dilettano delle cose di cosmografia, che non è mai giorno che d'alcuno non sia con molto piacere veduto e considerato, e fra gli altri miracoli di questa divina città, nell'andare de' forestieri a vedere i lavori di vetro a Murano, non sia per bella e rara cosa mostrato. E ancor che quivi si vegghino molte cose essere fatte alquanto confusamente e senza ordine, grado o misura (il che si deve attribuire a colui che 'l dipinse e forní), vi si comprendono per ciò di molto belle e degne particularità, non sapute ancora né conosciute meno dagli antichi: come che verso l'antartico, ove Tolomeo e tutti gli altri cosmografi mettono terra incognita senza mare, in questo di San Michele di Murano già tanti anni fatto si vede che 'l mare circonda l'Africa e che vi si può navicare verso ponente, il che al tempo di messer Marco si sapeva, ancor che a quel capo non vi sia posto nome alcuno, qual fu per Portughesi poi a' nostri tempi l'anno 1500 chiamato di Buona Speranza. Vi si vede appresso l'isola di Magastar, ora detta di San Lorenzo, e quella di Zinzibar, delle quali messer Marco parla nÈ capitoli 35 e 36 del terzo libro, e molte altre particularità nelli nomi dell'isole orientali, che dapoi per Portughesi a' tempi nostri sono state scoperte. Dalla parte poi di sotto la nostra tramontana, che ciascuno scrittore e cosmografo di questi e dÈ passati tempi fin ora vi ha messo e mette mare congelato, e che la terra corra continuatamente fin a 90 gradi verso il polo, sopra questo mappamondo, all'incontro, si vede che la terra va solamente un poco sopra la Norvega e Svezia, e voltando corre poi greco e levante nel paese della Moscovia e Rossia e va diritto al Cataio. E che ciò sia la verità, le navigazioni che hanno fatte gl'Inglesi con le loro navi volendo andare a scoprire il Cataio al tempo del re Odoardo sesto d'Inghilterra, questi anni passati, ne possono far vera testimonianza: perché nel mezo del loro viaggio, capitate per fortuna ai liti di Moscovia, dove trovarono allora regnare Giovanni Vaschelluich, imperatore della Rossia e granduca di Moscovia, il quale con molto piacere e maraviglia vedutogli fece grandissime carezze, hanno trovato quel mare essere navigabile e non agghiacciato. La qual navigazione (ancor che con l'esito fin ora non sia stata bene intesa), se col spesso frequentarla e col lungo uso e cognizione di quÈ mari si continuerà, è per fare grandissima mutazione e rivolgimento nelle cose di questa nostra parte del mondo. E tutte queste particolarità senza dubio alcuno furono cavate dalle carte e mappamondo del Cataio, perché messer Marco non fu mai nel seno Arabico né verso l'isole quivi vicine, e gran parte dell'informazione del terzo libro è da credere che gli fusse data da marinari di quelli mari d'India, li quali grossamente gli dicevano per arbitrio loro quanto era da un'isola all'altra (e mille e duemila miglia a loro non pareva troppo gran cosa); e anche per qual vento vi s'andasse non sapevano cosí chiaramente come al presente si sa, per le carte sí diligentemente e con tanta misura fatte e con li venti e con li gradi. E vi sono anco dÈ nomi di una medesima provincia duplicati, di che il lettore non piglierà ammirazione; e alcuna volta in cambio d'isole dice regni: come nella Zava minore, al capitolo decimo del terzo libro, mette otto regni, li quali a giudicio d'uomini pratichi sono isole, come saria dire che il regno di Samatra (chiamata da lui Samara) è quella grandissima isola di Sumatra, e così di molte altre le quali al presente ci sono incognite, che nell'avenire, col tempo e per la navigazione de' Portughesi, facilmente si saperanno. Si conosce ancora come al suo tempo non v'era el bussolo e la calamita a' nostri tempi ritrovata, cosa tanto maravigliosa e rara, né si sapeva la elevazione del polo con li gradi come ora si sa, ma grossamente guardandolo dicevano: la stella tramontana può essere tanti cubiti o braccia alta dal mare. Il fabricare delle navi, nel principio del terzo libro, è simile a quello che usano nell'isole delle Moluche e della China. Ultimamente nel fine del terzo libro, ove parla della Rossia e del regno delle Tenebre, come quello che in varii mappamondi antichi è posto per fine del nostro abitabile sotto la tramontana non s'inganna punto del sito del detto regno, nelli mesi però ch'egli scrive dell'inverno. E questo basti per ora per dichiarazione d'alcuni luoghi del libro di messer Marco Polo".

Doc. 15. António Galvão, *Tratado que compôs o nobre & notauel capitão António Galuão, dos diuersos & desuayrados caminhos, por onde nos tempos passados a pimenta & especearia veyo da India às nossas partes, & assi de todos os descobrimentos antigos & modernos, que são feitos até a era de mil & quinhentos & cincoenta. Com os nomes particulares das pessoas que os fizeram: & em que tempos & as suas alturas, obra certo muy notauel & copiosa*, [Lisbon], Rua de Sã Mamede em casa de Ioam da Barreira, 15 Dezembro 1563, f. 18r.

No anno de 1428. diz que foy o Infante dom Pedro a Inglaterra, França, Alemanha à casa sancta, & a outras de aquella bãda, tornou por Itália, esteue em Roma, & Veneza, trouxe de lá hü Mapamundo que tinha todo o âmbito da terra, & o estreito do Magalhães se chamaua, Cola do dragam, o cabo de Boa esperança, fronteira de África, & que deste pedram se ajudara o Infante dom Anrrique em seu descobrimento, Francisco de Sousa Tauarez me disse que no anno de 1528 o Infante dom Fernando lhe amostrara hüa Mapa que se achara no cartório d'Alcobaça que auia mais de cento & vinte annos que era feito, o qual tinha toda nauegaçam da índia, com ho cabo de Boa esperança [...].

Doc. 16. Agostino Fortunio, *Historiarum Camaldulensium pars posterior. Ubi ab origine templi, atque coenobij classis, et sanctissimi patris Romualdi conuersione, describuntur reliquorum monasteriorum exordia, et incrementa: priuilegia, et monachi sanctitate, dignitate, vel doctrina praestantes. Itemhabetur vita beati Petri Damiani cardinalis ostiensis, cum multis aliis, quae antiquitatem illustrant, et ostendunt. Augustino monacho Angelorum Florentiae auctore*, Venetijs ex typographia Guerraea, Venetijs, apud Dominicum, et Io. Baptistam Guerraeos fratres, 1579 mense Septembris, Lib. IV, ch. 23, Pars Posterior, pp. 194-195.

The chronicler and historian of the Camaldolese order, Don Agostino Fortunio (m. 1595), summarized Ramusio's description of the map in his history of the order. This confirms the archetypal role inherited from Ramusio's work in the development of a tradition of studies on Fra Mauro's *mappa mundi*. Following is the transcription of Fortunio's brief description.

"Insignis habetur ad Sancti Michaelis Murani Carta, universam Orbis situationes descriptionemque exprimens; quae, ut scribit Ioan. Bapt. Ramusius ex relatione Pauli arlandini, à quodam converso loci, ab insigni lomgem carta marina desumpta est; quae olim ex Cataio per marcum polum Venetias delata fuerat. Qui quidem, sicut ipse & pater per provicias magni Canis itinera faciebant; ita fuis cartis Civitates, & loca diligentissime addebat. Sed Pictor, seu miniator, à iustis eius mensuris ordine, & situationibus non paucis in locis deflexit plus fatis. Verum tamen nil pulchrius, aut venustius in eo genere desiderari posse videtur".

Doc. 17. December 20, 1655. Autograph note of Francesco Gherardo Erizzo, monk at San Michele, posted at mid point on the right side of the map.

The note points out that the *mappa mundi*, displayed since 1460 in the sacristy of the church of San Michele, was moved to the library of the monastery in 1655.

"Haec Tabula Geographica cum per centum et nonaginta quinque annorum curricula partim in Ecclesia partim vero in Aula, quae suo nomine dicta erat et dicebatur *Mappamondo*, fuisset appensa, tandem iussu Rev.mi P. D. Francisci Gherardi dicti Errici, Abbatis huius Monasterii, in hac Bibliotheca ab ipso instaurata, ditata et exomrnata translata et collocata fuit Anno 1655 Die 20 Decembris."

I. – Cosmographic legends placed in the four corners of the *mappa mundi*

Doc. 1. Del numero de hi cieli secondo l'auctorità de hi sacri theologi. *Mappamondo*, upper left corner (left).[1]

Del numero de hi cieli secondo l'auctorità de hi sacri theologi | Circa el numero dei cieli fra gli sacri doctori par che'l ne sia qualche diuersità et maxima | mente fra Basilio et Chrisostomo, el qual Chrisostomo dice che'l no(n) è altro che uno cielo. Et | benchè ne la sacra scriptura se dicha celi celorum, questo è per la proprietade de la lengua | ebraicha, ne la qual ae in consuetudine che'l cielo solo se nomina pluralmente, chome etiam | sono molti nomi in latin che no(n) à el proferir in singular. Ma Basilio (et) Damasceno che'l síegue di | chono esser molti cieli, ma questa diuersità è pliù ne la uoce che ne la chosa, cioè nel modo del | significar che nel significato. Et benchè Chrisostomo dica che'l sia uno cielo, per questo lui nomi | na uno cielo tuto quel corpo che è sopra la terra e l'aqua, p(er)chè etiam li oseli che uolano per | l'aiere sono dicti oseli del cielo. Ma perché in questo corpo sono molte distinctio(n), però Basilio me|te molti cieli. Et a saper la distinctio(n) de questi cieli, è da considerar che'l cielo se nomina a tre | modi ne la scriptura. Alguna uolta fi dito propria e naturalmente, (et) chusì el cielo fi dito [da] alguni | corpo sublime et luminoso per acto ouer p(er) potentia (et) i(n)corruptibile per natura, (et) seco(n)do questo p(r)i|mo modo sono posti tre cieli. El primo è totalme(n)te lucido, el qual se chiama empyreo, cioè igni|to over luminoso, el qual è ordinato al stato de la gloria (et) è deputado a la co(n)templatio(n) (no) n per | necessità ma per co(n)gruità, aciochè la chiateça exterior co(n)uegna a la interior. Unde dice Basilio ch(e)'l | spirito ministrator non podeua star in tenebre ma co(n)uegnia esser in luce (et) in letitia. Era etiam co(n) | ueniente che questo corpo fusse chreato luminoso per el stato de la gloria, chonciosia chossa chè | ne la futura remuneration se aspeta do' glorie, cioè la spiritual et la corporal, no(n) solame(n)te ne li corpi | humani per esser glorifichadi ma etiam in innouatio(n) de tuto el mundo. Ma nel principio del | mundo fo principiato la gloria spiritual ne la beatitudine de li angieli, la equalità de li quali fo promessa | a li sancti, unde fo co(n)ueniente che anchora nel principio fosse començada la gloria corporal | i(n) qualche corpo, el qual fusse sença corruptio(n) e mutabilità et totalmente lucido, cussì chome | aspeta esser in tuta la creatura corporal da poi la resurrection. Et perché quel cielo fi dito empyreo | cioè igneo, no(n) per ardor ma per splendor. El sego(n)do cielo è totalmente diaphano (et) è dito cielo aqueo | ouer christalino. El tercio parte è diaphano et parte lucido p(er) acto, el qual fi dito sydereo | ouer stellato, nel qual sono le stelle fixe. Et questo fi diuiso in octo spere, cioè la spera de le stelle | fixe et septe spere dei pianeti, i qual se pò dir octo cieli. El sego(n)do modo fi dito el cielo per partecipa | tion de qualche proprietà de corpo celestial, cioè de sublimità et de luminosità per acto ouer | per potentia. Et cussì tuto questo spatio che è da le aque fina al circulo de la luna Damasceno | mete uno cielo nominandolo aereo (et) per questo secondo lui sono tre cieli, cioè aereo | sydereo e l'altro de sopra del qual lui intende che l'apostolo Paulo fo rato al terçio cielo. | Ma perché questo spatio co(n)tien doi elementi, cioè el fuogo e l'aiere, (et) in chadauni de questi | do' fi notado la superior et inferior region, però Rabano ae distincto questo cielo in quatro | cioè la superior regio(n) del fuogo, nominando quel ciel igneo, ma la inferior region cie|lo olimpo che uol dir tuto lucido, ma la suprema regio(n) de l'aiere l'ae nominado cielo | ethereo per la i(n)flamatio(n) e la inferior regio(n) cielo aereo. Et siando questi quatro | cieli co(n)numeradi cho(n) li tre superior, sono i(n) tuto seco(n)do Rabano sete cieli corporei. | El tercio modo fi dito el cielo

[1] "The authority of the holy theologians on the number of heavens." Cf. *Mappamondo* (ed. 1956), p. 19. For the English translation see *Mappamondo* (ed. 2006), pp. 720-722.

metaphisicame(n)te cioè sopranatural et cussì | [40]algune fiade la sancta trinità fi dita cielo p(er) la sua spiritual sublimità e luce. | Del qual cielo se expone che lucifero dicesse : Io ascenderò in cielo, cioè a la equa|lità de Dio. Alguna fiade anchora significha li beni spirituali ne li quali | è la remuneratio(n) dei sancti, i qual per la sua emine(n)tia sono chiamati | cieli. Algune uolte significha le tre cho(n)ditio(n) de le uisio(n) sopranatural | [45]cioè de la corporal, de la ymaginaria et de la intellectual, sono nomina|de tre cieli, chome expone s(an)c(t)o Augustino. Ma cho(n)cluda(n)do dei cieli | segu(n)do la chomuna opinion, cho(m)me(n)çando dal cielo de la luna so|no diece cieli computando el cielo empyreo, chome apar | qui ne la presente pictura.

Doc. 1a : *S. Thomae Aquinatis Summma theologiae*, Pars I, Q. LXVIII, art. 4 'Utrum sit unum caelum tantum'.[2]

Ad quartum sic proceditur. Videtur quod sit unum caelum tantum.

1. Caelum enim contra terram dividitur, cum dicitur : In principio creavit Deus caelum et terram. Sed terra est una tantum. Ergo et caelum est unum tantum.

2. Praeterea, omne quod constat ex tota sua materia, est unum tantum. Sed caelum est huiusmodi, ut probat Philosophus in I De caelo. Ergo caelum est unum tantum.

3. Praeterea, quidquid dicitur de pluribus univoce, dicitur de eis secundum unam rationem communem. Sed si sunt plures caeli, caelum dicitur univoce de pluribus : quia si aequivoce, non proprie dicerentur plures caeli. Oportet ergo, si dicuntur plures caeli, quod sit aliqua ratio communis, secundum quam caeli dicantur. Hanc autem non est assignare. Non est ergo dicendum quod sint plures caeli.

Sed contra est quod dicitur in Psalmo 148, 4 : Laudate eum, caeli caelorum.

Respondeo dicendum quod circa hoc videtur esse quaedam diversitas inter Basilium et Chrysostomum. Dicit enim Chrysostomus non esse nisi unum caelum ; et quod pluraliter dicitur, caeli caelorum, hoc est propter proprietatem linguae hebraeae, in qua consuetum est ut caelum solum pluraliter significetur ; sicut sunt etiam multa nomina in latino quae singulari carent. Basilius autem, et Damascenus sequens eum, dicunt plures esse caelos. Sed haec diversitas magis est in voce quam in re. Nam Chrysostomus unum caelum nominat totum corpus quod est supra terram et aquam : nam etiam aves, quae volant in aere, dicuntur propter hoc volucres caeli. Sed quia in isto corpore sunt multae distinctiones, propter hoc Basilius posuit plures caelos.

Ad distinctionem ergo caelorum sciendam, considerandum est quod caelum tripliciter dicitur in Scripturis. Quandoque enim dicitur proprie et naturaliter. Et sic dicitur caelum corpus aliquod sublime, et luminosum actu vel potentia, et incorruptibile per naturam. Et secundum hoc, ponuntur tres caeli. Primum totaliter lucidum, quod vocant empyreum. Secundum totaliter diaphanum, quod vocant caelum aqueum vel crystallinum. Tertium partim diaphanum et partim lucidum actu, quod vocant caelum sidereum ; et dividitur in octo sphaeras, scilicet in sphaeram stellarum fixarum, et septem sphaeras planetarum ; quae possunt dici octo caeli.

Secundo dicitur caelum per participationem alicuius proprietatis caelestis corporis, scilicet sublimitatis et luminositatis actu vel potentia. Et sic totum illud spatium quod est ab aquis usque ad orbem lunae, Damascenus ponit unum caelum, nominans illud aereum. Et sic, secundum eum, sunt tres caeli : aereum, sidereum, et aliud superius, de quo intelligit quod Apostolus legitur [*II Cor.* 12, 2] *raptus usque ad tertium caelum. Sed quia istud spatium continet duo elementa, scilicet ignis et aeris, et in utroque*

[2] *S. Thomae Aquinatis Summma theologiae*, cura et studio sac. Petri Caramello cum textu ex recensione leonina, pars prima et prima secundae, Turin : Marietti, 1952, p. 334. The part of the article translated in the legend in the *mappa mundi* is in italics.

eorum vocatur superior et inferior regio; ideo istud caelum Rabanus distinguit in quatuor, supremam regionem ignis nominans caelum igneum; inferiorem vero regionem, caelum olympium, ab altitudine cuiusdam montis qui vocatur Olympus; supremam vero regionem aeris vocavit caelum aethereum, propter inflammationem; inferiorem vero regionem, caelum aereum. Et sic, cum isti quatuor caeli tribus superioribus connumerantur, fiunt in universo, secundum Rabanum, septem caeli corporei.

Tertio dicitur caelum metaphorice. Et sic quandoque ipsa sancta Trinitas dicitur caelum, propter eius spiritualem sublimitatem et lucem. De quo caelo exponitur diabolum dixisse [*Is. 14, 13*]: *Ascendam in caelum, idest ad aequalitatem Dei. Quandoque etiam spiritualia bona in quibus est Sanctorum remuneratio, propter eorum eminentiam, caeli nominantur; ubi dicitur* [*Mt. 5, 12; Lc. 6, 23*]: *Merces vestra multa est in caelis, ut Augustinus exponit. Quandoque vero tria genera supernaturalium visionum, scilicet corporalis, imaginariae e intellectualis tres caeli nominantur. De quibus Augustinus exponit quod Paulus est raptus usque ad tertium caelum.*

Ad primum ergo dicendum quod terra se habet ad caelum, ut centrum ad circumferentiam. Circa unum autem centrum possunt esse multae circumferentiae. Unde una terra existente, multi caeli ponuntur. Ad secundum dicendum quod ratio illa procedit de caelo, secundum quod importat universitatem creaturarum corporalium. Sic enim est unum caelum tantum. Ad tertium dicendum quod in omnibus caelis invenitur communiter sublimitas et aliqua luminositas, ut ex dictis patet.

Doc. 2. De la distantia de hi cieli. Rubrica. *Mappamondo*, upper left corner (right).[3]

De la distantia de hi
cieli. Rubrica
Dal centro del mondo a la superficie de la terra sono miglia 3245 5/11
Dal dicto centro a la superficie inferior del cielo de la luna sono miglia 107936 20/33
Tuto el diametro de la luna è miglia 1896 26/33
Dal dicto centro del mondo a la superficie inferior del cielo de Mercurio sono miglia 209198 26/33
Tuto el diametro de Mercurio sono miglia. 230 26/33
Dal centro del mondo a la superficie inferior del cielo de Venere sono miglia 579320 560/660
Tuto el diametro de Venere sono miglia 2884 560/660
Dal centro del mondo a la superficie inferior del cielo del sol sono miglia 3892866 820/660
Tuto el diametro del Sol è miglia 35700
Dal centro del mondo a la superficie inferior del cielo de Marte sono miglia 4268629
Tuto el diametro de Marte è miglia 75724 480/660
Dal centro del mondo a la superficie inferior del cielo de Joue son miglia 323520 480/660
Tuto el diametro de Jove è miglia 29641 540/660
Dal centro del mondo a la superficie inferior del cielo de Saturno sono miglia 52544702 280/660
Tuto el diametro de Saturno è miglia 29202
Dal centro del mondo a la superficie inferior de la octaua spiera, çoè del cielo
de le stelle fixe, sono miglia. 73387747 180/660
Queste miglia sono cadauno de lor de mesura de cubiti .400.
Quanto sia la grossezza de questa octava spiera ne la qual sono le stelle fixe
non è possibile per alguna demonstrativa raxon provar per che le predicte
stelle fixe se muoveno sopra el centro del mondo e non come li sette

3 "On the Distance of the Heavens. Rubric." Cf. *Mappamondo* (ed. 1956), pp. 19-20. For the English translation see *Mappamondo* (ed. 2006), pp. 722-723.

planeti i qual se muoveno sopra altri centri per la diversità de li qual
centri, dal centro del mondo se mesura la quantità de hi suo
circuli. E per qyùuesta medesima raxon riman ignota
a nui la quantità de la spera nona
e quela etiam dio de cielo cristalin.

Doc. 3. C De hi elementi, cioè quanto uno elemento | excieda l'altro in quantità. Rubrica. *Mappamondo*, lower right corner (right).[4]

C De hi elementi, cioè quanto uno elemento | excieda l'altro in quantità. Rubrica | Comu(n) et uulgar parlar è che la qua(n)tità de uno elemento excieda la qua(n)|tità del subseque(n)te menor elemento diexe fiade, cioè che lo elemento de l'a|qua sia mazor diexe fiade de q(ue)llo de la terra e simel sia de l'aiere a l'aqua e del | fuogo a la terra. Al qual parlar par co(n)sentir el dicto del phylosopho, che dixe | nel libro de le generatio(n) e corruptio(n) che de uno pugno de t(er)ra se genera diexe pugni de aqua | e cussì m(u)ltiplicando fin al fuogo. Ma la raxon geometricha a q(ue)sto parlar co(n)tradice co(n)ciosiachè | quella proportio(n) è de uno circulo a l'altro, che è dal quadrato del diametro de uno de quelli circ(u)li | al quadrato del diametro de l'altro circulo, come chiaramente è prouado nel duodecimo libro de | Euclides a la propositio(n) secu(n)da. Onde per q(ue)sta regula, trouati quatro circ(u)li tra i quali fosse q(ue)sta decu|pla proportio(n) co(n)tinua, exe(m)pli gr(ati)a che'l seco(n)do circ(u)lo cioè l'aqua fosse diexe fiade mazor che'l primo cir|culo cioè la t(er)ra, et el tergo circulo cioè l'aere fosse mazor del seco(n)do cioè l'aqua diexe fiade, et el quar|to circulo cioè el fuogo fusse diexe fiade mazor del terzo circ(u)lo cioè de l'aere, seguita che questo quar|to circ(u)lo cioè el fuogo fosse mazor del primo cioè de la t(er)ra mille fiade p(er) la propositio(n) XVI del qui(n)to | libro de Euclides. E più seguita che'l quadrato del diametro de q(ue)sto quadrato circ(u)lo cioè del fuogo sia mille | fiate mazor del quadrato del diametro de la t(er)ra che poneuemo esser el primo. La qual sequela co(n)tradice a la | mesura de la terra e de li circuli celesti p(er) instrume(n)ti e mesure geometrice tolta, p(er) la qual se uede che'l quadra|to de tuto el diametro de lo elemento del fuogo, cioè del circ(u)lo co(n)cauo cioè inferior del cielo de la luna el q(u)al | è conuexo cioè superior circulo del fuogo, q(ue)sto quadrato dico è mazor del quadrato de tuto el diametro de la | t(er)ra 1107 fiade. Onde no(n) è possibile che lo elemento del fuogo etia(m) sia mille fiade mazor de l'elemento de la | terra, p(er)chè la proportio(n) de li circuli i(n) prima dicta presuppone che infra tuto el quarto circulo no(n) sia altro cha | fuogo. Onde se dal cielo de la luna in zoso no(n) fosse altro cha | fuogo, quel fuogo seria 1107 fiade mazor de qua(n)tità | de la terra. Ma perché no(n) è cussì, anci infra la ci(r)co(n)ferentia del circulo del fuogo e aere, lo qual seco(n)do el primo co|mu(n) dicto è cento fiade mazor de la terra, resteria el fuogo mazor de la t(er)ra 1007 fiade se fra l'aere et centro no(n) fos|se l'aqua e la t(er)ra e p(er) el parlar dicto l'aqua è mazor de la terra diexe fiade computando la t(er)ra seria da substrar 11 del | dicto numero 1007 e resteria el fuogo 996 fiade solamente mazor de la t(er)ra. E benchè la propositio(n) del duodeci(m)o | de Euclides parla de li circuli, niente di me(n) quello medemo è ne li corpi sferici. Ancora la mesura de la terra cioè de la | sup(er)ficie inferior è stada cussì limitada no(n) p(er)chè li cosmographi, ouer mesuradori del mo(n)do, habiano a parte a parte tuta la | terra mesurada, anci auendo i(n) questa n(ost)ra parte habitada mesurado qua(n)ta parte de t(er)ra corresponde ad uno grado in cielo, m(u)l|tiplicando tal parte de terra p(er) 360 che sono i gradi del cielo ano limitado la mesura de la circu(m)ferentia ouer sup(er)ficie de la | terra. Per la qual cossa noto è che tal dime(n)sio(n) presuppone che tuta la p(re)dicta circu(m)fere(n)tia de la t(er)ra habi equal distante

[4] "Of the elements. That is, how much one element exceeds another in quantity." Cf. *Mappamondo* (ed. 1956), pp. 22-23. For the English translation see *Mappamondo* (ed. 2006), pp. 731-732.

dal meço | del mondo a questa parte ne la qual nui habitemo, la qual essendo eleuada da la sup(er)ficie de l'aqua seguita che'l dicto nu(mer)o de me|sura de la t(er)ra sia mazor ouer alme(n) equal cum la dime(n)sio(n) de la circu(m)fere(n)tia de l'aqua; e più oltra co(n)cludendo, che lo elemento | del fuogo è solamente mazor de la terra 100 fiade, la qual co(n)clusio(n) manifestamente contradice a la regola geometrica sopra|dicta et al parlar uulgar prescripto.

Doc. 4. Come per diuina prouidentia la terra habitabile è subleuada da l'aqua. *Mappamondo,* upper right corner (right).[5]

Come per diuina prouidentia la terra habitabile è subleuada da l'aqua | Essendo la t(er)ra luogo de la generatio(n) di a(n)i(m)ali e de le altre cosse a la uita humana neccessarie, come sono li arbori e piante ouer herbe, bixogna | q(ue)lla t(er)ra auer diuersità ne le sue parte, zoè che i(n) alguni luogi de quella sia più rara et habia manifeste et sensibile porosità e co(n)cauità, come | uedemo esser ne li luogi de le minere de li metali et etia(m) de le pietre preciose e no(n) p(re)ciose. Onde, se'l se pesasse do' parte de terra equal de | qua(n)tità de mole ouer de grandeza ma desp(ar) i(n) esser dense e rare, ouer cu(m) ma(n)ifeste porosità e (con)cauità e ssenza alcuna porosità e co(n)cauità, quella parte no(n) porosa più | peseria de l'altra. Se adoncha tal differencia se troua ma(n)ifesta e sensibelme(n)te ne le parte de la terra e quella parte ch'è descoperta da le aque sia luogo | de la generatio(n) de li a(n)i(m)ali e de le cosse p(re) dicte, senza alguna dubietà q(ue)sta tal parte è più rara e porosa che l'altra parte de la t(er)ra coperta da l'aque, priuada de nec|cessità de le cosse predicte. Co(n)clude adoncha la natural raxo(n), che se p(er) imaginaria diuisio(n) fosse partita la t(er)ra i(n) do' parte equal qua(n)to a la mole ouer qua(n)tità i(n) | tal modo che una mità e parte co(m)prehe(n)desse q(ue)lla portio(n) de la terra scoperta da le aque che è uerso nui e rara e l'altra mitade q(ue)lla che è cop(er)ta et | densa, la prima mità rara men peseria de la densa. E più oltra seguita che quella parte più densa sia dep(re)sa zoè più bassa et propi(n)qua al mezo del mo(n)do | de l'altra, perché essendo la mità de la terra q(uan)to a la mole cum gra(n)deça in peso equal, el mezo tuta la terra seria pu(n)ctalmen(n)te nel centro del | mondo, ma perché le cosse più graue naturalme(n)te uano al dito centro etiamdio cu(m) uiolentia, chazando dal dito centro le cosse me(n) | graue, no(n) può ess(er) de meno che tal p(ar)te e mità de la terra non sia più uicina al centro, dico zoè l'è mezo el mondo et e più spatio da tal | centro a la sup(er)fitie de la p(ar)te de la tera da nui habitada che dal dicto centro a la sup(er)fitie da l'altra p(ar)te a questa nostra opposita, la qual cossa il guber|natore sumo Idio cu(m) mirabel prouide(n)tia ha p(er) tal forma ordinado, p(er)chè se tuta la sup(er)fitie de la tera fosse equalme(n)te distante dal cent(r)o ouer | mezo del mondo, l'aqua da ogni p(ar)te circu(m)daria e copriria equalmente tuta la terra no(n) auendo lo eleme(n)to de la qual tanta dinsità ne le suo parte | qua(n)ta se truoua ne la p(ar)te de la terra la qual equalmente graue e così equame(n)te remota dal dito mezo q(uan)to de la p(ar)te sup(er)ior zoè sup(er)fitie | e p(er) (con) seque(n)te copra(n)do una p(ar)te de la t(er)ra tute le coure(r)ia, la qual cossa seria i(n) detrime(n)to de la uita de li animali terestri i qual sopra l'aqua no(n) se poso|no sustentar p(er) la lor grauità nè sotto l'aqua posono uiuer p(er) la respiration et i(n)spiratio(n) de l'aere p(er) lo anelito ouer fiador, le qual sono | necessarie a la uita animal p(er) refrigerio del cor e mu(n)dification sua. Ma essendo la terra eleuada piui i(n) una cha ne l'altra p(ar)te, l'aqua | che equalmente p(er) tuto el centro ouer mezo el mondo p(er) natural apetito circunda, auegna che la p(ar)te de la terra dep(re)ssa et al dito | centro più propinqua da tal aqua sia cop(er)ta, no(n) p(er)ò q(ue)sta aq(u)a può coprir i(n) tuto la terra zoè q(uan) to a la parte eleuada che p(er) diuino con|seio a cons(er)uatio(n) de la uita de li animali p(re)dicti cussì

[5] "How by Divine Providence the [element] earth is raised above the [element] water." Cf. *Mappamondo* (ed. 1956), p. 21. For the English translation see *Mappamondo* (ed. 2006), pp. 726-729.

è disposita chomo è p(re)dicto. | Mirauegliosa cossa par a le humane mente che senza difficoltà no(n) possono (con)p(re)hender che sustentame(n)to habiano li elleme(n)|ti, e precipue la terra i(n) mezo de tuti situada e posta, e maxime che p(er) qua(n)te cosse che a li nostri sentime(n)ti occore se de|uede niuna cossa star senza sustentame(n)to, il qual qua(n)do se rimoue cade la cossa sustentada. Questa admiration in tuto | cessa qua(n)do se considera e per experime(n)to se cognose che a tanto se de' extender la operation et effecto de chadauna | cossa qua(n)to inclina la natura prop(r)ia. Onde la intellectiua cognition ne l'omo, la qual a lui è natural, non è maravegli|oxa, che seria degna de ogni admiratio(n) se i(n) una piera ouer altra cossa se trouase, la qual a questo no(n) auesse natural | inclinatio(n). Perché adoncha la terra p(er) natural apetito è inclinada ad ess(er) in mezo dei mo(n)do, el qual apetito lei | haueua et(iam) se niuna altra cossa corporal fosse, no bixogna che la habi alchun sustentame(n)to el qual la tegni | nel dito sito e maxime p(er)chè a quele solle cosse che possono ess(er) in diu(er)si stati e dispositio(n) e de uno stato o|uer dispositio(n) p(er) sè medemi possono hauer mutatio(n), in queli q(ua)n(do) uolemo che q(ue)ste tal remagnano i(n) una dis|position li metemo reparo che in q(ue)la disposition la (con)s(er)ua. Uerbi gr(ati)a, uole(n)do che una piera rimagni in lo|go alto, la qual per sì medema cadeua, bixogna meterli repparo açochè la no(n) posi cazer. Ma se quella | piera remagnisse nel logo doue la se uolese meter no(n) bixognaria a questo algun sustentame(n)to ouer | reparo. Adoncha essendo la natural inclinatio(n) de la grauità e per (con)seque(n)te de la terra, la qual tra tu|ti li corpi è grauissima, de descender zoè andar al centro ouer mezo del mo(n)do e contra la sua | natura sia ascender zoè andar uerso el zielo, se la terra como è uerità è nel mezo del mondo, lei | no(n) può per sì auer altra dispositio(n), p(er)chè se la se p(ar)tise dal mezo lei anderia uerso el cielo e per conse|que(n)te asce(n)deria contra la sua natural inclinatio(n). No(n) bixogna adoncha alguno sustenta|mento a la terra ma li basta el suo natural apetito, al qual sequita l'ordine de la propor|tione e (con)nexion insita da Dio ne li eleme(n)ti, chomo dice Boetio nel libro de la phylosophi|ca consolatione.

Doc. 5 : C [Questa rubrica de sopra dechiara come per la uirtù atractiua de la luna le aque cresce e descresce]. *Mappamondo*, upper right corner (left).[6]

Non picola cossa par tra quelle che nui uedemo, che le aque marine do' fiade i(n) uinti quatro hore cresca et in quel medemo spacio de tempo do' fiade descresca ouer minuisse. Et benchè per molti se diuulgi che el mouer de la luna induga tanta mutation | ne le aque, no(n) però se co(m)prehende la particular caxo(n) de tanta uarietà. Donde per alguna particular informatio(n) nui diremo che el sol, auendo efficacia de un caldo digestiuo et co(n)sumatiuo de humidità et de quelle resolutiuo, molti uapori de le aque lieua, quelle | in parte conuertendo ouer resoluendo in uapori. I qual uapori no(n) solamente se genera ne la exterior et apparente sup(er)ficie de le aque, ma etia(m) nel profondo de quelle. Et se le aque, nel fondo de le quale sono tai uapori geneti, ano in sì subtilità, quelli uapori, | no li facendo resistentia la subtilità de le aque, esceno a pocho a pocho, secondo chomo quelli uapoti sono geniti. Et pertanto in questo exito no(n) se fa apparente movimento i(n) quelle aque. Ma doue le aque sono grosse et ponderose e maxime essendo profonde, li uapori per el dicto | modo geniti nel fondo de quelle aque no(n) possono insire senza notabile adiutorio, facendoli ressistentia a la dicta eleuatio(n) la grosseza, ponderosità et quantità de le aque. Et perché la luna ha in sì uirtù attractiua de ogni humidità e per consequente tira ad sì etia(m) li ua|pori de queste humidità geniti e maxime se quelli sono corrupti, como se dichiara nel primo libro del quadripartito, pertanto qua(n)do la luna come(n)za cu(m) li razi suoi far inpressio(n) ne le aque grosse et profonde, adora quelli uapori grossi se moueno a la superficie de le | aque. Et in

[6] "The above rubric describes how waters rise and subside due to the attractive properties of the moon." Cf. *Mappamondo* (ed. 1956), p. 20. For the English translation see *Mappamondo* (ed. 2006), pp. 724-726.

questo mouimento et subleuatio(n) è in quelle aque uno mouime(n)to de bolimento, cioè cossì como ne l'olla al fuogo per li uapori leuadi dal fuogo in fondi de la olla l'aqua bolie, cossì per la eleuatio(n) de tal uapori leuadi dal fondo de le aque essa aqua boglie. E | però in questo boglimento le aque eleuade nel mezo de la sua congregatio(n) se spande uerso le sue estremità, çoè uerso le riue, e tanto se multiplica l'aque uerso le riue quanto dura questo bulimento e però l'aque uerso tal riue in questo tempo cresce. El qual bulimento | dura finchè li uapoti sono da la luna eleuadi, che è dal tempo che la luna se lieua sopra de l'emisperio finchè lei uie(n) al circulo meridiano, nel qual luogo essendo la luna, mancha lo bulimento perché tuti li uapori ça sono insidi e la luna descendando continua|mente ha più debeli hi suo razi finchè lei uie(n) a l'occaso et in tuto questo tempo le aque che sono uerso le riue torna al suo luogo e tien compresi quelli uapori che in questo tempo se genera ut supra. Unde manchando da le riue le aque discresceno. Et qua(n)do | la luna è ne l'occaso, zoè sopra l'oriçonte occidentale, li suo raçi comenza auer quella uirtù che lei auea quando la luna se leuaua sopra lo emisperio, perché li raçi sono in questa parte de occidente directi in una linea cum quelli primi. Et in però li uapori | che sono ingeneradi dal tempo che la luna se parte da la linea meridional fin a l'occaso, questi tal uapori aiutati dal uigor de li raçi de la luna che comenza a crescer qua(n)do la luna è in occaso, se muoue e fano ci segondo bulimento et ci segondo crescer, | che dura da po' che la luna è i(n) occaso finchè lei p(er)uie(n) a la linea | che meça lo emisp(er)io i(n)ferior. | e in questo tempo le aque cresceno per la raxon predicta. Et dal fin del predicto te(m)po finchè la è in oriente essendo za exaladi e insidi i uapori de l'aqua e declinando la uirtù de li razi de la luna, l'aqua descresce como nel tempo che la luna se | parte da la linea meridiana fi(n)chè lei uien a l'orizonte occidental, perché li razi de queste do' sono directi in una linea. Ma bench'el sia como è predicto ampuo ad uno tempo cresce più le aque che a l'altro e cussì ad un tempo più decresce che a l'al|tro *per molte caxon.* | *La prima* perché qua(n)do la luna è più propi(n)qua al zenith de le aque ouer al polo de l'orizonte, perché in questo tempo la luna ha più uigor che qua(n)do la no(n) uie(n) cussì propi(n)qua. In exemplo : in queste nostre parte qua(n)do la luna è i(n) | cancro la uie(n) più propinqua al nostro zenith e per co(n)seque(n)te a le aque nostre che quando la è in leon, e cussì più qua(n)do la è in leon che qua(n)do la è in uirgine. Onde, essendo par le altre caxon subseque(n)te, le aque più cresce qua(n)do la luna è in cancro. *La* | *segonda* caxon è la propinquità de la luna a le aque, zoè qua(n)do la luna è in la parte inferior del suo epicido, questa propi(n)quità fa trar più le aque. *Terza* qua(n)do el tempo che la luna se muoue da l'oriçonte oriental a la linea meridiana è mazor | de te(m)po che quela luna se muoue da la predicta linea meridiana a l'orizonte occidentali le | aque cresceno più perché auanti che l'aqua sia partida da le riue descresce(n)do, la qual cresce(n)do era mossa da le riue, un'altra fiada crescendo torna a quelle riue. *Quarta* se sono uenti i qual muouano le aque uerso le riue. Item el crescer de le aque | aiuta el lume del sol, zoè qua(n)do la luna è cu(m) el sol conzonta, ouer quando lei è piena de lume zoè nel qui(n)todecimo, alora per el uigor el qual la luna receue dal sol in queste due situation, lei ha più forza de trar i sopraditi uapori, e però | che questa ultima caxon è più potente de le altre, semp(re) a questi do' tempi le aque sono colme e una fiada più che l'altra, segondo como una ouer più de le altre caxon dicte aiuta oue le contrarie minuisse. E nel tempo che la luna | è meza piena del lume del sol, lei no(n) ha pien uigor de mouer i uapor del profondo e per tanto i(n) questo te(m)po le aque no(n) cresceno molto, nè ancor molto discresceno, saluo se una ouer più de le predicte caxon no(n) aiuta el crescer | ouer el discrescer. Et perché le aque de questi nostri mari sono grosse, essendo permixte cu(m) molte parte terrestre, et etia(m)dio sono profonde, pertanto quele aque crescono e discrescono. Ma le aque dei fiumi, le qual no(n) sono | profonde e se le sono profonde per esser sotil e no(n) grosse no(n) possono tenir reclusi insieme molti uapori, i qual | nel suo exito possi far mouime(n)to de bulicion ma cusì presto como sono generati li uapori escono, pertanto queste tal aque no(n) cresceno nì discresceno come le aque del mar.

Doc. 6. C Come la terra supposita a l'equinoctial e a | la torrida çona è abitabile. Rubrica. *Mappamondo*, lower right corner (left).[7]

C Come la terra supposita a l'equinoctial e a | la torrida çona è abitabile. Rubrica | La terra che è suposita al circulo equinoctial se può habitar, zoè in quella parte de la terra no(n) è tanto calor che i(m) paci | la humana habitatio(n) e questo se conclude e prouasse per raxo(n) natural. E prima perché hi phylosophi hano diuiso la t(er)ra | habitabile a nui nota e manifesta in septe clima, el primo di qual è ne la torrida zona, zoè tra el tropico de cancro | et el circulo equinocial. Onde essendo in quella parte de la torrida zona mazor caldo de quello è sotto el circulo equinotial, seguita de | neccessità che se esso primo clima che è ne la dicta parte de la torrida zona se habita, nè i(n) quello no(n) i(m)paci gra(n)deza de calor che no(n) se posseno ha|bitar, etia(m)dio sotto l'equinotial doue è men caldo se può habitar. Ma che sotto el dicto equinotial sia men caldo cha nel p(r)imo clima se proua p(er)ch(è) | le caxo(n) ch(e) fano excesso de caldo ne l'aiere e la terra sono: p(r)ima, q(ua)n(do) el sol passa de directo sopra quel luogo, onde i razi suo sono perpe(n)dicular e p(er) co(n)se|que(n)te essendo reflexi se unisse cu(m) i primi directi e perché la uirtù unida è più forte che la uirtù dispersa, p(er)tanto in quel luogo i razi del sol sono più ef|ficaci a scaldar. La sego(n)da caxo(n) si è la longeza di zorni, perch(è) qua(n)to el zorno è più lo(n)go tanto el sol sta più sopra la terra e p(er) co(n)seque(n)te più scalda. La terza | si è la lo(n)geza del te(m)po de l'instate sopra q(ue)llo de l'i(n)uerno, zoè che'l te(m)po de l'instate sia mazor cha q(ue)llo de l'inuerno. La quarta caxo(n) è la co(n)tinuatio(n) del caldo, zoè | q(ua)n(do) el te(m)po del caldo no(n) è interoto dal fredo. Le qual caxo(n) ouer più d'esse se troua nel p(r)imo clima e no(n) tute ouer tante se troua soto l'equinotial, perché cussì passa el | sol de directo sopra la terra del primo clima do' uolte a l'a(n)no, come sopra a la terra suposita a l'equinotial e p(er)ò qua(n)to a la prima caxo(n) le dicte do' habitatio(n) sono e|qual. Ma sotto lo equinotial mai i zorni no(n) excede 12 hore, cussì de instade come de inuerno, et i(n) p(er)ò ta(n)to è el te(m)po del scaldar del sol de zorno qua(n)to è el te(m) po del refri|gerio de la nocte. Nel primo uerame(n)te clima i zorni de instade sono molto più longi cha la nocte e p(er)ò el refrigerio de la nocte no(n) è sufficie(n)te a mitigar el cal|do inducto el zorno dal sol. Item ben che'l tempo del caldo del p(r)imo clima che è per do' instade sia equal al tempo del caldo che è p(er) do' instade sotto l'equino|tial, tame(n) quel te(m)po del caldo nel primo clima è co(n)tinuo e no(n) i(n)teroto da algu(n) tempo fredo, perché le do' instade sono co(n)tinue, ma sotto l'equinotial el dicto | te(m)po de do' instade è interoto da do' inuerni. V(er)bi gratia, quelli che sotto la mità del segno de thauro hano l'instade de tre mexi come hauemo nui, onde uien | a finir l'instade a loro q(ua)n(do) el sol è nel mezo del segno de leon, e alora perché el sol passa de directo sopra del capo alora come(n)za la sego(n)da i(n)stade, la qual uie(n) | esser co(n)tinua. Ma sotto l'equinotial come(n)za l'instade q(ua) n(do) el sol è nel principio de ariete e durando tre mexi uien a finir q(ua)n(do) el sol è in cancro, la s(econd)a instade | come(n)za q(ua)n(do) el sol è nel principio de libra e durando simelme(n)te tre mexi finisse q(ua)n(do) el sol è nel principio de capricorno e p(er)ò tra la prima i(n)stade e la sego(n)da | cade de mezo tre mexi de inuerno ouer de altro men caldo tempo e simelme(n)te tuta la secu(n)da i(n)stade e l'altra prima seque(n)te cade tre altri mexi de i(n)uerno ouer de | altro men caldo te(m)po. Et a questa raxo(n) co(n)sona l'auctorità del phylosopho i(n) diu(er)si luogi ma più exp(re)ssa nel libro de le proprietà e n(atur)a di eleme(n)ti e Alberto Mag(n)o | nel suo simel libro et Auerois come(n)tator nel come(n)to del libro del cielo e del mondo et Auicena nel suo libro di canoni de medicina. Vero è che alguni, come | l'auctor de la spera, mete co(n)traria opinione, la qual etia(m) fo de Socrates e de alcuni altri autori, mossi a questa opinio(n) p(er) l'andar del sol sopra el capo de li habita(n)ti. | Onde credeuano

7 "How the Earth which is under the [celestial] Equator and the zone may be habitable." Cf. *Mappamondo* (ed. 1956), pp. 19-20. For the English translation, though with a different interpretation of the title and the contents of the legend, see *Mappamondo* (ed. 2006), pp. 733-735.

q(ue)sti tal che'l sol q(ua)n(do) el passa de directo sopra algu(n) luogo i(n)duce sì excessiuo caldo che quel luogo no(n) se può habitar, et ben che'l transito del | sol p(re)dicto i(n)duga gran caldo, no(n) però è tanto che'l remoua i(n) tuto la habitatio(n) humana, p(er)chè nel primo clima che è situado i(n) luogo che'l sol i passa sopra de di|recto no(n) è tanto caldo che'l dicto clima no(n) se habiti. E pertanto se nel primo clima el transito del sol de directo sopra el capo no(n) remuoue la habitation p(er) exces|[30]so de caldo, neanche el dicto transito sopra la t(er)ra supposita a l'equinotial remuove la habitatio(n) p(er) excesso de caldo e maxime essendo prima i zorni dei predicto | transito nel primo clima mazor cha i zorni proporcional sotto l'equinotial, et etia(m) sta(n)do più tempo el sol propi(n)quo al zenith ouer al capo de li habitatori del | p(r)imo clima di quello el sta quelli habita sotto lo equinotial. Onde p(er) le do'p(re) dicte raxo(n) el sol nel circulo equinotial, passando de directo sopra el capo de li ha|bitatori de quelle parte, no(n) scalda tanto qua(n)to lui fa q(ua)n(do) el passa sopra el capo de li habitatori del p(r)imo clima Se adoncha cu(m) tuto questo nel p(r)imo clima se | può habitar come concorda tuti I phylosophi, p(er) co(n)seque(n)te etia(m)soto l'equinotial se può habitar, zoè no(n) è impazada la habitatio(n) humana p(er) excesso de | [35]caldo. E però se può co(n)cluder che tuti I zona se possono habitar, perché el zona septe(n)trional, come è noto a nui, è habitado e p(er) lo simile el zona nostro, situ|ado e posto tra el circulo de cancro et el parallelo septe(n)trional, ite(m) el zona caldo, come di sopra è argume(n)tado. Et perche el zona austral, zoè quello che | è tra el tropico de capricorno et el parallelo austral, et etia(m) quello è infra el dicto parallelo sono correspo(n)denti a q(ue)sti do' septe(n)trional habitadi, seguita. | che anchora quelli do' austral siano habitadi. Et a questa raxo(n) consente Ptholomeo nel libro de la dispositio(n) de la spera et el Magno Alberto come(n)tador | nel suo libro del la n(atur)a de i luogi e Auerois nel suo comento del libro del cielo e del mondo co(n)clude questa esser sta' ferma sente(n)tia de Aristotile phylosopho.

II. – The cosmographic lexicon of Fra Mauro

As the conclusion to the study of Fra Mauro's cosmographic conceptions developed in chapter II, I offer a vocabulary of the terms he used to define and describe the earth, the waters, and the celestial realms. The definitions of these terms refer exclusively to the way the Camaldolese friar employs them in the legends on the *mappa mundi*, and are not meant as general definitions within the context of the language of his contemporaries. On the contrary, it is presumed here that an appreciation of the specificity of Fra Mauro's cosmographic vocabulary is essential to an understanding of how he conceived, imagined, and described the celestial and sublunar worlds. The explanations summarize and organize the principal qualities that Fra Mauro attributed to the "world." The superscript numbers placed immediately after the entries refer to the legends in which the terms are found, according to the following system of numeration that corresponds to the transcriptions of such documents provided above : Doc. 1 : "Del numero de hi cieli secondo l'auctorità de hi sacri theologi" ; Doc. 2 : "De la distantia de hi cieli" ; Doc. 3 : "Come per diuina prouidentia la terra habitabile è subleuada da l'aqua" ; Doc. 4 : "De hi elementi, cioè quanto uno elemento excieda l'altro in quantità" ; Doc. 5 : "Come per la uirtù atractiua de la luna le aque cresce e descresce."

Acqua (Water)

Aque dense (dense waters) (*Mappamondo* (ed. 1956) Pl. VIII, 1) : waters located in proximity to the border between the inhabited world (which in the *mappa mundi* includes the navigable seas) and the *regione delle Tenebre* (*s.v.*). Perhaps Fra Mauro wished to allude to the way in which these waters, due to their coldness, become icy and therefore difficult to navigate, even to the point of causing the death of anyone who should venture through them.

Aque tegnente (sticky waters) (*Mappamondo* (ed. 1956) Pl. VIII, 1) : waters located in proximity to the border between the inhabited world (which in the *mappa mundi* includes the navigable seas) and the *regione delle Tenebre* (*s.v.*). With their stickiness, these waters cling to vessels and, like the *aque dense* (*s.v.*), they impede navigation, causing certain death to anyone who should venture through them.

Aque, boglimento[5] (waters, boiling of) : the phenomenon of the tides. In Fra Mauro's mind the rising of the tidal waters in the seas and the ocean was caused by the combined effects of the sun and moon. The evaporation of the *aque grosse et ponderose* (*s.v.*), that is, those waters lying in the depths of the seas, in response to the heating action of the sun and the attractive force exerted by the moon upon watery elements (*s.v. Luna, virtù atractiua*[5]) provoked the elevation of the waters. The *boglimento* (boiling) is due to the fact that the *vapori grossi* (*s.v.*) (thick vapors) are not able to escape either because of the weight of the mass of oceanic water above them, or because of their excessive intrinsic weight. To explain this phenomenon Fra Mauro made use of an example from everyday experience, namely, the effect of heat upon the water in a cauldron : "cioè cossì como ne l'olla al fuogo per li uapori leuadi dal fuogo in fondi de la olla l'aqua bolie, cossì per la eleuation de tal uapori leuadi dal fondo de le aque essa aqua boglie".[5] The probable source, albeit perhaps indirect, for Fra Mauro's concept is Albert the Great's *De causis proprietatum elementorum*.

Aque, dei fiumi[5] (river waters) : waters that are shallow and light, that is *subtili* (*s.v.*).

Aque, grosse et ponderose[5] (thick and heavy waters) : deep waters, mixed with terrestrial matter, constituting a type of muddy waters ("permixte cum molte parte terrestre") that do not evaporate easily.

Aque, subtili[5] (light water) : superficial waters, fresh or minimally saline waters, such as those in rivers, from which the *vapori* (*s.v.*) produced by solar heating easily escape.

Aque, zenith de le[5] (waters, zenith of the) : the parte of the sphere of the element of water located at the Arctic Pole.

Cielo (Heaven)

Cieli[1] (the heavens) : the entirety that is constituted by the *cielo empyreo* (*s.v.*), the *cielo acqueo* (*s.v.*) and the *cielo sydereo* (*s.v.*) including the eight planetary spheres and that of the fixed stars (a concept derived from Thomas Aquinas, *Summa theologica*, pars I, q. 68, a. IV).

Cielo[1] (heaven) : "tuto quel corpo che è sopra la terra e l'aqua" ("that entire body which is above the earth and the water").

Cielo aqueo[1] (aqueous heaven) : the second heaven, crystalline and diaphonous.

Cielo empyreo[1](empyrean heaven) : the first heaven, fiery "no(n) per ardor ma per splendor" ("not with heat, but with shimmering light").

Cielo sidereo[1] (sidereal heaven) : the starry third heaven, where the fixed stars abide ; it is divided into eight spheres of which one is the sphere of the fixed stars, and the seven others each contain one of the seven planets, giving rise to the expression "octo cieli."

Cielo, primo[1](heaven, first) : synonymous with *cielo empyreo* (*s.v.*).

Cielo, secondo[1] (heaven, second) : synonymous with *cielo acqueo* (*s.v.*).

Cielo, terzo[1] (heaven, third) : synonymous with *cielo sydereo* (*s.v.*).

Circulo[1-2] (Celestial sphere)

The sphere of the Moon, i.e., the first celestial sphere ; the sphere of Venus, i.e., the second celestial sphere ; the sphere of Mercury, i.e., the third celestial sphere ; the sphere of Mars, i.e., the fourth celestial sphere ; the sphere of the Sun, i.e., the fifth celestial sphere ; the sphere of Jupiter, i.e., the sixth celestial sphere ; and the sphere of Saturn, i.e., the seventh celestial sphere.

Elementi (the Elements)

Elementi, ci(r)co(n)ferentia del circulo del aere[4] (Elements, circumference of the circle of Air) : external dimensions of the sphere of the element Air.

Elementi, circonferentia del circulo del fuogo[4] (Elements, circumference of the circle of Fire) : maximum external dimensions of the sphere of the element Fire.

Elementi, ordine e proporzione[3] (Elements, order and proportion of) : an inherent quality in each of the five elements (Earth, Water, Air, Fire, Aether) that determines its position and relative dimensions (Fra Mauro cites Boethius, *De consolatione philosophiae*, III, 9, 3 ; *Auct. Arist.*, p. 290, n. 47).

Elementi, sustentamento[3] (Elements, sustaining of) : an intrinsic property of the elements that causes them to align themselves in the order that was conferred upon them *ab origine* at the moment of Creation, without need for the continuous intervention of God. The weight and specific qualities of earth, water, air, and fire determine the structure of the sublunary order and regulate the constant process of generation and corruption.

Elemento de la terra[4] (the element of Earth) : the first sphere of the sublunary world. Placed at the center of the world with respect to its center of gravity, but placed eccentrically with respect to its geometric center. By means of this distinction, Fra Mauro explains the emergence of the element of Earth beyond the sphere of the element of Water. He probably derive this concept from a commentary by John Buridan on Aristotle's *De caelo* (ch. VII).

Elemento del fuogo[4] (the element of Fire) : the fourth sphere of the sublunary world, which, according to Fra Mauro, is one hundred times greater (presumably in terms of volume, although the text is not clear on this point) than the sphere of the Earth.

Elemento, quantità de uno[4] (element, quantity of an) : the relative dimensions of one element with respect to the others. Fra Mauro claims that Earth and Water are equivalent ; he does not define the dimensions of the sphere of Air, and he defines the element of Fire to be one hundred times greater than that of Earth. He considers to be false the "comun et vulgar parlar" which interprets literally a passage from Aristotle's *De generatione et corruptione* (II, 6, 333a, 24-27) where the dimensions of the elements are determined by an incremental proportion of 1 to 10, beginning with the earth (E=1 ; W=10 ; A=100 ; F=1000).

Inclination natural[3] (natural inclination)

Intrinsic property of an element which determines its motion, weight, and position with respect to the other elements. In the case of mixed substances, the *inclination natural* is determined by the element of greatest proportion. Synonymous with *apetito natural* (*s.v.*).

Isole perse (lost islands) (*Mappamondo* (ed. 1956), Pl. VIII, 1)

Numerous islands located along the southern limit of the known world, designating the edge of the world's inhabitable lands and waters. These are laid out in an arc on the *mappa mundi*.

Luna (the Moon)

Luna, razi della[5] (Moon rays) : the rays through which the moon exercises an influence upon terrestrial bodies.

Luna, virtù atractiua[5] (Moon, attractive force) : attractive force exerted by the Moon on the watery and earthly elements due to the humidity in its own composition.

Mondo (the World)

Mondo[3] (world) : the universe ; the sum total of the sublunary world and the celestial world.
Mondo, centro del[3] (world, center of the) : geometric center of the universe.

Raxon (Proof)

Raxon geometricha[4] (geometric proof) : geometric demonstrations involving calculation and, in particular, the application of Euclidian theorems.

Raxon natural[3] (natural proof) : an explanation based on a reasoned argument that propounds concepts of *philosofia naturalis*.

Tenebre, regione delle (Shadows, region of the)

Tenebre, regione delle (Shadows, region of the) (VIII, 1) : the parts of the Earth that are situated beyond the *Isole perse* (*s.v.*), uninhabited and unexplored. Near the Shadows navigation is both impossible and deadly due to the extreme darkness, the *aque tegnente* (*s.v.*) and the *aque dense* (*s.v.*). A red circle, drawn about a centimeter within the perimeter of the wooden frame that encircles the *mappa mundi* appears to indicate the border between the inhabited world and the region of the Shadows.

Tenebre, densità (Shadows, density of) (VIII, 1) : Fra Mauro seems to be referring to the profound darkness that characterizes the region of the Shadows.

Terra (Earth)

Terra[3] (Earth) : the element of Earth ; a sphere situated at the *centro del mondo* (*s.v.*). The sphere of the element of Earth is not internally homogeneous. Some parts are denser and thus heavier, while the subtler and cavernous parts are lighter. The heterogeneity is the reason that the sphere of the element of Earth is located at the *centro del mondo* (*s.v.*) with respect to its center of gravity (*centro di gravità*) rather than at its geometric center (*centro di grandeza*). This also explains why parts of the sphere of the earth project beyond the *elemento acqua* (*s.v.*). Reference to this theory could have been found in John Buridan's commentary on the *De caelo*, chapt. VII.

Terra coperta[3] (covered Earth) : the part of the sphere of the element of Earth that is submerged beneath that of Water ; this part, being heavier and more compact than those parts of the element of Earth that project above the element of water is also called *terra densa* (*s.v.*).

Terra densa[3] (dense Earth) : the densest, most concentrated, part of the sphere of the element of Earth, having neither cavities nor porosity. Due to its greater weight this part of the element of Earth remains submerged beneath the element of Water. The distance between the surface of the *terra densa* and the *centro del mondo* (*s.v.*) is less than the distance between the surface of the *terra rara* (*s.v.*) and the *centro del mondo* (although Fra Mauro does not specify how much less).

Terra discoperta[3] (exposed Earth) : the part of the sphere of the element of Earth that projects beyond the sphere of the element of Water ; porous and riddled with cavities, it is composed primarily of what Fra Mauro calls *terra rara* (*s.v.*), which is lighter and less compact that the part of the element of Earth submerged below the element of Water.

Terra rara[3] (rarefied Earth) : the part of the sphere of the element of Earth that is porous and full of cavities, and therefore lighter than *terra densa* (*s.v.*) which in contrast, remains "covered" or submerged beneath the sphere of the element of Water (*s.v.*). The distance between the surface of the *terra rara* and the center of the world is greater than the distance between the surface of the *terra densa* and the center of the world (but Fra Mauro does not specify how much greater). The sphere of the element of Earth is eccentric with repect to the center of the world from a geometric perspective.

Terra, centro[3a] (Earth, center of the) : (demographic-anthropic) defined according to population densities. In the *mappa mundi* it is located at Jerusalem which, although not situated at the geometric center of the world, nevertheless defines the center of the human and created world. In this particular context *terra* signifies the inhabited world, known to man and able to support life, which projects beyond the element of Water. This part includes the navigable seas and the navigal part of the ocean. It remains within the (metaphysical) border established by the *Isole perse* (*s.v.*). Fra Mauro, aware of the spherical quality of the elements, tends to conceive of their third dimension, in other words their volume, as if it were flattened out and irrelevant to his representational project.

Terra, centro di gravità[3b] (Earth, gravitational center of) : the center of gravity of the sphere of the element of Earth which, due to its internal heterogeneity, does not coincide with its geomentric center.

Terra, centro geografico[3c] (Earth, geographic center of) : geographic center of the part of the superficial of the element of Earth's sphere that projects beyond the sphere of the element of Water. In Fra Mauro's cosmographic conception this is a place with no symbolic significance, but is simply designated by an iron pinion, from which Fra Mauro may have traced the red circle that is set one centimeter in from the wooden frame encircling his *mappa mundi*.

Terra, [centro geometrico][3d] (Earth, geometric center of) : the geometric center of the sphere of the element of Earth which, due to the uneven distribution of the weight of this element, does not coincide with its gravitational center (*s.v. terra, centro di gravità*).

Terra, circu(m)ferentia ouer sup(er)ficie de la[4] (Earth, circumference or surface of) : the astronomical measurement of the entire collective body of Water and Earth, conceived as a sphere whose center coincides with the center of the world.

Terra, parte ne la qual nui habitemo[4] (Earth, the part of which we inhabit) : the part of the sphere of the element of Earth that projects above water ; it appears that Fra Mauro thought of this part as a kind of swelling, or *tumor*. Cfr. "la qual essendo eleuada da la sup(er)ficie de l'aqua seguita che'l dicto nu(mer)o de mesura de la t(er)ra sia mazor ouer alme(n) equal cum la dime(n)sio(n) de la circu(m) fere(n)tia de l'aqua" ("which being elevated above the surface of the water, it follows that the aforementioned measurement of the earth is greater, or at least equal to, the dimensions of the surface of the water"), being a literal translation of Averroes' commentary on Aristotle's *Meteorologica* : see Aristotle / Averroes, *Comm. Met.* 2.2 'De Ventis,' 1562, ff. 438-441. "Et. cum hoc positum sit ita, sequitur ut sit maior pars terrae discooperta : nec esset diameter aquae maior diametro terrae, sed erit minor, aut aequalis eis" (f. 439 I).

Vapori (Vapors)

Vapori, corrupti[5] (Vapors, corrupt) : vapors mixed with parts of the elements of Earth and Water.

Vapori, grossi[5] (Vapors, thick) : vapors mixed with parts of the elements of Earth and Water.

CHAPTER III – DOUMENTARY APPENDIX

THE LEGEND DESCRIBING THE EARTHLY PARADISE

Doc. 1. Del sito del paradiso terrestro. Rubrica. *Mappamondo,* lower left corner.[1]

Del sito del paradiso terrestro. Rubrica. | El paradiso de le delicie no(n) solamente ha sentimento spiritual ma etia(m) quello esser uno luogo ne la terra si|tuado mette s(an)c(t)o Augustino sopra el Genesis et ancora nel libro De ciuitate Dei el qual luogo è molto remoto da la habi|tatio(n) e cog(ni)tion humana, posto ne le parte oriental, segu(n)do la doctrina del sacro doctor Beda p(er) la cui auctorità el mai|stro da le sente(n)cie tal oppinio(n) aferma, auegna ch'el come(n)tator Alberto Magno nel libro de la na(tura) di luogi metta quello oltra el circ(u)lo | equinotia(l), pur ne la regio(n) oriental. Nel qual paradiso fo posto el primo nostro progenitor Adam nel stato de la sua innocentia, e | nel mezo de quello è uno fonte p(er) irrigatio(n) del dicto luogo, dal qual poi procede i quatro grandi e principal fiumi, et etia(m)dio | in questo paradiso Idio su(m)mo auea piantado nel mezo do' arbori, uno di qualli era l'arboro de la scientia del ben e del mal, zoè | quello del fructo del qual Dio fece comandame(n)to ha Adam che no(n) ne ma(n)zasse dal qual coma(n)dame(n)to parte(n)dosse Adam per inobe|die(n)tia, zoè ma(n)zando del fructo, oltra el bene che prima lui auea ueduto e sentito, lui p(er) esperimento cognoscete el male, cadendo | ne li diffecti che seguitano el peccato dal qual aduenime(n)to quello arboro è stato nominato de scientia de bene e de male. L'altro fo l'ar|boro de uita, el qual ma(n)zato hauea proprietà de co(n)seruar l'omo in uno stato per alqua(n)to tempo. Onde essendo sta' continuato el man|zar de questo arboro l'omo seria co(n)tinuame(n)te uiuesto, come dechiara s(an)c(t)o Augustino nel libro De ciuitate Dei e anchor nel libro de le q(ue)|stio(n) del Nuouo et Vechio Testame(n)to. In questo paradiso se crede esser posto Enoch q(ua)n(do) fo translatado, de la cui translatio(n) è sc(ri)pto nel li|bro de Genesis al quarto cap(itu)lo, e ancora nel libro del Ecclesiastico. E simelme(n)te se crede esser el proph(et)a Helia da poi la sua tra(n)slatio(n), de la | qual se fa memoria nel quarto libro de li Re. Qui ancora pietosame(n)te se crede fosseno le anime de li s(an)c(t)i padri uisitade dal n(ost)ro Rede(m)ptor | el uenerdì de la sua passio(n) da poi la morte sua, cu(m) li qual se crede l'anima del nostro Redemptor unita a la diuinità esser stata nel limbo fina al zorno de la sua resurrection, dal qual luogo quelli padri liberadi dal uinculo del peccato original, come parla sancto Augustino nel sermon de la passion, fono posti in questo paradiso, nel qual fin al zorno de la ascension lor steteno, cumciosiaché Christo fosse el primo che nel cielo e paradisi de li beati intrasse monstrando a li homini la uia de andar a quello nel dì de la sua ascension, come predisse el propheta Michea. Et auegna che'l Redemptor nostro per la passion e morte sua meritasse a la humana natura lo introito del paradiso de li beati, ampuo li piaque per la ascension indur le anime predicte in possession del dicto luogo de beatitudine za a li homini aperto per li meriti de la sua passion. E però da questo paradiso de le delicie se crede le anime sancte predicte ascen-desse in cielo el zorno de la ascension seguendo Christo ampuo uisibilmente dal monte Oliueto ascese in cielo, ne la costa del qual monte era situado el castello ouer uila de Betania, come narra el sacro doctor Beda. Perché la sacra scriptura fa mention del fonte del paradiso et ancora de queli quatro fiumi che nasceno da quello, però ne la presente pictura se demostra questo, e perché sono molti che se merauegliano come sia possibile che questi quatro fiumi nassando da questo luogo remotissimo habi li suo fonti distantissimi l'uno da l'altro, respondo secondo sancto Augustino sopra el Genesi, che questi fiumi, hi fonti di qual ne sono noti, habino luogi subterranei e passando molte region nassano in diuerse parte e che uno cioè Ganges nassa in India, Tygris in Armenia dal monte Charabach, Eufrates similiter in Armenia apresso la cità Arçeron, et Gen ouer Nilo in Ethyopia in Abassia, in la prouincia de Meroa.

[1] "The location of the earthly paradise." Cf. *Mappamondo* (ed. 1956), p. 22. For the English translation see *Mappamondo* (ed. 2006), pp. 729-731.

Partial transcription of MS. It. VI, 24 (= 6111) of the Biblioteca Nazionale Marciana in Venice

For their documentary value it is worth to transcribe the list of the provinces in Europe, Africa and Asia described and represented in the MS. It. VI, 24 (= 6111). The original form of the toponyms from the incipit of the brief descriptions placed inside each individual map is used (ff. 17vb-46ra). Following is the transcription of the procedure to draw the Ptolemaic projections in the Venetian vernacular (ff. 90va-92vb ; f. 87va).

List of the provinces

Europa

Ing[h]elterra isola de Britania (f. 17v, col. b) ;
Spagna, divi-|sa in tre parte principale, c[i]oè Belgiticha, Lusitania e Taraconense (f. 18r, col. a) ;
Aragon, Catalogna e reame de Valenza (f. 18r, col. b) ;
Franc[i]a chiamata Galia (f. 18v, col. a) ;
Seconda parte de Galia si è Narbonense (f. 18v, col. b) ;
Galia Belgicha (f. 19r, col. a) ;
Carta di Francia (f. 19v) ;
Germania la granda (f. 20r, col. a) ;
Rezia e Vindelizia (f. 20r, col. b) ;
Prussio che verso la Frisa Menore iuxta el fiume | Seguita Pannonia Superiore (f. 20v, col. a) ;
Illiris (f. 21r, col. a) ;
Dalmatia (f. 21r, col. b) ;
Italia, vincitrice de ogni altra gente (f. 21v, col. a) ;
Sardegnia insola marina (f. 22r, col. b) ;
Insola appellata Cirno (f. 22r, col. b) ;
Sicilia insola notabelissima (f. 22v, col. a) ;
Missia Superiore nel reame de Boemia (f. 23r, col. a) ;
Missina Inferior (f. 23r, col. b) ;
Jacige Metanasti è propria quella parte che anchora | Ungaria appellemo (f. 23v, col. a) ;
Dacia nel mar Occeano(f. 23v, col. b) ;
Chersoneso isola è nel mar de Prussia ditta Gotlant (f. 24r, col. a) ;
Sarmatia (f. 24r, col. b) ;
Tauricha Chersoneso è quella parte che nui | appellemo el mar de la Tana (f. 24v, col. a) ;
Tracia (f. 24v, col. b) ;
Macedonia (f. 25r, col. a) ;
Epiro parte de Grecia sopra el mar Adria|ticho (f. 25v, col. a) ;
Achaia si è la prima parte de Grecia (f. 26r, col. a) ;
Euboyca si è una insola posta nel mar Egeo (f. 26r, col. b) ;
Peloponeso (f. 26v, col. a) ; Candia isola notabile (f. 26v, col. b).

Africa

Mauritania Tingi-|tana, (f. 27v, col. a) ;
Mauritania Cesariense (f. 27v, col. b) ;
Numidia (f. 28 r, col. a) ;
Affrica (f. 28r, col. b) ;
Cirenaica (f. 28v, col. a) ;
Marmarica (f. 28v, col. b) ;
Libia Cirenense (f. 29r, col. a) ;
Egipto (f. 29r, col. b) ;
Egipto Thebais, c[i]oè Superior (f. 29v, col. a) ;
Libia sopra de Affrica o Libia de Ethiopia (f. 29v, col. b) ;
Ethiopia sopra Egipto (f. 30r, col. a) ;
Ethiopia (f. 30r, col. b).

Asia

Bitimia (f. 32r, col. b) ;
Cine ‹overo Phrigia› (f. 32v, col. a) ;
Phrigia Granda (f. 32v, col. b) ;
Licia posta nel mar Egeo e Cretico (f. 33r, col. a) ;
Panfilia sul Mar Nostro posta (f. 33r, col. b) ;
Galacia (f. 33v, col. a) ;
Paflagonia over Pissidia (f. 33v, col. b) ;
Capadocia (f. 34r, col. a) ;
Armenia Menore (f. 34r, col. b) ;
Cilicia (f. 34v, col. a) ;
Sarmatia, contigua al fonte de Tana-|is, (f. 34v, col. b) ;
Cholchos reame de re Octa (f. 35r, col. a) ;
Iberia fra Colchos et Albania (f. 35r, col. b) ;
Albania (f. 35v, col. a) ;
Armenia Magiore (f. 35v, col. b) ;
Cipro (f. 36r, col. a) ;
Prima Siria (f. 36r, col. b) ;
Seconda Siria, ditta Fenicea (f. 36v, col. b) ;
Terza Siria, che Judea overo Pa-|lestina è nomatta (f. 36v, col. b) ;
Arabia Petrea (f. 37r, col. a) ;
Mesopotania (f. 37r, col. b) ;
El mar Persico (f. 37v, col. a) ;
Babilonia antiqua (f. 37v, col. b)
Assiria contigua a Eufrate, Tigre e Gorgon (f. 38r, col. a) ;
Susiana propinqua al mar Persico (f. 38r, col. b) ;
Media (f. 38v, col. a) ;
Persia (f. 38v, col. b) ;
Hircania (f. 39r, col. a) ;
Margiana (f. 39r, col. b) ;
Bactriana provintia (f. 39v, col. a) ;
Sogdiana (f. 39v, col. b) ;
Partia (f. 40r, col. a) ;

Carmania Deserta (f. 40r, col. b) ;
Carmania posta fra i mon-|ti Persici e i monti ditti Semiramis posta | in sul mar Indico (f. 40v, col. a) ;
Arabia Felice (f. 40v, col. b) ;
Aria regione (f. 41r, col. a) ;
Dragiana (f. 41r, col. b) ;
Gedrosia (f. 41v, col. a) ;
Aracosia (f. 41v, col. b) ;
Trapobane (f. 42r, col. a) ;
Sachea regione (f. 42v, col. a) ;
Sitia intra el monte Ymao (f. 43r, col. a) ;
Paropansada (f. 43v, col. a) ;
India intra el fiume Gange (f. 44r, col. a) ;
Sitia for del monte Imao ultima provintia | habitabile de l'Asia (f. 44v, col. a) ;
India oltra el Gange (f. 45r, col. a) ;
Serica regione (f. 45v, col. a) ;
Sinay regione, ultima habitabile ne l' | oriente (f. 46r, col. a).

First projection (plate VI.1)

(f. 90v, col. a)

Per voler venir hora a la descriptione in | forma piana, quella dechiarerò quanto meglio | porò, a ciò che niuna cossa inexperta remanga | de la predicta descriptione. Dico adunque formata | per sé una linea diritta che sia lunga parte | clxv, overo divisa in tante parte ; quella se | divida in dui parte e l'una sia de xxxxiiii parte, | l'altra sia ‹cento e una› cxxxi, et a la fine de quella de | le xxxiiii parte formaremo una linea per | contario, distante ingualmente da l'una e | l'altra parte, e cussì una altra equale a la | fine de la linea divisa per cxxxi ; e quelle due | linee se vol congiunger insieme con una altra | linea per metà a le ditte due linee. Verbi | gratia sia ab la linea de le clxv parte, | cd ef le due linee interseccante gh per | la mittà de cd ef serà quella che ‹per la› | ligarà quelle dui in ‹modo› quadrangulo, e cussì | serà ‹v› ii quadranguli recti ; po' se prenda | ‹per cons› compasso da l'a al b, e tirisse una linea | flexa per fino al quadro che fia ik, e quella | fia la linea per Siene principio del primo | clima che vede el nostro polo. Unde perché | la distantia è clxxx gradi da uno po-|lo a l'altro, e questa de la forma in piano | è clxv, devemo comprender manifesta-|mente quindese gradi oltra verso l'austro | ogni uncha parte habitabile non vedere | el nostro polo, e cussì in figura piana | non poter meter regula de quello ; e ducta | la predicta linea da l'i a l'a e dal k al a, | protraheremo due linee in forma de triangulo, | el qual triangulo mostrerà el nostro habitabile | da b fino l ; puo' per la dicta linea da l' | i verso a in distantia de gradi xvi 1/3 1/12 | traremo una altra linea, intersecando la | linea alb verso la linea ak, e quella | fia la linea equinociale da la qualle prende-|remo ogni mesura per distantia dal punto | occidentale, che sarà mn, e da questa linea | mn, lassatto el spacio de gradi xxxvii, | una altra linea in similitudine de mn tra-|remo, che serà la linea per Rodo notata | da op. Un'altra distante linea da op | verso la ‹tiramento› gradi xxvii, tiraremo | una linea che fia per la isola de Tille, che fia | qr per fina a la linea cd che da questa dista |

(f. 90v, col. b)

gradi xviii, et in questa figura porre-|mo tutto el nostro habitabile comune. | Ponere ordenando la distantia da l'equi-|nociale punto l'alteza da l'a e volse | alora dividere da l'a a l'i e da l'i al | k e dal k a l'a in ccclx parte, le quale | po' ne indica l'alteza dal polo ; et azò | che più chiara vengha la cossa, quella po-|rò per figura, po' la cui positione an-|cora la dechiarerò altramente. |

Né però questa figura fia tutta habitabile, | onde per vedere la vera parte de lie habitabile | se convien formar una altra figura | quatrangula como questa, e far in quella el | triangulo ‹che comenci da l'equinociale›, se per | altra ‹p› rasone non se volesse procedere, | como tocherò de sotto ; e per meglio dechi[a]-|rare porrò la figura. |

(f. 91r, col. a)

Vedemo questa dopea figura sopra scripta | menar una linea da l'a al b da l'una | e l'altra figura e dal c al d, benché | dicha alchuni doverse menare a la | sumità de la linea dividente l'uno e l'al-|tro quatrangulo, perché la parte australle | molto meno habitabile è che la parte | septentrionale. O veramente se volemo per al-|tra rasione fare la ditta figura | triangulare ‹per una altra rasio› for-|maremo una linea tuta diritta in | parte clxv divisa, de la quale faremo | quinque parte. La prima parte continerà in sé | xxxiiii parte, la seconda ‹di› xviii, la | terza xxvii, la quarta xxxvi, la quinta | xvi 1/3 1/12,[1] como de fora per exemplo è ; nel ca-|po de sotto de la linea del xxxvi tra-|heremo una figura distante da quella | da una parte e l'altra do parte e meza, | sì che tutta integra fia v parte, et a la | fine de cadauna de le parte sì redure-|mo una linea da la sumità de la | linea de 34 da intrambe le parte, e ne | la fine de la linea xvi 1/3 1/12 quatro parte | e meza ‹da quelle v› e una terza de le v | sopraditte, reducendo però una linea | picola ‹da› che congionga queste due ; a la | linea del xxvii daremo parte iiii in equa-|le mesura de le v, sì che quella linea sia | una parte magiore de questa ; a la linea | del 18 daremo do parte e una quarta de | le v parte ; e tutte queste linee da l'una | parte a l'altra termenerà in le due linee | che tendeno in zoso da la linea del 34 | a la linea sotto el 36 inarcata per dritta | misura. E per questa rasione, vogliando ‹nu› | nui formar uno universo in figura | piana, non ne fia besogno formar | el quatrangulo ; o veramente forma-|mo la equinociale[2] ‹prima› seconda linea gradi xxx, la | sotto vol ‹e›essere xxviiii, la sopra la | equinociale lv, overo xi[3] siando la equinociale lx | overo xii, la ultima sopra sia gradi viiii, | siando la equinociale gradi xx ; e viene la |

(f. 91r, col. b)

linea equinociale tanto fora de la line-|a de clxv, per redure le due linee | che ha a farre lo triangulo quanto è alta | la linea del cxxxi, como in la presente | figura se demostra.

Questa si è la figura de l'univer-|so nostro habitabile in forma piana | dedutta con le misure sue, de la quale | non altramente dico ‹che de s› de la | situatione sua che ho ditto de | sopra nella figura emisperia, | salvo che questa prendi la sua alteza | dal polo, ove la emisperiale la | prendi da l'equinociale e partisse el dri-|tto triangulo de questa che è abc in | ccclx ‹parte› gradi, i quai se attribu-|isse ai ‹dì› giorni ; e dividesse in parte | xii, che sono xxx per parte, che ànno | a formare xii mesi de l'anno a | xxx dì l'uno, e vedemo pure al-|chuno averne xxxi, uno xxviii | et a la fi[n]a xxix, i quai tutti computa-|ti fano ccclxv e hore e ponti, e ne | la spera ‹non sono›, overo nel tondo | celeste, non sono più che ccclx ; dico | esser vero, ‹b› perché a uno continuo, con | equal tempo se duseno le hore, che sono | a l'anno in questo modo : viii^{m} vii^{c} lxxxx | e ponti xii ; ma le hore equinociale |

(f. 91v, col. a)

che più portano de tempo che le altre, se-|condo el corso loro, hanno retiratto al | diriettto hore cl e punti xii, che fano | vi dì e vi hore, como se pol mani-|festamente vedere nella spiera. E per | questa ragione

[1] Il numero de le parti contenute nelle cinque parti è ripetuto a margine in numeri arabi : *34 / 18 / 27 / 36 / 16 1/3 1/12*.

[2] Aggiunta a margine : *equinociale*.

[3] Aggiunta a margine : *overo XI*.

consideroreno li mensura-|tori del ciello el spacio de le hore equi-|nociale più longo esser che le altre, | per tanto posero solo ‹di› ccclx gradi, | benché ‹'l con› el contrario vediamo, per effetto | el sole levarse ccclxvi fiatte a l'‹ano› | anno. Nella qual rasone, perché tocha | altre parte che 'l proposito mio, non me | extenderò più oltra, vogliando che | basti questa dechiaratione quanto a la parte | predicta. Vogliando omai dar fine | a l'opera mia quanto a la descriptione de l' | universo, po' el seguente capitolo, per venire | ‹a la› a una certa descriptione de nuo-|ve gente de aspri costumi e varie | in tutto da nui, como per alchuni mo-|derni scriptori ho trovatto, benché | per auctorità alchuna non siano proba-|ti, ‹p› ma perché pure alchuna vag[h]e-|za serà ad odire, tocherò quello più | me parerà de tochar. | | Vidutte et examinatte tutte le soprascripte | figure, potemo comprendere manifesta-|mente ogni termene del nostro universo | habi-tabile redutto in figura piana esser | quello paralello, overo linea sotto lo | equinociale, più verso l'austro, secondo alch-|uni gradi xvi 1/3 1/12, altri sono che xvii | compiti pongono ; e questi gradi sono de tan-|ta quantità, qaunta è quelli del cerchio grando | che prendeno circa xxv e mezo de quelli | de ‹l'equ› la linea equinociale, che se parti in | clxxx ; e l'altro paralello per Meroe ‹cotan› | cotanti gradi più septentrionale che 'l equinociale, e 'l | paralello più septentrionale termenase ‹el paralello | più› da l'equinociale verso septentrione gradi | lxiii ; e questi gradi non sono equali a li | gradi del cerchio, perché verreberro

(f. 91v, col. b)

fuora del cerchio, ma sono de quelli de la | linea equinociale non archata, dutta da occi-|dente in oriente, divisa per clxxx parte ‹e | que›, perché questo paralello se mena per la isola | Tille, che non serebbe ai primi gradi del | cerchio. Unde se comprende l'abitabile comune | del ‹nostro› sito a nui senza dificultà noto | sia gradi 80, che fano stadii xl^{m}, co-|mo ho ditto de sopra de ‹a› la rasone dei | stadi per grado ; el circuito de l'habitabile | nostro si è stadi clxxx^{m}, che fa ccclx | gradi ; la ultima oriental parte del | nostro habitabile noto a nui termena el | merediano ducto per ‹Sina regione› Ti-|ne, metropoli de la regione de' Sini, di-|stante dal merediano descripto per Alexandria | gradi 119 1/2 ; e ponino alchuni questi 119 | gradi e 1/2 far octo hore equinociale. | La parte occidentale termena el mere-|diano dutto per le Insole Fortunatte | e questo dista dal merediano dutto per | Alexandria gradi lx 1/2, che valeno hore | quatro equinociale, dal semicirculo | orientale gradi clxxvi, contro quelli che di-|cono clxxx, perché Barbario promontorio | de Portugalia, ultimo termene habitabile | nostro verso occidente, dista dal punto ‹g› | gradi iiii. Onde se conchiude la longitudine | de l'habitabile per la linea equinociale ultima sia | gradi clxx, contro la opinione de quelli che | dicono esser clxxx, ma non excipi-|no la parte de oriente, né de occidente | inhabitabile.‹Ancora nel paralello per Rodo | nel qual spesso se prendeno le mesure›.

Second projection

‹Tutto quello che è scripto de sopra del modo | e dispositione universale del sito de la terra | habi-tabile a nui noto ànose con iusta men-|sura, ma adurò qui una altra raso-|ne : in che modo lo emisperio de la terra | contento sotto spera circulare se possa | descrivere in plano. Descrivasse in pla-|no una+ spera circulare, et dentro ve sia la | terra compresa ; l'aspetto de l'ochio averà› |

(f. 92r, col. a)

Avanti che ad altra parte io pervegna, farò | una repilogatione de tutto l'abitabile, | reducendo la terra in spera circulare, de | la qual de sopra assai evidentemente | è demostrato. Proponasse una spera | circulare in plano, che comprenda in sé la | terra ; l'ispetto del circulo habia el suo | sito, per lo qual sia fatta una linea drita | in parte comune merediane, el qual | aspetto passi per li segni topici, e che | parti la longitudine del nostro universo habitabile | el paralello notato in terra per Siene, che | divide la latitu-

+ Addition in the margin : *una*.

dine. Cussì ancho se con-|vien avere le rasone de la grandeza | de la spera circulare e de la terra e de l'a-|bundantia de quello medesimo aspetto, che da | la distantia che ‹è intra› se coglie iuxta lo equinociale e 'l tropi-|co extivo verso septentrione, tutta quella parte | del noto sito sia manifesta al semicirculo | australe, per mezo el zodiacho constitu-|to sopra la terra, azò che oltra quello | non se faci adicione veruna de l'habitabile | nostro a l'emisperio boreale protratto da | lì. Questi suppositi li circuli merediani | de una dritta linea, secondo quello axile, | presteno fantasia, como cadente in quello | aspetto per quelli merediani in plano, | e 'l paralello per Siene diritto sia a quella li-|nea con simile rasone ; li altri se descriva | flexi in forma de archo e tendano a quelle | due diritte linee, li merediani alla line-|a che passa per li poli, li paralelli a quella | che passa per Siene, e più questi che da | quelle linee distano, che lingieramente | se demostrerà. Sia el merediano ‹i› | che passa per li segni equinociali nella circula-|re spera abcd circa el mezo e 'l di-|ametro sia ae(f. Intendemo a nel | polo articho, c ne l'antarticho ; tol-|lemo bf et gd et kh et dr de la cir-|cumferentia nelle distantie dei tropici | da l'equinociale a et l et am et nc et go | nelle distantie dei poli artico et antartico, | e dividasse el diametro extivo ag iuxta | | (f. 92r, col. b) : | | p, convinendo esser posto el paralello per Siene intra | e et p ; e la rasone della circunferentia | dal paralello per Siene a l'equinociale como quatro | a xv, la mità de ep ad ea de quelli 4 | sia quasi ad xv, serà ea epiciclo de la | linea dirita stante nel centro. Prendasse | eq de tale tre, quale quatro sono ‹quelle› | ea, et in lo centro e descrivasse le distantie | de ep et in quello[5] plano circundi la terra uno | circulo qr e divisa ‹una› la diritta li-|nea e in lxxxx equale portione a uno | tetartunoro [...] ; se lasserano | ha portione de ep, ha e rs portione | de xvi 1/3 1/12 e rt de quelli medesimi lxiii, | e producasse ‹s› xsy sopra la cadente sopra | la diritta linea, secondo el paralello per Sie-|ne. Adunque s serà quel per lo quale se scriverà | el paralello seccante el fine australle | habitabile e l'opposito al paralello per Meroe | v, per lo qual se descriverà el paralello terme-|nante el fine articho descripto per Tille ; | e prendasse alchuna cossa più au-|strale che z, e giungasse za, e pro-|dutte ‹za› sy et zd cadano ‹in quelle› in ll. | Se adunque nui intenderemo li expositi | circuli in piano per li segni dei tropici | e dei poli e l'aspetto da ll ns dk | et oe produtte ad ac farano le sectione | de quella, per la quale se descriverà a l'aspe-|to de l'ochio sectione de ‹v› cinque | paralelli ad z, per la qual se descriverà quello | che ne in ne l'equinociale circa el d, quella che | se lung[h]erà da ll ad d et b et f et h et | g ‹g› farano quelle medesime sezzione, | zoè ad ae, per le qual se designerà la | fine de la terra e le portione de quelli paralelli. | Ma se nei paralelli che si descriverano in | terra torremo q et prime absentie de l' | equinociale, como z et t, le sectione facte | da quelli juncti diritti da ll de quello se-|micirculo qyr e le opposite de quelli | secondo li siti equidistanti, averemo sectione | da descrivere a l'aspetto de l'ochio dei | v paralelli ad ll, per lo qual se descriverà | lo equinociale che è cercha d ; ma quella | | (f. 92v, col. a) ; | | da ll ad d et b et f et h s se giungeno, | farano quelle medexime sictione che ac | portione dei predicti paralelli v a t b et | c t d a, da le quale, togliando le distan-|tie dai merediani ordenati da l'u-|na e l'altra parte de a v et xy dritta | nelle prime rasone dei tre paralelli, scriveremo | per simile tre punti, portione dei subiecti | merediani, como terminanti la longitudine | de e y f g ; la multitudine dei paralelli | da scrivere in terra serà desiderata a la | grandezza de la descriptione. Nella de-|ductione dei circuli ‹quello› è da observare | che cadauno passi per quatro assumpti et | in pictura, overo in accuto, nella sec-|tione a l'extivo circulo finiscano, a | ciò che non induca la fantasia contra | l'ordine. Ancora qui è da servar equa-|lità per quello che seguiterano, benché fo-|ra del circulo continente quella pictura cada-|no curvità ‹fini› che finisca quello de-|fecto con lo circulo, perché in quella vera | imagine parerà cadere. Ancora è | de considerare ‹qu› che quelle sole linee sia-|no quelli medesimi circuli con alchuna | commensurabile latitudine e colore deferente, | ma azò che le portione che oltra la | terra se porrano habiano più obscu-|ri colori che le portione ‹ingualmente› | insieme

[5] Addition in the margin : *et*.

incidente, da le più propinque se ta-|ino per le aditione equalmente de le | vere imagine ‹ne› in quei circuli e quella terra, | salvo questo che 'l cerchio dei segni so-|pra la terra ‹cada› passi[6] iuxta lo semicirculo | hiemale. ‹Intercidasse da quella secon-|do la borealle e per quello extivo tropico | ascriviamo e le proprie demostratione a | questi nei opportuni luochi et in quelli | circuli de la terra numeri de le distantie | e de le hore de quella habitabile, cercha lo | cerchio exteriore dei venti ; ingualmente | darase lì la demostratione, como nella | spera circulare ‹ad› ai adiacenti cinque | paralelli e poli›.

Third projection

(f. 92v, col. b)

Epilogo de questa prefata extensione secondo Ptholomeo | apparerà la descriptione de ‹circulare› la spera circula-|re in plano ‹con› la terra compresa in quello avendose. | Se gli ochii nostri a l'aspetto figuri in lo si-|to secondo le quale ne le comune sectione el | merediano circulo, passando per li punti dei | tropici, serà la linea diritta, el qual circulo | parte la longitudine de l'habitabile nostro. Ma | el paralello descripto per Siene, el qual circulo quasi | partisse la latitudine de quello habitabile equalmente | se se averano le rasone de le grandeze | de la spera e de quella terra e de le distantie | de quello aspetto che nel spacio che la | cade intra el circulo iuxta lo equinotiale | e lo circulo iuxta lo tropico extivo tu-|ta nota a nui parerà habitabile ‹el›. Constituto el | cerchio più australe per mezo del circulo | significherò sopra la terra, a ciò che so-|pra questo non se faci alchuna aditione | de la nostra habitabile, che se extenda a l'arti-|co emisperio, dal quale li mezi secondo quello | axille de una diritta linea faciano | fantasia, como aspetto cadente in piano | per quelli paralelli e' l paralello che se descriverà di-|ritto per Siene ancora per simile casione | apparà ; tutto lo resto dei circuli da notare | versi per curvitate apparano a quelle dri-|tte linee ; li merediani a quella che pa-|ssa per li polli ; li paralelli a quella che per | Siene e quelli più che da l'una e l'altra | parte da quella più distante, como el circulo | articho più che l'extivo che 'l tropicho | più chini al septentrione che 'l hiemale tropico | e l'equinociale più al mezodì sia in-|chinato e l'antartico più che quello hie-|male tropico ; e per più chiarezza redu-|rò questa figura in forma, a ziò che la | sua obscura traditione per la figura | sia compresa, et in quella se fia fine quanto | a la descriptione de l'universo.

First projection "in forma sferica"

(f. 87v, col. a)

Secondo li diversi scriptori diversamente è dechi-|aratta la figura precedente, perché alchuni | voleno le linee meredional che sono | da l'uno polo a l'altro xxxvi, cominc[i]a | dal septentrione e venga flectendosse verso l'austro ; | alchuni voleno ‹e contro› e contro. Alchuni vo-|leno che cominci dal punto occidenta-|le verso lo oriente ; alchuni è contro. E puo' | li paralelli traversar quelle con la loro | distantia sopraditta dal punto mezo de l'|equinociale, cominc[i]ata da uno punto | de l'uno dei poli e finisca sempre ne l' | altro, aducendo che quando con suo mesura | è adutto doveunque se incrosa con | la linea meredionale, lì vengha luo-|cho sotto mesura ordinato, la quale | de lieve se puol vedere. E chiamano | le linee meredionale, perché tutte se | conflecteno verso el mezodì ; e perché in questa | forma, dutte le linee da qual canto se vu-|ole con li paralelli, non cade difficultà, per in-|tanto non curo de soprastare, perché da | sé instessa se dechiara, et è questa una | facile forma a la descriptione de l'‹universo› | universo. Alchuni sono che, non ponendo | linee merediane tutto lo circulo impie | de' paralelli da ogni punto cominc[i]ati e fi-|nitti ‹per› ne lo suo contrario tra uno tropico e | l'altro, dicono in ogni + cader luocho per | li men-

[6] Correction in the space between lines : *passi* instead of *cada*.

suratori notato ‹Né te inganasse | però le line meredionale poste in tanta di-|stantia, ma con cadeno fra cadauna | de quelle do altre›. E dà a vedere questa | ‹mesura› figura[7] non essere differentia alchuna | de ‹qu› lie a quella sopra posta in forma triangulare, | perché questa in forma sperica, quella in forma | plana. Sono etiamdio alchuni che | formando l'universo quello in circulo sperico | forma quadri da l'uno tropico a l'altro ;| questi pono vignire partendo per xxxvi, co-|mo sono le linee meredionale ; e quelle | linee cussì quadre dicono esser linee me-|redionale, et in cadauno tondo figura-|no o cità o provintia o altra parte datta | per li mensori ; la qual figura, per più evidentia

Alchuni prendeno le | ditte linee meredio-|nale da l'axille fra | uno e l'altro polo, co-|mo nella predicta figura | appare, e per quelle reduse-|no li paralelli como | è ditto.

(f. 87v, col. b)

porrò qui de sotto per maor speculatione | de chiunque a tal misure vorà pervenire.

[7] Correction in the space between lines : *figura* instead of ‹*mesura*›.

CHAPTER VI – DOCUMENTARY APPENDIX

I. – List in alphabetical order of the authors and works cited explicitly by Fra Mauro with the transcription of the legends in which they are quoted

Here below is a list in alphabetical order of the authors and works cited explicitly by Fra Mauro. In those cases where he noted only the author without also citing a title, or, inversely, when he cited a title without naming the author, these missing authors and titles have been supplied here between parentheses whenever it was possible to identify the lacking element with reasonable probability.

Albertus Magnus (1200-1280), *De causis et proprietatibus elementorum.*

Et a questa raxo(n) co(n)sona l'auctorità del philosopho i(n) div(er)si luogi ma più exp(re)ssa nel libro de le proprietà e n(atur)a di eleme(n)ti e Alberto Mag(n)o | nel suo simel libro et Averois come(n)tator nel comento del libro del cielo e del mondo et Avicena nel suo libro di canoni de medicina. Vero è che alguni, come | l'auctor de la spera, mete co(n)traria opinione la qual etiam fo de Socrates e de alcuni altri autori.[1]

Albertus Magnus (1200-1280), *De natura loci.*

Et a questa raxo(n) consente Ptholomeo nel libro de la dispositio(n) de la spera et el Magno Alberto come(n)tador | nel suo libro de la n(atura) de i luogi e Averois nel suo comento del libro del cielo e del mondo co(n)clude questa esser sta' ferma sente(n)tia de Aristotile philosopho.[2]

El paradiso de le delicie no(n) solamente ha sentimento spiritual ma etia(m) quello esser uno luogo ne la terra si|tuado mette s(an)cto Augustino sopra el Genesis et ancora nel libro De Civitate Dei, el qual luogo è molto remoto da la habi|tation e cog(n)ition humana, posto ne le parte oriental, sego(n)do la doctrina del sacro doctor Beda p(er) la cui auctorità el mai|stro da le sente(n)cie tal oppinion aferma, avegna ch'el come(n)tator Alberto

[1] "Come la terra supposita a l'equinoctial e a | la torrida çona è abitabile. Rubrica," *Mappamondo*, lower right corner (left). *Mappamondo* (ed. 1956), p. 23. "All of this agrees with what Aristotle says in various places, but above all in his *De proprietatibus elementorum*; what Albertus Magnus says in his book on the same subject [*De causis proprietatum elementorum*]; what Averroës says in his commentary on [Aristotle's] *De Caelo*; what Avicenna says in his *Canon medicinae*. It is true that some, such as the author of The Sphere [Johannes de Sacrobosco, *Tractatus de sphaera*] are of the opposite opinion, as was Socrates and other writers." The English translation derives from *Mappamondo* (ed. 2006), p. 734.

[2] "Come la terra supposita a l'equinoctial e a la torrida çona è abitabile. Rubrica" *Mappamondo*, lower right corner (left). *Mappamondo* (ed. 1956), p. 24. "This is affirmed by Ptolemy in the "De la dispositione de la spera" and Albertus Magnus, the commentator, in his *De natura loci*. And in his commentary to *De caelo*, Averroës concludes that this was Aristotle's firm opinion". This is my English translation. For a different version, see *Mappamondo* (ed. 2006), p. 735. Falchetta's identification of Ptolemy's *De la dispositione de la spera* with the *Geography* is wrong.

Magno nel libro de la n(atur)a di luogi metta quello oltra el circ(u)lo | equinotial, pur ne la regio(n) oriental.[3]

Arrianus, Lucius Flavius (*c.* 86-*post* 146), *Anabasis Alexandri*: Book VIII (*Indica*).

Nota che alguni istoriographi, di quali sono Ariano e Soli(n) et etiam Dionisio, el qual fo mandado da Re Philadelpho ad in|vestigar de l'india, questi dicono de le indie molte cosse nota|bile, e de mo(n)ti e de fiumi che sono ganges, el qual è largo dove | men viij dove più xx mia et ha pessi gra(n) dissimi de più sor|te, maximame(n)te anguile, le q(u)al, come dice Statio, | Thebaidas, sono longe 300 piè [...].[4]

San Hieronymo nel prohemio | de la Bibia dice che Apolonio phy|osopho passò q(ue)sto fiume phison | e vene ad alguni pop(o)li diti bragma|ni p(er) aldir Archan [...]. Q(ue)sto dico p(er)ché | questi p(o)p(u)li [...] a nui no(n) sono noti mol|to, p(er)hò de li lor costumi et habiti | discrepatissimi chi ne vol haver notitia | leça Ariano e Strabo, i qual de questa india | scriveno copiosamente.[5]

Aristoteles, *De generatione et corruptione*.

Al qual parlar par co(n)sentir el dicto del philosopho, che dixe | nel libro de le generatio(n) e corruptio(n) che de uno pugno de t(er)ra se genera diexe pugni de aqua | e cussì m(u)ltiplicando fin al fuogo.[6]

[3] "Del sito del paradiso terrestro. Rubrica" *Mappamondo*, lower left corner. *Mappamondo* (ed. 1956), p. 22. "The location of the Earthly Paradise. Rubric. The Paradise of Delights does not only have a spiritual meaning; it is also a real place on the earth, as St. Augustine says in his *De Genesi* and in his book *De Civitate Dei*. This place is very far from all human settlements and knowledge; and according to the teachings of the holy doctor Bede, whose authority is also followed by the Master of the Sentences, it is to be found in the East. In his book *De natura loci*, Albertus Magnus also puts it in the East, beyond the circle of the equinox," *Mappamondo* (ed. 2006), p. 729.

[4] *Mappamondo* (ed. 1956) Pl. XXI, 57. "Note that some historiographers, such as Arrian, Solinus and even Dionysius (who was sent by King Philadelphus to investigate India) say many notable things of the Indies, of its mountains and of its rivers. There is the Ganges, which at its narrowest is eight miles wide and at its widest twenty; this contains very large fish of various sorts, including eels which Statius, in the *Thebaidas*, says are 300 feet long." This translation differs slightly from the one in *Mappamondo* (ed. 2006) *779.

[5] *Mappamondo* (ed. 1956) Pl. XXVI, 18. "In the Proem to his Bible, St. Jerome says that Apollonius the Philosopher crossed this river Phison to travel to certain peoples called *Bragmani* and hear Archan [...] I say this because these peoples [...] are not very well known to us. Those who want information on their very different customs and practices should read Arrianus and Strabo, who write extensively about this India," *Mappamondo* (ed. 2006) *1380.

[6] "De hi elementi, cioè quanto uno elemento excieda l'altro in quantità. Rubrica," *Mappamondo*, right lower corner (right). *Mappamondo* (ed. 1956), p. 22. "Of the Elements. That is, how much one element exceeds another in quantity. [...] This opinion would seem to be agreed with by a comment made by the Philosopher [Aristotle], who in his book *On Generation and Corruption* says that from a handful of earth ten handfuls of water are generated, multiplying by tens in this way up to fire," *Mappamondo* (ed. 2006), p. 731.

Aristoteles, *Meteorologica*.

[...] e queli che de q(ue)ste cosse mirabile e de al|tre mo(n)struose desiderano esser copiosi, leçano | Julio Soli(n) polistoro, Po(m)ponio Mela, s(an)c(t)o Agustino, Alber|to Magno, s(an)c(t)o Thomaso de Aqui(n), nel libro co(n)tra i cu|riosi, ite(m) le Metaure d'Aristotile e Plinio de le me|raveie del mo(n)do e vederano che de mile cosse non ne ho dite una.[7]

Aristoteles (Pseudo), *De proprietatibus elementorum*.

Et a questa raxo(n) co(n)sona l'auctorità del philosopho i(n) div(er)si luogi ma più exp(re)ssa nel libro de le proprietà e n(atur)a di eleme(n)ti e Alberto Mag(n)o | nel suo simel libro et Averois come(n)tator nel comento del libro del cielo e del mondo et Avicena nel suo libro di canoni de medicina. Vero è che alguni, come | l'auctor de la spera, mete co(n)traria opinione.[8]

Averroes (1126-1198), *Commentarium in Aristotelis De caelo et mundo*.

Et a questa raxo(n) co(n)sona l'auctorità del philosopho i(n) div(er)si luogi ma più exp(re)ssa nel libro de le proprietà e n(atur)a di eleme(n)ti e Alberto Mag(n)o | nel suo simel libro et Averois come(n)tator nel comento del libro del cielo e del mondo et Avicena nel suo libro di canoni de medicina. Vero è che alguni, come | l'auctor de la spera, mete co(n)traria opinione.[9]

Et a questa raxo(n) consente Ptholomeo nel libro de la dispositio(n) de la spera et el Magno Alberto come(n)tador | nel suo libro de la n(atura) de i luogi e Averois nel suo comento del libro del cielo e del mondo co(n)clude questa esser sta' ferma sente(n)tia de Aristotile philosopho.[10]

Augustinus (354-430), *De civitate Dei*.

El paradiso de le delicie no(n) solamente ha sentimento spiritual ma etia(m) quello esser uno luogo ne la terra si|tuado mette s(an)cto Augustino sopra el Genesis et ancora nel libro De Civitate Dei, el qual luogo è molto remoto da la habi|tation e cog(n)ition humana, posto ne le parte oriental, sego(n)do la doctrina del sacro doctor Beda p(er) la cui auctorità el mai|stro da le sente(n)cie tal oppinion

[7] *Mappamondo* (ed. 1956) Pl. XXX, 120: "[...] Those who wish to have plentiful information on these wonderful and these monstrous things should read Julius Solinus's *Polyhistor*, Pomponius Mela, St. Augustine, Albertus Magnus and St. Thomas Aquinas in the *Book against the curious*. Similarly, they can read Aristotle's *Meteorology* and Pliny on the wonders of the world, and they will see thousands of things of which I have not mentioned one." This is my translation of the legend. For a different version, see *Mappamondo* (ed. 2006) *2212.

[8] "Come la terra supposita a l'equinoctial e a | la torrida çona è abitabile. Rubrica." *Mappamondo*, lower right corner (left). *Mappamondo* (ed. 1956) p. 23; *Mappamondo* (ed. 2006) p. 734. *Vd. supra* n. 2.

[9] "Come la terra supposita a l'equinoctial e a | la torrida çona è abitabile. Rubrica." *Mappamondo*, lower right corner (left). *Mappamondo* (ed. 1956) p. 23; *Mappamondo* (ed. 2006) p. 734. *Vd. supra* n. 2.

[10] "Come la terra supposita a l'equinoctial e a la torrida çona è abitabile. Rubrica" *Mappamondo*, lower right corner (left). *Mappamondo* (ed. 1956) p. 24. For the English translation *vd. supra* n. 1015. For a different version, see *Mappamondo* (ed. 2006) p. 735.

aferma, avegna ch'el come(n)tator Alberto Magno nel libro de la n(atur)a di luogi metta quello oltra el circ(u)lo | equinotial, pur ne la regio(n) oriental.[11]

Onde essendo sta' continuato el man|zar de questo arboro l'omo seria co(n)tinuame(n)te vivesto, come dechiara s(an)c(t)o Augustino nel libro De Civitate Dei e anchor nel libro de le q(ue)|stio(n) del Nuovo et Vechio Testame(n)to.[12]

Alguni scrive che a le radice del monte caspio over pocho lutan sono queli populi i qual co|me se leçe sono seradi per Alexandro Macedo. Ma certo questa opinio(n) manifestamente è erronea | e da non esser sostenuta p(er) algun modo [...] P(er)hò io me acosto a la auctorità de sanc|to Augustino, el q(u)al nel suo De Civitate Dei reprova la opini|on de queli che dicono che gog e magog | significa q(ue)li populi che darano favor | ad antichristo et i(n) q(ue)sta sente(n)cia etia(m) co(n)corda Nicolò de Lira, el q(u)al | expone questi do' nomi secondo la he|braica verità.[13]

Augustinus (354-430), *Quaestiones evangeliorum or Quaestiones in Heptateuchum?*

Onde essendo sta' continuato el man|zar de questo arboro l'omo seria co(n)tinuame(n)te vivesto, come dechiara s(an)c(t)o Augustino nel libro De Civitate Dei e anchor nel libro de le q(ue)|stio(n) del Nuovo et Vechio Testame(n)to.[14]

Augustinus (354-430), *De passione Domini.*

[...] dal qual luogo q(ue)lli padri liberadi dal vinculo del peccato original, come parla s(an) c(t)o Augusti(n)o nel sermo(n) de la pas|sion, fono posti i(n) questo paradiso, nel qual fin al zorno de la asce(n)sion lor steteno.[15]

[11] "Del sito del paradiso terrestro. Rubrica," *Mappamondo*, lower left corner. *Mappamondo* (ed. 1956) p. 22; *Mappamondo* (ed. 2006) p. 729. *Vd. supra* n. 5.

[12] "Del sito del paradiso terrestro. Rubrica," *Mappamondo*, lower left corner. *Mappamondo* (ed. 1956) p. 22. "In fact, if this had been his constant nourishment, man would have lived forever, as St. Augustine says in the book *De civitate Dei* and also in the book *Le question del Nuovo et Vechio Testamento*," *Mappamondo* (ed. 2006) pp. 729-730. Falchetta identifies "Le question del Nuovo et Vechio Testamento" with *De diversis quaestionibus*. I do not agree on this choice. See the entry "Augustinus (354-430), *Quaestiones evangeliorum* or *Quaestiones in Heptateuchum?*".

[13] *Mappamondo* (ed. 1956) Pl. XXXIII, 77: "Some write that on the slopes of Mount Caspian, or not far from there, live those peoples who, as one reads, were shut in by Alexander the Macedonian. But this opinion is certainly and clearly mistaken and cannot be upheld in any way [...] So, I am not differing here from the authority of St. Augustine, who in his *De Civitate Dei* rejects all the opinions of those who claim that Gog and Magog are the peoples that will support the Antichrist. And Nicholas of Lyra agrees with this claim, explaining the two names by their Hebrew origin [...]," *Mappamondo* (ed. 2006) *2403.

[14] "Del sito del paradiso terrestro. Rubrica," *Mappamondo*, lower left corner. *Mappamondo* (ed. 1956) p. 22. *Quaestiones evangeliorum cum appendice quaestionum 16. in Matthaeum*, edidit Almut Mutzenbecher. Turnhaut: Brepols 1980; Id., *Questionum in Heptateuchum libri septem*, in Id., *Opera Omnia*, editio latina, PL 34.

[15] "Del sito del paradiso terrestro. Rubrica," *Mappamondo*, lower left corner. *Mappamondo* (ed. 1956) p. 22; "[...] and that the Fathers were freed from the bonds of Original Sin and placed in this Paradise until the day of the Ascension, as St. Augustine says in his sermon *De passione Domini*," *Mappamondo* (ed. 2006) p. 730.

Augustinus (354-430), *De Genesi ad litteram.*

El paradiso de le delicie no(n) solamente ha sentimento spiritual ma etia(m) quello esser uno luogo ne la terra si|tuado mette s(an)cto Augustino sopra el Genesis et ancora nel libro De Civitate Dei, el qual luogo è molto remoto da la habi|tation e cog(n)ition humana, posto ne le parte orien-tal, sego(n)do la doctrina del sacro doctor Beda p(er) la cui auctorità el mai|stro da le sente(n)cie tal oppinion aferma, avegna ch'el come(n)tator Alberto Magno nel libro de la n(atur)a di luogi metta quello oltra el circ(u)lo | equinotial, pur ne la regio(n) oriental.[16]

Avicenna (Abu 'Ali al-Husayn ibn 'Abd Allah ibn Sina, 980-1037), *Canon medicinae.*

Et a questa raxo(n) co(n)sona l'auctorità del philosopho i(n) div(er)si luogi ma più exp(re)ssa nel libro de le proprietà e n(atur)a di eleme(n)ti e Alberto Mag(n)o | nel suo simel libro et Averois come(n)tator nel comento del libro del cielo e del mondo et Avicena nel suo libro di canoni de medicina. Vero è che alguni, come | l'auctor de la spera, mete co(n)traria opinione.[17]

Basilius Caesariensis (Magnus) (330-379), [*Hexaemeron*].[18]

Del numero de hi cieli secondo l'auctorità de hi sacri theologi. | Circa el numero dei cieli fra gli sacri doctori par che'l ne sia qualche diversità et maximamente fra Basilio et Chrisostomo, el qual Chrisostomo dice che'l non è altro che uno cielo. [...] Ma Basilio et Damasceno che'l siegue dichono esser molti cieli, ma questa diversità è più ne la voce che ne la chosa, cioè nel modo del significar che nel signficato. Et benchè Chrisostomo dica che'l sia uno cielo, per questo lui nomina uno cielo tuto quel corpo che è sopra la terra e l'aqua, perché etiam li oseli che volano per l'aiere sono dicti oseli del cielo. Ma perchè in questo corpo sono molte distinction, però Basilio mete molti cieli. [...] Ma perchè questo spatio contien doi elementi, cioè el fuogo e l'aiere, et in cha-dauni de questi do' fi notado la superior et inferior regioni però Rabano ae distincto questo cielo in quatro [...].[19]

16 "Del sito del paradiso terrestro. Rubrica." *Mappamondo*, lower left corner. *Mappamondo* (ed. 1956) p. 22 ; *Mappamondo* (ed. 2006) p. 729. *Vd. supra* n. 5.

17 *Vd. supra* n. 2.

18 Though the legend refers to the *Hexaemeron*, Fra Mauro is not quoting Basilius Magnus : he is quoting and translating the *articulus IV* of the *quaestio* 68 of the *pars prima* of St. Thomas Aquinas' *Summa theologiae*.

19 "Del numero de hi cieli secondo l'auctorità de hi sacri theologi."*Mappamondo*, upper left corner (left). *Mappamondo* (ed. 1956) p. 19. "The authority of the holy theologians as to the number of the heavens. With regard to the number of the heavens there is some difference between the holy doctors, above all between Basil the Great and John Chrysostom, who himself says the heavens are one and single. Although the Scriptures write *coeli coelorum*, this is because of the characteristics of the Hebrew language, which refers to the heavens and sky in the plural ; just as there are a number of words in Latin that do not have a singular. Basil and John Damascene, who follows him, claim that there are many heavens. However, this difference is more a question of words than substance – that is, more of language than of meaning. John Chrysostom does say that the heavens are one and single, but he uses this name for everything above the earth and water, for the same reason that the birds [of the sky] are said to be the birds of the heavens. [...] But as this space contains two elements – that is, fire and air – and each of these belongs to a distinct region (the upper and the lower), Rabanus Maurus divides the heavens into four [...]." *Mappamondo* (ed. 2006) p. 720.

Beda Venerabilis (673?-735), [*Super Genesim*]?

El paradiso de le delicie no(n) solamente ha sentimento spiritual ma etia(m) quello esser uno luogo ne la terra si|tuado mette s(an)cto Augustino sopra el Genesis et ancora nel libro De Civitate Dei, el qual luogo è molto remoto da la habi|tation e cog(n)ition humana, posto ne le parte oriental, sego(n)do la doctrina del sacro doctor Beda p(er) la cui auctorità el mai|stro da le sente(n)cie tal oppinion aferma, avegna ch'el come(n)tator Alberto Magno nel libro de la n(atur)a di luogi metta quello oltra el circ(u)lo | equinotial, pur ne la regio(n) oriental.[20]

Boethius, Anicius Manlius Severinus (*c.* 480-524), *De consolatione philosophiae.*

[…] No(n) bixogna adoncha alguno sustenta|mento a la terra ma li basta el suo natural apetito, al qual sequita l'ordine de la propor|tione e (con)nexion insita da Dio ne li eleme(n)ti, chomo dice Boetio nel libro de la philosophi|ca consolatione.[21]

Dionysius Periegetes (sec. II), *De situ habitabilis orbis.*

Nota che alguni istoriographi, di quali sono Ariano e Soli(n) et etiam Dionisio, el qual fo mandado da Re Philadelpho ad in|vestigar de l'india, questi dicono de le indie molte cosse nota|bile, e de mo(n)ti e de fiumi che sono ganges, el qual è largo dove | men viij dove più xx mia et ha pessi gra(n)dissimi de più sor|te, maximame(n)te anguile, le q(u)al, come dice Statio, | Thebaidas, sono longe 300 piè […].[22]

Euclides (*c.* 300 bc), *Elementa.*

Ma la raxon geometricha a q(ue)sto parlar co(n)tradice […] come chiaramente è provado nel duodecimo libro | de Euclides a la propositio(n) secu(n)da. Onde per q(ue)sta regula, trovati quatro circ(u)li […] seguita che questo quar|to circulo cioè el fuogo fosse mazor del primo cioè de la t(er)ra mille fiade p(er) la propositio(n) XVI del qui(n)to | libro de Euclides.[23]

20 "Del sito del paradiso terrestro. Rubrica," *Mappamondo*, lower left corner. *Mappamondo* (ed. 1956) p. 22; *Mappamondo* (ed. 2006) p. 729. *Vd. supra* note 5.

21 "Come per divina providentia la terra habitabile è sublevada da l'aqua," in *Mappamondo*, upper right corner (right); *Mappamondo* (ed. 1956) p. 21. "How by Divine Providence the earth is raised above the water. […] All it requires is its natural inclination, which results from the order of proportion and relation that God has instilled in the elements, as Boethius claims in his book *De consolatione philosophiae*," *Mappamondo* (ed. 2006) p. 726.

22 *Mappamondo* (ed. 1956) Pl. XXI, 57; "Note that some historiographers, such as Arrian, Solinus and even Dionysius (who was sent by King Philadelphus to investigate India) say many notable things of the Indies, of its mountains and of its rivers. There is the Ganges, which at its narrowest is eight miles wide and at its widest twenty; this contains very large fish of various sorts, including eels which Statius, in the *Thebaidas*, says are 300 feet long. This translation differs slightly from the one in *Mappamondo* (ed. 2006) *779.

23 "De hi elementi, cioè quanto uno elemento excieda l'altro in quantità. Rubrica," *Mappamondo*, right lower corner (right). *Mappamondo* (ed. 1956) pp. 22-23. "Of the Elements. That is, how much one element exceeds another in quantity. […] But this argument is contradicted by geometrical calculation […] This is clearly demonstrated in the second proposition of Book Twelve of Euclid [*The Elements*]. On the basis of this rule, if you have four spheres between which there is this tenfold ration […] the consequence is that the fourth sphere, fire, is one thousand times

Fazio degli Uberti (1305-*post* 1367), *Il Dittamondo*.

Alguni autori sc(r)ive del mar d'i(n)dia che'l sia serado come un stagnon | e che'l mar occean no(n) li entri, ma Solin vol che'l sia occea(n) e ch(e) q(ue)la p(ar)te austral | e del garbi(n) sia navigabile, et io affermo che algune nave açira e vol|ta quel cami(n). E questo a(n)chor (con)ferma Plinio qua(n)do el dice che al te(m)po | suo do' nave se mosse del mar de arabia, e dice la cason la qual q(ui) lasso, | ma charge de spetie ziroe q(ue)le p(ar)te perfin in spagna e a cibelter descar|gò. Ancora Facio afferma q(ue)sto et simelme(n)te hi exp(er)imentadori de quel | cami(n), ho(min)i de gra(n) prudentia i qual co(n)corda co(n) queli autori.[24]

Hieronymus (*c.* 347-420), *Epistola ad Paulinum presbyterum*.

San Hieronymo nel prohemio | de la Bibia dice che Apolonio phy|osopho passò q(ue)sto fiume phison | e vene ad alguni pop(o)li diti bragma|ni p(er) aldir Archan [...] Q(ue)sto dico p(er) ché | questi p(o)p(u)li [...] a nui no(n) sono noti mol|to, p(er)hò de li lor costumi et habiti | discrepatissimi chi ne vol haver notitia | leça Ariano e Strabo, i qual de questa india | scriveno copiosamente.[25]

Hrabanus Maurus (780-856), *De universo*.[26]

Del numero de hi cieli secondo l'auctorità de hi sacri theologi. | Circa el numero dei cieli fra gli sacri doctori par che'l ne sia qualche diuersità et maxima|mente fra Basilio et Chrisostomo, el qual Chrisostomo dice che'l no(n) è altro che uno cielo. [...] Ma Basilio (et) Damasceno che'l siegue di|chono esser molti cieli, ma questa diuersità è più ne la voce che ne la chosa, cioè nel modo del | significar che nel significato. Et benchè Chrisostomo dica che'l sia uno cielo, per questo lui nomi|na uno cielo tuto quel corpo che è sopra la terra e l'aqua, p(er)chè etiam li oseli che uolano per l'aiere sono dicti oseli del cielo. Ma perchè in questo corpo sono molte distinctio(n), però Basilio me|te molti cieli. [...]. Ma perchè questo spatio co(n)tien doi elementi, cioè el fuogo e l'aiere, (et) in

greater than the first sphere. This by the sixteenth proposition in Book Five of Euclid [*The Elements*]," *Mappamondo* (ed. 2006) p. 731.

[24] *Mappamondo* (ed. 1956) Pl. XI, 26 ; "Some authors write that the Sea of India is enclosed like a pond and does not communicate with the ocean. However, Solinus claims that it is itself part of the ocean and that it is navigable in the southern and south-western parts. And I myself say that some ships have sailed it along that route. This is confirmed by Pliny when he says that in his day two ships loaded with spices coming from the Sea of Arabia sailed around these regions to Spain and unloaded their cargo at Gibraltar (he gives the reason for this choice of route, but I omit it here). Fazio [Degli Uberti] says the same ; and those who have taken this route, men of great prudence, agree with these writers," *Mappamondo* (ed. 2006) *53.

[25] *Mappamondo* (ed. 1956) Pl. XXVI, 18 ; "In the Proem to his Bible, St. Jerome says that Apollonius the Philosopher crossed this river Phison to travel to certain peoples called *Bragmani* and hear Archan [...] I say this because these peoples [...] are not very well known to us. Those who want information on their very different customs and practices should read Arrianus and Strabo, who write extensively about this India," *Mappamondo* (ed. 2006) *1380.

[26] Though the legend refers to the *De universo*, Fra Mauro is not quoting Hrabanus Maurus : he is quoting and translating the *articulus IV* of the *quaestio* 68 of the *pars prima* of St. Thomas Aquinas' *Summa theologiae*.

chadauni de questi | do' fi notado la superior et inferior regioni però Rabano ae distincto questo cielo in quatro [...].[27]

Iohannes Chrysostomus (*c.* 347-407), [*Homiliae super Genesim*].

Del numero de hi cieli secondo l'auctorità de hi sacri theologi. | Circa el numero dei cieli fra gli sacri doctori par che'l ne sia qualche diuersità et maxima|mente fra Basilio et Chrisostomo, el qual Chrisostomo dice che'l no(n) è altro che uno cielo. [...] Ma Basilio (et) Damasceno che'l siegue di|chono esser molti cieli, ma questa diuersità è più ne la voce che ne la chosa, cioè nel modo del | significar che nel significato. Et benchè Chrisostomo dica che'l sia uno cielo, per questo lui nomi|na uno cielo tuto quel corpo che è sopra la terra e l'aqua, p(er)chè etiam li oseli che uolano per l'aiere sono dicti oseli del cielo. Ma perchè in questo corpo sono molte distinctio(n), però Basilio me|te molti cieli. [...]. Ma perchè questo spatio co(n)tien doi elementi, cioè el fuogo e l'aiere, (et) in chadauni de questi | do' fi notado la superior et inferior regioni però Rabano ae distincto questo cielo in quatro [...].[28]

Iohannes Damascenus (*c.* 675-749), *De fide orthodoxa*.[29]

Del numero de hi cieli secondo l'auctorità de hi sacri theologi. | Circa el numero dei cieli fra gli sacri doctori par che'l ne sia qualche diuersità et maxima|mente fra Basilio et Chrisostomo, el qual Chrisostomo dice che'l no(n) è altro che uno cielo. [...] Ma Basilio (et) Damasceno che'l siegue di|chono esser molti cieli, ma questa diuersità è più ne la voce che ne la chosa, cioè nel modo del | significar che nel significato. Et benchè Chrisostomo dica che'l sia uno cielo, per questo lui nomi|na uno cielo tuto quel corpo che è sopra la terra e l'aqua, p(er)chè etiam li oseli che uolano per l'aiere sono dicti oseli del cielo. Ma perchè in questo corpo sono molte distinctio(n), però Basilio me|te molti cieli. [...]. Ma perchè questo spatio co(n)tien doi elementi, cioè el fuogo e l'aiere, (et) in chadauni de questi | do' fi notado la superior et inferior regioni però Rabano ae distincto questo cielo in quatro [...].[30]

Iohannes de Sacrobosco († 1256), *De sphaera*.

Et a questa raxo(n) co(n)sona l'auctorità del philosopho i(n) div(er)si luogi ma più exp(re)ssa nel libro de le proprietà e n(atur)a di eleme(n)ti e Alberto Mag(n)o | nel suo simel libro et Averois come(n)tator nel comento del libro del cielo e del mondo et Avicena nel suo libro di canoni de medicina. Vero è che alguni, come | l'auctor de la spera, mete co(n)traria opinione.[31]

[27] "Del numero de hi cieli secondo l'auctorità de hi sacri theologi." *Mappamondo*, upper left corner (left). *Mappamondo* (ed. 1956), p. 19. For the English translation see *Mappamondo* (ed. 2006), pp. 720-722. *Vd. supra* n. 1029.

[28] *Vd. supra* n. 18.

[29] Though the legend refers to the *De fide orthodoxa*, Fra Mauro is not quoting Iohannes Damascenus: he is quoting and translating the *articulus IV* of the *quaestio* 68 of the *pars prima* of St. Thomas Aquinas' *Summa theologiae*.

[30] *Vd. supra* n. 18.

[31] *Vd. supra* n. 1.

Pseudo-Messalla, *De progenie Augusti Caesaris* (*recte saec.* xv).

Circa la division de la terra, çoè de la affrica da l'a|sia e similiter de la europa da l'asia, ne trovo apresso | cosmographi et istoriographi diverse opinio(n) de le qual se poria parlar diffusamente, ma per esser materia | tediosa a demorar i(n) q(ue)sta controversia farò qui un poco de no|ta de le opinio(n) de questi e q(ue)lo se de' tignir lasserò eleçer ai pru|denti. Alguni ch(e) siegue li antichi, di quali son Messala orator | che sc(r)ive la p(ro)genie de Otavia(n) Augusto e Po(m) ponio Mela e q(ue)li che'l siegue, vuol che'l nilo divida l'asia da l'affrica, et thanai la euro|pa. Alguni dice che Tolomeo vuol che q(ue)la costa de mo(n)ti de arabia, | che sono da ladi de nubia e tirano p(er) abassia e oltra q(ue)la ethyopia austral, | faça la division de l'affrica [...].[32]

Petrus Lombardus (*c.* 1095-*c.* 1160), *Sententiae in IV libris distinctae.*

El paradiso de le delicie no(n) solamente ha sentimento spiritual ma etia(m) quello esser uno luogo ne la terra si|tuado mette s(an)cto Augustino sopra el Genesis et ancora nel libro De Civitate Dei, el qual luogo è molto remoto da la habi|tation e cog(n)ition humana, posto ne le parte oriental, sego(n)do la doctrina del sacro doctor Beda p(er) la cui auctorità el mai|stro da le sente(n) cie tal oppinion aferma, avegna ch'el come(n)tator Alberto Magno nel libro de la n(atur)a di luogi metta quello oltra el circ(u)lo | equinotial, pur ne la regio(n) oriental.[33]

Pomponius Mela (I *saec.* ce), *Cosmographia.*

[...] e queli che de q(ue)ste cosse mirabile e de al|tre mo(n)struose desiderano esser copiosi, leçano | Julio Soli(n) polistoro, Po(m)ponio Mela, s(an)c(t)o Agustino, Alber|to Magno, s(an) c(t)o Thomaso de Aqui(n) nel libro co(n)tra i cu|riosi, ite(m) le Metaure d'Aristotile e Plinio de le me|raveie del mo(n)do e vederano che de mile cosse non ne ho dite una.[34]

Circa la division de la terra, çoè de la affrica da l'a|sia e similiter de la europa da l'asia, ne trovo apresso | cosmographi et istoriographi diverse opinio(n) de le qual se poria parlar diffusamente, ma per esser materia | tediosa a demorar i(n) q(ue)sta controversia farò qui un poco de no|ta de le opinio(n) de questi e q(ue)lo se de' tignir lasserò eleçer ai pru|denti. Alguni ch(e) siegue li antichi, di quali son Messala orator | che sc(r)ive la p(ro)genie de Otavia(n) Augusto e Po(m)ponio Mela e q(ue)li che'l siegue, vuol che'l nilo divida l'asia da l'affrica, et thanai la euro|pa. Alguni dice che

[32] *Mappamondo* (ed. 1956) Pl. XXXIV, 79. "Regarding the divisions of the world – that is, of Asia from Africa and of Europe from Asia – cosmographers and historiographers give various opinions. Of these one could discuss at length, but because it is boring to dwell on this controversy, I will make a brief note with their opinions and leave the prudent to decide which one they should hold as best. Some follow the Ancients – for example, Messala, the orator who wrote of the family of Octavian Augustus, and Pomponius Mela and those who followed him. They claim that it is the Nile which divides Asia from Africa and the Thanai Asia from Europe. Some claim that Ptolemy says that the chain of the Arabian Mountains, which lies to the side of Nubia and extends through Abassia and beyond southern Ethiopia, constitutes the division between Asia and Africa. [...]." *Mappamondo* (ed. 2006) *2489.

[33] "Del sito del paradiso terrestro. Rubrica." *Mappamondo*, lower left corner. *Mappamondo* (ed. 1956) p. 22; *Mappamondo* (ed. 2006) p. 729. *Vd. supra* n. 5.

[34] *Mappamondo* (ed. 1956) Pl. XXX, 120; for the English translation *vd. supra* n. 12. For a different version see *Mappamondo* (ed. 2006) *2212.

Tolomeo vuol che q(ue)la costa de mo(n)ti de arabia, | che sono da ladi de nubia e tirano p(er) abassia e oltra q(ue)la ethyopia austral, | faça la division de l'affrica [...].[35]

Nicolaus de Lyra (1270-1339) *Postilla super totam Bibliam.*

Alguni scrive che a le radice del monte caspio over pocho lutan sono queli populi i qual co|me se leçe sono seradi per Alexandro Macedo. Ma certo questa opinio(n) manifestamente è erronea | e da non esser sostenuta p(er) algun modo [...] P(er)hò io me acosto a la auctorità de sanc|to Augustino, el q(u)al nel suo De Civitate Dei reprova la opini|on de queli che dicono che gog e magog | significa q(ue)li populi che darano favor | ad antichristo et i(n) q(ue)sta sente(n)cia etia(m) co(n) corda Nicolò de Lira, el q(u)al | expone questi do' nomi secondo la he|braica verità.[36]

Plinius Secundus, Gaius (23?-79) *Historia naturalis.*

[...] e queli che de q(ue)ste cosse mirabile e de al|tre mo(n)struose desiderano esser copiosi, leçano | Julio Soli(n) polistoro, Po(m)ponio Mela, s(an)c(t)o Agustino, Alber|to Magno, s(an)c(t) o Thomaso de Aqui(n), nel libro co(n)tra i cu|riosi, ite(m) le Metaure d'Aristotile e Plinio de le me|raveie del mo(n)do e vederano che de mile cosse non ne ho dite una.[37]

Alguni autori sc(r)ive del mar d'i(n)dia che'l sia serado come un stagnon | e che'l mar occean no(n) li entri, ma Solin vol che'l sia occea(n) e ch(e) q(ue)la p(ar)te austral | e del garbi(n) sia navigabile, et io affermo che algune nave açira e vol|ta quel cami(n). E questo a(n)chor (con)ferma Plinio qua(n)do el dice che al te(m)po | suo do' nave se mosse del mar de arabia, e dice la cason la qual q(ui) lasso, | ma charge de spetie ziroe q(ue)le p(ar)te perfin in spagna e a cibelter descar|gò. Ancora Facio afferma q(ue)sto et simelme(n)te hi exp(er)imentadori de quel | cami(n), ho(min)i de gra(n) prudentia i qual co(n)corda co(n) queli autori.[38]

[35] *Mappamondo* (ed. 1956) Pl. XXXIV, 79. "Regarding the divisions of the world – that is, of Asia from Africa and of Europe from Asia – cosmographers and historiographers give various opinions. Of these one could discuss at length, but because it is boring to dwell on this controversy, I will make a brief note with their opinions and leave the prudent to decide which one they should hold as best. Some follow the Ancients – for example, Messala, the orator who wrote of the family of Octavian Augustus, and Pomponius Mela and those who followed him. They claim that it is the Nile which divides Asia from Africa and the Thanai Asia from Europe. Some claim that Ptolemy says that the chain of the Arabian Mountains, which lies to the side of Nubia and extends through Abassia and beyond southern Ethiopia, constitutes the division between Asia and Africa. [...]," *Mappamondo* (ed. 2006) *2489.

[36] *Mappamondo* (ed. 1956) Pl. XXXIII, 77; "Some write that on the slopes of Mount Caspian, or not far from there, live those peoples who, as one reads, were shut in by Alexander the Macedonian. But this opinion is certainly and clearly mistaken and cannot be upheld in any way [...] So, I am not differing here from the authority of St. Augustine, who in his *De Civitate Dei* rejects all the opinions of those who claim that Gog and Magog are the peoples that will support the Antichrist. And Nicholas of Lyra agrees with this claim, explaining the two names by their Hebrew origin [...]," *Mappamondo* (ed. 2006)*2403.

[37] *Vd. supra* n. 7.

[38] *Mappamondo* (ed. 1956) Pl. XI, 26; "Some authors write that the Sea of India is enclosed like a pond and does not communicate with the ocean. However, Solinus claims that it is itself part of the ocean and that it is navigable in the southern and south-western parts. And I myself say that some ships have sailed it along that route. This is confirmed by Pliny when he says that in his day two ships loaded with spices coming from the Sea of Arabia sailed

Ptolemaeus, Claudius (*fl.* 127-148), [*Almagestum* ?].

Et a questa raxo(n) consente Ptholomeo nel libro de la dispositio(n) de la spera[39] et el Magno Alberto come(n)tador | nel suo libro de la n(atura) de i luogi e Averois nel suo comento del libro del cielo e del mondo co(n)clude questa esser sta' ferma sente(n)tia de Aristotile philosopho.[40]

Ptolemaeus, Claudius (*fl.* 127-148), *Cosmographia.*

Alguni autori sc(r)ive del mar d'i(n)dia che'l sia serado come un stagnon | e che'l mar occean no(n) li entri, ma Solin vol che'l sia occea(n) e ch(e) q(ue)la p(ar)te austral | e del garbi(n) sia navigabile, et io affermo che algune nave açira e vol|ta quel cami(n). E questo a(n) chor (con)ferma Plinio qua(n)do el dice che al te(m)po | suo do' nave se mosse del mar de arabia, e dice la cason la qual q(ui) lasso, | ma charge de spetie ziroe q(ue)le p(ar)te perfin in spagna e a cibelter descar|gò. Ancora Facio afferma q(ue)sto et simelme(n)te hi exp(er)imentadori de quel | cami(n), ho(min)i de gra(n) prudentia i qual co(n)corda co(n) queli autori.[41]

Circa la division de la terra, çoè de la affrica da l'a|sia e similiter de la europa da l'asia, ne trovo apresso | cosmographi et istoriographi diverse opinio(n) de le qual se poria parlar diffusamente, ma per esser materia | tediosa a demorar i(n) q(ue)sta controversia farò qui un poco de no|ta de le opinio(n) de questi e q(ue)lo se de' tignir lasserò eleçer ai pru|denti. Alguni ch(e) siegue li antichi, di quali son Messala orator | che sc(r)ive la p(ro)genie de Otavia(n) Augusto e Po(m)ponio Mela e q(ue)li che'l siegue, vuol che'l nilo divida l'asia da l'affrica, et thanai la euro|pa. Alguni dice che Tolomeo vuol che q(ue)la costa de mo(n)ti de arabia, | che sono da ladi de nubia

around these regions to Spain and unloaded their cargo at Gibraltar (he gives the reason for this choice of route, but I omit it here). Fazio [Degli Uberti] says the same; and those who have taken this route, men of great prudence, agree with these writers." *Mappamondo* (ed. 2006) *53.

[39] *De dispositione sphaerae* of Ptolemy "qui est introductorius ad Almagesti," according to Albert the Great, *De natura loci*, Tract. I, chapter 7, p. 13, line 52; Avicenne; and Averroes "Averroes enim in Commento super quartum librum caeli et mundi dicit Aristotelis et suam fuisse opinionem, quod loca illa sint habitabilia, et rationem adducit id necessario probantem, ut mihi videtur: dicit enim inter calidissimum et frigidissimum esse necessario temperatum." see Albertus Magnus, *De natura loci*, Tract. I, chapter 7, *Utrum habitabilis sit quarta terrae, quae est ab aequinoctiali usque in polum australem*, p. 13, lines 1-7.

[40] "Come la terra supposita a l'equinoctial e a la torrida çona è abitabile. Rubrica." *Mappamondo*, lower right corner (left). *Mappamondo* (ed. 1956) p. 24; "This is affirmed by Ptolemy in the "De la dispositione de la spera" and Albertus Magnus, the commentator, in his *De natura loci*. And in his commentary to *De caelo*, Averroës concludes that this was Aristotle's firm opinion. This is my English translation. For a different version, see *Mappamondo* (ed. 2006) pp. 733-734.

[41] *Mappamondo* (ed. 1956) Pl. XI, 26; "Some authors write that the Sea of India is enclosed like a pond and does not communicate with the ocean. However, Solinus claims that it is itself part of the ocean and that it is navigable in the southern and south-western parts. And I myself say that some ships have sailed it along that route. This is confirmed by Pliny when he says that in his day two ships loaded with spices coming from the Sea of Arabia sailed around these regions to Spain and unloaded their cargo at Gibraltar (he gives the reason for this choice of route, but I omit it here). Fazio [Degli Uberti] says the same; and those who have taken this route, men of great prudence, agree with these writers." *Mappamondo* (ed. 2006) *53.

e tirano p(er) abassia e oltra q(ue)la ethyopia austral, | faça la division de l'affrica [...].[42]

Unde se algun contradirà a questa p(er)ché no(n) ho seguito | Claudio Tolomeo, sì ne la forma come etia(m) ne le sue mesure p(er) longeça e p(er) | largeça, no(n) vogli più curiosamente defenderlo de quel che lui | proprio non se defende, el qual nel seco(n)do libro capitulo primo | dice che quele parte de le qual se ne ha co(n)tinua pratica se | ne può p(ar)lar corretam(en)te, ma de q(ue) le che no(n) sono cussì fre|que(n)tade no(n) pe(n)si algun se ne possi p(ar)lar cussì correctam(en)te.[43]

Da questa provi(n)cia seri|ca in çoso Tholomeo fa ter|ra ignota.[44]

Questo colfo, el qual | Tolomeo no(n) mete, ha piusor | nomi, e si dito Lubech, Pru|sico, Sarmatico, Germanico, e p(er)ché questo ultimo | nome | è più chiaro, p(er)ò | ho notado Colfo Ger|manico.[45]

Ptolemaeus, Claudius (*fl.* 127-148), *Quadripartitum*.

Et perchè la luna ha in sí uirtù attractiua de ogni humidità e per consequente tira ad si etia(m) li ua|pori de queste humidità geniti e maxime se quelli sono corrupti, como se dichiara nel primo libro del quadripartito, pertanto qua(n)do la luna come(n)za cu(m) li razi suoi far inpressio(n) ne le aque grosse et profonde.[46]

[42] *Mappamondo* (ed. 1956) Pl. XXXIV, 79. "Regarding the divisions of the world – that is, of Asia from Africa and of Europe from Asia – cosmographers and historiographers give various opinions. Of these one could discuss at length, but because it is boring to dwell on this controversy, I will make a brief note with their opinions and leave the prudent to decide which one they should hold as best. Some follow the Ancients – for example, Messala, the orator who wrote of the family of Octavian Augustus, and Pomponius Mela and those who followed him. They claim that it is the Nile which divides Asia from Africa and the Thanai Asia from Europe. Some claim that Ptolemy says that the chain of the Arabian Mountains, which lies to the side of Nubia and extends through Abassia and beyond southern Ethiopia, constitutes the division between Asia and Africa. [...]." *Mappamondo* (ed. 2006) *2489.

[43] *Mappamondo* (ed. 1956) Pl. XL, 49. This is my translation of the legend: "So if someone contests this work because I have not followed Claudius Ptolemy either in the form of the world or in the measurement of latitude and longitude, I do not want to defend this map in any other way than that in which Ptolemy defends himself when, in the first chapter of the second book [of his *Geography*], he says that one can only speak correctly of regions that are visited continually: of those which are less frequented no-one should think himself capable of speaking with equal accuracy." For a different version, see *Mappamondo* (ed. 2006) *2834.

[44] *Mappamondo* (ed. 1956) Pl. XXXI, 9. "Southwards from this province of Serica, Ptolemy labels this land as *terra ignota*." *Mappamondo* (ed. 2006) *2243.

[45] *Mappamondo* (ed. 1956) Pl. XL, 23. "This gulf, which Ptolemy has not shown, has various names, and is called the Gulf of Lübeck, the Prussian Gulf, the Sarmatic Gulf, the Germanic Gulf, and because this last name is the clearest of them all, I have therefore noted it as the Germanic Gulf." *Mappamondo* (ed. 2006) *2862.

[46] "Questa rubrica de sopra dechiara come per la virtù atractiva de la luna le aque cresce e descresce." *Mappamondo*, upper right corner (left). *Mappamondo* (ed. 1956), p. 20. "Given that the moon has the property of attracting all humidity, it consequently attracts to itself also the vapours that are generated from that humidity, above all if they are impure, as it is stated in the first book of the *Tetrabiblos*." *Mappamondo* (ed. 2006), p. 724.

Servasanto da Faenza OFM (Servasacntus de Faventia, *c.* 1220/30-*c.* 1300), *Liber naturalium rerum* sive *Summa de exemplis contra curiosos*.[47]

> [...] e queli che de q(ue)ste cosse mirabile e de al|tre mo(n)struose desiderano esser copiosi. leçano | Julio Soli(n) polistoro, Po(m)ponio Mela, s(an)c(t)o Agustino, Alber|to Magno, s(an)c(t)o Thomaso de Aqui(n), nel libro co(n)tra i cu|riosi, ite(m) le Metaure d'Aristotile e Plinio de le me|raveie del mo(n)do e vederano che de mile cosse non ne ho dite una.[48]

Solinus, Caius Julius (*c.* 250), *Collectanea rerum memorabilium sive Polyhistor.*

> Nota che alguni istoriographi, di quali sono Ariano e Soli(n) et etiam Dionisio, el qual fo mandado da Re Philadelpho ad in|vestigar de l'india, questi dicono de le indie molte cosse nota|bile, e de mo(n)ti e de fiumi che sono ganges, el qual è largo dove | men viij dove più xx mia et ha pessi gra(n)dissimi de più sor|te, maximame(n)te anguile, le q(u)al, come dice Statio, | Thebaidas, sono longe 300 piè [...].[49]

> [...] e queli che de q(ue)ste cosse mirabile e de al|tre mo(n)struose desiderano esser copiosi. leçano | Julio Soli(n) polistoro, Po(m)ponio Mela, s(an)c(t)o Agustino, Alber|to Magno, s(an)c(t)o Thomaso de Aqui(n), nel libro co(n)tra i cu|riosi, ite(m) le Metaure d'Aristotile e Plinio de le me|raveie del mo(n)do e vederano che de mile cosse non ne ho dite una.[50]
> Alguni autori sc(r)ive del mar d'i(n)dia che'l sia serado come un stagnon | e che'l mar occean no(n) li entri, ma Solin vol che'l sia occea(n) e ch(e) q(ue)la p(ar)te austral | e del garbi(n) sia navigabile, et io affermo che algune nave açira e vol|ta quel cami(n). E questo a(n)chor (con)ferma Plinio qua(n)do el dice che al te(m)po | suo do' nave se mosse del mar de arabia, e dice la cason la qual q(ui) lasso, | ma charge de spetie ziroe q(ue)le p(ar)te perfin in spagna e a cibelter descar|gò. Ancora Facio afferma q(ue)sto et simelme(n)te hi exp(er)imentadori de quel | cami(n), ho(min)i de gra(n) prudentia i qual co(n)corda co(n) queli autori.[51]

[47] Piero Falchetta did not recognize that the "libro co(n)tra i cu|riosi" is a clear reference to one of Servasanto's da Faenza OFM most disseminated work. The *Liber naturalium rerum* is a collection of exempla, legends, visions and miracle stories, divided in three books. See M. Grabmann, "Der *Liber de exemplis naturalibus* des Franziskanertheologen Servasanctus,"*Franziskanische Studien* 7 (1920), pp. 83-117 (including the edition of the prologue and the table of contents) and L. Gargan, *Cultura e arte nel Veneto al tempo del Petrarca*, Padua : Antenore, 1978, p. 145. The Franciscan Servasanto da Faenza from the Province of Bologna had expecially preached in Tuscany and Florence, was in the vanguard of the invasion of preaching by philosophical arguments which he integrated into his sermons. He extensively employed secular knowledge in his sermons, expecially Aristotle and his commentators. J. Ziegler, *Medicine and Religion c. 1300. The Case of Arnau de Vilanova*, Oxford : Oxford University Press, 1998, pp. 165-174.

[48] See note 7 above.

[49] *Mappamondo* (ed. 1956) Pl. XXI, 57 : "Note that some historiographers, such as Arrian, Solinus and even Dionysius (who was sent by King Philadelphus to investigate India) say many notable things of the Indies, of its mountains and of its rivers. There is the Ganges, which at its narrowest is eight miles wide and at its widest twenty ; this contains very large fish of various sorts, including eels which Statius, in the *Thebaidas*, says are 300 feet long". This translation differs slightly from the one in *Mappamondo* (ed. 2006) *779.

[50] *Mappamondo* (ed. 1956) Pl. XXX, 120 ; for the English translation *vd. supra* note 1023. For a different version see *Mappamondo* (ed. 2006) *2212.

[51] *Mappamondo* (ed. 1956) Pl. XI, 26 : "Some authors write that the Sea of India is enclosed like a pond and does not communicate with the ocean. However, Solinus claims that it is itself part of the ocean and that it is navi-

Statius, Publius Papinius (*c.* 45-96), *Thebaidos.*

Nota che alguni istoriographi, di quali sono Ariano e Solin et etiam Dionisio, el qual fo mandado da Re Philadelpho ad investigar de l'india, questi dicono de le indie molte cosse notabile, e de monti e de fiumi che sono ganges, el qual è largo dove men viij dove più xx mia et ha pessi grandissimi de più sorte, maximamente anguile, le qual, come dice Statio Thebaidas, sono longe 300 piè. Item del fiume indo, dal qual è dita india, e del nobilissimo fiume hipano, el qual se dice terminasse la via de Alexandro. Item de citade, castelli, inumerabel populi uarietà, condition, costumi, gran potentie de signori, gran numero de elephanti, diversità de monstri quasi incredibili e de homeni e de animali, zoè serpenti et altre orribel bestie, maxime l'euchrota velocissima de tuti li animali, e de varie forme. Item de fructi pretiosi e legni et herbe e radice virtuose e diversità de çoie, id est diamanti, lichiniti, berrilli, chrisoberilli, chrisopassi, iacincti e molte altre cosse che qui dir non posso.[52]

Strabo (*c.* 63 bc-*c.* 24 ce), *Geographia.*

San Hieronymo nel prohemio | de la Bibia dice che Apolonio phy|osopho passò q(ue)sto fiume phison | e vene ad alguni pop(o)li diti bragma|ni p(er) aldir Archan [...] Q(ue)sto dico p(er)ché | questi p(o)p(u)li [...] a nui no(n) sono noti mol|to, p(er)hò de li lor costumi et habiti | discrepatissimi chi ne vol haver notitia | leça Ariano e Strabo, i qual de questa india | scriveno copiosamente.[53]

II. – The *mappa mundi*'s legends written in first person

Provided here are the transcriptions of the legends in which Fra Mauro constructs a dialogue between the cosmographer, his sources and his readers in first person. Since the goal of this transcription is uniquely hermeneutical, the legends are deliberately reported without diacritical signs.

Affrica è nominata da uno de i posteri de Abraam dito Affer. Ma Iulio Solin dice altramente, la qual auctorità qui lasso. *Mappamondo* (ed. 1956) Pl. XXIII, 45.[54]

gable in the southern and south-western parts. And I myself say that some ships have sailed it along that route. This is confirmed by Pliny when he says that in his day two ships loaded with spices coming from the Sea of Arabia sailed around these regions to Spain and unloaded their cargo at Gibraltar (he gives the reason for this choice of route, but I omit it here). Fazio [Degli Uberti] says the same; and those who have taken this route, men of great prudence, agree with these writers." *Mappamondo* (ed. 2006) *53.

52 *Mappamondo* (ed. 1956) Pl. XXI, 57; *Mappamondo* (ed. 2006) *779. *Vd. supra* n. 4.

53 *Mappamondo* (ed. 1956) Pl. XXVI, 18; "In the Proem to his Bible, St. Jerome says that Apollonius the Philosopher crossed this river Phison to travel to certain peoples called *Bragmani* and hear Archan [...] I say this because these peoples [...] are not very well known to us. Those who want information on their very different customs and practices should read Arrianus and Strabo, who write extensively about this India." *Mappamondo* (ed. 2006) *1380.

54 "Africa is named after one of the descendants of Abraham, called Affer. But Julius Solinus says otherwise, whose authority here I do not cite." *Mappamondo* (ed. 2006) *1067.

Alguni autori scriue del mar d'india che'l sia serado come un stagnon e che'l mar occean non li entri, ma Solin uol che'l sia occean e che quela parte austral e del garbin sia nauigabile, et io affermo che algune naue açira e uolta quel camin. E questo anchor conferma Plinio quando el dice che al tempo suo do' naue se mosse del mar de arabia, e dice la cason la qual qui lasso, ma charge de spetie ziroe quele parte perfin in spagna e a çibelter descargò. Ancora Facio afferma questo et simelmente hi experimentadori de quel camin, homini de gran prudentia i qual concorda con queli autori. *Mappamondo* (ed. 1956) Pl. IX, 26.[55]

Alguni scriue che a le radice del monte caspio ouer pocho lutan sono queli populi i qual come se leçe fono seradi per Alexandro Macedo. Ma certo questa opinion manifestamente è erronea e da non esser sostenuta per algun modo, perchè certo le si notò la diuersità de le nation che habitano circa quel monte che'l non è possibile che tanta numerosità de populi ne fosse ignoti cum sit che tute quele parte sono assai domestege per esser frequentade sì da nostri come da altre nation che sono çorçani, mingresi, armini, cercassi e tartari, e molte altre generation de populi, i qual fano continuamente quel camin. Vnde se questi populi fosse de li rechiusi credo che se queli ne hauesse notiçia anchora seriano a nui noti. Ma essendo questi tal populi ne la extremità de la terra come ne son certissimamente informato, adeuien che ancor tute queste nation de sopra nominate non ne ha maçor noticia de nui. Perhò concludo che questi populi siano molto lutani dal monte caspio e siano come ho dito ne la extremità de la terra tra grieco e tramontana e sono circundadi da monti asperimi e dal mar occean quasi da tre bande, e questi sono soto el regno de tenduch e sono chiamati vng e mongul, i qual el uulgo dice gog e magog extimando che questi sia queli che diebano usir de lì al tempo de antichristo. Ma certo questo error è adeuenuto per alguni che tirano la sacra scriptura al suo sentimento. Perhò io me acosto a la auctorità de sancto Augustino, el qual nel suo De Ciuitate Dei reproua la opinion de queli che dicono che gog e magog significa queli populi che darano fauor ad antichristo et in questa sentencia etiam concorda Nicolò de Lira, el qual expone questi do' nomi secondo la hebraica uerità, ma de questa materia in altro luogo se ne parlerà cum mior modo quelo se douerà tignir. *Mappamondo* (ed. 1956) Pl. XXXIII, 77.[56]

[55] "Some authors write that the Sea of India is enclosed like a pond and does not communicate with the ocean. However, Solinus claims that it is itself part of the ocean and that it is navigable in the southern and south-western parts. And I myself say that some ships have sailed it along that route. This is confirmed by Pliny when he says that in his day two ships loaded with spices coming from the Sea of Arabia sailed around these regions to Spain and unloaded their cargo at Gibraltar (he gives the reason for this choice of route, but I omit it here). Fazio [Degli Uberti] says the same; and those who have taken this route, men of great prudence, agree with these writers." *Mappamondo* (ed. 2006) *53.

[56] "Some write that on the slopes of Mount Caspian, or not far from there, live those peoples who, as one reads, were shut in by Alexander the Macedonian. But this opinion is certainly and clearly mistaken and cannot be upheld in any way because the diversity of the peoples who live around that mountain would certainly have been noticed: it is not possible that such a large number of peoples should remain unknown given that these regions are fairly well known to us: they have been frequented not only by our own peoples but also by the likes of the Georgians, the Mingrelians, the Armenians, the Circassians, the Tartars and many other populations that continually travel along this route. Hence, if those peoples had been enclosed there, I think that others would have had notice of it, and the fact would be known to us. But given that those peoples are at the limit of the earth – something of which I have information that is certain – this explains why all the peoples I listed above know no more about them than we do. Hence I conclude that these peoples are very far from Mount Caspian and are, as I said, at the extreme limit of the world, between the north-east and the north, and they are enclosed by craggy mountains and ocean on three sides. They are to the north of the kingdom of Tenduch and are called the Ung and the Mongul, which people know as Gog and Magog and believe that they will emerge at the time of the Antichrist. But certainly this mistake is due to the way some force the Sacred Scriptures to mean what they want them to mean. So, I am not differing here from the

Alguni scriue che la sithia è de qua e de là dal monte imao, ma certo se i hauesse ueduto ad ochio i haueriano altramente ordinato e dilatado i suo confini, perchè certo soto questo nome sithia se può dir che gran parte de questi populi che sono tra griego e leuante e griego e tramontana se contegna, i qual sono innumerabeli e de gran potentie e regni et imperij. De i qual perhò nomi credo che li antichi non ne hano possuto far bon çudisio, in perhò che la diuersità de li interpreti come assai error in exponer li nomi. Vnde ancora mi non me persuado molto in dimostrar tuta la uerità de li nomi perchè non è possibile acordar tanta uarietà de lengue le qual secondo el suo idioma uaria e confonde li nomi, i qual conuien siano cambiadi e corrupti e per le lengue e per longeça de tempo et error de i librarij Alguni scriueno. *Mappamondo* (ed. 1956) Pl. XXXIII, 109.[57]

Alguni scriueno che in queste Indie sono molte diuersità de monstri sì de homeni come de animali, ma perchè a queste cosse pochi dano fede qui non ne façо nota, saluo che pur à certo de alguni animali (…). *Mappamondo* (ed. 1956) Pl. XX, 68.[58]

Benchè in questa proxima rubrica io habia dito che le spetie uano fina al mar de ponto, ahora sono le strade sì rote che le non uano tanto auanti. *Mappamondo* (ed. 1956) Pl. XXI, 34.[59]

Benchè io habi seruato ogni diligencia in meter le staree de questo mar secondo la più iusta carta ho possudo, nondimen queli che sono experti non faça gran caso se io me discordo in qual cossa, perochè non è possibile meter tuto a ponto. *Mappamondo* (ed. 1956) Pl. XXIII, 153.[60]

Credo che qui molti se meraueierà perchè io meto el nascimento del nilo, ma certo si se mouerano cum raxion et uorano inuestigar quanto ho fato mi e cum quela diligentia che qui dir non posso, i uederano che io me mouo a demostrar questo per euidentissime chiareçe ho habuto. *Mappamondo* (ed. 1956) Pl. XVI, 58.[61]

authority of St. Augustine, who in his *De Civitate Dei* rejects all the opinions of those who claim that Gog and Magog are the peoples that will support the Antichrist. And Nicholas of Lyra agrees with this claim, explaining the two names by their Hebrew origin. But elsewhere one can speak more fully of this matter and of how it is to be understood." *Mappamondo* (ed. 2006) *2403.

[57] "Some write that Sithia extends on the near and far side of Mount Imao. But certainly if they had seen with their own eyes, they would have laid it out differently and enlarged its borders, because one can say that within Sithia are most of the peoples who live between the North-East and the East, between the North-East and the North. These peoples are innumerable and their kingdoms and empires of great power. I do not think, however, that the Ancients were in a position to judge their names appropriately, as their interpreters made many errors in reporting them. That is why I am not very convinced that one can find the real names : it is not possible to establish agreement between such a variety of languages, which change and confuse names in accordance with their own idioms. These names are now necessarily changed and in corrupt form due to the variety of languages, the long time that has passed and the errors made by copyists." *Mappamondo* (ed. 2006) *2435.

[58] "Some write that in these Indies there are many types of human and animal monster, but because few people believe these things, here I have made no note of them, except for certain animals (…)." *Mappamondo* (ed. 2006) *707.

[59] "Although in the nearby note I say that the spices travel as far as the Black Sea, today the roads are in such a poor state that they no longer reach that far." *Mappamondo* (ed. 2006) *749.

[60] "Though I have been most diligent in trying to put all the coastlines of this sea in accordance with the most accurate map that I possess, those who are experts should not take it amiss if I am not always consistent. Because it is not possible to put everything accurately." *Mappamondo* (ed. 2006) *1188.

[61] "I think that many will be amazed that here I put the source of the Nile. But certainly if they approach the question rationally and undertake the same investigations that I have – and with the diligence that I cannot here describe – they will see that here I am undertaking to demonstrate this thanks to the very clear evidence I have had." *Mappamondo* (ed. 2006) *389.

De qui è uulgo che questi populi rechiusi per Alexandro in questo paexe de hung e mongul deriua el nome suo da hi do' paexi ditti, i qual tra nui se chiama gog e magog, a la qual opinion io non credo. *Mappamondo* (ed. 1956) Pl. XXXVIII, 52.[62]

De sopra ho notado che'l thanay non nasce da j monti ryphey, uoglio intender per questo che'l non nasce tanto lutan come edil, çoè da questi monti, i qual per proprio uocabulo sono diti buletul e anchora ryphey per el continuo impeto di uenti che regnano. *Mappamondo* (ed. 1956) Pl. XL, 53.[63]

Del numero de le prouincie del mondo non troui mai algun cosmographo acordarse e questo perchè tuti hano abuto diuersa information e chi più e chi mancho. Perhò non dico qui el parer mio, ho però notado quasi per tuto etiam ne i luogi minimi uno p in luogo de prouincia solo per dar forma ad descriuer de le region e diuersità de populi. Ma a queli che non li piace tal nota oltra quele che son notade da Tolomeo in luogo de prouincie e'l resto intendi ouer tal populi ouer tal region, e se qui non son posti tuti queli nomi de le prouincie che mete Tolomeo le qual secondo lui sono 94, intendi che nì tuto ho possudo meter nì anche seruar hi suo propri nomi che al presente sono cambiati, credo però soto altri nomi hauer posto tute le sue et ancora algune altre che a lui non sono sta' note. Similiter de questa circumferentia troi uarie opinion, perhò non è possibile uerificarla benchè el se dica che la sia 22.500 ouer 24.000 ouer più ouer manco, secundo diuersa consideration ouer opinion cb'è non molto autentica per non esser experimentada. E benchè in diuersi tempi alguni habiano nauigato ne le parte austral e de septentrion, non di men non hano habuto tempo de mesurar ouer pur considerar questa distantia perchè el suo nauegar è stato casual e non determinato a tal nauigation. Perhò a l'Eterno Dio lasso la mesura de la sua opera la qual lui solo intende a ponto, non de men de questa materia se ne parlerà nel luogo debito. *Mappamondo* (ed. 1956) Pl. XXXIX, 88.[64]

In questa opera per necessità ho conuenuto usar nomi moderni e uulgari perchè al uero se io hauesse fato altramente pochi me haueria inteso saluo che qualche literato, auegna

[62] "Here it is said that these peoples were enclosed by Alexander in these lands of Hung and Mongul and that they derive their name from those two lands, which amongst us are called Gog and Magog. But I do not believe this opinion." *Mappamondo* (ed. 2006) *2752.

[63] "Above, I have noted how the river Thanay does not arise in the Riphei mountains. By this I mean that it does not arise as far away as does the Edil, which does arise in these mountains. These latter are properly called the Buletul mountains, or the Riphei because of the powerful winds that there prevail." *Mappamondo* (ed. 2006) *2884.

[64] "The cosmosgraphers never agree on the number of provinces because they received different information; some give more, some give less. Therefore, I do not give my opinion; almost everywhere – except in the smallest places – I have annotated a 'p' to indicate a province, but only to give a form to the description of regions and the diversity of peoples. But those who are not happy that more provinces are given than indicated by Ptolemy should accept the meaning – that is, of such and such a region or such and such a province. Here there are not all the names of provinces given by Ptolemy, who says there are 94. However, I recall that I have not been able to put everything, and not even to respect the names he uses, as they are different today. Yet I do think that, under different names, I have indicated all his provinces, and some that were not known to him. Similarly, there are various opinions regarding the circumference of the Earth, but it is not possible to verify them. It is said that it is 22,500 or 24,000 miles, more or less, according to the different estimates and opinions, which are not very reliable because they have not be verified by experience. And even if some have repeatedly sailed in the southern and northern regions, they have not had time to measure, or even estimate, these distances, given that they navigated as events dictated and not to measure the navigation itself. So I leave to eternal God the measurement of his work, which only He can understand in full. However, I will speak again of this matter in due place." *Mappamondo* (ed. 2006) *2828.

che ancora lor non possa acordar hi autori cum quel che hora se pratica. *Mappamondo* (ed. 1956) Pl. XXX, 89.[65]

Io ho lassato amplissimi desegni de tute queste parte, zoè armenia, mesopotamia, siria, capadocia, cilicia, pamphylia, licia, asia propria e menor, bitinia, galatia e tute le altre che lì sono meio distincte et ordinate. *Mappamondo* (ed. 1956) Pl. XXII, 177.[66]

Io ho notado de sopra che'l nilo nasce in abassia tra do' prouincie, zoè marora ouer meroa e salgu. Ma j libri punici dicono che nasce in mauritania, la qual cossa io non credo tuta esser uera che'l nilo habi qui el suo origine per le information ho habuto, ma che questo sia uno ramo del nilo io affermo, perchè se truoua queli simili animali che se truoua nel nilo. *Mappamondo* (ed. 1956) Pl. XVII, 41.[67]

Io ho più uolte aldido da molti che qui è una colona cum una man che dimostra cum scriptura che de qui non se uadi più auanti. Ma qui uoglio che portogalesi che nauegano questo mar dicano se l'è uero quel che ho audito perchè io non ardiso affermarlo. *Mappamondo* (ed. 1956) Pl. XVIII, 6.[68]

Io non credo derogar a Tolomeo se io non seguito la sua cosmographia, perchè se hauesse uoluto obseruar i sui meridiani ouer paralleli ouer gradi era necessario quanto a lademonstration de le parte note de questa circumferentia lassar molte prouincie de le qual Tolomeo non ne fa mention, ma per tuto maxime in latitudine çoè tra ostro e tramontana dice terra incognita, e questo perchè alsuo tempo non li era nota. *Mappamondo* (ed. 1956) Pl. XLI, 21.[69]

(...) Anchora io ho parlato cum persona digna de fede, che afferma hauer scorso cum una naue de india per rabia de fortuna de trauersà per zorni 40 fuora del mar d'india oltra el cauo de soffala e de le insule uerde e qui pur al garbin e al ponente e per lo arbitrar de i suo astrologi i qual son lor guida i scorse circa 2000 mia. Unde certamente el se può affermar e creder cussì a questi come a queli i qual uien hauer scorso mia 4000 (...). *Mappamondo* (ed. 1956) Pl. XI, 2.[70]

[65] "In this work I have of necessity decided to use the modern names of common speech because, to tell the truth, if I had done otherwise only a few learned men would have understood me. However, even they cannot always reconcile what the *auctores* say and what is current practice." *Mappamondo* (ed. 2006) *2202.

[66] "I have left ample drawings of all these parts – that is, Armenia, Mesoptomia, Siria, Capadocia, Cilicia, Pamphylia, Licia, Asia proper and Asia Minor, Bitinia, Galitia and all the others – in which they are shown more distinctly and in order." *Mappamondo* (ed. 2006) *1014.

[67] "I have noted above that the Nile arises in Abassia between two provinces – that is, Marora or Meroa and Salgu. But the Punic books say that it arises in Mauritania, which I do not think can be entirely true – that the Nile has its origins here – because of the information that I have had. But I say that this is a branch of the Nile, because within it one finds animals similar to those in the Nile." *Mappamondo* (ed. 2006) *480.

[68] "I have often heard many say that here there is a column with a hand and inscription that informs one that one cannot go beyond this point. But here I would like the Portuguese that sail this sea to say if what I have heard is true, because I am not so bold as to affirm it." *Mappamondo* (ed. 2006) *560.

[69] "I do not think that I have belittled Ptolemy by not having followed his Cosmography, because if I had wanted to observe his meridians, parallels and degrees, I would have had to omit from this depiction of the world [circumferentia] many provinces that are known today that Ptolemy does not mention, especially in the extreme latitudes, for to the north and south, he labels these [regions] *terra incognita*, because in his age they were not known." This is my translation. A slightly different one also appears in *Mappamondo* (ed. 2006) *2892.

[70] "(...) What is more, I have spoken with a person worthy of trust, who says that he sailed in an Indian ship caught in the fury of a tempest for 40 days out in the Sea of India, beyond the Cape of Soffala and the Green Islands towards west-southwest; and according to the astrologers who act as their guides, they had advanced almost 2,000

Non parà da nuouo a queli che uedeno in questa europa cussì picole citade e in asia maçor cussì grande, perchè doue ho habuto campo ho fato luogi grandi e doue son streto picoli, per queli che uede habi patientia se non li satisfazo in tuto quel i uol. *Mappamondo* (ed. 1956) Pl. XXIV, 140.[71]

Nota che alguni istoriographi, di quali sono Ariano e Solin et etiam Dionisio, el qual fo mandado da Re Philadelpho ad inuestigar de l'india, questi dicono de le indie molte cosse notabile, e de monti e de fiumi che sono ganges, el qual, è largo doue men viij doue più xx mia et ha pessi grandissimi de più sorte, maximamente anguile, le qual, come dice Statio, Thebaidas, sono longe 300 piè. Item del fiume indo, dal qual è dita india, e del nobilissimo fiume hipano, el qual se dice terminasse la uia de Alexandro. Item de citade, castelli, inumerabel populi uarietà, condition, costumi, gran potentie de signori, gran numero de elephanti, diuersità de monstri quasi incredibili e de homeni e de animali, zoè serpenti et altre orribel bestie, maxime l'euchrota uelocissima de tuti li animali, e de uarie forme. Item de fructi pretiosi e legni et herbe e radice uirtuose e diuersità de çoie, id est diamanti, lichiniti, berrilli, chrisoberilli, chrisopassi, iacincti e molte altre cosse che qui dir non posso. *Mappamondo* (ed. 1956) Pl. XXI, 57.[72]

Nota che i cosmographi fano division de le mauritanie e chiamano una cesariensis e l'altra stifensis e la terça tingintanea, e questi nomi sono derivadi da cesaria, stifi e tingi, i qual erano casteli cussi nominati, ma ahora questi nomi a nui non sono noti molto, perhò non ne faço questa distintion. *Mappamondo* (ed. 1956) Pl. XXIV, 32.[73]

Nota che in questa asia maçor sono molti Regni e molte prouincie, le qual io non meto per non hauer luogo, unde ho conuenudo lassar molte cosse de meço e tuor quele me ha parso più note et etiam non ho fato mention de molti fiumi e monti e deserti in diuerse parte e maxime uerso l'ostro in arabia e uerso tramontana in le parte de permia e de tartaria e de russia e uerso leuante e in molti altri luogi in diuerse parte de questa asia. E similiter ho conuenuto lassar de

miles. Thus one can believe and confirm what is said by both these and those, and that they had therefore sailed 4,000 miles (…)." *Mappamondo* (ed. 2006) *149.

[71] "Let people not be surprised that in Europe I have shown cities so small and in Asia so big. Where I have had space I have made the places big; where I have been short of space, I have made them small. Let those who see them bear with me if they do not find them totally satisfactory and fully as they would want," *Mappamondo* (ed. 2006) *1312.

[72] "Note that some historiographers, such as Arrian, Solinus and even Dionysius (who was sent by King Philadelphus to investigate India) say many notable things of the Indies, of its mountains and of its rivers. There is the Ganges, which at its narrowest is eight miles wide and at its widest twenty; this contains very large fish of various sorts, including eels which Statius Thebaidas says are 300 feet long. Then there is the river Indus, which gives its name to India, and the most noble river Hipano, which is said to mark the limit of Alexander's advance. As well, there are cities, castles and innumerable peoples of different varieties, standing and customs. There are powerful lords, great numbers of elephants and a diversity of almost incredible monsters, both human and animals. For example, there are serpents and other horrible beasts – especially the *euchrota*, the fastest of all animals. These writers also talk of the precious fruits, timber and herbs and of the roots with virtuous properties. There there is the variety of gemstones – diamonds, lichnites, beryls, chrysoberyls, chrysoprases, jacinths – and many other things that I cannot mention here," *Mappamondo* (ed. 2006) *779.

[73] "Note that the cosmosgraphers divide the Mauritanias, calling one *Cesariense*, one *Stifensis* and the third *Tingintanea*. These names derived from those of three fortified towns: Cesaria, Stifi and Tingi. However, to us, they are not very well known, so I have not made this distinction." *Mappamondo* (ed. 2006) *1219.

dir de le nouità e de costumi e condition de populi e magnificentie e potentie de signori e gran diuersità de animal e de altre cosse sono infinite. *Mappamondo* (ed. 1956) Pl. XXI, 75.[74]

Nota che in questo mar come è noto sono piusor insule de le qual hi autori ne parla molto, ma qui non ho fato se non le principal, la nobeltà de le qual io non posso dir per non hauer luogo, perhò queli che sono studiosi suplisca a questo manchamento. *Mappamondo* (ed. 1956) Pl. XXIII, 104.[75]

Nota che per tuta questa opera sono fati alguni segni uerdi e algune piantade de arboseli, i qual non demostra altro cha diuision e confini de le prouincie, ma queli uuol ben intender è necessario habino ben uisto ad ochio ouer ben leto e intendi li uenti et habi bona gieometria e bona intelligentia de desegno, perchè altramente ne cauerano pocho fructo e non porano far bon çudicio de quelo j uederano designato qui suso. *Mappamondo* (ed. 1956) Pl. XL, 19.[76]

Nota che Tolomeo mete algune prouincie in questa asia, çoè albania, iberia, bactriana, paropanisades, dragiana, Arachosia, Gedrosia et oltra ganges le Sine, de le qual tute non ne faço nota perchè sono cambiati e corropti queli nomi, perhò può bastar che ho notado altre prouincie de le qual Tolomeo non ne parla. *Mappamondo* (ed. 1956) Pl. XXVI, 58.[77]

Parme che Tolomeo ne la quarta taola de europa nomina questa scandinaria e dice che'l à 18 hore de zorno, unde me meraueio che_questa li sia sta' nota e tuta questa parte de noruegia e suetia li sia sta' ignota. *Mappamondo* (ed. 1956) Pl. XLI, 27.[78]

74 "Note that in this Asia Major there are many kingdoms and provinces which I have not put because of lack of space. Thus, I have decided to omit many things and take out those that seemed to be to be best known. And I have not even made mention of many rivers, mountains and deserts in various parts – especially towards the south in Arabia, towards the north in Permia, Tartary and Russia, and towards the east and in many other places in different parts of this Asia. Similarly, I have decided to say nothing about the novelties, customs and standing of the various peoples, about the magnificent and powerful seignories, about the great diversity of animals, or about an infinite number of other things." *Mappamondo* (ed. 2006) *793.

75 "Note that in this sea there are numerous islands that the *auctores* speak of a lot. But here I have only given the main ones, the noble character of which I cannot describe because of lack of space. Those who are scholars may make good this omission." *Mappamondo* (ed. 2006) *1179.

76 "Note that throughout this work there are some green marks or some rows of small trees which serve to indicate the divisions and borders between provinces. But those who want to understand [these signs] well must have seen with their own eyes or else have read well. And they must have a good sense of proportions and the ability to interpret the drawing; otherwise, they will draw little fruit from these signs and will not be able to form a good idea of what they see depicted above." *Mappamondo* (ed. 2006) *2847.

77 "Note that Ptolemy gives various provinces in Asia – for example, Albania, Iberia, Bactriana, Paropanisades, Dragiana, Arachosia, Gedrosia and, beyond the Ganges, Sine – which I do not mention in full because those names have been changed and corrupted. But one should also note that I give other provinces that Ptolemy does not mention." *Mappamondo* (ed. 2006) *1405.

78 "It seems to me that in the fourth plate of his *Europe* Ptolemy mentions this Scandinavia and says the day lasts 18 hours. But it amazes me that he knew this, when all this area of Norway and Sweden was unknown to him." *Mappamondo* (ed. 2006) *2904.

Perchè ad alguni par da nuouo che io parli de questa parte meridional, la qual quasi està incognita a li antichi, perhò io respondo che tuto questo desegno da sayto in suso io l' ò habuto da queli proprij che sono nasudi qui, che son sta' religiosi, i qual cum le suo man me hano desegnato tute queste prouincie e citade e fiumi e monti cum li suo nomi, le qual tute cosse non le ho possudo meter cum el debito ordine per non esserui logo. *Mappamondo* (ed. 1956) Pl. X, 30.[79]

Perchè io no ho habuto loco in europa de dir la uera sententia de Tolomeo circa la diuision de l'affrica da la asia, perhò qui dico che lui fa quasi do' diuision : la prima el commença da la ponta de ethiopia e uien zoso per la costiera del colfo de arabia, poi el dice che per non diuider l'egypto el mete el in de l'affrica nel nilo e cu(m) el modo suo io me acordo. *Mappamondo* (ed. 1956) Pl. XXIII, 51.[80]

Perchè sono molti cosmographi e doctissimi homeni i qual scriueno che in questa affrica, maxime ne le mauritanie, esserui molti monstruosi homeni e animali, parme neccessario qui notar el parer mio, non perhò che io uogli contradir a le autorità de tanti, ma per dir la diligentia ho habuta in inquirir tute le nouità se à possudo inuestigar per molti anni de questa affrica, commençando da libia, barbaria e tute le mauritanie perfina al fiume daloro e da i 7 monti atrauerso per terra de negri oltra el primo clima e de soto commençando da binimagra, marocho, fessa, siçilmensa e per quela costiera de monti e uerso el garbin per garamantia, saramantia, almaona, benichileb, cetoschamar e dolcarmin e dafur e più uerso l'ostro per el Regno de goçam e uerso la austral e in abassia e ne li suo Regni, che sono barara, saba, hamara, e più de soto uerso nuba per el regno de organa e per l'insula meroes e per tuti queli regni de negri nontroui mai alguno me ne sapesse dar auiso de quelo io trouo scripto da queli ; vunde nonne sapiando altro nonne posso testificar, lasso a çerchar a queli che sono curiosi de intender tal nouitade. *Mappamondo* (ed. 1956) Pl. XXIII, 23.[81]

[79] Because to some it will appear as a novelty that I should speak of these southern parts, which were almost unknown to the Ancients, I will reply that this entire drawing, from Sayto upwards, I have had from those who were born there. These people are clerics who, with their own hands, drew for me these provinces and cities and rivers and mountains with their names ; all these things I have not been able to put in due order for lack of space. *Mappamondo* (ed. 2006) *98.

[80] Because there was not space in Europe to give the full verdict of Ptolemy with regard to the division of Africa and Asia, I will observe here that he practically makes two divisions : the first begins from the tip of Ethiopia and runs down the coasts of the Gulf of Arabia. [But then] he says that, in order not to divide Egypt, he places the limit of Africa at the Nile – and I agree with him here. *Mappamondo* (ed. 2006) *1077.

[81] Because there are many cosmographers and most learned men whowrite that in this Africa – and, above all, in the Mauritanias – there are human and animal monsters, I think it necessary to give my opinion. Not because I want to contradict the authority of these men but because of the care I have taken in all these years in studying all possible information concerning Africa. [I have studied] : from Libia, Barbaria and all the Mauritanias to the river of gold ; from the Seven Mountains through the land of the negroes down beyond the first *clima* to Binimagra, Marocho, Fessa and Siçilmensa along the chain of mountains ; south-west through Garamantia, Saramantia, Almaona, Benichileb, Cetoschamar, Dolcarmin and Dafur ; southwards through the kingdom of Goçam towards southern Ethiopia : Abassia and its various kingdoms of Barara, Sabi and Hamara ; and even lower, through the kingdom of Organa, to Nuba and the island of Meroes. And in all these kingdoms of the negroes I have never found anyone who could give me information on what those men have written. Thus, not knowing anything, I cannot bear witness to anything ; and I leave research in the matter to those who are curious about such things. *Mappamondo* (ed. 2006) *1043.

Queli che sono experti suplisca in questa idumea e palestina e galilea quele che io non meto, zoè el fiume zordan, el mar tyberiadis, el mar morto et altri luogi, i qual cum rason non se hano possudo meter. *Mappamondo* (ed. 1956) Pl. XXII, 118.[82]

Questa opera, fata a contemplation de questa illustrissima signoria, non ha in sì quel compimento che la doueria, perchè certo non è possibile a l'intellecto human senza qualche superna demostration uerificar in tuto questa cosmographia ouer mapamundi, de la qual se può hauer qualche noticia più a degustation cha a supplimento del desiderio. Unde se algun contradirà a questa perchè non ho seguito Claudio Tolomeo, sì ne la forma come etiam ne le sue mesure per longeça e per largeça, non uogli più curiosamente defenderlo de quel che lui proprio non se defende, el qual nel secondo libro capitulo primo dice che quele parte de le qual se ne ha continua pratica se ne può parlar corretamente, ma de quele che non sono cussì frequentade non pensi algun se ne possi parlar cussì correctamente. Però intendando lui non hauer possudo in tuto uerificar la sua cosmographia, sì per la cossa longa e diffcile e per la uita brieue e l'experimento fallace, resta che'l conciede che cum longença de tempo tal opera se possi meglio descriuer ouer hauerne più certa noticia de qual habuto lui. Per tanto dico che io nel tempo mio ho solicitado uerificar la scriptura cum la experientia, inuestigando per molti anni e praticando cum persone degne de fede, le qual hano ueduto ad ochio quelo che qui suso fedelmente demostro. *Mappamondo* (ed. 1956) Pl. XL, 49.[83]

Sel parerà ad alguno incredibile de qualche inaudita cossa io ho notado qui suso, non conferisca quela cum el suo inçegno ma tribuisca a hi secreti de la natura, la qual adopera cosse inumerabile de le qual quele che sauemo son la minor parte de quele che ignoremo, e quele che sauemo per el suo continuo uso non sono extimade, etiam essenndo admirabile, e quele che ne pareno inusitade non li demo fede e questo adeuien perchè i la natura exciede l'intellecto e queli che non l'ano suleuado nol può accomodar non tanto a le cosse insolite ma etiam a quele che assiduamente se pratica, e perhò queli che uol intender prima creda azò le intenda. *Mappamondo* (ed. 1956) Pl. XXX, 99.[84]

[82] Those who are expert, let them complete this Idumea and Palestina and Galilea with what I do not put – that is, the river Jordan, the Tiberian Sea, the Dead Sea and other places which it has not been possible to include. *Mappamondo* (ed. 2006) *960.

[83] This work, created as an act of homage to this most illustrious Seignory, is not as complete as it should be because it is not possible for the human intellect, without the help of some higher demonstration, to verify completely this cosmography or that *mappamundi* ; from these, one gets what is more like a sample of information rather than a full satisfaction of one's desire. So if someone contests the work because I have not followed Claudius Ptolemy either in the form of the world or in the measurement of latitude and longitude, I do not want to defend this map in any other way than that in which Ptolemy defends himself when, in the first chapter of the second book {of his *Geography*}, he says that one can only speak correctly of regions that are visited continually ; of those which are less frequented no-one should think himself capable of speaking with equal accuracy. But if he meant by this that he had not been able to verify his cosmography thoroughly, because it is a long and difficult thing, and life is short and experience fallible, he is actually admitting that with the passage of time the work could be improved – that is, one could have more certain information than that available to him. So I say that in my own day I have been careful to verify the texts by practical experience, investigating for many years and frequenting persons worthy of faith, who have seen with their own eyes what I faithfully report above. *Mappamondo* (ed. 2006) *2834.

[84] If someone finds incredible certain of the previously-unheard things which I have noted above, he should not submit them to the judgement of his own reason but rather list them amongst the secrets of Nature. We know only a small part of the innumerable things that Nature does, and those that we know by constant experience are not really held in consideration – even if they are admirable – whilst those which seem to us to be unusual we do not believe. This occurs because Nature goes beyond the human intellect ; and those who do not have an elevated intellect cannot grasp even the things of constant experience, let alone those which are unusual. Thus, those who want to understand must first believe in order to then understand. *Mappamondo* (ed. 2006) *2193.

Se'l parerà ad algun che io non habi ben posto babilonia per hauerla descrita sopra tygris e non eufrates, come scriueno li autori, piaquali prima considerar questo desegno e da poi domandi queli hano ueduto ad ochio e intenderano che io non me parto da la uerità. *Mappamondo* (ed. 1956) Pl. XXII, 93.[85]

Sono alguni istoriographi i qual dice_del fonte de j garamanti el qual de note è tanto caldo che chi li metesse le man se scoteria, e similiter el çorno è tanto fredo che'l non se poria patu[?]. Dicono ancora de questi ethyopi molte nouitade maxime de i panphagi, agriophagi, antropophagi e cinomolgi e de li lor bestial costumi. Item de animali monstruosi, zoè serpenti dragoni, basilischi et altre nouità le qual dir non posso. *Mappamondo* (ed. 1956) Pl. XVII, 16.[86]

[85] If to some it seems that I have not located Babylon well, because I have shown it on the Tigris and not the Euphrates, as the Authors write, may it please them to look first at the drawing and then ask those who have seen with their own eyes; they will thus understand that I do not stray from the truth. *Mappamondo* (ed. 2006) *957.

[86] Various historiographers write of the source of the Garamantes, which is so hot at night that anyone putting their hand in the water would be scolded; whereas during the day, the water is so cold one cannot stand it. These writers give information on the Ethiopians, above all, the *Panphagi*, the *Agriophagi*, the *Antropophagi* and the *Cinomolgi* and their bestial customs. They also talk about certain monstruous animals – such as serpents, dragons and basilisks – and give other information I cannot mention here. *Mappamondo* (ed. 2006) *460.

II

CHAPTER I — DOCUMENTARY APPENDIX

Calculations of the costs of fifteenth-century cartographic production

Florence, Biblioteca Medicea Laurenziana, Pluteo XXX.2						
Map		*ff.*	*H*	*W*	*Surface (mm²)*	*Surface (m²)*
1	*Oikumene*	68-69	560	830	464800	0,46
2	I Europae	70-71	495	664	328680	0,33
3	II Europae	72-73	536	639	342504	0,34
4	III Europae	74	433	346	149818	0,15
5	IV Europae	75	405	324	131220	0,13
6	V Europae	76-77	373	623	232379	0,23
7	VI Europae	78-79	480	704	337920	0,34
8	VII Europae	80-81	333	628	209124	0,21
9	VIII Europae	82-83	502	503	252506	0,25
10	IX Europae	84-85	508	687	348996	0,35
11	X Europae	86-87	502	590	296180	0,30
12	XI Europae	88-89	494	688	339872	0,34
13	XII Europae	90-91	462	600	277200	0,28
14	I Africae	92-93	485	784	380240	0,38
15	II Africae	94-95	458	746	341668	0,34
16	III Africae	96-97	528	743	392304	0,39
17	IV Africae	98-99	530	553	293090	0,29
18	I Asiae	100-101	494	650	321100	0,32
19	II Asiae	102-103	501	700	350700	0,35
20	III Asiae	104-105	469	648	303912	0,30
21	IV Asiae	106-107	455	733	333515	0,33
22	V Asiae	108-109	458	705	322890	0,32
23	VI Asiae	110	363	350	127050	0,13
24	VII Asiae	111	430	330	141900	0,14
25	VIII Asiae	112-113	536	694	371984	0,37
26	IX Asiae	114-115	528	542	286176	0,29
27	X Asiae	116	366	351	128466	0,13
	Average dimensions		470	606	7806194	7,81

Vatican City, Biblioteca Apostolica Vaticana, Urb. Lat. 277						
Map		*ff.*	*H*	*W*	*Surface (mm²)*	*Surface (m²)*
1	*Oikumene*	72-73	446	830	370180	0.37
2	I Europae	75-76	494	659	325546	0.33
3	II Europae	77-78	520	633	329160	0.33
4	III Europae	79	415	350	145250	0,15
5	IV Europae	80	390	340	132600	0.13
6	V Europae	81-82	356	613	218228	0.22
7	VI Europae	83-84	466	710	330860	0.33
8	VII Europae	85-86	320	628	200960	0.20
9	VIII Europae	87-88	491	502	246482	0.25
10	IX Europae	89-90	500	682	341000	0.34
11	X Europae	91-92	492	592	291264	0.29
12	XI Europae	93-94	485	685	332225	0.33
13	XII Europae	95-96	454	600	272400	0.27
14	I Africae	97-98	478	781	373318	0.37
15	II Africae	99-100	451	747	336897	0.34
16	III Africae	101-102	518	752	389536	0.39
17	IV Africae	103-104	520	558	290160	0.29
18	I Asiae	105-106	482	653	314746	0.31
19	II Asiae	107-108	492	707	347844	0.35
20	III Asiae	109-110	460	649	298540	0.30
21	IV Asiae	111-112	450	697	313650	0.31
22	V Asiae	113-114	444	693	307692	0.31
23	VI Asiae	115	343	339	116277	0,12
24	VII Asiae	116	417	322	134274	0,13
25	VIII Asiae	117-118	520	682	354640	0.35
26	IX Asiae	119-120	520	535	278200	0,28
27	X Asiae	121	355	346	122830	0,12
	Average dimensions		455	603	7514759	7.51

	Average dimensions (cm) of the two manuscripts	46	60		7,66

Normalization in *fiorini* and *ducati a oro* of the costs of the maps analyzed (with treferene to 1461 exchange rates)						
Mappa mundi of Fra Mauro, 1457-59	*Ducati*	*Soldi*	*Ducati/Soldi*	Total in *soldi* (1 D = 124 s)	Normalization in *fiorini* and *ducati a oro*	*Soldi*
8 February 1457	28	0	124	3472	28	
Manuscripts of the *Geography*, 1460-61	*Fiorini di conto*	*Soldi*	*Fiorini/Soldi*	Total in *soldi* (1 F = 88 s)	Normalization in *fiorini* and *ducati a oro*	*Soldi*
2 December 1460	23	17				
11 February 1461	14	8				
7 March 1461	18					
16 July 1461	8	8				
Total	63	33	88	5577	51	71
Catalan *mappae mundi*, 1398-1400	*Florí d'or d'Aragó*		Quantity	*Florí d'or d'Aragó / Ducato*	Normalization in *fiorini* and *ducati a oro*	
Small *mappa mundi* (1399)	60		2	75%	45	
Large *mappa mundi* (1399)	100		2	75%	75	
Total	320			75%	240	
Amount paid						
Acconto	192			75%	144	
Payment to Cresques	73			75%	54 12/16	
Payment to Becaria	38			75%	28 8/16	
Total	303			75%	227 4/16	

Monetary equivalences

1 *Fiorino di conto* = 0.78 *Fiorino a oro*. Cf. R. A. Goldthwaite, *Studi sulla moneta fiorentina : secoli XIII-XVI*, Florence : Olschki, 1994, pp. 94-95, 176, 180-182

1 *Fiorino a oro* = 1 *Ducato d'oro*. Cf. F. C. Lans and R. C. Mueller, *Money and Banking in Medieval and Renaissance Venice*, Baltimore and London : Johns Hopkins University Press, 1985, pp. 174-176, 308.

1 *Florí d'or d'Aragó* = 0.75 *Ducato or Fiorino a oro*. Cf. J. Hamilton, *Money, Prices, and Wages in Valencia, Aragon, and Navarre, 1351-1500*, Philadelphia : Porcupine Press, 1975, pp. 12-23

Average cost per square meter of the maps analyzed, normalized to gold ducats and florins (*fiorino* and *ducato a oro*)									
	Height (cm)	Width (cm)	Maps	Area (cm²)	Area (m²)	Average cost (m²) *Fiorini d'Aragona*	Average cost (m²) *Soldi di piccioli*	Average cost (m²) *Ducati and Fiorini*	
Mappa mundi of Fra Mauro	200	200	1	40000	4.00		868	7	Average cost of one m² of the *mappa-mundi*
Geography of Ptolemy	46	60	27	74520	7.45		748 4/10	6 9/10	Average cost of one m² of a manuscript with 27 maps
	46	60	34	93840	9.38		594 3/10	5 5/10	Average cost of one m² of a manuscript with 34 maps
							671 3/10	6 2/10	Average
Catalan *mappae mundi*	155	310	2	96100	9.61	12 5/10		9 4/10	Average cost of one m² of the two small *mappae mundi*
Specifica-tions in the contract	368	368	2	270848	27,08	7 4/10		5 5/10	Average cost of one m² of the two large *mappae mundi*
				366948	36,69	8 7/10		6 5/10	Average cost of one m² of the four *mappae mundi*
Specifica-tions of final maps				366948	36,69	8 3/10		6 2/10	Average cost of one m² of the four *mappae mundi*
Overall average								5 9/10	Average cost of one m² of all of the documents considered

Monetery equivalences

1 *Fiorino di conto* = 0.78 *Fiorino a oro*. Cf. R. A. Goldthwaite, *Studi sulla moneta fiorentina : secoli XIII-XVI*, Florence : Olschki, 1994 1994, pp. 94-95, 176, 180-182

1 *Fiorino a oro* = 1 *Ducato d'oro*. Cf. F. C. Lans and R. C. Mueller, *Money and Banking in Medieval and Renaissance Venice*, Baltimore and London : Johns Hopkins University Press, 1985, pp. 174-176, 308.

1 *Florí d'or d'Aragó* = 0.75 *Ducato or Fiorino a oro*. Cf. J. Hamilton, *Money, Prices, and Wages in Valencia, Aragon, and Navarre, 1351-1500*, Philadelphia : Porcupine Press, 19751975, pp. 12-23

CHAPTER II – DOCUMENTARY APPENDIX

Selection of documents attesting to the reception of the *Mappa mundi* in the eighteenth and nineteenth centuries

Doc. 1. Giovanni Battista Mittarelli, *Bibliotheca codicum manuscriptorum S. Michaelis prope Murianum una cum Appendice librorum impressorum saeculi XV, opus posthumum Johannis-Benedicti Mittarelli Veneti, abbatis ex-generalis benedictino-Camaldulensis*, Venice : Typographia Fentiana, 1779, col. 756-760.

(Col. 756) Mauro Monaco Camaldolese. Planisferio con le sue annotazioni. Ext. in Codd. 607 & 1112. Excusum fuit in ejus honorem aereum numisma cum titulo in peripheria : Frater Maurus Monachus Camaldulensis sancti Michaelis de Muriano, Cosmographus incomparabilis. Floruit medio seculo XV. Memorie per servir ad una dissertazione da farsi sopra il di lui antico Planisferio. Ext. in Cod. 626. Adnotationes, quas fecit suo Planisphaerio. In Cod. 1112. Cogitabamus has Mauri adnotationes producere ; verum, quoniam longiores sunt, (col. 757) & potissimum versantur super numerum coelorum & planetarum juxta auctoritatem factorum Theologorum, circa situm Paradisi Terrestris, circa distantiam planetarum a centro terrae, circa elementa variasque ipsorum qualitates, circa demum descriptionem partium orbis terraquei, Regnorum, Provinciarum & illustrium insularum, ideo eas libenter omittimus, contenti sequentes observationes, notulasque nonnullas ab eo confectas excerpere. Planisphaerium hocce per quinque pedes & octo uncias Venetas protenditur, pictumque est variis coloribus, & auro illitum super subtilissimas membranas, cum adnotationibus, nunc prolixioribus, nunc brevioribus, locatis ad regna vel urbes, licet illud fateatur Maurus, se plures adnotationes praetermisisse, quia situ carebat, quo eas super membranas describeret. V. gr : ad Norvegiam inferiorem scribit : In questo Oceano sono molte insule, le qual non hò notado, per non haver luogo. Ita ad mare Orientale super oras regni Fuguy inter Eogin & Wichen : In questo mar Orientale sono molte insule grande & famose, le qual non hò posto, per non haver luogo. Affirmat se Ptolomaei systema non fuisse sequutum in omnibus ; ipso tamen duce utitur in pluribus locis, praesertim in divisione Africae ab Asia. Signis arborum virentium provincias a provinciis dividit ; Europae urbes describit minori charactere, e contra majori regna& urbes Asiae & Africae. Loca nominat, uti ea suo tempore appellabantur. Laboris autem sui rationem reddit his verbis : Questa opera facta a contemplation di questa illustrissima Signoria, non hà in si quel compimento, che la doveria, perche certo non è possibile al inteleto humano verificar in tutto, senza qualche superna demostration questa Cosmografia over Mappamundi, de la qual se può haver qualche notitia più a degustation, che a suplimento del desiderio ; unde se algun contradirà a questa, perche non hò seguito Claudio Tolomeo, si ne la forma, come etiam ne le sue mensure per lunghezza, non voglj più curiosamente difenderlo di quel, che lui proprio non se deffende, el qual nel segondo libro capitolo rpimo dice, che quele parte, de le qual sene hà continua practica, se ne può parlar correctamente, ma de quele, che non sono cussi frequentade, non pensi algun se ne possi parlar cussi correctamente, però intendando lui non haver possudo in tucto verificar la sua Cosmografia, si per la cosa longa & difficile, & per la vita brieve & l'experimento falace, resta chel conciede, che cum lungheza de tempo l'opera se possi meglio descriver, over haverne più certa notitia, de quel hà habuto lui. Per tanto dico nel tempo mio ho solicitado verificar cum l'experientia, investigando per molti anni, & practicando cum persone degne di fede, le qual hano veduto ad ochio quelo, che qui suso fedelmente demonstro. Revera ope virorum, qui ea loca viderunt, tabulam suam expressit Maurus. Memorat Petrum Quirinum Patritium Venetum ad Norvegiam inferiorem. In questa provincia di Norvegia scorse missier Piero Quirini, come è noto. Oculatorum testium assertione pariter Africam Meridionalem describit a Sayto prope Nilum inter Nuba & Gion usque ad caput Diab, vocatum subinde Capo de buona Speranza : Perchè ad alghuni par da nuovo, che io parli de questa parte Meridional, la qual quasi è stà incognita a li antichi, però io respondo, che tucto questo desegno da Sayto in suso io hò habuto da queli propri, che sono nassudi qui, che sono stà religiosi, i qual cum le so man mi hanno desegnato tucte queste provincie e citade e fiumi e monti con li suo nomi. Le qual tutte chose non le hò possudo metter cum el debito ordene, per non esservi luogo. Ad idem caput Diab adnotat : oltra questo cavo non impazano i naviganti. Refert etiam navim Indicam anno 1420. praetergressam dictum Caput, sed post longam navigationem reversam. Ad ipsum caput Diab exsurgere scribit magnam insulam medio canali : Demum de ea regione differens ait : Questa region fertilissima è stà conquistà nuovamente per el gran Rè de Abbassia circa l'anno 1430. qui comenza el mar scuro. Terminos suo planisphaerio facit Maurus, ad Orientem Java majoren, qua statuit inter Meridiem & Septentrionem, non inter Orientem & Occidentem, sicut Java minor non respicit ad Septentrionalem plagam Javae majoris, sed ad Meridiem inter Java majorem & Sumatram : Ad Eurum confine statuit Catajum. Ad (col. 759) Quare emendandus est Ramusius, qui tomo II. suae collectionis Navigationum, Maurus affirmat Planisphaerium fuum concinnasse ex Mappamundo, quod ex Catajo in Italiam detulerat Marcus Polus Patritius Venetus, qui una cum patre suo provincias remotissimas perlustraverat ; licet enim non dissentiamus Maurum aliquid profecisse ex Itineribus Marci Poli, praesertim quoad Catajum, exploratissimum tamen est ipsum plura addidisse, & de novo expressisse ex auditu eorum, qui ea loca incoluerant, & ex aliis relationibus, quae post tempora Poli audierat. Praeterea Polus in suis itineribus nunquam memorat se tabulam universalem orbis confecisse ; ipse Maurus nunquam memoriam ingerit in suis fusis notis Marci Poli, ut memorat ceteros indigenas locorum, cum quibus colloquutus fuerat. Postremo non hocce solum Planisphaerium Maurus elaboravit, sed amplissimos etiam typos aliorum regnorum & provinciarum : Io hò lassato, inquit, amplissimi desegni de tutte queste parte, zoe Armenia, Mesopotamia, Siria, Cappadocia, Cilicia, Pamphilia, Licia, Asia propria menor, Bitinia, Galacia, e tutte le altre, che si sono mezo distincte & ordinate : ut observat noster Abundius Collina pag. 83 suarum Considerationum historicarum super Pixide nautica. Alterum pariter simillimum exemplar Planisphaerii confecit Maurus, servatum nunc in monasterio Alcobazae in Lusitania. Ex codicibus Monasterii nostri, qui continent rationes dati & accepti, exscriptis tempore Maphaei Gherardi Abbatis Coenobii S. Michaelis de Muriano, sub quo postremis annis vitae suae degit Maurus, Veneto idiomate leguntur expensae, quae fuere necessariae pro eo conficendo ; pecuniae quoque, quas Alphonsus IV, Rex Portugalliae Venetias ad Maurum transmisit ; neque dubitamus Maurum plures topographicos typos concinnasse, cum tempore obitus ipsius multi reperti fuerit, qui depositi fuere Monasterio S. Iohannis-baptistae de Judaica Venetiarum. Ceterum Petrus Delphinus Abbas generalis Camaldulensis omnia adnotata vulgari sermone a Mauro in pictura orbis, Latina fecit, ut scribit die XXVI martii anni 1494 ad Bernardinum Gadolum, eaque se deposuisse scribit asservanda in domo Petri Medicei, sed in eam rerum perturbatione quae tunc (col. 760) temporis Petrum Mediceum divexabat, aegre posse reperiri. Ipsa ut deperdita dolemus, quae si existerent, magnam in partem pepercissemus labori, quem subivimus in his describendis ex tabula, quae ob sui antiquitataem fugientes pluribus in locis charcteres modo retinet. Consulantur Annales Camaldulenses T. VII p. 252 ad annum 1457. Aliqua etiam ex hoc Planispherio derivarunt in suas epistolas collectores Delle Memorie per servir alla Storia Letteraria T. VII P. I. pag. 56 praeter superius laudatum Abundium Collinam, exterosque pariter scriptores & geographos. Nostris temporibus nullus fere est eruditorum virorum, qui Venetias pertransiens, Bibliothecam nostram non adeat, ut Planisphaerium hocce observet & consulat, ut usuveniebat etiam temporibus Ramusii, qui de eo verba facit, ut diximus, Tomo II. suarum Navigationum.

Doc. 2. London, British Library, MS Add. 76092, Earl George John Spencer – General Correspondence (January-June 1804), Letters sent by William Vincent.

February 24, 1804

My Lord

By the abominable folley of my servants and the mistake of my son I missed the opportunity of seeing your Lordship this morning which I regard above all things.

I send your Lordship enclosed the paper I had fro Lord Hobart, by which it appears that the Map of St. Michael at Murano, thus not Marco Polo's, has been perpetually mistaken for it. And as it is the Map copied for the King of Portugal on which the Portuguese discoveries depended, a copy of it may be worth bringing to England. The India Company has subscribed £ 50, and by accepting £ 20 as your Lordship's subscription, I have now £ 150. And should I obtain your opinion in favour of procuring the Copy, be the map what it may, I would solicit to other no other judgements but Lord Macartney's to determine me in pursuing the research always submitting the truth to the other subscribers before I take their money.

Could you favour me with a quarter of an hour conversation, I would wait on you any hour on any morning you may appoint as a few minutes conversation would do more than much writing.

Your Lordship's
Most obedient
And faithful servant.
W. Vincent

March 3, 1804

My Lord

I do not vouch for the accuracy of the translation, but it is not far different from the original, and if your Lordship still think I ought to proceed I shall be happy to receive your subscription whenever you please to pay it. Perhaps we may meet on Tuesday as I shall be at the club. Believe me

Your Lordship's
Most obedient
And faithful servant.
W. Vincent

March 4, 1804

My Lord

I return to you my best thanks for the subscription, and entertain the same opinion of the of the map as your Lordship, still thinking is worth the expense and trouble of procuring, if it be only to say what was and what was not known in 1430. The translation is intended for your Lordship's use to keep as a memorandum and a similar copy has been or will be sent to every subscriber.

I remain
Your Lordship's
Most faithful
And obedient servant.
W. Vincent (Dean of Westminster)

Doc. 3. London, British Library, MS Add. 76092, Earl George John Spencer – General Correspondence (January-June 1804), *Translation of a paper communicated by Mr Stuart (d) to Lord Hobart relative to the Map now existing in the Church of St. Michael di Murano at Venice*. This manuscript report follows the letter sent by Vincent to Earl Spencer on March 4, 1804.

(F. 1r) From the Library of St Michael at Venice in the Account of Manuscripts preserved there, with the Appendix, drawn up by Jo. Benet. Mirtarelli (*sic*), 1779 p. 756 et segg.

Mauro a monk of the Order of Camaldulenses. A planisphere of his with his Annotationes, is still extant in the Collections marked 607 & 1112. A medal was struck in honour of Mauro with this inscription in the circumference : Frater Maurus Monachus Camaldulensis sancti Michaelis di Murano Cosmographus incomparabilis. Mauro flourished in the middle of the fifteenth century. "A Memoir to serve as a Dissertation upon his ancient planisphere" is extant in the Codex 626. The Annotationes made by him, in the Codex 1112 – it was our intention to give these Annotations at full length ; but as they are very large, and contain much upon the Heavens and the number of Planets according to the authority of theologians, as well as concerning the situation of the Terrestrial Paradise ; concerning the distance of the Planet the centre of the Earth ; concerning the Elements, and their various qualities and lastly concerning the description of the parts of the (f. 1v) terraqueous Globe, the kingdoms Provinces and most remarkable Islands – these therefore we omit, confining ourselves to extract only the following observations and some of his annotations.

The Planisphere itself is five feet eight inches of Venice measure, it is delineated in a variety of colours illuminated with gold upon, the finest vellum, added to which are observations of different extent, affixed to different kingdoms and cities. But Mauro confessed he has omitted [in] many observations, for which his sheet did not allow space sufficient. For example at the lower part of Norway he writes, "In this sea are many islands which I have not noted for want of room." He asserts likewise that he has not followed Ptolemy in everything, but he takes him for his guide in most instances, particularly in the limit between Asia and Africa. He devides his Provices by the fugures of tree [Add. : this is conformable to the Arabian Geographers]. He describes the Cities of Europe in a smaller character, and those of Asia and Africa in a larger, he gives the names of places as they were called in his own age. The Account he gives of his own work is in these words :

"This work prepared for the inspection of our most illustrious City (Venice) is by no means so complete as it ought to be, because certainly it is not possible for the powers of Man fully to ascertain every particular in this Cosmography [Planisphere] or Map of the World, so far as it is discovered, bu that something must remain which will rather (f. 2r) offend than satisfy the curious. There may likewise appear something like contradiction, because I have not always followed Ptolemy either in the form of Countries or his degrees of Longitude, neither am I willing to defend him anxiously where he cannot defend himself ; as when he says in his second Book Chap 1. : "that in those parts with which we have not continual intercourse it is impossible to speak with precision". "For it is evident, that it was not in his power to verify his whole Cosmography, as the endeavour is of the greatest difficulty while life is short and information fallacious. He in himself also confesses that the whole attempt must be matured by time and by more certain accounts than he he had been able to obtain." "For my own part I can say that I have taken great pains to inform myself by the study of many years and by enquires of all persons worthy of credit who had personally visited the countries I have attempted to describe."

So far Fra Mauro. And he does in reality appear toh ave made use of all travellers who had seen the Countries contained in his Map. He mentions Peter Quirinus a Venetian nobleman in describing Norway. "Here, says he, it is well known that Piero Quirino travelled.". He likewise appeals to ocular testimony in regard to the whole of Africa from the [said] upon the Nile between Nuba and Gion, even to Cape Diab which was afterwards called the Cape of Good Hope. His words are these. Now because it may appear extraordinary to some, that I speak of these southern parts which were in a manner unknown (f. 2v) to the Ancients, to this I answer that the whole of this delineation from the Said upwards [Add.: That is towards the South as appears by the context. And whether not rather towards the North as no missionaries in that age had been farther south than EgyptI have received from particular person who were either atives of the Country or from Missionaries, who with their own hands traced out for me all these Provinces, Cities, and rivers, the mountains, with their names, all which things I have not been able to arrange in due order, for want of room. He than adds at the point where he [Add. This seems as if Mauro had written Diab himself] has marked Cape Diab: "Beyond this Cape navigators do not pass." But he observes, "that an Indian vessel had passed this Cape in the year 1420. And after a long course of Navigation it had returned." He mentions also that at the very Cape a large island appears in the middle of the channel, and speaking of he country he says, "this tract is extremely fertile and has been lately conquered by the great King of Abassia (Habez or Abyssinia) about the year 1430 [Add.: Who this king of Abassia is, does not very well appear from Bruce, but perhaps Theodorus who died about 1412. But we are by this brought from the Cape to the country of Mogadoxo (*sic*), e.g. about 35 degrees at one step, see Bruce, vol. II, p. 84]. And here commences (el Mar obscuro) [Add.: The mare Tenebrorum of Arabian geography] the Dark Sea.

Fra Mauro likewise fixed the limits of his planisphere, East, at Java Mayor which he lays down North and South, and not East and West [Add.: In Marco Polo Java Minor is Sumatra]. Also his Java Minor does not face north to Java Mayor, but South between Java Mayor and Sumatra. To the North he lays down Catai (China) as his termination, and to the North Permia immediately after the Russies. On the North West his limit is Scandinavia and Iceland, on the West the Kingdom of Spain ì; and on the South West the island of Dragons and Aethiopia but his termination south is Cape Diab, beyond which he plas islands without a name; the principal of which however he calls Mahal [Add.: I think he means the Mal-dives and if this should seem extraordinary I have an Arabian map which I think places the great Pepper Island [east ?] of Africa i:e: Ceylon or Sumatra] (f. 3r) meaning perhaps Madagasiar [sic] finally his limit on the South East is the Island of Columbus.

From these observations of Fra Mauro and the limits of his Planisphere it appears clearly that he is not a mere copier of others but the real Composer and Author of this map and upon these grounds the assertion of Ramusio vol. II ought to be corrected, when he says that Fra Mauro composed his planisphere from a Map which Marco Polo the Venetian Nobleman brought out of Catai (China) into Italy: who, in company with his Father, had visited the most distant countries of the East. For this we may allow that Fra Mauro made some use of the travels of Marco Polo, particularly in regard to Catai, yet it is most evident that he added much and had acquired fresh information from the accounts of these who were inhabitants of the countries and other reports which he received of a later date than Marco Polo. But beside this, Marco Polo never says in his own Account of his travels that he had composed a general Map of the World; and Fra Mauro never mentions the name of Marco Polo in his observations which are very numerous, and notice several natives of the countries he describes with whom he had conversed.

We ought likewise to remember that this Planisphere was not the only work of Fra Mauro for he produced several other large Maps of different kingdoms and provinces. "I have left – says he – very large description of all these Countries, (f. 3v) that is of Armenia, Syria, Cappadocia, Cilicia, Pamphilia, Lycia, Asia Minor, Bythinia, Galatia, and all the rest nearb[l]y [Add.: mezzo] composed and put in order." See Abundius Collina p. 83 in his treatise on the history of the compass. Another copy of this Planisphere was made by Fra Mauro still preserved in the Monastry of Alcobaça (Add.: See *Periplus* p. 199) in Portugal, as appears by an Account in our Convent (St. Michael di Murano) which contains a detail of the Receipit of Expense written in the time of Maffei Gerardi Abbot of St. Michael under whom Fra Mauro past the latter years of his life.

The account is written in the Venetian dialect with all the sums necessary for completing the Work and all that Alfonso IV [Add.: Alfonso V] paid, or remitted from Portugal to Fra Mauro. Neither can there be any doubt but that he made many printing types as many were found after his death, which were deposited in the Convent of St. John Baptist de Judaica at Venice. And Peter Delphinus Abbot and General of the Camaldulenses, translated all the observations of Fra Mauro upon his map from the Venetian into Latin, as he writes on the 26th day of March 1494 to Bernardinus Gadolo; and adds that he deposited them in the house of Peter di Medicis. These papers however perished perhaps in the trobles that beset the Family of Medicis soon after, or at least cannot now be found. There is much reason to lament this loss as these papers would have saved us much trouble in taking out these particular from the Map itself, which in several places are (f. 4r) decaying from the estreme antiquity of the work. See Annali of the camaldulenses, T. VII. 252 Anno 1457. The Collectors of the memoirs for forming the Literary History also copied something from this planisphere, T. VI. Part I, p. 56 as well as Abundius Collina aforementioned, with several writers and geographers. In the present day scarce any man of learning visits Venice without coming to out convent to see and make his observations on this Planisphere, as was the custom in the time of Ramusio [Add.: This is apparently the origin of the error in attributing this Planisphere to Marco Polo for more than the last 200 years] who mentions this Work in his second volume.

Doc. 4. London, British Library, MS Add. 76094 Earl George John Spencer – General Correspondence (January-June 1805), Letter sent by William Vincent.

April 19, 1805

My Lord

I have the pleasure to acquaint your Lordship that by a letter I have received from Mr. B. Bathcroft, who is in Venice, I have been informed that the copy of Fra Mauro's map is completed, and is now in Vienna in the possession of Mr. [G]enkinson, who will send it to England by the first safe conveyance. I have also received the attestation of its fidelity and correctness.

I have the honour to remain
Your Lordship's
Most faithful
And obedient servant,
William Vincent (Dean of Westminster)

Doc. 5. London, British Library, MS Add. 76096 Earl George John Spencer – General Correspondence (January-June 1806), Letter sent by William Vincent.

27 June 1806

My Lord I have ordered the Memoir on Fra Mauro's Map to be sent to your Lordship's house, and you will receive it in he course of few days. I only lament that your engagements of a public nature have prevented you from seeing he original, as the inspection of it, I think, would have persuaded you that I am not unreasonable in intrading farther on your liberality; the artist complains that after his engagement, he was obliged to raise a scaffold and [escape] the water twice a day at the expense of nearly five shillings to reach the Monastery of St. Michael. I asked him if £ 30 would satisfy him; and he said, most amply. I have taken £ 10 from the Bishop of Durham and Mr. [Strahan]; and if your Lordship will make up the remaining £ 10. I will take upon myself the conveyance and mounting, which are short of £ 20, and then the whole expence

Is provided for. Nothing remains but for the subscribers to determine where the copy shall finally be deposited. I propose the British Museum that shall chearfully [sic] submit to the majority.

I remain
Your Lordship's
Most faithful
and obedient servant
W. Vincent

Doc. 6. London, British Library, India Office Record, B/137, Court Book (13 April 1803 – 28 September 1803), "Index," *s.v.* "Publication," p. 1214" : "50 Guineas subscribed for map of Marco Polo."

Doc. 7. London, British Library, India Office Record, B/138, Court Book (5 October 1803 – 11 April 1804), p. 1214.

February 22, 1804

Present
Jacob Busanquet – Chairman
John Roberts – Deputy Chairman
On a motion
Resolved that this Court subscribe the sum of Fifty Guineas towards a fund now raising for the purpose of obtaining a Copy of the original Map of Marco Polo which has been discovered in Venice.

Doc. 8. M. F. Santarém de Barros e Sousa, "Note sur la mappemonde de Fra Mauro," *Bulletin de la Société de Géographie de Paris*, 1846, pp. 251-252.

Ceux qui se sont consacrés à l'étude de l'histoire de la géographie connaissent les rapports qui eurent lieu entre le célèbre infant D. Pedro frère du prince Henri, Alphonse V, roi de Portugal, et le fameux cosmographe vénitien Fra-Mauro. Tous savent que le roi fit construire par ce cosmographe une mappemonde que ce même cosmographe envoya à Lisbonne. Il parait que cette mappemonde se trouvait encore aux archives d'Alcobaça en 1528 ; mais malheureusement, quelques années après, elle disparut, sans que personne, pendant l'espace de plus de deux siècles, ait pu la découvrir en Portugal. Lorsque Jean V, au commencement du dernier siècle a fait copier tous les documents concernant le Portugal qui se trouvent dans les bibliothèques de l'Italie, compilation précieuse qui, sous le titre de *Symmicta Lusitana*, se composent de 200 énormes volumes in-folio, ce grand roi eut l'idée de faire exécuter une copie de la mappamonde de Fra-Mauro. Ce dernier projet n'eut pas d'effet. Lorsque j'ai redigé mes *Recherches sur la priorité des découvertes des Portugais*, j'ai tout d'abord compris que, parmi les pièces justificatives, une des plus positives et des plus précieuses était la mappemonde de Fra-Mauro où l'on trouve des légendes relatives aux découvertes des Portugais. Mais la copie d'un document de si grande dimensione exigeant beaucoup de temps, je pensais que la publication de mon ouvrage et de mon atlas serait retardée de longtemps si j'attendais (p. 252) la copie entière du monument en question. Cependant, dès lors, j'ai entamé une correspondance pour obtenir une copie fidèle de cette grande mappemonde ; copie que je viens d'obtenir, et que j'ai fait graver en cette moment. Toutesfois, en attendant qu'elle soit entièrement terminée, je me crois heureux de pouvoir non seulement annoncer cette nouvelle à mes honorables confrères, mais aussi de leurs dire que ce même monument est entre mes mains, à la disposition de tous ceux qui voudront le consulter. Visconte de Santarém.

SOURCES

Apographon descriptionis Orbis terrae, figuris et narratiunculis distinctae, Manu Germanica opere nigelliari discolorio circa medium Saec. XV. Tabulae aeneae Musei Borgiani Velitris consignatae, Quod Camillus Joh. Paulli F. Borgia, Cruce Hieros. ornatus, ab intimo cubiculo Electoris Bavarici, Patrui Cardinalis exempla imitatus Summa fide, maximoque artificio expressum, recognitumque Eruditis spectandum proponit A. C. MDCCXCVII. (1797) ; Arnold Hermann Heeren, "Explicatio planiglobii orbis terrarum faciem exhibentis, ante medium seculum XV. summa arte confecti, Musei Borgiani Velitris," *Commentationes Societatis Regiae Scientarum Gottingensis* 16 (1808), pp. 250-284.

Agostino Fortunio, *Historiarum Camaldulensium pars posterior…*, Venetijs ex typographia Guerraea, Venetijs, apud Dominicum, et Io. Baptistam Guerraeos fratres, 1579 mense Septembris, pp. 174-175 ; see also the reprint in *Annales Camaldulenses*, vol. VII, pp. 299-302.

Albertus Magnus, *De causis proprietatum elementorum*, in *Alberti Magni OFM episcopi Opera Omnia*, P. Hossfeld (ed.), t. V, 2, Monasterii Westfalorum in aedibus Aschendorff, 1980, pp. 47-106.

—, *De natura loci*, in *Alberti Magni OFM episcopi Opera Omnia*, P. Hossfeld (ed.), t. V, 2, Monasterii Westfalorum in aedibus Aschendorff, 1980, pp. 1-46.

—, *Meteora*, in *Alberti Magni OFM episcopi Opera Omnia*, P. Hossfeld (ed.), t. VI, Monasterii Westfalorum in aedibus Aschendorff, 2003.

Albrecht Dürer, [Descriptio spherae armillaris una cum tota habitabili in plano], in *Claudii Ptolomaei Geographicae enarrationis libri octo Bilibaldo Pirckeymhero interprete. Annotationes Joannis De Regio Monte, in errores commissos a Jacobo Angelo in traslatione sua*, Argentorati, Johannes Grieningerus, communibus Johannis Koberger Impensis excudebat. Anno Christi nativitate MCXXV. Tertio Kal. Apriles, f. 69v.

Alexander Neckam, *Alexandri Neckam De naturis rerum libri duo with the poem of the same author, De laudibus divinae sapientiae*, T. Wright (ed.), London : Longman & Co., 1863.

Ambrogio di Lorenzo de' Rocchi, *Una pratica di mercatura in formazione : 1394-1395*, B. Dini. (ed.), Florence : Le Monnier, 1980.

St. Ambrose, *De paradiso*, I, 5-6 (Ambrosius, *Il giardino piantato a Oriente : De paradiso* ; introduction by U. Mattioli ; translation and notes by C. Mazza, Rome : Edizioni Paoline, 1981.

Andrea Bianco, *Atlante nautico (1436)*, P. Falchetta (ed.), Venice : Arsenale Editrice, 1993.

Antonio di Marco Morosini, *Cronicha de Veniexia*. A. Nanetti (ed.), PhD thesis. Bologna : Università degli Studi di Bologna, 2000.

António Galvão, *Summary of the Discoveries of the World, from Their First Original to the Year 1555*, in R. Kerr (ed.), *A General History and Collections of Voyages and Travels*, 18 vols., London : 1811.

—, *Tratado que compôs o nobre & notauel capitão António Galuão, dos diuersos & desuayrados caminhos, por onde nos tempos passados a pimenta & especearia veyo da India às nossas partes, & assi de todos os descobrimentos antigos & modernos, que são feitos até a era de mil & quinhentos & cincoenta. Com os nomes particulares das pessoas que os fizeram : & em que tempos & as suas alturas, obra certo muy notauel & copiosa*, [Lisbon], Rua de Sã Mamede : em casa de Ioam da Barreira, 15 Dezembro 1563.

Aristotle, *On the Heavens*, trans. W. K. C. Guthrie, London : Heinemann ; Cambridge, MA : Harvard University Press, 1939.

—, *The Physics*, with an English translation by P. H. Wicksteed and F. M. Cornford, London and Cambridge, MA : Harvard University Press, 1970.

St. Augustine, *The Literal Meaning of Genesis*, transl. and annotated by J. H. Taylor, 2 vols., NewYork-Ramsey : Newman Press, 1982.

—, *La Genesi alla lettera : testo latino dell'edizione maurina confrontato con il Corpus scriptorum ecclesiasticorum Latinorum* ; translation, notes and indices by L. Carrozzi, Rome : Nuova biblioteca agostiniana, Città nuova, 1989.

Avicenna, *Il poema della medicina*, A. Borruso (ed.), Turin : Zamorani, 1996.

Bartholomaeus Anglicus, *De rerum proprietatibus*, Frankfurt : 1606.

—, *On the Properties of Things. John Trevisa's Translation... of de Proprietatibus Rerum : A Critical Text*, Oxford : Oxford University Press, 1988.

Basilio di Cesarea, *Sulla Genesi*, M. Naldini (ed.), Milan : Fondazione Lorenzo Valla and Arnoldo Mondadori Editore, 1990.

Bede, *Libri quatuor in principium Genesis usque ad nativitatum Isaac et electionem Ismahelis adnotationum*, C. W. Jones (ed. and studies), Turnhout : Brepols, 1967.

Benedetto Cotrugli, *Il libro dell'arte di mercatura*, U. Tucci (ed.), Venice : Arsenale, 1990.

Bernardino Gadolo, *Biblia latina. cum glossa ordinaria Walafridi Strabonis ; aliorumque et interlineari Anselmi Laudunensis ; et cum postillis Nicolay de Lyra ; expositionibusque Guillelmi Britonis in omnes prologos S. Hieronymi ; Nicolaus de Lyra, Contra perfidiam Judaeorum*, (Venice) : (Paganino de' Paganini), 1495.

Bernardino Zendrini, *Memorie storiche dello stato antico e moderno delle lagune di Venezia*, Venice : 1726.

Bernardo Gomes de Brito, *História trágico-marítima, em que se escrevem chronologicamente os naufragios que tiverão as naos de Portugal, depois que se pozem exercicio a navegação da India*, Tom. 1-2, Lisboa Occidental, Na Officina da Congregação do Oratorio, 1735-36.

—, *História Trágico-Marítima*, Alexei Bueno (ed.), Rio de Janeiro : Lacerda Editores/Contraponto, 1998.

Brunetto Latini, *Li Livres dou Trésor*, F. O. Carmody (ed.), Berkeley : University of California Publications in Modern Philosophy 22, 1948.

Bullarium Patronatus Portugalliae Regum in ecclesiis Africae, Asiae atque Oceanie : bullas, brevia, epistolas, decreta actaque sedis ab Alexandro III ad hoc usque tempus amplectens, 5 vols., Jordão, L. M. (ed.), Lisbon : Typographia Nationali, 1868-1879.

Caius Iulius Solinus, *C. Iulii Solini collectanea rerum memorabilium iterum*, T. Mommsen (ed.), Berlin : Weidmann, 1895.

Campanus Novariensis, *Campanus of Novara and Medieval Planetary Theory. Theorica Planetarum*, edition with an introduction and English translation and commentary by F. S. Benjamin and G. J. Toomer, Madison : University of Wisconsin Press, 1971.

Christoph Gottlieb von Murr, *Diplomatische Geschichte des portugesichen beruehmten Ritters Martin Behaims. Aus Originalurkunden*, Nuerenberg : Johann Eberhard Zeh, 1778 (French translation : *Histoire diplomatique du chevalier portugais Martin Behaim de Nuremberg avec la description de son globe terrestre*, traduite de l'allemand par le citoyen H.-J. Jansen. 3rd edition, Strasbourg & Paris : Treuttel et Würtz, 1802).

Christophorus Buondelmontius, *Librum Insularum Archipelagi a codicibus Parisinis regiis nunc primum totum adidit*, G. R. L. von Sinner (ed.), Leipzig, 1824.

—, *Liber insularum archipelagi: Transkription des Exemplars Universitäts und Landesbibliothek Düsseldorf Ms. G 13, Übersetzung und Kommentar*, K. Bayer (ed.), 2 vols., Wiesbaden: L. Reichert, 2005 and 2007.

Ciriaco di Ancona (Kiriaki Anconitanis), *Itinerarium*, ed. L. Mehus (Editionem recensuit, animadversionibus, ac praefatione illustravit, nonnullisque ejusdem Kyriaci epistolis partim editis, partim ineditis locupletavit Laurentius Mehus), Florence: 1742 (reprint, Bologna: Forni, 1968).

—, *Later Travels*, E. W. Bodnar and C. Foss (eds), Cambridge, MA and London: I Tatti Renaissance Library and Harvard University Press, 2003.

Claudius Ptolemaeus, *Claudii Ptolemaei Geographiae codex Urbinas graecus 82*, phototypice depictus consilio et opera curatorum Bibliothecae Vaticanae, [Vatican City], Biblioteca Apostolica Vaticana, 1932: Tomus prodromus: J. Fischer S.J., *De Cl. Ptolemaei vita operibus Geographia praesertim eiusque fatis*. Pars prior. *Commentatio*, Vatican City: Biblioteca Apostolica Vaticana, 1932.

—, *Cosmographia: Bologna, 1477*. Introduction by R. A. Skelton, Amsterdam: N. Israel, 1963.

—, *Ptolemy's Geography: An Annotated Translation of the Theoretical Chapters*, A. Jones and J. L. Berggren (eds), Princeton: Princeton University Press, 2000.

—, *The Almagest*, G. Toomer (ed.), Princeton: Princeton University Press, 1984.

Dante Alighieri, *Quaestio de aqua et terra*, in *Opere minori*, tomo II, F. Mazzoni (ed.), Naples: Ricciardi, 1979, pp. 693-880.

Der 'Deutsche Ptolemäus' aus dem Ende des XV. Jahrhunderts (um 1490), in Faksimile-druck herausgegeben mit einer Einleitung von J. Fischer S.J., Strassburg, J. H. Ed. Heitz, 1910, ff. 2v-3r (before the chart marked "a2," pp. 52-53, unnumbered, in the facsimile edition).

Domenico Silvestri, *De insulis et earum proprietatibus*, C. Pecoraro (ed.), Atti dell'Accademia di Scienze Lettere e Arti di Palermo, s. IV, vol. XI, 1954, pp. 5-270.

Edmé-François Jomard, *Les Monuments de la Géographie, ou Recueil d'anciens cartes Européennes et Orientales accompagnés de Sphéres Terrestres et Célestes, de Mappemondes et Tables Cosmographiques, d'Astrolabes et autres instruments d'observation, depuis les temps les plus reculés jusqu'à l'époque d'Ortelius et de Gérard Mercator, publiés en facsimile de la grandeur des originaux…*, Paris: 1854.

Francesco Balducci Pegalotti, *Nomi di spezierie*, in *La pratica della mercatura*; A. Evans (ed.), Cambridge, MA: Harvard University Press, 1936, pp. 293-297 (reprint, New York, 1970).

Fazio degli Uberti, *Il Dittamondo e le Rime*, G. Corsi (ed.), Bari: Laterza, 1952, 2 vols.

Federico Crisogono, *Tractatus de occulta causa fluxux et refluxus maris* in *Federici Chrisogoni nobilis Jadertini Artium et Medicine doctoris Subtilissimi et Astrologi excellentissimi de modo Collegiandi Prognosticandi et curandi Febris Necnon de humana Felicitate ac denique de Fluxu et Refluxu Maris* …, Venetiis: impressum a Joan. Anto. De Sabbio et fratribus, 1528.

Federico Delfino, *De fluxu et refluxu aquae maris*, Venice: Accademia Veneta, 1528.

Francesco Berlinghieri, *In questo uolume si contengono septe giornate della geographia di Francesco Berlingeri fiorentino allo illustrissimo Federigo duca durbino*. Florence: Niccolò di Lorenzo della Magna, [1482]

Francesco Grisellini, *Succinta descrittione delle bellissime tele geografiche ora rinnovate ed accresciute nella sala del palazo ducale di san Marco detta dello Scudo ed esposte alla pubblica vista il dì 24 dicembre MDCCXII*, Venice: Modesto Fenzo, 1763, (reprinted under the same title as "Pubblicazione per Nozze Zajotti-Antonini," Venice, 1880).

Francesco Poggio Bracciolini, *India Recognita. Christoforus bullatus ducis Isubrium senator Petro Carae ducis alobrogum Senatoris. Poggii Florentini de uarietate Fortunae*, Milan : Uldericus Scinzenzeler, 1492.

—, *De varietate fortunae*, critical edition with an introduction and commentary edited by O. Merisalo, Helsinki : Suomalainen Tiedeakatemia, 1993 (Annales Academiae Scientiarum Fennicae, Ser. B).

—, L'India di Nicolò de' Conti : un manoscritto del Libro IV del De varietate fortunae di Francesco Poggio Bracciolini da Terranova (Marc. 2560), A. Grossato (ed.), Padua : Editoriale Programma, 1994.

—, *De l'Inde. Les voyages en Asie de Niccolò de' Conti. De varietate fortunae livre IV*, M. Guéret-Laferté (ed.), Turnhout : Brepols, 2004.

—, *The Travels of Nicolò Conti in the East in the Early Part of the Fifteenth Century*, translated from the original of Poggio Bracciolini, with notes, by J. Winter Jones. [Esq. FSA, Keeper of Printed Books, British Museum], in R. H. Major (ed.), *India in the Fifteenth Century. Being a Collection of Narratives of Voyages to India in the Century Preceding the… Discovery of the Cape of Good Hope ; from Latin, Persian, Russian and Italian sources. Now First translated into English*, London : Hakluyt Society, 1857, Part 2, pp. 1-39.

Gaius Plinius Secundus, *Storia naturale. Cosmologia e geografia : Libri 1-6*, introductory essay by G. B. Conte ; bibliographic note by A. Barchiesi, C. Frugoni, and G. Ranucci ; translation and notes by A. Barchiesi [et al.], Turin : Einaudi, 1982.

Geofroy de Tory de Bourges *Cosmographia Pii Papae, in Asiae et Europae eleganti descriptione Asia Historias rerum ubique gestarum cum locorum descriptione complectitur. Europa temporum authoris varias continet historias. Impressa per Henricum Stephanum impressorem diligentissimum Parisiis e regione schole decretorum sumptibus eius Henr. et Ioh. Hogonti VI Id. Octobris 1509.*

Giacomo Gastaldi (cartographer) and Antonio Lafrery (engraver ? and publisher), *Il disegno della terza parte dell'Asia. All' illmo. Sigr. il Sr. Marcho Fucharo Barone di Kirchberg e' d' Waisenhoren ; Giacomo di Castaldi Piamotese Cosmographo in Venetia* ; engraved and printed on paper, 70 × 71.8 cm, 1562. Plate 75 of Antonio Lafrery, [Tavole moderne di Geografia de la maggior parte del mondo di diversi autori raccolte…], Rome : A. Lafrery, 1545-1566.

Giorgio Vasari, *Le vite de' più eccellenti pittori, scultori, e architettori… di nuovo dal medesimo riviste et ampliate con i ritratti loro et con l'aggiunta delle vite de' vivi, & de' morti dall'anno 1550 infino al 1567…*, 3 vols., Florence : Giunti, 1568.

—, *Le vite dei più eccellenti pittori, scultori e architetti*. Rome : Newton Compton, 2007.

Giovan Battista Ramusio, *Dichiarazione d'alcuni luoghi ne' libri di messer Marco Polo, con l'istoria del reubarbaro*, in *Di messer Giovanbattista Ramusio prefazione sopra il principio del libro del magnifico messer Marco Polo. All'eccellente messer Ieronimo Fracastoro*, in *Navigationi et viaggi*, in Venetia, appresso i Giunti, 1559 (Modern edition ed. by M. Milanesi, Giovanni Battista Ramusio, "Dichiarazione d'alcuni luoghi nè libri di messer Marco Polo, con l'istoria del reubarbaro," in Giovanni Battista Ramusio, *Navigazioni e viaggi*, Turin : Einaudi, vol. III, 1980).

—, *Secondo volume delle nauigationi et viaggi, raccolto gia da m. Gio. Ramusio, et hora in questa nuoua editione accresciuto : nel quale si contengono l'Historia delle cose de Tartari, & diuersi fatti de' loro imperatori, descritta da m. Marco Polo gentil'huomo venetiano & da Hayton Armeno. Varie descrittioni di diuersi auttori,… et il viaggio della Tana. Con la descrittione de' nomi de' popoli,… Aggiuntoui in questa vltima editione la Descrittione dell'vna e dell'altra Sarmatia,… Con l'indice…*, In Venetia appresso i Giunti, 1583.

Giovanni Battista Mittarelli, Anselmo Costadoni, *Annales camaldulenses ordinis Sancti Benedicti quibus plura interseruntur tum ceteras Italico-monasticas res, tum historiam ecclesiasticam remque diplomaticam illustrantia*, 9 vols., Venice : Giambattista Pasquali, 1755-1773.

Giovanni Battista Mittarelli, *Bibliotheca codicum manuscriptorum S. Michaelis prope Murianum una cum Appendice librorum impressorum saeculi XV, opus posthumum Johannis-Benedicti Mittarelli Veneti, abbatis ex-generalis benedictino-Camaldulensis*, Venice : Typographia Fentiana [Modesto Fenzo editore], 1779.

Girolamo Zanetti, *Descrizione delle antiche Tavole geografiche collocate già nella Sala detta dello Scudo del Ducal Palagio di Venezia e rinnovate per decreto dell'Ecc.mo Senato nel presente Anno MDCCLXII. alla quale sono aggiunte moltissime memorie intorno alle Vite e Fatti de' principali e più celebri Viaggiatori Veneziani*, published as an appendix to G. L. Bertolini's "Relazione sul restauro delle Carte geografiche del Palazzo Ducale di Venice (1762) by Girolamo Zanetti," *Bollettino della Società Geografica Italiana* 52 (1925), pp. 326-340.

Hieronymus Münzer, *Viaje por España y Portugal (1494-1495)*. Prólogo de Manuel Gómez-Moreno : Traducción de José López Toro. Madrid : Almenara, 1951.

—, *Viaje por España y Portugal (1494-1495)*, Madrid : Polifemo, 1991.

Ioannes Buridanus, *Il cielo e il mondo. Commento al trattato 'Del cielo' di Aristotele*, A. Ghisalberti (ed.), Milan : Rusconi, 1983.

—, *Expositio et questiones in Aristotelis De Caelo*, B. Patar (ed.), Louvain-la-Neuve : Éditions de l'Institut supérieur de philosophie : Louvain and Paris : Éditions Peeters, 1996.

Iosephus Flavius, *Antichità giudaiche*, L. Moraldi (ed.), 2 vols., Turin : UTET, 1998.

Isidorus Hispalensis, *Etimologie, o Origini di Isidoro di Siviglia* ; 2 vols., A. Valastro Canale (ed.), Turin : UTET, 2004.

Isidorus of Kiev, "Lettere a Nicolò V, al Cardinal Bessarione, ai fedeli di Cristo, al Doge di Venezia, a Filippo di Brogogna," in A. Pertusi, *La caduta di Costantinopoli. Le testimonianze dei contemporanei*, Milan : Fondazione Lorenzo Valla and Arnoldo Mondadori Editore, 1997, pp. 52-111.

Jacobus de Voragine, *Le lege(n)de de tutti li sancti & le sancte dalla romana sedia acceptati & honorati* [Venice] : Nicolas Jenson, [1475, after 1 July] (IGI 5037).

Jacopo Dondi, *Il trattato della marea*, introduction, Latin text, translation and appendix by P. Revelli, Florence : Ricci, 1911.

Jean de Mandeville, *Le livre des merveilles du monde*, C. Deluz (ed.), Paris : CNRS, 2000.

Johannes de Sacrobosco, *Tractatus de sphaera*, Thorndike, *The Sphere of Sacrobosco.*

Jordanus Catalani, *Mirabilia descripta per fratrem Jordanum, Ordinis Praedicatorum, oriendum de Severaco, in India Majori episcopum columbrensem*, in *Recueil des voyages et mémoires, publié para la Société de Géographie*, Tome quatrième, Paris : Arthus-Bertrand, 1839.

—, *Mirabilia descripta, VII.6 Mirabilia descripta : The Wonders of the East, by Friar Jordanus, of the Order of Preachers, and Bishop of Columbum in India the Greater*, translated from the Latin original, as published at Paris in 1839, in the *Recueil de voyages et de mémoires*, of the Society of Geography, with the addition of a commentary, by Colonel Henry Yule, London : Hakluyt Society, 1863.

Jørgen Zoëga, *Catalogus codicum Copticorum manuscriptorum qui in Museo Borgiano Velitris adservantur*... Rome : typis Sacrae Congregationis de Propaganda Fide, 1810.

Leonardo da Vinci, *Codice Atlantico*, ff. 323 and 327v (formerly 117 r.b. and 119 v.a.). See also A. Marinoni, *Introduzione* in Leonardo da Vinci, *Codice Trivulziano*, Milan : Arcadia-Electa, 1980, p. xiv.

Lorenzo de' Medici, *Opere*, T. Zanato (ed.), Turin : Einaudi, 1992.

Libro intitulato Stimulo de amore : composto per il seraphico doctore sancto Bonauentura. In Venetia per Ioanne Antonio & Fratelli da Sabbio, 1521 dil mese de septembrio.

Ma Huan, *Ying-yai sheng-lan 'The Overall Survey of the Ocean's Shores' (1433)*, trans. J. V. G. Mills, Cambridge, MA : Hakluyt Society at the Cambridge University Press, 1970.

Marc'Antonio de Dominis, *Euripus, seu de fluxu et refluxu maris sententia*, Rome : Andreas Phaeus, 1624.

Marco Antonio Sabellico, M. A., *Enneades*, in *Opera omnia*, t. II, Basel, 1560.

Marco Polo, *De consuetudinibus et condicionibus orientalium regionum*, Gouda : Gerard Leeu, between 1483 and 11 June 1484 (IGI 7976). Modern edition : Marka Pavlova z Benátek, *Milion. Dle jediného rukopisu spolu s příslušným základem latinským*, Prague : Justin V. Prášek, 1902.

—, *The Book of Ser Marco Polo the Venetian Concerning the Kingdoms and Marvels of the East*, translated and edited, with notes, by Sir Henry Yule. Third edition revised by H. Cordier, 2 vols., London : John Murray, 1903.

—, *Il Milione. Prima edizione integrale a cura di L. F. Benedetto*, Florence : Olschki, 1928 (English translation London : Routledge, 1931).

—, *The Most Noble and Famous Travels of Marco Polo together with the Travels of Nicolò de Conti. Edited from the Elisabethan Translation* [1579] *of John Frampton with introduction, notes and appendixes by N. M. Penzer*, London : Argonaut Press, 1929 (repr. Amsterdam and New York 1971).

—, *The Description of the World*, edited, translated, and with an introduction by A. C. Moule and P. Pelliot, 2 vols., London : Routledge, 1938, (facsimile New York : AMS Press, 1976).

—, *Milione ; Le divisament dou monde. Milione nelle redazioni toscana e franco-italiana* ; G. Ronchi (ed.) ; introduction by C. Segre, Milan : Mondadori, 1982 (2000[4]).

—, *Il Milione* : introduction, edition of the Tuscan text (Ottimo), illustrative notes, exegetics, linguistics, onomastic repertoires and lexicons edited by R. M. Ruggieri, Florence : Olschki, 1986.

—, *Milione ; versione toscana del Trecento*, V. Bertolucci Pizzorusso (ed.) ; annotated index by G. R. Cardona, Milan, Adelphi, 1994.

—, *Milione. Redazione latina del manoscritto Z*. A. Barbieri (ed.), Parma : Fondazione Pietro Bembo, Guanda, 1998.

—, *Il 'Milione' veneto. Ms. CM 211 della Biblioteca Civica di Padova*, A. Barbieri and A. Andreose (eds), Venice : Marsilio, 1999.

—, *Le Devisement du monde*, P. Ménard (ed.), 6 vols., Genève : Droz, 2001-2009.

—, *Milione. Edizione toscana del Trecento*, V. Bertolucci Pizzorusso (ed.), annotated index G. R. Cardona, Milan : Adelphi, 2003.

—, *"Milione", redazione VB. Edizione critica commentata*. P. Gennari, Ph.D. dissertation in Italianistica e filologia romanza, 22° ciclo. Settore scientifico-disciplinare di afferenza : L-FIL-LET/09 : Dir. E. Burgio. Venice : Universita Ca' Foscari, 2010.

Marin Sanudo il Vecchio, *Liber secretorum fidelium Crucis super Terrae Sanctae recuperatione et conservatione, quo et Terrae Sanctae historia ab origine et eiusdem vicinarumque provinciarum geographica descriptio continetur…*, Jerusalem : Massada Press, 1972. (Partial reprint of the edition from Hannover, Typis Wechelianis apud heredes Ioannis Aubrii, 1611).

Monumenta Henricina, XIV (1460-1469), edited by Comissão Executiva dos Comemorações do V Centenário da Morte do Infante D. Henrique, Coimbra : 1973.

Neri di Bicci, *Le ricordanze : 10 marzo 1453 - 24 aprile* 1475, B. Santi (ed.), Pisa : Marlin, 1976.

Origenes Alexandrinus, *De principiis*, IV.3.1, Manlio Simonetti (ed.), Turin : UTET, 1968.

Paolo da Canal, *Zibaldone manoscritto mercantile del sec. XIV*, A. Stussi (ed.), with contributions by F. C. Lane, Th. E. Marston, O. Ore, Venice : Committee for the publications of sources relative to the history of Venice, 1967.

—, *Merchant culture in fourteenth century Venice. The Zibaldone da Canal*, translated with an introduction and notes by J. E. Dotson, Binghamton, New York : 1994.

Peter Lombard, *Sententiae in IV libris distictae*, t. 1.2, Grottaferrata : Editiones Collegii S. Bonaventurae ad Claras Aquas, 1971.

Petri Delphini Epistolae, *Veterum Scriptorum et monumentorum ecclesiasticorum et dogmaticorum amplissima collectio*, Martène, E., Durand, U. (eds), vol. III, Paris, 1724.

Philo Alexandrinus, *Le Allegorie delle leggi*, in *Tutti i trattati del Commentario allegorico alla Bibbia*, R. Radice (ed.), in collaboration with G. Reale, C. Kraus and C. Mazzarelli, Milan : Rusconi, 1994.

Pierre d'Ailly, *Ymago mundi : texte latin et traduction française des quatre traités cosmographiques de d'Ailly et de notes marginales de Cristophe Colomb. Etude sur les sources de l'auteur*, E. Buron (ed.), 3 vols., Paris : Maisonneuve, 1930.

Pietro Bembo, *Prose della volgar lingua : l'editio princeps del 1525 riscontrata con l'autografo Vaticano latino 3210*, C. Vela (ed.), Bologna : CLUEB, 2001.

Pietro Querini, Nicolò de Michiele, Cristofalo Fioravante, *Il naufragio della Querina : Veneziani nel circolo polare artico* ; P. Nelli (ed.), Rome : Nutrimenti, 2007.

Placido Zurla, *Il mappamondo di Fra Mauro camaldolese descritto ed illustrato da D. Placido Zurla dello stess'ordine*, Venice : s.n., 1806.

—, *Di Marco Polo e degli altri viaggiatori veneziani*, 2 vols., Venice : G. Fuchs, 1819.

—, *Dei vantaggi della cattolica religione derivati alla geografia e scienze annesse Dissertazione letta nell'Accademia di religione cattolica nel giorno 23 di maggio 1822*, Rome : Tipografia della Sagra Congregazione de Propaganda Fide, 1822.

Pomponius Mela, *De chorographia. Libri tres*, P. Parroni (ed.), Rome : Edizioni di Storia e Letteratura, 1984.

—, *Chorographie* ; Latin text, translation and notes by A. Silberman, Paris : Les Belles Lettres, 1988.

—, *Un antico volgarizzamento veneziano della Chorographia di Pomponio Mela*, L. Beltramo (ed.), Alessandria : Edizioni dell'Orso, 2002 (printed in 2003).

Prosdocimus de Beldemandis, "Super tractatu sphaerico commentaria," in Johannes de Sacro Bosco, *Sphaere tractatus Joannis de Sacro Busto. Gerardi Cremonensis theoricæ planetarum veteres. G. Purbachii theoricae planetarum novae. Prosdocimi de Beldomando super tractatu sphaerico commentaria, nup in lucem diducta per L. Ga*[*uricum*] *nunquam amplius impressa. Joannis Baptistae capuani expositio in sphaera & theoricis…*, Venetiis : in aedibus Luceantonii Iunte Florentini, 1531.

Pseudo-Aristotle, *De causis proprietatum et elementorum*, critical edition and study by S. L. Vodraska, Ph.D. Dissertation, University of London, 1969.

Rabanus Maurus. *De universo*, in *Patrologia Latina*, vol. III, col. 9-614.

Robertus Anglicus, *Commentus in Tractatum de spera Iohannis de Sacrobosco*, in L. Thorndike (ed.), *The Sphere of Sacrobosco and Its Commentators*, Chicago : The University of Chicago Press, 1949.

Roger Bacon, *Opus maius*, Pars IV, d. IV, c. 4 (edited, with introduction and analytical table, by J. H. Bridges, Oxford : Clarendon Press, 1897-1900).

Ruy González de Clavijo, *Embajada a Tamorlán. Estudio y edición de un manuscrito del siglo XV*, F. López Estrada (ed.), Madrid : C.S.I.C., 1943 ;

—, *Viaggio a Samarcanda 1403-1406. Un ambasciatore spagnolo alla corte di Tamerlano*, P. Boccardi Storoni (ed.), Rome : Viella, 1999.

Serto di documenti attenenti alle Reali Case di Savoja e di Braganza per le auspicatissime nozze di Sua A. R. la Principessa Pia di Savoja con Sua Maestá Don Luigi I Re di Portogallo, Florence : Stamperia Reale di Firenze, Francesco Cambiagi, 1862.

Strabo, *Geographia*, trans. H. L. Jones, Cambridge MA : Harvard University Press, 1917.

St. Thomas Aquinas, *The Summa Theologica of St. Thomas Aquinas*. Second and Revised Edition, 1920, literally translated by Fathers of the English Dominican Province.

—, *Summa theologiae*, cura et studio sac. Petri Caramello cum textu ex recensione leonina, pars prima et prima secundae, Turin : Marietti, 1952.

Thomas of Cantimpré, *Liber de natura rerum*, H. Boese (ed.), Berlin : de Gruyter, 1973.

Traité élémentaire de géographie, suivi d'un aperçu de la géographie ancienne… par Malte-Brun ; terminé… par ses collaborateurs MM. Larenaudière, Balbi et Huot… Tome second, Paris : Aimé-André ; Bruxelles : à la Librairie parisienne, 1831.

Vincenzo Terrarossa, *Riflessioni geografiche circa le terre incognite distese in ossequio perpetuo della nobilta veneziana… dal… Vitale Terra Rossa…*, Padua : Giacomo Cadorin, 1686.

Vincent of Beauvais, M. Paulmier-Foucart (ed.), *Vincent de Beauvais et le Grand miroir du monde*, Turnhout : Brepols, 2004.

Visconde de Santarém, *Essai sur l'histoire de la cosmographie et de la cartographie pendant le moyen-âge et sur les progrès de la géographie après les grandes découvertes du XVe siècle, pour servir d'introduction et d'explication à l'Atlas composé de mappemondes et de portulans, et d'autres monuments géographiques, depuis le VIe siècle de notre ère jusqu'au XVIIe*, Paris : Imprimerie Maulde et Renou, 1849, 1850 and 1852.

—, "Note sur le mappemonde de Fra Mauro," *Bulletin de la Société de Géographie de Paris*, 1849, pp. 251-252.

—, *Atlas composé de mappemondes, de portulans et de cartes hydrographiques et historiques depuis le VIe jusqu'au XVIIe siècle*, Paris : Maulde et Renou, 1849.

—, *Mappemonde dressée en 1459, par Fra Mauro, cosmographe vénetien, par ordre d'Alphonse V roi de Portugal : publiée pour la première fois de la grandeur de l'original avec toutes les légendes par le Vicomte de Santarém*, Lith. J. Feuquières, Paris : Kaeppelin, Lemercier, 1854.

—, *Algumas cartas ineditas do Visconde de Santarém*, introdução e notas de Vicente Almeida d'Eça, Lisbon : Sociedade de Geographia de Lisboa, 1906.

—, *Atlas du Vicomte de Santarém*, org. et préf. par Martim de Albuquerque, Lisbon : Administração do Porto de Lisboa, 1989.

—, *Correspondência do 2° Visconde de Santarém, colligida, coordenada e com anotações de Rocha Martins; publicada pelo 3° Visconde de Santarém*, 8 vols., Lisbon, Alfredo Lamas & Motta, 1918-1919.

Wilhelm Beer and Johann Heinrich von Mädler, *Mappa Selenographica totam Lunae hemisphaeram visibilem complectens observationibus propriis secundum projectionem orthographicam quatuor sectionibus constructa et delineate…*, Berlin, 1834.

—, *Der Mond nach seinen kosmischen und individuellen Verhaltnissen oder Allgemeine vergleichende Selenographie mit besondrer Beziehung auf die von den Verfassern herausgegebene Mappa selenographica*, Berlin: Schropp & Comp., 1837.

William Vincent, *The Commerce and Navigation of the Ancients in the Indian Ocean*, 2 vols., London: Printed for T. Cadell and W. Davies, 1807.

BIBLIOGRAPHY

Abu-Lughod, J. L., *Before European Hegemony: The World System A.D. 1250-1350*, New York and Oxford : Oxford University Press, 1989.

Adams, P. G., *Travellers and Travel Liars, 1660-1800*, Berkeley, CA : California University Press, 1962. Adravanti, F., *Tamerlano*, Milan : Fabbri, 2001.

Alegria, M. F., Daveau, S., Garcia, J. C., "Portuguese Cartography in the Renaissance," in D. Woodward (ed.), *The History of Cartography*, III, *Cartography in the European Renaissance*, Chicago : Chicago University Press, 2007, pp. 975-1068.

Alexander, J. J. G., "Notes on Some Veneto-Paduan Illuminated Books of the Renaissance," *Arte Veneta* 23 (1969), pp. 9-20.

—, "The Limbourg Brothers and Italian Art : A New York Source," *Zeitschrift für Kunstgeschichte* 46 (1983), pp. 425-435.

Allerston, P., "Wedding Finery in Sixteenth-Century Venice," in T. Dean and K. J. P. Lowe (eds), *Marriage in Italy, 1300-1650*, Cambridge : Cambridge University Press, 1998, pp. 25-40.

Almagià, R., "La dottrina della marea nell'antichità classica e nel Medio Evo," *Atti della R. Accademia dei Lincei* 301, *Memorie della classe di scienze fisiche, matematiche e naturali*, s. V, 5 (1904), pp. 377-513.

—, "Il Mappamondo di Albertin de Virga," *Rivista Geografica Italiana* 21 (1914), pp. 92-96.

—, "Gli italiani primi esploratori dell'America," in *L'opera del genio italiano all'estero*, Rome : Libreria dello Stato, 1937, pp. 393-395.

—, "Carta nautica con elementi corografici di Anonimo Veneziano (1450 circa)," in *Monumenta Cartographica Vaticana*, vol. I, Rome : Vatican City, 1944, pp. 32-39.

—, "*Osservazione sull'opera geografica di Francesco Berlinghieri*," *Archivio della R. Deputazione Romana di Storia Patria* 68 (1945), pp. 211-255.

—, "Presentazione" in T. Gasparrini Leporace, *Il Mappamondo di Fra Mauro camaldolese*, Rome : Poligrafico della Zecca dello Stato, 1956, pp. 5-9.

—, "Uno sconosciuto geografo umanista : Sebastiano Compagni," in *Miscellanea in onore di Giovanni Mercati*, IV, Vatican City : 1946, pp. 442-473, reprinted in his *Scritti geografici (1905-1957)*, Rome : Cremonese, 1961, pp. 527-552.

—, "Intorno ad alcune carte nautiche italiane conservate negli Stati Uniti," *Atti della Accademia Nazionale dei Lincei, Classe di scienze morali, storiche e filologiche*, Series 8, 7 (1952), pp. 356-366.

Ames-Lewis, F. (ed.), *Cosimo 'il Vecchio' de' Medici, 1389-1464*, Oxford : Clarendon Press, 1992.

André J., Filliozat, J., *L'Inde vue de Rome. Textes latins de l'Antiquité relatifs à l'Inde*, Paris : Les Belles Lettres, 1986.

Andreose, A., "Tra ricezione e riscrittura : la fortuna romanza della 'Relatio' di Odorico da Pordenone," in G. Carbonaro, M. Cassarino, E. Creazzo and G. Lalomia (eds), *Medioevo romanzo e orientale. Il viaggio nelle letterature romanze e orientali*, Genoa : Rubbettino, 2007, pp. 5-21.

Andresen Leitão, R., *Diário de viagem de França del rei Dom Pedro V (1855)*, Paris : Fundação Calouste Gulbenkian, 1970.

Archer, M., *British Drawings in the India Office Library*, 2 vols., London : Foreign and Commonwealth Office, 1969.

Ashtor, E., *Levant Trade in the Late Middle Ages*, Princeton : Princeton University Press, 1983.

Auer, J., *Il mondo come creazione*, Assisi : Cittadella Editrice, 1977.

Aujac, G., *Strabon et la science de son temps*, Paris : Les Belles Lettres, 1966.

—, "Le peintre florentin Piero del Massaio et la 'Cosmographia' de Ptolémée," *Geographia Antiqua* 3-4 (1994-1995), pp. 187-204.

Avril, F., "Das Buch der Wunder der Welt. Ms. fr. 2810 der Bibliothèque nationale de France," in *Das Buch der Wunder : aus "Le Livre des Merveilles du monde", Ms. fr. 2810 der Bibliothèque Nationale de France*. Übersetzung der Handschrift M-H. Tesnière. Kritischer Essay F. Avril. Bilderäuterungen M.-T. Gousset, Munich : Hirmen Verlag, 1999, pp. 197-216.

Bacci, P. (ed.), *Documenti toscani per la storia dell'arte*, Florence : Gonnelli, 2 vols., 1910-12.

Bagnoli, A. (ed.), *Duccio : alle origini della pittura senese*, Cinisello Balsamo : Silvana, 2003.

Balard, M., "Les Génois en Asie centrale et en Extrême-Orient au xive siècle : un cas exceptionnel ?" in *Mélanges offerts à Edouard Perroy*, Paris, 1973, pp. 681-689.

—, "Precursori di Cristoforo Colombo. I genovesi in Estremo Oriente nel XIV secolo," in *Atti del convegno internazionale di studi colombani, 13-14 ottobre 1973*, Genoa, 1974, pp. 149-164.

—, "Génois et Pisans en Orient (fin du xiiie - début du xive siècle)," in *Genova, Pisa e il Mediterraneo tra Due e Trecento. Atti della Società ligure di Storia Patria*, Genoa, 1984, pp. 179-209.

Barbato, M., *Il libro VIII del Plinio napoletano di Giovanni Brancati*, Naples : Liguori, 2001.

Barber, P., "Visual Encyclopedias. The Hereford map and Other Mappae Mundi," *The Map Collector* 48 (1989), pp. 2-8.

—, Harvey, P. D. A. (eds), *The Hereford Mappa Mundi : Proceedings of the Mappa Mundi Conference*, London : British Library, 2006.

—, Harper, T., *Magnificent Maps. Power, Propaganda and Art*, London : The British Library, 2010.

Barbieri, A., "Un Veneziano nel Catai : sull'autenticità del viaggio di Marco Polo," *Critica del testo* 3.3 (2000), pp. 993-1022 (reprinted in Id., *Dal viaggio al libro. Studi sul Milione*, Verona : Fiorini, 2004, pp. 9-43).

—, "Marco, Rustichello, il "patto", il libro : genesi e statuto testuale del Devisement dou monde alla luce degli studi recenti," in G. Carbonaro et al. (eds), *Medioevo romanzo e orientale. Il viaggio nelle letterature romanze e orientali*, Genoa : Rubbettino, 2006, pp. 23-42.

Barbieri, E., *Le bibbie italiane del Quattrocento e del Cinquecento : storia e bibliografia ragionata delle edizioni in lingua italiana dal 1471 al 1600*, 2 vols., Milan : Bibliografica, 1992.

—, *Il libro nella storia : tre percorsi*, Milan : CUSL, 1999.

Barbieri, G., Fiaccadori G. (eds), *Nigra sum sed Formosa. Sacro e bellezza dell'Etiopia cristiana*, Treviso : Terra-Ferma, 2009.

Barocchi, P., Gaeta Bertelà, G., *Collezionismo mediceo e storia artistica 1545-1743*, 2 vols., Florence : S.P.E.S, 2002-2010.

Battaglia, S., Bàrberi Squarotti, G. (eds), *Il grande Dizionario della lingua italiana*, 21 vols., Turin : UTET, 1961-2002.

Bauer-Eberhardt, U., "Die Rothschild-Miscellanea in Jerusalem Haupwerk des Leonardo Bellini," *Pantheon* 42 (1984), pp. 229-237.

Baxandall, M., *Pittura ed esperienze sociali nell'Italia del Quattrocento*, Turin : Eiunaudi, 1978.

Bazàn, B. C., Wippel, J. W., Fransen, G., Jacquart, D., *Les questions disputées et les questions quodlibétiques dans les facultés de théologie, de droit et de médecine*, Turnhout : Brepols, 1985.

Bec, C., *Les marchands écrivains. Affaires et humanisme à Florence, 1375-1434*, Paris : Mouton, 1967.

Beck, H. G., Manoussacas, M., Pertusi, A. (eds), *Venezia centro di mediazione fra Oriente e Occidente (secoli XV-XVI). Aspetti e problemi*, Florence : Olschki, 1977.

Belozerskaya, M., "Jan van Eyck's Lost *Mappa mundi* : A Token of Fifteenth-Century Power Politics," *Journal of Early Modern History* 4. 1 (2000), pp. 45-84.

Berchet, G., *Il planisfero di Giovanni Leardo dell'anno 1452. Fac-simile nella grandezza originale*, Venice : Ferdinando Ongania Editore, 1880.

Bernardinello, S., "Le carte dell'Africa nella Geographia di Tolomeo. Lettura dal codice Marciano gr. Z. 516," *Atti e memorie dell'Accademia Patavina di Scienze, Lettere ed Arti. Parte III* 109 (1996-97), pp. 48-95.

Bernocchi, M., *Le monete della Repubblica fiorentina*, Florence : Olschki, 1974-78.

Bertelli, S., Rubinstein, N., Smyth, C. H.(eds), *Florence and Venice, Comparisons and Relations*. Acts of Two Conferences at Villa I Tatti in 1976-1977, Florence : La Nuova Italia, 1979.

Bertolucci Pizzorusso, V., "Enunciazione e produzione del testo nel Milione," *Studi mediolatini e volgari* 25 (1977), pp. 5-43 (reprinted in Ead., *Morfologie del testo medievale*, Bologna : Il Mulino, 1989, pp. 209-241).

Bethencourt, F., Chaudhuri, K. N. (eds), *História da expansão portuguesa*, Lisbon : Temas e Debates, 5 vols., 1998-2000.

Bevilacqua, E., "Geografi e cosmografi," in G. Arnaldi and M. Pastore Stocchi (eds), *Storia della cultura veneta*, vol. 3.2, *Dal primo Quattrocento al Concilio di Trento*, Vicenza : Neri Pozza, 1980, pp. 355-374.

Bianchi, R., "Notizie del cartografo veneziano Antonio Leonardi. Con una appendice su Daniele Emigli (o Emilei) e la sua laurea padovana," in V. Fera and G. Ferraù (eds), *Filologia umanistica. Per Gianvito Resta*, Padua : Antenore, 1997, pp. 165-211.

Billanovich, E., "Per la storia della pittura nel Veneto," *Italia medioevale e umanistica* 16 (1973), pp. 362-367.

Billanovich, G., "Dall'antica Ravenna alle biblioteche umanistiche," *Aevum* 30 (1956), pp. 319-353 (reprinted under the same title but as a revised version in *Italia medioevale e umanistica* 36 (1993), pp. 107-174).

Blackmore, J., *Manifest Perdition : Shipwreck Narrative and the Disruption of Empire*, Minneapolis, MN : Minnesota University Press, 2002.

Blair, A., Grafton, A., "Reassessing Humanism and Science," *Journal of the History of Ideas* 53. 4 (1992), pp. 535-540.

Boitani, P., Bologna, C., Cipolla, A., Liborio, M. (eds), *Alessandro nel Medioevo occidentale*, introduction by P. Dronke, Rome : Fondazione Lorenzo Valla, 1997.

Borghini, G., Vasco Rocca, S. (eds), *Giovanni V di Portogallo (1707-1750) e la cultura romana del suo tempo*, Rome : Argos, 1995.

Borlandi, F., "All'origine del libro di Marco Polo," in *Studi in onore di Amintore Fanfani*, vol. I, Milan : Giuffrè, 1962, pp. 107-147.

Boschetto, L.,"Una nuova lettera di Giannozzo Manetti a Vespasiano da Bisticci. Con alcune considerazioni sul commercio librario tra Firenze e Napoli nei decenni centrali del Quattrocento," *Medioevo e Rinascimento* 27 (2004), pp. 175-205.

Boskovits, M., "Per Jacopo Bellini pittore (postilla ad un colloquio)," *Paragone. Arte* 36, nn. 419-423 (1985), pp. 113-123.

Bouloux, N., *Culture et savoirs géographiques en Italie au XIVe siècle*, Turnhout : Brepols, 2002.

Boxer, C. R., "An Introduction to the História Trágico-Marítima," in *Miscelânea de Estudos em Honra do Prof. Hernâni Cidade*, Lisbon : Publicações da Faculdade de Letras, 1957, pp. 48-99.

—, "An Introduction to the História Trágico-Marítima. Some Corrections and Clarifications," *Quaderni portoghesi* 5 (1979), pp. 99-112.

Branca, V., "Dal favoloso al realistico : esotismo fra pellegrini mercanti e Boccaccio," in R. Zorzi (ed.), *L'Europa delle scoperte*, Florence : Olschki, 1994, pp. 39-60.

—, *La sapienza civile. Studi sull'Umanesimo a Venezia*, Florence : Olschki, 1998.

Braudel, F., *La Méditerranée et le monde méditerranéen à l'époque de Philippe II*, Paris : Armand Colin, 1949.

—, *Civilisation matérielle, économie et capitalisme XVe-XVIIIe siècle*, 3 vols., Paris : Armand Colin, 1979.

Bray, N., Sturlese, L., (eds), *Filosofia in volgare nel medioevo*. Atti del Convegno della Società Italiana per lo Studio del Pensiero Medievale (SISPM), Lecce, 27-29September 2002, Turnhout : Brepols, 2003.

Brenk, B., "Originalità e innovazione nell'arte medievale," in E. Castelnuovo and G. Sergi (eds), *Arti e storia nel Medioevo*, vol. 1, Tempi Spazi Istituzioni, Turin : Einaudi, 2002, pp. 3-69.

Briant, T., *Les plus beaux textes sur la Mer*, Paris : La Colombre, 1950.

Brincken, A-D von den, "Mappa Mundi und Chronographia : Studien zur imago mundi des abendländlischen Mittelalters," *Deutsches Archiv für Erforschung des Mittelalters* 24 (1968), pp. 118-186.

Briquet, C. M., *Les filigranes : dictionnaire historique des marques du papier dès leur apparition vers 1282 jusqu'en 1600 : a facsimile of the 1907 edition with supplementary material contributed by a number of scholars*, 4 vols., A. Stevenson (ed.), Amsterdam : Paper Publication Society, 1968.

Brunello, F., *Marco Polo e le merci dell'Oriente*, Vicenza : Neri Pozza, 1986.

Burgio, E., "Marco Polo e gli 'idolatri'," in N. Pasero and S. Barillari (eds), *Le voci del Medioevo. Testi, immagini, tradizioni*, Alessandria : Edizioni dell'Orso, 2006, pp. 31-62.

—, "'Cartografie'" di viaggio. Sulle relazioni fra la *Mappamundi* di Fra Mauro e il *Milione*," *Critica del testo* 12.1 (2009), pp. 59-106.

Burke, P., "Early Modern Venice as a Center of Information and Communication," in J. Martin and D. Romano (eds), *Venice Reconsidered : the History and Civilization of an Italian City-State, 1297-1797*, Baltimore and London : Johns Hopkins University Press, 2000, pp. 389-419.

Busi, G., *Qabbalah visiva*, Turin : Einaudi, 2005.

Caby, C., "Culte monastique et fortune humaniste : Ambrogio Traversari, vir illuster de l'ordre camaldule," in Mélanges de l'École française de Rome. Moyen Âge 108 [1], Rome : École française de Rome, 1996, pp. 321-354.

—, "Bernardino Gadolo ou les débuts de l'historiographie camaldule," *Mélanges de l'Ecole française de Rome. Moyen-Age* 109, 1 (1997) pp. 225-268.

—, *De l'érémitisme rural au monachisme urbain. Les Camaldules en Italie à la fin du Moyen Âge*, Rome : École française de Rome, 1999.

—, "Entre observance et humanisme. Définition et pratiques d'une orthodoxie culturelle dans l'Ordre Camaldule," in *Orthodoxie, christianisme, histoire*, Collection de l'École française de Rome 270, Rome : École française de Rome, 2000, pp. 3-22.

Campbell, T., "Portolan Charts from the Late Thirteenth Century to 1500," in J. B. Harley and D. Woodward (eds), *The History of Cartography*, vol. 1, *Cartography in Prehistoric, Ancient and Medieval Europe and the Mediterranean*. Chicago and London : University of Chicago Press, 1987, pp. 415-421.

—, *The Earliest Printed Maps, 1472-1500*, London : British Library ; Berkeley, CA : University of California, 1987.

Capacci, A., "Planisfero detto 'genovese'," in G. Cavallo (ed.), *Cristoforo Colombo e l'apertura degli spazi*, Rome : Istituto Poligrafico e Zecca dello Stato, 1992, 2 vols., vol. 1, entry III.18, pp. 491-494.

Caraci, G., "Mappamondo anonimo del secolo XV," *Enciclopedia Italiana*, vol. XI, plate LXXI, fig. 13, Rome : Istituto Giovanni Treccani, 1931, pp. 230-250.

Caraci, G. "The Italian Cartographers of the Benincasa and Freducci families and the So-Called Borgiana Map of the Vatican Library," *Imago Mundi* 10 (1953), pp. 23-45.

Cardona, G. R., "Indice ragionato," in Marco Polo, *Milione. Edizione toscana del Trecento*, V. Bertolucci Pizzorusso (ed.), Milan : Adelphi, 2003, pp. 489-761.

Carnesecchi, C., "Paolo Toscanelli e gli ambasciatori del re di Portogallo nel 1459," *Archivio Storico Italiano* 5, 21 (1898), pp. 316-318.

Caroti, S., "Filosofia e scienza della natura nel Medioevo e nel Rinascimento," in W. R. Shea (ed.) *Storia delle scienze. Le scienze fisiche e astronomiche*, Turin : Einaudi, 1992, pp. 110-147.

Carpo, M., *Metodo ed ordini nella teoria architettonica dei primi moderni* : *Alberti, Raffaello, Serlio e Camillo*, Geneva : Droz, 1993.

Carrère, C., *Barcelone, centre économique à l'époque des difficultés, 1380-1462*, 2 vols., Paris and La Haye : Mouton, 1967.

Casanova, E., "Un anno della vita privata di Pio II," *Bullettino senese di storia patria* 2 (1931).

Casapullo, R., "Segmentazione del testo e modalità d'uso delle enciclopedie tra latino e volgare," in R. Gualdo (ed.), *Le parole della scienza : scritture tecniche e scientifiche in volgare, secoli XIII-XV*, Galatina : Congedo, 2001, pp. 153-185.

—, Policardo, M. R., "Tecniche della divulgazione scientifica nel volgarizzamento mantovano del "De proprietatibus rerum" di Bartolomeo Anglico," *Lingua e Stile* 2 (2003), pp. 139-176.

Casella, N., "Pio II tra geografia e storia : la 'Cosmographia'," *Archivio della Società Romana di storia patria*, s. III, 26 (1972), pp. 35-112.

Cassandro, C., Giovè, N., Marchioli, C., Massalin, P., Zamponi, S. (eds), *I manoscritti datati della provincia di Vicenza e della Biblioteca Antoniana di Padova*, Tavarnelle (Florence) : Sismel-Edizioni del Galluzzo, 2000.

Castellani, F., *Quaternuccio e giornale B, 1459-1485*, Florence : Olschki, 1995.

Cattaneo, A., "Fra Mauro Cosmographus Incomparabilis and His *Mappamundi* : Documents, Sources, and Protocols for Mapping," in A. Cattaneo, A. Ferrand Almeida, D. Ramada Curto (eds), *La cartografia europea tra primo Rinascimento e fine dell'Illuminismo*, Florence : Olschki, 2003, pp. 19-48.

—, "God in His World. Leonardo Bellini Illuminator of the Earthly Paradise in Fra Mauro's *Mappa mundi*," *Imago Mundi* 40, (2003), pp. 97-102.

—, "A Forgotten Mid-Fifteenth Century Venetian Manuscript Codex with Sixty Maps and also the First Vernacular Adaptation of Ptolemy's Geography," in *ICHC 2003 Program, Abstracts, and List of Participants*, Cambridge and Portland, 2003.

—, "La 'Ptolomei Cosmographia' di Henricus Martellus Germanus," in *Ptolomei Cosmographia*, Florence : Vallecchi, 2004, pp. 23-55.

—."Scritture di viaggio e scrittura cartografica. La *mappamundi* di Fra Mauro e i racconti di Marco Polo e Niccolò de' Conti." *Itineraria* 3-4 (2004-2005), pp. 157-202.

—, "Letture e lettori della Geografia di Tolomeo a Venezia intorno alla metà del Quattrocento," *Geographia Antiqua* 4-5 (2005), pp. 41-65.

—, "From the Mediterranean to the Atlantic. Venice, Florence and Lisbon. Commercial Routes and Networks of Knowledge, 1300-1550," in J. A. Levenson (ed.), *Encompassing the Globe : Portugal and the World in the 16th & 17th Centuries*. Essays [Vol. 3], Washington, D.C. : Arthur M. Sackler Gallery, 2007, pp. 13-21.

—, "La cosmografia del Rinascimento : una storia polifonica," in G. Ernst and A. Clericuzio, *Le scienze nel Rinascimento italiano*, vol. 4. G. L. Fontana and L. Molà (eds), *Il Rinascimento italiano e l'Europa*, 12 vols., Verona : Fondazione Cassamarca, Angelo Colla Editore, 2008, pp. 551-569.

—, "Cosmographie et prédication médiévale et renaissante," *Cahiers V.-L. Saulnier* 26 (Paris, Editions Rue d'Ulm, 2008) pp. 57-70.

—, *Mappa mundi 1457 (Biblioteca nazionale centrale di Firenze, Port. 1). Analisi, trascrizione e commentario*, Rome : Istituto dell'Enciclopedia Italiana, 2008.

—, "La cosmografia di Cosimo," in F. Camerota and M. Miniati, *I Medici e le scienze. Strumenti e macchine nelle collezioni granducali*. Catalogo della mostra (Firenze, maggio 2008-gennaio 2009), Florence : Giunti, pp. 147-151.

—, "Orb and Sceptre : Cosmography and World Cartography in Portugal and Italian Cities in the Fifteenth Century," *Archives Internationales d'Histoire des Science* 59 (2009), pp. 531-555.

—, "Découvertes littéraires et géographiques au xve siècle. Le 'Portolano 1' de la Bibliothèque nationale Centrale de Florence," *Médiévales* 58 (2010), pp. 79-98.

—, "Venedig, 1450. Ozen – Meer – Seefahrt – Welthandelsrouten – Schiffbrüche," in H. Baader, G. Wolf (eds), *Das Meer, der Tausch und die Grenzen der Repräsentation*, Zürich and Berlin : Diaphanes, 2010, pp. 263-292.

—."Il lungo viaggio della cartografia universale attraverso il mondo islamico, la Cina, la Corea e il Giappone (sec. XIV-XVI). Studio dell carta universale *Honil kangni yŏktae kukto chi to* (Corea, c. 1479-85) in una prospettiva comparata," in M. Angelillo (ed.), *Culture, religioni e diritto nelle società dell'Asia Orientale*, Rome : Bulzoni, 2010, pp. 177-200 (*Asiatica Ambrosiana, Saggi e ricerche di cultura religioni e società dell'Asia* 2. Atti del convegno "Culture, religioni e diritto nelle società dell'Asia Orientale" organizzato dalla Accademia Ambrosiana, Classe di Studi sull'Estremo Oriente, Milano).

—, "Inventare musei per ordinare e rappresentare il mondo. La Guardaroba nuova di Palazzo Vecchio e le Sale della cosmografia e delle matematiche agli Uffizi a Firenze," *Revista de História da Arte* 8 (2010), in press.

Cattani, A., *Da Venezia in viaggio con la Posta. Pagine e documenti di storia veneta*, Padua : Editrice Elzeviro, 2002.

Cavallo, G., "Dallo *scriptorium* senza biblioteca alla biblioteca senza *scriptorium*," in *Dall'eremo al cenobio. La civiltà monastica in Italia dalle origini all'età di Dante*, Milan : Scheiwiller (Antica Madre X), 1987, pp. 331-421.

—, (ed.), *Cristoforo Colombo e l'apertura degli spazi*, 2 vols., Rome : Istituto Poligrafico e Zecca dello Stato, 1992.

Cerulli, E., "Fonti arabe del mappamondo di Fra Mauro," *Orientalia : Commentarii periodici Pontifici Istituti Biblici* 4 (1935) pp. 336-338.

Chartier, R., *Au bord de la falaise. L'histoire entre certitudes et inquiétudes*, Paris : Albin Michel, 1998.

Chaudhuri, K. N., *The Trading World of Asia and the English East India Company, 1600-1760*, Cambridge : Cambridge University Press, 1978.

—, *Trade and Civilisation in the Indian Ocean : an Economic History from the Rise of Islam to 1750*, Cambridge : Cambridge University Press, 1985.

Ciano, C., *La pratica di mercatura datiniana secolo XIV*, introd. F. Melis, Milan : Giuffré, 1964.

Ciappelli, G., *Una famiglia e le sue ricordanze. I Castellani di Firenze nel Tre-Quattrocento*, Florence : Olschki, 1995.

—, "Carte geografiche e politica nei rapporti tra Firenze e il Portogallo nel Quattrocento," *Annali dell'Istituto storico italo-germanico in Trento* 32 (2006), pp. 47-70.

Cieli e terre nei secoli XI-XII. Orizzonti, percezioni, rapporti. Atti della tredicesima Settimana internazionale di studio, Mendola, 22-26 agosto 1995, Milan : Vita e pensiero, 1998.

Cigni, F. (ed.), *Il romanzo arturiano di Rustichello da Pisa*, foreword by V. Bertolucci Pizzorusso, Pisa : Pacini and Cassa di Risparmio di Pisa, 1994.

Clagett, M., "Some General Aspects of Physics in the Middle Ages," *Isis* 39, 1-2 (1948), pp. 29-44.

—, "The Life and Works of Giovanni Fontana," *Annali dell'Istituto e Museo di Storia della scienza di Firenze* 1 (1976), pp. 5-28.

Classen, A., "Die iberische Halbinsel aus der Sicht eines humanistischen Nürnberger Gelehrten. Hieronymus Münzer, *Itinerarium Hispanicum* (1494-1495)," *Mitteilungen des Instituts für Österreichische Geschichtsforschung* 111 (2003), pp. 317-340.

Coli, E., *Il Paradiso terrestre dantesco*, Florence : Carnesecchi, 1897.

Colish, M., *Peter Lombard*, 2 vols., Leiden : Brill, 1994.

—, *Medieval Foundations of the Western Intellectual Tradition, 400-1400*, New Haven : Yale University Press, 1997.

Coluccia, G. L., *Basilio Bessarione. Lo spirito greco e l'occidente*. Florence : Olschki, 2009.

Conterio, A., "L'arte del navegar : cultura, formazione professionale ed esperienze dell'uomo di mare veneziano nel XV secolo," in *L'uomo e il mare nella civiltà occidentale : da Ulisse a Cristoforo Colombo*, "Atti della Società Ligure di Storia Patria", new series, 32 (1992), pp. 187-225.

Cook, A. S., "Alexander Dalrymple's 'A Collection of Plans of Ports in the East Indies' (1774-1775) : A Preliminary Examination," *Imago Mundi* 33 (1981), pp. 46-64.

—, *Alexander Dalrymple (1737-1808), hydrographer to the East India Company and to the admiralty, as publisher*, Ph.D. dissertation., St. Andrews : University of St. Andrews, 1992.

—, "Dalrymple, Alexander (1737-1808)," in H. C. G. Matthew and B. Harrison (eds), *Oxford Dictionary of National Biography*, Oxford : Oxford University Press, 2004 (on-line edition).

Cordier, H., *Ser Marco Polo. Notes and Addenda to Sir Henry Yule's Edition, Containing the Results of Recent Research Discovery*, London : Murray, 1920.

Corner, F., *Notizie storiche delle chiese e monasteri di Venezia e di Torcello*, Bologna : Forni, 1990.

Cortelazzo, M., Paccagnella, I., "Il Veneto," in F. Bruni (ed.), *L'italiano nelle regioni*, Turin : UTET, 1992, pp. 220-280.

Cortesão, A., *Cartografia e cartógrafos portugues dos séculos XV e XVI (Contribudo para um estudo completo)*, 2 vols., Lisbon : Seara Nova, 1935.

—, Teixeira da Mota, A., *Portugaliae Monumenta Cartographica*, Lisbon : 1960, 6 vols.

Cortesão, J., *A politica de sigilo nos descobrimentos nos tempos do infante D. Enrique e de D. João II*, Lisbon : Comissão Executiva das Comemorações do V Centenário da Morte do Infante D. Henrique, 1960.

Cortesi, M, "Umanisti alla ricerca di Padri greci," in S. Gentile (ed.), *Umanesimo e Padri della Chiesa. Manoscritti e incunaboli di testi patristici da Francesco Petrarca al primo Cinquecento*, Rome : Rose, 1997, pp. 63-75.

—, Leonardi, C., *Tradizioni patristiche nell'Umanesimo*, Florence : Sismel, 2000.

—, "Gli umanisti lettori di s. Basilio tra proposte pedagogiche, motivi ascetici e dottrina teologica," in *Basilio tra Oriente e Occidente*. Convegno internazionale 'Basilio il Grande e il monachesimo orientale', Cappadocia, 5-7 ottobre 1999. Magnano : Qiqajon, 2001, pp. 253-278.

Costa, J. P. Oliveira, *Henrique, O Infante*, Lisbon : Esfera dos Livros, 2009.

Crawford, O. G. S. (ed.), *Ethiopian Itineraries circa 1400-1524. Including those collected by Alessandro Zorzi at Venice in the years 1519-24*, Cambridge, MA : Hakluyt Society at the Cambridge University Press, 1958.

Crinò, S., *Come fu scoperta l'America*, Milan : Hoepli, 1943.

Crivellari, G., *Alcuni cimeli della cartografia medievale esistenti a Verona*, Florence : Bernardo Seeber, 1903.

Crombie, A. C., "Intuizioni storiche della scienza medievale," in P. Tourbert and A. Paravicini Bagliani (eds), *Federico II e le scienze*, Palermo : Sellerio, 1994, pp. 15-24.

—, *Styles of Scientific Thinking in the European Tradition*, London : G. Duckworth, 1994.

Curtius, E. R., *Letteratura europea e Medio Evo latino*, Florence : La Nuova Italia, 1992.

D' Ancona, P., *La miniatura florentina (Secoli XI-XVI)*, Florence : Olschki, 1914.

D' Arienzo, L., "La presenza italiana in Portogallo e nella Spagna meridionale all'epoca di Cristoforo Colombo," in G. Cavallo (ed.), *Cristoforo Colombo e l'apertura degli spazi*, 2 vols., Rome : Istituto Poligrafico e Zecca dello Stato, 1992, vol. I, pp. 535-565.

—, *La presenza degli italiani in Portogallo al tempo di Colombo*, Ministero per i Beni Culturali e Ambientali, Comitato nazionale per le celebrazioni del V Centenario della scoperta dell'America, Nuova Raccolta Colombiana, Rome : Istituto Poligrafico e Zecca dello Stato, 2004.

Da Mosto, A., *L'Archivio di Stato di Venezia. Indice generale, storico, descrittivo ed analitico*, Venice : Archivi dell'amministrazione centrale, 1937.

Da Silva Marques J. M., Iria, A. (eds), *Descobrimentos portugueses : documentos para a sua história*, 3 vols., Lisbon : Instituto para a Alta Cultura, 1944-1971.

Das Gupta, A., *The World of the Indian Ocean Merchant, 1500-1800*, New Delhi : Oxford University Press, 2001.

De Albuquerque, L., *Os descobrimentos portugueses*, Lisbon : Alfa, 1989.

De la Mare, A., "Notes and Documents. Observations on Two Italian Manuscripts from Madrid Recently Exhibited in the Bodleian," *Bodleian Library Review* 12 (1985-1988) pp. 242-247.

De Nicolò Salmazo, A., "Padova," in M. Lucco (ed.), *La pittura nel Veneto. Il Quattrocento*, vol. 2, Milan : Electa, 1990, pp. 481-540.

—, *Francesco Squarcione 'pictorum gymnasiarcha singularis'*, Padua : Il Poligrafo, 1999.

De Reparaz, G. Jr., *Mestre Jacome de Malhorca, cartógrafo do Infante*, Coimbra : Imprensa da Universidade, 1930.

De Roover, R., *The Rise and Decline of the Medici Bank, 1397-1494*. New York : W. W. Norton, 1966.

De Smet, A., "Les géographes de la Renaissance et la cosmographie," in *Album Antoine de Smet*, Brussels : Centre national d'Histoire des Sciences, 1974, pp. 149-160.

De Toni, G. B., "Francesco Griselini [*sic*] viaggiatore e naturalista veneziano del secolo XVIII," *Archivio di storia della scienza* 1 (1919), pp. 1-21.

Degenhart, B., Schmitt, A., *Corpus der Italienischen Zeichnungen 1300-1450*, Teil II : *Venedig. Jacopo Bellin*i, Berlin : Mann, 1990 : V : Text ; VI : Katalog ; VII, figs 1-119 (Drawing book at the Louvre) ; VIII, Tab. 120-319 (Drawing book at the British Museum).

Del Badia, J., *Egnazio Danti cosmografo e matematico e le sue opere in Firenze*, Florence : M. Cellini, 1881.

Delumeau, J., *Une histoire du paradis. Le jardine des délices*, Paris : Fayard, 1992.

—, *Que reste-t-il du paradis ?*, Paris : Fayard, 2000.

Desmond, R. G. C., *The India Museum, 1801-1879*, London : HMSO, 1982.

Destombes, M., *Mappemondes A.D. 1200-1500. Catalogue préparé par la Commission des Cartes Anciennes de l'Union Géographique Internationale*, Amsterdam : N. Israel, 1964.

Devos, P., "Le miracle postume de S. Thomas l'Apôtre," *Analecta Bollandiana* 66 (1948), pp. 270-272.

Dias, P., *Importação de esculturas de Itália no séculos XV e XVI*, Porto : Paisagem, 1982.

Diffie, W., "Foreigners in Portugal and the 'Policy of Silence'," *Terrae Incognitae* 1 (1969), pp. 23-34.

Dilke, O. A. W., (with additional material supplied by the editors), "The Culmination of Greek Cartography in Ptolemy," in J. B. Harley and D. Woodward (eds), *History of Cartography*, vol. I : *Cartography in Prehistoric, Ancient, and Medieval Europe and the Mediterranean*, Chicago and London : University of Chicago Press, 1987, pp. 177-200.

Diller, A., "A Geographical Treatise by Georgius Gemistus Pletho," *Isis* 27, 3, (1937), pp. 441-451.

—, Kristeller, P. O., "Strabo," in F. E. Cranz, P. O. Kristeller and V. Brown (eds), *Catalogus Translationum et Commentariorum. Medieval and Renaissance Latin Translations and Commentaries : Annotated Lists and Guides*, Washington D.C. : Catholic University of America Press, 1960-1986.

—, "The Library of Francesco and Ermolao Barbaro," *Italia Medioevale e Umanistica* 6 (1963), pp. 253-262.

—, *The Textual Tradition of Strabo's Geography*, Amsterdam : Adolf M. Hakkert, 1975.

Dini, B., "La vita e l'attività professionale di Ambrogio di messer Lorenzo de' Rocchi," in Ambrogio di Lorenzo de' Rocchi, *Una pratica di mercatura in formazione : 1394-1395*, Florence : Le Monnier, 1980, pp. 5-33.

—, *Saggi su una economia-mondo : Firenze e l'Italia fra Mediterraneo ed Europa, secc. XIII-XVI*, Ospedaletto, Pisa : Pacini, 1995.

Dionisotti, C., "Tradizione classica e volgarizzamenti," in Id., *Geografia e storia della letteratura italiana*, Turin : Einaudi, 1983, pp. 124-178.

Donatello, A., Florio, G. M., Giovè, N., Granata, L., Mariani Canova, G., Massalin, P., Mazzon, A., Toniolo, F., Zamponi S.(eds), *I manoscritti della Biblioteca del Seminario vescovile di Padova*, Florence : Sismel, Edizioni del Galluzzo, 1998.

Donattini, M., *Spazio e modernità : libri, carte, isolari nell'età delle scoperte*, Bologna : CLUEB, 2000.

Doumerc, B., "Il dominio del mare," in A. Tenenti and U. Tucci (eds), *Storia di Venezia. Dalle origini alla caduta della Serenissima*, vol. IV, *Il Rinascimento. Politica e cultura*, Rome : Istituto della Enciclopedia Italiana, 1996, pp. 113-180.

Dubourg-Glatigny, P., "Egnazio Danti O.P. (1536-1586). Itinéraire d'un mathematicien parmi les artistes," *Mèlanges de l'École Française de Rome* 114, 2 (2002), pp. 543-605.

Duffy, J., *Shipwreck & Empire : Being an Account of Portuguese Disasters in a Century of Decline*, Cambridge, MA : Harvard University Press, 1955.

Duhem, P., *Le système du monde : histoire des doctrines cosmologiques de Platon à Copernic*, Paris : Libr. scientifique A. Hermann, 1913-1917, 5 vols. (reprinted Paris : Hermann, 1954-1965, 10 vols.).

Dupuigrenet Desroussiles, F., "L'Università di Padova dal 1405 al Concilio di Trento," in G. Arnaldi and M. Pastore Stocchi (eds), *Storia della cultura veneta*, vol. 3.2, *Dal primo Quattrocento al Concilio di Trento*, Vicenza : Neri Pozza, 1980, pp. 607-647.

Durand, D. B., "Tradition and Innovation in Fifteenth Century Italy. 'Il primato dell'Italia' in the Field of Science," *Journal of the History of Ideas* 4 (1943), pp. 1-20.

—, *The Vienna-Klosterneuburg Map Corpus of the Fifteenth Century : A Study in the Transition from Medieval to Modern Science*, Leiden : Brill, 1952.

Durel, A., *L'imaginaire des épices. Italie médiévale, Orient lointain XIIe-XVIe siècle*, Paris : Harmattan, 2006.

Dürst, A., "Die Weltkarte von Albertin de Virga von 1411 oder 1415," *Cartographia Helvetica* 13 (1996), pp. 18-21.

Dutschke, C. W., *Francesco Pipino and the Manuscripts of Marco Polo's "Travels"*, Ph.D. dissertation, Los Angeles : University of California, 1993.

Duval-Arnould, L., "Les manuscrits de la Géographie de Ptolémée issus de l'atelier de Piero del Massaio (Florence, 1469-vers 1478)," in D. Marcotte (ed.), *Humanisme et culture géographique à l'époque du Concile de Constance : autour de Guillaume Fillastre ; actes du Colloque de l'Université de Reims 18-19 novembre 1999*, Turnhout : Brepols, 2002, pp. 227-243.

Duyvendak, J. J. L., "The True Dates of the Chinese Maritime Expeditions in the Early Fifteenth Century," *T'oung Pao – Archives concernant l'histoire, les langues, la géographie, l'ethnographie et les arts de l'Asie orientale* 34 (1939), pp. 341-412.

Edson, E., *Mapping Time and Space. How Medieval Mapmakers Viewed Their World*, London : British Library, 1999.

Eelsalu, H., *Johann Heinrich von Mädler, 1794-1874 : eine dokumentarische Biographie*, Berlin : Akademie-Verlag, 1985.

Eisler, C., *The Genius of Jacopo Bellini. The Complete Paintings and Drawings*, New York : Abrams, 1989.

Eisler, C., Rubiés, J-P., *Voyages and visions : towards a Cultural History of Travel*, London : Reaktion Books, 1999.

Emiliani, M., "Le carte nautiche dei Benincasa cartografi anconitani," *Bollettino della Reale Società Geografica Italiana* 7, 1 (8-9), 1936, pp. 485-510.

Erdö, P., *Storia della scienza del diritto canonico : una introduzione*, Rome : Editrice Pontificia Università Gregoriana, 1999.

Erlande-Brandenburg, A., *La Dame à la licorne*, Paris : M. Aveline, 1993.

Esch, A., Frommel, C. L. (eds), *Arte, committenza ed economia a Roma e nelle corti del Rinascimento (1420-1530)*, Turin : Einaudi, 1995.

Falchetta, P., "Marinai, mercanti, cartografi, pittori. Ricerche sulla cartografia nautica a Venezia (sec. XIV-XV)," *Ateneo Veneto* 182 (1995), pp. 40-45.

—, "Maps ans Mapmakers in Venice in Cabot's Time," in R. Mamoli Zorzi (ed.), *Attraversare gli oceani. Da Giovanni Caboto al Canada multiculturale*. Venice : Marsilio, 1999, pp. 73-82.

—, "Carte veneziane dell'Asia da Fra Paolino a Giacomo Gastaldi," in G. Curatola (ed.), *Sciamani e dervisci dalle steppe del Prete Gianni : religiosità del Kazakhistan e percezione del fantastico a Venezia*, Venice : Multigraf, 2000, pp. 39-50.

—, *Fra Mauro's World Map*, Turnhout : Brepols, 2006.

Fantoni, M., Matthews-Grieco, S. (eds), *The Art Market in Italy, 15th-17th Centuries*, Ferrara : Istituto di Studi Rinascimentali, Modena : Panini, 2003.

Farmer, E. L., *Early Ming Government : The Evolution of Dual Capitals*, Cambridge, MA : Harvard University Press, 1976.

Favi, A., *Ambrogio Traversari. Lettere a Niccolò Niccoli. Edizione critica e commento*, Ph.D. dissertation, Florence : Università degli Studi di Firenze, 2003-2004.

Fernández-Armesto, F., *Before Columbus : Exploration and Colonization from the Mediterranean to the Atlantic, 1229-1492*, Philadelphia : University of Pennsylvania Press, 1987.

Fiorani, F., *The Marvel of Maps. Art, Cartography and Politics in Renaissance Italy*, New Haven and London : Yale University Press, 2005.

Fischer, T., *Sammlung mittelalterlicher Welt und Seekarten italienischen Ursprungs und aus italienischen Bibliotheken und Archiven*, Venice : Ongania 1886.

Folena, G., *Lingue e culture nel Veneto medievale*, Padua : Editoriale Programma, 1990.

Formisano, L. (ed.), *'Iddio ci dia buon viaggio e guadagno'. Biblioteca Riccardiana, Florence, Ms. 1910 (Codice Vaglienti)*, Florence : Edizioni Polistampa, 2006.

Fragnito, G., *La Bibbia al rogo : la censura ecclesiastica e i volgarizzamenti della Scrittura, 1471-1605*, Bologna : Il Mulino, 1997.

Franchi, S., *L'itinerario di Marco Polo in Persia*, Turin : Tipografia Collegio Artigianelli, 1941.

Frick, C. C., *Dressing Renaissance Florence : Families, Fortunes, & Fine Clothing*, Baltimore and London : Johns Hopkins University Press, 2002.

Fubini, R., Caroti, S. (eds), *Poggio Bracciolini nel VI Centenario della nascita. Codici e documenti fiorentini*, Florence : Biemme, 1980.

—, *Il "teatro del mondo" nelle prospettive morali e storico politiche di Poggio Bracciolini*, Florence : INSR, 1982.

—, *Storiografia dell'umanesimo in Italia da Leonardo Bruni ad Annio da Viterbo*, Rome : Edizioni di storia e letteratura, 2003.

—, "Umanesimo e Scolastica. Saggio per una definizione," in *Medioevo e Rinascimento* 17/n.s. 15 (2004), pp. 165-174 (English translation,"Humanism and Scholasticism. Toward an Historical Definition," in A. Mazzocco (ed.), *Interpretations of Renaissance Humanism*, Leiden : Brill, 2006, pp. 127-136.).

Fumaroli, M., *L'età dell'eloquenza : retorica e 'res literaria' dal Rinascimento alle soglie dell'epoca classica*, Milan : Adelphi, 2002 (original French edition, Geneva : Droz, 1980).

Gadrat, C., "Le rôle de Venise dans la diffusion du livre de Marco Polo (xive-début xvie siècle)," *Médiévales* 58 (2010), pp. 63-78.

Gallo, R., "Le mappe geografiche del Palazzo ducale di Venezia," *Archivio Veneto* 5, 32-33 (1943), pp. 47-113.

García Espada, A., "Marco Polo, Odorico of Pordenone, the Crusades, and the Role of the Vernacular in the First Descriptions of the Indies," *Viator* 40 (2009), pp. 201-222.

Garcia J. C., Feijão, M. J. (eds), *O 2º Visconde de Santarém e a História da Cartografia*, Lisbon : Biblioteca Nacional de Portugal, 2006.

—, "Um Castelo de Cartas Antigas : construir e comemorar o Império," in T. Pinto Coelho (ed.), *Os Descobrimentos no mundo de língua inglesa, 1880-1972*, Lisbon : Colibri, 2006, pp. 167-174.

Gargan, L., *Lo studio teologico e la biblioteca dei domenicani a Padova nel Tre e Quattrocento*. Padua : Antenore, 1971.

—, *Cultura e arte nel Veneto al tempo del Petrarca*, Padua : Antenore, 1978.

Garin, E., "Cultura filosofica toscana e veneta nel Quattrocento," in V. Branca (ed.), *Umanesimo europeo e Umanesimo veneziano*. Florence : Sansoni, 1963, pp. 11-33.

—, *Ritratti di umanisti*, Florence : Sansoni, 1967.

—, *Umanisti, artisti, scienziati. Studi sul Rinascimento italiano*, Rome : Editori Riuniti, 1989.

Garzelli, A., *Miniatura fiorentina del Rinascimento, 1440-1525 : un primo censimento*, Florence : Giunta regionale toscana ; Scandicci : La nuova Italia, 2 vols., 1985.

Gasparrini Leporace, T., *Il Mappamondo di Fra Mauro camaldolese*, con la presentazione di R. Almagià. Rome : Poligrafico della Zecca dello Stato, 1956.

Gautier Dalché, P., *La 'Descriptio mappe mundi' de Hugues de Saint-Victor*, Paris : Études augustiniennes, 1988.

—, "Modes de connaissance des *fines orbis terrarum* du Nord-Ouest (de l'Antiquité au XIIIe siècle)," in *L'Europe et l'Océan au Moyen Age. Contribution à l'Histoire de la Navigation*, S.H.M.E.S.-Cid édition, 1988, pp. 217-233.

—, "De la glose à la contemplation. Place et fonction de la carte dans les manuscrits du Haut Moyen Age," in *Testo e immagine nell'Alto medioevo, Settimane di studio del Centro italiano di studi sull'Alto Medioevo* 51, Spoleto, 1994, pp. 693-771.

—, "Saperi geografici del Mediterraneo cristiano," in P. Toubert and A. Paravicini Bagliani (eds), *Federico II e le scienze*, Palermo : Sellerio, 1994, pp. 162-184.

—, "Pour une histoire du regard géographique. Conception et usage de la carte au XVe siècle," *Micrologus* 4 (1996), pp. 77-103.

—, "Le souvenir de la 'Géographie' de Ptolémée dans le monde latin médiéval (VIe-XIVe siècles)," *Euphrosyne. Revista de Filologia clássica* 24 (1999) pp. 80-106.

—, "Décrire le monde et situer les lieux au XIIe siècle. L'*Expositio mappe mundi* et la généalogie de la mappemonde de Hereford." *Mélanges de l'École française de Rome. Antiquité-Moyen Âge* 113 (2001). Rome : École française de Rome, pp. 343-410.

—, "Jean Fusoris et la géographie. Un astronome, auteur d'un globe terrestre, à la découverte de Ptolémée," in D. Marcotte (ed.), *Humanisme et culture géographique à l'époque du Concile de Constance : autour de Guillaume Fillastre* ; Proceedings of the Colloquium of the University of Reims, November 18-19, 1999, Turnhout : Brepols, 2002, pp. 161-176.

—, "L'œuvre géographique du Cardinal Fillastre († 1428). Représentation du monde et perception de la carte à l'aube des découvertes," in D. Marcotte (ed.), *Humanisme et culture géographique à l'époque du Concile de Constance : autour de Guillaume Fillastre* ; Proceedings of the Colloquium of the University of Reims, November 18-19, 1999, Turnhout : Brepols, 2002, pp. 293-355 (originally published in *Archives d'histoire doctrinale et littéraire du Moyen Âge* 59, 1992, pp. 319-383).

—, "Sur l'"originalité" de la "géographie" médiévale," in M. Zimmermann (ed.), *Auctor et auctoritas : invention et conformisme dans l'écriture médiévale*, Paris : École des Chartes, 2001, pp. 130-143.

—, "Le paradis aux antipodes ? Une *Distinctio divisionis terre et paradisi delitiarum* (xive siècle)," in D. Barthélemy and J.-M. Martin (eds), *Liber largitorius. Études d'histoire médiévale offertes à Pierre Toubert par ses élèves*, Geneva : Droz, 2003, pp. 615-637.

—, "Weltdarstellung und Selbsterfahrung : Der Kartograph Fra Mauro," in H.-D. Heimann, P. Monnet (eds), *Kommunication mit dem Ich*, Bochum : 2004, pp. 39-51.

—, "Les représentations de l'espace de l'Antiquité tardive au xvie siècle", in *École pratique des hautes études. Sciences historiques et philologiques. Annuaire. Résumé des conférences et travaux*, 139e année, 2006-2007, pp. 119-123 ; 140e année, 2007-2008, pp. 142-127

—, "The Reception of Ptolemy's Geography (End of the Fourteenth to Beginning of the Sixteenth Century)," in D. Woodward (ed.), *The History of Cartography*, vol. III, 1, *Cartography in the European Renaissance*, Chicago : University of Chicago Press, 2007, pp. 285-364.

—, "A propos de la mappemonde d'Ebstorf," Médiévales 55 (2008), pp. 163-170.

—, "L'héritage antique de la cartographie médiévale. Les problèmes et les acquis," in R. J. A. Talbert and R. W. Unger (eds), *Cartography in Antiquity and the Middle Ages : Fresh Perspectives, New Methods*, Leiden and Boston : Brill, 2008, pp. 29-66.

—, *La Géographie de Ptolémée en Occident (ive-xvie siècle)*, Turnhout : Brepols, 2009.

—, "Usages critiques et scientifiques de la carte marine au xive siècle : Pétrarque, Boccace, Paolo dell'Abbaco," in T. Michalsky, F. Schmieder and G. Engel (eds), *Aufsicht – Ansicht – Einsicht. Neue Perspektiven auf die Kartographie an der Schwelle zur Frühen Neuzeit*, Frankfurt : 2009, pp. 81-92.

Geanakoplos, D. J., *Greek Scholars in Venice*. Cambridge : Harvard University Press, 1962.

Gentile, S., "L'ambiente umanistico fiorentino e lo studio della geografia nel secolo XV," in L. Formisano et al. (eds), *Amerigo Vespucci. La vita e i viaggi*, Florence : Banca Toscana, 1991, pp. 9-63.

—, *Firenze e la scoperta dell'America. Umanesimo e geografia nel '400 fiorentino*, Florence : Olschki, 1992.

—, "Emanuele Crisolora e la 'Geographia' di Tolomeo," in M. Cortesi and E. V. Maltese (eds), *Dotti bizantini e libri greci nell'Italia del secolo XV*, Naples : D'Auria, 1992, pp. 291-308.

—, "Toscanelli, Traversari, Niccoli e la geografia," *Rivista geografica italiana. Annata del Centenario* 100 (1993), p. 115 ;

—, "Parentucelli e l'ambiente fiorerentino : Niccoli e Traversari," in F. Bonatti and A. Manfredi (eds), *Niccolò V nel sesto centenario della nascita*, Vatican City : Biblioteca Apostolica Vaticana, 2000, pp. 217-235.

—, "Umanesimo e cartografia : Tolomeo nel secolo XV," in D. Ramada Curto, A. Cattaneo, and A. Ferrand Almeida (eds), *La cartografia europea tra primo Rinascimento e fine dell'Illuminismo*, Florence : Olschki, 2003.

Gernet, J., *A History of Chinese Civilization*, J. R. Foster trans., Cambridge : Cambridge University Press, 1982 (transl. of *Le Monde Chinois*, Paris : Armand Colin, 1972).

Ghisalberti, A. (ed.), *Il cielo e il mondo. Commento al trattato 'Del cielo' di Aristotele*, Milan : Rusconi, 1983.

Gilbert, F., "Humanism in Venice," in S. Bertelli, N. Rubinstein and C. H. Smyth (eds), *Florence and Venice, Comparisons and Relations*, 2 vols., Florence : La Nuova Italia, 1979, vol. 1, pp. 13-26.

Gilson, É., *La filosofia nel Medioevo*, Bari : Laterza, 1998.

Ginzburg, C., *Il formaggio e i vermi : il cosmo di un mugnaio del '500*, Turin : Einaudi, 1976.

—, *The Cheese and the Worms. The Cosmos of a Sixteenth-Century Miller*, trans. J. and A. Ledeschi, London and Henley : Routledge & Kegan Paul, 1980.

—, "Microstoria : due o tre cose che so di lei," *Quaderni storici* 86 (1994), pp. 511-539.

—, *Rapporti di forza : storia, retorica, prova*, Milan : Feltrinelli, 2000.

—, *Nessuna isola è un'isola. Quattro sguardi sulla letteratura inglese*, Milan : Feltrinelli, 2002.

—, *Il filo e le tracce : vero, falso, finto*, Milan : Feltrinelli, 2006.

Goffen, R., *Giovanni Bellini*, New Haven : Yale University Press, 1989.

Goitein, S. D., "From the Mediterranean to India : Documents on the Trade to India, South Arabia, and East Africa from the Eleventh and Twelfth Centuries," *Speculum* 29, 2 (1954), pp. 181-197.

Goldthwaite, R. A., *La costruzione della Firenze rinascimentale : una storia economica e sociale*, Bologna : Il Mulino, 1984.

—, *Wealth and the Demand for Art in Italy, 1300-1600*, Baltimore and London : Johns Hopkins University Press, 1993.

—, *Studi sulla moneta fiorentina : secoli XIII-XVI*, Florence : Olschki, 1994.

Gomes, P. F., "Alessandro Zorzi e l'invenzione dei Tropici," in A. Cattaneo, D. Ramada Curto and A. Ferrand Almeida (eds), *La cartografia europea tra primo Rinascimento e fine dell'Illuminismo*, Florence : Olschki, 2003, pp. 109-132.

Gomes, S. A., *D. Afonso V o Africano*, Lisbon : Temas e Debates, 2009.

Gonçalves, J., *O Infante D. Pedro as sete partidas e a génese dos descobrimentos*, Lisbon : Agência Geral do Ultramar, 1955.

Gorshenina, S. M., *De la Tartarie à l'Asie centrale. Le cœur d'un continent dans l'histoire des idées entre la cartographie et la géopolitique*, Ph.D. dissertation, Paris : Université Panthéon-Sorbonne, and Lausanne : Université de Lausanne, 2007.

Graf, A., *La leggenda del Paradiso terrestre : lettura fatta nella R. università di Torino, addì 11 novembre 1878*, Turin : Loescher, 1878 (reprinted Rome : Edizioni del Graal, 1982).

—, *Miti, leggende e superstizioni del Medio Evo*, C. Allasia and W. Meliga (eds) ; introduction by M. Guglielminetti ; critical essays by E. Artifoni and C. Allasia, Milan : Mondadori, 2002.

Grant, E., "Late Medieval Thought, Copernicus, and the Scientific Revolution," *Journal of the History of Ideas* 23, 2 (1962), pp. 197-220.

—, "A New Look at Medieval Cosmology, 1200-1687," *Proceedings of the American Philosophical Society* 129, 4 (1985), pp. 417-432.

—, *Planets, Stars, and Orbs : the Medieval Cosmos, 1200-1687*, Cambridge and New York : Cambridge University Press, 1994.

—, *The Foundations of Modern Science in the Middle Ages : Their Religious, Institutional, and Intellectual Contexts*, Cambridge and New York : Cambridge University Press, 1996.

Greenblatt, S., *Marvellous Possessions. The Wonder of the New World*, Oxford and New York : Clarendon Press, 1991.

Gregorio XVI. Miscellanea commemorativa, Rome : Università Gregoriana, 2 vols., 1948.

Grendi, E., "Micro-analisi e storia sociale," *Quaderni storici* 35 (1977), pp. 506-520.

—, *Il Cervo e la repubblica. Il modello ligure di antico regime*, Turin : Einaudi, 1993.

—, "Ripensare la microstoria ?," *Quaderni storici* 86 (1994), pp. 539-549.

Grieco, A. J., Redon, O., Tongiorgi-Tomasi, L. (eds), *Le monde végétal (XIIe-XVIIe siècles), savoirs et usages sociaux*, Saint-Denis : Presses Universitaires de Vincennes, 1993.

Grierson, P., "The Weight of the Gold Florin in the Fifteenth Century," *Quaderni ticinesi di numismatica e antichità classiche* 10 (1981), pp. 421-431.

Griffante, C., "L'umanesimo," in A. Tenenti and U. Tucci (eds), *Storia di Venezia. Dalle origini alla caduta della Serenissima*, vol. IV, *Il Rinascimento. Politica e cultura*, Rome : Istituto della Enciclopedia Italiana, 1996 pp. 723-755.

Grosjean, G., *Mapamundi : der katalanische Weltatlas vom Jahre 1375*, herausgegeben und kommentiert von G. Grosjean, Dietikon : Urs Graf Verlag, 1977.

Gualdo, R. (ed.), *Le parole della scienza : scritture tecniche e scientifiche in volgare, secoli XIII-XV*, Galatina : Congedo, 2001.

Hair, P. E. H., "Portuguese Documents on Africa and Some Problems of Translation," *History in Africa* 27 (2000), pp. 91-97.

Halikowski Smith, S., *Portugal and the European Spice Trade, 1480-1580*, Ph.D. dissertation, Florence : European University Institute, 2001.

Hallberg, I., *L'Extrême-Orient dans la litérature et la cartographie de l'Occident des* XIIIe, XIVe, XVe *siècles*, Göteborg : W. Zachrissons Boktryckeri, 1906.

Hamesse, J., *Les auctoritates Aristotelis : un florilège médiéval. Étude historique et édition critique*, Louvain : Publications Universitaires ; Paris : Béatrice-Nauwelaerts, 1974.

Hamilton, E. J., *Money, Prices, and Wages in Valencia, Aragon, and Navarre*, 1351-1500, Philadelphia : Porcupine Press, 1975 (reprint of the edition published in Cambridge, MA, Harvard University, 1936.

Hankey, T., "Domenico di Bandino of Arezzo," *Italian Studies* 12 (1957), pp. 100-128.

—, "The Library of Domenico di Bandino," *Rinascimento* 8 (1957), pp. 177-207.

Harley, J. B., Woodward, D. (eds), *The History of Cartography*, vol. 1 : *Cartography in Prehistoric, Ancient and Medieval Europe and the Mediterranean*. Chicago and London : University of Chicago Press, 1987.

—, Woodward D., "Greek Cartography in the Early Roman World," (prepared from materials supplied by G. Aujac), in J. B. Harley and D. Woodward (eds), *History of Cartography*, vol. 1 : *Cartography in Prehistoric, Ancient, and Medieval Europe and the Mediterranean*, Chicago and London : University of Chicago Press, 1987, pp. 148-175.

Haskell, F., *Mecenati e pittori : studio sui rapporti tra arte e società italiana nell'età barocca*, Florence : Sansoni, 1966.

Haussmann, F. R., "Guglielmo Capello," in *Dizionario Biografico degli Italiani*, vol. 28, Rome, 1975, pp. 494-495.

Heeren, A. H., "Explicatio Planiglobi Orbis terrarum faciem exhibentis ante medium saeculum XV summa arte confecti Musei Borgiani Velitrensi," *Commentarius Soc. Regiae Scientiarum Gottingensis* 16 (1808), pp. 250-284.

Helden, A. van, *Measuring the Universe : Cosmic Dimensions from Aristarchus to Halley*, Chicago and London : University of Chicago Press, 1985.

Herbert F., "Jacob Gråberg af Hemsö, the Royal Geographical Society, the Foreign Office, and Italian Portolan Charts for the British Museum," in G. Bäärnhielm, F. Sandgren, and A. Burius (eds), *Accurata descriptio : Papers in Cartography, Numismatics, Oriental Studies and Librarianship Presented to Ulla Ehrensvärd*, Stockholm : Kungliga Biblioteket, 2003, pp. 269-314.

Herriott, J. H., "The 'Lost' Toledo Manuscript of Marco Polo," *Speculum* 12, 4 (1937), pp. 456-463.

Herrmann, A., "Taprobane," in *Paulys Real-Encyclopädie der Classischen Altertumswissenschaft. Neue Bearbeitung begonnen von Georg Wissowa* […], 2nd ser., vol. IV [Stuttgart, 1932], col. 2260-2272.

Heyd, W. von, *Histoire du commerce du Levant au Moyen-Age*, 2 vols., Leipzig : O. Harrassowitz, 1885-1886 (reprint Amsterdam : Hakkert, 1967).

Hiatt, A., *Terra Incognita. Mapping the Antipodes before 1600*, Chicago : University of Chicago Press, 2008.

Hofmann, C., "Publishing and the Map Trade in France, 1470-1670," in D. Woodward (ed.) *The History of Cartography*, vol. 3 : *Cartography in the European Renaissance*, Chicago and London : University of Chicago Press, 2007, pp. 1569-1588.

Hole, R., "Horsley, Samuel (1733-1806)," in H. C. G. Matthew and B. Harrison (eds), *Oxford Dictionary of National Biography*, Oxford : Oxford University Press, 2004 (on-line edition).

Hollingsworth, M., *Patronage in Renaissance Italy from 1400 to the Early Sixteenth Century*, London : John Murray, 1994.

Hoogvliet, M., "Mappae Mundi and Medieval Encyclopedias : Image versus Text," in P. Binkley (ed.), *Pre-Modern Encyclopaedic Texts*, Proceedings of the Second COMERS Congress, Groningen, 1-4 July 1996, Leiden and New York and Köln : Brill, 1997, pp. 63-74.

—, *Pictura et scriptura : textes, images et herméneutique des mappae mundi (XIIIe-XVe siècles)*, Turnhout : Brepols, 2007.

Humboldt, A. von, *Examen critique de l'histoire de la géographie du nouveau continent et des progrès de l'astronomie nautique aux quinzième et seizième siècles*, Paris : Librairie de Gide, 5 vols., 1836-1839.

Humphreys, K. W., *The Book Provisions of the Mediaeval Friars, 1215-1400*, Amsterdam : Erasmus Booksellers, 1964.

—, *The Library of the Franciscans of the Convent of St. Anthony, Padua, at the Beginning of the Fifteenth Century*, Amsterdam : Erasmus Booksellers, 1966.

Iaria, S., *Ambrogio Traversari, Hodoeporicon : edizione critica e commento*, Università degli Studi di Firenze, Ph.D. dissertation, Florence : Università degli Studi di Firenze, 2003-2004.

Il cielo : intorno alla Cosmologia di Plinio. Capolavori delle rappresentazioni miniate nei codici delle principali biblioteche del mondo, Turin : Allemandi, 1994.

Irwin, R., *For Lust of Knowing : The Orientalists and Their Enemies*, London : Penguin and Allen Lane, 2006.

Iwanczak, W., "Entre l'espace ptolémaïque et l'empirie : les cartes de Fra Mauro," *Médiévales* 18 (1990), pp. 53-68.

Jacob, C., *L'empire des cartes. Approche théorique de la cartographie à travers l'histoire*, Paris : Albin Michel, 1992.

Jaeger, W., *Early Christianity and Greek Paideia*, Cambridge, MA : Harvard University Press, 1961.

Jammer, M., *Storia del concetto di spazio*, Milan : Feltrinelli, 1963.

Jones A., Berggren, J. L., (eds), *Ptolemy's Geography. An Annotated Translation of the Theoretical Chapters*, Princeton : Princeton University Press, 2000.

Jones, E. M. (ed.), "Landing at Fra Mauro," 1995 (last revised 12 February 2009) : ‹http ://www.hq.nasa.gov/office/pao/History/alsj/a14/a14.html›.

Kaiser, O., *Die mythische Bedeutung des Meeres in Ägypten*, Berlin : Ugarit und Israel, 1959 (second edition 1962).

Kamal, Y., *Monumenta Cartographica Africae et Aegypti*, 5 vols., Cairo, 1926-1951.

Keay, J., *The Honourable Company: A History of the English East India Company*, London: Harper Collins, 1991.

Kimble, G. H. T., *Geography in the Middle Ages*, London: Methuen, 1938.

King, M. L., *Venetian Ideology and Reconstruction of Knowledge: Giovanni Caldiera (c. 1400-1474)*, Ph.D. dissertation, Palo Alto: Stanford University, 1972.

—, *Venetian Humanism in an Age of Patrician Dominance*, Princeton: Princeton University Press, 1986.

—, *Umanesimo e patriziato a Venezia nel Quattrocento*, 2 vols., Rome: Il Veltro, 1989.

Koyrè, A., *From the Closed World to the Infinite Universe*, Baltimore: Johns Hopkins University Press, 1957.

Kretschmer, K., "Eine neue mittelalterliche Weltkarte der vatikanischen Bibliothek," *Zeitschrift der Gesellschaft für Erdkunde* 26 (1891), pp. 371-406.

Kristeller, P. O., *Renaissance Thought and Its Sources*, New York: Columbia University Press, 1979.

Kugler, H., *Die Ebstorfer Weltkarte: Kommentierte Neuausgabe in zwei Bänden*. Band 1: *Atlas*. Band 2: *Untersuchungen und Kommentar*, Berlin: Akademie Verlag, 2007.

La Bibbia di Borso d'Este: ms. lat. 422-423, Biblioteca estense e universitaria, Modena – Ferrara, seconda metà del XV secolo, 3 vols., Rome: Istituto della Enciclopedia italiana, 1997.

Labowsky, L., "Bessarione," in *Dizionario Biografico degli italiani*. Rome: Istituto della Enciclopedia italiana, 1967, vol. 11, pp. 686-696.

—, *Bessarion's Library and the Bibliotheca Marciana*, Rome: Edizioni Storia e Letteratura, 1979.

Laird Clowes, G. S., "Ships of Early Explorers," *The Geographical Journal* 69, 3 (1927), pp. 216-230.

Laissus, Y., *Jomard – le dernier Égyptien*, Paris: Fayard, 2004.

Lanciani, G. (ed.), *Viaggi e naufragi portoghesi sulla via delle Indie*, Naples: Liguori, 2002.

Lane, F. C., "The Spice Trade," in Id., *Venice a Maritime Republic*, Baltimore and London: Johns Hopkins University Press, 1973, pp. 285-294.

—, Mueller, R. C., *Money and Banking in Medieval and Renaissance Venice*, Baltimore and London: Johns Hopkins University Press, 1985.

Lazzi, G., "Mondi celesti e terrestri nei libri medicei," in A. Cecchi and P. Pacetti (eds), *La Sala delle Carte geografiche in Palazzo Vecchio: 'capriccio et invenzione nata dal Duca Cosimo'*. Florence: Polistampa, 2008, pp. 41-65.

Leed, E. J., *La mente del viaggiatore. Dall'Odissea al turismo globale*, Bologna: Il Mulino, 1992.

Lelewel, J., *Géographie du Moyen Âge*, 7 vols., Bruxelles: V. Pilliet and J. Pilliet, 1852-1857.

Lepori, F., "La scuola di Rialto dalla fondazione alla metà del Cinquecento," in G. Arnaldi and M. Pastore Stocchi (eds), *Storia della cultura veneta*, vol. 3.2, *Dal primo Quattrocento al Concilio di Trento*, Vicenza: Neri Pozza, 1980, pp. 539-605.

Lester, M. "Spencer, George John, second Earl Spencer (1758-1834)," in the *Oxford Dictionary of National Biography*, ed. H. C. G. Matthew and B. Harrison, Oxford: Oxford University Press, 2004 (on-line edition).

Levi, G., "On Microhistory," in P. Burke, *New Perspectives on Historical Writing*, London: Polity Press, 1991, pp. 93-113.

Levi d'Ancona, M., *Miniatura e miniatori a Firenze dal XIV al XVI secolo: documenti per la storia della miniatura*, with a preface by M. Salmi, Florence: Olschki, 1962.

Levi-Donati, G. R. (ed.), *Egnatio Danti. Le trentacinque cartelle della Guardaroba medicea di Palazzo Vecchio in Firenze*, Perugia : Benucci, 2002 (first edition 1995).

Libois, A., "À propos des modes de preuve et plus spécialement de la preuve par témoins dans la juridiction de Léau au xve siècle," in *Hommage au Professeur Paul Bonenfant* (1899-1965), pp. 332-546.

Librandi, R., *La Metaura d'Aristotile. Volgarizzamento fiorentino anonimo del XIV secolo*, 2 vols., Naples : Liguori, 1995.

—, "Auctoritas e testualità nella descrizione dei fenomeni fisici," in R. Gualdo (ed.), *Le parole della scienza : scritture tecniche e scientifiche in volgare, secoli XIII-XV*, Galatina : Congedo, 2001, pp. 99-126.

—, "Il lettore di testi scientifici in volgare," in *Lo spazio letterario del medioevo*, vol. 3 : P. Boitani, M. Mancini and A. Varvaro (eds), *La ricezione del testo*, Rome : Salerno, 2003, pp. 125-154.

—, "Tratti sintattico-testuali e tipologia di testi : la trattatistica scientifica," in M. Dardano and G. Frenguelli (eds), *La sintassi dell'italiano antico*, Atti del Convegno internazionale (Roma, 18-21 settembre 2002), Rome : Aracne, 2004, pp. 271-292.

Lightbown, R., *Mantegna : With a Complete Catalogue of the Paintings, Drawings and Prints*, Oxford : Phaidon-Christie's, 1986.

Lindberg, D. C., *The Beginnings of Western Science : the European Scientific Tradition in Philosophical, Religious, and Institutional Context, 600 B.C. to A.D. 1450*, Chicago : University of Chicago Press, 1992.

Livi, G., *Dall'Archivio di Francesco Datini mercante pratese*, Florence : F. Lumachi, 1910.

Longhena, M. (ed.), *Viaggi in Persia, India e Giava di Niccolò de' Conti, Girolamo Adorno e Girolamo da Santo Stefano*, Milan : Alpes, 1929.

Lopez, R. S., *Venezia e le grandi linee dell'espansione commerciale nel sec. XIII*, in *La civiltà veneziana del secolo di Marco Polo*, Florence : Sansoni, 1955.

Lusignan, S., *Parler vulgairement : les intellectuels et la langue française aux XIIIe et XIVe siècles*, Paris : Vrin, 1987.

Maccagni, C. (ed.), *La scienza a Venezia tra Quattrocento e Cinquecento. Opere manoscritte e a stampa*, Venice : Comune di Venezia and Biblioteca Nazionale Marciana, 1985.

—, "Leggere, scrivere e disegnare la 'scienza volgare' nel Rinascimento," *Annali della Scuola normale superiore di Pisa. Classe di lettere e filosofia* 23 (1993), pp. 631-676.

—, "Sapere e cultura dei tecnici nel Rinascimento," in M. Dalai Emiliani and V. Curzi (eds), *Piero della Francesca tra arte e scienza*, Venice : Marsilio, 1996.

Mandich, G., "Per una ricostruzione delle operazioni mercantili e bancarie della compagnia dei Covoni," in A. Sapori (ed.), *Libro giallo della compagnia dei Covoni*, Milan : Ist. Editoriale Cisalpino, 1970, pp. clxxviii-cxciii.

Mandosio, J.-M., "Alberti et l'esprit du caméléon," in F. Choay and M. Paoli (eds), *Alberti humaniste, architecte*, Paris : ENSBA, 2006, pp. 128-141.

—, *D'or et de sable*, Paris : Éditions de l'Encyclopédie des nuisances, 2008.

Manodori, A., (ed.), *La preghiera del marinaio : la fede e il mare nei secoli della Chiesa e nelle tradizioni marinare*, Rome : Istituto Poligrafico e Zecca dello Stato, 1992.

Manzoni, L., "Frate Francesco Pipino di Bologna dei PP. Predicatori, geografo, storico e viaggiatore," *Atti e memorie della R. Deputazione di Storia Patria per le Provincie di Romagna*, s. 3, 13 (1895), pp. 257-334.

Maqbul Ahmad, S., "Cartography of al-Sharif al-Idrisi," in D. Woodward (ed), *The History of Cartography*, Vol. 2, Book 1, *Cartography in the Traditional Islamic and South Asian Societies*, pp. 156-173.

Maracchi Bigiarelli, B., "Niccolò Tedesco e le carte della Geografia di Francesco Berlinghieri autore-editore," in B. Maracchi Bigiarelli and D. E. Rhodes (eds), *Studi offerti a Roberto Ridolfi direttore de 'La bibliofilia'*, Florence : Olschki, 1973, pp. 377-397.

Marcolin, M., "I cartigli delle tavole della sala della Guardaroba," in A. Cecchi and P. Pacetti (eds), *La Sala delle Carte geografiche in Palazzo Vecchio. 'Capriccio et invenzione nata dal Duca Cosimo'*, Florence : Polistampa, 2009, pp. 107-133.

Marcon, S., "Cardinal Bessarion's Ptolemy Codex : A Book of Inestimable Worth (Il prezioso Tolomeo di Bessarione)," in Klaudios Ptolemaios, *Geographike yphegesis. O ellenikos kodikas 388 tes Markianes Bibliothekes tes Benetias*. Athens : Militos, 1999, pp. 11-32.

—, "Il Mappamondo di Fra Mauro e Leonardo Bellini," in M. Piantoni and L. de Rossi (eds), *Per l'arte da Venezia all'Europa. Studi in onore di Giuseppe Maria Pilo*, Venice : Edizioni della Laguna, 2001, pp. 103-108.

—, "Bellini, Leonardo," in M. Bollati (ed.), *Dizionario biografico dei miniatori italiani*, Milan : Sylvestre Bonnard, 2004, pp. 76-78.

—, "Il Mappamondo di Fra Mauro e Leonardo Bellini," in P. Falchetta, *Fra Mauro's World Map*, Turnhout : Brepols, 2006, pp. 135-168.

Marcotte, D. (ed.), *Humanisme et culture géographique a l'époque du Concile de Constance : autour de Guillaume Fillastre*, Turnhout : Brepols, 2002.

Mariani Canova, G., "Per Leonardo Bellini," *Arte veneta* 22 (1968), pp. 9-20.

—, *La miniatura veneta del Rinascimento*, Venice : Alfieri, 1969.

—, "Leonardo Bellini," in *Storia della miniatura veneta*, Padua : 1969, pp. 103-104 and figures 10-19.

—, "Riflessioni su Jacopo Bellini e sul libro dei disegni del Louvre," *Arte Veneta* 26 (1972), pp. 9-30.

—, "Miniatura e pittura in età tardogotica (1400-1440), in *La pittura nel Veneto. Il Quattrocento*, vol. 1, Milan : Electa, 1989, pp. 193-222.

—, "La miniatura a Venezia dal Medioevo al Rinascimento," in R. Pallucchini (ed.), *Storia di Venezia. Dalle origini alla caduta della Serenissima*, vol. II, *Temi. L'arte*, Rome : Istituto della Enciclopedia italiana, 1995, pp. 769-843.

Marinelli, G., *Scritti minori. Metodo e storia della geografia*, vol. 1, Florence : Le Monnier, 1878, pp. 304-308.

Marshall, D. W., "A List of Manuscript Editions of Ptolemy's Geography," *Bulletin of the Geography and Maps Division. Special Libraries Association* 87 (1982), pp. 17-38.

Martineau, J., *Andrea Mantegna*, London : Thames and Hudson, 1992.

Martines, L., "Review of 'Una Famiglia e le sue Ricordanze : I Castellani di Firenze nel Tre-Quattrocento'," *Speculum* 71, 3 (July, 1996), pp. 707-708.

Massa, E. (ed.), *Paolo Giustiniani. Trattati, lettere e frammenti dai manoscritti originali del beato Paolo Giustiniani dell'Archivio dei Camaldolesi di Monte Corona nell' Eremo di Frascati*, Rome : Edizioni di Storia e Letteratura, 1967.

Mather, R. G., "Documents Mostly New Relating to Florentine Painters and Sculptors of the Fifteenth Century," *The Art Bulletin* 30, 1 (1948), pp. 20-65.

Maula, E., *Studies in Eudoxus' Homocentric Spheres*, Helsinki : Societas Scientiarum Fennica, 1974.

Mazzatinti, G., *Inventari dei manoscritti delle biblioteche d'Italia*, vol. LXXVII, Venice-Marciana Mss. Italiani – Classe VI (Storia e Geografia), P. Zorzanello (ed.), Florence : Olschki, 1950.

Mazzoni, F., *Contributi di filologia dantesca. Prima serie*, Florence : Sansoni, 1966.

Melis, F., *Documenti per la storia economica dei secoli XIII-XVI*, Florence : Olschki, 1972.

—, "La diffusione dell'informazione economica nel Mediterraneo," in E. Le Roy Ladurie (ed.), *Histoire économique du monde Méditerranéen 1450-1650, Mélanges en l'honneur de Fernand Braudel*, 2 vols., Toulouse : Privat, 1973, vol. 1, pp. 389-424 ;

—, "Di alcune figure di operatori economici fiorentini attivi nel Portogallo nel XV secolo," in L. Frangioni (ed.) *I mercanti italiani nell'Europa medievale e rinascimentale*, intro. H. Kellenbenz, Florence : Le Monnier, 1990, pp. 1-18.

Ménard, P., "L'illustration du Devisement du monde de Marco Polo : étude d'iconographie comparée," in *Métamorphoses du récit de voyage*, Paris : Champion, 1986, p. 17-31.

—, "Réflexions sur l'illustration du texte de Marco Polo dans le manuscrit français 2810 de la Bibliothèque nationale," *Mélanges in memoriam Takeshi Shimmura*, Tokyo, 1998, p. 81-93.

Meneghin, V., *San Michele in Isola di Venezia*, 2 vols., Venice : Stamp. di Venezia, 1962.

Merolla, L., "Un corale di San Michele di Murano," *Nuovi Annali della Scuola speciale per archivisti e bibliotecari* 9 (1997), pp. 111-128.

—, *La biblioteca di San Michele di Murano all'epoca dell'abate Giovanni Benedetto Mittarelli. I codici ritrovati*, Manziana : Vecchiarelli, 2009.

Meyer, P., *Alexandre le Grand dans la littérature française du Moyen Age*, 2 vols., Paris : F. Vieweg, 1886.

Milanesi, G., *Nuovi documenti per la storia dell'arte toscana dal XII al XV secolo per servire d'aggiunta all'edizione del Vasari edita da Sansoni nel 1885 raccolti e annotati da G. Milanesi*, Florence : G. Dotti, 1901.

Milanesi, M., "Il 'De insulis et earum proprietatibus' di Domenico Silvestri (1385-1406)," *Geographia Antiqua* 2 (1993), pp. 133-146.

—, "La Cartografia italiana nel Medio Evo e nel Rinascimento," in M. Milanesi, V. Valerio, E. Casti Moreschi, and L. Rombai, *La Cartografia italiana. Cicle de conferències sobre Història de la Cartografia. Tercer curs, 1991*, Barcelona : Institut Cartogràfic de Catalunya, 1993, pp. 15-80.

—, "G. B. Ramusio e le 'Navigazioni e viaggi' (1550-59)," in R. Zorzi (ed.), *L'Europa delle scoperte*, Florence : Olschki, 1994, pp. 75-101.

—, "A Forgotten Ptolemy : Harley Codex 3686 in the British Library," *Imago Mundi* 48 (1996), pp. 43-64.

—, "G. B. Ramusio e i piloti portoghesi," in L. Cusati (ed.), *Atti del Congresso Internazionale 'Portogallo e i mari : un incontro tra culture'*, Naples : Guida, 1998, pp. 231-248.

—, "Terza parte dell'Asia," in A. Dentoni Litta, I. Massabò Ricci (eds), *Architettura militare. Luoghi, città, fortezze, territori in età moderna*, Rome : Ministero per i beni e le attività culturali. Direzione generale per gli archivi, 2003, pp. 245-246.

—,"El mundo visto desde Venecia en el siglo XVI. Ramusio y Gastaldi," *El mundo de los mapas, Quadernos de la Fundaciòn M. Botin* 7 (2005), pp. 131-160.

—, "Antico e moderno nella cartografia umanistica. Le grandi carte d'Italia nel Quattrocento," *Geographia Antiqua* 26 (2007-2008), pp 153-175.

—, "Cartografia per un principe senza corte : Venezia nel Quattrocento," *Micrologus. Natura, scienze e società Medievali / Nature, Sciences and Medieval Societies* 16 (2008), pp. 189-216.

—, "Le regard de la postérité. L'âge des découvertes vu un siècle après," *Médiévales* 58 (2010), pp. 11-26.

—, "Nelle stanze di palazzo. Venezia, Firenze, Roma : qualche confronto." Proceedings of the Symposium "Cristoforo Sorte e il suo tempo," Verona, 31 October 2008, in press.

Milano, E (ed.), *Commentario al facsimile Mappamondo Catalano Estense*, supplemented by a transcription by A. Battini, Dietikon-Zürich, 1995.

—, *Mapamundi Catalán Estense. Escuela Cartográfica Mallorquina*. Transcription of the original text by A. Battini. [Modena, Bibl. Estense, C.G.A.1]. Barcelona : M. Moleiro, 1996.

Miller, K., *Mappae mundi : Die ältesten Weltkarten. Vol. V : Die Ebstorfkarte. Mit dem Facsimile der Karte in den Farben des Originals*, Stuttgart : 1896.

Minnis, A. J., *Medieval Theory of Authorship : Scholastic Literary Attitudes in the Later Middle Ages*, London : Scolar Press, 1984.

Mioni, E., *Bibliothecae Divi Marci Venetiarum codices Graeci manuscripti. Thesaurus antiquus*, vol. 2, *codices 300-625*, Rome, 1985.

Mitjià, M., "Albandò de les Illes Canàries per Joan I d'Aragò," *Anuario de Estudios Atlánticos* 8 (1962), pp. 325-353.

Molho, A., *Florentine Public Finances in the Early Renaissance, 1400-1433*, Cambridge, MA : Harvard University Press, 1971.

—, Ramada Curto, D., "Les réseaux marchands à l'époque moderne," *Annales* 58, 3, (2003).

Monumenta Henricina, 15 vols. Coimbra : Comissão Executiva das Comemorações do V Centenário da Morte do Infante D. Henrique, 1960-1978.

Moody, E. A., "John Buridan on the Habitability of the Earth," *Speculum* 16, 4 (1941), pp. 415-425.

Moretti, G., *Gli antipodi : avventure letterarie di un mito scientifico*, Parma : Pratiche, 1994.

Moretti, L., "Di Leonardo Bellini, pittore e miniatore," *Paragone. Arte* 99 (1958) pp. 58-96.

—, "Bellini, Leonardo," in *Dizionario biografico degli italiani*, vol. VII, Rome : Istituto della Enciclopedia italiana, 1965, pp. 712-713.

Morozzo della Rocca, R., Lombardo, A., *Documenti del commercio veneziano nei sec. XI-XIII*, 2 vols., Turin : Libreria italiana, 1940.

Morpurgo, S., Papa, P., Maracchi Bigiarelli, B. (eds), *Catalogo dei manoscritti Panciatichiani della Biblioteca Nazionale Centrale di Firenze*, Rome, 1887-1962.

Muccillo M., "Giovanni Fontana," *Dizionario biografico degli Italiani* 48 (1997), pp. 672-675.

Mueller, R. C., "'Chome l'ucciello di passagio' : la demande saisonnière des espèces et le marché des changes à Venise au Moyen Age," in J. Day (ed.), *Etudes d'histoire monétaire, XIIe-XIXe siècles*, Lille : Lavoisier, 1984, pp. 195-219.

—, "Mercanti e imprenditori fiorentini a Venezia nel tardo medioevo," *Società e storia* 55 (1992).

Muir, E., Weissman, R. F. E., "Social and Symbolic Places in Renaissance Venice and Florence," in J. A. Agnew and J. S. Duncan (eds), *The Power of Place : Bringing Together Geographical and Sociological Imaginations*, Boston : Unwin Hyman, 1989, pp. 81-103.

Muntz, E., Fabre, P., *La Bibliothèque du Vatican au XVe siècle d'après des documents inédits*, Paris : E. Thorin, 1887.

Murschel, A., "The Structure and Function of Ptolemy's Physical Hypotheses of Planetary Motion," *Journal for the History of Astronomy* 26 (1995), pp. 33-61.

Muzzarelli, M. G., *Guardaroba medievale. Vesti e società dal XII al XVI secolo*, Bologna : Il Mulino, 1999.

Nanetti, A., *Antonio di Marco Morosini : Cronicha de Veniexia (1094-1434) : edizione critica*, Ph.D. dissertation, Bologna : Università degli Studi di Bologna, 2000.

—, *Il Codice Morosini. Il mondo visto da Venezia (1094-1433)*, 4 vols., Spoleto : Fondazione CISAM, 2009.

—, *La caduta di Lucifero e l'autenticità della 'Quaestio de aqua et terra'*, Turin-Genoa-Milan : SEI, 1959.

Nardi, B., "Letteratura e cultura veneziana del Quattrocento," in *La civiltà veneziana del Quattrocento*, Venice : Sansoni, 1957, pp. 99-145.

Nigro, G., *Mercanti in Maiorca. Il carteggio datiniano dall'isola (1387-1396)*, vol. II, *Documenti* (Tome 1), Florence : Le Monnier, 2003.

Novembri, V., *La geografia dei Padri della Chiesa : il caso di Basilio di Cesarea*, Ph.D. dissertation, Florence : University of Florence, 2005.

Olivieri, L. (ed.), *Aristotelismo veneto e scienza moderna*, Atti del XXV anno accademico del Centro per la storia della tradizione aristotelica nel Veneto, Padua : Antenore, 1983.

Olschki, L., *L'Asia di Marco Polo. Introduzione alla lettura e allo studio del Milione*, Venice : Fondazione Giorgio Cini, 1957.

Origo, I., *The merchant of Prato : Francesco di Marco Datini*, London : J. Cape, 1957 (Italian translation I. Origo, *Il mercante di Prato : Francesco di Marco Datini*, preface by L. Einaudi ; transl. by N. Ruffini, Milan : BUR, 1988).

Ortalli, G., Puncuh, D. (eds), *Genova, Venezia, il Levante nei secoli XII-XIV : atti del Convegno internazionale di studi, Genova-Venezia, 10-14 marzo 2000*, Genoa : Società ligure di storia patria ; Venice : Istituto veneto di scienze, lettere ed arti, 2001.

Ortigão, R., "A Symmicta Lusitanica. Inédito de Alexandre Herculano. Relatório apresentado a D. Manuel II," *Archivo historico portuguez*, vol. I, n.° II, Lisboa, Novembro de 1903, pp. 369-370.

Paci, G., Sconocchia, S. (eds), *Ciriaco d'Ancona e la cultura antiquaria dell'Umanesimo*, Reggio Emilia : Diabasis, 1998.

Padovani, S., *Il cenacolo del Perugino detto 'di Fuligno'*, Ministero per i beni culturali ambientali, Soprintendenza per i beni artistici e storici di Firenze, Florence : Tipografia Coppini, 1990.

Paoletti, P., *Raccolta di documenti indediti per servire alla storia della pittura veneziana nei secoli XV e XVI*, Fascicolo I : *I Bellini*, Padua : R. Stabilimento P. Prosperini, 1894.

Pastor, L. von, *Storia dei papi dal Medio Evo*, 16 vols., Rome : Desclee, 1958-64.

Pastore Stocchi, M., "Quaestio de aqua et terra," in *Enciclopedia Dantesca*, Rome : Istituto della enciclopedia italiana, vol. IV (1973), pp. 761-765.

—, "Scuola e cultura umanistica fra due secoli," in G. Arnaldi and M. Pastore Stocchi (eds), *Storia della cultura veneta*, vol. 3.1, *Dal primo Quattrocento al Concilio di Trento*, Vicenza : Neri Pozza, 1980, pp. 93-122.

Pearson, M. N., *The Indian Ocean*, New York : Routledge, 2003.

Pélissier, V. R., *História da Guiné : Portugueses e Africanos na Senegâmbia (1841-1936)*, Lisbon : Editorial Estampa, 1989.

Pelliot, P., "Les grands voyages maritimes chinois au début du xve siècle," *T'oung Pao* 30 (1933), pp. 237-452.

—, *Notes on Marco Polo*, 2 vols. and *Index*, Paris : Imprimerie nationale, Librairie A. Maisonneuve, 1959-1963.

Perocco, D., "Giacomo Gastaldi e la Universale Descrittione del mondo," in S. Ballo Alagna (ed.), *Esplorazioni geografiche e immagine del mondo nei secoli XV e XVI*, Messina : Grafo, 1994.

Pertusi, A., *La caduta di Costantinopoli. Le testimonianze dei contemporanei*, Milan : Fondazione Lorenzo Valla and Arnoldo Mondadori Editore, 1997.

Picone, M. (ed.), *L'enciclopedismo medievale. Atti del convegno di San Gimignano del 1992*, Ravenna : Longo, 1994.

Pinheiro Marques, A., *A maldição da memória do Infante Dom Pedro*, Figueira da Foz : Centro de Estudos do Mar e das Navegações, 1994.

—, *Vida e obra do Príncipe Perfeito Dom João II : Rei de Portugal, Senhor de Coimbra* etc., Figueira da Foz : Centro de Estudos do Mar e das Navegações ; Mira : Câmara Municipal, 1997.

—, "The Portuguese Prince Pedro's Purchase of the Fra Mauro Map from Venice," *The Globe* 48 (1999), pp. 1-34.

Pittaluga, S. (ed.), *Relazioni di viaggio e conoscenza del mondo fra Medioevo e Umanesimo*, Genoa : 1993.

Po, G., "La collaborazione italo-portoghese alle grandi esplorazioni geografiche e alla cartografia nautica," in *Relazioni storiche fra l'Italia e il Portogallo*, XVIII, Rome : Reale Accademia d'Italia, 1940, pp. 276-350.

Polidori Calamandrei, E., *Le vesti delle donne fiorentine nel Quattrocento*, Florence : La Voce, 1924, reprinted Rome : Multigrafica, 1973.

Poncelet, A., "Le légendier de Pier Calo," Analecta Bollandiana 29 (1910), pp. 30-34 and 44-116.

Poppi (ed.), A., *Scienza e filosofia all'Università di Padova nel Quattrocento*, Padua : Lint, 1983.

Portugal : Dicionário Histórico, Corográfico, Heráldico, Biográfico, Bibliográfico, Numismático e Artístico, Lisbon : João Romano Torres, 1904-1915.

Possedoni, G. A. (ed.), *Ciriaco d'Ancona e il suo tempo*, Ancona : Canonici, 2002.

Preto, P., "Grisellini Francesco," *Dizionario biografico degli italiani* 59 (2002), pp. 691-696.

Prontera, F.,"Prima di Strabone : materiali per uno studio della geografia come genere letterario," in Id. (ed.) *Strabone : contributi allo studio della personalità e dell'opera*, vol. 1, Istituto di Storia Antica, Medievale e Moderna, Facoltà di Lettere e Filosofia, Università di Perugia, 1984, pp. 187-256.

Pujades, R. J., *Les cartes portolanes. La representació medieval d'una mar solcada*, Barcelona : Istitut d'Estudis Catalans, 2007.

Pullè, F. L., "Illustrazione del Mappamondo catalano della Biblioteca Estense di Modena" *Atti del Sesto Congresso Geografico Italiano*, Venice, 1908, pp. 341-397.

—, *Cartografia antica dell'India*, Florence : Carnesecchi, 1932.

Randles, W. G. L., *L'image du Sud-Est Africain dans la littérature européenne du XVIe siècle*, pref. A. da Silva Rego, Lisbon : Centro de Estudos Históricos Ultramarinos, 1959.

—, *La signification cosmographique du passage du cap Bojador*, Lisbon : Centro de Estudos Históricos Ultramarinos, 1961 (Sep. Studia, 8).

—, *La représentation de l'Atlantique dans la conscience européenne au Moyen Age et à la Renaissance*, Funchal : Secretaria Regional do Turismo, Cultura e Emigração, Centro de Estudos de História do Atlântico, 1989.

Ratti, A., "A Lost Map of Fra Mauro Found in a Sixteenth-Century Copy," *Imago Mundi* 40 (1988), pp. 77-85.

Relaño, F., *The Shaping of Africa. Cosmographic Discourse and Cartographic Science in Late Medieval and Early Modern Europe*, Aldershot and Burlington : Ashgate, 2002.

—, "Paradise in Africa. The History of a Geographical Myth from Its Origins in Medieval Thought to Its Gradual Demise in Early Modern Europe," *Terrae Incognitae* 36 (2004), pp 1-11.

Revel, J., "Microanalisi e costruzione del sociale," *Quaderni storici* 86 (1994), pp. 549-575.

Rey Pastor, J., García Camarero, E., *La cartografía mallorquina*, Madrid : CSIC, 1960.

Ribemont, B., *De natura rerum : études sur les encyclopédies médiévales*, Orleans : Paradigme, 1995.

—, *Les origines des encyclopédies médiévales, D'Isidore de Séville aux Carolingiens*, Paris : Champion, 2001.

Ricciotti, G., "St. Agostino come interprete del Genesi," *Didaskaleion* 10 (1931).

Richard, J., "Les navigations des occidentaux sur l'océan Indien et la mer Caspienne (XIIe-XVe siècles)," in *Sociétés et compagnies de commerce dans l'Océan Indien*, M. Mollat (ed.), Paris, S.E.V.P.E.N, 1970, pp. 353-363 (reprinted in *Orient et Occident au Moyen Âge. Contacts et relations*, Aldershot : Ashgate, 1976, n. XXI).

Robertson, G., "The Earlier Work of Giovanni Bellini" *Journal of the Warburg and Courtauld Institutes* 23 (1960), pp. 45-59.

—, *Giovanni Bellini*, Oxford : Clarendon Press, 1968.

Romano, D., *Patricians and 'Popolani': The Social Foundations of the Venetian Renaissance State*, Baltimore : Johns Hopkins University Press, 1987, pp. 120-148.

—, *The Likeness of Venice : A Life of Doge Francesco Foscari, 1373-1457*, New Haven and London : Yale University Press, 2007.

Romm, J. S., *The Edges of the Earth in Ancient Thought. Geography, Exploration, and Fiction*, Princeton : Princeton University Press, 1992.

Rosenberg, C., "Arte e politica alle corti di Leonello e Borso d'Este," in *Le Muse e il Principe. Arte di corte nel Rinascimento pagano*, 1991, 2 vols., Modena : Panini, vol. 1, pp. 39-52.

Rosenthal, E., "The German Ptolemy and Its World Map," *Bulletin of the New York Public Library*, 1944, pp. 1-15.

Runia, D. T., "Philo of Alexandria and the Beginnings of Christian Thought. Alexandrian and Jew," *Studia Philonica Annual* 7 (1995), pp. 143-160.

Runni Anderson, A., "Caspian Gates," *Transactions and Proceedings of the American Philological Association* 59 (1928), pp. 130-163.

Russo, L., *Flussi e riflussi. Indagine sull'origine di una teoria scientifica*, Milan : Feltrinelli, 2003.

Sabbadini, R., Guariniana. *1, Vita di Guarino veronese, 2, La scuola e gli studi di Guarino veronese*, M. Sancipriano (ed.), Turin : Bottega d'Erasmo, 1964 (reprint of the Genoa edition : Tipografia del R. Istituto sordo-muti, 1891 ; Catania, 1896).

Sabra, A. "Farghānī, al-," *Dictionary of Scientific Biography*, vol. IV, New York : Charles Scribner's Sons, 1971, pp. 541-545.

Said, E., *Orientalism*, New York : Pantheon Books, 1978.

Sardella, P., *Nouvelles et spéculations à Venise au début du XVIe siècle*, Paris : Armand Colin, 1948.

Scafi, A., "Il Paradiso Terrestre di Fra Mauro," *Storia dell'Arte* 93/94 (1998), pp. 411-419.

—, *The Notion of the Earthly Paradise from the Patristic Era to the Fifteenth Century*, Ph.D. dissertation, London : University of London, 1999.

—, "Mapping Eden," in D. Cosgrove (ed.), *Mappings*, London : Reaktion Books, 1999, pp. 51-69.

—, *Mapping Paradise : A History of Heaven on Earth*, London : British Library ; Chicago : University of Chicago Press, 2006.

Schiapparelli, G., "Le sfere omocentriche di Eudosso, di Callippo e di Aristotele," *Memorie del Reale Istituto Lombardo, Classe di Scienze matematiche e naturali* 13 (1875) (reprinted in *Scritti sulla storia dell'astronomia antica*, II, Bologna, 1926).

Schlosser, J. von, "Die Werkstatt der Embriachi in Venedig," *Jahrbuch der kunsthistorischen Sammlungen [...] in Wien* 20 (1899), pp. 220-282.

Shirley, R., *The Mapping of the World. Early Printed Maps 1472-1700*, London : Holland Press, 4th edition, 2003.

Schönberger, R., *La Scolastica medievale : cenni per una definizione*, Milan : Vita e pensiero, 1997.

Schulz-Forberg, H. (ed.), *Unravelling Civilisation : European Travel and Travel Writing*, Bruxelles : Peter Lang, 2005.

Seaver, K., "Albertin de Virga and the Far North," *Mercator's World* 2.6 (1997), pp. 58-62.

Segre, C., "Marco Polo : filologia e industria culturale," in *Il menante filologo. Bullettino dell'Istituto di filologia moderna*, Parma : Università degli Studi di Parma – Edizioni Zara, 1983, pp. 9-20.

Sinica franciscana, Vol. I : *Itinera et relationes fratrum minorum saeculi XIII et XIV*, Collegit, ad fidem codicum redegit e adnotavit p. Anastasius Van Denert O.F.M., Florence : Apud Collegium S. Bonaventurae, 1929 (Barbera, Alfani e Venturi).

Siniscalco, P., "Due opere a confronto sulla creazione dell' uomo : il De Genesi ad litteram libri XII di Agostino e i Libri in principium Genesis di Beda," *Augustinianum* 25 (1985), pp. 435-452 (Miscellanea di studi in onore di P. Agostino Trapé).

Siraisi, N. G., *Avicenna in Renaissance Italy : the Canon and Medical Teaching in Italian Universities after 1500*, Princeton : Princeton University Press, 1987.

Skelton, R. A., "Bibliographical Note," in Francesco Berlinghieri, *Geographia. Florence 1482*, Amsterdam : Theatrum Orbis Terrarum, 1966, pp. v-xiii.

—, "A Contract for World Maps in Barcelona, 1399-1400, " *Imago Mundi* 22 (1968), pp. 107-113.

Sommerbrodt, E., *Die Ebstorfer Weltkarte, nebst Atlas von 25 Lichtdrucktafeln*, Hannover : 1891.

Sousa Viterbo, F. M., "A livraria real specialmente no reinado de D. Manuel," *Memoria da Academia Real das Sciencias de Lisboa*, classe II, vol. 9, pp. 1, and 23, n. 95.

Spallanzani, M., *Ceramiche orientali a Firenze nel Rinascimento*, Florence : Cassa di Risparmio di Firenze, 1978.

—, Gaeta Bertelà, G. (eds), *Libro d'inventario dei beni di Lorenzo il Magnifico*, Florence : SPES, 1992.

—, *Inventari medicei, 1417-1465 : Giovanni di Bicci, Cosimo e Lorenzo di Giovanni, Piero di Cosimo*, Florence : Associazione Amici del Bargello, 1996.

Spufford, P., *Handbook of Medieval Exchange* ; with the assistance of W. Wilkinson and S. Tolley, London : Royal Historical Society ; Woodbridge, Suffolk ; Wolfeboro, N.H., distributed by Boydell & Brewer, 1986.

Standish, J. F., "*The Caspian Gates*," *Greece & Rome*, 2nd Ser., 17, 1 (1970), pp. 17-24.

Stevenson, E. L., *Genoese World Map 1457. Fac-simile and Critical Text Incorporating, in Free Translation, the Studies of Professor Theobald Fisher*, New York : American Geographical Society and Hispanic Society of America, 1912.

Storia della cultura veneta, 6 vols., Vicenza : Neri Pozza, 1976-1986.

Storia di Venezia. Dalle origini alla caduta della Serenissima, 11 vols., Rome : Istituto della Enciclopedia italiana, 1992-1998.

Stussi, A., "Lingua, dialetto e letteratura", in *Storia d'Italia*, vol. I, *I caratteri originali*, Turin : UTET, 1972, pp. 677-728.

Surdich, F., "Niccolò de' Conti", in *Dizionario biografico degli italiani*, vol. XIII (1971), pp. 640-646.

Tanturli, G., "Volgarizzamenti e ricostruzione dell'antico. I casi della terza e quarta Deca di Livio e di Valerio Massimo, la parte del Boccaccio (a proposito di un'attribuzione)," *Studi medievali* 27 (1986), pp. 811-888.

—, "La cultura fiorentina volgare del Quattrocento davanti ai nuovi testi greci," *Medioevo e Rinascimento* 2 (1988), pp. 217-243.

—, "Umanesimo civile, Umanesimo volgare : i sonetti di Coluccio Salutati," in M. Bendinelli Predelli (ed.), *Firenze alla vigilia del Rinascimento. Antonio Pucci e i suoi contemporanei*, Fiesole : Edizioni Cadmo, 2006, pp. 333-378.

—, "Coluccio Salutati e i letterati del suo tempo," in T. De Robertis, G. Tanturli, and S. Zamponi (eds), *Coluccio Salutati e l'invenzione dell'Umanesimo*, Florence : Mandragora, 2008, pp. 41-47.

Targioni Tozzetti, G., "Del fiorino di sigillo," G. A. Zanetti (ed.), *Nuova raccolta delle monete e zecche d'Italia*, vol. 1, Bologna : Istituto delle Scienze, 1775, pp. 249-274.

Tedeschi, S., "Etiopi e Copti al concilio di Firenze," *Annuarium historiae Conciliorum* 21, 2 (1989), pp. 380-407.

Teeuwen, M., *The Vocabulary of Intellectual Life in the Middle Ages*, Turnhout : Brepols, 2003.

Teixeira da Mota, A., *Topónimos de origem Portuguesa na costa ocidental de Africa desde o Cabo Bojador ao Cabo de Santa Caterina*, Bissau : Centro de Estudos da Guiné Portuguesa, 1950.

—, "Reflexos do Tratado de Tordesilhas na cartografia nautica do seculo XVI," in *El Tratado de Tordesillas y su projección*, Serie Americanista del Seminario de Historia de América, Universidad de Valladolid, Valladolid, 1973, vol. 3, pp. 119-136.

Tenenti A., Vivanti, C., "Le film d'un grand système de navigation : les galères vénitiennes, xive-xvie siècles," *Annales : histoire, sciences sociales* 16, 1 (1961), pp. 83-86.

Testa, E., "Il paradiso dell'Eden secondo i SS. Padri (Contributo alla storia dell'esegesi)," *Studi Biblici Franciscani*, Liber Annus 18 (1968), pp. 94-102.

Thomaz, L. F., *Le Portugal et l'Afrique au xve siècle : les débuts de l'expansion*, Lisbon : Inst. de Investigação Científica Tropical, Sep. Arquivos do Centro Cultural Português da Fundação Calouste Gulbenkian – Volume XXVI, Série Separatas 221, 1989.

Thorndike, L., "An unidentified work by Giovanni Fontana : 'Liber de omnibus rebus naturalibus'," *Isis* 15 (1931), pp. 31-46.

Thorne, R., "Hobart, Robert, Fourth Earl of Buckinghamshire (1760-1816)," in *Oxford Dictionary of National Biography*, ed. H. C. G. Matthew and B. Harrison, Oxford : Oxford University Press, 2004 (on-line edition).

—, "Macartney, George, Earl Macartney (1737-1806)," in *Oxford Dictionary of National Biography*, ed. H. C. G. Matthew and B. Harrison, Oxford : Oxford University Press, 2004 (on-line edition).

Thornson, D., *Renaissance Architecture : Critics, Patrons, Luxury*, Manchester : Manchester University Press, 1993.

Tognetti, S., *Il banco Cambini : affari e mercati di una compagnia mercantile-bancaria nella Firenze del XV secolo*, Florence : Olschki, 1999.

Toniolo, F., "Marco dell'Avogaro e la decorazione all'antica," in *Le Muse e il Principe. Arte di corte nel Rinascimento pagano*, 2 vols., Panini : Modena, vol. II, pp. 133-140.

Toomer, G. J., "Campanus of Novara," in *Dictionary of Scientific Biography* 3 (1981), pp. 23-29.

Tramontin, S., "Un programma di riforma della Chiesa per il Concilio Laterano V : il Libellus ad Leonem X dei veneziani Paolo Giustiniani e Pietro Quirini," in *Venezia e i Concili*, Venice : Fondazione Giorgio Cini, 1962.

Trexler, R. C., "The Magi Enter Florence : The Ubriachi of Florence and Venice," *Studies in Medieval and Renaissance History* 1 (1978), pp. 127-218.

Trifogli, C., "The Place of the Last Sphere in Late-Ancient and Medieval Commentaries," in S. Knuttila, R. Työrinoja, S. Ebbesen (eds), *Knowledge and the Sciences in Medieval Philosophy. Proceedings of the Eighth International Congress of Medieval Philosophy (SIEPM)*, II, Helsinki, pp. 342-350.

Trowles, T., "Vincent, William (1739-1815)," in *Oxford Dictionary of National Biography*, ed. H. C. G. Matthew and Brian Harrison, Oxford : Oxford University Press, 2004 (on-line edition).

Tucci, U., "Tariffe veneziane e libri toscani di mercatura," *Studi veneziani* 10 (1968), pp. 65-108.

—, "I primi viaggiatori e l'opera di Marco Polo," in G. Arnaldi e M. Pastore Stocchi (eds), *Storia della cultura veneta*, vol. 1, *Dalle origini al Trecento*, Vicenza : Neri Pozza, 1976, pp. 633-670.

—, "Manuali di mercatura e pratica degli affari nel Medioevo," in *Fatti e idee di storia economica nei secoli XII-XX. Studi dedicati a Franco Borlandi*, Bologna : Il Mulino, 1977, pp. 215-231.

—, "Mercanti, viaggiatori, pellegrini nel Quattrocento," in G. Arnaldi and M. Pastore Stocchi (eds), *Storia della cultura veneta*, vol. 3.2, *Dal primo Quattrocento al Concilio di Trento*, Vicenza : Neri Pozza, 1980, pp. 317-353.

—, "I servizi marittimi veneziani per il pellegrinaggio in Terrasanta nel medioevo," in *Studi veneziani* 9, 1985, pp. 43-55.

—, "Mercanti veneziani in Asia lungo l'itinerario poliano," in L. Lanciotti (ed.), *Venezia e l'Oriente*, Florence : Olschki, 1987, pp. 307-321.

Tzanaki, R., *Mandeville's Medieval Audiences : a Study on the Reception of the Book of Sir John Mandeville (1371-1550)*, Aldershot : Ashgate, 2003.

Uzielli, G., *La vita e i tempi di Paolo dal Pozzo Toscanelli. Ricerche e studi con un capitolo (VI) sui lavori astronomici del Toscanelli di Giovanni Celoria*, Rome, 1894.

Vasoli, C., "A proposito di umanisti fiorentini e umanisti veneziani," in *Continuità e discontinuità nella storia politica, economica e religiosa. Studi in onore di Aldo Stella*, Vicenza : Neri Pozza, 1993, pp. 89-116.

Vicentini, E., "Il Milione di Marco Polo come portolano," *Italica* 71, 2 (1994), pp. 145-152.

Virtual Encyclopaedia of Portuguese Expansion edited by the CHAM, the Centre for Overseas History, of the Faculty of Social and Human Sciences of the New University of Lisbon : http ://www.fcsh.unl.pt/cham/eve/index.php ?lang=en.

Vila Vilar, E., Acosta Rodríguez, A., González Rodríguez, A. L.(eds), *La Casa de Contratación y navegación entre España y las Indias*, Seville : Universidad de Sevilla – Consejo Superior de Investigaciones Científicas (CSIC), 2004.

Vogel, K. A., *Sphaera terrae : Das mittelalterliche Bild der Erde und die kosmographische Revolution*, Ph.D. dissertation, Göttingen, 1995.

Wallerstein, I., *The Modern World-System I: Capitalist Agriculture and the Origins of the European World-Economy in the Sixteenth Century*, New York : Academic Press, 1974.

Waquet, F., *Latin, or, The Empire of the Sign : from the Sixteenth to the Twentieth Century*, trans. J. Howe, London and New York : Verso, 2001, (original ed., *Le latin ou l'empire d'un signe, XVIe-XXe siècle*, Paris : Albin Michel, 1999).

Weerakoddy, D. P. M., "Greek and Roman Notices of Sri Lanka and their Historical Context," *Sri Lanka Journal of Humanities* 20 (1994), pp. 65-86.

Weiss, R., "Cristoforo Buondelmonti," in *Dizionario biografico degli italiani*, vol. XV, pp. 198-200.

Westrem, S. D., *The Hereford Map. A Transcription and Translation of the Legends with Commentaries*. Turnhout : Brepols, 2001.

Wey-Gómez, N., *The Tropics of Empire. Why Columbus Sailed South to the Indies*, Cambridge MA, and London : MIT Press, 2008.

Whitfield, P., *The Image of the World : Twenty Centuries of World Maps*, San Francisco : Pomegranate Artbooks, 1994.

Wieser, F. von, *Die Weltkarte des Albertin de Virga aus dem Anfange des XV. Jahrhunderts in der Sammlung Figdor in Wien*, Innsbruck : Heinrich Schurick, 1912.

Winter, H., "The Fra Mauro Portolan Chart in the Vatican," *Imago Mundi* 16 (1962), pp. 17-28.

Wittkover, R., "Marco Polo and the Pictorial Tradition of the Marvels of the East," in *Oriente poliano, Studi e conferenze tenute all'ISMEO in occasione del vii centenario della nascita di Marco Polo (1254-1954)*, Rome : Ismeo, 1957, pp. 155-72.

—, "Marvels of the East. A Study in the History of Monsters," *Journal of the Warburg and Courtauld Institutes* 5 (1942), pp. 159-197, reprinted in Id., *Allegory and the Migration of Symbols*, London : Thames and Hudson, 1977, pp. 75-92.

Wolkan, R. (ed.), "Der Briefwwhsel des Eneas Silvius Piccolomini," *Fontes Rerum Austriacarum* 2, 68 (1918), Epist. 109, p. 19.

Wolters, W., *Storia e politica nei dipinti di Palazzo Ducale. Aspetti dell'autocelebrazione della Repubblica di Venezia nel Cinquecento*, Venice : Arsenale, 1987.

Wood, F., *Did Marco Polo Go to China ?*, London : Secker & Warburg, 1995.

Woodward, D.,"Medieval *Mappae mundi*," in J. B. Harley and D. Woodward (eds), *The History of Cartography*, vol. 1, Chicago and London : Chicago University Press, 1987, pp. 286-370.

—, *Maps as Prints in the Italian Renaissance. Makers, Distributors & Consumer*, The Panizzi Lectures, London : British Library, 1996.

—, "Il ritratto della Terra," in F. Camerota (ed.), *Nel segno di Masaccio. L'invenzione della prospettiva*, exhibit catalog, Firenze, Uffizi, maggio 2001-febbraio 2002, Florence : Giunti, 2001, pp. 259-261.

—, "The 'Two Cultures' of Map History : Scientific and Humanistic Traditions : a Plea for Reintegration," in D. Woodward, C. Delano-Smith and C. D. K. Yee (eds), *Plantejaments i objectius d'una història universal de la cartografia – Approaches and challenges in a worldwide history of cartography*, Barcelona : Institut Cartogràfic de Catalunya, 2001, pp. 49-67.

—, "The Italian Map Trade, 1480-1650," in Id (ed.) *The History of Cartography*, Vol. 3 : *Cartography in the European Renaissance*, Chicago and London : University of Chicago Press, 2007, pp. 773-803 ;

Yule, H., *The Book of Ser Marco Polo the Venetian Concerning the Kingdoms and Marvels of the East*. Translated and edited, with notes, by Colonel Sir H. Yule. Third edition, revised throughout in light of recent discoveries by Henry Cordier, 2 vols., London : Murray, 1921.

Zaccagnini, G., "Francesco Pipino traduttore del Milione, cronista e viaggiatore in Oriente nel secolo XIV," *Atti e memorie della R. Deputazione di Storia Patria per le Provincie di Romagna*, s. 5, 1 (1936), pp. 70-95.

Zaccaria, R., "Dolfin Pietro," in *Dizionario biografico degli italiani*. Rome : Istituto della Enciclopedia italiana, vol. 40 (1991), pp. 565-571.

Zippel, G., "Cosmografi al servizio dei papi nel Quattrocento," *Bollettino della Società geografica italiana* 4, 2 (1910).

Zorzi, M., "Dal manoscritto al libro," in A. Tenenti and U. Tucci (eds), *Storia di Venezia. Dalle origini alla caduta della Serenissima*, vol. IV, *Il Rinascimento. Politica e cultura*, Rome: Istituto della Enciclopedia Italiana, 1996, pp. 817-958.

MANUSCRIPT SOURCES

INDEX NOMINUM ET OPERUM

I.1. Fra Mauro's *mappa mundi* : Junks (*çoncho da India*) in the Indian Ocean,
a realistic visual citation of Marco Polo's *Milione*.
(Venice, Biblioteca Nazionale Marciana)

I.2. Fra Mauro's *mappa mundi* : The Gran Kahn's hunting tents, the Palace of Cambaluc, and the Polisanchin bridge. (Venice, Biblioteca Nazionale Marciana)

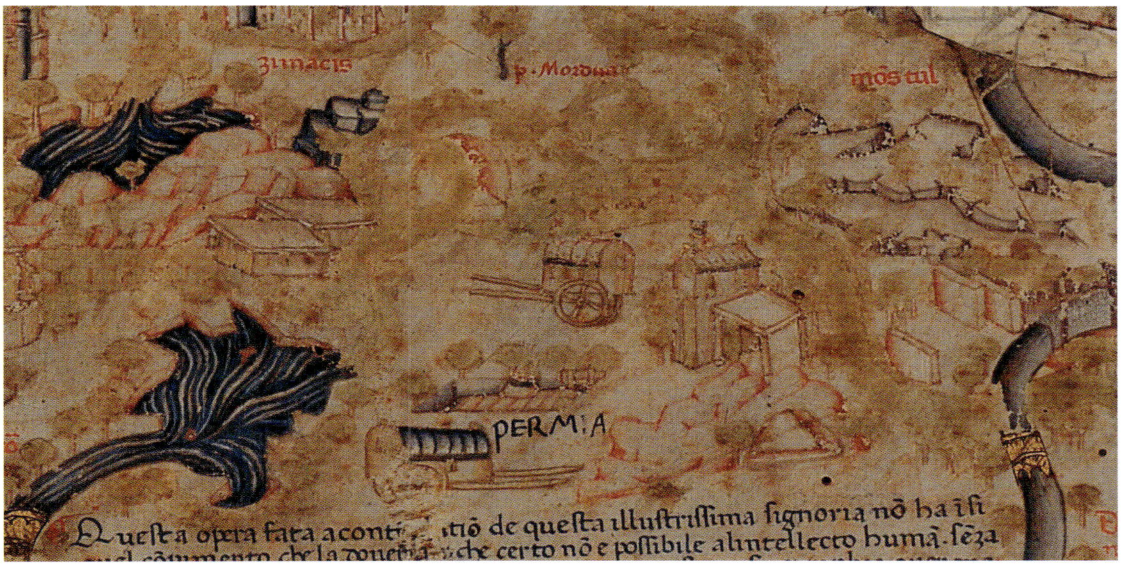

I.3. Fra Mauro's *mappa mundi*: The felt-covered carriages of the Tartars,
a faithful visual citation of Marco Polo's *Milione*.
(Venice, Biblioteca Nazionale Marciana)

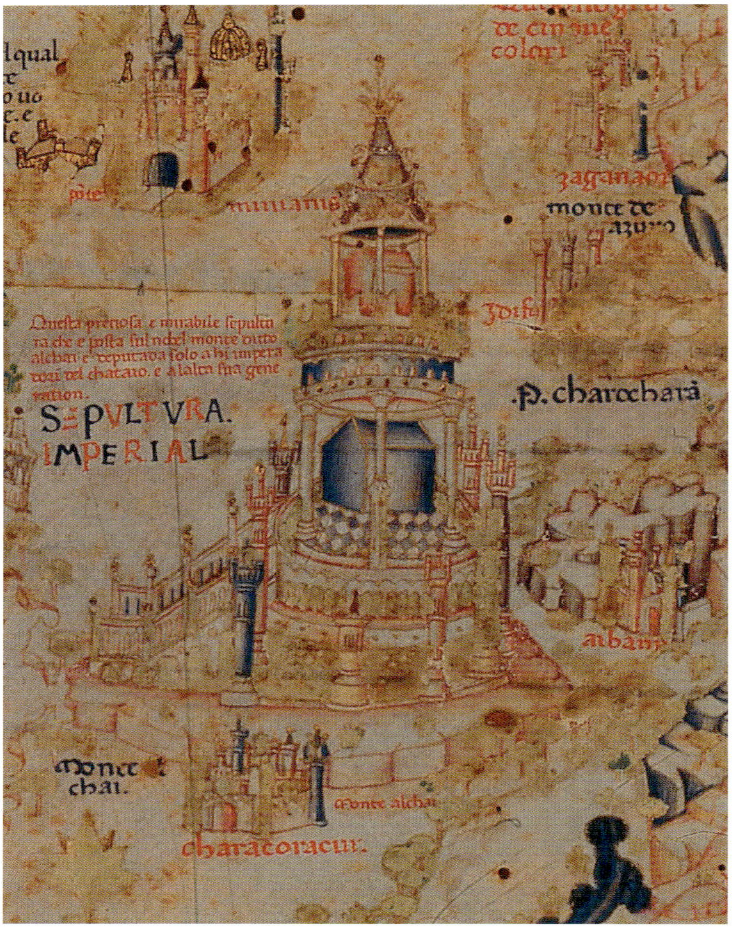

I.4. Fra Mauro's *mappa mundi* : Mount Alchai and the sepulchral shrine of the Khans, an imaginative visual citation of Marco Polo's *Milione.*
(Venice, Biblioteca Nazionale Marciana)

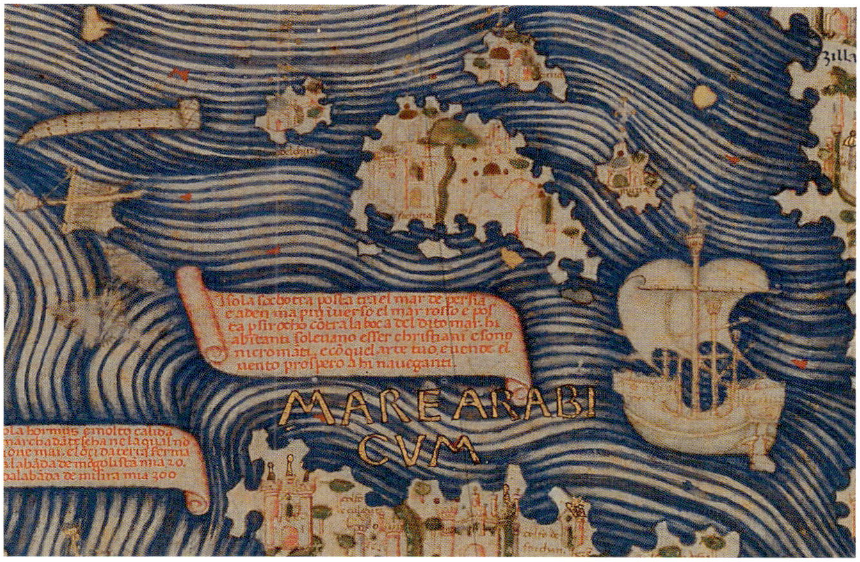

I.5. Fra Mauro's *mappa mundi*: Shipwrecks in the Indian Ocean and off the coast of Gibraltar.
(Venice, Biblioteca Nazionale Marciana)

INDIA PRIMA
PERSIA
MEDIA

II. This map is contemporary with, and very probably derived from, a prototype also used to create Fra Mauro's *mappa mundi*. (Vatican City, Biblioteca apostolica vaticana, Carta Borgiana V)

Libya inferior
Sinus affrica
Felix Arabia

III. Andrea Bianco drew this first Venetian Ptolemaic planisphere in 1436 as part of a composite nautical atlas. (Venice, Biblioteca Nazionale Marciana, Ms. It. Z. 76 (=4783), ff. 12v-13r)

F. 260 Inf
N° 1
(cf. Revelli pp 59-60)

IV. A Marine chart drawn by Andrea Bianco in London in 1448 attesting to contemporary Portuguese navigations. (Milan, Biblioteca Ambrosiana, F. 260, inf. 1)

V. *Mappa mundi* designed in 1457 which, like the *mappa mundi* of Fra Mauro, is a synthesis of all of the cartographic idioms of the middle of the fifteenth century. (Florence, Biblioteca Nazionale Centrale, Portolano 1)

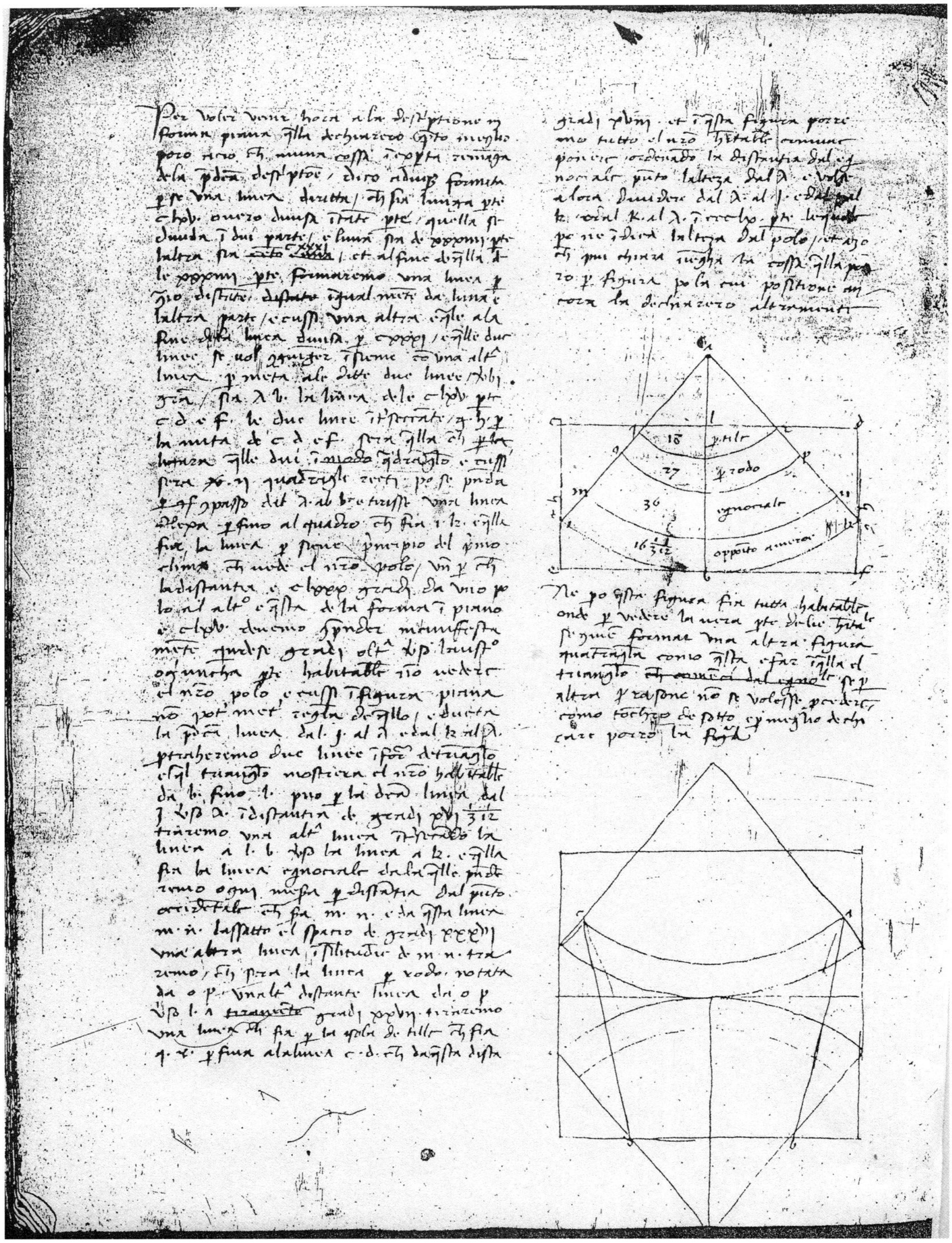

VI.1. [Double figure], *c.*1430-50: the unique example of a "double" first Ptolemaic projection. (Venice, Biblioteca Nazionale Marciana, Ms. It. Cl. VI, 24 (=6111), f. 90v)

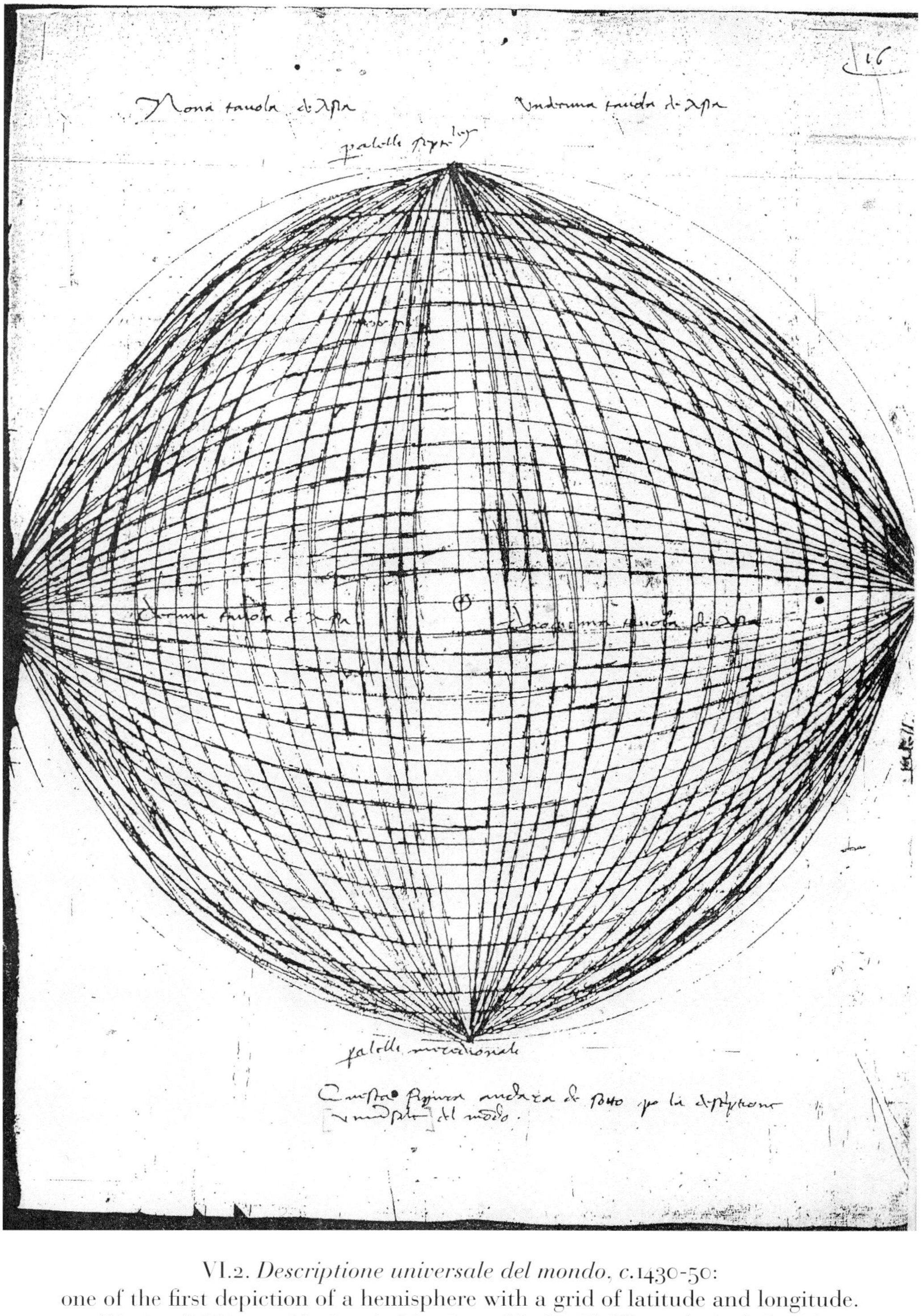

VI.2. *Descriptione universale del mondo*, *c.*1430-50:
one of the first depiction of a hemisphere with a grid of latitude and longitude.
(Venice, Biblioteca Nazionale Marciana, Ms. It. Cl. VI, 24 (=6111), f. 16r)

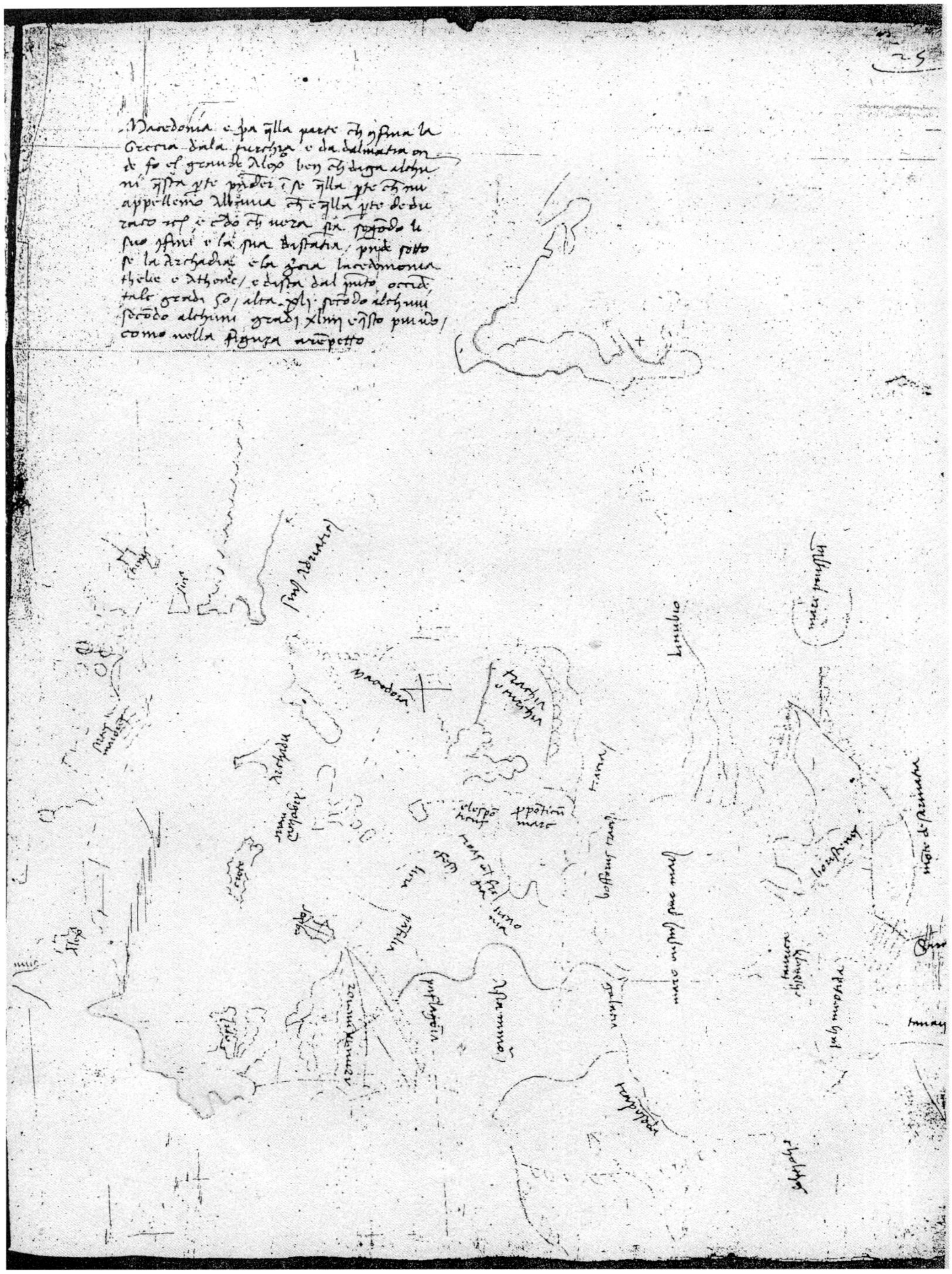

VI.3. [Regional map of Macedonia] *c.*1430-50:
an innovative re-elaboration of Ptolemaic cartography for the creation of regional maps.
(Venice, Biblioteca Nazionale Marciana, Ms. It. Cl. VI, 24(=6111), f. 25r)

VII. Leonardo Bellini's illuminations in the *Promissione dogale di Niccolò Marcello*, 1473, which shows stylistic similarities with the image of paradise on Fra Mauro's *mappa mundi*. (Venice, Museo Correr, Ms. l. III, 322, f. Ir)

VIII. Andrea Mantegna's *Agony in the Garden*, *c*.1455-1460: the shape of the rocky landscape is very similar to that of paradise on Fra Mauro's *mappa mundi*. (London, National Gallery, inv. 1417)

ETHYOPIA AVSTRAL
ABASSIA
ABASSIA
ETHIOPIA
NILO
NUBA
MARE RUBRUM
ARABIA
DESERTO
DESERTO
ARABIA
TIBAIDA
EGYPTO
EGYPTO
LIBIA
SIRIA

IX. William Frazer, *Copy of the Large Planisphere or Mappa mondo by Fra Mauro*, c.1804. (London, British Library, Add. MS 11267)

AVSTER
ETHYOP IA
AVS TRAL
ABASSIA
ETYOPIA
FATAGA
HAMARA
ABASSIA
ADEL
AFRIC
ETHIOPIA
SINVS ARABICVS
NILO
NUBA
MARE RUBRUM
ARABIA
ARABIA FELIX
DESER TO

X. The published version of the manuscript copy of Fra Mauro's map made by William Frazer and commissioned by the Viscount de Santarém, 1854 (cf. plate IX). (Lisbon, Biblioteca Nacional de Portugal, C.C. 770 R)

XI.1. Map of central Asia made by Angelo Freducci "Anconitanus" in 1556, which shows the influence of the Borgia V map (cfr. plate II).
(Mantua, Biblioteca Comunale Teresiana, Ms. 646)

XI.2. Map of southeast Asia made by Angelo Freducci "Anconitanus" in 1556, which shows the influence of the cartography of Fra Mauro (cf. plates I and II).
(Mantua, Biblioteca Comunale Teresiana, Ms. 646)

PHOTO CREDITS

Images are reproduced by permission of the following institutions:

Città del Vaticano, Biblioteca Apostolica Vaticana (pl. II)
Firenze, Biblioteca Nazionale Centrale (pl. V)
Lisboa, Biblioteca Nacional de Portugal (pl. X)
London, British Library (pl. IX) (By Permission of The British Library)
London, National Gallery (pl. VIII)
Mantova, Biblioteca Comunale Teresiana (pl. XI.1-2)
Milano, Biblioteca Ambrosiana (pl. IV)
Venezia, Biblioteca Nazionale Marciana (pl. I.1-5, III, VI.1-3)
Venezia, Museo Correr (pl. VII)